T0189559

Lecture Notes in Computer Science   9516

Commenced Publication in 1973
Founding and Former Series Editors:
Gerhard Goos, Juris Hartmanis, and Jan van Leeuwen

More information about this series at http://www.springer.com/series/7409

Qi Tian · Nicu Sebe
Guo-Jun Qi · Benoit Huet
Richang Hong · Xueliang Liu (Eds.)

MultiMedia Modeling

22nd International Conference, MMM 2016
Miami, FL, USA, January 4–6, 2016
Proceedings, Part I

Springer

Editors
Qi Tian
University of Texas at San Antonio
San Antonio, TX
USA

Nicu Sebe
Department of Information Engineering
University of Trento
Povo, Trento
Italy

Guo-Jun Qi
EECS
University of Central Florida
Orlando, FL
USA

Benoit Huet
EURECOM
Sophia-Antipolis
France

Richang Hong
Hefei University of Technology
Hefei, Anhui
China

Xueliang Liu
School of Computing and Information
Hefei University of Technology
Hefei, Anhui
China

ISSN 0302-9743   ISSN 1611-3349 (electronic)
Lecture Notes in Computer Science
ISBN 978-3-319-27670-0   ISBN 978-3-319-27671-7 (eBook)
DOI 10.1007/978-3-319-27671-7

Library of Congress Control Number: 2015957238

LNCS Sublibrary: SL3 – Information Systems and Applications, incl. Internet/Web, and HCI

This Springer imprint is published by SpringerNature
The registered company is Springer International Publishing AG Switzerland

Preface

The 22nd International Conference on Multimedia Modeling (MMM 2016) was held in Miami, USA, January 4–6, 2016, and was hosted by the University of Central Florida at Orlando, USA. MMM is a leading international conference for researchers and industry practitioners to share their new ideas, original research results, and practical development experiences from all multimedia-related areas. University of Central Florida is a Space-Grant university and has made noted research contributions to digital media, engineering, and computer science.

MMM 2016 featured a comprehensive program including three keynote talks, eight oral presentation sessions, two poster sessions, one demo session, five special sessions, and the Video Browser Showdown (VBS). The 168 submissions from authors of 20 countries included a large number of high-quality papers in multimedia content analysis, multimedia signal processing and communications, and multimedia applications and services. We thank our 130 Technical Program Committee members who spent many hours reviewing papers and providing valuable feedback to the authors. From the total of 117 submissions to the main conference and based on at least three reviews per submission, the program chairs decided to accept 32 regular papers (27.8 %) and 30 poster papers (25.6 %). In total, 38 papers were received for 5 special sessions, with 20 being selected, and 11 submissions were received for a demo session, with 7 being selected. Video browsing systems of nine teams were selected for participation in the VBS. The authors of accepted papers come from 17 countries. This volume of the conference proceedings contains the abstracts of three invited talks and all the regular, poster, special session, and demo papers, as well as special demo papers of the VBS. MMM 2016 included the following awards: the Best Paper Award, the Best Student Paper Award, and the winner of the VBS competition.

The technical program is an important aspect but only provides its full impact if complemented by challenging keynotes. We were extremely pleased and grateful to have three exceptional keynote speakers, Wen Gao (ACM/IEEE Fellow), Chang Wen Chen (IEEE Fellow), and Changsheng Xu (IEEE Fellow), accept our invitation and present interesting ideas and insights at MMM 2016.

We are heavily indebted to many individuals for their significant contributions. We thank the MMM Steering Committee for their invaluable input and guidance on crucial decisions. We wish to acknowledge and express our deepest appreciation to the organizing chairs, Xueliang Liu and Luming Zhang, the special session chairs, Wen-Huang Chen, Haojie Li and Rongrong Ji, the panel chair, Tat-Seng Chua, the demo chairs, Cathal Gurrin and Björn Þór Jónsson, the VBS chairs, Klaus Schöffmann and Werner Bailer, the publicity chairs, Yu-Gang Jiang, Shuicheng Yan, Hengtao Shen, Zhengjun Zha, and Sheng Wu, the publication chairs, Na Zhao and Zechao Li, and last but not least the Webmaster, Jun He. Without their efforts and enthusiasm, MMM 2016 would not have become a reality. Moreover, we want to thank our sponsor the

University of Central Florida. Finally, we wish to thank all committee members, reviewers, session chairs, student volunteers, and supporters. Their contributions are much appreciated.

January 2016

Guo-Jun Qi
Benoit Huet
Richang Hong
Nicu Sebe
Qi Tian

Organization

MMM 2016 was organized by the University of Central Florida, USA.

MMM 2016 Steering Committee

Phoebe Chen	La Trobe University, Australia
Tat-Seng Chua	National University of Singapore
Shiqiang Yang	Tsinghua University, China
Kiyoharu Aizawa	University of Tokyo, Japan
Noel E. O'Connor	Dublin City University, Ireland
Cess G.M. Snoek	University of Amsterdam, The Netherlands
Meng Wang	Hefei University of Technology, China
R. Manmatha	University of Massachusetts, USA
Cathal Gurrin	Dublin City University, Ireland
Klaus Schoeffmann	Klagenfurt University, Austria
Benoit Huet	Eurecom, France

MMM 2016 Organizing Committee

General Co-chairs

Qi Tian	University of Texas at San Antonio, USA
Nicu Sebe	University of Trento, Italy

Program Co-chairs

Guojun Qi	University of Central Florida, USA
Benoit Huet	Eurecom, France
Richang Hong	Hefei University of Technology, China

Organizing Co-chairs

Xueliang Liu	Hefei University of Technology, China
Luming, Zhang	National University of Singapore, Singapore

Special Session Co-chairs

Wen-Huang Cheng	Academia Sinica, Taiwan
Haojie Li	Dalian University of Technology, China
Rongrong Ji	Xiamen University, China

Demo Session Co-chairs

Cathal Gurrin	Dublin City University, Ireland
Björn Þór Jónsson	Reykjavík University, Iceland

Publication Co-chairs

Na Zhao   National University of Singapore, Singapore
Zechao Li   Nanjing University of Science and Technology, China

Publicity Co-chairs

Yu-Gang Jiang   Fudan University, China
Shuicheng Yan   National University of Singapore, Singapore
Hengtao Shen   University of Queensland, Australia
Zhengjun Zha   Chinese Academy of Sciences, China
Sheng Wu   Google, USA

Panel Chair

Tat-Seng Chua   National University of Singapore, Singapore

Video Search Showcase Co-chairs

Werner Bailer   Joanneum Research, Graz, Austria
Klaus Schoeffmann   Klagenfurt University, Austria

Web Master

Jun He   Hefei University of Technology, China

Technical Program Committee

Selim Balcisoy   Sabanci University, Turkey
Yingbo Li   Ecole Normale Superieure, France
Lifeng Sun   Tsinghua University, China
Cathal Gurrin   Dublin City University, Ireland
Haojie Li   Dalian University of Technology, China
Rainer Lienhart   University of Augsburg, Germany
Rossana Damiano   University of Turin, Italy
Zheng-Jun Zha   Institute of Intelligent Machines, CAS, China
Vincent Charvillat   University of Toulouse, France
Liqiang Nie   National University of Singapore, Singapore
Wolfgang Hurst   Utrecht University, The Netherlands
Wei-Guang Teng   National Cheng Kung University, Taiwan
Bo Yan   Fudan University, China
Werner Bailer   Joanneum Research, Austria
Mei-Ling Shyu   University of Miami, USA
Luiz Fernando Gomes   Catholic University of Rio de Janeiro, Brazil
    Soares
Joemon Jose   University of Glasgow, UK
Mylene Farias   University of Brasilia, Brazil
Wolfgang Huerst   Utrecht University, The Netherlands

Sponsors

University of Central Florida

Google

FX Palo Alto Laboratory

Springer Publishing

Springer

Contents – Part I

Regular Papers

Special Session Poster Papers

Contents – Part II

Demo Session Papers

Video Browser Showdown

Regular Papers

Video Event Detection Using Kernel Support Vector Machine with Isotropic Gaussian Sample Uncertainty (KSVM-iGSU)

Christos Tzelepis[1,2]([⊠]), Vasileios Mezaris[1], and Ioannis Patras[2]

[1] Information Technologies Institute (ITI), CERTH, 57001 Thermi, Greece
{tzelepis,bmezaris}@iti.gr
[2] Queen Mary University of London, Mile End Campus, London E14NS, UK
i.patras@qmul.ac.uk

Abstract. In this paper, we propose an algorithm that learns from uncertain data and exploits related videos for the problem of event detection; related videos are those that are closely associated, though not fully depicting the event of interest. In particular, two extensions of the linear SVM with Gaussian Sample Uncertainty are presented, which (a) lead to non-linear decision boundaries and (b) incorporate related class samples in the optimization problem. The resulting learning methods are especially useful in problems where only a limited number of positive and related training observations are provided, e.g., for the 10Ex subtask of TRECVID MED, where only ten positive and five related samples are provided for the training of a complex event detector. Experimental results on the TRECVID MED 2014 dataset verify the effectiveness of the proposed methods.

Keywords: Video event detection · Very few positive samples · Related samples · Learning with uncertainty · Kernel methods · Relevance degree SVMs

1 Introduction

High-level video event detection is concerned with determining whether a certain video depicts a given event or not. Typically, a high-level (or complex) event is defined as an interaction among humans, or between humans and physical objects [16]. Some typical examples of complex events are those provided in the Multimedia Event Detection (MED) task of the TRECVID benchmarking activity [22]. For instance, indicative complex events defined in MED 2014 include "Attempting a bike trick", "Cleaning an appliance", or "Beekeeping", to name a few.

There are numerous challenges associated with building effective video event detectors. One of them is that often there is only a limited number of positive video examples available for training. Another challenge is that video representation techniques usually introduce uncertainty in the input that is fed to

© Springer International Publishing Switzerland 2016
Q. Tian et al. (Eds.): MMM 2016, Part I, LNCS 9516, pp. 3–15, 2016.
DOI: 10.1007/978-3-319-27671-7_1

the classifiers, and this also needs to be taken into consideration during classifier training. In this work we deal with the problem of learning video event detectors when a limited number of positive and related (i.e., videos that are closely related with the event, but do not meet the exact requirements for being a positive event instance [22]) event videos are provided. For this, we exploit the uncertainty of the training videos by extending the linear Support Vector Machine with Gaussian Sample Uncertainty (LSVM-GSU), presented in [27], in order to arrive at non-linear decision functions. Specifically, we extend this version of LSVM-GSU that assumes isotropic uncertainty (hereafter denoted LSVM-iGSU) into a new kernel-based algorithm, which we call KSVM-iGSU. We also further extend KSVM-iGSU, drawing inspiration from the Relevance Degree kernel SVM (RD-KSVM) proposed in [28], such that related samples can be effectively exploited as positive or negative examples with automatic weighting. We refer to this algorithm as RD-KSVM-iGSU. We show that the RD-KSVM-iGSU algorithm results in more accurate event detectors than the state of the art techniques used in related works, such as the standard kernel SVM and RD-KSVM.

The paper is organized as follows. In Sect. 2 we review related work. In Sect. 3 the two proposed SVM extensions are presented. Video event detection results, by application of the proposed KSVM-iGSU and RD-KSVM-iGSU to the TRECVID MED 2014 dataset, are provided in Sect. 4, while conclusions are drawn and future work is discussed in Sect. 5.

2   Related Work

There are many works dealing with event detection in video (e.g. [2,5,7,9,11–15,19,21]), several of them being in the context of the TRECVID MED task. Despite the attention that video event detection has received, though, there is only a limited number of studies that have explicitly examined the problem of learning event detectors from very few (e.g. 10) positive training examples [13,28], and developed methods for addressing this exact problem. In [13], for instance, the authors present VideoStory, a video representation scheme for learning event detectors from a few training examples by exploiting freely available Web videos together with their textual descriptions. Several other works (e.g. [2]) treat the few-example problem in the same way that they deal with event detection when more examples are available (e.g. training standard kernel SVMs). Learning video event detectors from a few examples is a problem that is simulated in the TRECVID MED task [22] by the 10Ex subtask, where only 10 positive samples are available for training.

In the case of learning from very few positive samples, it is of high interest to further exploit video samples that do not exactly meet the requirements for being characterized as true positive examples of an event, but nevertheless are closely related to an event class and can be seen as "related" examples of it. This is simulated in the TRECVID MED task [22] by the "near-miss" video examples provided for each target event class. Except for [28], none of the above works takes full advantage of these related videos for learning from few positive samples; instead, the "related" samples are either excluded from the training procedure [2,11], or they are mistreated as true positive or true negative instances [7]. In contrast, in [28] the authors exploit related samples by handling them as weighted positive or negative ones, applying an automatic weighting technique during the training stage. To this end, a relevance degree in $(0, 1]$ is automatically assigned to all the related samples, indicating the degree of relevance of these observations with the class they are related to. It was shown that this weighting resulted in learning more accurate event detectors.

Regardless of whether the above works address the problem of learning from a few positive examples or assume that an abundance of such examples is available, they all treat the training video representations as noise-free observations in the SVM input space. Looking beyond the event detection applications, though, assuming uncertainty in input under the SVM paradigm is not unusual and has been shown to lead to better learning. Lanckriet et al. [18] considered a binary classification problem where the mean and covariance matrix of each class are assumed to be known. Xu et al. [29,30] considered the robust classification problem for a class of non-box-typed uncertainty sets, in contrast to [1,18,25], who robustified regularized classification using box-type uncertainty. Finally, in [27], Tzelepis et al. proposed a linear maximum-margin classifier, called SVM with Gaussian Sample Uncertainty, dealing with uncertain input data. The uncertainty in [27] can be modeled either isotropically or anisotropically, arriving at a convex optimization problem that is solved using a gradient descent approach.

To the best of our knowledge, there has been no study dealing with uncertainty in the video event detection problem, except for [27]. However, [27] introduces linear classifiers, which in the event detection problem are not expected to perform in par with traditional kernel SVMs that are typically used (e.g. [11,31]), despite the advantages of considering data uncertainty in the learning process. In this work, we extend the above study and kernelize the LSVM-iGSU of [27], under the assumption of isotropic sample uncertainty. We apply the resulting KSVM-iGSU to the event detection problem when only a few positive samples are available for training. Moreover, we propose a further extension of KSVM-iGSU, namely Relevance Degree KSVM-iGSU (RD-KSVM-iGSU), inspired by [28], such that related samples can also be exploited as weighted positive or negative ones, based on an automatic weighting scheme.

3 Kernel SVM-iGSU

3.1 Overview of LSVM-iGSU

LSVM-iGSU [27] is a classifier that takes a input training data that are described not solely by a set of feature representations, i.e. a set of vectors $\mathbf{x}_i$ in some n-dimensional space, but rather by a set of multivariate isotropic Gaussian distributions which model the uncertainty of each training example. That is, every training datum is characterized by a mean vector $\mathbf{x}_i \in \mathbb{R}^n$ and an isotropic covariance matrix, i.e. a scalar multiple of the identity matrix, $\Sigma_i = \sigma_i^2 I_n \in \mathbb{S}_{++}^n$[1]. LSVM-iGSU is obtained by minimizing, with respect to $\mathbf{w}, b$, the objective function $\mathcal{J} \colon \mathbb{R}^n \times \mathbb{R} \to \mathbb{R}$ given by

$$\mathcal{J}(\mathbf{w}, b) = \frac{1}{2}\|\mathbf{w}\|_2^2 + C\sum_{i=1}^{l}\mathcal{L}(\mathbf{w}, b, \mathbf{x}_i, \sigma_i^2 I_n, y_i), \tag{1}$$

where l is the number of training data, $\mathbf{w} \cdot \mathbf{x} + b = 0$ denotes the separating hyperplane, and the loss $\mathcal{L} \colon (\mathbb{R}^n \times \mathbb{R}) \times (\mathbb{R}^n \times \mathbb{S}_{++}^n \times \{\pm 1\}) \to \mathbb{R}$ is given by

$$\mathcal{L}(\mathbf{w}, b, \mathbf{x}_i, \sigma_i^2 I_n, y_i) = \frac{y_i - \mathbf{w} \cdot \mathbf{x}_i - b}{2}\left(\operatorname{erf}\left(\frac{y_i - \mathbf{w} \cdot \mathbf{x}_i - b}{\sqrt{2\sigma_i^2\|\mathbf{w}\|_2^2}}\right) + y_i\right)$$
$$+ \frac{\sqrt{\sigma_i^2\|\mathbf{w}\|_2^2}}{\sqrt{2\pi}}\exp\left(-\frac{(y_i - \mathbf{w} \cdot \mathbf{x}_i - b)^2}{2\sigma_i^2\|\mathbf{w}\|_2^2}\right), \tag{2}$$

where $\mathbf{x}_i$ and $\sigma_i^2 I_n$ denote the mean vector and the covariance matrix of the i-th input entity (Gaussian distribution), respectively, y_i denotes its ground-truth label, and $\operatorname{erf}(x) = \frac{2}{\sqrt{\pi}}\int_0^x e^{-t^2}\,dt$ denotes the error function.

As discussed in [27], (1) is convex and thus a (global) optimal solution $(\mathbf{w}, b)$ can be obtained using a gradient descent algorithm. The resulting (linear) decision function $f(\mathbf{x}) = \mathbf{w} \cdot \mathbf{x} + b$ is used in the testing phase for classifying an unseen sample similarly to the standard linear SVM algorithm [4]; that is, according to the distance between the testing sample and the separating hyperplane, without taking into account any uncertainty estimates that could be made for the testing sample representation.

3.2   Kernelizing LSVM-iGSU (KSVM-iGSU)

The optimization problem discussed in the previous section can be recast as a variational calculus problem of finding the function f that minimizes the functional $\Phi[f]$:

$$\min_{f \in \mathcal{H}} \Phi[f], \tag{3}$$

where the functional $\Phi[f]$ is given by

$$\Phi[f] = \frac{1}{2}\lambda\|f\|_{\mathcal{H}}^2 + \sum_{i=1}^{l}\left[\frac{y_i - f(\mathbf{x}_i) - b}{2}\left(\operatorname{erf}\left(\frac{y_i - f(\mathbf{x}_i) - b}{\sqrt{2\sigma_i^2\|f\|_{\mathcal{H}}^2}}\right) + y_i\right)\right.$$
$$\left. + \frac{\sqrt{\sigma_i^2\|f\|_{\mathcal{H}}^2}}{\sqrt{2\pi}}\exp\left(\frac{(y_i - f(\mathbf{x}_i) - b)^2}{2\sigma_i^2\|f\|_{\mathcal{H}}^2}\right)\right], \tag{4}$$

---

[1] $\mathbb{S}_{++}^n$ denotes the convex cone of all symmetric positive definite $n \times n$ matrices with entries in $\mathbb{R}$. I_n denotes the identity matrix of order n.

where $\lambda = 1/C$ is a regularization parameter and f belongs to a Reproducing Kernel Hilbert Space (RKHS), $\mathcal{H}$, with associated kernel k. Using a generalized semi-parametric version [24] of the representer theorem [17], it can be shown that the minimizer of the above functional admits a solution of the form

$$f(\mathbf{x}) = \sum_{i=1}^{l} \alpha_i k(\mathbf{x}, \mathbf{x}_i) - b, \tag{5}$$

where $b \in \mathbb{R}$, $\alpha_i \in \mathbb{R}$, $\forall i$.

Using the reproducing property, we have

$$\|f\|_{\mathcal{H}}^2 = \langle f, f \rangle_{\mathcal{H}} = \left\langle \sum_{i=1}^{l} \alpha_i k(\cdot, \mathbf{x}_i), \sum_{j=1}^{l} \alpha_j k(\cdot, \mathbf{x}_j) \right\rangle_{\mathcal{H}} = \boldsymbol{\alpha}^\top K \boldsymbol{\alpha}, \tag{6}$$

where K is the kernel matrix, i.e. the symmetric positive definite $l \times l$ matrix defined as $K = (k(\mathbf{x}_i, \mathbf{x}_j))_{i,j=1}^{l}$, and $\boldsymbol{\alpha} = (\alpha_1, \cdots, \alpha_l)^\top$. Moreover, we observe that $f(\mathbf{x}_i) = \sum_{j=1}^{l} \alpha_j k(\mathbf{x}_i, \mathbf{x}_j) = \mathbf{K}_i \cdot \boldsymbol{\alpha}$, where $\mathbf{K}_i$ denotes the i-th column of the kernel matrix K. Then, the objective function $\mathcal{J}_{\mathcal{H}} \colon \mathbb{R}^l \times \mathbb{R} \to \mathbb{R}$ is given by

$$\mathcal{J}_{\mathcal{H}}(\boldsymbol{\alpha}, b) = \frac{1}{2}\lambda\boldsymbol{\alpha}^\top K \boldsymbol{\alpha} + \sum_{i=1}^{l} \left[\frac{y_i - \mathbf{K}_i \cdot \boldsymbol{\alpha} - b}{2} \left(\mathrm{erf}\left(\frac{y_i - \mathbf{K}_i \cdot \boldsymbol{\alpha} - b}{\sqrt{2\sigma_i^2 \boldsymbol{\alpha}^\top K \boldsymbol{\alpha}}} \right) + y_i \right) \right.$$
$$\left. + \frac{\sqrt{\sigma_i^2 \boldsymbol{\alpha}^\top K \boldsymbol{\alpha}}}{\sqrt{2\pi}} \exp\left(-\frac{(y_i - \mathbf{K}_i \cdot \boldsymbol{\alpha} - b)^2}{2\sigma_i^2 \boldsymbol{\alpha}^\top K \boldsymbol{\alpha}} \right) \right], \tag{7}$$

where the above sum gives the total loss. We (jointly) minimize the above convex[2] objective function with respect to $\boldsymbol{\alpha}$, b similarly to [27] using the Limited-memory BFGS (L-BFGS) algorithm [20]. L-BFGS is a quasi-Newton optimization algorithm that approximates the Broyden-Fletcher-Goldfarb-Shanno (BFGS) [3] algorithm using a limited amount of computer memory. It requires the first order derivatives of the objective function with respect to the optimization variables $\boldsymbol{\alpha}$, b. They are given[3], respectively, as follows

$$\frac{\partial \mathcal{J}_{\mathcal{H}}}{\partial \boldsymbol{\alpha}} = \lambda K \boldsymbol{\alpha} + \sum_{i=1}^{l} \left[\frac{\sigma_i^2 \exp\left(-\frac{(y_i - \mathbf{K}_i \cdot \boldsymbol{\alpha} - b)^2}{2\sigma_i^2 \boldsymbol{\alpha}^\top K \boldsymbol{\alpha}} \right)}{\sqrt{2\pi \sigma_i^2 \boldsymbol{\alpha}^\top K \boldsymbol{\alpha}}} K \boldsymbol{\alpha} \right.$$
$$\left. - \frac{1}{2} \mathrm{erf}\left(\frac{y_i - \mathbf{K}_i \cdot \boldsymbol{\alpha} - b}{\sqrt{2\sigma_i^2 \boldsymbol{\alpha}^\top K \boldsymbol{\alpha}}} \right) \mathbf{K}_i - \frac{y_i}{2}\mathbf{K}_i \right], \tag{8}$$

and

$$\frac{\partial \mathcal{J}_{\mathcal{H}}}{\partial b} = -\frac{1}{2} \sum_{i=1}^{l} \left[\mathrm{erf}\left(\frac{y_i - \mathbf{K}_i \cdot \boldsymbol{\alpha} - b}{\sqrt{2\sigma_i^2 \boldsymbol{\alpha}^\top K \boldsymbol{\alpha}}} \right) + y_i \right]. \tag{9}$$

---

[2] Convexity can be shown using Theorem 2 proved in [27].
[3] Their derivation is omitted, as it is technical but straightforward.

Since J is a convex function on $\mathbb{R}^l \times \mathbb{R}$, L-BFGS leads to a global optimal solution; that is, at a pair $(\boldsymbol{\alpha}, b)$ such that the decision function given in the form of (5) minimizes the functional (4). We call this classifier kernel SVM-iGSU (KSVM-iGSU).

3.3 Relevance Degree KSVM-iGSU

Motivated by [28], we reformulate the optimization problem in (3)-(4) such that a different penalty parameter $c_i \in (0, 1]$ (hereafter called as relevance degree) is introduced to each input datum. That is, the functional $\Phi[f]$ of (4) is now given by

$$\Phi[f] = \frac{1}{2}\lambda\|f\|_{\mathcal{H}}^2 + \sum_{i=1}^{l} c_i \left[\frac{y_i - f(\mathbf{x}_i) - b}{2} \left(\text{erf}\left(\frac{y_i - f(\mathbf{x}_i) - b}{\sqrt{2\sigma_i^2\|f\|_{\mathcal{H}}^2}} \right) + y_i \right) \right.$$
$$\left. + \frac{\sqrt{\sigma_i^2\|f\|_{\mathcal{H}}^2}}{\sqrt{2\pi}} \exp\left(\frac{(y_i - f(\mathbf{x}_i) - b)^2}{2\sigma_i^2\|f\|_{\mathcal{H}}^2} \right) \right]. \tag{10}$$

To solve $\min_{f \in \mathcal{H}} \Phi[f]$, following a similar path as in the Sect. 3.2, we arrive at the following convex objective function

$$\mathcal{J}_{\mathcal{H}}(\boldsymbol{\alpha}, b) = \frac{1}{2}\lambda\boldsymbol{\alpha}^\top K\boldsymbol{\alpha} + \sum_{i=1}^{l} c_i \left[\frac{y_i - \mathbf{K}_i \cdot \boldsymbol{\alpha} - b}{2} \left(\text{erf}\left(\frac{y_i - \mathbf{K}_i \cdot \boldsymbol{\alpha} - b}{\sqrt{2\sigma_i^2\boldsymbol{\alpha}^\top K\boldsymbol{\alpha}}} \right) + y_i \right) \right.$$
$$\left. + \frac{\sqrt{\sigma_i^2\boldsymbol{\alpha}^\top K\boldsymbol{\alpha}}}{\sqrt{2\pi}} \exp\left(-\frac{(y_i - \mathbf{K}_i \cdot \boldsymbol{\alpha} - b)^2}{2\sigma_i^2\boldsymbol{\alpha}^\top K\boldsymbol{\alpha}} \right) \right], \tag{11}$$

which we again minimize using L-BFGS. The (global) optimal solution $(\boldsymbol{\alpha}, b)$ determines the decision function given in the form of (5). The new extension of KSVM-iGSU obtained in this way is hereafter referred to as a Relevance Degree KSVM-iGSU (RD-KSVM-iGSU).

Furthermore, following the approach presented in [28], we solely assign a single relevance degree $c \in (0, 1]$ only to related samples, keeping the relevance degrees for the rest of the training set equal to 1. The above training parameter needs to be optimized, using a cross-validation procedure.

4   Experiments and Results

4.1   Dataset and Evaluation Measures

The proposed algorithms are applied in the problem of video event detection and are tested on a subset of the large-scale video dataset of the TRECVID Multimedia Event Detection (MED) 2014 benchmarking activity [22]. Similarly to [27], we use only the training portion of the TRECVID MED 2014 task dataset, which provides ground-truth information for 30 complex event classes, since for the corresponding evaluation set of the original TRECVID task there is no ground-truth data available. Hereafter, we refer to the aforementioned ground-truth-annotated dataset as MED14 and we divide it into a training subset, consisting of 50 positive and 25 related (near-miss) samples per event class, together with 2496 background samples (i.e. videos that are negative examples for all the event classes), and an evaluation subset consisting of approximately 50 positive and 25 related samples per event class, along with another 2496 background samples.

For assessing the detection performance of each trained event detector, the average precision (AP) [23] measure is utilized, while for measuring the detection performance of a classifier across all the event classes we use the mean average precision (MAP), as is typically the case in the video event detection literature, e.g. [8, 22, 28].

4.2 Video Representation and Uncertainty

For video representation, 2 keyframes per second are extracted at regular time intervals from each video. Each keyframe is represented using the last hidden layer of a pre-trained Deep Convolutional Neural Network (DCNN). More specifically, a 16-layer pre-trained deep ConvNet network provided in [26] is used. This network had been trained on the ImageNet data [6], providing scores for 1000 ImageNet concepts; thus, each keyframe has a 1000-element vector representation. Then, the typical procedure followed in state of the art event detection systems includes the computation of a video-level representation for each video by taking the average of the corresponding keyframe-level representations [2, 5, 11, 31].

In contrast to the existing event detection literature, in the case of RD-SVM-iGSU (or also KSVM-iGSU and the original LSVM-iGSU), the aforementioned keyframe-level video representations can be seen as observations of the input Gaussian distributions that describe the training videos. That is, let $\mathcal{X}$ be a set of l annotated random vectors representing the aforementioned video-level model vectors. We assume that each random vector is distributed normally; i.e., for the random vector representing the i-th video, $\mathbf{X}_i$, we have $\mathbf{X}_i \sim \mathcal{N}(\mathbf{x}_i, \Sigma_i)$. Also, for each random vector $\mathbf{X}_i$, a number, N_i, of observations, $\{\mathbf{x}_i^t \in \mathbb{R}^n : t = 1, \ldots, N_i\}$ is available; these are the keyframe-level model vectors that have been computed. Then, the mean vector and the covariance matrix of $\mathbf{X}_i$ are computed respectively as follows

$$\mathbf{x}_i = \frac{1}{N_i} \sum_{t=1}^{N_i} \mathbf{x}_i^t, \quad \Sigma_i = \sum_{t=1}^{N_i} (\mathbf{x}_i^t - \mathbf{x}_i)(\mathbf{x}_i^t - \mathbf{x}_i)^\top. \tag{12}$$

Now, due to the assumption for isotropic covariance matrices, we approximate the above covariance matrices as multiples of the identity matrix, i.e. $\widehat{\Sigma_i} = \sigma_i^2 I_n$ by minimizing the squared Frobenious norm of the difference $\Sigma_i - \widehat{\Sigma_i}$ with respect to σ_i^2. It can be shown (by using simple matrix algebra [10]) that for this it suffices to set σ_i^2 equal to the mean value of the elements of the main diagonal of Σ_i.

4.3   Experimental Results and Discussion

The proposed kernel extensions of LSVM-iGSU [27] (KSVM-iGSU, RD-KSVM-iGSU) are tested on the MED14 dataset, and compared to standard kernel SVM (KSVM), LSVM-iGSU [27] and RD-KSVM [28]. We note here that for the problem of video event detection (and especially when only a few positive training samples are available), kernel SVM is the state-of-the-art approach [2,5], while, when also a few related samples are available, RD-KSVM leads to state-of-the-art detection performance [28]. We experimented on the problem of learning from 10 positive examples per each event class, together with 5 related samples, that are drawn from the set of 25 related samples provided for each event class following the method presented in [28]; i.e., the 5 nearest to the median of all 25 related samples were kept for training both RD-KSVM and RD-SVM-iGSU. Also, we randomly chose 70 negative samples for each event class, while we repeated each experiment 10 times. That is, for each different experimental scenario, the obtained performance of each classifier (KSVM, RD-KSVM, LSVM-iGSU, KSVM-iGSU, and RD-SVM-iGSU) is averaged over 10 iterations, for each of which 10 positive samples have been randomly selected from the pool of 50 positive samples that are available in our training dataset for each target event class.

For all the above experimental scenarios where a kernel classifier is used, the radial basis function (RBF) kernel has been used. Training parameters (C for LSVM-iGSU; C, γ for KSVM, KSVM-iGSU; and C, γ, and c for RD-KSVM, RD-KSVM-iGSU) are obtained via cross-validation. For C, γ, a 10-fold cross-validation procedure (grid search) is performed with C, γ being searched in the range $\{2^{-16}, 2^{-15}, \ldots, 2^2, 2^3\}$. For c, an approach similar to that presented in [28] is followed. That is, related samples are initially treated as true positive and true negative ones (in two separate cross-validation processes) and C, γ are optimized as described above; then, by examining the minimum cross-validation errors of the two above processes, we automatically choose whether to treat the related samples as weighted positive or weighted negative ones, and also fix the value of C to the corresponding optimal value. Using this C, we proceed with a new cross-validation process (again grid search) for finding the optimal γ, c pair (where c is searched in the range $[0.01, 1.00]$ with a step of 0.05).

Table 1 shows the performance of the proposed KSVM-iGSU and RD-KSVM-iGSU, compared to LSVM-iGSU [27], the standard KSVM, and the RD-KSVM [28], respectively, in terms of average precision (AP), for each target event, and mean AP (MAP), across all target events. Bold-faced values indicate the best performance for each event class. We can see that LSVM-iGSU, whose improved performance over the standard linear SVM was studied extensively in [27], cannot outperform the kernel methods that are typically used for the video event detection problem, achieving a MAP of 0.1761. Without using any related samples, KSVM-iGSU that takes into account the input uncertainty, outperformed the standard kernel SVM for 25 out of 30 target event classes, achieving a MAP of 0.2527 in comparison to KSVM's 0.2128 (achieving a relative boost of 18.75 %). Moreover, when related samples were used for training, the proposed RD-KSVM-iGSU

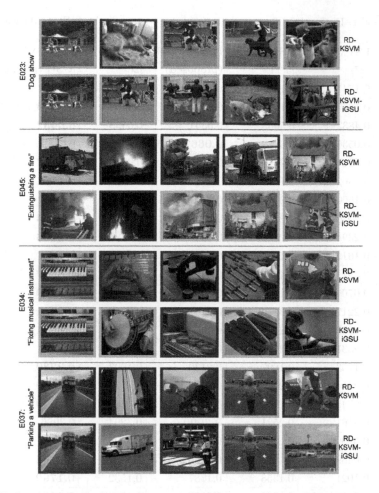

Fig. 1. Indicative results (top-5 returned shots) for comparing RD-KSVM-iGSU with RD-KSVM, for four event classes.

outperformed the baseline RD-KSVM for 27 out of 30 target event classes, achieving a MAP of 0.2730, in comparison to RD-KSVM's 0.2218 (i.e. a relative boost of 23.08 %). This RD-KSVM-iGSU result also represents a 8 % relative improvement (MAP of 0.2730 versus 0.2527) in comparison to KSVM-iGSU, which does not take advantage of related video samples during training. The above results suggest that using uncertainty for training video event detectors leads to promising results, while the additional exploitation of related samples can further improve event detection performance.

Finally, in Fig. 1 we present indicative results of the proposed RD-KSVM-iGSU in comparison with the baseline RD-KSVM [28] for four event classes, showing the top-5 videos each classifier retrieved. Green borders around frames indicate correct detection results, while red ones indicate false detection. These

Table 1. Evaluation of event detection approaches on the MED14 dataset.

Event class	LSVM-iGSU [27]	KSVM (e.g. [5,11])	KSVM-iGSU (proposed)	RD-KSVM [28]	RD-KSVM-iGSU (proposed)
E021	0.1741	0.1763	0.1923	0.1823	**0.2167**
E022	0.1847	0.1903	0.2495	0.2009	**0.2604**
E023	0.4832	0.5665	0.6361	0.5435	**0.6432**
E024	0.0536	0.0482	**0.0667**	0.0489	0.0549
E025	0.0117	0.0210	0.0257	0.0200	**0.0287**
E026	0.1002	0.1388	0.1530	0.1385	**0.1701**
E027	0.1600	0.2882	**0.4162**	0.2899	0.4002
E028	0.2030	0.2234	0.2338	0.2250	**0.2495**
E029	0.2394	0.2321	0.2948	0.2521	**0.3106**
E030	0.1612	**0.2464**	0.2220	0.2398	0.2451
E031	0.4911	0.4595	0.6122	0.4762	**0.6497**
E032	0.0706	0.1278	0.1490	0.1301	**0.1729**
E033	0.2217	0.3170	0.3731	0.3265	**0.3971**
E034	0.1658	0.2129	0.3302	0.2231	**0.6541**
E035	0.2331	0.2650	0.3580	0.2874	**0.3771**
E036	0.1753	0.1897	0.2139	0.1923	**0.2230**
E037	0.2454	0.2928	0.3325	0.3133	**0.3569**
E038	0.0745	0.1127	0.1231	0.1187	**0.1259**
E039	0.2161	0.2531	**0.3990**	0.3294	0.3986
E040	0.5809	**0.3205**	0.3157	0.3095	0.3021
E041	0.0489	0.1589	0.2166	0.1782	**0.2254**
E042	0.1021	0.1358	0.1787	0.1532	**0.1799**
E043	0.0967	0.1568	0.2037	0.1890	**0.2101**
E044	0.0732	**0.2697**	0.2087	0.2543	0.1968
E045	0.1307	0.2315	0.2517	0.2385	**0.2786**
E046	0.1952	0.2457	0.2668	0.2412	**0.2721**
E047	0.0531	0.0837	0.1796	0.1187	**0.1865**
E048	0.0672	0.0642	0.0672	0.0654	**0.0674**
E049	0.0641	0.1250	0.1245	0.1189	**0.1329**
E050	0.2076	0.2321	0.1867	**0.2489**	0.2039
MAP	0.1761	0.2128	0.2527	0.2218	**0.2730**

indicative results illustrate the practical importance of the AP and MAP differences between these two methods that are observed in Table 1.

5   Conclusions and Future Work

Two extensions of LSVM-iGSU, which is a linear classifier that takes input uncertainty into consideration, were proposed in this paper. The first one (KSVM-iGSU) results in non-linear decision boundaries, while the second one (RD-KSVM-iGSU), which is proposed especially for the problem of video event detection, exploits related class observations. The applicability of the aforementioned methods was verified using the TRECVID MED 2014 dataset, where solely a limited number of positive and related samples were used during training.

In the future, we plan to extend KSVM-iGSU such that the uncertainty of the input data is taken into consideration anisotropically. Also, we plan to exploit related samples in a more elaborate way; for instance, by clustering them into subclasses and assigning a different relevance degree to each subclass.

Acknowledgment. This work was supported by the European Commission under contract FP7-600826 ForgetIT.

References

1. Bhattacharyya, C., Pannagadatta, K., Smola, A.J.: A second order cone programming formulation for classifying missing data. In: Neural Information Processing Systems (NIPS), pp. 153–160 (2005)
2. Bolles, R., Burns, B., Herson, J., et al.: The 2014 SESAME multimedia event detection and recounting system. In: Proceedings of the TRECVID Workshop (2014)
3. Broyden, C.G.: The convergence of a class of double-rank minimization algorithms 1. general considerations. IMA J. Appl. Math. **6**(1), 76–90 (1970)
4. Chang, C.C., Lin, C.J.: LIBSVM: a library for support vector machines. ACM Trans. Intell. Syst. Technol. **2**, 27:1–27:27 (2011). http://www.csie.ntu.edu.tw/cjlin/libsvm
5. Cheng, H., Liu, J., Chakraborty, I., Chen, G., Liu, Q., Elhoseiny, M., Gan, G., Divakaran, A., Sawhney, H., Allan, J., Foley, J., Shah, M., Dehghan, A., Witbrock, M., Curtis, J.: SRI-Sarnoff AURORA system at TRECVID 2014 multimedia event detection and recounting. In: Proceedings of the TRECVID Workshop (2014)
6. Deng, J., Dong, W., Socher, R., Li, L.J., Li, K., Fei-Fei, L.: Imagenet: a large-scale hierarchical image database. In: IEEE Conference on Computer Vision and Pattern Recognition CVPR 2009, pp. 248–255. IEEE (2009)
7. Douze, M., Oneata, D., Paulin, M., Leray, C., Chesneau, N., Potapov, D., Verbeek, J., Alahari, K., Harchaoui, Z., Lamel, L., Gauvain, J.L., Schmidt, C.A., Schmid, C.: The INRIA-LIM-VocR and AXES submissions to TRECVID 2014 multimedia event detection (2014)
8. Gkalelis, N., Markatopoulou, F., Moumtzidou, A., Galanopoulos, D., Avgerinakis, K., Pittaras, N., Vrochidis, S., Mezaris, V., Kompatsiaris, I., Patras, I.: ITI-CERTH participation to TRECVID 2014. In: Proceedings of the TRECVID Workshop (2014)
9. Gkalelis, N., Mezaris, V.: Video event detection using generalized subclass discriminant analysis and linear support vector machines. In: Proceedings of International Conference on Multimedia Retrieval, p. 25. ACM (2014)

10. Golub, G.H., Van Loan, C.F.: Matrix Comput., vol. 3. JHU Press, Baltimore (2012)
11. Guangnan, Y., Dong, L., Shih-Fu, C., Ruslan, S., Vlad, M., Larry, D., Abhinav, G., Ismail, H., Sadiye, G., Ashutosh, M.: BBN VISER TRECVID 2014 multimedia event detection and multimedia event recounting systems. In: Proceedings of the TRECVID Workshop (2014)
12. Habibian, A., van de Sande, K.E., Snoek, C.G.: Recommendations for video event recognition using concept vocabularies. In: Proceedings of the 3rd ACM Conference on International Conference on Multimedia Retrieval, pp. 89–96. ACM (2013)
13. Habibian, A., Mensink, T., Snoek, C.G.: Videostory: A new multimedia embedding for few-example recognition and translation of events. In: Proceedings of the ACM International Conference on Multimedia, pp. 17–26. ACM (2014)
14. Jiang, L., Meng, D., Mitamura, T., Hauptmann, A.G.: Easy samples first: self-paced reranking for zero-example multimedia search. In: Proceedings of the ACM International Conference on Multimedia, pp. 547–556. ACM (2014)
15. Jiang, L., Yu, S.I., Meng, D., Mitamura, T., Hauptmann, A.G.: Bridging the ultimate semantic gap: a semantic search engine for internet videos. In: ACM International Conference on Multimedia Retrieval (2015)
16. Jiang, Y.G., Bhattacharya, S., Chang, S.F., Shah, M.: High-level event recognition in unconstrained videos. Int. J. Multimedia Inf. Retrieval 2(2), 73–101 (2013)
17. Kimeldorf, G., Wahba, G.: Some results on Tchebycheffian spline functions. J. Math. Anal. Appl. 33(1), 82–95 (1971)
18. Lanckriet, G.R., Ghaoui, L.E., Bhattacharyya, C., Jordan, M.I.: A robust minimax approach to classification. J. Mach. Learn. Res. 3, 555–582 (2003)
19. Liang, Z., Inoue, N., Shinoda, K.: Event Detection by Velocity Pyramid. In: Gurrin, C., Hopfgartner, F., Hurst, W., Johansen, H., Lee, H., O'Connor, N. (eds.) MMM 2014, Part I. LNCS, vol. 8325, pp. 353–364. Springer, Heidelberg (2014)
20. Liu, D.C., Nocedal, J.: On the limited memory BFGS method for large scale optimization. Mathematical prog. 45(1–3), 503–528 (1989)
21. Mazloom, M., Habibian, A., Liu, D., Snoek, C.G., Chang, S.F.: Encoding concept prototypes for video event detection and summarization (2015)
22. Over, P., Awad, G., Michel, M., Fiscus, J., Sanders, G., Kraaij, W., Smeaton, A.F., Quenot, G.: An overview of the goals, tasks, data, evaluation mechanisms and metrics. In: Proceedings of the TRECVID 2014. NIST, USA (2014)
23. Robertson, S.: A new interpretation of average precision. In: Proceedings of the 31st Annual International ACM SIGIR Conference on Research and Development in Information Retrieval, pp. 689–690. ACM (2008)
24. Schölkopf, B., Herbrich, R., Smola, A.J.: A generalized representer theorem. In: Helmbold, D.P., Williamson, B. (eds.) COLT 2001 and EuroCOLT 2001. LNCS (LNAI), vol. 2111, pp. 416–426. Springer, Heidelberg (2001)
25. Shivaswamy, P.K., Bhattacharyya, C., Smola, A.J.: Second order cone programming approaches for handling missing and uncertain data. J. Mach. Learn. Res. 7, 1283–1314 (2006)
26. Simonyan, K., Zisserman, A.: Very deep convolutional networks for large-scale image recognition (2014). arXiv preprint arXiv:1409.1556
27. Tzelepis, C., Mezaris, V., Patras, I.: Linear maximum margin classifier for learning from uncertain data (2015). arXiv preprint arXiv:1504.03892
28. Tzelepis, C., Gkalelis, N., Mezaris, V., Kompatsiaris, I.: Improving event detection using related videos and relevance degree support vector machines. In: Proceedings of the 21st ACM International Conference on Multimedia, pp. 673–676. ACM (2013)

29. Xu, H., Caramanis, C., Mannor, S.: Robustness and regularization of support vector machines. J. Mach. Learn. Res. **10**, 1485–1510 (2009)
30. Xu, H., Mannor, S.: Robustness and generalization. Mach. Learn. **86**(3), 391–423 (2012)
31. Yu, S.I., Jiang, L., Mao, Z., Chang, X., Du, X., Gan, C., Lan, Z., Xu, Z., Li, X., Cai, Y., et al.: Informedia at TRECVID 2014 MED and MER. In: NIST TRECVID Video Retrieval Evaluation Workshop (2014)

Video Content Representation Using Recurring Regions Detection

Lukas Diem[1]([✉]) and Maia Zaharieva[1,2]

[1] Multimedia Information Systems Group, University of Vienna, Vienna, Austria
diem@cs.univie.ac.at
[2] Interactive Media Systems Group, Vienna University of Technology,
Vienna, Austria

Abstract. In this work we present an approach for video content representation based on the detection of recurring visual elements or regions. We hypothesize that such elements play a potentially central role in the underlying video sequence. The approach makes use of fundamental intrinsic properties of a video and, thus, it does not make any assumptions about the video content itself. Furthermore, our approach does not require for any training or prior knowledge about the general settings and video domain. Preliminary experiments with a small and heterogeneous dataset of web videos demonstrate the potential of the approach to be employed as a compact summary of the video content with focus on its central visual elements. Additionally, resulting representations enable the retrieval of video sequences sharing common visual elements.

Keywords: Video content-based analysis · Recurring regions · Video representation

1 Introduction

Video content representation plays a crucial role for consumers in assessing its relevance to the personal interests and needs. Automated generation of content representation is a non-trivial task. The core challenge considers the selection of central and relevant aspects of the underlying content while preserving simplicity and low redundancy.

The assessment of relevance and significance is a high-level task that usually requires for additional knowledge about the content or its settings. For example, face detection can help to identify the main characters in a movie and to provide a compact overview of the playing actors. However, such an overview only presents a single aspect of the video, which does not allow for the assessment of the content itself. In a general setting, where there is no any prior knowledge about the explored video, a common way of content representation is by means of keyframes. The selection of keyframes is usually based on uniform sampling or on some visual criteria, e.g. large motion differences between consecutive frames often indicate substantial content change. However, this may still result in a large

© Springer International Publishing Switzerland 2016
Q. Tian et al. (Eds.): MMM 2016, Part I, LNCS 9516, pp. 16–28, 2016.
DOI: 10.1007/978-3-319-27671-7_2

amount of both irrelevant and redundant information for the video consumers that is not directly interpretable.

In this work we make use of a very fundamental characteristic of human perception. Humans tend to memorize things they are repeatedly confronted with. We, therefore, hypothesize, that elements recurring throughout a video sequence bear potential importance for the video content. Such elements can be characters or objects playing a central role. We propose an approach for video content representation based on the detection of recurring regions. The proposed approach makes use of the intrinsic properties of a video in terms of visual and motion coherence between consecutive frames. Therefore, we define a region as a visually and/or motion coherent element and, this, detected regions can represent a character (usually the main actors), an object, or a part of it. As a result, the approach does not require for any training in order to detect potential objects. Furthermore, it does not make any assumptions about the general settings and video domain (e.g. that objects of interest should be moving). The resulting video representation can be employed as a compact and interpretable video overview with the focus on central recurring elements. Furthermore, detected regions can be seen as an analogy to detected terms in a conventional text document. Thus, recurring regions enable novel, higher level similarity measure (e.g. in terms of term frequency - inverse document frequency) and retrieval approaches such as the search for videos sharing common visual elements.

In previous work, we transformed recurring element detection into the problem of matching and grouping local image descriptors [24]. In order to reduce complexity, we recently presented an approach that employs region segmentation and tracking [4]. In this paper, we extend our previous work on recurring region detection and introduce improvements of the underlying tracking and matching system.

This paper is organized as follows. Section 2 outlines related work in the context of video representations. We describe the proposed approach for the detection of recurring regions in a video sequence in Sect. 3. Section 4 presents the performed quantitative and qualitative analysis. Section 5 concludes the paper and provides an outlook for future work.

2   Related Work

Related work on video representations can be categorized into *visual-based*, *feature-based*, and *object-based* approaches. *Visual-based* summarizations provide an overview of the video content for the end-users. A commonly employed approach considers the use of key frames, e.g. by means of clustering all frames and selecting one or more frames per cluster for the final visual summary [1,15,21]. The number of key frames is a crucial parameter in this context since it directly influences the quality of the representation. Additionally, such visual summaries may contain several key frames depicting the same recurring elements in different environments that could be combined in a single representation. Another approach for video summarization visualizes a video as a graph. For example,

Zhang et al. group video shots into scenes and build a graph that models the scene structure of the underlying video [26]. A drawback of this representation is that a user does not get a visual summarization of the content (such as key frames) but only a set of nodes representing the detected video scenes. In contrast, Huang et al. depict a whole video collection as a graph of key frames and model the relationships between video shots [11]. Thus, the user is able to browse through the video collection and to retrieve videos containing similar shots.

Feature-based video representations are commonly employed for tasks such as near duplicate detection [13,14], video retrieval, and activity recognition [20,22]. The goal of a feature-based representation is to abstract the video content for further analysis rather than for visual summarization. Usually, frame- and shot-based features are aggregated to get a more compact video representation, e.g. by means of bag-of-words sampling and feature hashing [13,14,22]. Due to their nature extracted features are not directly interpretable by non-experts.

Object-based approaches aim at the detection and tracking of objects of interest within a video. This category can be further divided into supervised and unsupervised methods. Supervised object detection is commonly employed when the object of interest is known in advance, e.g. for surveillance tasks. In contrast, video segmentation methods usually have little assumptions on object categories. For example, several approaches employ long range point trajectories in combination with different grouping strategies (e.g. motion-based grouping) to segment video objects [6,17,18]. These methods report state-of-the-art results on segmentation benchmarks such as the VSB100 [8] and or FBMS [17]. A common limitation of such approaches is the assumption that objects of interest are moving within the video. Other methods employ a segmentation as graph-based grouping by building spatio-temporal graph representations [7,10,23]. Such approaches are often computationally complex since the graphs for a video representation easily become demanding in terms of space and computational power. Some recent approaches employ multiple segmentation proposals per frame that are temporally linked by similarity measures within a video sequence [2,5,12]. In general, the selection of the right hierarchy layer is highly depending to the video content and, thus, not feasible in a generic application.

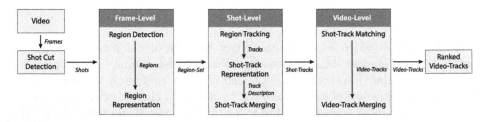

Fig. 1. Overview of the proposed approach for recurring region detection.

3 Recurring Region Detection

Figure 1 outlines the proposed approach for the detection of recurring regions in a video sequence. We start with the detection of shot boundaries employing the method presented by Zeppelzauer et al. [25]. Within each shot, we detect and track regions using simple and efficient image-based techniques. Resulting shot-tracks are matched across the video in order to identify regions that recur repeatedly throughout the whole video sequence. The resulting video-tracks are sorted by their visibility score and represent the final set of detected recurring regions.

3.1 Region Detection and Representation

We employ Statistical Region Merging (SRM) [16] to detect regions in each frame of the video. Core advantage of SRM is that it results in near object level segmentation. We describe each region by a color histogram with 72 bins (12 for achromatic colors, 12 for hue, and 5 for saturation) extracted from the HSV color space [4,19]. Additionally, since very small regions are difficult to interpret and track over time, we merge regions with an area below a predefined threshold with the nearest neighboring region in terms of visual similarity.

3.2 Tracking, Shot-Track Representation, and Merging

In order to track detected regions within each shot, we first initialize a *shot-track pool* with the region representations of the first frame in a shot. Following, we process all subsequent frames of the shot iteratively. For each region of a frame we consider all shot-tracks as possible assignments which are (1) within a predefined radius r of the current region position and (2) visually similar by means of the χ^2-distance between the corresponding regions' descriptors, d_{χ^2}. For all regions within the radius and below the maximum descriptor distance we compute a similarity score that penalizes distant assignments:

$$S_m = (1 - d_{\chi^2}) \cdot f(\frac{d_c}{r}, \mu, \sigma), \tag{1}$$

where f is a Gaussian weighting function, d_c the distance between the regions' centroids, r the predefined radius, and μ and σ the parameters of the Gaussian distribution. The resulting list of potential assignments between previously tracked regions and the regions of the current frame is processed in decreasing order according to the achieved similarity score. Processed regions and shot-tracks are removed from the list to avoid multiple assignments. Eventually, for all regions, which are not assigned to an existing shot-track, we initialize a new shot-track and proceed with the next frame until all frames are processed.

Tracked regions (shot-tracks) are represented using the Color and Edge Directivity Descriptor (CEDD) [3]. Additionally, we employ the medoid region (i.e. the region with the minimum distance to all other regions within the track) as a representative for the shot track. The representative region reduces the complexity of the following steps and is further employed for visualization purposes.

Objects within a shot are often represented by several shot-tracks due to occlusions and color variations within the object (e.g. a head is commonly split into a hair region and a face region). To account for such splits we introduce two refinement steps. First, we merge shot-tracks with similar medoid regions if the underlying tracked regions are immediate neighbors in the corresponding frames. Second, we additionally merge neighboring tracks if they carry similar motion patterns. Since the motion vectors defined by the region centroids of a shot-track are partly strongly jittering due to variations of the segmentation, we smooth the motion trajectories using an optimized version of a penalized least squares regression for discrete data [9]. To avoid merging stationary tracks we estimate the average motion per shot-track and allow for merges with a significant motion in comparison to the average motion of all shot-tracks of a shot. The average shot-track motion is estimated by

$$\bar{m} = \frac{1}{N-1} \sum_{t=1}^{N-1} \|c_{t+1} - c_t\|_2, \tag{2}$$

where N is the number of region centroids, $\|.\|_2$ is the Euclidean distance, and c_t is the smoothed region centroid at time t. For all neighboring shot-tracks, we compute the motion distance as the median of the (normalized) cosine distance between the motion vectors of consecutive frames and we merge shot-tracks with a distances below a predefined threshold. As a result, tracks that belong to the same rigid object but with a different visual appearance are merged.

3.3   Shot-Track Matching and Video-Track Merging

Recurring elements of a video usually appear in several shots. Therefore, we match detected shot-tracks across different video shots. The matching approach is split into two steps. First, we match shot-tracks similarly to tracking regions. We start by setting all shot-tracks of the first shot as initial video-tracks. For the following shot, we compute the χ^2-distance between the medoid region of the video-tracks and the current shot-tracks and accept a match if the distance is below a predefined matching threshold t_m. Accepted matches are optionally refined to reduce the number of false positives. The refinement step considers the compactness of a shot-track before and after the merging step. We measure compactness in terms of average distance between all region descriptions of a shot-track. If the compactness changes significantly, i.e. the difference exceeds a predefined threshold t_c, the match is disregarded as it results in a visually inhomogeneous shot-track. If a shot-track is assigned to an existing video-track, the representative region of the track is updated otherwise a new video-track is initialized and we proceed with the next shot.

The final refinement step merges video-tracks that are visually similar and that fulfill one of the following requirements: (1) regions of the video-tracks are

Fig. 2. An overview of the employed dataset (each video is represented by a keyframe).

neighbors in all frames in which the video-tracks are visible, or (2) the video-tracks are never visible in a frame together. These requirements prevent merges of video-tracks that are spatially discontinuous.

Eventually, the final visibility score for each video-track corresponds to the fraction of time the video-track is visible (tracked) in the video sequence (e.g. a score of 0.75 means that a recurring element is visible in 75 % of all frames).

4 Evaluation

In this section we present the results of quantitative and qualitative experiments performed to evaluate the proposed approach.

4.1 Dataset

We employ the same dataset as presented in [4]. It consists of 12 YouTube videos with a duration between 1 and 6 min and an average duration of approx. 3 min ($\sigma = 1.28$). Each video has 53 shots on average ($\sigma = 29.8$) and the shots consist of 107 frames on average ($\sigma = 44.3$). The genres of the videos are strongly varying, e.g. advertisement, music, sport (see Fig. 2). Additionally, all videos contain both static and moving cameras and objects.

4.2 Quantitative Evaluation

We first investigate the length of time a region is successfully traced within a shot, which corresponds to the visibility score of shot-tracks. Figure 3a visualizes the distribution of the visibility scores of the shot-tracks for all videos of the employed dataset. The distribution shows two major peaks, one for regions that are only visible in a few frames (visibility score below 0.05) and one for regions that are traced in nearly all frames of a shot (visibility score above 0.95). Overall, about 44 % of the regions are traced in less than 10 % of the frames. Such short shot-tracks often emerge from segmentation inaccuracies (e.g. region splits), or they represent visual elements (and potentially objects) which are only visible for a short period of time during the shot (e.g. due to camera or object motion). Additionally, about 33 % of the regions are traced and, thus, visible in more than 40 % of the shot. We assume that such longer lasting regions are the primary candidates for long-term recurring regions that can be employed to represent the content of the underlying video sequence. Therefore, we explore the traceability

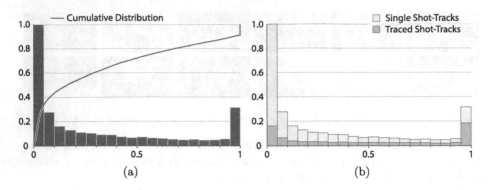

Fig. 3. (a) Distribution of the visibility scores of detected shot-tracks for all videos. (b) Distribution of the visibility scores of detected shot-tracks with respect to their traceability across different shots.

of shot-tracks with respect to their visibility score. Figure 3b shows the results of this evaluation. Long shot-tracks are commonly traced across several shots. This implies that visual elements that last in time in one shot will most probably reappear in the video sequence. On the opposite, short shot-tracks tend to remain unmatched throughout the underlying video sequence. Thus, such short shot-tracks can be considered for removal in order to improve the overall efficiency of the approach.

Next, we evaluate the influence of the proposed video-track merging step on the visibility scores of the complete video-tracks. As discussed in Sect. 3.3, we merge two visually similar video-tracks if either the underlying regions are neighbors in all frames they coincide or the underlying regions temporally complement each other. Figure 4a shows the distribution of the visibility scores of video-tracks with and without considering the merging step. The introduction of the step results in a merge of 21 % of all video-tracks. As a result, the visibility scores of the merged video-tracks increases, which is indicated by the shift of their distribution to the right. Overall, approximately 95 % of all video-tracks are visible in less than 10 % of the total video length. This confirms our assumption that there are only few central visual elements that recur throughout a complete video sequence. The quality of the detected regions in terms of content representation is discussed in Sect. 4.3.

Eventually, we compare the precision of tracing (matching) shot-tracks across different video shots with the approach presented in [4]. In this experiment, we match the shot-tracks of one shot with the shot-tracks of all other shots of the same video. All resulting matches are manually evaluated as true positives and false positives. Figure 4b shows the results for both approaches and for different parameter settings of the optional refinement step considering the compactness of shot-tracks: $t_c = \{0.02, 0.06, \text{and } No\ Refinement\}$. Note that, the parameter $t_c = 0.06$ is not evaluated by the previous approach. The average precision

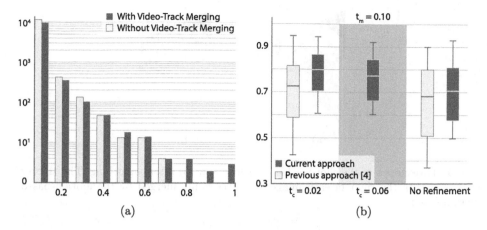

(a)   (b)

Fig. 4. (a) Distribution of the visibility scores of the video-tracks for all videos. (b) Precision of the matching performance of shot-tracks for different parameter settings.

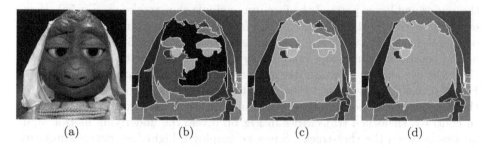

(a)   (b)   (c)   (d)

Fig. 5. Example shot-tracks: (a) (a detail of an) input frame, (b) tracked regions, (c) shot-tracks after merging by similarity, and (d) shot-tracks after merging by motion. Each shot-track is visualized by a pseudo-color (Color figure online).

increases by more than 7 % for the best parameter settings ($t_m = 0.10, t_c = 0.02$) as a result of the improved region tracking and representation.

4.3   Qualitative Evaluation

We fix the matching parameters to $t_m = 0.10$ and $t_c = 0.06$ for all experiments. All examples are based on a video that shows the comedian Sascha Grammel talking to his turtle puppet Josie about marriage. The setting of the video is a gala night and the comedian is performing in front of the audience on a stage. The shots of the video show the puppet and the comedian from different perspectives as well as the audience.

We first analyze the quality of the proposed steps for merging shot-tracks (merging by similarity and merging by motion) and their impact on the final shot-tracks. Figure 5 shows detected regions in a frame and their alteration as a results of the merging steps. Originally, the face of the turtle is represented by

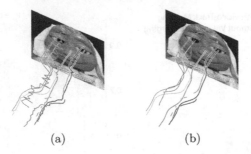

(a)   (b)

Fig. 6. Trajectories of the shot-tracks (a) before and (b) after smoothing. The thicker trajectories on the right in each image are successfully merged by the motion-based merging step.

several regions (and in following by several shot-tracks) due to the underlying over-segmentation. The first merging step, *merging by similarity*, is able to link the upper and lower parts of the face into a single shot-track (see Fig. 5c). However, the eyes and the top part of the face fail to merge since the descriptors of these shot-tracks achieve a low similarity score. Figure 5d shows the effect of the second merging step, *merging by motion*. The shot-tracks representing the right eye and eyebrow are merged into the shot-track representing the face. Figure 6 visualizes the trajectories of the tracked regions. Figure 6a demonstrates the necessity for smoothing trajectories in order to account for jittering regions. In contrast, smoothed trajectories in Fig. 6b reveal the underlying motion similarities between the shot-tracks. Since we employ a highly restrictive similarity estimation between trajectories to account for the heterogeneity of the underlying data, only a subset of the potential tracks is merged (cp. Fig. 5d).

Figure 7 shows the top 10 detected recurring regions for the investigated video sequence. The regions are sorted by the corresponding visibility score. This top 10 list demonstrates different aspects of our approach. Most of the top regions (8 out of 10) are part of the recurring regions identified by a human observer: the turtle puppet and the comedian. The regions, that belong to the turtle, are: 1 - left hand, 2 - turtle's face, 3 - wedding dress, 5 - right hand, and 7 - necklace. Regions, that belong to the comedian, are: 4 - head, 6 - shirt, and 10 - detail of the hair. The regions 8 and 9 show background elements. Since we only rank the regions according to their visibility, we cannot avoid that there are background regions within the top set of video-tracks. To remove such tracks from the results, additional information about the objects of interest is needed, which contradicts to the unsupervised methodology. Additionally, recurring background parts can be of interest during video analysis, for example when searching for videos sharing the same environment. We only visualize the top 10 regions since this limited amount of regions is already meaningful for the investigated video. However, a dynamic threshold can be employed to estimate the final number of regions from the visibility scores.

| 1 / 0.83 | 2 / 0.83 | 3 / 0.80 | 4 / 0.78 | 5 / 0.65 |

| 6 / 0.60 | 7 / 0.59 | 8 / 0.56 | 9 / 0.54 | 10 / 0.49 |

Fig. 7. Top 10 detected recurring regions for the *Sascha Grammel* video. Labels below the regions indicate the corresponding rank/visibility score.

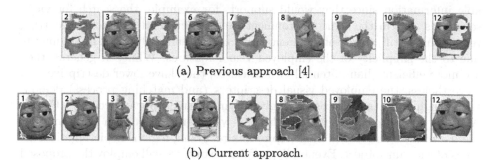

(a) Previous approach [4].

(b) Current approach.

Fig. 8. Representative regions of all shot-tracks of the turtles' face. The shot number is noted in the left upper corner. Orange (thin) and green (thick) borders indicate missed and successfully linked regions respectively (Color figure online).

Eventually, we investigate the shot-tracks that represent the face of the turtle, since this region lead to problems in the previous approach. Despite significant visual differences, we are able to match the turtle's face in all shots it is present. Figure 8 compares the results of both approaches. The green (thick) borders mark representative regions of the shot-tracks that are successfully matched into the same video-track. The representative regions of shots 2, 5, and 12 better represent the face due to the improved shot-track merging steps. The shot-tracks 7 and 9 of the current approach are part of a separate video track before the final video-track merging step. In this example the additional merging step worked as intended and significantly improved the result.

5 Conclusion

In this paper we presented a generic approach for the detection of recurring regions in a video sequence. This work is motivated by the fact that humans usually memorize elements that are temporally persistent. As a result, recurrence in a video sequence can be seen as an indicator for importance. The result of our approach is a ranked list of detected recurring regions, which represents the content of the video sequence in a compact way. In addition to the visual representation, detected regions can be further employed for the retrieval of videos sharing common visual elements.

Performed experiments demonstrate both the potentials and the limitations of the proposed approach. *First*, achieved results show that detected recurring regions effectively capture visual elements that mostly play a central role for the underlying video sequence. However, a few recurring regions tend to capture background elements if the video sequence is shot in a single environment. The main reason is the aim of the approach to account for an arbitrary video data. As a result, both stationary and moving elements may capture a potential object of interest. However, the differentiation between a stationary object and the background is not feasible without any additional assumptions or a prior knowledge about the video settings. Nevertheless, background elements may still hold valuable information since they would support, for example, the search for video sequences shot in the same environment. *Second*, currently, detected recurring regions may still represent different parts of the same object. This is mainly due to the simplicity of the employed visual features. While, in general, they are more efficient than potential local features, they have lower descriptiveness. Nevertheless, the employed visual descriptors (and matching process) demonstrate high robustness across strongly varying visual appearances (cp. Fig. 8). A potential approach to improve the final visual representation is to consider the use of local features as final step to link recurring regions depicting different parts of the same object. Eventually, in future work, we will employ the proposed video content representation for the retrieval of videos sharing common visual elements. A core challenge in this context is the acquisition and annotation of a dataset of publicly available videos.

Acknowledgment. This work has been partly funded by the Vienna Science and Technology Fund (WWTF) through project ICT12-010.

References

1. de Avila, S.E.F., Lopes, A.P.B., da Luz, A., de Albuquerque Araújo, A.: VSUMM: a mechanism designed to produce static video summaries and a novel evaluation method. Pattern Recognit. Lett. **32**(1), 56–68 (2011)
2. Banica, D., Agape, A., Ion, A., Sminchisescu, C.: Video object segmentation by salient segment chain composition. In: IEEE International Conference on Computer Vision Workshops, pp. 283–290 (2013)
3. Chatzichristofis, S.A., Boutalis, Y.S.: CEDD: color and edge directivity descriptor: a compact descriptor for image indexing and retrieval. In: Gasteratos, A., Vincze, M., Tsotsos, J.K. (eds.) ICVS 2008. LNCS, vol. 5008, pp. 312–322. Springer, Heidelberg (2008)

4. Diem, L., Zaharieva, M.: Interpretable video representation. In: International Workshop on Content-based Multimedia Indexing, pp. 1–6 (2015)
5. Fragkiadaki, K., Arbelaez, P., Felsen, P., Malik, J.: Learning to segment moving objects in videos. In: IEEE Conference on Computer Vision and Pattern Recognition (2015)
6. Fragkiadaki, K., Zhang, G., Shi, J.: Video segmentation by tracing discontinuities in a trajectory embedding. In: IEEE Conference on Computer Vision and Pattern Recognition, pp. 1846–1853 (2012)
7. Galasso, F., Keuper, M., Brox, T., Schiele, B.: Spectral graph reduction for efficient image and streaming video segmentation. In: IEEE Conference on Computer Vision and Pattern Recognition, pp. 49–56 (2014)
8. Galasso, F., Nagaraja, N.S., Cardenas, T.J., Brox, T., Schiele, B.: A unified video segmentation benchmark: annotation, metrics and analysis. In: IEEE International Conference on Computer Vision, pp. 3527–3534 (2013)
9. Garcia, D.: Robust smoothing of gridded data in one and higher dimensions with missing values. Comput. Stat. Data Anal. **54**(4), 1167–1178 (2010)
10. Grundmann, M., Kwatra, V., Han, M., Essa, I.A.: Efficient hierarchical graph-based video segmentation. In: IEEE Conference on Computer Vision and Pattern Recognition (2010)
11. Huang, H., Liu, H., Zhang, L.: Videoweb: space-time aware presentation of a video-clip collection. IEEE J. Emerg. Sel. Top. Circ. Syst. **4**(1), 142–152 (2014)
12. Li, F., Kim, T., Humayun, A., Tsai, D., Rehg, J.M.: Video segmentation by tracking many figure-ground segments. In: IEEE International Conference on Computer Vision, pp. 2192–2199 (2013)
13. Liu, D., Yu, Z.: A computationally efficient algorithm for large scale near-duplicate video detection. In: He, X., Luo, S., Tao, D., Xu, C., Yang, J., Hasan, M.A. (eds.) MMM 2015, Part II. LNCS, vol. 8936, pp. 481–490. Springer, Heidelberg (2015)
14. Liu, J., Huang, Z., Cai, H., Shen, H.T., Ngo, C., Wang, W.: Near-duplicate video retrieval: current research and future trends. ACM Comput. Surv. **45**(4), 44:1–44:23 (2013)
15. Mahmoud, K.M., Ghanem, N.M., Ismail, M.A.: Unsupervised video summarization via dynamic modeling-based hierarchical clustering. Int. Conf. Mach. Learn. Appl. **2**, 303–308 (2013)
16. Nock, R., Nielsen, F.: Statistical region merging. IEEE Trans. Pattern Anal. Mach. Intell. **26**(11), 1452–1458 (2004)
17. Ochs, P., Malik, J., Brox, T.: Segmentation of moving objects by long term video analysis. IEEE Trans. Pattern Anal. Mach. Intell. **36**(6), 1187–1200 (2014)
18. Ommer, B., Mader, T., Buhmann, J.M.: Seeing the objects behind the dots: recognition in videos from a moving camera. Int. J. Comp. Vis. **83**(1), 57–71 (2009)
19. Phan, R., Chia, J., Androutsos, D.: Unconstrained logo and trademark retrieval in general color image databases using color edge gradient co-occurrence histograms. IEEE Int. Conf. Acoust. Speech, Sign. Proces. **114**(1), 1221–1224 (2008)
20. Sadanand, S., Corso, J.J.: Action bank: a high-level representation of activity in video. In: IEEE Conference on Computer Vision and Pattern Recognition, pp. 1234–1241 (2012)
21. Truong, B.T., Venkatesh, S.: Video abstraction: a systematic review and classification. ACM Trans. Multimedia Comput. Commun. Appl. **3**(1), 1–37 (2007)
22. Wang, H., Kläser, A., Schmid, C., Liu, C.: Dense trajectories and motion boundary descriptors for action recognition. Int. J. Computer Vision **103**(1), 60–79 (2013)

23. Xu, C., Xiong, C., Corso, J.J.: Streaming hierarchical video segmentation. In: Fitzgibbon, A., Lazebnik, S., Perona, P., Sato, Y., Schmid, C. (eds.) ECCV 2012, Part VI. LNCS, vol. 7577, pp. 626–639. Springer, Heidelberg (2012)
24. Zaharieva, M., Breiteneder, C.: Recurring element detection in movies. In: Schoeffmann, K., Merialdo, B., Hauptmann, A.G., Ngo, C.-W., Andreopoulos, Y., Breiteneder, C. (eds.) MMM 2012. LNCS, vol. 7131, pp. 222–232. Springer, Heidelberg (2012)
25. Zeppelzauer, M., Mitrovic, D., Breiteneder, C.: Analysis of historical artistic documentaries. In: International Workshop on Image Analysis for Multimedia Interactive Services, pp. 201–206 (2008)
26. Zhang, L., Xu, Q., Nie, L., Huang, H.: Videograph: A non-linear video representation for efficient exploration. Vis. Comput. **30**(10), 1123–1132 (2014)

Group Feature Selection for Audio-Based Video Genre Classification

Gerhard Sageder[1,2]([✉]), Maia Zaharieva[1,2]([✉]), and Christian Breiteneder[2]

[1] Multimedia Information Systems Group, University of Vienna, Vienna, Austria
gerhard.sageder@unvie.ac.at, maia.zaharieva@tuwien.ac.at
[2] Interactive Media Systems Group, Vienna University of Technology,
Vienna, Austria

Abstract. The performance of video genre classification approaches strongly depends on the selected feature set. Feature selection requires for expert knowledge and is commonly driven by the underlying data, investigated video genres, and previous experience in related application scenarios. An alteration of the genres of interest results in reconsideration of the employed features by an expert. In this work, we introduce an unsupervised method for the selection of features that efficiently represent the underlying data. Performed experiments in the context of audio-based video genre classification demonstrate the outstanding performance of the proposed approach and its robustness across different video datasets and genres.

Keywords: Genre classification · Group feature selection · Audio features

1 Introduction

Video genres commonly represent a first coarse categorization of large media collections. Although partly subjective, such a categorization enables end users to efficiently access and retrieve media of potential interest. As a result, automated video genre classification is subject to active research in the context of e.g. sport events [24], TV programs [5,10,13], web videos [5,21,23]. The selection of an appropriate feature set is crucial for any approach for video genre classification. In general, feature selection strongly depends on both the underlying data and on the application scenario (genres). As a result, an alteration of the genres of interest leads to a reconsideration of the employed features. This process requires for the intervention of an expert in order to assess potential feature candidates. We facilitate and support this process by proposing a generic and unsupervised approach for the automated selection of features that best represent the underlying data. In a next step, we employ the selected features for training a classifier and performing video genre classification.

Existing approaches for video genre classification usually consider the combination of different modalities, such as visual, acoustic and textual information, in order to differentiate between video genres [5,8,13,15]. For thorough

© Springer International Publishing Switzerland 2016
Q. Tian et al. (Eds.): MMM 2016, Part I, LNCS 9516, pp. 29–41, 2016.
DOI: 10.1007/978-3-319-27671-7_3

reviews on current research, please refer to [2,18]. In this work, we focus on audio features only. Recently, audio features demonstrate competitive performance to multimodal approaches [10]. Furthermore, audio features are less computational expensive and allow for the efficient analysis of large media collections. Existing audio-based approaches for video genre classification commonly use well-established features from the temporal and frequency domains, e.g. mel filter cepstral coefficients (MFCC) [6,16], wavelet coefficients [4], short-term cepstral analysis by means of MFCC, perceptual linear prediction (PLP) and Rasta-PLP [17], acoustic topic models [10], background acoustic features [20]. Employed audio-based features often originate from the task of speech/non-speech discrimination [14].

Feature selection in the context of audio-based genre classification is usually based on previous successful experiments in similar application scenarios [14,18]. The major drawback of such a strategy is the use of prior information about the audio content, such as the existence of various content elements (e.g. speech, music). The inflexibility and resource-demanding calculations of features, that may not even contribute to a classification performance are current core limitations intended to overcome by the proposed method.

This work is organized as follows. Section 2 outlines our approach for feature selection, which we employ in order to automatically select those audio features that best describe a given dataset. Section 3 presents the evaluation setup for the performed experiments including employed features, datasets, classifiers, and performance metrics. Section 4 discusses the experimental results in detail. Eventually, Sect. 5 concludes the paper.

2   Feature Selection

We propose an unsupervised group feature selection approach, which makes use of canonical correlation analysis (CCA) in order to identify low-correlated and, thus, complementary and relevant features that efficiently describe the underlying data. In general, CCA is a multivariate regression method that measures the relationship between multidimensional variables in a linear manner [7]. CCA calculates correlations between features of different dimensions and provides canonical correlation coefficients for all pairs of features. We employ the term *group features* to emphasize the fact that we select multidimensional features as a whole in contrast to conventional feature selection methods, which typically ignore existing groupings of feature components.

Figure 1 illustrates the main components and basic steps of our feature selection workflow. First, we calculate the canonical correlations for all feature combinations of the input data (feature matrix). The result of this step is a symmetrical canonical correlation matrix (CCM). We employ the pairwise correlation coefficients as a measure of redundancy. Feature pairs with high correlation coefficients are considered redundant, whereas low-correlated features provide complementary and, therefore, additional information. We prune the CCM and reduce it to the upper triangular matrix. Additionally, we remove correlations that exceed a

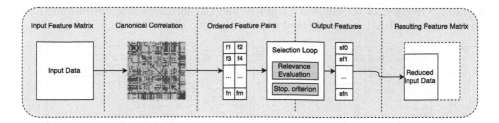

Fig. 1. Feature selection workflow.

certain threshold. The purpose of this threshold is to remove highly-correlated features since they are considered too redundant and, thus, non-expressive for the underlying data. The remaining pairs of features are sorted in decreasing order according to their correlation coefficients. This constitutes an initial feature ranking that is iteratively and sequentially processed. In every iteration, a candidate feature pair is evaluated whether or not to be included in the target feature set based on an internal relevance measure. In our implementation, this relevance criterion is designed in a flexible, modular manner and thus can be exchanged easily. Depending on the actual relevance evaluation, the feature selection process can terminate autonomously in every iteration. One possible relevance criterion is the entropy-based information gain (IG), which has been applied in [19]. In this work, we apply CCA directly as relevance measure. Therefore, we calculate the canonical correlation between the current candidate feature pair and the already constructed feature set. We measure the significance of a correlation following the principle of significant modes [9]. If the correlation is significant, it is considered to provide additional, descriptive, and low-redundant information to the target feature set and is added to the previously selected feature set. Eventually, an additional, optional stopping criterion can be employed to terminate the feature selection process if, for example, a feature set of a certain size is desired. In this work, we do not employ such stopping criterion but investigate all feature pairs to autonomously identify the optimal set of features for the given data.

In our experiments, we split the input data into 10 % training set and 90 % test set. The feature selection process is applied on the training data only.

3   Evaluation Setup

3.1   Audio Features

We conduct our experiments using a set of 50 high-dimensional audio features that consists of 679 feature components in total (see Table 1). This selection incorporates representative and comprehensive audio features from the temporal and frequency domains, that cover various audio aspects, such as harmonics, beat and rhythm, pitch, timbre, and loudness. For more details on the audio features please refer to [11].

Table 1. Overview of the employed features and the corresponding dimensions (D). The features are listed in alphabetical order.

	Feature	Feature Name	D		Feature	Feature Name	D
1	AD	Amplitude Descriptor	40	26	M7_LAT	MPEG-7 Log Attack Time	1
2	BFCC	Bark-scale Frequency Cepstral Coeff.	40	27	M7_SC	MPEG-7 Spectral Centroid	1
3	BTHI	Beat Histogram	7	28	MFCC	Mel-scale Frequency Cepstral Coeff.	40
4	CRMA	Chroma CENS Features	24	29	PLP	Perceptual Linear Prediction	38
5	E4Hz	4 Hz Modulation Energy	2	30	PTCH	Pitch	2
6	GPD	Group Delay	40	31	PTCT	Pitch Contour	2
7	HMDV	Harmonic Derivate	16	32	PTVB	Pitch Vibration	1
8	HZCR	High Zero Crossing Rate	1	33	R_ZC	Range of Zero Crossing Rate	1
9	LPC	Linear Predictive Coding	40	34	RMS	Root Mean Square	2
10	LPCC	Linear Prediction Cepstral Coefficients	40	35	ROFF	Spectral Rolloff	2
11	LPZC	Linear Prediction ZCR	2	36	RPLP	Raster PLP	38
12	LSP	Line Spectral Pairs	40	37	RYPT	Rhythm Patterns	20
13	M7_AFF	MPEG-7 Audio Fundamental Frequency	4	38	SBER	Subband Energy Ratio	10
14	M7_AH	MPEG-7 Audio Harmonicity	4	39	SF	Spectral Flux	2
15	M7_AP	MPEG-7 Audio Power	2	40	SONE	Loudness	40
16	M7_ASB	MPEG-7 Audio Spectrum Basis	72	41	SPCR	Spectral Crest	8
17	M7_ASC	MPEG-7 Audio Spectrum Centroid	2	42	SPCT	Spectral Center	2
18	M7_ASF	MPEG-7 Audio Spectrum Flatness	34	43	SPDI	Spectral Dispersion	2
19	M7_ASP	MPEG-7 Audio Spectrum Projection	16	44	SPEY	Spectral Entropy	8
20	M7_ASS	MPEG-7 Audio Spectrum Spread	2	45	SPPS	Spectral Peak Structure	2
21	M7_AW	MPEG-7 Audio Waveform	4	46	SPRE	Spectral Renyi Entropy	8
22	M7_HSC	MPEG-7 Harmonic Spectral Centroid	1	47	SPSL	Spectral Slope	8
23	M7_HSD	MPEG-7 Harmonic Spectral Deviation	1	48	STE	Short Time Energy	2
24	M7_HSS	MPEG-7 Harmonic Spectral Spread	1	49	VDR	Volume Dynamic Range	1
25	M7_HSV	MPEG-7 Harmonic Spectral Variation	1	50	ZCR	Zero Crossing Rate	2
						Total dimensionality	679

3.2 Datasets

We investigate two video datasets in our experiments: BBC documentaries and RAI TV broadcasts. Table 2 provides an overview of the employed datasets and their characteristics.

The **BBC documentaries** dataset is a self-collected set of videos from the BBC's YouTube channel[1]. It covers three sub-genres: *technical, nature,* and *music*. Although the semantic focus of the three sub-genres is strongly varying, all videos in this set are composed, edited, and post-processed in a very similar way, at least from a technical point of view.

The **RAI TV broadcasts** dataset contains more than 100 hours of complete broadcasted programmes of RAI television [12, 13]. The data is divided into subsets of different sizes and with partly different genres, which can be investigated separately. For our experiments we employ two subsets. The first one, *RAI-6*, compromises 6 genres: *commercials, football, music, news, talk shows,* and *weather forecasts*. The second one, *RAI-3*, is a subset of RAI-6 and covers 3 genres: *commercials, football,* and *music*. In contrast to the BBC documentaries, the RAI broadcasts exhibit strongly varying structures and no explicit regularities among the different genres. As a result, this heterogeneous corpus corresponds to a conventional genre classification task, whereas the BBC documentaries allow for the investigation of a sub-genre classification scenario.

3.3 Data Preprocessing

Since the different datasets are available as different video container files, we first extract the audio tracks and convert them to PCM audio files. Next, the

---

[1] https://www.youtube.com/user/BBC/.

Table 2. Overview of the employed datasets.

Dataset	Videos	Total Duration	Classes	Segment Size	Total Samples
BBC	9	4.5 h	3	2 s	16,140
				10 s	3,225
				30 s	1,070
RAI-3	49	19.4 h	3	2 s	45,935
				10 s	9,067
				30 s	2,998
RAI-6	93	30.9 h	6	2 s	85,517
				10 s	17,041
				30 s	5,636

audio tracks are segmented into chunks of 2, 10, and 30 s. This subdivision is carried out with an overlap of 50 % in order to maintain acoustic information near the segmentation boundaries. Especially when considering small segments of the audio signals, passages of constant silence may appear. These segments do not have any expressiveness and may cause errors in the feature extraction process. Therefore, we perform silence detection by means of a noise threshold of $-60dB$ and remove detected silent segments from the dataset. This step has a low impact on the following analysis: none of the segments from the two RAI TV datasets and only 0.31 % of the 2 s segments from the BBC documentaries are identified as silence and removed.

3.4   Classification

We employ three, in the audio domain well-established classifiers: K-Nearest Neighbor (KNN) [3], Support Vector Machine (SVM) [22], and Random Forest (RF) [1]. The parameter settings for the different classifiers have been selected based on preliminary experiments with respect to classification performance. We employ KNN with $k = 2$ and the Euclidean distance as distance measure without any additional weighting. The SVM implementation uses a polynomial inhomogeneous kernel in order to support non-linear hyperplane separation: $K(x_i, x_j) = (x_i \cdot x_j + 1)^e$, with $e = 2$, a complexity factor of $c = 1$. RF as a tree-based classifier generates a forest of random trees having unlimited depth $mD = \infty$ and a maximum number of $nT = 10$ trees. Although many works employ RF with more trees (e.g. 500 trees by default in some implementations), we could not identify any significant increase in the classification performance while the runtime increased notably (about 50 times in our preliminary experiments). Therefore, we chose $nT = 10$ for all experiments in this work.

All classifications are randomly initialized, 10-fold cross-validated with respect to the underlying class distribution, and run 10 times independently.

3.5 Performance Metrics

We employ the *weighted F-score* to measure the accuracy of the performed classifications. The weighted F-score takes into account the varying class distribution of the datasets:

$$F_\beta^w = \frac{1}{n} \sum_{c \in C} F_\beta(c) \times n_c, \tag{1}$$

where n_c denotes the number of instances per class c, n the number of instances in total, and F_β the standard F-score:

$$F_\beta = (1 + \beta^2) \frac{precision \times recall}{\beta^2 \times precision + recall} \tag{2}$$

In addition to the quantitative performance evaluation, the selected groups of features are investigated in terms of *semantic expressiveness* for the underlying data and *robustness* of the feature selection method. We measure the robustness of the feature selection method by considering the occurrences of the selected features averaged over the 10 independent and randomly initialized runs.

4 Experimental Results

4.1 Classification Performance

In this experiment, we focus on the classification performance in terms of F1-score for the different segment sizes for all three datasets (BBC, RAI-3, and RAI-6) and the three classifiers (KNN, SVM, and RF). Table 3 summarizes the achieved results. All three datasets and all three segment sizes achieve an outstanding performance in terms of F1-score given the notable reduction of dimensionality of the selected feature set. For the BBC data and a segment size of 30 s, for example, only 21 features covering 6 % of the full feature set are selected achieving 95 % F1-score with the KNN classifier. For all three datasets, a decrease of the segment size tends to result in a feature set of higher dimensionality. The reason for this trend is that, in general, smaller segments bear more details that need to be described and, thus, they require for more precise features. On the opposite, larger segments tend to blur details (primarily due to feature averaging) and need fewer features for their representation.

The results show a notable difference between the performance across the different segment sizes for the different datasets. The BBC dataset performs better for smaller segments, e.g. F1-score of 99 % for segments of size 2 s vs. F1-score of 94 % for segments of size 30 s using the RF classifier. On the opposite, both RAI datasets perform slightly better for increasing segment sizes, e.g. F1-score of 97 % for segments of size 2 s vs. F1-score of 99 % for segments of size 30 s using again the RF classifier on the RAI-3 dataset. This inverse tendency is primarily due to the substantial difference in the nature of the underlying data resulting in different feature selections. While the BBC dataset is very homogeneous (all

Table 3. Performance results for the three datasets in terms of weighted F1-scores. N: number of selected features, D: dimensionality of the corresponding feature set. Classification of the full feature set is conducted using the best performing classifier for the corresponding dataset and segment size.

Dataset	Classifier	Segment size											
		30 s				10 s				2 s			
		N	D		F1	N	D		F1	N	D		F1
BBC	KNN	21	38	(6 %)	0.948	25	82	(12 %)	0.966	27	159	(23 %)	0.990
	SVM				0.906				0.984				0.996
	RF				0.940				0.971				0.993
	Full feature set	50	679	(100 %)	0.993	50	679	(100 %)	0.995	50	679	(100 %)	0.996
RAI-3	KNN	19	89	(13 %)	0.986	22	117	(17 %)	0.985	28	221	(33 %)	0.953
	SVM				0.995				0.982				0.976
	RF				0.990				0.958				0.974
	Full feature set	50	679	(100 %)	0.998	50	679	(100 %)	0.997	50	679	(100 %)	0.989
RAI-6	KNN	18	84	(12 %)	0.951	22	117	(17 %)	0.937	20	114	(17 %)	0.939
	SVM				0.969				0.972				0.954
	RF				0.993				0.991				0.975
	Full feature set	50	679	(100 %)	0.996	50	679	(100 %)	0.994	50	679	(100 %)	0.972
Average over all datasets	KNN	19	80	(12 %)	0.961	22	113	(16 %)	0.955	23	152	(23 %)	0.949
	SVM				0.970				0.976				0.965
	RF				0.986				0.979				0.977
	Full feature set	50	679	(100 %)	0.996	50	679	(100 %)	0.995	50	679	(100 %)	0.980

documentaries have similar structure and share common elements), the RAI data is distinctive to a certain degree (cp. discussion in the performed case study on full video classification). As a result, the BBC dataset requires for higher granulated segments in order to capture more descriptive information and, thus, to better distinguish across the different sub-genres of documentaries.

The average performance of the employed classifiers over all datasets and different segment sizes shows that the RF classifier achieves best performance over all datasets followed by SVM and lastly KNN.

Eventually, we compare our approach with related work reporting evaluation results on the RAI dataset (see Table 4). Please note, that our experiments are conducted on a subset of the dataset employed by the compared approaches and does not include the *cartoon* genre. The results indicate the outstanding performance of the selected features. The performance achieved on the employed data (F1-score of 99 %) demonstrate strong competitiveness to the top reported performance by Ekenel et al. [5]. In addition to some acoustic features, Ekenel et al. consider visual, structural, and cognitive features. The features are selected in a way to reflect the editor's process in TV production and cannot be applied for arbitrary data. In contrast, our approach autonomously selects the features that are relevant for the provided data set. The achieved performance demonstrates the quality of the selected features (see Sect. 4.2 for a detailed analysis) while at the same time the exploration of a single modality notably reduces the computational effort.

Table 4. Comparison with related works on the RAI dataset, used modalities (A=audio, V=video, S=structural, C=cognitive), number of genres, dataset size, and achieved accuracy in terms of F1-score.

Authors	Modalities	# Genres	Dataset size	F1
Montagnuolo et al. [12]	A, V, S, C	7	6,690 min	0.924
Montagnuolo et al. [13]	A, V, S, C	7	6,690 min	0.949
Ekenel et al. [5]	A, V, S, C	7	6,600 min	0.992
Ekenel et al. [5]	A	7	6,600 min	0.957
Kim et al. [10]	A	7	4,167 min	0.943
This work	A	6	1,850 min	**0.993**

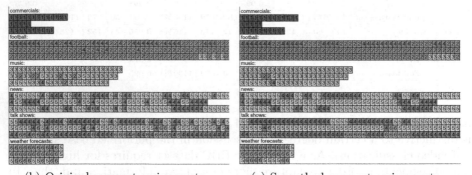

(a) Color legend for predicted genres.

(b) Original segment assignments.   (c) Smoothed segment assignments.

Fig. 2. Segment assignments of three randomly selected video sequences for each genre. Groupings in 2(b) and 2(c) correspond to the ground truth. Colors and numbers in the corresponding segments represent the predicted assignments. The timeline is cut to the right due to space limitations (Color figure online).

Case Study: Full Video Classification. In our first case study, we investigate the question: *Can we successfully detect the genre of a full video sequence based on the classification of its segments?* For this case study we employ the RAI-6 dataset, 30 s segments. For each genre, one of the videos is used for training and the remaining videos for testing. Figure 2 shows examples for segment assignments for videos of different genres. Since the underlying video segments have an overlap of 50 %, we smooth the predicted assignments using a sliding window of size 3 in order to remove single outliers. Figure 2 additionally indicates that while some genres such as *commercials*, *football*, and *music* can be clearly identified, segments of the remaining genres, *news*, *talk shows*, and *weather forecasts*, are often misclassified.

We employ majority voting as classification strategy for the assignment of a genre to a video sequence. Majority voting is a simple decision rule that selects

Table 5. Confusion matrix for the full video classification task. Rows correspond to the ground truth and columns to the predicted genre. Blank cells represent zero values.

	commercials	football	music	news	talk shows	weather
commercials	18 (100 %)					
football		17 (85 %)		2 (10 %)	1 (5 %)	
music			3 (100 %)			
news				3 (30 %)	7 (70 %)	
talk shows				1 (9 %)	10 (91 %)	
weather forecasts					6 (35 %)	11 (65 %)

the genre that is in the majority in the genre assignments of the underlying video segments. The overall classification performance of the full video sequences achieves a F1-score of 80 % , which is significantly lower than the classification accuracy of single audio segments. A crucial difference between the two experiments is the amount of available data (both training and test data) which significantly influences the quality of the underlying models. Table 5 shows the confusion matrix of the classification. Due to the low number of video sequences, a single misclassification has a notable influence on the overall classification rate. For example, one video sequence from *talk shows* has been misclassified as news. As a result, the retrieval performance for *talk shows* decreases to 91 %. Furthermore, *news, talk shows*, and *weather forecasts* bear similar audio characteristics. Hence, multiple video sequences from these genres are incorrectly assigned within this group (predominantly as *talk shows*, cp. Fig. 2).

In general, video durations vary strongly. For example, the RAI dataset consists of video sequences between 90 s and 53 min. The analysis of longer video sequences can easily become computationally expensive. Therefore, in a next experiment we investigate the question if it is feasible to classify a full video sequence based on the analysis of a small subsequence only. For this experiment, we select varying numbers of segments, starting with 3 segments and iteratively increasing the number of segments (step of 2) until the full video sequence is considered. Figure 3 compares the performance of segments originating from different parts of the underlying video sequences: from the beginning of a video, from the mid part, as well as a combination from the beginning, mid, and ending. Due to space limitations we only show the results for the first 30 segments, which correspond to a video subsequence of 10 min. The results demonstrate, that two genres, *commercials* and *football*, achieve an outstanding performance independently of the length of subsequences analyzed or the part of the video it is contained. On the opposite, *news, talk shows*, and *weather forecasts* perform poorly in general since the audio models are less discriminative in this context and the employment of visual features will definitely help to better distinguish between the three genres. Additionally, more training data can significantly improve the quality of the underlying audio models as proved by the experiments on single audio segments. Finally, *music* is well identified if either the full length of the

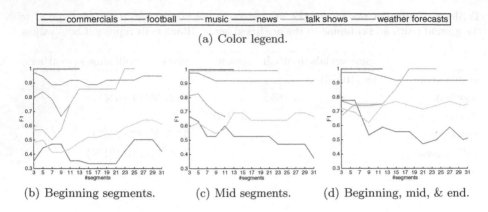

(a) Color legend.

(b) Beginning segments.   (c) Mid segments.   (d) Beginning, mid, & end.

Fig. 3. Performance of the full video classification task using a subset of segments taken from different video parts. 3(b): from the beginning of a video sequence; 3(c): from the mid part of a video sequence; and 3(d): from the beginning, middle, and the end part of a video sequence.

video or a subsequence from its mid part is analyzed. These results confirm the previous analyses in this case study. Genres, which are in general well recognized, require for the analysis of a small subsequence of the full video only. On the opposite, genres, which are commonly confused by the employed features, do not improve or worsen significantly with varying length of analyzed segments.

4.2   Feature Analysis

Core advantage of the proposed unsupervised feature selection approach is that it selects complete features independently of their dimensionality. Especially in the audio domain, where features usually carry a higher level of semantics as e.g. visual features, group features show an advantage in terms of interpreting the data. The purpose of this section is a brief discussion of the robustness of selected feature sets across different runs within a dataset and across the different datasets.

Figure 4(b) depicts the amount of intersection of feature sets (robustness) computed in different test runs for the different datasets and segment sizes. It can be seen that different test runs show very little differences in the computed feature sets. The overlapping of selected features sets between different datasets is illustrated in Fig. 4(c). Here, the amount of intersection is about 0.1 points lower for all combinations. One might argue that this indicates that the selected features are less depended on the datasets than expected or argued so far.

Therefore, in a small experiment we investigate the features actually selected for the RAI-3 data for a classification of the BBC dataset (30 s segments). While the original 21 BBC features achieve an F1-score of 95 % using the KNN classifier (see Table 3), the RAI-3 features obtain an F1-score of only 86 %. This notable drop in the performance stresses strengths of a high-quality feature selection

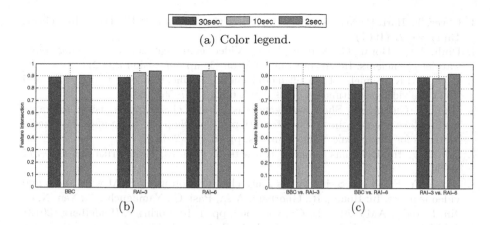

(a) Color legend.

(b)   (c)

Fig. 4. (a) Robustness of selected feature sets for the different datasets and segment sizes across 10 test runs. (b) Overlapping of selected feature sets between different datasets.

with respect to the provided data. Even if the percentage of overlap between datasets is rather high (about 85 %) it is the remaining data-specific feature groups that contribute to excellent results.

5   Conclusion

This paper addressed a core question in the context of video genre classification concerning the selection process of an appropriate feature set. We proposed a generic approach for feature selection, which does not make any assumptions about the underlying data or investigated video genres, but it autonomously selects a feature set that efficiently describes the data. Performed experiments demonstrated the outstanding performance of the approach for different datasets and video genres. The analysis of the selected features showed the robustness of the approach across different runs on the same data. Additionally, the analysis demonstrated the necessity of selecting different feature sets for varying datasets, which is a core argument why a generic feature selection process is required.

Acknowledgments. This work has been partly funded by the Vienna Science and Technology Fund (WWTF) through project ICT12-010. The authors are thankful to Marcus Hudec for pointing our interest towards CCA. The authors would also like to thank Maurizio Montagnuolo from RAI Centre for Research and Technological Innovation for providing the RAI TV dataset.

References

1. Breiman, L.: Random forests. Mach. Learn. **45**(1), 5–32 (2001)
2. Brezeale, D., Cook, D.: Automatic video classification: a survey of the literature. IEEE Trans. Syst. Man Cybern. **38**(3), 416–430 (2008)

3. Cover, T., Hart, P.: Nearest neighbor pattern classification. IEEE Trans. Inf. Theor. **13**(1), 21–27 (1967)
4. Dinh, P.Q., Dorai, C., Venkatesh, S.: Video genre categorization using audio wavelet coefficients. In: Asian Conference on Computer Vision (2002)
5. Ekenel, H.K., Semela, T.: Multimodal genre classification of TV programs and YouTube videos. Multimedia Tools Appl. **63**(2), 547–567 (2013)
6. Guo, J., Gurrin, C.: Short user-generated videos classification using accompanied audio categories. In: ACM International Workshop on Audio and Multimedia Methods for Large-scale Video Analysis, pp. 15–20 (2012)
7. Hotelling, H.: Relations between two sets of variates. Biometrika **28**, 321–377 (1936)
8. Huang, Y.-F., Wang, S.-H.: Movie genre classification using SVM with audio and video features. In: Huang, R., Ghorbani, A.A., Pasi, G., Yamaguchi, T., Yen, N.Y., Jin, B. (eds.) AMT 2012. LNCS, vol. 7669, pp. 1–10. Springer, Heidelberg (2012)
9. Jolliffe, I.: Principal Component Analysis. Springer Series in Statistics. Springer, Heidelberg (2002)
10. Kim, S., Georgiou, P., Narayanan, S.: On-line genre classification of TV programs using audio content. In: IEEE International Conference on Acoustics, Speech and Signal Processing, pp. 798–802 (2013)
11. Mitrovic, D., Zeppelzauer, M., Breiteneder, C.: Features for content-based audio retrieval. Adv. Comput.: Improving Web **78**, 71–150 (2010)
12. Montagnuolo, M., Messina, A.: TV genre classification using multimodal information and multilayer perceptrons. In: Basili, R., Pazienza, M.T. (eds.) AI*IA 2007. LNCS (LNAI), vol. 4733, pp. 730–741. Springer, Heidelberg (2007)
13. Montagnuolo, M., Messina, A.: Parallel neural networks for multimodal video genre classification. Multimedia Tools Appl. **41**(1), 125–159 (2009)
14. Natarajan, R., Chandrakala, S.: Audio-based event detection in videos - a comprehensive survey. Int. J. Eng. Technol. **6**(4), 1663–1674 (2014)
15. Roach, M., Mason, J., Xu, L.Q.: Video genre verification using both acoustic and visual modes. In: IEEE Workshop on Multimedia Signal Processing, pp. 157–160 (2002)
16. Roach, M., Mason, J.: Classification of video genre using audio. Eurospeech **4**, 2693–2696 (2001)
17. Rouvier, M., Linares, G., Matrouf, D.: On-the-fly video genre classification by combination of audio features. In: IEEE International Conference on Acoustics Speech and Signal Processing, pp. 45–48 (2010)
18. Rouvier, M., Oger, S., Linares, G., Matrouf, D., Merialdo, B., Li, Y.: Audio-based video genre identification. IEEE/ACM Audio, Speech, Lang. Process. **23**(6), 1031–1041 (2015)
19. Sageder, G., Zaharieva, M., Zeppelzauer, M.: Unsupervised selection of robust audio feature subsets. In: SIAM International Conference on Data Mining, pp. 686–694 (2014)
20. Saz, O., Doulaty, M., Hain, T.: Background-tracking acoustic features for genre identification of broadcast shows. In: IEEE Spoken Language Technology Workshop, pp. 118–123 (2014)
21. Song, Y., Zhao, M., Yagnik, J., Wu, X.: Taxonomic classification for web-based videos. In: IEEE Conference on Computer Vision and Pattern Recognition, pp. 871–878 (2010)
22. Vapnik, V.: Nature Statistical Learning Theory. Springer, Heidelberg (1995)

23. Wu, X., Zhao, W.L., Ngo, C.W.: Towards google challenge: Combining contextual and social information for web video categorization. In: ACM International Conference on Multimedia, pp. 1109–1110 (2009)
24. Zhang, N., Duan, L.Y., Li, L., Huang, Q., Du, J., Gao, W., Guan, L.: A generic approach for systematic analysis of sports videos. ACM Trans. Intell. Syst. Technol. 3(3), 46:1–46:29 (2012)

Computational Cartoonist: A Comic-Style Video Summarization System for Anime Films

Tsukasa Fukusato[1]([envelope]), Tatsunori Hirai[1], Shunya Kawamura[1], and Shigeo Morishima[2,3]

[1] Waseda University, Tokyo, Japan
tsukasa@moegi.waseda.jp
[2] Waseda Research Institute for Science and Engineering, Tokyo, Japan
[3] JST CREST, Tokyo, Japan
shigeo@waseda.jp

Abstract. This paper presents Computational Cartoonist, a comic-style anime summarization system that detects key frame and generates comic layout automatically. In contract to previous studies, we define evaluation criteria based on the correspondence between anime films and original comics to determine whether the result of comic-style summarization is relevant. To detect key frame detection for anime films, the proposed system segments the input video into a series of basic temporal units, and computes frame importance using image characteristics such as motion. Subsequently, comic-style layouts are decided on the basis of pre-defined templates stored in a database. Several results demonstrate the efficiency of our key frame detection over previous methods by evaluating the matching accuracy between key frames and original comic panels.

Keywords: Comic generation · Shot clustering · Shot boundary detection

1 Introduction

Comic has grown to become one of the most efficient storytelling mediums across the world, with many cartoonists creating their own compositions and imagery [10]. Recently, cartoonists extend comic techniques to cartoon animation (or anime films), and this has resulted in the increasing general population in producing and viewing many anime films. However, the number of anime episodes is very large. When people choose their favorite anime contents from the large number of anime films, to read comics for understanding all episodes of the anime is more effective than to watch the films.

To facilitate intuitive access to video archives, the main challenge for video summarization and browsing systems is achieving a satisfactory balance between removing redundant sections and maintaining representative coverage of the video. Some works developed comic-style browsing tools, to display thumbnails

© Springer International Publishing Switzerland 2016
Q. Tian et al. (Eds.): MMM 2016, Part I, LNCS 9516, pp. 42–50, 2016.
DOI: 10.1007/978-3-319-27671-7_4

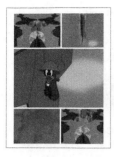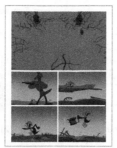

Fig. 1. Two comic-style summarizations generated by our method. Left: "Daffy The Commando"(1943, public domain). Right: "A Wild Hare"(1941, public domain).

appropriately indicating the contents of input video by computing scene importance. For example, comic-style video summarization methods based on image features or annotation features (e.g., subtitles or game play logs) have been proposed [1,12,14,16,19]. However, previous methods did not provide evaluation criteria to determine whether the key frames could be appropriately used in comic-style summarization. The domain-specific approach of cartoonists cannot be easily extended to generate comic-style summaries such as a film comic.

Our goal is to generate a comic-style video summary for anime films (referred to as comicalization) by computing the importance of shots and frames from anime sequence (Fig. 1). In this paper, to quantify the rule of key frame, we focus on Japanese anime films that are generally created by interpolating the panels of original comics. There is the correspondence between the keyframes of the anime film and the original comic panels. The main idea of our approach is to measure the matching accuracy between key frames and original comic panels. This system detects key frames based on image features and then automatically generate the comic layout. The proposed method's overall process is as follows:

1. Segment an input anime film into shots using image features.
2. Determine the cost function based on the correspondence between anime films and original comics.
3. Extract key frames of each shot using image features.
4. Compute the paneling score according to the importance of key frames and shots.
5. Generate the comic layout from a template layout stored in a database.

The summary should clearly present the meaningful content throughout the scenes of the anime film. The proposed method enables automatic and unsupervised key frame detection using various image features from an input anime film. By calculating the frame importance, we improve the appropriateness of detected key frames for comic-style summaries. Using the calculated importance values, the system automatically generates comic paneling from database of template layout. As the result, we can reduce the manual labor required for creating comics from anime films (Fig. 2).

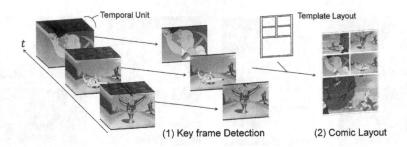

Fig. 2. System overview.

The remainder of this paper is organized as follows. Related works are reviewed in Sect. 2. We discuss the main ideas underlying the algorithms used in the proposed method in Sect. 3. Section 4 presents experimental results for some anime films, and we conclude this paper and discuss limitations and future work in Sect. 5.

2   Related Work

There are some researches of comic-style video summarization, which determine key frames in video sequence. Cho et al. [4] use Lifelog data on mobile devices, and Thawonmas [15] proposes comic summaries based on Gamelog, i.e., the playing data in an online game. Furthermore, Shamir's method [13] can be configured to accommodate user preferences. By contrast, these methods require recorded content information.

In image-based methods, Zhuang's [20] method provides an unsupervised clustering method based on hue-saturation (HS) color features (16×8). Calic's [2] method generates a comic-style summarization system based on HSV features and clustering. These methods allow users to visualize key frames in videos easily; however, it does not focus on shot importance.

Uchihashi et al. [18] propose Video Manga, which computes the shot importance (using the total length of each shot cluster) and layouts key frames based on a film comic format. However, they do not consider the frame importance for key frame detection. Kasamatsu et al. [7] propose a method to detect key frames using the central frames of video shots, and determine panel sizes using YCbCr color clustering. With these methods, it is possible to visualize video content with relative ease. However, their methods does not consider the temporal information in each shot. Hence, to generate more effective summaries, computing the frame and shot importance for key frame detection and comic layout is essential.

These researches cannot assign optimum key frames since they do not consider a large number of the scene structure in anime films. Therefore, we propose an evaluation criteria based on the correspondence between anime films and original comics.

3   Comic-Style Video Summarization

3.1   Shot Transition Detection

A shot is a series of interrelated consecutive pictures taken continuously by a single camera, representing a continuous motion in time and space. in particular, each anime shot (e.g., character motion) is created on the vasis of the composition of original comic panels. First, we investigate whether each shot in 20 Japanese anime films matches the original comic panel, and confirm that approximately 88.7 % of the anime shots include key frames matched with the original comic panels. Therefore, we assume that each anime shot has one key frame, and segment an anime film into basic temporal units (shots). Kasamatsu [7] uses shot segmentation based on the sum of absolute differences (SAD) between two consecutive frames. However, the shot transitions of the anime films include both abrupt transitions between two consecutive frames and gradual transitions, e.g., white/black fades and various camera techniques. Therefore, to determine abrupt shot transitions, we use the edge of orientation histograms [8] and Lian's temporal segmentation method [9]. Frame differences based on pixel differences, RGB color histograms, and a block-matching motion estimation algorithm are performed. These measurements are robust against camera operation (e.g., zoom in/out, pan, and tilt) and object motions. In this study, the minimum shot sequence length is defined as ten frames.

We have verified the results of our shot segmentation with an accuracy evaluation using precision and recall rate. In the experiments, 20 Japanese anime films are used. The mean accuracy of the shot segmentation are $Precision = 92.34\,\%$, $Recall = 86.09\,\%$. These results show that this approach enables the classification of abrupt transitions, white-fades, black-fades, pans, and zooms in anime films.

3.2   Key Frame Detection

In this section, we describe a method to detect key frames in anime films. Ideally, key frames, which are matched with original comic panels, should capture the semantics of a shot. However, current techniques in Computer Vision techniques are not advanced enough to automatically generate such key frames. Instead, we have to rely on low level visual features, such as color, motion of the object in a shot. Therefore, we detect the key frames based on (1) color transitions, e.g., black and white reversal, (2) characters' motion, and (3) frame composition.

In the color features, Zhuang [20] proposed an unsupervised clustering method based on HS color features. The frame closest to the cluster center is detected as the representative key frame for a given shot. However, capturing consecutive and similar key frames is problematic, because a substantial number of shots are reused in anime films. Therefore, we focus on the cluster outliers such as cutaways and establishing shots. To determine the centroid features for HSV histograms in scene clusters, we perform the k-means clustering algorithm.

The cost function E_1 of the i th frame $(i = 1, \cdots, N)$ is defined as follows

$$E_1(i) = \alpha \cdot \left\{ 1.0 - \sum_k \beta_k \exp\left(\frac{-D_k(i)^{\mathrm{T}} \sum^{-1} D_k(i)}{2} \right) \right\} \tag{1}$$

$$\beta_k = \frac{\omega_k}{\sqrt{(2\pi)^k | \sum |}}$$

$$D_k(i) = H(i) - \hat{H}_k$$

where $\sum$ is a covariance matrix and ω_k is a weight value for the kth cluster. $E_1(i)$ is normalized to the maximum row height α, $H(i)$ is the HSV histogram of ith frame, and $\hat{H}_k$ is the HSV histogram's centroid of the kth cluster. As a result, the cluster outliers are presented as more important and attract user's notice compared to key frames concentrated around the cluster center. This grouping around the cluster centers is caused by common repetitions of similar content in the video, often adjacent in time.

The color-based term (Eq. (1)) does not place constraints on object motion. In the original comics, cartoonists draw the special motions of characters, such as punch motions. Then, we compute the moving object area, i.e., the number of grid flow vectors E_2 having large flow vectors, based on Farneback's [5] optical flow algorithm. The second cost function is expressed as follows:

$$E_2(i) = \iint_{(x,y) \in f} \phi_1(\boldsymbol{v}_f(i, x, y)) dx dy \tag{2}$$

$$\phi_1(\boldsymbol{v}_f(i, x, y)) = \begin{cases} 1.0 \ |\boldsymbol{v}_f| > threshold \\ 0.0 \ else \end{cases}$$

In this function, $E_2(i)$ is normalized to the maximum height E_{max} and $\boldsymbol{v}_f$ is a grid flow vector based on optical flow. However, this energy is not strong enough to control the frame comparison (motion of the object). Then we assume that the edge intensity of important frame is high. A regularization term based on the edge intensity, which is computed by Canny's edge detection, is added. The third cost function E_3 is defined as follows:

$$E_3(i) = \iint \phi_2(i, x, y) dx dy \tag{3}$$

$$\phi_2(i, x, y) = \begin{cases} 1.0 \ if (x, y) \in edge \\ 0.0 \ else \end{cases}$$

In this function, $E_3(i)$ is normalized to the maximum height E_{max}.

By integrating all the energy terms, the key frame detection of the jth shot is formulated as:

$$\max_{i \in j} \lambda_1 E_1(i) + \lambda_2 E_2(i) + \lambda_3 E_3(i) \tag{4}$$

Generally, the weights are set as $\lambda_1 = 2.0$ and $\lambda_2 = \lambda_3 = 1.0$ to balance the contribution of different terms.

We note that when our key frame detection is applied to anime films (total length of approximately 30 minutes), the number of key frames is typically lower than 100.

3.3 Comic Layout

We describe a method to generate comic-style layouts using key frame importance (as described in Sect. 3.2). This system is mainly inspired by Uchihashi's [17] Video Manga, Cao's [3] and Myodo's [11] layout method. In this paper, we design various layout templates, which contains one, five, and six panels per page. We assume that large panels are very important frames that help the reader understand the comic content, and define the displayed function F_j, which relates the panel size to the layout template (according to the paneling score) for paneling key frames.

To determine paneling size (represented by the paneling score), we use the key frame importance of each shot (Eq. (4)) and the shot importance. Given C clusters in an anime film, a measure of normalized weight W_k for kth cluster (as described in Sect. 3.2) is computed as

$$W_k = \frac{S_k}{\sum_{l=1}^{C} S_l} \tag{5}$$

where S_k is the total length of all shots in the kth cluster, computed by summing the length of all shots in the cluster. W_k is the proportion of shots from the entire anime film that are in the kth cluster. We assume that a shot is important if it is both long and unique, i.e., it does not resemble other shots. Thus, weighting the shot length with the inverse of the cluster's weight yields a measure of shot importance. The importance I of the jth shot is

$$I_j = \frac{L_j}{L_{max}} \log \frac{1}{W_k} \tag{6}$$

where L_j is the length of the jth shot and L_{max} is the maximum length of the kth cluster shot. By combining Eq. (4) and the shot importance (Eq. (6)), the paneling score for the key frames of the jth shot is defined as follows:

$$F_j = \lambda_4 I_j + \lambda_5 E_j \tag{7}$$

where the weights are set as $\lambda_4 = 0.6$ and $\lambda_5 = 0.4$ to balance the contribution of different terms.

To determine the page label associated with key frames, we utilize the Euclidean distance from the Eq. (7) to the paneling score of the layout template.

$$\min_{n} \sum_{j} |F_j - \Omega_n(j)|^2 \tag{8}$$

where $\Omega_n(j)$ is the jth paneling score of the nth template comic page (*large panel* = 3.0, *middle panel* = 2.0, *small panel* = 1.0).

In addition, we consider the number of comic panels for the summary. Realizing the level of detail control for users, Kasamatsu [7] reduces the displayed key frames by eliminating less important frames. This technique automatically manages the process, by comparing adjacent two key frames in time, and eliminating the frame with the lower importance score. However, this method does not focus on the key frame's positions in anime films, and there is a high probability that this method will lead to the deflection of key frames. Thus, to prevent key frame deflections, we divide the input anime film into four categories of comic-based composition: introduction, development, turn, and conclusion. In each category, we reduce the number of comic panels (displayed key frames) using Eq. (7).

4   Evaluation

We verify the effectiveness of our key frame detection method using an accuracy evaluation. To evaluate anime film summarization, we propose a criteria that measures the matching accuracy of key frames and original comic panels (precision rate, recall rate and F-measure). In the experiments, we use 20 different types of Japanese anime films, including martial arts, romantic comedy, mystery, and fantasy. These anime films are among the top 20 films in annual sales of DVD and comic book. The ground truths of key frames are manually labeled according to the original comic panels.

For comparison, we use Kasamatsu's [7] method, wherein key frames are the center frame in each shot sequence (SC), and the equal interval method, wherein the key frames are detected by equal time intervals in an anime film (EQL). The results of key frame detection are shown in Table 1. These results indicate that our key frame detection method provides higher accuracy than the previous methods. In contrast to the previous methods, our precision rate is higher than our recall rate. Consequently, false-positive detection of key frames is significantly reduced. Moreover, the recall rate for our method is higher than that of SC; this occurred because the SC method does not consider the frame importance of each shot. The results indicate that our method successfully improves the accuracy of key frame detection.

Table 1. Key frame detection result.

	Precision (%)	Recall (%)	F-measure (%)
Our method	83.47	80.17	81.17
SC	41.95	76.76	53.15
EQL	56.58	58.46	56.99

In addition, we apply our key frame detection to original anime films *'Spirited Away'* and *'Puella Magi Madoka Magica,'* which were released as movies before serialized in a comic magazine. These mean scores of key frame detection were $Precision = 77.62\%$, $Recall = 88.59\%$, and $F-measure = 82.74\%$. These results show that our method can create highly accurate summaries for any anime films without original comics. This evaluation measure represents how well the method generates video summaries in an aspect of completion of an original comic from the anime film.

5   Conclusions and Future Work

We have presented Computational Cartoonist, a video summarization system that generates key-frame-based video summaries from anime films. We define the correspondence between the anime's key frames and the original comic panels. In the future, we are planning to take high-level image features such as character faces to understand the specific character's movement in anime films. In addition, we append to add speech bubbles and sound effects (e.g., onomatopoeia) using acoustic features or physics parameters [6]. Furthermore, we are planning to help users review the information they need, and create richer comic summaries by using subtitle features.

Acknowledgements. This research is supported in part by OngaCREST, CREST, JST.and by Research Fellowship for Young Scientists of Japan Society for the Promotion of Science (JSPS).

References

1. Boreczky, J., Girgensohn, A., Golovchinsky, G., Uchihashi, S.: An interactive comic book presentation for exploring video. In: Proceedings of the SIGCHI Conference on Human Factors in Computing Systems, pp. 185–192. ACM (2000)
2. Calic, J., Gibson, D.P., Campbell, N.W.: Efficient layout of comic-like video summaries. IEEE Trans. Circ. Syst. Video Technol. **17**(7), 931–936 (2007)
3. Cao, Y., Chan, A.B., Lau, R.W.: Automatic stylistic manga layout. ACM Trans. Graph. (TOG) **31**(6), 141 (2012)
4. Cho, S.-B., Kim, K.-J., Hwang, K.-S.: Generating cartoon-style summary of daily life with multimedia mobile devices. In: Okuno, H.G., Ali, M. (eds.) IEA/AIE 2007. LNCS (LNAI), vol. 4570, pp. 135–144. Springer, Heidelberg (2007)
5. Farneback, G.: Very high accuracy velocity estimation using orientation tensors, parametric motion, and simultaneous segmentation of the motion field. In: Eigth IEEE International Conference on Computer Vision, ICCV 2001, Proceedings, vol. 1, pp. 171–177. IEEE (2001)
6. Fukusato, T., Morishima, S.: Automatic depiction of onomatopoeia in animation considering physical phenomena. In: Proceedings of the Seventh International Conference on Motion in Games, pp. 161–169. ACM (2014)
7. Kasamatsu, S., Itoh, T.: A browser for summarized multiple videos. In: Proceedings of the 8th NICOGRAPH International (2009)
8. Levi, K., Weiss, Y.: Learning object detection from a small number of examples: the importance of good features. In: Proceedings of the 2004 IEEE Computer Society Conference on Computer Vision and Pattern Recognition, CVPR 2004, vol. 2, pp. II-53. IEEE (2004)

9. Lian, S., Dong, Y., Wang, H.: Efficient temporal segmentation for sports programs with special cases. In: Qiu, G., Lam, K.M., Kiya, H., Xue, X.-Y., Kuo, C.-C.J., Lew, M.S. (eds.) PCM 2010, Part I. LNCS, vol. 6297, pp. 381–391. Springer, Heidelberg (2010)
10. McCloud, S.: Making Comics: Storytelling Secrets of Comics, Manga and Graphic Novels Author. William Morrow Paperbacks, New York (2006)
11. Myodo, E., Ueno, S., Takagi, K., Sakazawa, S.: Automatic comic-like image layout system preserving image order and important regions. In: Proceedings of the 19th ACM International Conference on Multimedia. pp. 795–796. ACM (2011)
12. Ryu, D.S., Park, S.H., Lee, J.w., Lee, D.H., Cho, H.G.: Cinetoon: A semi-automated system for rendering black/white comic books from video streams. In: IEEE 8th International Conference on Computer and Information Technology Workshops, CIT Workshops 2008, pp. 336–341. IEEE (2008)
13. Shamir, A., Rubinstein, M., Levinboim, T.: Generating comics from 3d interactive computer graphics. IEEE Comput. Graph. Appl. **26**(3), 53–61 (2006)
14. Taniguchi, Y., Akutsu, A., Tonomura, Y.: Panoramaexcerpts: extracting and packing panoramas for video browsing. In: Proceedings of the fifth ACM International Conference on Multimedia, pp. 427–436. ACM (1997)
15. Thawonmas, R., Shuda, T.: Comic layout for automatic comic generation from game log. In: Ciancarini, P., Nakatsu, R., Rauterberg, M., Roccetti, M. (eds.) New Frontiers for Entertainment Computing. IFIP International Federation for Information Processing, vol. 279, pp. 105–115. Springer, Heidelberg (2008)
16. Tobita, H.: Comic engine: interactive system for creating and browsing comic books with attention cuing. In: Proceedings of the International Conference on Advanced Visual Interfaces, pp. 281–288. ACM (2010)
17. Uchihashi, S., Foote, J.: Summarizing video using a shot importance measure and a frame-packing algorithm. In: Proceedings of IEEE International Conference on Acoustics, Speech, and Signal Processing, vol. 6, pp. 3041–3044. IEEE (1999)
18. Uchihashi, S., Foote, J., Girgensohn, A., Boreczky, J.: Video manga: generating semantically meaningful video summaries. In: Proceedings of the Seventh ACM International Conference on Multimedia (Part 1), pp. 383–392. ACM (1999)
19. Wang, M., Hong, R., Yuan, X.T., Yan, S., Chua, T.S.: Movie2comics: towards a lively video content presentation. IEEE Trans. Multimedia **14**(3), 858–870 (2012)
20. Zhuang, Y., Rui, Y., Huang, T.S., Mehrotra, S.: Adaptive key frame extraction using unsupervised clustering. In: 1998 International Conference on Image Processing, ICIP 1998, Proceedings, vol. 1, pp. 866–870. IEEE (1998)

Exploring the Long Tail of Social Media Tags

Svetlana Kordumova$^{(\boxtimes)}$, Jan van Gemert, and Cees G.M. Snoek

University of Amsterdam, Amsterdam, The Netherlands
{s.kordumova,j.c.vanGemert,cgmsnoek}@uva.nl

Abstract. There are millions of users who tag multimedia content, generating a large vocabulary of tags. Some tags are frequent, while other tags are rarely used following a long tail distribution. For frequent tags, most of the multimedia methods that aim to automatically understand audio-visual content, give excellent results. It is not clear, however, how these methods will perform on rare tags. In this paper we investigate what social tags constitute the long tail and how they perform on two multimedia retrieval scenarios, tag relevance and detector learning. We show common valuable tags within the long tail, and by augmenting them with semantic knowledge, the performance of tag relevance and detector learning improves substantially.

1 Introduction

In this paper we focus on the long tail frequency distribution of social tags. It is well known that tag frequencies in social media form a long tail distribution [17,24]. While some tags are frequent, like *snow, beach, coffee*, there is a large number of tags which are rare, with only few example images per tag, like *mierkat, tank suit, dyippy*, see Fig. 1. Current works note that many tags from the long tail are "misspelled" or "meaningless" words [24], or only useful for "exceptional cases" [17]. It seems this observation has been accepted in the community as such, and no further investigation has been performed so far. We believe that there are also meaningful tags within the long tail which have been overlooked. Since the tags from the long tail make up a significant portion of the data, they deserve more detailed analysis. On that account, we pose the question: *What tags constitute the long tail?*.

We believe that the long tail distribution should not be just accepted as such, but looked at as a challenge. The challenge is to augment the frequencies of rare tags, so that the long tail distribution will change its shape. By augmenting the rare tags, they become a richer source for many multimedia algorithms. Motivated from common approaches in the literature of enriching tags with semantic knowledge [3,26], we investigate the effect of augmenting the examples of rare tags with semantically related tagged images. We question *What happens when rare tags are augmented?*.

Social tags come for free and are a valuable resource for many multimedia methods that aim to automatically understand visual content. Examples include automatic concept detection [6,9,21], user profiling [4,14], sentiment analysis [1,23,25],

© Springer International Publishing Switzerland 2016
Q. Tian et al. (Eds.): MMM 2016, Part I, LNCS 9516, pp. 51–62, 2016.
DOI: 10.1007/978-3-319-27671-7_5

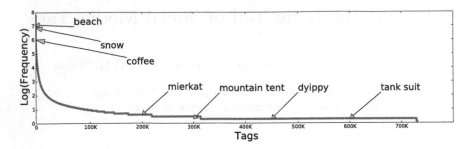

Fig. 1. The log-frequency of Flickr images for 730 K tags. Some tags like *beach*, *snow*, *coffee* have millions of example images. We investigate the majority of rare tags such as *miercat*, *mountain tent*, and *tanksuit*.

and assessing tag relevance [10, 20]. These works give excellent results when ample training examples are available per tag. It is not clear, however, how these methods will perform for rare tags with fewer examples. Since there are many tags which fall on the long tail, we argue that it is of great importance for the tags from the long tail to also be successful on these methods. Therefore, we question *What is the effect of rare tags on multimedia retrieval scenarios?*, and evaluate the effect when rare tags are augmented with semantics.

The contributions of this paper are three fold. First, we analyze the type of tags that occur in the long tail distribution of social tags. We base the analysis on a representative snapshot of Flickr containing 1 million photos with 700 K diverse tags, and additional 38 K tags from three categories *objects, scenes* and *fine-grain animals*. Second, we investigate augmenting the rare tags with semantic knowledge. Third, we exploit two multimedia retrieval scenarios on sampled tags from the long tail distribution, and analyze their performance on both rare and augmented tags.

2 Related Work

Many works in social media are focused on frequent tags with many example images per tag. This is quite understandable, since those tags are popular and evidently important for many users. In Fig. 2 we show the trend of social media works over the frequency distribution, which confirms the tendency towards frequent tags. In this paper we analyze the non frequent tags.

Frequent Tags. One widely recognized problem is that tags are noisy, ambiguous and often not directly related to the visual content [10, 26]. For that reason, Li *et al.* [10] defined a *tag relevance* metric, calculated by counting neighborhood votes from visually similar images, with the intuition that visually similar images should share the same tags. Their method has shown impressive performance when evaluated on frequent tags. For rare tags we believe that this method could be problematic. For example if a tag is relevant to an image but occurs very rarely or not at all in the corpus, it will not get enough votes from

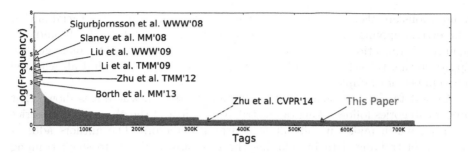

Fig. 2. The trend of related works over tag frequency distribution, confirming the emphasis in the literature towards frequent tags.

its neighbors and will be falsely considered as irrelevant. Zhu *et al.* [26] use the related tags to handle imprecise and incomplete image tags. They calculate the relevance of an image tag by collectively analyzing statistics from Flickr and the WordNet hierarchy, so that a tag receives examples from its child nodes. It might be troublesome if a rare tag is a leaf node in the WordNet hierarchy, without child nodes to be enriched with. Having this in mind, we investigate how calculating tag relevance on rare tags performs, and if including semantic knowledge would help improving the relevance of a tag.

Many other diverse topics exist when using frequent tags. The authors of [1] have created a visual sentiment ontology by sending queries of adjective noun pairs (ANP) to Flickr. In their work ANP candidates (about 320 k) are ranked by occurrence frequency, with the goal to remove extremely rare constructions and to keep only frequent ones. While the frequent ANPs show valuable performance when learning detectors, it is unclear what will be the performance of the rare ANPs. The tag ranking method of [12] automatically ranks tags associated with an image using the neighborhood of images containing a specific tag. Finding tagged neighbors of rare tags is problematic, which might lead to assigning low scores for relevant tags. In other works, like tag ambiguity [18], learning detectors to tag [24] or tag recommendation [17], the authors use co-occurrence statistics. If a tag is rare, its co-occurrence statistics will be unreliable, which might result in erroneous scores. In [6,9,21] social tags are used to collect training data and learn concept detectors. For rare tags there are less images to be used as training data. Therefore, we believe it is interesting to also investigate the learning performance of rare tags. We choose to evaluate detectors learning, since it is a popular topic, and the performance can be evaluated on a standard benchmark dataset [2].

Rare Tags. To the best of our knowledge there is no related work investigating rare tags in social media. The problem of rare tags is somewhat similar to few- or zero-shot learning [8,19], designed for problems where there is a lack of training samples. Most of these works use textual or attribute descriptions of known concepts and compare with the textual description or attribute scores of the zero-shot concept. These textual or attribute detectors are learned with manual annotations, which limits the type of concepts that can be detected. The manual annotations are also quite precise compared to the noisy tags from social media. Therefore, applying these techniques to multimedia methods is difficult, and also requires modifications in the existing algorithms. We investigate a more generic approach that takes the noise of social tags into account, and still deals with the low number of examples.

Closest to our work is a recent study on objects [27], where it is shown that object categories follow a long tail distribution. Most of the object subcategories are rare, which makes it difficult for learning detectors. The authors address the lack of training data by allowing the rare subcategories to share training examples with dominant ones in their learning model. This is possible since the sematic relationships of object categories-subcategories are known, and the images have precise manual annotations. However, social media does not have known semantic relationships between tags. Tagged images are easy to obtain, but they come with noise as additional complication. The method of [27] does not address these challenges, thus, it can not be directly applied on multimedia applications that use social media tags.

Semantic Relationships. Relationships between tags have been calculated using co-occurrence statistics [17,18,24], or semantic knowledge from external sources [3,17,26]. Co-occurrence statistics can be unreliable for calculating tag relationships of rare tags, since they occur rarely or not at all. In the work of Fergus *et al.* [3] an ontology from an external source is adopted to expand query sets for label propagation. In [26] Zhu *et al.* use an ontology to expand the training data. For example, a training set of a non-leaf concept (e.g., building) is enriched by including representative examples from its child nodes (e.g., church). In this paper we investigate external semantic knowledge for a different purpose, to augment the frequency of rare tags.

Although there are many works investigating the possibilities of socially tagged images, none has so far looked into the tags from the long tail of the frequency distribution. In this paper we do a first attempt to analyze rare tags.

3   What Tags Constitute the Long Tail?

Tag Vocabulary. We analyze tags from Flickr as one of the most popular social media platform. We consider the *Flickr 1M* dataset from [10] as a representative sample. This dataset has 1 million images, downloaded by randomly generating photo ids as queries, making it unbiased towards any tags. The images come with about 700K diverse tags. Since its unpractical to analyze all these tags manually, we consider tags from three categories of existing image recognition datasets: *objects, scenes* and *fine-grained* animals. Many object tags are available in ImageNet [2], 22K classes resulting in 40K tags, since some classes come with multiple tags, like *sea cow, sirenian mammal, sirenian* represent one object class. The SUN Attribute dataset [15] has the highest number of scene tags so far, 717 classes like *amusement park, coast, squash court*. Tags from fine-grained animal categories are available in the 120 dog tag from Stanford Dogs [5] like *pekinese, irish terrier, chihuahua*, and 200 bird tags from Caltech Birds [22] like *shiny*

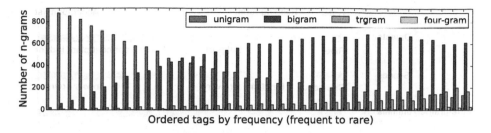

Fig. 3. Occurrence of n-grams, following the long tail frequency distribution.

cowbird, bobolink, blue jay. All tags together form a vocabulary of 730K unique tags. We call this tag set the *Tag vocabulary*, and we analyze it in this experiment.

Tag Composition. It is interesting to investigate the occurrence of n-grams, following the long tail frequency distribution. We aim to find if there is a correlation between the number of words a tag is composed of, and its occurrence frequency. It has been noted that tags from the long tail are mostly complex phrases [17], we investigate if this is indeed the case. We merge the tags of objects, scenes and fine-grained animal categories, in total 38K. We send queries with the tag names to Flickr with the Flickr api, and count the occurrence of each tag. We order the tags by their frequency and we count the unigrams, bigrams, trigrams and four-grams in 40 steps. We visualize the histograms in Fig. 3. It can be observed that the frequent tags are mostly unigrams. As the frequency goes down, more and more bigrams appear, and towards the end trigrams and four-grams start occurring. Some frequent bigrams are *christmas tree* with frequency 250K, or *polar bear* with frequency 160K. The bigrams *zebrawood tree* or *kaffir cat* have frequency zero. The histograms shows that most of the rare tags are bigrams, and less uni-/tri-/four-grams. We believe this is so because users often use general one word tags to describe an image, and do not try to be precise. We also manually analyzed sampled Flickr tags from the Tag vocabulary, not necessarily only coming from the preselected categories. We looked into 50 tags around 10 steps over the long tail distribution. Interestingly, among the frequent tags we found tags like *iphoneography* and *instagramapp*, which are added from popular mobile applications. We expect that with thousands of images being uploaded in Flickr daily, the frequency of the tags changes, new tags appear, and old ones become more frequent. However, at any given time, when a snapshot of Flickr is taken, like at 2008 [17], at 2009 [24] and ours, the distribution stays heavy tailed. Among the sampled tags from the long tail there were attribute and noun pairs like *scratching post, showing work*, named entities like *saint petersburg russia, arnold aragon*, phrases like *what color is your time, i enjoyed all types of outdoor adventures as a child*, number and word compositions like *photo domino 357, hp850* which represent models of products, and also tags in witch we could not find meaning like *wo0, ell:mcc = 222*. Moreover, we summarize, the long tail contains attribute noun pairs, entities of less famous people or geographic places, as well as phrases. The bigrams occur most often in the long tail, and there are less unigrams, trigrams and fourgrams.

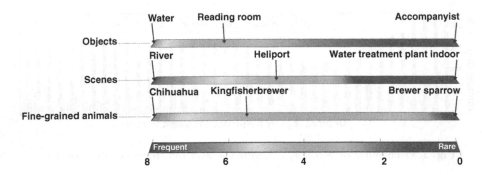

Fig. 4. Heat bars of tag occurrence frequency in Flickr for objects, scenes and fine-grained animal tags. We observe there are many object and scene tags that appear rarely in Flickr, whereas most of the fine-grained animal tags occur reasonably often.

Tag Categories. We visualize in heat bars the tag occurrence of object, scenes, and fine-grained animal categories in Fig. 4. Many **object** tags are colored in blue and appear rarely in Flickr. In numbers, 67 % of the object tags appear in less then 1K images. The rare objects are mostly specific tags like *lingberry* which so far occurs in only 3 Flickr images, *marsh tea* in 28 images or *space vehicle* in 67 images. From the **scene** category, 46 % of the tags occur in less then 1K images. The strong blue response in the heat bar shows that a great portion of scene tags have low frequency. These rare tags mostly represent fine-grained scenes like *artists loft, canal urban, bakery kitchen.* From the **fine-grained** animal categories, green is the dominant color in the hear bar, and red and blue take small proportions at the ends of the bar. This shows that images of fine grained animal categories occur reasonably often in Flickr. In numbers, 41 % of the fine-grained animal tags appear in 1K–10K images, and 36 % appear in less then 1K images. This hints that fine-grained classification can be made even more challenging than the classes suggested in the existing fine-grained datasets. Within the fine-grained dogs and birds classes, some unpopular ones are *brabancon griffon* with frequency 0, *brewer blackbird* with frequency 7 and *blenheim spaniel* with frequency 36. Overall, our analysis go against the prevailing norm in the literature where tags from the long tail have been considered unimportant. We show that there are meaningful tags of objects, scenes and fine-grained categories that occur rarely in Flickr and should not be overlooked.

4   Utilizing the Long Tail

4.1   Augmenting Rare Tags

We investigate augmenting rare tags from the long tail with semantically similar tags. We believe that augmenting the rare tags is important for good performance of multimedia retrieval methods. Therefore we analyze the performance of both rare and augmented tags on two multimedia retrieval scenarios in Sects. 4.2 and 4.3. Motivated from common approaches in the literature of using semantics from external sources like Wikipedia or WordNet [3, 26], we investigate if rare tags can be augmented with semantics. From WordNet we consider synonyms and child nodes of a tag in the hierarchy. From Wikipedia we consider titles of redirect pages, as commonly used for semantic linking in information retrieval [13].

Datasets. Since it is not possible to evaluate all 730 K tags from the Tag vocabulary, we obtain a representative sample by uniformly sampling each 2,000th tag from the distribution shown in Fig. 1. We make sure that the sampled tags have ground truth annotations in some dataset, so that we can evaluate their performance on tag relevance and learning detectors. We consider ground truth classes from ImageNet, as one of the largest available image dataset, with 50 validation images per tag for 1000 classes. If the 2000th tag does not appear in one of the 1000 ImageNet classes, we move to the 2001th tag, and so on. In this manner we select 81 representative tags. These 81 tags contain frequent tags like *light, marmot, blue jean* and rare tags like *whiskey jug, bottle screw, rock snake*. For each tag we download up to 2,000 images from Flickr if available, otherwise as much as we can. This resulted in a new *LongTail* dataset with 13K Flickr images. Additionally, we also downloaded images tagged with semantically similar tags found from WordNet and Wikipedia, forming a new *LongTailAugmented* dataset with 160K images.

Analysis. In Fig. 6(a) we show the frequency of the sampled 81 tags, as well as their frequency when augmented with images of semantically related tags. For most of the rare tags we could find images tagged with their synonym tags, magnifying the frequency when joined. Some rare tags have synonyms from WordNet which are quite frequent, like for example tag 18:*patrol wagon* (id:tag, where id is the position of the tag on the horizontal axis in Fig. 6) appears in only 41 images, whereas its WordNet synonyms appear in much more: *police van* in 2310, *paddy wagon* in 3736, *wagon* in 200K images. For tag 56:*Chlamydosaurus kingi* which so far appears in one image, its WordNet synonym *frilled lizard* has 250 images, and Wikipedia finds more synonyms, *Chlamydosaurus* in 278, *frilled dragon* in 150 and *frillnecked* in 150 images. For some tags like 17:*english foxhound* and 67:*mountain tent*, we did not find semantically similar tags, and for tag 39:*plumbers helper* its synonym *plunger* has 0 tagged images. We expect that with more sophisticated language processing more semantically related tags can be found, and the rare tags will be even better augmented. Overall we conclude, that rare tags from the long tail can be augmented with simple synonyms.

4.2   Tag Relevance

In this experiment we investigate the performance of calculating tag relevance for tags which fall on the long tail of the frequency distribution. We investigate the performance of the most popular tag relevance method proposed be Li *et al.* [10], due to its good performance [20].

Augmented Tag Relevance. We simply extend the tag-relevance method of [10] to take semantically related tags into account. Instead of just counting votes

58   S. Kordumova et al.

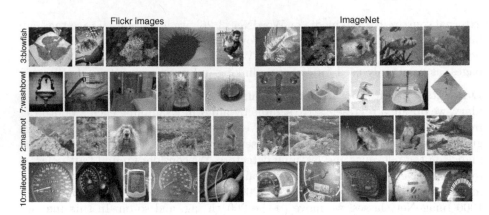

Fig. 5. Example images from Flickr and ImageNet. Tags 3:*blowfish* and 7:*washbowl* have visually very diverse appearance. Tags 2:*marmot* and 10:*mileometer* have more consistent visual features, thus show better performance.

of the same tag, we also count votes from its synonyms. The augmented tag relevance of a rare tag is computed as follows.

We denote a rare tag with r and $S_r = \{s_1, s_2, ..., s_n\}$ is its set of n synonyms. For an image I_r with tags $\{t | t \in \text{tags}(I)\}$, the set of images tagged with semantically similar tags within its k visual nearest neighbors is

$$N(I_r, S_r, k) = \{I | I \in \text{NN}(I_r, k) \wedge \exists t (t \in \text{tags}(I) \wedge t \in S_r)\}. \tag{1}$$

We calculate the rare tag relevance R for an image I_r as

$$R(r, I_r, k) = |N(I_r, S_r, k)| - P(S_r, k), \tag{2}$$

where $P(S_r, k)$ is the prior tag distribution, which in our case denotes the average prior of the synonym set

$$P(S_r, k) = \frac{1}{|S_r|} \sum_{s \in S_r} k \frac{|L_s|}{|L|}, \tag{3}$$

where k is the number of visual neighbors, $|L_s|$ the number of all images labeled with s, and $|L|$ the size of the entire collection. The difference in this formulation from [10], is adding the set of synonyms S_r, on places where only one tag t was used in the voting.

The tag relevance method is developed for tags which are composed of only one word. If a tag is an n-gram, composed of two or more words, we follow the recommendation from [11], and average the tag relevances or the augmented tag relevances for each word.

Features. As visual features for tag relevance and augmented tag relevance we use the same settings as the multi-feature color and texture variant of [11].

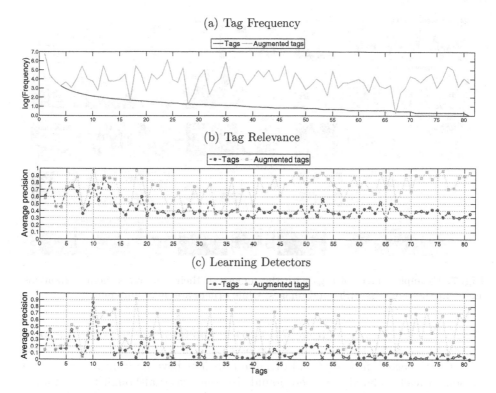

Fig. 6. Frequency and AP of 81 tags, without and with synonym augmentation.

Results and Analysis. To evaluate, we use the ImageNet validation images of the same 81 classes as the selected tags in the LongTail dataset. For each tag, there are 50 positive images in the ImageNet validation set. To evaluate tag relevance we also need images with noisy tags. Thus, for each tag we additionally sample 100 random images from the other classes of ImageNet, resulting in 150 images for evaluation of each tag. We evaluate with average precision (AP) per tag and mean average precision (MAP) for overall score.

We plot the AP per tag in Fig. 6(b), for both regular and augmented tag relevance. A negative effect of the long tail can be clearly seen for tag relevance, where most of the the rare tags obtain low average precision. One reason for this is that when there are few or even zero images in Flickr which have the tag, it is unlikely for the visual neighbors to contain the tag resulting in no votes and failing to learn the relevance of the tag. We also notice some outliers, like for 3:*blowfish*, 4:*bluejean* and id 7:*washbowl* which have lower tag relevance performance compared to their neighbors in the frequency distribution. In Fig. 5 we show how visually diverse and ambiguous these concepts are, compared to for example 2:*marmot* and 5:*easternfox squirrel*, and 10:*mileometer*.

The augmented tag relevance has a positive effect on the rare tags, since for most of the rare tags the average precision improves. For example, the AP of

Fig. 7. Example images of tags with the images of their synonyms. Some synonyms augment the rare tags with good images, while others are more ambiguous.

73:*lycaenid butterfly*, grows from 37 % to 64 %, and for 18:*patrol wagon* improves from 49 % to 86 %. For some tags, there is none or a small improvement. For example 17:*english foxhound* and 67:*mountain tent* have no improvement since no semantically related tags were found. In some cases although the synonym tags are frequent, the results do not improve. For example tag 24:*woodworking plane* has a synonym *plane* which is not as specific and contains many diverse and visually different images, see Fig. 7. In cases like this, we expect a more sophisticated semantic method for augmenting the rare tags would help.

Overall, when only few tagged images are present, the tag relevance is determined to fail upfront. When we use augmented tag relevance with synonyms, the mean average precision grows from 43 % to 73 %. We conclude the tag relevance of the rare tags can be better calculated with augmented semantics.

4.3 Learning Detectors

We investigate learning concept detectors from tagged images, and analyze their performance in correlation with the tag frequency occupance. As a training set we use the LongTail and LongTailAugmented datasets. From the LongTail dataset we select the top ranked images based on their tag relevance score, and from the LongTailAugmented dataset we select the top ranked images based on their augmented tag relevance score. We select the top 1,300 ranked images per tag, as the settings of ImageNet, or less if not as many available. We evaluate on validation images from ImageNet, with AP per tag, or MAP overall.

Features. We recognize that deep learning has shown a great improvement in image classification. The features used from the last layer of a Convolutional Neural Network (CNN), or one layer before the last have become popular and widely used [7]. The CNN is mostly trained on images from ImageNet. Thus, these features have already seen all the classes of ImageNet. Since we evaluate on ImageNet classes, and we want to see the performance of using only few images from a rare tag to learn a concept detector, we do not use the CNN features to keep the evaluation fair. Instead as features we employ the once popular Fisher vector encoding [16] with a GMM of size 1,024 and a spatial pyramid of 1×1 and 1×3. We extract SIFT descriptors with dense sampling at every 6 pixels at two scales, PCA reduced to 80D. As a classifier we use a one-vs-all linear SVM.

Results and Analysis. We visualize the results in Fig. 6 (c). As expected, the less frequent the tags are, the lower the average precision is. Similarly as in tag relevance, some tags have low AP, even though they are frequent. For example tag 1:*light* has low score since its meaning is ambiguous, as well as tags 3:*blowfish*, 4:*bluejean* and 7:*washbowl*, see Fig. 5. For most rare tags, the results improve when we augment the training data with images tagged with their synonyms. We show augmented examples of few tags in Fig. 7. For tag 26:*chrysanthemum dog* and 53:*galeocerdo cuvieri* the augmented images are quite relevant, improving the tag annotation result from 55 % to 74 %, and 2 % to 56 % respectfully. For tags 33:*barracouta* and 34:*sleuthhound* the results do not improve. For example *snoek* is a synonym of *sleuthhound*, and also a name of a car model, which adds noise to the training data of *sleuthhound*.

Overall, learning detectors from tagged images with rare tags gives poor performance since there are not enough images to learn reliable detectors. When simply augmenting the training data with images tagged with synonyms, the MAP improves from 53 % to 79 %. We conclude, learning detectors for rare tags from the long tail can be improved by augmenting the training data with images tagged with their synonyms.

5   Conclusions

We have looked into the long tail of social tags, and analyzed three questions: *What tags constitute the long tail?*, *What happens when rare tags are augmented?* and *What is the effect of rare tags on multimedia retrieval scenarios?*. We uncover that the long tail has valuable tags of objects, scenes and fine-grained animal categories. We show that by augmenting the rare tags with simple semantics, the performance of tag relevance and detector learning improves considerably. Thus, we conclude the rare tags from the long tail are valuable and perform better when augmented with semantic knowledge.

Acknowledgments. This research is supported by the STW STORY project and the Dutch national program COMMIT.

References

1. Borth, D., Ji, R., Chen, T., Breuel, T., Chang, S.-F.: Large-scale visual sentiment ontology and detectors using adjective noun pairs. In: MM (2013)

2. Deng, J., Dong, W., Socher, R., Li, L.-J., Li, K., Fei-Fei, L.: ImageNet: a large-scale hierarchical image database. In: CVPR (2009)
3. Fergus, R., Bernal, H., Weiss, Y., Torralba, A.: Semantic label sharing for learning with many categories. In: Daniilidis, K., Maragos, P., Paragios, N. (eds.) ECCV 2010, Part I. LNCS, vol. 6311, pp. 762–775. Springer, Heidelberg (2010)
4. Ginsca, A.L., Popescu, A., Ionescu, B., Armagan, A., Kanellos, I.: Toward an estimation of user tagging credibility for social image retrieval. In: MM (2014)
5. Khosla, A., Jayadevaprakash, N., Yao, B., Fei-Fei, L.: Novel dataset for fine-grained image categorization. In: CVPR (2011)
6. Kordumova, S., Li, X., Snoek, C.: Best practices for learning video concept detectors from social media examples. MTAP **74**, 1291–1315 (2014)
7. Krizhevsky, A., Sutskever, I., Hinton, G.E.: Imagenet classification with deep convolutional neural networks. In: NIPS (2012)
8. Lampert, C., Nickisch, H., Harmeling, S.: Attribute-based classification for zero shot visual object categorization. TPAMI **36**, 453–465 (2014)
9. Li, G., Wang, M., Zheng, Y.-T., Li, H., Zha, Z.-J., Chua, T.-S.: Shottagger: tag location for internet videos. In: ICMR (2011)
10. Li, X., Snoek, C.G.M., Worring, M.: Learning social tag relevance by neighbor voting. TMM **11**, 1310–1322 (2009)
11. Li, X., Snoek, C.G.M., Worring, M., Smeulders, A.W.M.: Harvesting social images for bi-concept search. TMM **14**, 1091–1104 (2012)
12. Liu, D., Hua, X.-S., Yang, L., Wang, M., Zhang, H.-J.: Tag ranking. In: WWW (2009)
13. Meij, E., Weerkamp, W., de Rijke, M.: Adding semantics to microblog posts. In: WSDM (2012)
14. Ni, Y., Zheng, M., Bu, J., Chen, C., Wang, D.: Personalized automatic image annotation based on reinforcement learning. In: ICME (2013)
15. Patterson, G., Xu, C., Su, H., Hays, J.: The sun attribute database: beyond categories for deeper scene understanding. IJCV **108**, 59–81 (2014)
16. Sánchez, J., Perronnin, F., Mensink, T., Verbeek, J.: Image classification with the fisher vector: theory and practice. IJCV **105**, 222–245 (2013)
17. Sigurbjörnsson, B., van Zwol, R.: Flickr tag recommendation based on collective knowledge. In: WWW (2008)
18. Slaney, M., Weinberger, K., van Zwol, R.: Resolving tag ambiguity. In: MM (2008)
19. Socher, R., Ganjoo, M., Manning, C.D., Ng, A.Y.: Zero-shot learning through cross-modal transfer. In: NIPS (2013)
20. Truong, B., Sun, A., Bhowmick, S.: Content is still king: the effect of neighbor voting schemes on tag relevance for social image retrieval. In: ICMR (2012)
21. Ulges, A., Koch, M., Borth, D.: Linking visual concept detection with viewer demographics. In: ICMR (2012)
22. Wah, C., Branson, S., Welinder, P., Perona, P., Belongie, S.: The Caltech-UCSD Birds-200-2011 Dataset. Technical report (2011)
23. Wang, H., Wu, F., Li, X., Tang, S., Shao, J., Zhuang, Y.: Jointly discovering fine-grained and coarse-grained sentiments via topic modeling. In: MM (2014)
24. Wu, L., Yang, L., Yu, N., Hua, X.-S.: Learning to tag. In: WWW (2009)
25. Yang, Y., Cui, P., Zhu, W., Zhao, H.V., Shi, Y., Yang, S.: Emotionally representative image discovery for social events. In: ICMR (2014)
26. Zhu, S., Ngo, C.-W., Jiang, Y.-G.: Sampling and ontologically pooling web images for visual concept learning. TMM **14**, 1068–1078 (2012)
27. Zhu, X., Anguelov, D., Ramanan, D.: Capturing long-tail distributions of object subcategories. In: CVPR (2014)

Visual Analyses of Music Download History: User Studies

Dong Liu[1]([⊠]) and Jingxian Zhang[2]

[1] CAS Key Laboratory of Technology in Geo-spatial Information Processing
and Application System, University of Science and Technology of China, Hefei, China
dongeliu@ustc.edu.cn
[2] Department of Computer Science,
University of Illinois at Urbana-Champaign, Champaign, IL, USA
jzhng144@illinois.edu

Abstract. Users' download history is a primary data source for analyzing user interests. Recent work has shown that user interests are indeed time varying, and accurate profiling of user interest drifts requires the temporal dynamic analyses. We have proposed a visualization approach to analyzing user interest drifts from the download history, taking music as an example, and studied how to depict the underlying relevances among the downloaded music items to identify the drifts. We designed three new kinds of plots to display the music download history of one user, namely Bean plot, Transitional Pie plot, and Instrument plot. In this paper, we report our conducted user studies that ask normal users to visually analyze the download history of other users in a given real-world data set. User studies are performed in a learning-practice-test workflow. The results demonstrate the feasibility of our visualization design.

Keywords: Download history · Time varying · User interest drifts · User studies · Visualization

1 Introduction

Downloading digital content from online content distribution services has been a common practice for Internet users. To enhance user satisfaction and loyalty, service providers often analyze the users' download history to mine the interests and needs of users, and utilize the *recommender systems* to provide personalized recommendations for content that suit a user's interest [2].

Modern recommender systems have been developed in two dimensions. The first is content-based filtering that finds content items similar to what the user has liked before, where content similarity can be defined on the features of content items. The second and more successful one is collaborative filtering (CF) [11], which utilizes not only the past behavior of one user but also of other users. The content-based filtering and CF approaches have been combined in existing recommender systems [2].

Q. Tian et al. (Eds.): MMM 2016, Part I, LNCS 9516, pp. 63–75, 2016.
DOI: 10.1007/978-3-319-27671-7_6

Recently, recommender systems have been further empowered by taking into account the temporal dynamics. Indeed, user interests are often not static but rather time varying. Some work has been done to detect and to adapt to the user interest drifts in making recommendations. For example, Koren studied a series of temporal dynamic models within the framework of CF [13]; Cao *et al.* proposed four patterns to describe user interest drifts, including Single Interest Pattern, Multiple Interests Pattern, Interests Drift Pattern, and Casual Noise Pattern [9]; Abel *et al.* investigated temporal dynamic user modeling from user behavior on the Social Web such as Twitter [1]. All such work reported more accurate recommendations compared to traditional systems that adopted time invariant user models.

Though automated recommender systems have achieved remarkable success, enabling user interactive recommendations is nontheless important to improve user experience, where visualization is key to engage user's participation [7,8,16]. Visually analyzing user behavior also provides great help to service providers, as it capably supports the open-ended exploration and flexible questions that analysts may generate [4]. One question that interests service providers, is whether visualization of users' data can be understandable by users themselves, and thus recommendations based on this visualization are also understandable and more acceptable?

However, there are some challenges in visualizing user behavior for analyses as well as for making recommendations. In view of the possible user interest drifts, it is crucial to display the relevances between the content items that one user had accessed, but how to quantify and visualize the relevances is a difficulty. Displaying temporal and drifting data remains a challenge in visualization, especially taken into account the interpretability and perceivability. Moreover, how to evaluate the visualization design, both objectively and subjectively, is not well studied before.

In our previous work [19], we have studied a visualization approach to analyzing user interest drifts from users' music download history. Our main purpose was to depict the underlying relevances among the downloaded music tracks so as to identify the user interest drifts. To that end, we considered feature-based relevances and *collaborative relevances* in accordance with the existing recommender systems. For feature-based relevances, we utilized the metadata of music and selected genre and release year to represent categorical and numerical features, respectively. For collaborative relevances, we had been inspired by the CF approach and define relevance between music tracks as co-occurrence in all users' download history. Moreover, we designed three new kinds of plots to display the music download history of one user, namely Bean plot, Transitional Pie plot, and Instrument plot.

In this paper, we report our conducted user studies to evaluate the usability of the visualization design, during which we focus on the capability of the visualization in assisting analyses, as well as the ease of learning and use and the analyst's experience. Such user studies remain largely explorative rather than quantitative in the literature, and how to perform user studies is also an open problem. Our studies try to deliver a learning-practice-test workflow to observe how normal users could leverage the visualization tools to perform data analyses tasks as if they were professional analysts, which may be inspiring for further research.

The remainder of this paper is organized as follows. Section 2 describes some related work. Section 3 summarizes our visualization design. The conducted user studies are discussed in detail in Sect. 4. Concluding remarks are presented in Sect. 5.

2   Related Work

Our work is closely related to visual recommendations that help user discover interesting items and help the service provider interpret the reason of recommendations. Visualization of music itself is also of great interests recently. How to evaluate the visualization design has been studied mostly in an empirical manner. On the above aspects will we briefly review some related work.

As mentioned before, recommender systems can be roughly classified into three categories: content-based filtering [17], CF [6], and hybrid approaches [3]. Visual recommendations also fall into one of the three categories. For example, Bogdanov et al. [7] proposed a content-based recommendation that infers high-level semantic descriptors from the music tracks of one user, and then utilizes the semantic descriptions to perform recommendations or visualization of user's preferences. PeerChooser [15] is a collaborative recommender system with an interactive graphical explanation interface, which enables user to select "similar users" in her own mind. Hybrid visual recommendations are more attended, for example the recommendations from multiple social and semantic web resources [8], and the visual user-controllable interface that encourages user to manually control the recommendation strategies [16]. Regarding the type of content, the work in [5] may be the most similar one to our work, as it proposed several visualizations for music download history, also for recommendations. However, all the above-mentioned work did not consider the underlying user interest drifts in the user's behavior data.

Music-related visualization attracts the attention of researchers in a wide range. Earlier work was done to visualize music collections such as personal music libraries [12,18]. The visualization of music download history was presented in [5], which added the temporal dimension into consideration. Based on the user's music library, a humanoid cartoon-like character called *Musical Avatar* was generated to visualize the user's interests [7]. In such work, the implicit relations among music tracks are less studied, and the temporal dynamics are not taken into account.

Evaluation of information visualizations has always been an important part of related research. Carpendale discussed different types of evaluations as well as their pros and cons [10]. Lam et al. [14] focused on empirical studies in information visualization and summarized seven scenarios to discuss what might be the most effective evaluation of a given information visualization. Basole et al. [4] designed a three-phase user study including tutorial, practice, and evaluation for assessing their visualization design. In this paper, we design user studies as a learning-practice-test workflow, similar to that in [4] but enables users to utilize the visualization tools as if they were professional analysts.

3   Visualization Design

The raw data recording the music download history of one user can be described as pairs of downloading timestamp and the identity of music track. Raw data are pre-processed in two steps. First, one user's download history is divided into *sessions*, where each session consists of a series of consecutive downloads that have short intervals, and the intervals between sessions are usually much longer. Second, from the metadata of each music track, genre and release year are extracted as features. Moreover, we calculate collaborative relevances within any pair of two tracks, the relevance is evaluated by the number of users who downloaded both tracks *within a short period*. Then, we designed three kinds of visualization plots, namely Bean plot, Transitional Pie plot, and Instrument plot, to display the music download history that indeed imply the dynamics of user interests. Please refer to Figs. 1, 2, and 3 for the plots and user interactions. For more details please refer to our previous paper [19].

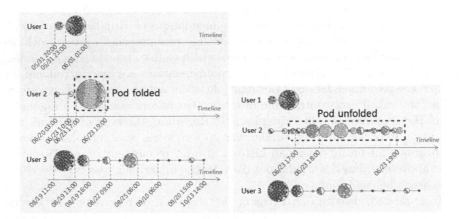

Fig. 1. (Left) Bean plot showing the download history of three users. Each small, color-filled circle (named a bean) represents a music track and each larger disc (named a pod) represents a download session. Colors of beans stand for genres. Pods are arranged in chronological order. (Right) Bean plot provides interactive display that one pod being clicked will unfold to multiple smaller pods to represent subsessions, each of which has a single genre (Colour figure online).

4   User Studies

We have conducted user studies to evaluate the usability of our proposed visualization. In the studies, participants are required to first learn the design of the

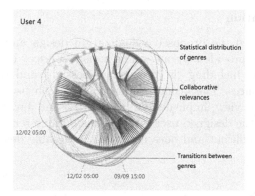

Fig. 2. Transitional Pie plot showing the download history of one user. Similar to pie chart, the disc shows the proportions of different genres. Within each genre, tracks are arranged in chronological order so that each downloaded track has a corresponding position on the disc. Bezier curves inside the disc display the collaborative relevances among music tracks. Bezier curves outside the disc show the transitions between genres, two successively downloaded tracks of different genres are connected by a outer-disc curve.

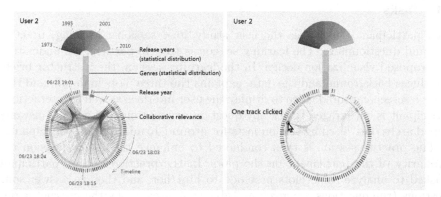

Fig. 3. (Left) Instrument plot showing the download history of one user. The timeline is represented by a disc as the body of the instrument, where the music tracks are arranged in chronological order. The gray bars alongside the tracks indicate release years. Bezier curves inside the disc represent the collaborative relevances among music tracks. The distributions of release year and genre are shown as the headstock and the neck of the instrument, respectively. (Right) Instrument plot provides interactive display, once a track is clicked, all its related tracks will be highlighted.

three plots; then after practice, they are asked to analyze the plots of new users and to answer both specific questions and open-ended questions regarding the user interest drifts; finally, questionnaire and survey are performed to collect the participants' feedback about the experience of using the visualization plots.

4.1 Implementation

We have implemented the proposed visualization design and tested it with a real-world data set provided by an online music service. Data preprocessing is performed offline, including the division of sessions and the calculation of collaborative relevances. The visualization plots of each user are drawn upon request in a webpage view, implemented by HTML5 and JavaScript techniques, which also enables the designed user interactions. Online rendering of the plots is computationally efficient and does not incur noticeable delay in mainstream web browsers.

4.2 Participants

15 undergraduate students (6 females and 9 males) participate the user studies. Participants have different majors including science and engineering, with ages ranging from 20 to 23 (mean: 21.6, median: 22). Participants reported different levels of interests in online music services, but none of them had experience of visual analyses.

4.3 Tasks

Each participant undertakes the user study in 4 sessions: learning, practice, test, and questionnaire. The learning session is to help participants understand the proposed visualization design. In the learning session, the instructor briefly introduces background and the data, explains the three plots in detail, and then shows a screen-captured video to display the user interface as well as interactions. Participant is encouraged to ask any question at any time and will be answered immediately. The learning session lasts for around 15 min for one participant.

The practice session is then conducted to enhance the comprehension and familiarity of the participant on the plots. In the practice session, participant is asked to analyze some plots provided to him/her, and then to answer some questions (the questions are given in Table 1). For example, the instructor provides the Bean plots of user 7 and user 8, and then asks the participant: "whose download history is more consistent in music genre, user 7 or user 8?" (PQ3 in Table 1) Instructor will answer any question of the participant, and will explain the plots again to the participant, if necessary. The practice session lasts for around 19 min for one participant.

The test session is the most important part of the user study. In the test session, the visualization plots of 10 users[1] are provided to participant for visual analyses. Participant is asked to analyze all the plots and then to answer 3 specific questions and 2 open-ended questions (the questions are given in Table 2). In the test session, instructor does not offer any assistance to the participant, neither tell the participant which plot to look at, nor answer any question regarding the plots. The only instruction is to encourage the participant to answer the

---

[1] All the plots can be accessed from http://staff.ustc.edu.cn/%7Edongeliu/stuff/userStudyPlots.pdf.

Table 1. Questions in the practice session

Bean plot	PQ1: How many subsessions are there in the session A of user 2?
	PQ2: Whose download history is more continuous in time, user 2 or user 3?
	PQ3: Whose download history is more consistent in music genre, user 7 or user 8?
Transitional Pie plot	PQ4: Whose download history has more genres, user 12 or user 18?
	PQ5: Which genre is the most lasting in user 4's download history?
	PQ6: Combining the collaborative relevances and the transitions, whose interest lasts for longer time, user 2 or user 3?
Instrument plot	PQ7: Who prefers music tracks with older release year, user 2 or user 5?
	PQ8: List the music tracks that have the highest relevances in user 2's download history
	PQ9: Find the track that has relevances with the most kinds of genres in user 27's download history

Table 2. Questions in the test session

Specific questions	Q1-1: Whose download history is the most consistent in time?
	Q1-2: Whose download history is the most consistent in genre?
	Q2: Whose downloaded tracks have the highest relevances?
	Q3-1: Whose interest is the most consistent?
	Q3-2: Whose interest is the least consistent?
Open-ended questions	Q4: Write down your findings of user 12's download history as many as possible
	Q5: Write down your findings of the differences between the download history of user 2 and user 4 as many as possible

open-ended questions as comprehensive as possible. By this design, we hope to verify whether the participant has learnt the characteristics of different plots and can choose the plots for visual analyses, and to observe what the participant can find from the exploration of the plots. The test session lasts for around 37 min for one participant.

The last session is to ask the participant to finish a five-point Likert-scale questionnaire regarding the experience of using the plots for visual analyses. Also, a quick survey is conducted to collect the participant's feedback on the visualization design. In total, the user study for one participant lasts for 57 to 84 min (mean: 71, median: 70). All participants are instructed by the same investigator.

4.4   Results

Practice Session. In the practice session, we ask the participants to answer some specific questions (given in Table 1) regarding the visual analyses of the provided plots. Participants' answers show a good consistence: for 5 out of 9 questions (i.e. PQ1, PQ4, PQ5, PQ7, PQ9), all the 15 participants give the same answer; for the other 3 questions (PQ2, PQ3, PQ8), 14, 14, and 13 participants give the same answer, respectively. The only exception is PQ6: for this question, 7 out of 15 participants choose user 2 and the other 8 participants choose user 3. These results show that the three plots are not difficult to understand for the participants except for the collaborative relevances and transitions that are shown simultaneously in the Transitional Pie plot, which may lead to diverse understandings.

Test Session. In the test session, we ask the participants to answer both specific questions and open-ended questions (given in Table 2) after their analyses of the visualization plots of 10 users. These questions are believed to be much more difficult and more subjective compared to the questions in the practice session. On the one hand, the analyst shall compare the plots of 10 users to give out the answers to the specific questions. On the other, the open-ended questions indeed have diverse answers.

For the specific questions shown in Table 2, Q1-1, Q1-2, and Q2 all receive quite consistent answers from the participants. For Q1-1, 13 participants select user 6 and the other two select user 24, which can be verified by looking at the Bean plots of these two users. For Q1-2, 11 participants select user 2, two select user 3, and the other two select user 4, which has been made visible in the Bean plots and Instrument plots. For Q2, observing the Instrument plots or Transitional Pie plots, 10 participants select user 6, and the other five select user 24. Note that such consistence is not trivial, since the participants need to pick one user from the 10 users as the answer to these questions, and the interest patterns of several users are very similar. Therefore, we believe that the plots have depicted the characteristics of user interests that help participants make the correct analyses.

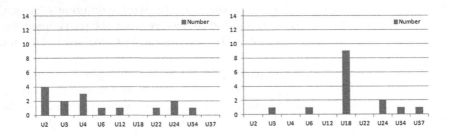

Fig. 4. The answers to Q3-1 (left) and Q3-2 (right) in the test session. Please refer to the questions shown in Table 2.

The answers to Q3-1 and Q3-2 (summarized in Fig. 4) show more diversity among participants. These two questions are not specific to time, genre, or collaborative relevance, but require the participants to present their own understandings of "user interest drifts," which is quite subjective. Participants report difficulty in considering genre and collaborative relevance together when looking for the user with the *most* or the *least* consistent interest. This is not surprising since genre and collaborative relevance are displayed separately in our designed plots. This issue shall be addressed in the future work.

The open-ended questions Q4 and Q5 (also shown in Table 2) ask the participants to write down as many as possible their findings of one user's download

Table 3. The answers to the open-ended questions in the test session

(a) Answers to Q4

Findings	Related Participants
The user prefers music tracks with modern release years.	1, 2, 3, 4, 6, 7, 8, 9, 10, 11, 12, 13, 14, 15
The release years of downloaded tracks are relatively concentrated.	5
The user prefers mandarin pop music.	1, 2, 3, 4, 5, 6, 7, 8, 10, 11, 12, 15
The user's interest is not consistent in genre.	1, 3, 7, 12
The user likes only a few genres.	1, 2, 13
The user's downloading actions are concentrated in time.	2, 3, 13, 14, 15
There are long intervals between the user's downloading sessions.	9, 10, 14
The user's interest is quite stable within each session.	4, 5
In different sessions, the numbers of tracks vary quite a lot.	9, 10
The user might have different moods in different sessions.	6
The downloaded tracks have high collaborative relevances.	1, 5, 7, 8, 9, 10, 11, 12, 13, 15
The downloaded tracks have low collaborative relevances.	14
Transitions between genres only happen among tracks that have collaborative relevances.	4, 7

(b) Answers to Q5

Findings	Related Participants
User 2 prefers music tracks with older release years.	1, 2, 3, 4, 6, 7, 8, 9, 10, 11, 12, 13, 14, 15
User 2 likes pop and soul music whilst user 4 likes mandarin pop.	3, 4, 5, 6, 7, 8, 10, 11, 12, 13, 15
User 4 likes more genres than user 2.	1, 2
User 2's interest is more consistent in time and in genre.	1, 2, 3, 4, 5, 7, 8, 9, 10, 11, 12, 13, 14, 15
User 2's downloading sessions contain more tracks.	3, 4, 7
User 4's downloaded tracks have higher collaborative relevances.	1, 7, 9, 10, 11, 12, 14, 15
User 2's downloaded tracks have higher collaborative relevances.	2
User 2 has more consistent interest than user 4.	2, 3, 4, 5, 11
User 4's taste seems to be of popular mass compared to user 2.	6

history and of comparison between two users'. Participants' findings are summarized in Table 3. Overall, these findings achieve a good consistence among different participants and at the same time exhibit subjective diversity. There are only two findings inconsistent with the others (typeset *italic* in Table 3), each approved by

Table 4. Questions and answers in the questionnaire and survey session

Questions		Strong disagree 1 2 3	Strong agree 4 5
Ease	Easy to learn.		
	Easy to use.		
Usability	Collaborative relevance is helpful in analyses.		
	The plots make user interests obvious.		
	The designed interactions are intuitive.		
Redundancy	Redundancy exists in Bean plot.		
	Redundancy exists in Instrument plot.		
	Redundancy exists in Transitional Pie plot.		
Fondness	I am willing to use the plots to analyze.		

only one participant, and both relate to collaborative relevances. It reveals that few participants have misunderstanding on the collaborative relevances, which seem a less straightforward concept for them. Moreover, participants have made findings at different aspects including the release year and genre of music, the download sessions, and the collaborative relevances. Interestingly, participants mention some findings that are beyond our previous analyses. For example, "transitions between genres only happen among tracks that have collaborative relevances," which is mentioned by two participants, was not easy to find at the first sight. Also, some findings are more *conjectured* than *analyzed*, e.g. "the user might have different moods in different sessions," "since the relevances between tracks within each session is high, but cross-session relevances are almost none," said the participant. Last but not the least, the findings of comparison between two users (answers to Q5) are more consistent among different participants compared to the answers to Q4, which implies that the visual comparison may be easier and more obvious than the visual analysis of single user.

Questionnaire and Survey Session. After practice and test sessions, we ask the participants to finish a five-point Likert-scale questionnaire that consists of 9 questions shown in Table 4, which also shows the average score of answers to each question. We also conduct a survey to collect participants' feedback.

Questionnaire. According to the questionnaire, participants feel the visualization interface is easy to learn and use. This is also verified by the fact that all participants, having no experience of visual analyses, can finish the entire user study with good performance in less than 90 min. Regarding the usability of the visualization, participants agree that collaborative relevance is helpful in analyzing user interests, and all of them indeed take the collaborative relevance into consideration during the test session. 13 out of 15 participants agree that the three plots make user interests obvious, and the other two are neutral to this question. Most participants think the designed interactions are intuitive, and we have observed that all participants learn the interactions in Instrument plot very quickly and use the highlighting feature effectively; the interactions in Bean plot are also not a difficulty for the participants, but three of them need some time to understand the concept of subsessions (unfolded pods), once understood, they all use the interactions well in the test session. Furthermore, on whether there is redundancy in the three plots, Bean plot is believed to have none, but Instrument plot and Transitional Pie plot are believed to have some redundancy to some extent by several participants. The reason may be that the Instrument plot displays the genre and release year of each track as well as the statistics of them, and the Transitional Pie plot uses gradient color for the transitions between genres. Finally, participants report willingness to use the visualization plots (9 agree, 5 neutral, and 1 disagree).

Survey. In the survey, we ask the participants for their opinions on the visualization plots. About the Bean plot, although it does not display collaborative relevance, its distinctive layout is acknowledged by the participants and it is utilized frequently in the test session. Almost all participants believe that Instrument plot displays the *most* information and the *most important* information; participants express an overwhelming preference on the Instrument plot in the test session. The Transitional Pie plot is endorsed by some participants but disliked by some others. Several participants are fond of the Transitional Pie plot as "the inner- and outer-disc curves can be jointly considered," but some others think the Transitional Pie plot can be covered by the Instrument plot and thus is unnecessary. Moreover, participants provide comments and suggestions on the plots. For example, the shape of beans can be changed to square so that the layout of Bean plot may look more regular; the color coding of genres may be adaptive for each user to better distinguish different genres; and so on. These issues will be addressed in our future work.

4.5 Discussion

Comparing the three plots, Instrument plot receives the most preference due to its comprehensiveness. Bean plot is also approved when the analysts do not concern collaborative relevance. Transitional Pie plot is endorsed by several people but less utilized by some others. Collaborative relevance is believed to be helpful in the analyses, but people have difficulty in combining feature-based and collaborative relevances at the same time since the plots display them separately.

With the help of the proposed visualization design, non-expert people can learn and use the plots for visual analyses of user interests without much difficulty. As the visualization makes user interests obvious, there is a good consistence among the analyses of different people. At the same time, people also make diverse, sometimes subjective findings from the explorative analyses, which implies the visualization may inspire analysts to further investigate the user interests.

5 Conclusions

In this paper, we have reported our conducted user studies to evaluate our proposed visualization approach to analyzing user interest drifts from the music download history. We examined users' feedback on our designed three new kinds of plots, i.e. Bean plot, Transitional Pie plot, and Instrument plot, that display the music download history while making the user interest drifts visible to analysts. The results demonstrate the feasibility of our visualization design, and the user studies may be inspiring for further research on visual analyses tasks.

Acknowledgment. This work was supported by National Program on Key Basic Research Projects (973 Program) under No. 2015CB351800, by Natural Science Foundation of China (NSFC) under No. 61303149 and No. 61331017, and by the Fundamental Research Funds for the Central Universities.

References

1. Abel, F., Gao, Q., Houben, G.J., Tao, K.: Analyzing temporal dynamics in twitter profiles for personalized recommendations in the social web. In: Proceedings of the 3rd International Web Science Conference, p. 2. ACM (2011)
2. Adomavicius, G., Tuzhilin, A.: Toward the next generation of recommender systems: a survey of the state-of-the-art and possible extensions. IEEE Trans. Knowl. Data Eng. **17**(6), 734–749 (2005)
3. Balabanović, M., Shoham, Y.: Fab: content-based, collaborative recommendation. Commun. ACM **40**(3), 66–72 (1997)
4. Basole, R.C., Clear, T., Hu, M., Mehrotra, H., Stasko, J.: Understanding interfirm relationships in business ecosystems with interactive visualization. IEEE Trans. Visual. Comput. Graph. **19**(12), 2526–2533 (2013)
5. Baur, D., Butz, A.: Pulling strings from a tangle: visualizing a personal music listening history. In: Proceedings of the 14th International Conference on Intelligent User Interfaces, pp. 439–444 (2009)
6. Billsus, D., Pazzani, M.J.: Learning collaborative information filters. In: Proceedings of the International Conference on Machine Learning (ICML), vol. 98, pp. 46–54 (1998)
7. Bogdanov, D., Haro, M., Fuhrmann, F., Xambó, A., Gómez, E., Herrera, P.: Semantic audio content-based music recommendation and visualization based on user preference examples. Inf. Process. Manage. **49**(1), 13–33 (2013)
8. Bostandjiev, S., O'Donovan, J., Höllerer, T.: Tasteweights: A visual interactive hybrid recommender system. In: Proceedings of the Sixth ACM Conference on Recommender Systems, pp. 35–42. ACM
9. Cao, H., Chen, E., Yang, J., Xiong, H.: Enhancing recommender systems under volatile user interest drifts. In: Proceedings of the 18th ACM Conference on Information and Knowledge Management, pp. 1257–1266. ACM (2009)
10. Carpendale, S.: Evaluating information visualizations. In: Kerren, A., Stasko, J.T., Fekete, J.-D., North, C. (eds.) Information Visualization. LNCS, vol. 4950, pp. 19–45. Springer, Heidelberg (2008)
11. Goldberg, D., Nichols, D., Oki, B.M., Terry, D.: Using collaborative filtering to weave an information tapestry. Commun. ACM **35**(12), 61–70 (1992)
12. Knees, P., Schedl, M., Pohle, T., Widmer, G.: An innovative three-dimensional user interface for exploring music collections enriched. In: Proceedings of the 14th annual ACM international conference on Multimedia, pp. 17–24. ACM (2006)
13. Koren, Y.: Collaborative filtering with temporal dynamics. Commun. ACM **53**(4), 89–97 (2010)

14. Lam, H., Bertini, E., Isenberg, P., Plaisant, C., Carpendale, S.: Empirical studies in information visualization: seven scenarios. IEEE Trans. Visual. Comput. Graph. **18**(9), 1520–1536 (2012)
15. O'Donovan, J., Smyth, B., Gretarsson, B., Bostandjiev, S., Höllerer, T.: Peer-Chooser: visual interactive recommendation. In: Proceedings of the SIGCHI Conference on Human Factors in Computing Systems, pp. 1085–1088. ACM (2008)
16. Parra, D., Brusilovsky, P., Trattner, C.: See what you want to see: visual user-driven approach for hybrid recommendation. In: International Conference on Intelligent User Interfaces (IUI), pp. 235–240 (2014)
17. Pazzani, M., Billsus, D.: Learning and revising user profiles: the identification of interesting web sites. Mach. Learn. **27**(3), 313–331 (1997)
18. Torrens, M., Hertaog, P., Arcos, J.L.: Visualizing and exploring personal music libraries. In: Proceedings of 5th International Conference on Music Information Retrieval, pp. 421–424 (2004)
19. Zhang, J., Liu, D.: Visualization of user interests in online music services. In: 2014 IEEE International Conference on Multimedia and Expo Workshops. IEEE (2014)

Personalized Annotation for Mobile Photos Based on User's Social Circle

Yanhui Hong, Tiandi Chen, Kang Zhang, and Lifeng Sun[✉]

Tsinghua National Laboratory for Information Science and Technology,
Department of Computer Science and Technology,
Tsinghua University, Beijing, China
hongyh13@mails.tsinghua.edu.cn, dalianctd@163.com, zk54188@gmail.com,
sunlf@tsinghua.edu.cn

Abstract. For mobile photos annotation, users are more interested in the context information behind the photos. The user's social circle can provide valuable information for it. However, the accompanying textual information of social network is sparse and ambiguous in nature. In this paper, we propose a personalized annotation framework for mobile photos leveraging the user's social circle. To address the unreliability problem of social network, we present an algorithm to generate reliable tags for social photos before assigning tags to the user's unlabeled photos. In the tag generation stage, a multi-modality hierarchical clustering algorithm is performed to detect social events. Besides, we use "Album" instead of individual photo as the basic unit for clustering. Finally, we employ a weighted nearest neighbor model for label propagation. We evaluate our framework on a large-scale, real-world dataset from Renren, the largest Facebook-like social network in China. Our evaluation results show promising results of our proposed framework.

Keywords: Personalized annotation · Social network · Event detection

1 Introduction

With the prevalence of mobile devices, people can take photos at anytime and anywhere. And owing to the screen and memory limitations of mobile devices, people are used to transmitting and storing their taken photos via third-party cloud services, such as iCloud. How to organize and manage these personal photos has become a much more pressing issue for users due to the large data set size.

Image annotation is an effective and promising technique to solve this problem. However, full manual image annotation is labor-intensive and time-consuming. Thus, automatic image annotation is crucial and has received a lot of research interests. Different from general-purpose photo annotation algorithms [13,23], which try to assign general visual labels to the images, such as "cats" or "boys", mobile photos annotation is a highly personal and user-centric task, for example users may more concern about "My brother Jack" instead of "a boy", "my graduation trip to

© Springer International Publishing Switzerland 2016
Q. Tian et al. (Eds.): MMM 2016, Part I, LNCS 9516, pp. 76–87, 2016.
DOI: 10.1007/978-3-319-27671-7_7

Hawaii" instead of "beach". The question is then how to get the context information behind the photos? Undoubtedly, the user's social circle is an important source since there are millions of users upload and share their personal photos in the social media platforms, such as Flickr, Facebook and Twitter. And there is a significant overlap in their real world activities as the participants in the network are often family, friends or co-workers. For a photo to be labeled in the user's mobile device, it is probably that there are social friends who participate in the same activity and upload the related photos to the social network. The community-contributed photos with their associated social information will be of tremendous value for personalized annotation.

In this paper, we aim to provide personalized annotation for mobile photos based on social network. However, it is a very challenging task because: (1) The accompanying tags of social images are noisy and heterogeneous in nature since they are annotated by different users, and far from uniform in quality and might often be misleading or ambiguous; (2) Given the diversity of social network, the tags are more personal and the number of tags to be modeled in this situation is larger than that of predefined labels in typical annotation problem, which renders most of the traditionally approaches undesirable for our scenario.

Therefore, we would like to propose a new framework to address the mentioned problems. The proposed framework is based on two observations:

1. Event is one of the most important elements of people's life and memories. And most of the personal photos are taken during specific events such as birthday party, family trip, sport meeting etc. The same event should share the same event attribution labels, such as what, where, when and who.
2. Besides, event is highly time dependent. And unlike the images from web searching engine or commercial image banks, the photos in user's social circle are related to each other and the photos uploaded by the same user at the same time are most likely belong to the same event.

Base on observation 1, we can come to conclusion that although the associated social information of a photo in social network may be missing or ambiguous, combing all the social information of photos belong to the same event will provide more reliable labels. So we detect social events at first to generate reliable event tags for social photos. What's more, considering the sparsity and unreliability of individual social photo and being inspired by observation 2, we use "Album" as the basic unit for clustering in social event detection stage innovatively - it can be the album structure on some social platforms or a manually defined way that the photos uploaded by the same user at the same time.

Based on the above analyses, we propose a personalized annotation framework based on the user's social circle. Our system can be divided into two main parts, label generation and image annotation. In the label generation stage, the events in the user's social circle are detected by a novel multi-modality hierarchical clustering algorithm at first. Different from the previous works, we exploit the intrinsic properties of social network and events for event detection. In our multi-modality hierarchical clustering algorithm, a temporal-based clustering algorithm is performed after separating photos into albums. And, a multi-modality agglomerative hierarchical clustering algorithm is employed in each temporal cluster result respectively. Then the representative labels for each event are extracted from the textual in the same cluster. After doing it, each photo in the user's social circle will be associated with some reliable event labels and its initial tags given by the uploader. In the image annotation stage, personalized labels for the mobile photos are generated by a weighted K-nearest neighbor model similar to [6]. We improve the model by using both visual and date information to get the neighbors.

The contributions of this paper are manifold:

1. To tackle the unreliability issue of tags in social network, we exploit the characteristic of personal social circle and detect social events at first to generate reliable event tags.
2. We use "Album" as the basic unit for clustering and event detection, which not only eases the problem of large scale clustering, but also addresses the problem caused by the unreliability and sparsity of individual photo.
3. We propose a novel hierarchical clustering algorithm exploiting multi cues including the content, time, textual and social behavior information for social event detection.

The rest of the paper is organized as follows: In Sect. 2, the related works is introduced. In Sect. 3, we present our personalized annotation system, detail with the label generation and propagation. Data set analysis and experimental results are shown in Sect. 4. In Sect. 5, we conclude the paper and discuss future work.

2   Related Work

In this section, we briefly review some existing literature related to our work.

Content-Based Annotation. In recent years, many content-based annotation algorithms have been proposed and dramatically advanced this field [4]. They can be categorized into three main groups: (1) generative models [15], which try to estimate the joint probabilities between image visual features and labels; (2) discriminative models [7,21], which regard image annotation as a classification problem and consider each pre-defined label as an independent class; (3) graph-learning models [10,19], which use label diffusion over a similarity graph of labeled and unlabeled images. However, all of these content-based methods suffer from the well-known semantic gap.

Social Event Detection. As mentioned above, social event detection is performed to generate event tags for social photos in our label generation stage. Although there are some work focus on event detection from photos social metadata, most part of the methods regard it as an event classification or recognition problem in which the event ontology and number are pre-defined [2]. It is not suitable for our situation where there may be hundreds or thousands events in the user's social circle and the events are unknown before detection.

Recently, there are some works try to detect social event by clustering. In [1], the authors employ both ensemble and classification-based similarity learning techniques in conjunction with an incremental clustering algorithm to solve this problem, which is naive and only the textual, location and date information is used. In [16], the authors use pairwise similarities to predict a "same cluster" relationship. However, a known clusters from the same domain is required to adjust the weights and finally a K-mean or spectral cluster is performed, which requires a priori knowledge of the cluster number.

Personalized Annotation. To provide personalized annotation, the rich contextual information has been investigated. Some works leverage other mobile applications such as weather API, personal calendar or email context to get personalized labels [5]. For example, the calendar entry "Bob's birthday party on July 12, 2013" provides strong complementary information. In [3], the GPS location, compass direction and image visual are employed to find potential point of interest (POI) in a given area by clustering. A cross-entropy based learning algorithm to personalize a generic annotation model is proposed in [9]. Whereas in [12], a personalized tag recommendation system is proposed, which takes users' characteristics and tagging habits into consideration and gets the tag list by tags voting. [11] proposes a unified framework using subspace learning method to suggest personalized and geo-specific tags dynamically. The intuitive efforts have obtained a certain success in some extend, but the information they can exploit is limited and most of them are relying on the user's tagging history or geographical information.

Social Annotation. Leveraging social data for photos annotation has attracted significant attention recently. In [22], the authors propose a graph-learning based personalized annotation framework leveraging the friends social network photos as training dataset. There are also some works trying to exploit the social behavioral information, such as comments and likes. In [14], the authors employ social metadata (common galleries, locations, uploaders) to extend the SVM model to include relational features, with the intuition that images sharing common properties are likely to share labels. In [8], a common-interest model is presented, which studies the common interests between pairwise users from the social diffusion records of sharing content. However, they always ignore the sparsity and unreliability of social metadata and don't exploit the wealth of events' intrinsic properties, which is the highlight of our framework.

3 Proposed Framework

As mentioned above, for mobile photos annotation, users are more interested in the context information behind the photos, which makes it different from the previous content-based image annotation works. On the other hand, the user's social circle can provide valuable information. Intuitively, similar images are more likely to have the same labels. Starting from this intuition, we propagate the tags in the user's social circle to personal unlabeled images based on image similarity. However, as is known to all, the textual information contributed by

Label Generation

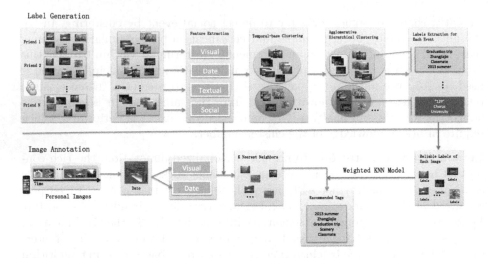

Fig. 1. Overview of our system

common users in the social network is sparse, ambiguous and unreliable. So, before label propagation, we should tackle this issue by generating reliable tags for social photos. Based on the observations and analyses in Sect. 1, we detect social events to generate high confidence event labels at first. For social event detection, we develop a multi-modality hierarchical clustering algorithm exploiting the intrinsic properties of social network and using "Album" as the basic clustering unit. Hence, our system can be divided into two main parts: label generation and image annotation. Figure 1 provides an overview of our system.

To summer, the process of our system is:

1. **Label Generation**
   1.1 Separate all photos in the user's social circle into albums.
   1.2 Extract visual, date, textual and social features, and learning multi-feature similarity metrics among albums.
   1.3 Hierarchical clustering based on albums, including density-based clustering according to the photos' taken date and multi-modality hierarchical agglomerative clustering in each temporal cluster result.
   1.4 Generate representative event labels in each cluster for all photos.
2. **Image Annotation**
   2.1 Train the weighted nearest neighbor model with discriminative metric learning from the user's social circle data as well as generated tags.
   2.2 Extract the visual and date feature of the given personal photos.
   2.3 Get k-nearest-neighbors of the given photo according to its visual and taken date information.
   2.4 Predict tags from the trained weighted k-nearest-neighbors neighbors by label propagation.

In the following, we provide more details on each of these components.

3.1 Label Generation

Album. The current social event detection algorithms suffer from two problems: the scalability problem and the unreliability of individual photo. How to address the two problems is a big challenge in our work.

Different from the images from web searching engine or commercial image banks, the photos in user's social circle are related to each other. And some social platforms even provide "Album" structure for users to organize their photos, such as "Renren". Experientially, the photos that a user uploads to the same album usually belong to the same event. What's more, even without the "Album" structure in social platforms, we can find that the photos uploaded by the same user at the same time are often closely correlated to each other, and likely to be taken at the same event. So, we introduce "Album" concept, it defined as a photo set, in which all the photos share the same event attribution and uploaded by the same user. Namely, the photos in the same "Album" belong to the same event, and an event may consist of one or more "Albums". Our experiments in Sect. 4 also indicate it.

So, instead of processing photo-by-photo, we use "Album" as basic unit for clustering to address the problem caused by the unreliability of individual photo and ease the pressure due to the large database.

In our paper, albums are generated by the following way: (1) If the social platforms have "Album" structure, then the photos in an album of social platform are regarded as belonging to the same album; (2) If the social platforms do not have "Album" structure, for each user in the social network, we get all the user's upload photos and sort them according to their upload dates. Then if the upload dates of two adjoining photos are within an hour, they are regarded as in the same album, otherwise they belong to different albums and we create a new album for the second photo. There are some works using the similar concept to us. However there are some fundamental differences between our "Album" concept and them. For example, in [20] the authors use Flickr Groups to provide more accurate annotation. They train specific annotation models for different Flickr groups (like Rome or Wedding), and chose the trained most appropriate Flickr group to generate labels for a given batch of images. They only use the common style of Flickr Groups, while we stress the concept of events.

Multi-modality Feature Extraction. As a distinctive characteristic, social networks include a variety of context features [1], which will help our clustering task. It is noteworthy that as stated above, we utilize "Album" as the base clustering unit, the similarity measures should be defined between two albums instead of two photos. In our work, the following features and corresponding similarity measures are used:

- **Visual Feature:** We use the 4096-dimensional visual feature vector extracted by Convolutional neural networks [7] to represent the image visual information, which has been widely used for different recognition problems. The visual similarity of two albums is defined as follow:

$$S_v(A, B) = \frac{1}{|A|} \sum_{i=1}^{|A|} \max_j \{v(a_i, b_j)\} \tag{1}$$

where A, B are two albums and a_i, b_j are the photos in A and B; $v(a_i, b_j)$ is the visual similarity of photo a_j and b_j defined as the cosine distance of the visual feature vectors of two images. Noted that $S_v(A, B)$ may be not equal to $S_v(B, A)$, so we use the average of $S_v(A, B)$ and $S_v(B, A)$ as the visual similarity score of albums A and B.

– **Date Feature:** We represent date as the number of minutes elapsed since the Unix epoch. Let t_a, t_b be the date value of image a and b, their similarity is defined as: $s_t = 1 - |t_a - t_b| / T$, where T equals $365 \times 24 \times 60$, namely T is the number of minutes in a year. If $s_t < \epsilon$, then $s_t = \epsilon$ (We use ϵ to avoid non-positive similarity score, and in practice we set $\epsilon = 10^{-6}$). The date similarity of two albums is defined as follow:

$$S_t(A, B) = \max(1 - \max(\frac{D_B^{min} - D_A^{max}}{T}, \frac{D_A^{min} - D_B^{max}}{T}, 0), 0) \tag{2}$$

where $D_A^{min} = \min_i\{t_{a_i}\}$, $D_A^{max} = \max_i\{t_{a_i}\}$, $D_B^{min} = \min_i\{t_{b_i}\}$, $D_B^{max} = \max_i\{t_{b_i}\}$. Intuitively speaking, $S_t(A, B)$ is the time span of album A and album B.

– **Textual Feature:** There are various textual features accompanying social photos, such as tag, title and description. They can be transformed into words by extracting nouns using natural language processing techniques. We defined a weighted similarity metric for different texts as they are unreliable at different level, for example title may provide strong complementary information than description. The weighted Jaccard similarity coefficient is employed to measure textual similarity. The textual feature similarity of two albums is defined as follow:

$$S_w(A, B) = w_{tag} * J_{tag} + w_{title} * J_{title} + w_{desc} * J_{desc} \tag{3}$$

where J_{tag} is the Jaccard similarity of tags in album A and album B, defined as follow:

$$J_{tag} = \frac{|Tag_A \cap Tag_B|}{|Tag_A \cup Tag_B|} \tag{4}$$

J_{title} and J_{desc} are defined the same as J_{tag}. And w_{tag}, w_{title} and w_{desc} are the weights and $w_{tag} + w_{title} + w_{desc} = 1$.

– **Social Feature:** We estimate social similarity according to multiple social factors such as friend relationship, comments, favorite images and share behavior. Let U_a, U_b be the owners of album A and B. The social similarity of two albums is defined as follow:

- If U_a and U_b are social friends, then their friend similarity is 1, otherwise is 0;
- If U_a comments photos in album B or U_b comments photos in album A, then their comment similarity is 1, otherwise is 0;

- If U_a favorites photos in album B or U_b favorites photos in album A, then their favorite similarity is 1, otherwise is 0;
- If U_a shares photos in album B or U_b shares photos in album A, then their share similarity is 1, otherwise is 0;

Having defined all these feature representation and corresponding similarity metrics, we combine all the features using a weighted similarity consensus function.

Hierarchical Clustering and Event Representation. As a key contribution, we propose a novel hierarchical clustering algorithm to detect social event.

For our scenario, the clustering algorithms should be scalable and not require a priori knowledge of the cluster number. So the traditional clustering algorithms required cluster numbers, such as K-means and spectral clustering, are not suitable in our situation.

Note that the events in our case are always small and there may be hundreds or thousands events in the user's social circle and many of them are hosted by few users. So the agglomerative hierarchical clustering is preferable for our clustering task, which is performed based on album similarity.

Considering that events are always time depended and do not last long, we employ a temeporal-based clustering at first and perform agglomerative hierarchical clustering algorithm in each date cluster result. By doing it, the data scale is reduced and clustering performance is improved due to the less noise.

For temeporal-based clustering, we exploit a density-based algorithm base on [18]. In this stage, we use photo instead of "Album" as the basic unit. The local density ρ_i is calculated by a Gaussian kernel based on date similarity for each photo. Then the minimum date distance between the photo i and any other photos with higher density δ_i is calculated. In our work, if the value of ρ or δ is larger than a pre-defined threshold, we then create a new cluster for it.

We exploit agglomerative hierarchical clustering on each temeporal-based cluster result, and merge them. For agglomerative hierarchical clustering, we combine all the feature similarities by a weighted function as a final similarity score between two albums. The process is as follow:

1. Each album is regarded as a separated cluster at first.
2. The two clusters with smallest distance are selected and merged into one cluster.
3. Calculate the similarity score between the new merged cluster and other clusters.
4. Repeat Step 2 and 3 until the smallest distance larger than a pre-defined threshold.

Then the representative labels for each event are extracted from the textual in the same cluster. And all the photos in the same cluster share the labels.

3.2 Label Propagation

As we mentioned, there are many approaches for annotation from labeled images. However, the discriminative models which should learn classifier for each label are unsuitable for our problem, since the labels of personalized annotation are heterogeneous and highly user-centric. Besides, generative models require a strong correlation relationship between images and tags, while the labels in our scenario are subjective. Intuitively, images have similar properties are likely to share labels. Graph-based label propagation algorithm is preferable for our annotation task, as it does not take the labels' inherent meaning into consideration. However, full graph method is time consuming due to the large data set. To tackle this issue, K-nearest-neighbor like methods have been introduced, which predict tags taking a combination of the tag absence/presence among neighbors.

So, we employ a weighted K-nearest-neighbor model similar to [6] to predict tags for given personal photos. Since both visual and date feature may provide complementary information for annotation, we get the K nearest neighbors based on both the visual and date similarity.

4 Experiment

4.1 DataSet

We employ both the public available dataset ReSEED [17] and real-world dataset for our experiments.

The ReSEED dataset consists of pictures collected from Flickr and the corresponding metadata such as user information, upload and capture time, geographic information, tags, title and description. And all the pictures are assigned to individual social events. We use a subset of the dataset with a capture time between January 1, 2012 and December 31, 2012, yielding a dataset of 15577 pictures assigned to 714 events in total. The dataset is employed to verify the assumption about "Album".

In this paper, we annotate a user's mobile photos exploiting his (or her) personal social circle, which is unavailable in public datasets. To evaluate our system, we crawl images together with their context from the user's social circle (Renren) from January 1, 2013 to December 31, 2013 and manually tag the event of each photo for evaluating our clustering algorithm. As a result, we construct a training data based on the user's social circle and give annotation for the user's personal images. Table 1 provides more details regarding the Renren dataset used in our experiments.

Table 1. Statistics of our real-world dataset

#Photos	#Friends	#Events	#Albums	#Min photos of a friend	#Max photos of a friend	#Min photos of an event	#Max photos of an event
33879	361	1174	1149	1	2663	1	1145

Table 2. The performance of Album

Dataset	Purity	Precision	Recall	F1
ReSEED	**0.9962**	**0.9992**	0.3986	0.5698
Renren	**0.9995**	**0.9998**	0.6241	0.7685

Table 3. Clustering performance comparison in terms of NMI and F1

Methods	NMI	F1
Incremental clustering [1]	0.7973	0.1983
Our algorithm	**0.9384**	**0.6381**

4.2 Evaluation of Album

The key hypothesis of this paper is that the photos belong to the same album are taken in the same event. Since the albums are generated by two different approaches as stated in Sect. 3.1, we analyze two datasets representing the two cases respectively.

For social media platforms with "Album" structure, we use the real-word dataset crawled from Renren. For social media platforms without "Album" structure, we use the ReSEED dataset from Flickr stated in Sect. 4.1. For the ReSEED dataset, we separate all the photos into albums according their taken time and get 1272 albums finally. To verify the proposed assumption and evaluate the performance of our album generation approach, we regard each album as a cluster and measure its performance using Precision and Purity. Table 2 shows the result. We can observe that the purity and precision scores are nearly 100 % in both the two datasets, which demonstrates our assumption. Besides, note that album is not equivalent to event, an event can consist of many albums, so the recall and F1 scores are low.

4.3 Evaluation of Tag Generation

To demonstrate the advantages of our proposed hierarchical clustering algorithm, we use NMI and F1 to measure the performance comparing with a single-pass incremental clustering algorithm used in [1]. Table 3 shows the results. As it indicates, our proposed hierarchical clustering algorithm is much better than the baseline for both NMI and F1 score.

4.4 Evaluation of Personalized Annotation

In this section, we present a experimental comparison between the performance of content-based image annotation system [7], employing original unreliable social accompanying tags directly on the KNN model and the performance of our proposed framework, which generates reliable tags by social event detection before performing the KNN model.

Table 4. Comparision of Recall, Precision and F1

Methods	Recall	Precision	F1
Conted-based [7]	0.0363	0.0205	0.0262
Without tags generation	0.8252	0.4615	0.5919
Our proposed algorithm	**0.8752**	**0.5026**	**0.6385**

In our experiments, we combine all the top 5 ranked labels generated by the three methods, and allow users to select their preferred labels. Recall, Precision and F1 are adopted to measure the performance. Table 4 shows the result. We can observe that our proposed framework obtains the best performance.

5 Conclusion

In this paper, we proposed a personalized annotation framework for mobile photos leveraging the user's social circle. To address the issue caused by the sparsity and unreliability of social photo tags, we generated reliable tags by detecting social events at first. An multi-modality hierarchical clustering algorithm using "Album" as the basic unit was proposed to detect social event by exploiting all the text, date, social behavior and visual features. By analyzing the characteristic of our scenario, a weighted KNN model was exploited to propagate the generated tags of social photos to the user's unlabeled photos. Experimental results show our system is effective. In the future work, we will use the additional information in personal photos as feedback to refine the tag generation stage.

Acknowledgments. This work was part-funded by 973 Program under Grant No. 2011CB302206, National Natural Science Foundation of China under Grant No. 61272231, 61472204, Beijing Key Laboratory of Networked Multimedia.

References

1. Becker, H., Naaman, M., Gravano, L.: Learning similarity metrics for event identification in social media. In: Proceedings of the Third ACM International Conference on Web Search and Data Mining, pp. 291–300. ACM (2010)
2. Cao, L., Luo, J., Kautz, H., Huang, T.S.: Image annotation within the context of personal photo collections using hierarchical event and scene models. IEEE Trans. Multimedia **11**(2), 208–219 (2009)
3. Cheng, A.J., Lin, F.E., Kuo, Y.H., Hsu, W.H.: Gps, compass, or camera? Investigating effective mobile sensors for automatic search-based image annotation. In: Proceedings of the International Conference on Multimedia, pp. 815–818. ACM (2010)
4. Datta, R., Joshi, D., Li, J., Wang, J.Z.: Image retrieval: Ideas, influences, and trends of the new age. ACM Comput. Surv. (CSUR) **40**(2), 5 (2008)
5. Gallagher, A.C., Neustaedter, C.G., Cao, L., Luo, J., Chen, T.: Image annotation using personal calendars as context. In: Proceedings of the 16th ACM International Conference on Multimedia, pp. 681–684. ACM (2008)

6. Guillaumin, M., Mensink, T., Verbeek, J., Schmid, C.: Tagprop: Discriminative metric learning in nearest neighbor models for image auto-annotation. In: 2009 IEEE 12th International Conference on Computer Vision, pp. 309–316. IEEE (2009)
7. Krizhevsky, A., Sutskever, I., Hinton, G.E.: Imagenet classification with deep convolutional neural networks. In: Advances in Neural Information Processing Systems, pp. 1097–1105 (2012)
8. Lei, C., Liu, D.: Image annotation via social diffusion analysis with common interests. In: 2014 IEEE International Conference on Multimedia and Expo Workshops (ICMEW), pp. 1–6. IEEE (2014)
9. Li, X., Gavves, E., Snoek, C.G., Worring, M., Smeulders, A.W.: Personalizing automated image annotation using cross-entropy. In: Proceedings of the 19th ACM International Conference on Multimedia, pp. 233–242. ACM (2011)
10. Liu, J., Li, M., Liu, Q., Lu, H., Ma, S.: Image annotation via graph learning. Pattern Recogn. **42**(2), 218–228 (2009)
11. Liu, J., Li, Z., Tang, J., Jiang, Y., Lu, H.: Personalized geo-specific tag recommendation for photos on social websites. IEEE Trans. Multimedia **16**(3), 588–600 (2014)
12. Liu, X., Qian, X., Lu, D., Hou, X., Wang, L.: Personalized tag recommendation for flickr users. In: 2014 IEEE International Conference on Multimedia and Expo (ICME), pp. 1–6. IEEE (2014)
13. Makadia, A., Pavlovic, V., Kumar, S.: A new baseline for image annotation. In: Forsyth, D., Torr, P., Zisserman, A. (eds.) ECCV 2008, Part III. LNCS, vol. 5304, pp. 316–329. Springer, Heidelberg (2008)
14. McAuley, J., Leskovec, J.: Image labeling on a network: using social-network metadata for image classification. In: Fitzgibbon, A., Lazebnik, S., Perona, P., Sato, Y., Schmid, C. (eds.) ECCV 2012, Part IV. LNCS, vol. 7575, pp. 828–841. Springer, Heidelberg (2012)
15. Niu, Z., Hua, G., Gao, X., Tian, Q.: Semi-supervised relational topic model for weakly annotated image recognition in social media. In: 2014 IEEE Conference on Computer Vision and Pattern Recognition (CVPR), pp. 4233–4240. IEEE (2014)
16. Petkos, G., Papadopoulos, S., Kompatsiaris, Y.: Social event detection using multimodal clustering and integrating supervisory signals. In: Proceedings of the 2nd ACM International Conference on Multimedia Retrieval, p. 23. ACM (2012)
17. Reuter, T., Papadopoulos, S., Mezaris, V., Cimiano, P.: Reseed: social event detection dataset. In: Proceedings of the 5th ACM Multimedia Systems Conference, pp. 35–40. ACM (2014)
18. Rodriguez, A., Laio, A.: Clustering by fast search and find of density peaks. Science **344**(6191), 1492–1496 (2014)
19. Tang, J., Hong, R., Yan, S., Chua, T.S., Qi, G.J., Jain, R.: Image annotation by k nn-sparse graph-based label propagation over noisily tagged web images. ACM Trans. Intell. Syst. Technol. (TIST) **2**(2), 14 (2011)
20. Ulges, A., Worring, M., Breuel, T.: Learning visual contexts for image annotation from flickr groups. IEEE Trans. Multimedia **13**(2), 330–341 (2011)
21. Wang, J., Yang, J., Yu, K., Lv, F., Huang, T., Gong, Y.: Locality-constrained linear coding for image classification. In: 2010 IEEE Conference on Computer Vision and Pattern Recognition (CVPR), pp. 3360–3367. IEEE (2010)
22. Wu, Z., Aizawa, K.: Building friend wall for local photo repository by using social attribute annotation. J. Multimedia **9**(1), 4–13 (2014)
23. Zhang, D., Islam, M.M., Lu, G.: A review on automatic image annotation techniques. Pattern Recogn. **45**(1), 346–362 (2012)

Utilizing Sensor-Social Cues to Localize Objects-of-Interest in Outdoor UGVs

Yingjie Xia[1], Luming Zhang[2]($\boxtimes$), Liqiang Nie[3], and Wenjing Geng[3]

[1] College of Computer Sciences, Zhejiang University, Hangzhou, China
[2] Department of CSIE, Hefei University of Technology, Hefei, China
zglumg@gmail.com
[3] School of Computing, National University of Singapore, Singapore, Singapore

Abstract. A huge number of outdoor user-generated videos (UGVs) are recorded daily due to the popularity of mobile intelligent devices. Managing these videos is a tough challenge in multimedia field. In this paper, we tackle this problem by performing object-of-interest (OOI) recognition in UGVs to identify semantically important regions. By leveraging geo-sensor and social data, we propose a novel framework for OOI recognition in outdoor UGVs. Firstly, the OOI acquisition is conducted to obtain an OOI frame set from UGVs. Simultaneously, the classified object set recommendation is performed to obtain a candidate category name set from social networks. Afterward, a spatial pyramid representation is deployed to describe social objects from images and OOIs from UGVs, respectively. Finally, OOIs with their annotated names are labeled in UGVs. Extensive experiments in outdoor UGVs from both Nanjing and Singapore demonstrated the competitiveness of our approach.

1 Introduction

Location-based services provided by social networks, such as Facebook and Twitter, remarkably enrich the quantity of multimedia content tagged by geo-sensor including latitude and longitude. Besides, the popularity of mobile devices with sensors makes capturing, uploading, and sharing of outdoor user-generated videos (UGVs) highly convenient. This motivates us to investigate effective techniques to manage these Internet-scale UGVs.

To handle the vast amount of UGVs on social networks, we focus on object-of-interest (OOI) recognition, *i.e.*, building an OOI recognition system by leveraging both visual features and sensor-social data. The major benefit is that not only it can localize OOIs, but also it is highly efficient and accurate by adopting sensor-social data. Such a recognition system would be of tremendous value and significance for a large body of multimedia applications. For example, Zheng *et al.* [1] proposed a web-scale landmark recognition engine by leveraging the vast amounts of multimedia data. However, most GPS-tagged recognition systems depend on a large collection of images to achieve accurate visual clusters. The existing techniques, however, are unsuitable for OOI recognition in UGVs because of two reasons: (1) OOI recognition in UGVs should be a lightweight

© Springer International Publishing Switzerland 2016
Q. Tian et al. (Eds.): MMM 2016, Part I, LNCS 9516, pp. 88–99, 2016.
DOI: 10.1007/978-3-319-27671-7_8

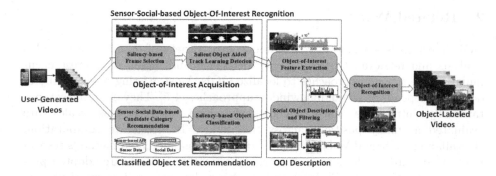

Fig. 1. The proposed OOI recognition pipeline using sensor-social data

application since UGVs are usually captured by mobile devices. Therefore, off-loading many recognition tasks onto cloud servers may increase latency and response time; (2) typical approaches which acquire a "complete" image dataset to handle object recognition may consume extra computation practically.

The explosive growth of UGVs leads to a significant challenge on how to efficiently organize large video repositories and make them searchable. Common approaches adopt content-based media analysis to extract visual features for similarity matching. However, due to the overwhelming amount of video materials, it is inappropriate to perform feature matching on a frame-by-frame level. In this work, we understand video content at object-level in a lightweight way. We propose to recognize OOIs in UGVs with user-intentionally captured objects. Similar to our work, Hao et al. [2] focused on point-of-interest detection in sensor-rich videos. It was achieved by analyzing a large number of sensor-rich videos automatically and comprehensively. This implies that the method is unsuitable for OOI recognition in a single UGV.

An overview of the proposed method is presented in Fig. 1. We focus on analyzing UGVs uploaded on social networks at object-level, by utilizing sensor-social data. Given a collection of UGVs, the OOI acquisition and the classified object set recommendation are conducted simultaneously. The former task can be formulated as salient objects extraction, where saliency indicates the informative/interesting regions within a scenery. To obtain the most representative frames, a saliency-guided selection algorithm is proposed to filter frames with similar saliency distributions. For the latter task, candidate categories are recommended by leveraging sensor-social data. Metadata including timestamps, GPS coordinates, accuracy, and visible distances are employed as sensor data. Afterward, salient objects are extracted in social images based on category classification. A spatial-pyramid architecture [3] is adopted to describe social objects and OOIs in UGVs for its robustness in scene modeling. And the Euclidean distance is employed to measure the similarity between the classified and labeled object sets. Finally, OOIs associated with their annotated names are labeled in UGVs frame-by-frame. Experiments on object-level video summarization and content-based video retrieval demonstrate the usefulness of our method.

2   Related Work

OOI detection is a widely used technique in a variety of domains, *e.g.*, video analysis and retrieval. Object/saliency detection and region of interest accumulation are typical approaches to localize OOIs. Most existing work on object detection depends on the sliding window approaches [4,5]. They might be computationally intractable since windows detection with various scales are evaluated at many positions across the image. To accelerate computation, Harzallah *et al.* [8] and Vedaldi *et al.* [9] designed cascade-based methods respectively to discard windows at each stage, where richer features are adopted progressively. Cinbis *et al.* [10] developed an object detection system by employing the Fisher vector representation. State-of-the-art performance was achieved for image and video categorization. Kim *et al.* [11] proposed an OOI detection algorithm based on the assumption that OOIs are usually located near the image centroid. Zhang *et al.* [12] introduced a novel approach to extract primary object segments in videos from multiple object proposals. Although the above methods performs well on object detection, they are not lightweight algorithms. Thus, they cannot effectively handle OOI recognition toward mobile devices.

Many recent OOI detection algorithms are based on visual saliency prediction [13,14]. It is generally accepted that OOIs are aroused by human perception and visual saliency can reflect the cognitive mechanism. Therefore, saliency prediction performance significantly influences these methods in detecting OOIs. Most of the existing saliency models are completely based on low-level visual features [15,16]. However, some high-level semantic cues [17,18] should also be integrated for saliency calculation [19]. Both biological and psychological studies [20] shown that, optimally fusing low-level and high-level visual features (including the location cue) can enhance saliency detection greatly. We employ the saliency detection by deploying the markov chain proposed by Jiang *et al.* [19]. One advantage of [19] is that both the appearance divergence and spatial distribution of foreground/background objects are integrated. It performs better on our multi-source location-aware dataset as compared with its competitors.

Many approaches have been proposed to predict where human perceives when viewing a scenery. The majority of the existing methods recognize OOIs based on the similarity of appearance features. Recently, the cheap availability of sensor-rich videos allows users to understand video semantics in a straightforward way [29]. For these different types of sensory metadata, we focus on the geo-attributes of sensor data throughout this paper. Associating GPS coordinates with digital photographs has becoming an active research domain over the last decade [30]. Toyama *et al.* [31] introduced a metadata-based image search algorithm and compiled a database which indexes photos using location and timestamp. Föckler *et al.* [32] developed a museum guidance system by utilizing camera-equipped mobile phones. Zheng *et al.* [1] constructed an efficient and effective landmark recognition engine, which organizes, models, and recognizes the landmarks on the world-scale. Gammeter *et al.* [33] introduced a fully functional and complete augmented reality system which can track both stationary and mobile objects. By utilizing geo-sensor data, a number of object recognition tasks are implemented based on GPS coordinates.

3   Sensor-Social-Based OOI Recognition

Given an outdoor UGV, we detect its OOIs and annotate them by utilizing a variety of multimedia features. The key to recognize OOIs in outdoor UGVs is to fuze video content, sensor data, and social factors optimally. Thereafter, video sequences with annotated OOIs can be generated.

3.1   OOI Acquisition from UGVs

Saliency-Based Frames Selection. Obviously, semantics between sequential video frames are highly correlated. Existing summarization algorithms typically detect key frames to alleviate computational burden. These techniques are popularly used in video editing and compression. Notably, two factors should be emphasized in our method: the computational efficiency and representative OOI sequences. This means that the conventional key frames selection algorithms may not be able to preserve the diverse OOI sequences. In order to select representative frames at OOI-level, we propose a novel saliency-based frame selection. First, saliency map of each UGV frame is calculated based on Jiang et $al.$'s algorithm [19]. We employ [19] because it jointly describes the appearance divergence and spatial distribution of foreground/background objects. By adopting the Markov chain theory [21], the saliency detection is conducted rapidly. Let $Sal_{c,s}$ denote the calculated saliency map based on color and spatial distributions, e index the transient graph nodes, and y_w be the normalized weighted absorbed time vector, then the saliency map is simply obtained as:

$$Sal_{c,s}(e) - y_w(e), \quad i = 1, 2, \cdots, t, \tag{1}$$

Afterward, region of OOI, denoted as $R_{bw}(\cdot)$, is binarized by an adaptive threshold τ_1. The criterion of saliency-based frames selection is:

$$decision(i) = \begin{cases} 1 & \text{if } ||Th(i) - Th(i+1)|| > \tau_2 \\ 0 & \text{otherwise} \end{cases}, \tag{2}$$

where τ_2 denotes the divergence of saliency values between neighboring frames; $Th(i)$ is the salient area in frame i and $R(i)$ the salient object region; $R(i, i+1)$ denotes the salient area intersection between frames i and $i+1$.

Salient-Object-Assisted Track Learning Detection. To recognize OOIs, it is necessary to generate a number of OOI candidates extracted from UGVs. To balance the efficiency and accuracy, we employ the track learning detection framework proposed by Kalal et $al.$ [22]. One advantage of [22] is that it can decompose a long-term tracking task into tracking, learning, and detection efficiently. Due to the complicated spatial context of a scenery, it is difficult to

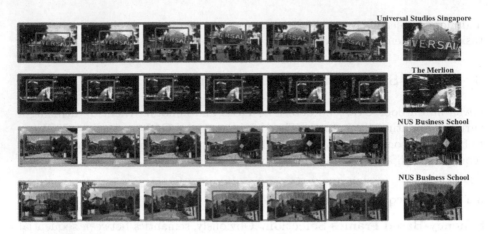

Fig. 2. Tracking recognition for OOIs in UGVs. The long box contains frames randomly selected from UGVs with a tracking box around OOIs. The right column displays the extracted objects marked by the annotated names.

detect all the objects in UGVs accurately. To solve this problem, we propose a salient-object-assisted track learning detection. It combines object and saliency detection when processing each UGV frame. If the object detection fails, saliency detection will be conducted and assists the similarity measure between patches. Based on the assisting scheme, OOI acquisition is conducted for each frame. Thereafter, the new object modeling can be formulated as:

$$M = \{p_1^+, p_2^+, \cdots, p_x^+, \cdots, p_m^+, p_1^+, p_2^-, \cdots, p_x^-, \cdots, p_m^-\}, \tag{3}$$

where p^+ and p^- denote the foreground and background patches respectively; p_x^+ and p_x^- are the saliency patches of object and background respectively. Example OOIs extracted from the UGVs are presented in Fig. 2. As can be seen, the proposed method not only detects those OOIs accurately, but also tracks them within the UGV frames. The tracking is performed by localizing a bounding box centered around each detected OOI.

3.2 Classified Object Set Recommendation

Assisted by human interactions, social data has become an intellective media conveying informative cues, e.g., tagged images, video clips, and user comments. It is worth emphasizing that social data also contains lots of noises. Thus, effectively exploiting social data is a challenging task.

Sensor data is recorded by sensory modules embedded in mobile devices. In this work, we model sensor data of UGVs as a frame-related feature vector, which can be specified as:

$$S = \{(t_i, lat_i, long_i, accur_i, visD_i) | t_i \in T, (lat_i; long_i) \in G,$$
$$accur_i \in A, visD_i \in V\}, \tag{4}$$

where T contains the capturing time of each frame; G is a set of GPS coordinates that describe the capturing location changes; A is a set of GPS location errors; V is a set of visible distances calculated by Arslan Ay *et al.* [28].

We constructed an image set containing candidate OOIs which are collected based on the category keywords. In particular, image retrieval is conducted by using different category names. Then, a collection of images are downloaded and classified from social networks. In order to compare at object-level, we calculate the saliency maps from these social images and then extract the salient objects as the OOIs adaptively. Saliency-based object classification minimizes the influence of noises resulted from the various backgrounds in social images. The classified OOIs from social images can be described as:

$$O_L = \{R_{bw}^1, R_{bw}^2, \cdots, R_{bw}^n | n \in \mathcal{N}_L\}, \tag{5}$$

where O_L is the OOI set labeled by L; and $\mathcal{N}$ is the candidate category set.

3.3   OOI Description and Recognition

We adopt a spatial-pyramid-based [3] feature to represent an image, since it combines the advantages of standard feature extraction method. Spatial pyramid is a simple and efficient extension of an orderless bag-of-features image representation. It exhibits significantly improved performance on challenging scene categorization tasks. More specifically, local visual descriptors are quantized into a D-sized dictionary. Then, the spatial pyramid feature for the c-th class and n-th object is calculated as:

$$F_n^c = \{[f_1^1, f_2^1, \cdots, f_t^1][f_1^2, f_2^2, \cdots, f_s^2], \cdots, [f_1^p, f_2^p, \cdots, f_q^p]\}, \tag{6}$$

where p represents the pyramid level; t, s, and q denote the feature dimensionality of each pyramid level. Examples of the above spatial pyramid representation are presented on the left of Fig. 3. Noticeably, to maximally eliminate the negative effects caused by the complicated scenic backgrounds, we introduce a salient object based image filtering scheme, as elaborated in Fig. 4. We perform k-means clustering of two subsets from each classified object image set to constitute two class feature samples. Generally, people tend to capture images with similar salient objects for a category. Thus, we discriminate positive and negative samples using the intra-class variance. Additionally, we extract features with the same spatial pyramid architecture for OOI in UGVs, toward a consistent feature description. A few examples are shown on the right of Fig. 3.

As the last step, we recognize objects in UGVs using a similarity metric to compare objects extracted from an image set. The similarity is calculated between the mean features of OOIs extracted from UGVs and those of salient objects extracted from social images.

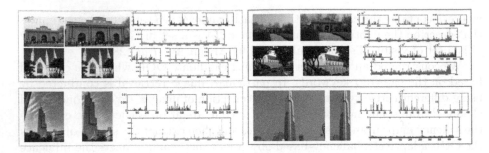

Fig. 3. Left: social objects and their three level spatial-pyramid features; right: OOIs of UGVs and their three level spatial-pyramid features

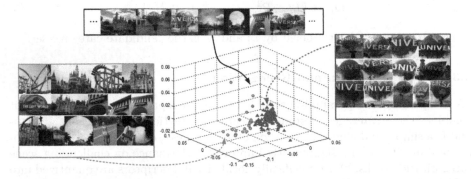

Fig. 4. Salient-object-based social images filtering

4   Experimental Results and Analysis

4.1   Dataset and Experimental Setup

The UGVs in our experiments consist of sensor-annotated videos captured from an Android/iOS device in Nanjing and Singapore. For the Nanjing dataset, five volunteers captured 676 UGVs using Sumsung Galaxy Note 3 and iPhone 6 respectively. Two resolutions 3840×2160 and 1920×1080 are employed. The Singapore dataset has 835 720×480 UGVs with complicated sceneries, *e.g.*, the Merlion, the Marina Bay, the Esplanade, and the Singapore Flyer.

Our approach is implemented on a desktop PC with an Intel i7-4770K CPU and 16 GB main memory. Java is adopted to parse the Json data collected from social servers. Matlab is used to implement the entire framework for its convenience in image/video processing. The location-based social network is implemented based on the Foursquare[1]. The threshold τ_1 for salient region detection is adaptively calculated by OTSU. The frame selection threshold τ_2 is set to 0.2 and the spatial-pyramid level p is set to 3.

---

[1] https://foursquare.com/.

4.2   Experimental Results and Analysis

The experiments are designed to evaluate: (1) whether the proposed frames selection method is capable to preserve OOIs from UGV in order to accelerate computation, (2) users' satisfaction about the proposed tracking detection for OOIs in UGVs, and (3) the recognition accuracy.

Efficiency of Frame Selection. Figure 5 presents some results of the saliency-based frames selection. To better elaborate our proposed frames selection, we design a PSNR-loss histogram to measure the quality of the selected frames. The PSNR measure is popularly used to evaluate the reconstruction quality of the loss compression codec between images. In our experiment, we construct a PSNR-loss histogram $H = \{P_{12}, \cdots, P_{ij}\}_L$ to calculate the PSNR difference between the i-th and j-th frames both in the original and the selected sequences. L is the frame number of the original UGVs. P denotes PSNR and is defined as:

$$P = 10 * \log_{10}\left(\frac{2^n - 1}{M_{SE}}\right),\qquad(7)$$

where $M_{SE} = \sum_{x=1}^{M} \sum_{y=1}^{M} (f(x,y) - g(x,y))/M * N$; n represents using n bits per sample, $f(x,y)$ and $g(x,y)$ are the grayscale of neighboring frames; $M \times N$ is the size of each frame.

The PSNR value of the selected sequences falls into the bin based on its frame number in the original videos. Therefore, the information loss can be compared at frame-to-frame level. Figure 6 presents the PSNR-loss histograms of one UGV, reflecting the information loss of the input UGV, and the selected sequences with $\tau_1 = 0.1$ and $\tau_2 = 0 : 2$. The red rectangle indicates that our method excludes the frames with very low information loss. It guarantees that the diversity changes of OOIs can be well preserved in the selected UGV frames.

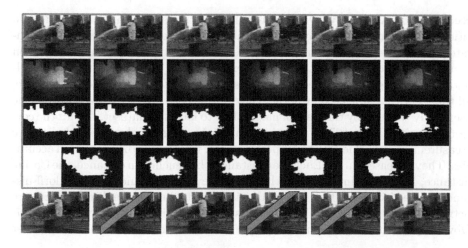

Fig. 5. Example frames of the saliency-based selection

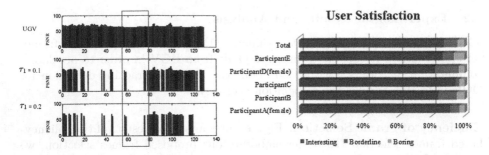

Fig. 6. Left: PSNR-loss histogram of the original UGV and the saliency-guided selected UGV frames; Right: user satisfaction with respect to the tracking detection

User Satisfaction. To evaluate the effectiveness of the proposed system, we invite five volunteers (two females and three males) whom are the photographers of the GeoVid[2] to participate our user study. As to the multi-source location-aware dataset in Singapore, we also invite them to rate the OOI tracking results generated by our system. Each volunteer rates the UGVs captured by himself/herself, and then randomly assigns one fifth part of the Singapore dataset. The participants are asked to choose from three feelings about the generated UGVs: "Interesting", "Borderline", and "Boring", which reflect their opinions after viewing the UGVs with the OOI tracking box. Noticeably, the five volunteers label each video to determine whether the OOIs are recognized successfully or not. Afterward, we accumulate the feedbacks from the five volunteers, as shown on the right of Fig. 6. We also explore the reasons why they feel boring about some UGVs. We observe that the reason is that the wrong trackings occurred on several frames. The borderline opinion primarily due to the size of bounding boxes. Some of them cannot fully contain the OOIs.

Recognition Accuracy. Our multi-source location-aware dataset contains two cities: Nanjing and Singapore. We first calculate the recognition accuracies separately on the two cities. Afterward, we average them to obtain a final recognition accuracy of our designed system. All the experimental UGVs are captured by volunteers spanning a long time, and there is no ground truth presented. Therefore, all the UGVs are labeled by whether they can be correctly recognized during the user study. We employ the traditional method to label the dataset, "1" for the correct recognition while "0" for the mistaken one. In order to validate which distance measurement can achieve the best performance, we calculate 6 recognition accuracies. They are based on the Euclidean distance, the Seuclidean distance, the Cosine distance, Histogram intersection, the Chebychev distance and the Hausdorff distance, respectively. The final recognition is calculated using the distance measure between feature vectors. All the accuracies on the two cities are presented in Fig. 7. Obviously, calculating the similarity by histogram intersection achieves the best

---

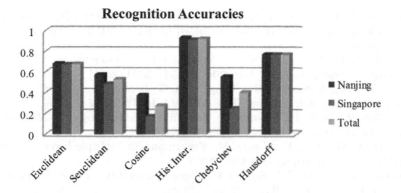

Fig. 7. OOI Recognition accuracies of UGVs extracted from Nanjing and Singapore

accuracy of 92.86 % on the Nanjing dataset, and 91.02 % on the Singapore dataset. Therefore, the average recognition accuracy of our system on the multi-sources dataset is 91.94 %.

5 Conclusions

OI recognition on UGVs is an important application in multimedia [24–27] and artificial intelligence [6,7,23,34]. This paper proposes an automatic system to achieve OOI recognition on UGVs by leveraging sensor-social data. The key contributions of this paper can summarized as follows. First, we propose a light-weight framework for recognizing OOIs in outdoor UGVs by leveraging geo-sensor data with the location-aware social networks. Second, we introduce a novel saliency-guided frame selection algorithm, which performs OOI recognition effectively and reduces the computational burden. Third, we compile a multi-source location-aware dataset containing two cities, Nanjing and Singapore, with three kinds of resolutions and two types of frame rates. Third, our system achieves an OOI recognition accuracy of 91.94 %, which demonstrated that it is useful in both mobile and desktop applications.

References

1. Zheng, Y.-T., Zhao, M., Song, Y., Adam, H.: Tour the world: building a web-scale landmark recognition engine. In: Proceedings of CVPR (2009)
2. Hao, J., Wang, G., Seo, B., Zimmermann, R.: Point of interest detection and visual distance estimation for sensor-rich video. IEEE T-MM **16**(7), 1929–1941 (2014)
3. Lazebnik, S., Schmid, C., Ponce, J.: Beyond bags of features: spatial pyramid matching for recognizing natural scene categories. In: Proceedings of CVPR (2006)
4. Dalal, N., Triggs, B.: Histograms of oriented gradients for human detection. In: Proceedings of CVPR (2005)

5. Felzenszwalb, P.F., Girshick, R.B., McAllester, D.A., Ramanan, D.: Object detection with discriminatively trained part-based models. IEEE T-PAMI **32**(9), 1627–1645 (2010)
6. Yang, K., Wang, M., Hua, X.-S., Yan, S., Zhang, H.-J.: Assemble new object detector with few examples. IEEE T-IP **20**(12), 3341–3349 (2011)
7. Wang, M., Hua, X.-S., Hong, R., Tang, J., Qi, G.-J., Song, Y.: Unified video annotation via multi-graph learning. IEEE T-CSVT **19**(5), 733–746 (2009)
8. Harzallah, H., Jurie, F., Schmid, C.: Combining efficient object localization and image classification. In: Proceedings of ICCV (2009)
9. Vedaldi, A., Gulshan, V., Varma, M., Zisserman, A.: Multiple kernels for object detection. In: Proceedings of ICCV (2009)
10. Cinbis, R.G., Verbeek, J.J., Schmid, C.: Segmentation driven object detection with fisher vectors. In: Proceedings of ICCV (2013)
11. Kim, S., Park, S., Kim, M.: Central object extraction for object-based image retrieval. In: Bakker, E.M., Lew, M., Huang, T.S., Sebe, N., Zhou, X.S. (eds.) CIVR 2003. LNCS, vol. 2728, pp. 39–49. Springer, Heidelberg (2003)
12. Zhang, D., Javed, O., Shah, M.: Video object segmentation through spatially accurate and temporally dense extraction of primary object regions. In: Proceedings of CIVR (2013)
13. Jiang, H., Wang, J., Yuan, Z., Liu, T., Zheng, N.: Automatic salient object segmentation based on context and shape prior. In: Proceedings of BMVC (2011)
14. Khuwuthyakorn, P., Robles-Kelly, A., Zhou, J.: Object of interest detection by saliency learning. In: Daniilidis, K., Maragos, P., Paragios, N. (eds.) ECCV 2010, Part II. LNCS, vol. 6312, pp. 636–649. Springer, Heidelberg (2010)
15. Margolin, R., Tal, A., Zelnik-Manor, L.: What makes a patch distinct? In: Proceedings of CVPR (2013)
16. Rosin, P.L.: A simple method for detecting salient regions. Pattern Recogn. **42**(11), 2363–2371 (2009)
17. Jia, Y., Han, M.: Category-independent object-level saliency detection. In: Proceedings of ICCV (2013)
18. Jiang, P., Ling, H., Yu, J., Peng, J.: Salient region detection by UFO: uniqueness, focusness and objectness. In: Proceedings of ICCV (2013)
19. Navalpakkam, V., Itti, L.: Modeling the influence of task on attention. Vision. Res. **45**(2), 205–231 (2005)
20. Borji, A.: Boosting bottom-up and top-down visual features for saliency estimation. In: Proceedings of CVPR (2012)
21. Bolch, G., Greiner, S., de Meer, H., Trivedi, K.S.: Queueing Networks and Markov Chains, 2nd edn. John Wiley, Hoboken (2006)
22. Kalal, Z., Mikolajczyk, K., Matas, J.: Tracking-learning-detection. IEEE T-PAMI **34**(7), 1409–1422 (2012)
23. Zhang, L., Bian, W., Song, M., Tao, D., Liu, X.: Integrating local features into discriminative graphlets for scene classification. In: Lu, B.-L., Zhang, L., Kwok, J. (eds.) ICONIP 2011, Part III. LNCS, vol. 7064, pp. 657–666. Springer, Heidelberg (2011)
24. Zhang, L., Song, M., Sun, L., Liu, X., Wang, Y., Tao, D., Bu, J., Chen, C.: Spatial graphlet matching kernel for recognizing aerial image categories. In: ICPR (2012)
25. Zhang, L., Gao, Y., Zimmermann, R., Tian, Q., Li, X.: Fusion of multichannel local and global structural cues for photo aesthetics evaluation. IEEE T-IP **23**(3), 1419–1429 (2014)
26. Zhang, L., Wang, M., Nie, L., Hong, L., Rui, Y., Tian, Q.: Retargeting semantically-rich photos. IEEE T-MM **17**(9), 1538–1549 (2015)

27. Zhang, L., Gao, Y., Hong, R., Hu, Y., Ji, R., Dai, Q.: Probabilistic skimlets fusion for summarizing multiple consumer landmark videos. IEEE T-MM **17**(1), 40–49 (2015)
28. Ay, S.A., Zimmermann, R., Kim, S.H.: Viewable scene modeling for geospatial video search. In: ACM Multimedia (2008)
29. Zheng, Y.-T., Zha, Z.-J., Chua, T.-S.: Research and applications on georeferenced multimedia. Multimedia Tools Appl. **51**(1), 77–98 (2011)
30. Rodden, K., Wood, K.R.: How do people manage their digital photographs? In: ACM SIGCHI (2003)
31. Kentaro, T., Logan, R., Roseway, A., Anandan, P.: Geographic location tags on digital images. In: ACM Multimedia (2003)
32. Föckler, P., Zeidler, T., Brombach, B., Bruns, E., Bimber, O.: PhoneGuide: museum guidance supported by on-device object recognition on mobile phones. In: Proceedings of Mobile and Ubiquitous Multimedia (2005)
33. Gammeter, S., Gassmann, A., Bossard, L.: Server-side object recognition and client-side object tracking for mobile augmented reality. In: Proceedings of CVPR (2010)
34. Wang, M., Gao, Y., Ke, L., Rui, Y.: View-based discriminative probabilistic modeling for 3D object retrieval and recognition. IEEE T-IP **22**(4), 1395–1407 (2013)

NEWSMAN: Uploading Videos over Adaptive Middleboxes to News Servers in Weak Network Infrastructures

Rajiv Ratn Shah[1]([✉]), Mohamed Hefeeda[2], Roger Zimmermann[1], Khaled Harras[3], Cheng-Hsin Hsu[4], and Yi Yu[5]

[1] National University of Singapore, Singapore, Singapore
{rajiv,rogerz}@comp.nus.edu.sg
[2] Qatar Computing Research Institute, Doha, Qatar
mhefeeda@qf.org.qa
[3] CMU in Qatar, Doha, Qatar
kharras@qatar.cmu.edu
[4] National Tsinghua University, Hsinchu, Taiwan
chsu@cs.nthu.edu.tw
[5] National Institute of Informatics, Tokyo, Japan
yiyu@nii.ac.jp

Abstract. An interesting recent trend, enabled by the ubiquitous availability of mobile devices, is that regular citizens report events which news providers then disseminate, *e.g.*, CNN iReport. Often such news are captured in places with very weak network infrastructures and it is imperative that a citizen journalist can quickly and reliably upload videos in the face of slow, unstable, and intermittent Internet access. We envision that some middleboxes are deployed to collect these videos over energy-efficient short-range wireless networks. Multiple videos may need to be prioritized, and then optimally transcoded and scheduled. In this study we introduce an adaptive middlebox design, called NEWSMAN, to support citizen journalists. NEWSMAN jointly considers two aspects under varying network conditions: (i) choosing the optimal transcoding parameters, and (ii) determining the uploading schedule for news videos. We design, implement, and evaluate an efficient scheduling algorithm to maximize a user-specified objective function. We conduct a series of experiments using trace-driven simulations, which confirm that our approach is practical and performs well. For instance, NEWSMAN outperforms the existing algorithms (i) by 12 times in terms of system utility (*i.e.*, sum of utilities of all uploaded videos), and (ii) by 4 times in terms of the number of videos uploaded before their deadline.

Keywords: News reporting · News scheduling · Adaptive transmission · Video transcoding · Multimedia information systems · Information systems applications

© Springer International Publishing Switzerland 2016
Q. Tian et al. (Eds.): MMM 2016, Part I, LNCS 9516, pp. 100–113, 2016.
DOI: 10.1007/978-3-319-27671-7_9

1   Introduction

Owing to technical advances in mobile devices and wireless communications, user-generated news videos have become popular since they can be easily captured using most modern smartphones and tablets in sufficiently high quality. Moreover, in the era of globalization, most news providers cover news from every part of the world, while on many occasions, reporters send news materials to editing rooms over the Internet. Therefore, in addition to traditional news reporting, the concept of citizen journalism, which allows people to play active roles in the process of collecting news reports, is also gaining much popularity. For instance, CNN allows citizens to report news using modern smartphones and tablets through its CNN iReport service. This service has more than one million citizen journalist users [2], who report news from places where traditional news reporters may not have access. Every month, it garners an average of 15,000 news reports and its content nets 2.6 million views [1]. It is, however, quite challenging for reporters to timely upload news videos, especially from developing countries, where Internet access is slow or even intermittent. Hence, it is crucial to deploy *adaptive middleboxes*, which upload news videos respecting the varying network conditions. Such middleboxes will allow citizen reporters to quickly *drop* the news videos over energy-efficient short-range wireless networks, and continue their daily life. Moreover, Short message service (SMS) is gaining much popularity due to its easy, fast, and cheap way of information retrieval in an area with weak network infrastructures [15,16]. This concept can be used in building an SMS-based news retrieval system in future.

Journalists can upload news videos to middleboxes or news providers either by using cellular or WiFi networks if available. Since an energy-efficient short-range wireless network between mobile devices and middleboxes can be leveraged using optimized mobile applications, we focus on a scheduling algorithm tuned for varying network conditions which can adaptively schedule the uploads of videos from the middleboxes to the server. Middleboxes can be placed in cloud servers or strategic places in towns such as city centers, coffee shops, train and bus stations, *etc.*, so that when reporters frequent these places then the short-range wireless communication can be leveraged for uploading videos. One can envision that an efficient smartphone application can further improve such communication among different reporters based on collaborative models. Shops at these places may host such middleboxes incentivized by the following reasons: (i) advertisement companies can sponsor the cost of resources (*e.g.*, several companies already sponsor Internet connectivity at airports), (ii) news providers can sponsor resources since they will receive news on time with less investment, (iii) more customers may be attracted to visit these shops, and (iv) a collaborative model of information sharing based on crowdsourcing is gaining popularity. Moreover, middleboxes can be used to decide whether reporters can directly upload videos to news providers based on current network conditions.

In designing the adaptive middlebox, we consider two categories of news videos, first, *breaking news* and, second, *traditional news*. Usually, the breaking news videos have stricter deadlines than those of the traditional news videos.

Table 1. Real world results of news uploading.

Location	India	Pakistan	Argentina	USA
Throughput	500 ~ 600 Kbps	300 ~ 500 Kbps	200 ~ 300 Kbps	20 ~ 23 Mbps
File Sizes	100 ~ 200 MB	50 ~ 100 MB	500 ~ 600 MB	100 ~ 200 MB
# of Interruptions	6	3	7	0

There is significant competition among news organizations to be the first to report breaking news. Hence, ubiquitous availability of mobile devices and the concept of citizen journalism help with fast reporting of news videos, using the mobile applications and the web sites of news providers. However, many times, the uploading of news videos is delayed due to reporters' slow Internet access and the big sizes of news videos. In pilot experiments among news reporters in early 2015, we noticed low throughput and non-trivial network interruptions in some of our test cases, as summarized in Table 1. Reporters tested uploading from a few locations in India, Pakistan, Argentina, and the USA, mostly through cellular networks. For example, when news reporters uploaded their videos over the Internet to an editing room in New York City for a leading news provider, they suffered from as many as 7 interrupts per upload. Without our proposed adaptive middleboxes, news reporters may be frustrated and eventually give up, because of long uploading times. This necessitates carefully designed adaptive middleboxes which run a scheduling algorithm to determine an uploading schedule for news videos considering factors such as optimal bitrates, videos deadlines, and network conditions.

In this study, we propose NEWSMAN, which maximizes the system utility by optimizing the number and quality of the videos uploaded before their deadlines from users to news editors under varying network conditions. We place middleboxes between reporters and news editors, to de-couple the local upload from the long-haul transmission to the editing room, in order to optimize both network *segments*, which have diverse characteristics. To optimize the system performance, we design an efficient scheduling algorithm in the middlebox to derive the uploading schedule and to transcode news videos (if required, to meet their deadlines) adaptively following a practical video quality model. The NEWSMAN scheduling process is described as follows: (i) reporters directly upload news videos to the news organizations if the Internet connectivity is good, otherwise (ii) reporters upload news videos to the middlebox, and (iii) the scheduler in the middlebox determines an uploading schedule and optimal bitrates for transcoding. Figure 1 presents the architecture of the NEWSMAN system.

The key contribution of this study is an efficient scheduling algorithm to upload news videos to a cloud server such that: (i) the system utility is maximized, (ii) the number of news videos uploaded before their deadlines is maximized, and (iii) news videos are delivered in the best possible video qualities under varying network conditions. We conducted extensive trace-driven simulations using real datasets of 130 online news videos. The results from the simulations show the merits of

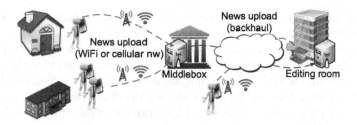

Fig. 1. Architecture of the proposed NEWSMAN system.

NEWSMAN as it outperforms the current algorithms (i) by 1,200 % in terms of system utility and (ii) by 400 % in terms of the number of videos uploaded before their deadlines. Furthermore, NEWSMAN achieves low average delay of the uploaded news videos. The remaining parts of this paper are organized as follows. In Sect. 2, we review the related literature, and we describe the NEWSMAN system in Sect. 3. We present and solve the upload scheduling problem in Sect. 4. The experiments and results are presented in Sect. 5. Finally, we conclude the paper with a summary in Sect. 6.

2   Related Work

The NEWSMAN scheduling process is described as follows: (i) reporters directly upload news videos to the news organizations if the Internet connectivity is good, otherwise (ii) reporters upload news videos to a middlebox, and (iii) the scheduler at the middlebox determines an uploading schedule and optimal bitrates for transcoding. In this section, we survey some recent related work.

In addition to traditional news reporting systems such as satellite news networks (SNN), the use of satellite news gathering (SNG) by local stations has also increased during recent years. However, SNG has not been adopted as widely as SNN due to reasons such as: (i) the high setup and maintenance costs of SNG, and (ii) the non-portability of SNG equipment to many locations due to its big size [8,13]. These constraints have popularized the citizen news reporting services such as CNN iReport [1].

Unlike significant efforts that have focused on systems supporting downloading applications such as video streaming and file sharing [11,19], little attention has been paid to systems that support uploading applications [4,20]. Media uploading with hard deadlines require an optimal deadline scheduling algorithm [3,5]. Abba et al. [3] proposed a prioritized scheduling algorithm using a project management technique for an efficient job execution with deadline constraints of jobs. Chen et al. [5] proposed an online preemptive scheduler which either accepts or declines a job immediately upon its sporadic arrival based on a contract where the scheduler looses the profit of the job and pays a penalty if the accepted job is not finished within its deadline.

Chen et al. [5] proposed an online preemptive scheduling of jobs with deadlines arriving sporadically. The scheduler either accepts or declines a job immediately upon arrival based on a contract where the scheduler looses the profit of the job and pays a penalty if the accepted job is not finished within its deadline. The objective of the online scheduler is to maximize the overall profit, *i.e.*, the total profit of completed jobs before their deadlines is more than the penalty paid for the jobs that missed their deadlines. Online scheduling algorithms such as earliest deadline first (EDF) [12] are often used for applications with deadlines. Since we consider jobs with diverse deadlines, we leverage the EDF concept in our system to determine the uploading schedule that will maximize the system utility.

Recent years have seen significant progress in the area of rate-distortion (R–D) optimized image and video coding [6,9]. In lossy compression, there is a tradeoff between the bitrate and the distortion. R–D models are functions that describe the relationship between the bitrate and expected level of distortion in the reconstructed video. In NEWSMAN, R–D models enable the optimization of the received video quality under different network conditions. To avoid unnecessary complexity of deriving R–D models of individual news videos, NEWSMAN categorizes news videos into a few classes using *temporal perceptual information* (TI) and *spatial perceptual information* (SI), which are the measures of temporal changes and spatial details, respectively [14,18]. Due to limited storage space, less powerful CPU, and constrained battery capacity, earlier works [7,10] suggested to perform transcoding at resourceful clouds (middleboxes in our case) instead of at mobile devices. In our work we follow this model.

3   System Overview

We refer to the uploading of a news video as a *job* in this study. NEWSMAN schedules jobs such that videos are uploaded before their deadlines in the highest possible qualities with optimally selected coding parameters for video transcoding.

NEWSMAN Scheduling Algorithm. Figure 2 shows the architecture of the scheduler. Reporters upload jobs to a middlebox. For every job arriving at the middlebox, the scheduler performs the following actions when the scheduling interval expires: (i) it computes the job's importance, (ii) it sorts all jobs based on news importance, and (iii) it estimates the job's uploading schedule and the

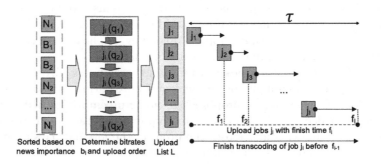

Fig. 2. Scheduler architecture in a middlebox.

optimal bitrate for transcoding. The scheduling algorithm is described in details in Sect. 4. As Fig. 2 shows, we consider χ video qualities for a job j_i and select the optimal bitrate for transcoding of j_i to meet its deadline, under current network conditions.

R-D Model. Traditional digital video transmission and storage systems either fully upload a news video to a news editor or not at all. The key idea for transcoding videos with optimal bitrates is to compress videos for transmission to adaptively transfer video contents before their deadlines, under varying network conditions. More motion in adjacent frames indicates higher TI values and scenes with minimal spatial detail result in low SI. For instance, a scene from a football game contains a large amount of motion (*i.e.*, high TI) as well as spatial detail (*i.e.*, high SI). Since two different scenes with the same TI/SI values produce similar perceived quality [18], news videos can be classified in G categories. Therefore, news videos can be categorized into different categories such as sport videos, interviews, *etc.*, based on their TI/SI values. It is important to determine the category (or TI/SI values) of a news video, so that we can select appropriate R–D models for these categories. A scene with little motion and limited spatial detail (such as a head and shoulders shot of a newscaster) may be compressed to 384 kbits/s and decompressed with relatively little distortion. Another scene (such as from a football game) which contains a large amount of motion as well as spatial detail will appear quite distorted at the same bit rate [18]. Therefore, it is important to consider different R–D models for all categories. Empirical piecewise linear R–D models can be constructed for individual TI/SI pairs. We encode online news videos with diverse content complexities and empirically analyze their R–D characteristics. We consider four categories in our experiments corresponding to *high TI/high SI*, *high TI/low SI*, *low TI/high SI*, and *low TI/low SI*. We adaptively determine the suitable coding bitrates for an editor-specified video quality for videos, using these piecewise linear R–D models.

4   Scheduling Problem and Solution

Formulation. Let there be B breaking news and N traditional news, with given arrival times A, deadlines D, and metadata M which consist of users' reputations and video metadata such as bitrates and fps (frames per second). For a job j_i, let $\xi(j_i)$ and $\lambda(j_i)$ be scores for video length and news-location, respectively. Let $\gamma(r)$ be the user reputation score for a reporter r. Let σ be the editor-specified minimum required video quality (say in PSNR, peak signal-to-noise ratio). Let $\bar{p}_i$, p_i, $\bar{b}_i$ and b_i be the original video quality, transcoded video quality, original bitrate, and transcoded bitrate for job j_i, respectively. Let $\omega(t_c)$ be the available disk size at some time t_c. Let $\bar{s}_i$ and s_i be the original and transcoded file sizes, respectively. Let $\eta(s_i)$ be the time required to transcode the job j_i with file size s_i and $\beta(t_1, t_2)$ the average throughput between time interval t_1 and t_2. Let $\delta(j_i)$ be the video length (in seconds) for j_i. Let τ be the time interval of running the scheduler in a middlebox.

The news importance u of a job j_i is defined as $u(j_i) = \mu(j_i) \cdot (w_1 \, \xi(j_i) + w_2 \, \lambda(j_i) + w_3 \, \gamma(r))$, where the multiplier $\mu(j_i)$ is a weight for boosting or ignoring the importance of any particular news type or category. E.g., in our experiments the value of $\mu(j_i)$ is 1 if job j_i is traditional news and 2 if job j_i is breaking news. By considering news categories such as sports a news provider can boost videos during a sports events such as the FIFA world cup. Moreover, the news decay function v is defined as:

$$
v(f_i) = \begin{cases} 1, & \text{if } f_i \leq d_i \\ e^{-\alpha(f_i - d_i)}, & \text{otherwise, where } f_i \text{ and } d_i \text{ are the finish time and} \\ & \text{deadline of job } j_i, \text{ respectively. } \alpha \text{ is an exponential} \\ & \text{decay constant.} \end{cases}
$$

The utility score of a news video j_i depends on the following factors: (i) the importance of j_i, (ii) how quickly the importance of j_i decays, and (iii) the delivered video quality of j_i. Thus, we define the news utility ρ for job j_i as $\rho(j_i) = u(j_i) \, v(f_i) \, p_i$. With the above notations and functions, we state the problem formulation as:

$$
\max \sum_{i=1}^{B+N} \rho(j_i) \tag{1a}
$$

$$
s.t. \quad \sigma \leq p_i \leq \bar{p}_i \;\; \forall \, 1 \leq i \leq B + N \tag{1b}
$$

$$
(f_i - f_{i-1}) \, \beta(f_{i-1}, f_i) \geq b_i \, \delta(j_i) \tag{1c}
$$

$$
\sum_{\forall j_k \in K} \eta(s_k) < f_i, \text{ where } K = \{ j_k \mid j_k \text{ is scheduled before } j_i \} \tag{1d}
$$

$$
\sum_{i=1}^{B+N} s_i \leq \omega(t_c) - \sum_{i=1}^{B+N} \bar{s}_i \tag{1e}
$$

$$
f_i \leq f_k, \;\; \forall \, 1 \leq i \leq k \leq B + N \tag{1f}
$$

$$
0 \leq f_i, \;\; \forall \, 1 \leq i \leq B + N \tag{1g}
$$

$$
0 < b_i \leq \bar{b}_i, \;\; \forall \, 1 \leq i \leq B + N \tag{1h}
$$

$$
j_i \in \{1, \ldots, B, B + 1, B + 2, \ldots, B + N\} \tag{1i}
$$

The objective function in Eq. (1a) maximizes the sum of news utility (*i.e.*, the product of importance, decay value and video quality) for all jobs. Equation (1b) makes sure that the video quality of the transcoded video is at-least the minimum video quality σ. Equation (1c) ensures bandwidth constraints for NEWSMAN. Equation (1d) enforces that the transcoding of a video completes before its uploading starts and Eq. (1e) ensures disk constraints of a middlebox. Equation (1f) ensures that the scheduler uploads jobs in the order scheduled by NEWSMAN. Equations (1g) and (1h) define the ranges of the decision variables. Finally, Eq. (1i) indicates that all jobs are either breaking news or traditional news.

Lemma. Let $j_{i\,i=1}^{\,n}$ be a set of n jobs in a middlebox at time t_c, and $d_{i\,i=1}^{\,n}$ their respective deadlines for uploading. The scheduler is executed when either the scheduling interval τ expires or when all jobs in the middlebox have been

---

Algorithm 1. A procedure for scheduling news videos

1: **procedure** SCHEDULEANDUPLOADNEWSREPORTS
2:     INPUT: $B + N$ jobs with given arrival times A and deadlines D
3:     OUTPUT: A uploading schedule for jobs at a middlebox
4:     Initialization: $t = t_c$, $\mathcal{U} = 0$ ▷ t_c is current time and $\mathcal{U}$ is total utility.
5:     Q = getAllJobs(t_c)   ▷ Q is the list of all jobs arrived till time t_c at the middlebox.
6:     compNewsImporatnce(Q)   ▷ Compute news importance for all jobs in Q.
7:     compEstmFinishTime(Q, χ)   ▷ Compute estimated finish time for χ video qualities.
8:     sortJobs(Q)   ▷ Sort all jobs in Q based on their news importance scores.
9:     **for** each job j_i in Q **do**   ▷ Read a job with the highest importance score first.
10:         t_d = getDeadline(j_i, A)   ▷ Get the deadline (maximum allowed finish time) for j_i.
11:         t_h = getEstmFinishTime(j_i, t)   ▷ Estimated time to upload j_i in original quality.
12:         **if** ($t_h \leq t_d$) **then**   ▷ Indicates j_i can be uploaded in original video before deadline.
13:             $\rho(j_i) = u(j_i)\, v(t_h)\, p_i$ ▷ Utility based on its importance, decay and quality.
14:             $\mathcal{U}\ +\ = \rho(j_i)$ ▷ Total utility for jobs scheduled so far.
15:             addJobToUploadList(j_i, L)   ▷ Added j_i to L in original video quality.
16:             $t\ +\ = t_h$ ▷ Update total scheduled time so that more jobs can be added to L.
17:         **else**   ▷ Try to add j_i to L in lower video quality, if possible.
18:             ϕ = addJobInLowerQuality(j_i, t_d, t, L, $\mathcal{U}$)   ▷ See Algorithm 2.
19:         **end if**   ▷ ϕ is $TRUE$ if j_i added to L after lowering qualities of some videos.
20:         **if** ($\phi == FALSE$) **then**   ▷ ϕ is $FALSE$ if j_i could not be added to L.
21:             addJobToScheduleLater(j_i, $Q\prime$)   ▷ Jobs in $Q\prime$ could not be scheduled.
22:         **end if**
23:     **end for** ▷ Output the uploading order and b_i (decision variables in formulation) of all j_i.
24:     transcodeAndUpload(L)   ▷ Set f_i (decision variable in formulation) for j_i after upload.
25: **end procedure**

---

uploaded before τ expires. Thus, the average throughput $\beta(t_c, t_c + \tau)$ (or β in short) during the scheduling interval is distributed among several jobs selected for parallel uploading[1], and as a consequence, the sequential upload of jobs has higher utility than parallel uploading.

Proof Sketch. Let k jobs $j_{i\,i=1}^{k}$ with transcoded sizes $s_{i\,i=1}^{k}$ be selected in parallel uploading. Let k_t of them $(j_{i\,i=1}^{k_t})$ require transcoding. Thus, it takes some time for their transcoding (i.e., $\eta_p(s_i)_{i=1}^{k_t} \geq 0$) before the actual uploading starts. Hence, uploading throughput is wasted during the transcoding of these jobs in parallel uploading. During sequential uploading, NEWSMAN ensures that transcoding of a job is finished (if required) before the uploading of the job is started. Thus, it results in a net transcoding time of zero (i.e., $\eta_s(s_i)_{i=1}^{k_t} = 0$) in sequential uploading, and it fully utilizes the uploading throughput β. Let t_u be the time (excluding transcoding time) to upload jobs $j_{i\,i=1}^{n}$. Thus, t_u is equal for both sequential and parallel uploading since the same uploading throughput is divided among parallel jobs. Let t_p (i.e., $t_u + \eta_p$) and t_s (i.e., $t_u + \eta_s$) be the uploading time for all jobs $(j_{i\,i=1}^{n})$ when the jobs are uploaded in parallel or sequential manner, respectively. Hence, the actual time required to upload in a parallel manner (i.e., t_p) is greater than the time required to upload in a sequential manner (i.e., t_s). Moreover, the uploading of important jobs is delayed in parallel uploading since throughput is divided among several other selected

---

[1] Some videos may require transcoding first before uploading to meet deadlines in NEWSMAN.

Algorithm 2. A procedure for determining transcoding parameters for jobs in L, if j is added to the uploading list

---

1: **procedure** ADDJOBINLOWERQUALITY
2:     Initialization: $\bar{t} = t, \bar{\mathcal{U}} = \mathcal{U}$ ▷ Initialize updated utility and scheduled time after adding j.
3:     addJobToUploadList(j, L)   ▷ Add j to L in original quality and reduce quality later.
4:     $n = $ getNumberOfJobs(L)   ▷ Check all existing jobs in L to determine scheduling list.
5:     **for** ($m = n$; $m >= 1$; $m - -$) **do**
6:         **for** each q_k of j_m **do**   ▷ q_k is k^{th} video quality of j_m.
7:             $t_k = $ getEstmFinishTimeLowQlty(j_m, t, q_k)   ▷ Time to upload j_m in k^{th} quality.
8:             $t_e = $ getEstmFinishTimeLowQlty(j_m, t, q_c)   ▷ Time to upload j_m in current qlty.
9:             $\rho_k = u(j_m) \, v(t_k) \, q_k$ ▷ Compute the utility of j_m for k^{th} quality.
10:            $\rho_c = u(j_m) \, v(t_e) \, q_c$ ▷ Compute the utility of j for current quality.
11:            $\bar{t} = $ getUpdatedScheduledTime(j_m, t_k, t_e)   ▷ Get the updated scheduled time.
12:            $\bar{\mathcal{U}} = $ getUpdatedUtility($j_m, \rho_c, \rho_k, \mathcal{U}, \bar{\mathcal{U}}$)   ▷ Updated total utility value.
13:            $\phi = $ isJobAccomodatedWihinDeadline($m, t_d, \bar{t}, t, L, \mathcal{U}, \bar{\mathcal{U}}$)   ▷ All jobs in L
        (including j) can be scheduled within deadline after lower qualities of jobs (from j_m to j_n).
14:            **if** ($\phi == TRUE$) **then**   ▷ Set the video quality of j_m to q_k.
15:                updateTranscodingParam(m, L, q_k)   ▷ Set quality of jobs (j_{m+1} to j_n) to σ.
16:                $\mathcal{U} = \bar{\mathcal{U}}$ ▷ Update total utility value after changing qualities of some jobs.
17:                $t = \bar{t}$ ▷ Update the total uploading time after scheduling all jobs in L.
18:                **return** $TRUE$
19:            **end if**
20:        **end for**
21:    **end for**
22:    removeJobFromUploadList(j, L)   ▷ j is removed from L since it could not be scheduled.
23:    **return** $FALSE$ ▷ j could not be added to L despite lowering qualities of jobs.
24: **end procedure**

---

jobs (β/k for each job). Therefore, the sequential uploading of jobs is better than the parallel uploading.

Upload Scheduling Algorithm. We design an efficient scheduling algorithm to solve the above formulation. Algorithm 1 shows the main procedure of scheduling a list of jobs at a middlebox. If it is not possible to upload any job within its deadline, NEWSMAN uploads the transcoded news videos to meet the deadline. Algorithm 2 shows the procedure of calculating the encoding parameters for transcoding under current network conditions and σ. Algorithm 2 is invoked on line 18 of Algorithm 1 whenever necessary. The NEWSMAN scheduler considers χ possible video qualities (hence, smaller video size and shorter upload time are possible) for a job. NEWSMAN considers σ as a threshold and divides a region between σ (minimum required video quality) and $\bar{p}_i$ (original video quality) among χ discrete qualities (say, $q_i{}_{i=1}^{\chi}$, with $q_1 = \sigma$ and $q_\chi = \bar{p}_i$). The scheduler keeps checking lower, but acceptable, video qualities starting with the least important job first, to accommodate j in L such that: (i) the total estimated system utility increases after adding j, and (ii) all jobs in L still meet their deadlines (maybe with lower video qualities), if they are estimated to meet deadlines earlier. However, if the scheduler is not able to add j in the uploading list, then this job is added to a missed-deadline list whose deadline can be modified later by news-editors based on news importance. Once the scheduling of all jobs is done, NEWSMAN starts uploading news videos from the middlebox to the editing room and transcodes (in parallel with uploading) the rest of the news videos (if required) in the uploading list L.

Algorithm 2 is invoked when it is not possible to add a job with the original video quality to L. This procedure keeps checking jobs at lower video qualities until all jobs in the list are added to L with estimated uploading times within their deadlines. The *isJobAccomodatedWihinDeadline()* method on line 13 of Algorithm 2 ensures that: (i) the selected video quality q_k is lower than the current video quality q_c (*i.e.*, $q_k \leq q_c$) since some jobs are already set to lower video qualities in earlier steps, (ii) the utility value is increased after adding the job (*i.e.*, $\bar{U} \geq U$), (iii) all jobs in L is completed (estimated) within their deadlines, and (iv) a job with higher importance comes first in L.

5   Simulations and Results

Real-Life Datasets. We collected 130 online news video sequences from Al Jazeera, CNN, and BBC YouTube channels during mid-February 2015. The shortest and longest duration of videos are 0.33 and 26 min, and the smallest and biggest news video sizes are 4 and 340 MB, respectively. We also collected network traces from different PCs across the globe, such as (Delhi and Hyderabad) India, and (Nanjing) China, which emulate middleboxes in our system. More specifically, we use IPERF [17] to collect throughput from the PCs to an Amazon EC2 server in Singapore (see Table 2). The news and network datasets are used to drive our simulator.

Simulator Implementation and Scenarios. We implemented a trace–driven simulator for NEWSMAN using Java. Our focus is on the proposed scheduling algorithm under varying network conditions. The scheduler runs once every scheduling interval τ (say, 5 min) and reads randomly generated new jobs following the Poisson process. We consider 0.1, 0.5, 1, 5, and 10 per min as mean job arrival rate and randomly mark a job as breaking news or traditional news in our experiments. In the computation of news importance for videos, we randomly generate a real number in [0,1] for user reputations and location importance in simulations. We set deadlines for news videos randomly in the following time intervals: (i) [1, 2] h for breaking news, and (ii) [2, 3] h for traditional news. We implemented two baseline algorithms: (i) earlier deadline first (EDF), and (ii) first in first out (FIFO) scheduling algorithms. For fair comparisons, we run the simulations for 24 h and repeat each simulation scenario 20 times. If not otherwise specified, we use the first–day

Table 2. Statistics of network traces.

Location	Dates	Avg. throughput
Delhi	2015-03-12 to 2015-03-14	409 Kbps
Hyderabad	2015-03-14 to 2015-03-18	297 Kbps
Nanjing	2015-03-23 to 2015-03-27	1138 Kbps

network trace to drive the simulator. We use the same set of jobs (with the same arrival times, deadlines, news types, user reputations, location importance, *etc.*) for three algorithms, in a simulation iteration. We report the average performance with 95 % confidence intervals whenever applicable.

Results. *NEWSMAN delivers the most news videos in time, and achieves the highest system utility.* Figures 3a, 4a and c show that NEWSMAN performs up to 1200 % better than baseline algorithms in terms of system utility. Figures 3b and c show that our system outperforms baselines (i) by up to 400 % in terms of number of videos uploaded before their deadlines, and (ii) by up to 150 % in terms of total number of uploaded videos. That is, NEWSMAN significantly outperforms the baselines either when news editors set hard deadlines (4X improvement) or soft deadlines (1.5X improvement).

NEWSMAN Achieves Low Average Lateness. Despite delivering the most news videos in time, and achieving the highest system utility for Delhi (see Figs. 3 and 4), NEWSMAN achieves fairly low average lateness (see Figs. 3d, 4b and d).

NEWSMAN Performs Well Under All Network Infrastructures. Fig. 4a shows that NEWSMAN outperforms baselines under all network conditions such as low average throughput in India, and higher average throughput in China (see Table 2).

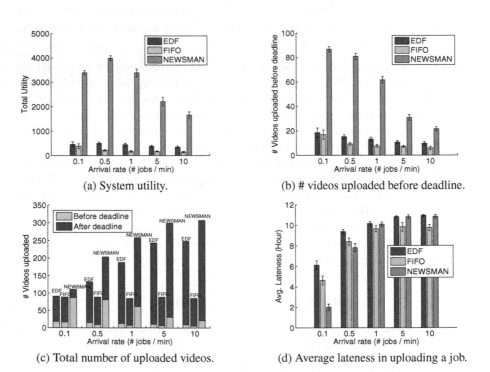

(a) System utility.

(b) # videos uploaded before deadline.

(c) Total number of uploaded videos.

(d) Average lateness in uploading a job.

Fig. 3. Results after running the simulator for 24 h using network traces from Delhi.

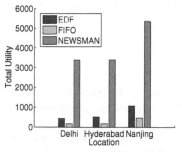

(a) System utility from different locations.

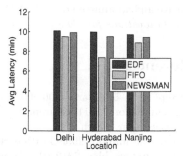

(b) Avg. lateness from different locations.

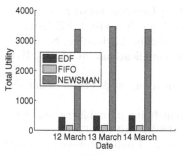

(c) System utility on different dates.

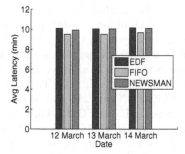

(d) Average lateness on different dates.

Fig. 4. Results after running the simulator for 24 h using network traces from different locations (a–b) on different dates (c–d).

6   Conclusions

We present an innovative design for efficient uploading of news videos with deadlines under weak network infrastructures. In our proposed news reporting system called NEWSMAN, we use middleboxes with a novel scheduling and transcoding selection algorithm for uploading news videos under varying network conditions. The system intelligently schedules news videos based on their characteristics and underlying network conditions such that: (i) it maximizes the system utility, (ii) it uploads news videos in the best possible qualities, and (iii) it achieves low average lateness of the uploaded videos. We formulated this scheduling problem into a mathematical optimization problem. Furthermore, we developed a trace-driven simulator to conduct a series of extensive experiments using real datasets and network traces collected between a Singapore EC2 server and different PCs in Asia. The simulation results indicate that our proposed scheduling algorithm improves system performance. We are currently deploying NEWSMAN in developing countries to demonstrate its practicality and efficiency in practice.

Acknowledgment. This research has been supported by the Singapore National Research Foundation under its International Research Centre @ Singapore Funding

Initiative and administered by the IDM Programme Office through the *Centre of Social Media Innovations for Communities* (COSMIC).

References

1. iReport at 5: Nearly 900,000 contributors worldwide, July 2012. http://www.niemanlab.org/2011/08/ireport-at-5-nearly-900000-contributors-worldwide/. Accessed on March 2015
2. Meet the million: 999,999 iReporters + you! July 2012. http://ireport.cnn.com/blogs/ireport-blog/2012/01/23/meet-the-million-999999-ireporters-you. Accessed on March 2015
3. Abba, H.A., Shah, S.N.M., Zakaria, N.B., Pal, A.J.: Deadline based performance evaluation of job scheduling algorithms. In: CyberC, pp. 106–110. IEEE (2012)
4. Bhattacharjee, S., Cheng, W.C., Chou, C.-F., Golubchik, L., Khuller, S.: Bistro: a framework for building scalable wide-area upload applications. ACM SIGMETRICS Perform. Eval. Rev. **28**(2), 29–35 (2000)
5. Chen, S., Tong, L., He, T.: Optimal deadline scheduling with commitment. In: ALLERTON, pp. 111–118. IEEE (2011)
6. Hefeeda, M., Hsu, C.-H.: On burst transmission scheduling in mobile tv broadcast networks. IEEE/ACM Trans. Networking (TON) **18**(2), 610–623 (2010)
7. Jokhio, F., Ashraf, A., Lafond, S., Porres, I., Lilius, J.: Prediction-based dynamic resource allocation for video transcoding in cloud computing. In: PDP, pp. 254–261. IEEE (2013)
8. Lacy, S., Atwater, T., Qin, X., Powers, A.: Cost and competition in the adoption of satellite news gathering technology. J. Media Econ. **1**(1), 51–59 (1988)
9. Lambert, P., De Neve, W., De Neve, P., Moerman, I., Demeester, P., Van de Walle, R.: Rate-distortion performance of H. 264/AVC compared to state-of-the-art video codecs. IEEE Trans. Circuits Syst. Video Technol. **16**(1), 134–140 (2006)
10. Li, Z., Huang, Y., Liu, G., Wang, F., Zhang, Z.-L., Dai, Y.: Cloud transcoder: bridging the format and resolution gap between internet videos and mobile devices. In: NOSSDAV, pp. 33–38. ACM (2012)
11. Liang, C., Guo, Y., Liu, Y.: Is random scheduling sufficient in P2P video streaming? In: International Conference on Distributed Computing Systems, pp. 53–60. IEEE (2008)
12. Liu, C.L., Layland, J.W.: Scheduling algorithms for multiprogramming in a hard-real-time environment. J. ACM (JACM) **20**(1), 46–61 (1973)
13. Livingston, S., Van Belle, D.A.: The effects of satellite technology on newsgathering from remote locations. Political Commun. **22**(1), 45–62 (2005)
14. Recommendation ITU-T P.910. Subjective video quality assessment methods for multimedia applications (2008)
15. Shaikh, A.D., Jain, M., Rawat, M., Shah, R.R., Kumar, M.: Improving accuracy of SMS based FAQ retrieval system. In: Majumder, P., Mitra, M., Bhattacharyya, P., Subramaniam, L.V., Contractor, D., Rosso, P. (eds.) FIRE 2010 and 2011. LNCS, vol. 7536, pp. 142–156. Springer, Heidelberg (2013)
16. Shaikh, A.D., Shah, R.R., Shaikh, R.: SMS based FAQ retrieval for Hindi, English and Malayalam. In: Forum on Information Retrieval Evaluation, p. 9. ACM (2013)
17. Tirumala, A., Qin, F., Dugan, J., Ferguson, J., Gibbs, K.: Iperf: the TCP/UDP bandwidth measurement tool (2005). http://dast.nlanr.net/Projects/Iperf/

18. Webster, A.A., Jones, C.T., Pinson, M.H., Voran, S.D., Wolf, S.: Objective video quality assessment system based on human perception. In: IS&T/SPIE's Symposium on Electronic Imaging: Science and Technology, pp. 15–26. SPIE (1993)
19. Xie, D., Qian, B., Peng, Y., Chen, T.: A model of job scheduling with deadline for video-on-demand system. In: WISM, pp. 661–668. IEEE (2009)
20. Zhang, M., Wong, J., Tavanapong, W., Oh, J., de Groen, P.: Deadline-constrained media uploading systems. Multimedia Tools Appl. **38**(1), 51–74 (2008)

Computational Face Reader

Xiangbo Shu[1,4(✉)], Liyan Zhang[2], Jinhui Tang[1], Guo-Sen Xie[3],
and Shuicheng Yan[4]

[1] School of Computer Science and Engineering,
Nanjing University of Science and Technology, Nanjing, China
shuxb104@gmail.com, jinhuitang@njust.edu.cn
[2] College of Computer Science and Technology,
Nanjing University of Aeronautics and Astronautics, Nanjing, China
zhangliyan.uci@gmail.com
[3] NLPR, Institute of Automation, Chinese Academy of Sciences, Beijing, China
guosen.xie@nlpr.ia.ac.cn
[4] Department of ECE, National University of Singapore, Singapore, Singapore
eleyans@nus.edu.sg

Abstract. The long-history Chinese anthroposcopy has demonstrated
the often satisfying capabilities to tell the characteristics (mostly exaggerated as fortune) of a person by reading his/her face, i.e. understanding
the fine-grained facial attributes (e.g. single/double-fold eyelid, position
of mole). In this paper, we study the face-reading problem from the
computer vision perspective and present a computational face reader to
automatically infer the characteristics of a person based on his/her face.
For example, it can estimate the attractive and easy-going characteristics of a Chinese person from his/her big eyes according to the Chinese
anthroposcopy literature. Specifically, to well estimate these fine-grained
facial attributes, we propose a novel deep convolutional network in which
a facial region pooling layer (FRP layer) is embedded, called FRP-net.
The FRP layer uses the searched facial region windows (locates these
facial attributes) instead of the commonly-used sliding windows. The
experiments on facial attribute estimation demonstrate the potential of
the automatic face reader framework, and qualitative and quantitative
evaluations from the attractive and smart perspectives of face reading
validate the excellence of the presented face reader framework.

Keywords: Chinese anthroposcopy · Facial attribute estimation · Deep
convolutional neural networks · Facial region pooling layer

1 Introduction

A Chinese proverb says, *"The face is the index of the mind."* By observing a person's
facial features, Chinese anthroposcopy, also called Chinese face reading, is able to
tell his/her characteristics (mostly exaggerated as the fortune of this person). Chinese anthroposcopy is a comprehensive expression of Chinese ancient philosophy

---

This work was performed when X. Shu was visiting National University of Singapore.

© Springer International Publishing Switzerland 2016
Q. Tian et al. (Eds.): MMM 2016, Part I, LNCS 9516, pp. 114–126, 2016.
DOI: 10.1007/978-3-319-27671-7_10

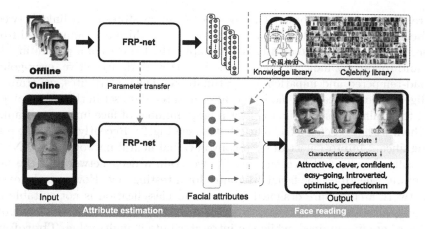

Fig. 1. Overview of our face reader framework. In the offline phase, a proposed FRP-net is trained on the collected database with facial landmarks and attribute annotations. In the online phase, for an input face, this system predicts the facial attributes via the trained FRP-net and translates them into the person's characteristics by referring to Chinese anthroposcopy literature. The output includes the text descriptions and the image show of celebrities from a celebrity library.

(e.g. I Ching [1], Yin and Yang [2], Confucianism [3], Taoism [4]). The outcome of a face reading is determined by many facial attributes, including not only the most obvious facial features (e.g. area of eyes, length of nose) but also some less conspicuous facial attributes (e.g. eyebrow density, shape of eyelids, shape of mouth corner).

Only some people who are very familiar with the Chinese anthroposcopy are good at reading faces. This appealing skill is apparently not so easy for most common ones. Recently, some web demos (e.g., Mianxiang Yuce[1]) and Apps (e.g., Face Reader+[2], and Mianxiang Dashi[3]) have provided the Chinese anthroposcopy function, and gained certain popularity. A smart face reader system should accurately estimate the fine-grained facial attributes. **Intuitively, the higher the estimation accuracy of these facial attributes is, the more popularity of the face reader system will gain.** However, existing facial attribute estimation systems (e.g. linkface[4]) only focus on the estimation of common facial attributes (e.g., accessories, hair color), and do not take the fine-grained facial attributes into consideration.

In this paper, we study the face-reading problem from the computer vision perspective and present a computational face reader framework, as shown in Fig. 1. Give an input face of a person, this presented framework can automatically estimate the facial attributes and then output his/her characteristics from a

[1] http://www.guabu.com/mxyc.

[2] https://itunes.apple.com/cn/app/face-reader-pro/id774539886?mt=8.

[3] https://itunes.apple.com/cn/app/mian-xiang-ce-suan-da-shi/id844999156?mt=8.

[4] http://www.linkface.cn.

face-reading knowledge library. The knowledge library stores many links between the facial attributes and the human characteristics, which are summarized from Chinese anthroposcopy literature. For example, the big eyes of a person in Chinese anthroposcopy indicate that he/she is extroverted, attractive, easy-going, yet bad-tempered and impatient. To estimate the facial attributes, we train an intelligent machine learning model on the annotated data set in the offline phase.

In the offline phase, we first collect a large number of face images and annotate them with pre-defined facial attributes (see Fig. 2). Recently, encouraged by the significant improvement of deep convolutional neural networks (DCNN) [5] in classification [6] and face analysis [7], we train a deep network on the face dataset and then infer the facial attributes for a testing face. However, the existing network architecture designed for common classification is not suitable for the fine-grained facial attribute estimation task. In Fig. 2, we can see that inter-attributes are fine-grained, while the intra-attribute is multi-value. **Therefore, our facial attribute estimation task is more challenging than the common facial feature estimation.**

Empirically, observing the pre-defined facial attributes in Fig. 2, we can see that only some local facial patterns are the interesting regions related to the facial attributes. In the max pooling layer of DCNN, if the pooling operation is based on these interesting regions instead of the commonly-used sliding windows, the learned features will be more representative and discriminative. Therefore, we propose a novel DCNN architecture with a facial region pooling layer (called FRP-net) to address the fine-grained facial attribute estimation challenge in this paper. Specifically, the facial region pooling (FRP) layer is added on top of the last convolutional layer, followed by the fully-connected layers. This proposed FRP-net is terminated at multiple softmax loss functions, of which one estimates a type of facial attribute values. The experiments for facial attribute estimation demonstrate the excellent performance of the proposed FRP-net compared with the existing DCNN models and other baselines. Figure 5 plots the architecture of FRP-net, which will be described in Sect. 5.

Main contributions in this paper can be summarized as follows.

- The first work targeting at the face-reading problem. To the best of our knowledge, this is the first work which studies the face-reading problem in the computer vision filed. Formally, we present a computational face reader framework to automatically infer the human characteristics for one input human face.
- A new fine-grained face dataset. We collect a Chinese face database, containing 5,562 faces and 19 fine-grained facial attributes. This Chinese face database contains complete facial attribute (the number is 19) annotations and can be used for fine-grained facial attribute estimation in the emerging applications.
- A new DCNN architecture. In our proposed FRP-net there is a specific FRP layer embedded, which can estimate the fine-grained facial attributes. According to the experimental results, the proposed FRP-net outperforms the baseline and other counterpart deep networks.

The rest of this paper is organized as follows. After the related work is reviewed, Sect. 3 introduces our face reader framework, followed by the dataset preparation and library construction in Sect. 4., i.e., attribute definition and dataset collection. The proposed FRP-net is described in Sect. 5. Experimental evaluations are presented in Sect. 6. The conclusions and future work will be given in the last section.

2 Related Work

2.1 Face Reading

To the best of our knowledge, this work is the first attempt to study the face-reading problem in computer vision. Recently, some demos and Apps have been proposed and developed, such as Mianxiang Yuce, Face Reader+, Mianxiang Dashi, etc., which can output face-reading results. Mianxiang Yuce requires the user to choose her/his facial attributes and then provides the face-reading results. Mianxiang Dashi requires well-aligned faces when the photo is taken. Face Reader+ can automatically generate the face-reading results, but the face attribute estimation is usually not accurate enough, which decreases the satisfaction. Overall, there has been no related literature on the estimation methods used for this problem.

2.2 Deep Convolutional Neural Network

Deep learning has been comprehensively reviewed and discussed in [8]. Recently, DCNN has been widely used for face recognition [9], image classification [10], object detection [11], etc. The classical stacked auto encoders linearly stack multiple layers of Auto Encoder together to learn higher-level representation. Some variants of the DCNN have also been proposed, such as network-in-network (NIN) [12], SPP-net [13], Fast r-CNN [14], etc. NIN is proposed to enhance model discriminability for local patches. The conventional convolutional layer uses the linear filters followed by a micro network (e.g. multilayer perceptron). To abolish the requirement of the fixed-size input image in DCNN, SPP-net equips the networks with a new spatial pyramid pooling layer. The performance of SPP-net is also superior in object detection. Fast r-CNN is the extension of the r-CNN [15]. The improvement lies in a region of interesting (ROI) layer proposed in the Fast r-CNN to only pool the interesting regions of feature maps on top of the convolutional layer. To improve the discriminability of the fine-grained facial attributes, we add a facial region pooling layer to pool the feature maps on top of the convolutional layer by leveraging the facial location information.

3 Overview of Our Framework

The presented face reader framework is shown in Fig. 1. Some preliminary work has been done. First, 19 fine-grained facial attributes are defined; second, we collect and annotate a Chinese face database containing 5,562 face photos; third, we construct a knowledge library which stores links between facial attributes and characteristics, and also a celebrity library of 200 annotated Chinese celebrities with characteristic tags. In the Offline phase, the proposed FRP-net is trained on the collected face database. In the Online phase, for an input face of the estimated gender[5], we first estimate the facial attributes on the trained FRP-net, and translate them to the person's characteristics based on the knowledge library. The characteristic descriptions for each attribute are integrated into a text, where the high-frequency words are put at the top, and the antonyms are given up. Besides, we further search for the similar celebrities of the same gender among Chinese celebrities by simultaneously matching the facial attributes and the characteristics. The final outputs are the text descriptions which combine the characteristics and the image show of the celebrities (with the character-similar coefficients) from a celebrity library.

4 Dataset Preparation and Library Construction

By referring to the Chinese anthroposcopy literature, we define 19 facial attributes, including *area of forehead*, *density of eyebrows*, *area of eyes*, *shape of eyelids*, etc., which can hint the human characteristics. For each attribute, we assign three labels, such as *large*, *medium*, and *small* for the attribute *area of forehead* and *thick*, *medium*, and *thin* for the attribute *density of eyebrows*. These descriptions from the Chinese anthroposcopy literature are used to construct the knowledge library which contains the links between facial attributes and characteristics. For example, *large area of forehead* hints "sensible, optimistic,

Attribute name	Attribute values	Attribute name	Attribute values	Attribute name	Attribute values
Area of forehead	Large Medium Small	**Shape of forehead**	Towering Medium Collapsing	**Length of eyebrow**	Long Medium Short
Density of eyebrows	Thick Medium Thin	**Distance of eyebrows**	Wide Medium Narrow	**Shape of eyebrows**	Eight One Inverted-eight
Area of eyes	Large Medium Small	**Distance of eyes**	Large Medium Small	**Depth of eyes**	Convex Medium Concave
Shape of eyelids	Single Medium Double	**Eye corners**	Bullish Horizontal Dropping	**Outline of eyes**	Linear Normal Round
Tip of nose	Towering Medium Collapsing	**Bridge of nose**	Towering Medium Collapsing	**Length of nose**	Long Medium Short
Width of nose	Wide Medium Narrow	**Area of mouth**	Large Medium Small	**Shape of lips**	Thick Medium Thin
Mouth corner	Bullish Horizontal Dropping				

Fig. 2. Detailed definition of facial attributes.

---

[5] It is estimated by our gender recognition system [7].

and clever" characteristics, and *thick density of eyebrows* hints "impatient, bad-tempered, simple" characteristics. Figure 2 lists the detailed definition of the facial attributes. As aforementioned, inter-attributes are fine-grained (e.g., *area of forehead* and *shape of forehead*), and the intra-attribute is multi-value (e.g., *convex, medium,* and *concave* in *depth of forehead*). Appendix 1 gives the whole knowledge library.

We download a large number of Chinese face images from Baidu image search and Bing image search by using queries such as "Chinese face", "Chinese portrait", etc. Ten students check these downloaded photos, and non-frontal faces, small-size faces and non-Chinese faces are removed. We crop and transform the faces to a common size (i.e., 245 × 196 × 3) via the face landmarks, which are detected by the face alignment algorithm [16]. The total number of the collected faces is 5,562. Several exemplar face photos are shown in Fig. 3. We employ 5 student annotators to annotate the facial attributes of the whole face database. For each face, the highest-frequency value of each attribute is set to its attribute value. The distribution of facial attributes in our collected Chinese face database is plotted in Fig. 4. In addition, to vividly show the face-reading output, we also consider showing some celebrity templates to an individual input from a Chinese celebrity library. Therefore, we collect the celebrity images with well-known characteristic tags from Sina Weibo[6] (e.g., the most beautiful Chinese people list can tell who are attractive; Forbes ranks China's largest charitarians who are

Fig. 3. Examples and their facial regions in the collected Chinese face database.

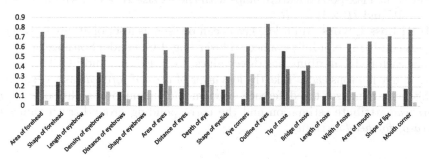

Fig. 4. Distribution for each facial attribute in the collected Chinese face database. Each attribute has three values.

---

[6] www.weibo.com.

kind-hearted; the top-N follower ranking on Weibo indicates who have more personal charm). The celebrity images are also annotated with the facial attributes by the student annotators. Appendix 2 shows some exemplar celebrities from the collected Chinese celebrity library.

5 Deep Networks with Facial Region Pooling

5.1 Architecture

Following the common DCNN architecture, the architecture of FPR-net is designed to consist of several convolutional (conv) layers and max pooling layers, which are alternately stacked into a Deep ConvNet. The facial region pooling (FRP) layer is embedded on top of the last conv layer. The FRP layer is followed by several stacked fully-connected layers (FCs). The R facial regions are searched by the box search on the facial landmarks. The top of FCs is terminated by the multiple loss (ML) layers, where each loss layer consists of a fully-connected layer and a Softmax loss function. Therefore, FRP-net outputs 19 probability estimates, each of which covers three attribute values. Figure 5 shows the FRP-net architecture.

- **Box Search.** The function of Box search is to locate the facial regions based on the facial landmarks, and all the 19 facial attributes are located. The face landmarks are detected by the face alignment algorithm [16]. The r-th region is defined by a duple $\{r, x_{min}, y_{min}, x_{max}, y_{max}\}$, where $\{x_{min}, y_{min}\}$ and $\{x_{min}, y_{min}\}$ denote its top-left location and right-down location, respectively. Figure 3 shows some exemplar facial regions.
- **FRP Layer.** The FRP layer takes as inputs the R facial regions and N feature maps with size $H \times W$, where H and W denote the number of rows and that of columns, respectively. The N feature maps are the outputs on top of the Deep ConvNet, while the R facial regions are obtained by Box search. For each feature map, FRP layer implements the max pooling on R facial regions instead of the commonly used sliding windows. Finally, the output of the FEP layer is the max-pooled feature maps with the fix size $H' \times W'$, where $H' < H$, $W' < W$ and $H' \times W' = R$.
- **Multi Loss Layers.** Multi loss layers can be seen as a set of attribute-specific layers consisting of a fully-connected layer (e.g. FC1) and a Softmax function. Since each facial attribute has three values, FC1, FC2, ... and FC19 are set to have three neurons each. Take the loss layer of *area-of-forehead* as an example. This layer penalizes the difference between the output probability of *area-of-forehead*-attribute estimation and the corresponding ground-truth values of *area-of-forehead*-attribute.

5.2 Training the Network

Generally, training a DCNN requires a large-scale database, e.g., ImageNet [17] and CIFAR-10 [18]. For example, the AlexNet model on 1,000 classes of ILSVRC

2012 [10] has 60 million parameters. Our collected Chinese face database contains no sufficient data to train the FRP-net with millions of parameters, which may result in the overfitting problem. Therefore, we consider manually enlarging the dataset through the commonly used data augmentation strategy. For the original image, we randomly shake the pixels of the facial landmarks within the range $[-6, +6]$, and then transform and crop the faces to a common size. For each original image, we generate a "standard-cropped" image and 11 "shaking-cropped" images. Thus the number of images in the face database is augmented to 66,744.

To optimize the parameters of the proposed FRP-net well, we consider using the model learned from the ImageNet dataset as the initializing parameters[7]. Two pre-trained ImageNet network models are used in this paper. One is the AlexNet model [10], and the other is the VGG16 model [19]. Both of these two

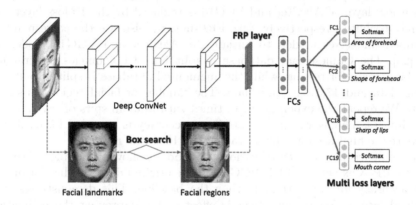

Fig. 5. FRP-net architecture. An image with facial landmarks is firstly input to a fully convolutional network (Deep ConvNet), and is terminated by multi loss layers. Box search is used to locate the coordinates of the facial regions based on the facial landmarks. On top of the Deep ConvNet, the embedded facial region pooling (FRP) layer with the input of facial regions pools the feature maps for each image, which is followed by the fully connected layers (FCs).

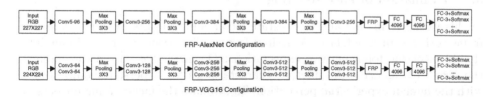

Fig. 6. FRP-net configuration. Conv5-96 denotes the kernel size and the number of channels which are 5 and 96, respectively. FC-3 denotes the three neurons in this FC. The ReLU activation function is not shown for brevity.

---

[7] Here, we also use the random Gaussian values as the parameter initialization. However, it produces less stable results and slower convergence.

pre-trained models (parameters) are available online[8]. Following the configurations of AlexNet and VGG16, the proposed FRP-net has two corresponding configurations as shown in Fig. 6, called FRP-AlexNet and FPR-VGG, respectively. For our FRP-net, we train two models, namely FRP-AlexNet and FPR-VGG on the Chinese face database.

6   Experiments

6.1   Evaluation of Facial Attribute Estimation

As aforementioned in Introduction, the facial attribute estimation is the most important part for the face reading. We conduct experiments to evaluate the performance of facial attribute estimation by our FRP-net (i.e., FEP-AlexNet and FRP-VGG16) and the baseline methods (i.e., AlexNet and VGG16). The common loss layer of AlexNet and VGG16 is replaced by the 19 loss layers for 19 facial attributes, respectively. Like FRP-net, we also use the corresponding pre-trained ImageNet models to initiate the AlexNet and VGG16 models in Caffe [20]. The Chinese face database is split into 4,114 images for training and 1,448 images for testing, while the augmented database is split into 49,368 training data and 17,376 testing data. The number of facial regions is set as $R = 10$. We repeat the experiments 10 times with random spits of training and testing images. The average of per-attribute accuracy is recorded for each run. We report the final accuracy by averaging the 10 times' results.

The comparison accuracy of various methods is shown in Table 1. We can see that FRP-AlexNet and FRP-VGG16 on the original database have achieved higher accuracy for each attribute than the baseline. This demonstrates that the added face region pooling layer is effective for improving the estimation accuracy of the fine-grained attributes. Besides, the FRP-VGG16 outperforms FRP-AlexNet, sine the former adopts the deeper network architecture. When we use the augmented data, the estimation accuracy of FRP-VGG16 and VGG16 is further improved.

6.2   Evaluation of Face Reading

In this work, we mainly aim to generate face-reading results by our face reader framework for an input face. A desired face-reading system for one input face should meet the following two expectations: (1) *attractive* – it should interest the users; (2) *smart* – the inferred characteristics for a person should accord with the human experiential perception. To evaluate the face-reading outputs of our approach, we compare our outputs with one popular App, Mianxiang Dashi (short for MD-App). We show the qualitative and quantitative evaluations on the collected photo sets. First, some outputs obtained by our approach and MD-App for different persons are shown. We investigate the properties of our face reader system by inviting volunteers to use this system and then do voting. Second, we

---

[8] http://dl.caffe.berkeleyvision.org/.

Table 1. Accuracy (%) of facial attributes for various methods. Superscript ∗ denote used database consists of original and augmented data.

Method	Area of forehead	Shape of forehead	Length of eyebrows	Density of eyebrows	Distance of eyebrows	Shape of eyebrows	Area of eyes
AlexNet [10]	76.65	70.58	56.32	60.98	74.24	71.75	61.25
VGG16 [19]	78.52	74.10	61.32	64.08	81.05	75.31	63.89
VGG16* [19]	82.40	75.27	67.81	69.44	87.01	78.49	69.63
FRP-AlexNet	81.45	78.74	66.81	65.00	81.38	75.89	76.57
FRP-VGG16	83.75	80.02	68.90	68.42	87.25	82.32	80.82
FRP-VGG16*	**90.66**	**85.58**	**76.39**	**74.40**	**95.67**	**85.37**	**85.45**

Method	Distance of eyes	Depth of eyes	Shape of eyelids	Eye corners	Outline of eyes	Tip of nose	Bridge of nose
AlexNet [10]	74.46	60.70	59.39	61.53	82.47	61.88	52.56
VGG16 [19]	77.97	62.00	60.38	65.12	84.11	62.57	55.46
VGG16* [19]	80.63	68.74	67.00	73.59	88.99	68.40	66.28
FRP-AlexNet	78.86	66.39	62.09	72.22	87.75	70.80	59.68
FRP-VGG16	84.94	72.97	68.83	74.68	89.34	72.96	60.77
FRP-VGG16*	**88.79**	**75.93**	**74.09**	**82.39**	**96.50**	**77.85**	**68.37**

Method	Length of nose	Width of nose	Area of mouth	Shape of lips	Mouth corner	Average accuracy
AlexNet [10]	75.01	64.22	66.78	67.13	79.63	67.24
VGG16 [19]	79.28	65.85	69.20	68.78	81.15	70.00
VGG16* [19]	83.23	68.90	70.82	69.79	82.91	74.70
FRP-AlexNet	80.88	70.14	72.61	79.73	82.32	74.17
FRP-VGG16	86.19	75.14	76.47	78.96	84.67	77.76
FRP-VGG16*	**89.24**	**83.96**	**78.82**	**81.39**	**92.36**	**83.33**

conduct the user study on face-reading outputs in two aspects: *attractive* and *smart*.

Qualitative Evaluation. Figure 7 shows the exemplar outputs by our approach and MD-App for different persons. The 1st and 4th columns are input face images, the 2nd and 5th columns are the output results of our approach, and the 3rd and 6th columns are the output results of MD-App. We can see that our results are more interesting in the image-text mixed show. We investigate the properties of the presented face reader system. 50 volunteers are invited to test our face reader system and are asked to answer two questions: (1) *Reasonable: Do you think the output results are reasonable?* (2) *Interesting: Do you like the face reader system?* In these two questions, we set five values for the subjects to choose: *excellent, good, ordinary, weak* and *poor.* The investigation results are given in Fig. 8. It can be seen that our face reader system has gained very high scores in this questionnaire.

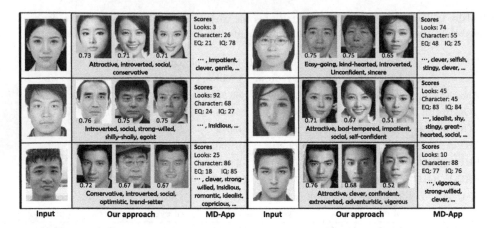

Fig. 7. Comparison results of characteristic estimation. The 2nd and 5th columns are our outputs, while the 3rd and 6th columns are outputs of MD-App. For space limitation, we only show some characteristic descriptions in the results of MD-App. Better view in ×3 size original color PDF file.

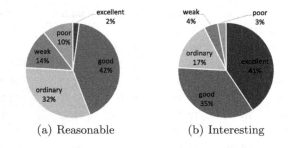

(a) Reasonable   (b) Interesting

Fig. 8. Investigation results of our face reader system.

Quantitative Evaluation. Although not everyone is familiar with the Chinese anthroposcopy, people are still able to accumulate some experience to infer someone's characteristics by observing her/his face. The characteristic output from a smart system should accord with the common perception of most people. Therefore, we employ 30 adults (13 females and 17 males who are students, staffs and businessmen ranged from 18 to 50 years old) to participate in the user study. They are asked to observe each comparison group including two output results of Ours and MD-App respectively. The order of the two output results in each comparison group is random. We provide one question to the volunteers: *Empirically, do you think the results are reasonable?* Based on this question, the users are asked to give comparison results using "≫", ">" and "=", which mean "much better","better" and "comparable", respectively. We assign a score of 1 to the worst result, and the other results are assigned with a score of 3, 2, or 1 if it is much better than, better than, or comparable to this one, respectively. Thus, for each comparison on one question, there are 30 ratings.

Table 2. The left part illustrates the average rating scores and standard deviation values from the user study on our Ours and MD-App. The right part shows the ANOVA test results.

Ours vs. MD-App		Factor of approaches		Factor of evaluators	
Ours	MD-App	F-statistic	p-value	F-statistic	p-value
2.0819 ± 0.0607	1.1158 ± 0.0342	49.3992	9.1136×10^{-7}	0.1970	0.9995

The two-way ANalysis Of VAriance (ANOVA) [21] test partitions the observed rating scores into components corresponding to different explanatory factors, and tests the significance levels of the rating differences with respect to the factors of approaches and users, respectively. The two comparison scores are illustrated in Table 2, respectively. We can see that the voting scores of ours are superior to MD-demo on both of two comparisons. The p-values of ANOVA test show that this superiority is statistically significant and the difference of the volunteers is insignificant, which further confirms the effectiveness of our approach.

7   Conclusions and Future Work

In this paper, we study the face-reading problem from the computer vision perspective and present a computational face reader to automatically infer the characteristics of a person based on his/her face. Specifically, the proposed FRP-net architecture can well estimate the fine-grained facial attributes by equipping a FRP layer, a box search and several multi-loss layers, which is trained on the collected Chinese face database in the offline phase. The experiments on facial attribute estimation demonstrated the potential of the automatic face reader framework, and user studies from automatic, attractive, and smart perspectives of face reading validated the excellence of the presented face reader framework. As an early exploration of Chinese anthroposcopy, we would like to devote more consideration to the age and gender influence of face reading in the future.

Acknowledgments. This work was partially supported by the 973 Program of China (Project No. 2014CB347600), the National Natural Science Foundation of China (Grant No. 61522203 and 61402228), and the Program for New Century Excellent Talents in University under Grant NCET-12-0632.

References

1. Wikipedia: I ching. https://en.wikipedia.org/wiki/I_Ching
2. Wikipedia: Yin and yang. https://en.wikipedia.org/wiki/Yin_and_yang
3. Wikipedia: Confucianism. https://en.wikipedia.org/wiki/Confucianism
4. Wikipedia: Taoism. https://en.wikipedia.org/wiki/Taoism
5. LeCun, Y., Boser, B., Denker, J.S., Henderson, D., Howard, R.E., Hubbard, W., Jackel, L.D.: Backpropagation applied to handwritten zip code recognition. Neural Comput. **1**, 541–551 (1989)

6. Chen, Q., Huang, J., Feris, R., Brown, L.M., Dong, J., Yan, S.: Deep domain adaptation for describing people based on fine-grained clothing attributes. In: CVPR (2015)
7. Li, S., Xing, J., Niu, Z., Shan, S., Yan, S.: Shape driven kernel adaptation in convolutional neural network for robust facial trait recognition. In: CVPR (2015)
8. Bengio, Y.: Learning deep architectures for ai. Found. Trends Mach. Learn. **2**, 1–127 (2009)
9. Sun, Y., Liang, D., Wang, X., Tang, X.: Deepid3: Face recognition with very deep neural networks (2015). arXiv preprint arXiv:1502.00873
10. Krizhevsky, A., Sutskever, I., Hinton, G.E.: Imagenet classification with deep convolutional neural networks. In: NIPS (2012)
11. Ren, S., He, K., Girshick, R., Sun, J.: Faster r-cnn: Towards real-time object detection with region proposal networks (2015). arXiv preprint arXiv:1506.01497
12. Lin, M., Chen, Q., Yan, S.: Network in network. In: ICLR (2014)
13. He, K., Zhang, X., Ren, S., Sun, J.: Spatial pyramid pooling in deep convolutional networks for visual recognition. In: Fleet, D., Pajdla, T., Schiele, B., Tuytelaars, T. (eds.) ECCV 2014, Part III. LNCS, vol. 8691, pp. 346–361. Springer, Heidelberg (2014)
14. Girshick, R.: Fast r-cnn (2015). arXiv preprint arXiv:1504.08083
15. Girshick, R., Donahue, J., Darrell, T., Malik, J.: Rich feature hierarchies for accurate object detection and semantic segmentation. In: CVPR (2014)
16. Viola, P., Jones, M.: Rapid object detection using a boosted cascade of simple features. In: Computer Vision and Pattern Recognition (2001)
17. Russakovsky, O., Deng, J., Su, H., Krause, J., Satheesh, S., Ma, S., Huang, Z., Karpathy, A., Khosla, A., Bernstein, M., et al.: Imagenet large scale visual recognition challenge. Int. J. Comput. Vis., 1–42 (2014)
18. Krizhevsky, A., Hinton, G.: Learning multiple layers of features from tiny images. Master's thesis, University of Toronto (2009)
19. Simonyan, K., Zisserman, A.: Very deep convolutional networks for large-scale image recognition (2014). arXiv preprint arXiv:1409.1556
20. Jia, Y., Shelhamer, E., Donahue, J., Karayev, S., Long, J., Girshick, R., Guadarrama, S., Darrell, T.: Caffe: convolutional architecture for fast feature embedding. In: ACM Multimedia (2014)
21. Minium, E.W., King, B.M., Bear, G.: Statistical reasoning in psychology and education (2003)

Posed and Spontaneous Expression Recognition Through Restricted Boltzmann Machine

Chongliang Wu and Shangfei Wang[✉]

School of Computer Science and Technology,
University of Science and Technology of China, Hefei, China
clwzkd@mail.ustc.edu.cn, sfwang@ustc.edu.cn

Abstract. This paper presents a new method to recognize posed and spontaneous expression through modeling their global spatial patterns in Restricted Boltzmann Machine (RBM). First, the displacements of facial feature points between apex and onset facial images are extracted as features, which capture spatial variations of facial points. Second, the point displacement related facial events are extracted from its displacements. Third, two RBM models are trained to capture spatial patterns embedded in posed and spontaneous expressions respectively. The recognition results on both USTC-NVIE and SPOS databases demonstrate the effectiveness of the proposed RBM approach in modeling complex spatial patterns embodied in posed and spontaneous expressions, and good performance on posed and spontaneous expression distinction.

Keywords: Posed and spontaneous · Expression recognition · Restricted Boltzmann Machine · Spatial pattern

1 Introduction

As we all know, spontaneous expressions convey one's true feelings, while posed expressions disguise one's real emotions. A method to distinguish posed expressions from spontaneous expressions can be used in many areas, including real-life human-robot communications, healthcare, and security. For example, robots can be more perceptual by analyzing users' true emotions through posed and spontaneous expressions differentiating technology. Doctors can make a more precise diagnosis by detecting patients' genuine pain. Deceptive facial expression recognition can be used by the police for lie detection.

Many nonverbal behavior researches presented the differences between posed and spontaneous expressions in both spatial and temporal patterns [4–7]. Spatial patterns mainly involve the movements of facial muscles. For example, when one smiles spontaneously, both the zygomatic major and the orbicularis oculi should be contracted. However, if a smile is posed, the contraction will only appear on zygomatic major, but not on orbicularis oculi [5]. The contraction of zygomatic major is more likely to occur asymmetrically in posed smiles than in spontaneous ones [6]. Ekman *et al.* [5,7] claimed that a good way to recognize

© Springer International Publishing Switzerland 2016
Q. Tian et al. (Eds.): MMM 2016, Part I, LNCS 9516, pp. 127–137, 2016.
DOI: 10.1007/978-3-319-27671-7_11

a posed expression from a spontaneous one is to analyze the absence of muscles movements, since some movements is difficult to make voluntarily [5,7]. Temporal patterns include the trajectory, speed, amplitude and total duration of onset and offset. Such as, for posed expressions, the total duration is usually longer and the onset is more abrupt than spontaneous expressions in most cases [4,5]. For spontaneous expressions, the trajectory appears often smoother than posed expressions [5].

Motivated by the properties revealed by behavior researches, researchers in computer vision have begun to pay attention to posed and spontaneous expressions recognition. The first research on posed and spontaneous expression recognition using machine learning method is presented by Cohn and Schmidt [1], which extracted temporal features, i.e. amplitude, duration, and the ratio of amplitude to duration, and applied a linear discriminant as classifier for posed and spontaneous smile recognition. Valstar [18] proposed a posed and spontaneous smile recognition method. They studied posed and spontaneous brow actions using velocity, duration, and the order of occurrence and fused head, face, and shoulder modalities for the recognition. Littlewort et al. [10] proposed a real and faked pain expression classification method by feeding the detected 20 facial action units into a classifier. Dibeklioglu et al. [2] used the dynamics of eyelid, check and lip corner movement to distinguish posed and spontaneous smile.

However, most computer vision researches only focus on one specific expression. To the best of our knowledge, only two works [15,20] considered all six basic expressions (i.e. happiness, disgust, fear, surprise, sadness and anger) for posed and spontaneous expressions recognition. Zhang et al. [20] used SIFT [3] and FAP [12] features to investigate the performance of a machine vision system for posed and spontaneous expressions recognition of six basic expression on USTC-NVIE database. Pfister et al. [15] proposed a spatiotemporal local texture descriptor (CLBP-TOP) and a generic facial expression recognition framework to differentiate posed from spontaneous expressions from both visible and infrared images on SPOS database.

Furthermore, most current works applied different classifiers for posed and spontaneous expression recognition. Few works captured the spatial patterns embedded in posed and spontaneous expressions explicitly. Thus, we proposed to use Restricted Boltzmann Machine (RBM) to explicitly model spatial patterns embedded in both posed and spontaneous expressions respectively. As a graphical model, RBM can model higher-order dependencies among random variables by introducing a layer of latent units [11]. It has been widely used to model complex joint distributions over structured variables such as image pixels.

In this paper, we first extract the facial point displacements between the apex and the onset facial images. Second, these displacements are discretized to extract facial point displacement related facial events which are used as inputs to RBMs. Third, two RBM models are trained from posed and spontaneous expression samples to capture the spatial patterns embedded in posed and spontaneous expressions explicitly. During testing, the label of an unknown sample is given by selecting the RBM model that is more likely to have generated its set of displacements. The recognition results on both USTC-NVIE and SPOS databases demonstrate the effectiveness of the proposed RBM approach in modeling complex spatial patterns embodied in posed and spontaneous expressions.

2 The Proposed Method

The procedure of our proposed approach is shown in Fig. 1, includes features extraction, and posed and spontaneous expression modeling by RBMs. The details are described as follows.

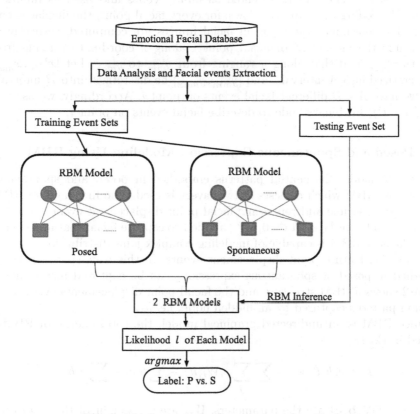

Fig. 1. The framework of our proposed method

2.1 Feature and Facial Events Extraction

First, 27 facial feature points, as shown in the bottom part of Fig. 2, are automatically detected on both the onset and apex expression frames using the algorithm introduced in [17]. The onset frame is the beginning of the onset phase, which is similar to the neutral frame here, and the apex frame is the most exaggerated expression frame during the apex phase. Both onset and apex frames are provided by the databases. Second, the facial region around each facial feature point is extracted and normalized to 100×100, in which the locations of the eyes and the tip of the nose are fixed. Through the face alignment and normalization, the facial feature points are robust to different subjects and to moderate face pose variation.

Then, the displacements of the feature points between the onset frame and the apex frame are calculated as features. After that, for each facial feature point, the displacements are discretized with unequal intervals. Each displacement interval represents a certain movement of facial feature points, called as a facial event here.

Last, facial events are represented as binary codes and used as inputs of RBMs. The coding rules are as follows: for every facial point, the displacements of x and y coordinate values of all facial images are computed, respectively. We assume the movement of facial point i along x coordinate ranging from x_{min}^i to x_{max}^i, and that along y ranging from y_{min}^i to y_{max}^i. Let $[x_{min}^i, x_{max}^i]$ be discretized into A intervals, and $[y_{min}^i, y_{max}^i]$ be discretized into B intervals, then we have $A \times B$ different facial events on point i. Accordingly, we use $t_i = \lceil \log_2(A \times B) \rceil$ bit binary code to describe facial events on point i.

2.2   Posed and Spontaneous Expression Modeling Using RBM

In order to model the spatial patterns embodied in posed and spontaneous expressions, RBM which consists of two layers is used. The first layer of RBM, $v \in \{0, 1\}^n$, is visible and represents facial point displacement events. The second layer is a latent layer, $h \in \{0, 1\}^m$, applied to capture facial spatial patterns. Latent layer of RBM is capable of modeling complex joint distribution over visible layer, i.e. feature point displacement events. In this way, spatial patterns embodied in posed or spontaneous expressions can be captured in the model. Figure 2 shows a RBM structure and the facial point displacements events combination patterns captured by an hidden node, i.e. h_1.

Since RBM is an undirected graphical model, the total energy of RBM is defined in Eq. 1.

$$E < v, h; \theta > = - \sum_i \sum_j v_i W_{ij} h_j - \sum_i b_i v_i - \sum_j c_j h_j \tag{1}$$

where $\theta = \{\mathbf{W}, \mathbf{b}, \mathbf{c}\}$ are the parameters. W_{ij} are the weight of the connection between visible node v_i and hidden node h_i which measures the compatibility between v_i and h_j. $\{b_i\}$ and $\{c_j\}$ are the biases of v_i and h_i respectively.

The distribution over visible units of RBM is calculated by marginalizing over all hidden units with Eq. 2, where $Z(\theta)$ is the partition function and $P(h, v; \theta)$ is joint distribution over hidden nodes and visible nodes. This allows RBM to capture global dependencies among the visible variables.

$$P(v; \theta) = \sum_h P(h, v; \theta) = \frac{\sum_h exp(-E(v, h; \theta))}{Z(\theta)} \tag{2}$$

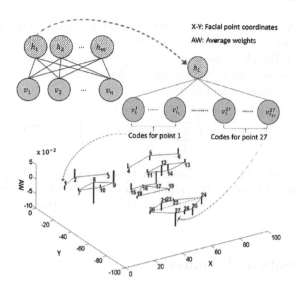

Fig. 2. RBM structure and the combination patterns of facial point events captured by h_1. t_i represents the length of codes for facial point i. At the bottom part, we drew a facial points distribution map at $x - y$ plane. z coordinate is the average weights (AW) of visible nodes for each facial point

According to Bayesian theorem, conditional probability of hidden nodes given visible nodes and of visible nodes given hidden nodes can be estimated as follow:

$$P(v|h, \theta) = \prod_i \delta(\sum_j w_{ij} h_j + b_i);$$

$$P(h|v, \theta) = \prod_j \delta(\sum_i v_i w_{ij} + c_j) \tag{3}$$

Given the training data $\{v_i\}_{i=1}^N$, where N indicates the number of the training samples, the parameters are learned by maximizing the log likelihood with Eq. 4.

$$\theta^* = argmax_\theta L(\theta); L(\theta) = \frac{1}{N} \sum_{i=1}^N log P(v; \theta) \tag{4}$$

The gradient with respect to θ can calculated with Eq. 5,

$$\frac{\partial log P(v; \theta)}{\partial \theta} = \langle \frac{\partial E}{\partial \theta} \rangle_{P(h|v, \theta)} - \langle \frac{\partial E}{\partial \theta} \rangle_{P(h, v|\theta)} \tag{5}$$

where $\langle \cdot \rangle_P$ represents the expectation over distribution P. In Eq. 5, inferring partition function $Z(\theta)$ in $P(h, v)$ analytically is intractable. However, Hinton *et al.* [9] proposed an very efficient way to estimate it approximately, namely contrastive divergence (CD). The basic idea of CD algorithm is to approximate $P(h, v)$ with one step sampling from the data. Gradient calculation is basic procedure of RBM training. With the gradients of all parameters, stochastic gradient descent is applied for RBM training. Meta-parameters such as the learning rate, the momentum, are decided by following the instruction of [8].

We train two RBMs for posed and spontaneous expression respectively. After training, for a test sample t, the log probability that RBM trained on class c assign to the test sample is as follows:

$$\log P(t;\theta_c) = \log\left(\sum_h exp(-E(t;\theta_c))\right) - \log Z(\theta_c) \qquad (6)$$

This is the logarithm of Eq. 2. The partition function $Z(\theta_c)$ can be approximately estimated through Annealed Importance Sampling (AIS) approach [16]. With these log probabilities, the label of the test sample is the class with greater value.

3   Experiments and Analysis

3.1   Experimental Conditions

As far as we know, there are several databases available for posed and spontaneous expressions recognition, such as, BBC Smile Dataset [13], MAHNOB-Laughter database [14], UvA-NEMO smile database [2], SPOS database [15], and USTC-NVIE database [19]. Among them, the BBC, MAHNOB-Laughter and UvA-NEMO databases only contain posed and spontaneous expressions for smiles, while the USTC-NVIE and SPOS databases consist of posed and spontaneous expressions for six basic expression categories. Thus, these two databases are adopted in our experiments.

The USTC-NVIE database [19] is a natural visible and thermal infrared facial expression database, which contains both spontaneous and posed expressions with six basic categories (i.e. happiness, disgust, fear, surprise, anger and sadness) of more than 100 subjects. The onset and apex frames are provided for both posed and spontaneous subsets. The SPOS database [15] is a visible and near infrared expression database, including both posed and spontaneous expressions with six basic categories from seven subjects (four males and three females). The image sequences in this database start from onset frame and end with apex frame.

For USTC-NVIE database, both the apex and onset frames of all posed and spontaneous expression samples, which come in pairs from the same subject, are selected. In this procedure, we discarded spontaneous samples which have no expressions, and finally select 1028 samples, including 514 posed and 514 spontaneous expression samples from 55 male and 25 female subjects. The distribution of posed and spontaneous expression samples is shown in Table 1. Our experimental results on the database are obtained by applying a 10-fold cross validation method on all samples according to the subjects.

For SPOS database, the first and the last frames of all the posed and spontaneous samples are selected, including 84 posed expression samples and 150

Table 1. The distribution of samples on USTC-NVIE database

	Happiness	Disgust	Fear	Surprise	Anger	Sadness
Posed	104	93	68	78	91	80
Spontaneous	104	93	68	78	91	80

spontaneous expression samples, as shown in Table 2. Since SPOS database consists of images from only seven subjects (4 males and 3 females), and it does not include all six expression images for certain subjects, we did not select samples in pairs from SPOS database as we did on USTC-NVIE database. In order to compare with [15], leave-one-subject-out cross validation is used.

Table 2. The distribution of samples on SPOS database

	Happiness	Disgust	Fear	Surprise	Anger	Sadness
Posed	14	14	14	14	14	14
Spontaneous	66	23	32	11	13	5

3.2 Experimental Results and Analysis

Figure 3 shows the histogram of the feature point displacements along x and y axis, From Fig. 3, we find that most displacements are at the middle, which means that the movements of feature points are small in most cases.

In order to make each facial event cover similar number of samples, the displacements are discretized into multiple intervals with unequal length. In our experiments, we let the number of samples fall into every interval as close as possible to 100 but no more than 100. After extracting point related facial events, these events are used to form binary codes which are the inputs of RBMs. According to the binary coding rule described in Sect. 2.1, we are able to generate 108-bit binary codes for samples in SPOS database and 216-bit for samples in USTC-NVIE database.

We trained two RBM models from posed and spontaneous samples using the generated binary codes, respectively. Then, the RBMs are used for distinguishing posed vs. spontaneous expression.

In order to analyze the ability of our proposed RBM for modeling spacial patterns in posed and spontaneous expressions, the global spacial pattern captured by both RBMs are showed in Fig. 4. As described in Sect. 2.2, parameters W_{ij} measures the compatibility between visible node v_i and latent node h_j. The greater the absolute value of W_{ij}, the more the point displacement facial events affect the captured spacial pattern. We first summated W over all hidden units for every visible unit, and then computed the average W for visible units represent facial events of the same facial point. Due to the unbalanced data in

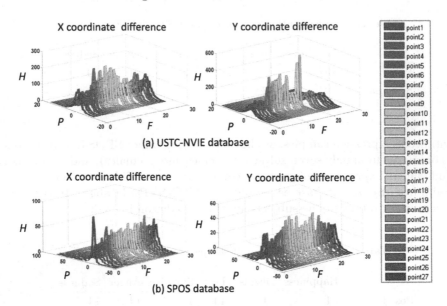

Fig. 3. The histogram of x and y coordinate value differences between apex and onset facial points for all samples on two database. The number of bins is 50 for all facial points. P: Position of bins; F: Facial points number; H: Height of bins

SPOS database, the weights are normalized by the number of samples for each RBM. From Fig. 4, we can obtain the following observations: first, the W values of RBM for posed expressions (red bars) and those for spontaneous expressions (blue bars) are different, proving that the spacial pattern for posed expressions and that for spontaneous expressions are different. This is consistent with current behavior research. Second, in most cases, the weights of posed RBM are larger than those of spontaneous RBM. It further confirms that posed expressions are more exaggerated than spontaneous one.

The recognition results on both databases are shown in Table 3. The recognition accuracy for all expressions reaches 81.23 % and the F1-score is 0.8012, on USTC-NVIE database. On SPOS database, the accuracy achieves 74.36 %, however, due to the unbalanced data the F1-score is only 0.5231.

3.3   Comparison with Other Methods

In order to further demonstrate the effectiveness of our proposed method, we compared our work with Zhang's [20] and Pfister's [15] works. In additional, as a baseline, we conducted experiments using support vector machine (SVM) with linear kernel under the same experimental conditions with our work.

Zhang *et al.* selected 3572 posed and 1472 spontaneous images. Since they did not explicitly state which images were selected, it is hard for us to select the same images as theirs. We can only compare the experimental results as a reference. The results of our work and the best results of [20] are shown in

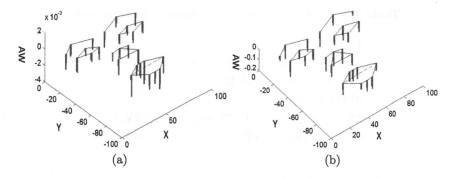

Fig. 4. Average W at every facial points from the trained RBMs, (a) and (b) are from the RBMs trained on USTC-NVIE database and SPOS database, respectively. We drew a facial points distribution map at $x-y$ plane. z coordinate is the average weights (AW). The red bars represent average W value from RBM for posed expressions, and the blue bars represent that for spontaneous ones (Color figure online)

Table 3. Experiment result on USTC-NVIE database and SPOS database

Confusion matrix		USTC-NVIE database		SPOS database	
		Posed	Spont.	Posed	Spont.
	Posed	389	125	35	49
	Spont.	68	446	11	139
Accuracy		81.23 %		74.36 %	
F1-score		0.8012		0.5385	

"Posed" represents posed expression.
"Spont." represents spontaneous expression.

Table 4. From Table 4, we can find that although the number of samples are smaller than Zhang *et al.*'s, our proposed models outperform their.

Pfister *et al.* [15] distinguished posed and spontaneous expression from both visible and near-infrared image sequences on SPOS database. Here, we only compare our work with their work on visible images, as shown in Table 4. From this table, we can find that the accuracy of our method is 2.36 % higher than Pfister *et al.*'s. Furthermore, they extracted CLBP-TOP texture features from facial expression sequence, while our features are extracted from the apex frames and onset frames. It indicates that, with less information, our proposed method achieve better results.

Table 4 also shows the results achieved by using SVM with linear kernel as classifier. The classification accuracy reaches 78.02 % on USTC-NVIE database, and 69.66 % on SPOS database. Our approach outperforms the baseline.

The above comparison demonstrates the performance of our approach is better than the state of the art. The most important reason that contributes the performance of our method is the use of the RBM to globally capture the relationships among the spatial movements of facial landmark points. This explains

Table 4. Comparison between our method and related work

USTC-NVIE database			
Method	ours	L. Zhang *et al.* [20]	SVM
Accuracy (%)	81.23	79.43	78.02
SPOS database			
Method	ours	T. Pfister *et al.* [15]	SVM
Accuracy (%)	74.36	72.0	69.66

why our method outperforms [15,20] in spite of their use of more powerful features and classifiers.

4 Conclusion and Future Works

In this paper, we proposed a new method to recognize posed and spontaneous expressions by capturing the global facial spatial patterns of posed and spontaneous expressions using RBM models. First, the displacements of facial feature points between apex and onset facial images are recorded in the form of coordinates value variation. We analyzed distributions of these displacements of all facial points by computing their histograms. Second, the coordinates value variation of all facial points are discretized as facial point displacement related facial events which are used to form binary codes as inputs of RBMs. Third, two different RBM models are trained to capture spatial patterns embedded in posed and spontaneous expressions respectively. The recognition results on both USTC-NVIE and SPOS databases demonstrated the effectiveness of our method in modeling spacial patterns of posed and spontaneous expressions, good performance on posed and spontaneous expression recognition.

In the future works, we will consider modeling the temporal patterns of posed and spontaneous expressions to improve recognition performance.

Acknowledgments. This paper was supported by the National Science Foundation of China (Grant No. 61175037, 61228304, 61473270), and the project from Anhui Science and Technology Agency (1106c0805008).

References

1. Cohn, J., Schmidt, K.: The timing of facial motion in posed and spontaneous smiles. Int. J. Wavelets Multiresolut. Inf. Process. **2**(02), 121–132 (2004)
2. Dibeklioğlu, H., Salah, A.A., Gevers, T.: Are you really smiling at me? spontaneous versus posed enjoyment smiles. In: Fitzgibbon, A., Lazebnik, S., Perona, P., Sato, Y., Schmid, C. (eds.) ECCV 2012, Part III. LNCS, vol. 7574, pp. 525–538. Springer, Heidelberg (2012)
3. Lowe, D.G.: Distinctive image features from scale-invariant keypoints. Int. J. Comput. Vis. **60**, 91–110 (2004)

4. Ekman, P.: Darwin, deception, and facial expression. Ann. NY Acad. Sci. **1000**(1), 205–221 (2003)
5. Ekman, P., Friesen, W.: Felt, false, and miserable smiles. J. Nonverbal Behav. **6**(4), 238–252 (1982)
6. Ekman, P., Hager, J., Friesen, W.V.: The symmetry of emotional and deliberate facial actions. Psychophysiology **18**(2), 101–106 (1981)
7. Ekman, P., Rosenberg, E.: What the Face Reveals: Basic and Applied Studies of Spontaneous Expression using the Facial Action Coding System (FACS). Oxford University Press, New York (1997)
8. Hinton, G.: A practical guide to training restricted boltzmann machines. Momentum **9**(1), 926 (2010)
9. Hinton, G.E.: Training products of experts by minimizing contrastive divergence. Neural Comput. **14**(8), 1771–1800 (2002)
10. Littlewort, G., Bartlett, M., Lee, K.: Automatic coding of facial expressions displayed during posed and genuine pain. Image Vis. Comput. **27**(12), 1797–1803 (2009)
11. Nie, S., Ji, Q.: Capturing global and local dynamics for human action recognition. In: ICPR (2014)
12. Pandzic, I.S.: Mpeg-4 Facial Animation: The Standard, Implementation and Applications. John Wiley & Sons Inc., New York (2003)
13. Paul, E.: Bbc-dataset. http://www.bbc.co.uk/science/humanbody/mind/surveys/smiles/
14. Petridis, S., Martinez, B., Pantic, M.: The mahnob laughter database. Image Vis. Comput. **31**, 186–202 (2013)
15. Pfister, T., Li, X., Zhao, G., Pietikainen, M.: Differentiating spontaneous from posed facial expressions within a generic facial expression recognition framework. In: ICCV Workshops, pp. 868–875. IEEE (2011)
16. Salakhutdinov, R., Murray, I.: On the quantitative analysis of deep belief networks. In: ICML, pp. 872–879. ACM (2008)
17. Tong, Y., Wang, Y., Zhu, Z., Ji, Q.: Robust facial feature tracking under varying face pose and facial expression. Pattern Recogn. **40**(11), 3195–3208 (2007)
18. Valstar, M., Gunes, H., Pantic, M.: How to distinguish posed from spontaneous smiles using geometric features. In: Proceedings of the 9th International Conference on Multimodal Interfaces, pp. 38–45. ACM (2007)
19. Wang, S., Liu, Z., Lv, S., Lv, Y., Wu, G., Peng, P., Chen, F., Wang, X.: A natural visible and infrared facial expression database for expression recognition and emotion inference. IEEE Trans. Multimedia **12**(7), 682–691 (2010)
20. Zhang, L., Tjondronegoro, D., Chandran, V.: Geometry vs. appearance for discriminating between posed and spontaneous emotions. In: Lu, B.-L., Zhang, L., Kwok, J. (eds.) ICONIP 2011, Part III. LNCS, vol. 7064, pp. 431–440. Springer, Heidelberg (2011)

DFRS: A Large-Scale Distributed Fingerprint Recognition System Based on Redis

Bing Li[1]($\boxtimes$), Zhen Huang[1], Jinbang Chen[2], Yifan Yuan[1], and Yuxing Peng[1]

[1] National University of Defence and Technology, Changsha, China
tanmu991331@163.com
[2] East China Normal University, Shanghai, China

Abstract. As the fast growth of users, matching a given fingerprint with the ones in a massive database precisely and efficiently becomes more and more difficult. To fight against this challenging issue in "big data" era, we have designed in this paper a novel large-scale distributed Redis-based fingerprint recognition system called DFRS that introduces an innovative framework for fingerprint processing while incorporating many key technologies for data compression and computing acceleration. By using Base64 compressive encoding method together with key-value pair storage structure, the space reduction can be achieved up to 40% in our experiments – which is particularly important as Redis is an in memory read-write NoSQL data storage system. To compensate the cost introduced by compressive encoding, the parallel decoding is adopted with the help of OpenMP, saving the time by above one third. Furthermore, the granularity-based division (RM+AM architecture) and the *Quick-Return* strategy bring significant improvement in matching time, making the whole system – DFRS feasible and efficient in large scale for massive data volume.

Keywords: Fingerprint recognition · Distributed computing · Redis · Big data

1 Introduction

In today's information society, the biologic features are popularly deployed for identity authentication, in particular in the area of public security, hospital and finance [1] – as these features are commonly recognizable, distinguishable and storable, thus technologically feasible. They are manifold, such as face, fingerprint, retina, *etc.*. Among all these features, the fingerprint recognition is the

Y. Peng—This work was supported by the National Basic Research Program of China (973) under Grant No. 2014CB340303, National Natural Science Foundation of China (NSF) under Grant No. 61402490, the Science and Technology Commission of Shanghai Municipathy under research Grant No. 14DZ2260800, China Postdoctoral Science Foundation under Grant No. 2014M561438 and Excellent Ph.D. Dissertation Foundation of Hunan.

© Springer International Publishing Switzerland 2016
Q. Tian et al. (Eds.): MMM 2016, Part I, LNCS 9516, pp. 138–149, 2016.
DOI: 10.1007/978-3-319-27671-7_12

most popular one due to its credibility, security and convenience, and its popularity brings the sharp increase in the volume of users. The fingerprint recognition technology is well rounded and broadly applied today. However, the stand-alone fingerprint recognition system can't satisfy the fast handling for billions of data – "big data", so it's necessary for a distributed fingerprint recognition system.

Facing with big data, we at first have to choose an I/O efficient storage. These years, NoSQL database is rising, and becoming the mainstream in data storage. According to the database rank announced by DB-engines in July 2015 [2], Redis has been ranked the 10th of all databases and 1st of key-value databases. It features easy-expansibility, high performance with big data, flexible data model and great usability. Among NoSQL [3] databases, Redis occupies an significantly important position. Now about 12 % of the Internet companies, such as Sina, GitHub and Stack Overflow, use Redis.

However, Redis is an in-memory database, and the memory is smaller and more expensive compared to the hard-disk. In our system, we encode the fingerprint information with compression to store the massive data. The encoding method increases the time expenses of decoding, we therefore design a parallel decoding strategy to address this time issue. To further enhance the efficiency, we present a new fingerprint-match model, which divides the whole matching process into two part: rough match (RM) and accurate match (AM). Finally we propose a matching strategy called *Quick-Return* to speed up the process.

To evaluate our design, we have implemented a distributed prototype system called DFRS based on Redis and have carried out extensive evaluation experiments in the cluster. The results reveal that our design achieves the following objects:

1. The new matching model makes the recognition process simpler and faster on condition to satisfy the matching accuracy.
2. Compressive encoding reduces the storage space by about 39 %, and millions of data can be well stored into Redis. Moreover, the disorder fingerprint features get regular and convenient to read and write. Parallel decoding attains the distribution effect and the time consumption is shortened by two-thirds.
3. *Quick-Return* matching strategy vastly shortens the recognition time consumption and guarantees the highly correct recognition rate. The optimal node can be accumulated by nearly 50 %.

The remaining of the paper is organized as follows: In Sect. 2, we briefly introduce the necessary background – Redis (a popular NoSQL database) and the basis for fingerprint recognition; as the significant part, we then present our system architecture and the key methods proposed for the purpose of optimization in Sect. 3; the setting of the experiments and the performance evaluation are given in Sect. 4; and finally Sect. 5 concludes the paper and suggests the future work.

2   Background

2.1   Redis

Redis means REmote DIctionary Server. It is a high-performance database using key-value (KV) stores. Different from other structured storage systems, Redis supports not only Strings, but also many other data types, consisting of Lists, Sets, Order-Sets and Hash tables [4]. Thus Redis maps keys to types of values. The read-write operation is fairly fast as Redis typically holds the whole database in memory, comparing to other disk-stored NoSQL database. In general, with Redis, a task can be fulfilled within 20 operations (called commands).

We use hash table as the data structure in our system. Every KV pair satisfies one-to-one correspondence. In Redis, hash object is encoded by *ziplist* or *hashtable*. Hash object encoded by *ziplist* uses compressive list as its low-level implementation. And the KV pair storage is constructed by pushing the key and its value together into the compressive list. The newest KV pair will be then added into the end of the compressive list. On the other hand, hash object encoded by *hashtable* uses dictionary as its low-level implementation. The encoding way can be selected when existing a key or a value with the length of 64 or having more than 512 nodes in the compressive list.

2.2   Fingerprint Feature

Conventionally, a single fingerprint is characterized by four features, namely the fingerprint's type *type*, the ridge distance *avlrd*, the minutiae group and the fingerprint's name. In our RM+AM architecture, The first two features are used in RM while the latter two are deployed in AM.

By Galton-Henry scheme [5], the fingerprints are divided into five classes: whorl, left loop, right loop, arch and tented arch. Research has shown that arch and tented arch account for less than 6 % [6] in total. For the sake of simplicity, we take arch, tented arch and others not included into any class mentioned above together as one class in this paper. As a result, we have four types for fingerprints. With the method proposed by En Zhu *et al.* [7], the ridge distance is set to be the integer values ranging from 0 and 255. Minutia means small detail. In the context of fingerprints, it refers to various ways that ridges can be discontinuous. It consists of four features [8]: the minutia's type, the position parameters (the horizontal coordinate X and the vertical coordinate Y) and the direction of the distance. Note that the minutia's type is usually difficult to distinguish because of the fingerprint impression [9]. Thus, we move it out from our consideration and focus on other three features of minutia in our study.

As a summary, we use a 4-tuple, consisting of the fingerprint's type, the ridge distance, the minutiae group and the fingerprint's name to characterize a fingerprint, and further use it to achieve the goal of fingerprint recognition in large scale.

2.3   Fingerprint Recognition

It is intuitive that fingerprint recognition can be conducted by comparing fingerprints' features. However, every time the obtained fingerprint images are different when collecting one's fingerprint using the sensor – which brings difficulty in image matching.

As one of the most popular technologies for identity authentication, fingerprint recognition attracts much attention in the research community. Today, many studies focus on the stand-alone fingerprint recognition system. Many optimization methods have been proposed. Ruggero Donida Labati *et al.* [10] have designed a neural approach to decrease the matching error rate from 3.04 % to 2.20 %. In addition, some other studies perform fingerprint recognition relying merely on the minutiae [11–13].

3   DFRS

3.1   System Architecture

In our design, DFRS mainly consists of three modules: client, Configure Server and Data Server. For the purpose of presentation, Fig. 1 provides the system architecture and the illustration of what each part is responsible for and how each module works.

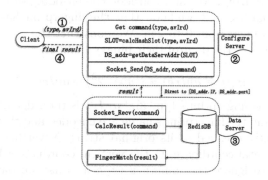

Fig. 1. Process of distribute system

1. The client inputs the user's fingerprint image with the help of fingerprint collector (sensor). Then we need to do normalization [14] on the image to remove the effects of sensor noise and the finger pressure difference. After that, the client extracts the features from the normalized image and sends the query request in the form of $(type, avIrd)$ to Configure Server.
2. The Configure Server receives the query requests from the client, directs it to the proper Data Server according to $(type, avIrd)$, and receives the matching results from the Data Server.

3. The Data Server receives the queries from the Configure Server, and then sends the query request to the native database. The database returns the values (the minutiae and the fingerprint's name) to the Data Server, and then the Data Server carries out the matching process with given matching algorithm. Finally, the Data Server sends the optimal fingerprints' information back to the Configure Server.
4. The Configure Server gets all results (more than one for each request) from the Data Server, selects the optimal among them, and sends it to the client.

3.2   Encoding and Decoding

The fingerprint information extracted from the fingerprint image can be not used directly for KV store in Redis due to their complex data formats. We therefore need to encode the fingerprint features into key-value pairs in some rules. In our design, we have taken four fingerprint features into consideration: fingerprint type *type*, ridge distance *avIrd*, minutiae group and fingerprint's name, which make up a tetrad. We take the first two features as one group and the latter two as the other one. The two groups are hashed to a new 2-tuple $[H(type, avIrd), H(minutiae, name)]$, which is stored in Redis as KV pair. We define encoding as the conversion from the tetrad to the two-tuple, inversely decoding as the conversion from the 2-tuple to the tetrad.

Compressive Encoding. We first consider the encoding of the type and the ridge distance. From Sect. 2.3, we know that the fingerprint has four types and its ridge distance ranges from 0 to 255. We construct the following hash code function on the basis of the digit of their binary number:

$$H(type, avIrd) = (type \cdot 2^8) + avIrd \tag{1}$$

where *type* is the fingerprint's type and *avIrd* is the ridge distance.

The minutiae and the fingerprint's name are encoded and then concatenated. The name is directly encoded as its original string. Nevertheless, the minutia includes the horizontal coordinate X, the vertical coordinates Y and direction field *Orin* (here we ignore the minutia's type as what we have explained in Sect. 2.3). The structure of a single minutia is shown in Table 1.

Table 1. The structure of a minutiae

Name	X	Y	$Orin$
Value	$-1024 \ldots +1024$	$-1024 \ldots +1024$	$0 \ldots 3599$

To store the chaotic minutiae information into the database, Table 2 shows how to encode a single minutia to a binary hash number with a *Tag* for X and Y's signs.

Table 2. Encoding of a minutia

Name	$|X|$	$|Y|$	$Orin$	Tag
Value	$0 \dots 1024$	$0 \dots 1024$	$0 \dots 3599$	$0 \dots 3$
Binary Digit	10	10	12	2

The value of Tag is shown in the Eq. 2:

$$Tag = \begin{cases} 0, & x \geq 0, y \geq 0; \\ 1, & x \leq 0, y \geq 0; \\ 2, & x \geq 0, y \leq 0; \\ 3, & x \leq 0, y \leq 0. \end{cases} \tag{2}$$

We then construct a hash function as shown in Eq. 3 by the rules given Table 2:

$$H(minutia) = x \cdot 2^{24} + y \cdot 2^{14} + orin \cdot 2^2 + tag \tag{3}$$

According to Eq. 3, a single minutia is encoded to a 11-digits string, all of which are linked to a minutiae group, as shown in Eq. 4:

$$H(minutiae, name) = S(H(minutia_0)) + \cdots + S(H(minutia_K)) + S(name) \tag{4}$$

Note that encoded number for the minutiae is too long. Considering the execution time and the algorithm efficiency, we use Base64 for compression, in which $0 \dots 9, a \dots z, A \dots Z, +, /$ correspond to $0 \dots 63$, so as to reduce the space expense.

Parallel Decoding. After obtaining the optimal results (KV pairs) from the database, we need to revert them to the tetrad – we call this process decoding.

Based on Eq. 1, the way to decode the fingerprint's type and the ridge distance is given as:

$$\begin{cases} type & = \lfloor \frac{H_0(type, avIrd)}{2^8} \rfloor; \\ avIrd & = H_0(type, avIrd) \mod 2^8. \end{cases} \tag{5}$$

For the encoding of the minutia and the name (*i.e.* $H_0(M, N)$) given in Eq. 4, we decode it in turn from the end of the string, as shown in Eq. 6

$$\begin{cases} x & = \lfloor \frac{H_0(minutia_i)}{2^{24}} \rfloor; \\ y & = \lfloor \frac{H_0(minutia_i)}{2^{14}} \rfloor \mod 2^{10}; \\ orin & = \lfloor \frac{H_0(minutia_i)}{2^2} \rfloor \mod 2^{12}; \\ tag & = H_0(minutia_i) \mod 4. \end{cases} \tag{6}$$

Moreover, we restore the sign of X and Y from Tag according to the rules defined in Table 3.

Table 3. The rule of sign of X and Y

Tag	0	1	2	3
sign(X,Y)	$(+,+)$	$(-,+)$	$(+,-)$	$(-,-)$

When decoding the similar fingerprint set from RM, it spends 3 s on 250 thousand fingerprints – which is time-consuming. To solve this problem, we apply OpenMP in the RM process to enhance the decoding efficiency. With OpenMP, we create N threads since the processor has N cores, and conduct the decoding operation in N threads in parallel. And finally we construct the decoding results from N threads in order.

3.3 Process of Fingerprint Recognition

Usually the fingerprint recognition system includes two stages: register stage and recognition stage. This paper is focused on the latter one.

Most past stand-alone fingerprint recognition systems adopt the way to match one by one, which is low-efficient when facing with big data. To fight for the challenge, we propose a stage treatment mechanism including RM and AM, as shown in Fig. 2.

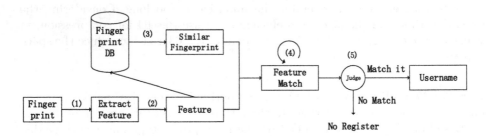

Fig. 2. Process of fingerprint recognition

RM is shown from Step (1) to Step (3) in Fig. 2. We enter the unknown fingerprint through the client and extract its feature information to get its features. By comparing the extracted features with the fingerprints' feature in the database, we get all similar fingerprints with the same type and the approximate ridge distance with the unknown one. The number of the similar fingerprints is about 3 %–5 % of the whole database. The percentage is controlled by a constant parameter *scale* in the query sentence. The constant *scale* means that in RM we will find all similar fingerprints with $avIrd$ ranging from $avIrd - scale$ to $avIrd + scale$.

AM, shown from Step (4) to Step (5) in Fig. 2, mainly gets one with the highest similarity by comparing the minutiae group between the unknown fingerprint

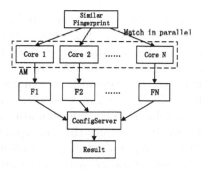

Fig. 3. Accurate match in parallel

and the similar fingerprint set. To get the result from the massive database more quickly, parallel coding can also be used to improve the system performance effectively, whose mechanism is shown in Fig. 3. We can create the equivalent threads with the number of processor's cores. Every thread handles $\lceil \frac{1}{N} \rceil$ of all similar fingerprints. When getting the results from threads, we can select to get all N results from N threads by OpenMP, and also can select to get one result from the thread which firstly gets a optimal fingerprint (refer to Sect. 3.4). Finally we will send the results to Configure Server. Configure Server selects the fingerprint with the highest similarity value as the matching result.

3.4 Match Strategy Based on *Quick-Return*

Based on the system architecture presented in Sect. 3.1, we have additionally brought a strategy called *Quick-Return* in the stage of AM, with the purpose of further improving the whole matching time.

Assuming that there is a cluster consisting of a Configure Server and N Data Servers, and we use this cluster to build Redis database. It is easy to know that the total time T needed for fingerprint recognition comprises three parts: the transport time T_{tran}, the RM time spent in the jth Data Server $T_{rou,j}$ and the AM time in the jth node $T_{accu,j}$. Every Data Server is supposed to have C cores and an AM task is run in parallel in C branches. Moreover, in the jth Data Server, every branch has $f_{i,j}$ fingerprints and the comparison time between every two entries (i.e. the unknown fingerprint with each in the subset obtained by RM) is ϵ_k. Then the AM time spent in the i^{st} branch is $t_{i,j} = \sum_{k=1}^{f_{i,j}} \epsilon_k$, and therefore the AM time in this Data Server (i.e. the j_{th} Data Server) is $T_{accu,j} = \max_{1 \le i \le c} t_{i,j}$. Since the deviation of the transmission time T_{tran} between two nodes is fairly small, for the sack of simplicity, we assume that they are equal to each other. Finally, we can obtain the complete format of the matching time as Eq. (7).

$$T = T_{tran} + \max_{1 \le j \le N}(T_{rou,j} + \max_{1 \le i \le C} t_{i,j}) \qquad (7)$$

In general, the amount of data a single fingerprint carries is quite small, resulting that the time needed for transmission T_{tran} is small. Besides, the value of $T_{rou,j}$ is small as well since it is closely related to hash operation in Redis which usually runs very fast. Therefore, from Eq. 7, the matching time in a branch gets longer when the number of similar fingerprints is large. From all the above analysis, we know that improving the AM process is necessary.

When two fingerprints match successfully, the similarity ranges in a very small interval. If not, the similarity lies far outside the interval. To accelerate the AM process, we define a value p. When the similarity in a matching process reaches p, we immediately stop it and return this fingerprint – which is the target we are looking for in the Data Server. This strategy, which is called *Quick-Return*, successfully shorten the residence time in each branch and further reduces the matching time.

Note that, in AM stage, the procedure with OpenMP can not be interrupted by the commands commonly used, such as *break*, *continue*, *goto* and *return*. For this reason, we adopt a parallel tool called *pthread*. With this tool, if one of multi-threads working in parallel finds a fitting fingerprint, the thread returns the results and all the others are killed at the same time. Thus, the matching time in a Data Server lies on the fastest thread, and the AM time turns to $T_{accu,j} = \min_{1 \leq i \leq c} t_{i,j}$.

4   Evaluation

4.1   Experiment Environment

The experimental cluster includes multi-nodes, each of which consists of 32cores and runs on the OS of Ubuntu 12.04. All the fingerprint images are obtained from FVC database [15], the features of which are extracted from images and stored into Redis cluster. The version of Redis we use is redis-3.0.0. The details of the experimental environment are given as follows:

1. One Configure Server and multiple Data Servers (4 to 8); 2. Set the number of fingerprints inserted into the database to different values; 3. Transact concurrent-queries (10, 20, 50, 100 and 200) from client.

We measure the following parameters to evaluate the performance of mechanisms of DFRS:

1. The space consumption S_e when encoding; 2. The time consumption T_d when decoding; 3. The matching time T_m^s when adopting different matching strategies.

4.2   Performance of Encoding and Decoding

We have measured the encoding space consumption and the decoding consumption when inserting different number of data into the database, and the result is shown in Figs. 4 and 5. In Fig. 4, the white columns represent space consumption with compressive encoding. In Fig. 5, the white columns represent time consumption with parallel decoding.

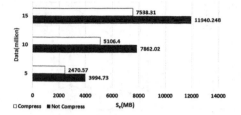

Fig. 4. Encoding space consumption

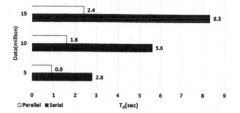

Fig. 5. Decoding time consumption

For space optimization, compressive encoding with Base64 is superior to that with decimal system. For a fingerprint with 80 minutiae, if using the decimal system, the length of the encoded minutiae group and fingerprint name is $11 * 80 + length(name)$ which consumes about 0.9 KB, while if using Base64, it's $6 * 80 + length(name)$ which consumes about 0.5 KB. We can find from Fig. 4 that the compression ratio is about 61 %.

For time optimization, we can find from Fig. 5 that parallel decoding precedes serial decoding. After applying OpenMP on data decoding, the RM efficiency is faster about 3.5 times than before. The results in Fig. 5 clearly expose that, when applying parallel decoding, it takes only 0.9 s (less than 1 s) by returning 5 % of data compressively encoded from the fingerprint database which has a scale of 5 million records in total.

4.3   Analytical Evaluation for Matching Method

The initial matching strategy is simply to compare the input fingerprint one by one with all in the database. It's pretty intractable when the number of fingerprints in the database is up to 10^8.

Our new matching strategy only need get 3 %–5 % of all fingerprints in Redis. We know from Sect. 2.2 that fingerprint has four types and ridge distance ranges from 0 to 255, so assuming that the probability of fingerprint with fingerprint's type $type$ is $P(type)$ and that of fingerprint with ridge distance $avlrd$ is $P(avlrd)$. The probability of a fingerprint with $type$ and $avlrd$ is about $P(type) \cdot P(avlrd)$. According to the constant $scale$, the probability of getting all similar fingerprints is $P(type) \cdot \sum_{i=avlrd-scale}^{avlrd+scale} P(i)$, the value of which can be controlled from 3 % to 5 % by changing the constant $scale$. After doing like this, we greatly reduce the number of the matching fingerprints. For the new matching model, further data analysis isn't necessary. Just the seeming data volume indicates that our new matching model enhances the system's performance by using less data.

4.4   Evaluation for *Quick-Return*

In order to evaluate the performance by adopting the strategy of *Quick-Return*, we have run the experiments by varying the number of concurrent queries and

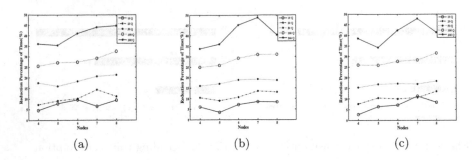

(a)   (b)   (c)

Fig. 6. The result of experiment. The number of data in the database is respectively 1 million, 2 million and 5 million in (a), (b) and (c), which shows the reduction percentage of time with different nodes and queries(Q).

the number of nodes in the cluster. The results under different scale of database are given in Fig. 6(a), (b) and (c), representing 1 million, 2 million and 5 million fingerprints in the database respectively.

In each sub-figure of Fig. 6, to highlight the improvement *Quick-Return* brings, we have plot the reduction of the time in percentage by changing two factors: the number of concurrent queries and the number of nodes in the cluster. Note that, on one hand, five values (10, 20, 50, 100, 200) are chosen for the number of concurrent queries in our experiments, denoted respectively from the bottom up in the figure by distinct combination of dot-lines; on the other hand, we range the number of nodes from 4 to 8.

In general, the results shown in Fig. 6(a), (b) and (c) are essentially similar. Furthermore, it is clear to observe that: (1) when fixing the number of nodes, the reduction of the time in percentage increases as the number of concurrent queries expands – this reduction can go up to nearly 50 % for the best case; (2) if the number of concurrent queries is the same, the more nodes the cluster consists of, the faster the matching process goes.

To sum up, *Quick-Return* strategy shows a high performance in DFRS. On one hand, the matching strategy can be good to handle larger scale of data; on the other hand, with the number of nodes changing, the system's performance embodies better with *Quick-Return*, which shows good scalability.

5   Conclusion and Future Work

In this paper, we have designed a large-scale distributed fingerprint recognition system based on Redis, with the purpose of improving the matching efficiency for "big data". In this system, we have proposed an innovative framework and incorporated several key technologies: (i) Compressive encoding and parallel decoding save the storage space and accumulate the computing; (ii) The division of RM+AM architecture significantly improves the efficiency; (iii) *Quick-Return* strategy reduces the matching time while achieving a high precision.

As the future work, we will: (1) compare the performance by adopting Redis and Mongo respectively as the database for our system; (2) test its applicability to real system and further evaluate its global performance.

References

1. Maltoni, D., Maio, D., Jain, A.K., et al.: Handbook of Fingerprint Recognition. Springer, London (2009)
2. http://db-engines.com/en/ranking
3. Bartholomew, D.: SQL vs. NoSQL. Linux J. 2010(195) (2010)
4. Carlson, J.L.: Redis in Action. Manning Publications Co., Greenwich (2013)
5. Ross, A.A., Shah, J., Jain, A.K.: Toward reconstructing fingerprints from minutiae points. In: Defense and Security. International Society for Optics and Photonics, pp. 68–80 (2005)
6. Kary, K., Jain, A.K.: Fingerprint classification. Pattern Recogn. **29**(3), 389–404 (1996)
7. Zhu, E., Yin, J., Hu, C., et al.: A systematic method for fingerprint ridge orientation estimation and image segmentation. Pattern Recogn. **39**(8), 1452–1472 (2006)
8. Prabhakar, S., Jain, A.K., Wang, J., et al.: Minutia verification and classification for fingerprint matching. In: Proceedings of the 15th International Conference on Pattern Recognition, vol. 1, pp. 25–29. IEEE (2000)
9. Jea, T.Y., Govindaraju, V.: A minutia-based partial fingerprint recognition system. Pattern Recogn. **38**(10), 1672–1684 (2005)
10. Labati, R.D., Genovese, A., Piuri, V., et al.: Contactless fingerprint recognition: a neural approach for perspective and rotation effects reduction. In: Proceedings of the IEEE Symposium on Computational Intelligence in Biometrics & Identity Management, pp. 22–30 (2013)
11. Kaur, M., Singh, M., Girdhar, A., et al.: Fingerprint verification system using minutiae extraction technique. Proc. World Acad. Sci. Eng. Technol. **46**, 497–502 (2008)
12. Bhargava, D.N., Bhargava, R., Narooka, P., et al.: Fingerprint recognition using minutia matching. Int. J. Comput. Trends Technol. **3**(4), 641–643 (2012)
13. Zhu, E., Yin, J., Zhang, G.: Fingerprint matching based on global alignment of multiple reference minutiae. Pattern Recogn. **38**(10), 1685–1694 (2005)
14. Wang, S., Zhang, W.W., Wang, Y.S.: Fingerprint classification by directional fields. In: Proceedings of the Fourth IEEE International Conference on Multimodal Interfaces, pp. 395–399. IEEE (2002)
15. http://bias.csr.unibo.it/fvc2004/

Logo Recognition via Improved Topological Constraint

Panpan Tang and Yuxin Peng[✉]

Institute of Computer Science and Technology, Peking University,
Beijing 100871, China
panpantangbjtu@gmail.com, pengyuxin@pku.edu.cn

Abstract. Real-world logo recognition is challenging mainly due to various viewpoints and different lighting conditions. Currently, the most popular approaches are usually based on bag-of-words model due to their good performance. However, their shortcomings lie in two main aspects: (1) wrong recognition results caused by mismatching of keypoints. (2) high computational complexity and extra noise caused by a large number of keypoints which are irrelevant to the target logo. To address these two problems, we propose a new approach which combines feature selection and topological constraint for logo recognition. Firstly, feature selection is applied to filter out most of the irrelevant keypoints. Secondly, an improved topological constraint, which considers the relative position between a keypoint and its neighboring points, is proposed to reduce the number of mismatched keypoints. It is proven in this paper that the proposed constraint can remove the keypoints which are not on the same planar surface with the others from the k nearest neighbors of a keypoint. This property is very important to logo recognition because logos are planar objects in real world. The proposed approach is evaluated on two challenging logo recognition benchmarks, FlickrLogos-32 and FlickrLogos-27, and the experimental results show its effectiveness compared to other popular methods.

Keywords: Logo recognition · Improved topological constraint · Feature selection

1 Introduction

Logo recognition is a sub-problem of object recognition, and has attracted increasing interests in recent years because of its commercial benefits such as measurement of brands' exposure. Given an image, the goal of logo recognition is to check whether it contains any logo and to determine where the logos are located. In real world, affine transformations such as rotation, shearing and scaling caused by variety of viewpoints make logo recognition a challenging task. Besides, occlusion, different lighting conditions as well as diverse appearance of logo itself increase the difficulty of logo recognition.

© Springer International Publishing Switzerland 2016
Q. Tian et al. (Eds.): MMM 2016, Part I, LNCS 9516, pp. 150–161, 2016.
DOI: 10.1007/978-3-319-27671-7_13

Recently, a number of works have tackled logo recognition using bag-of-words (BoW) model [1 7]. These methods firstly extract local features such as SIFT or SURF from images. Then these features of keypoints are clustered and quantized into individual integer numbers which are called visual words, and an image is represented as a collection of visual words, known as bag-of-words (BoW). Finally, the similarity between a logo image and a test image is measured based on these visual words. Compared with the original features, the quantized features are more suitable for large scale image retrieval/recognition systems.

However, approaches based on BoW model have two common shortcomings. The first one is that the discriminative ability of visual words decreases in some degree due to quantization, causing that two keypoints belonging to the same visual word may locate at different objects or different regions of the same object, which is called mismatching. To enhance the discriminative ability of visual words and reduce mismatching, Zhou et al. [8] proposed a novel scheme, named spatial coding, to encode the spatial relationships among local features in an image. Since it is specifically for partial-duplicate image search and is based on the assumption that the query image and matched images share the same or very similar spatial layout, it is not inherently suitable for logo recognition in real-world images. Romberg et al. [4] proposed a Bundle min-Hashing (BmH) approach, which aggregates individual local features with the features from their spatial neighborhood into bundles. However, it ignores the relative position between features in a bundle, which limits the discriminative ability in some degree. Kalantidis et al. [1,3] proposed to locally group features in triples and represent the triangles by the signatures capturing both visual appearance and local geometry. However, the geometric constraint in [1,3] is not strictly affine invariant and occasionally fails to detect logos especially with rotation changes which are typically the case for logos in real-world images. Wan et al. [5] proposed a Tree-based Shape Descriptor (TSD) to encode the shape of logos by depicting both the appearance and spatial information of four local key-points. Although the Tree-based Shape Descriptor is strictly invariant to affine transformation in real-world images, it is easily affected by the loss of keypoints which commonly happens in the case of occlusion. In this paper, we propose an Improved Topological Constraint (ITC) which considers the relative position between a keypoint and its neighboring points. Moreover, we apply the cyclic Longest Common Subsequence (LCS) in ITC for similarity measurement in order to be robust to the loss of keypoints.

The second problem is that a large number of keypoints in an image are irrelevant to the target logo, which increase computational complexity as well as bring some noise. An optional solution is by only considering keypoints in the mask region. However, not all keypoints in the mask region is relevant to the target logo. Additionally, the number of keypoints in different mask regions varies a lot, which significantly affects the result of recognition. In our approach, we apply feature selection based on Mutual Information (MI) to filter out most irrelevant keypoints.

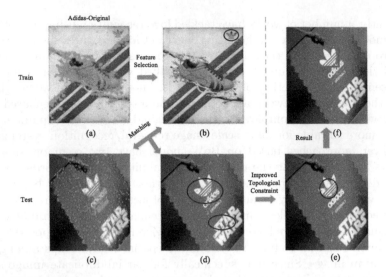

Fig. 1. Logo recognition process of our approach (taking Adidas-Original as an example). (a)→(b), most irrelevant keypoints are filtered out by feature selection; (b), (c)→(d), keypoints matching process, note that some mismatched keypoints appears in (d) (keypoints in the smaller ellipse); (d)→(e), mismatching is removed by the proposed improved topological constraint; (e)→(f), Adidas-Original logo is recognized correctly)

The contributions of this paper are twofold: (1) we propose an improved topological constraint to reduce the number of mismatched keypoints. (2) we propose a new approach for logo recognition by combining feature selection and the proposed topological constraint. Figure 1 shows the process of the proposed approach.

2   Improved Topological Constraint

Tell et al. [12] proposed a topological constraint which performs string matching to ensure one-to-one matching and to preserve cyclic order. Firstly, they represent the k nearest neighbors (kNN) of a point as a string according to their cyclic order, as shown in Fig. 2. Then for a pair of matched points, they get the cyclic Longest Common Subsequence (LCS) of their kNN using the algorithm proposed by Gregor et al. [10]. Finally, two points are truly matched if and only if they have longer cyclic LCS than any other pair of points that contain at least one of them. This constraint is invariant to affine transformation if the profile is on a planar surface [12].

The constraint proposed by Tell et al. [12] which combines local and semi-local feature comparison has good discriminative power. But it only considers the cyclic order of kNN and ignores the angular relationship between them. The following is a lemma proven by Wan et al. [5].

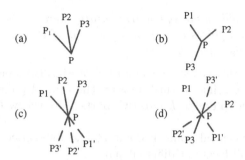

Fig. 2. The cyclic order of the points in (a) and (b) are both $P_1P_2P_3$. The difference between them can not be reflected in the cyclic order. After adding the symmetric points such as in (c) and (d), P_1, P_2, P_3 will not appear in the cyclic LCS together.

Lemma 1. *As shown in Fig. 2(a), if PP_2 lies in $\angle P_1PP_3$ and $\angle P_1PP_3$ is less than π, PP_2 is still in $\angle P_1PP_3$ after affine transformation.*

As shown in Fig. 2, the cyclic order of the points in Fig. 2(a) and (b) are both $P_1P_2P_3$, so their cyclic LCS is $P_1P_2P_3$ whose length is 3. However, in Fig. 2(b), PP_2 is not in the $\angle P_1PP_3$ that is less than π, which means that the obtained cyclic LCS does not satisfy Lemma 1, and P_1, P_2, P_3 can not be on the same planar surface.

To address this problem, we propose an Improved Topological Constraint (ITC). As shown in Fig. 2(c) and (d), for each neighboring point of the point P, we add an extra symmetric point. P_1', P_2', P_3' are the symmetric points of P_1, P_2, P_3 respectively. Then the cycle orders of the points are changed into $P_1P_2P_3P_1'P_2'P_3'$ and $P_1P_3'P_2P_1'P_3P_2'$ respectively. Then their cyclic LCS becomes $P_1P_2P_1'P_2'$. Finally, we remove the added points P_1', P_2' from the cyclic LCS and it becomes P_1P_2, whose length is 2. Compared with the original topological constraint, the proposed improved topological constraint removes point P_3 which is not on the same planar surface with the others. Furthermore, it can be proven that there exists at least one cyclic LCS obtained by ITC can satisfy that any three points of it meet Lemma 1.

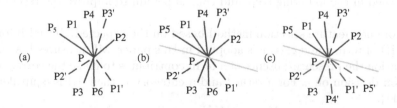

Fig. 3. (a) is the original original cyclic LCS; In (b), the yellow line which crosses P divide all points into two parts; In (c), points in the under part are replaced by the symmetry points of the other part (Color figure online).

Theorem 1. *After adding the symmetric points, there exists at least one cyclic LCS satisfying that all points appear in pairs, e.g., if P_i appears, its symmetric point P_i' appears in the cyclic LCS too.*

Proof. Suppose that there exists a cyclic LCS of Fig. 3(a), denoted as L.

1. We can always use a line s which crosses the central point P but does not cross any other point to divide L into two parts. The yellow line in Fig. 3(b) is one of these lines.

2. As point P is the central point, if point P_i and its symmetric point P_i' both appear in L, they will locate at different part.

3. If we replace all the points in the under part with the symmetric points of the other part, we can get another subsequence L', as shown in Fig. 3(c).

4. As $Length(L') \geq Length(L)$, L' must be a cyclic LCS too. And all points of L' appear in pairs.

5. As described above, we can always find a cyclic LCS which satisfies that all the points appear in pairs after adding the symmetric points.   □

Theorem 2. *If a cyclic LCS satisfies that all the points appear in pairs, any three points of it will satisfy Lemma 1.*

Proof. Suppose there are three points of a cyclic LCS which satisfies that all the points appear in pairs, cyclic LCS of these three points and their symmetric points will not exceed four points, as shown in Fig. 2(c) and (d). It signifies that these three points and their symmetric points can not appear in a cyclic LCS together. So the hypothesis is invalid and any three points will satisfy Lemma 1.   □

Through Theorems 1 and 2 we can prove that there exists at least one cyclic LCS obtained by ITC can satisfy that any three points of it meet Lemma 1.

3   Logo Recognition

3.1   Feature Selection

The first step of our approach for logo recognition is feature selection. It has two advantages: Firstly, it can filter out a lot of keypoints in the test images which are irrelevant to the target logo, thus helps to improve the recognition accuracy. Secondly, it can greatly reduce the number of keypoints which need to be matched in the following step, and thus is helpful to improve the recognition speed.

A common feature selection method based on the expected Mutual Information (MI) of term t and class c is adopted in this paper. MI measures how much information the presence/absence of a term contributes to make the correct classification decision on c. For a term t and a category c, their MI is equivalent to Eq. 1 [11].

$$
\begin{aligned}
I(t,c) = &\frac{N_{11}}{N} \log_2 \frac{N N_{11}}{N_{1.} N_{.1}} + \frac{N_{01}}{N} \log_2 \frac{N N_{01}}{N_{0.} N_{.1}} \\
&+ \frac{N_{10}}{N} \log_2 \frac{N N_{10}}{N_{1.} N_{.0}} + \frac{N_{00}}{N} \log_2 \frac{N N_{00}}{N_{0.} N_{.0}}
\end{aligned}
\tag{1}
$$

Algorithm 1. Feature selection for a positive sample containing logo c

---

Input:

   $I = \{P_i\} :=$ All BoWs in the positive sample

   $S = \{I_j\} :=$ All images in the training set

   $R = \{I_k\} :=$ All images in the training that contain logo c

   $k :=$ Number of BoWs to be selected

Output:

   $Q :=$ The top k features of I

1: $V \leftarrow []$

2: **for each** $P_i \in I$

3: **do** $N_{11} \leftarrow 0$, $N_{10} \leftarrow 0$, $N_{01} \leftarrow 0$, $N_{00} \leftarrow 0$

4:      **for each** $I_j \in S$

5:          **if** $P_i \in I_j$

6:             **then if** $I_j \in R$ **then** $N_{11} \leftarrow N_{11} + 1$

7:                 **else** $N_{10} \leftarrow N_{10} + 1$

8:          **else**

9:                 **if** $I_j \in R$ **then** $N_{01} \leftarrow N_{01} + 1$

10:                **else** $N_{00} \leftarrow N_{00} + 1$

11:      $N \leftarrow \#(S)$

12:      $MI \leftarrow$ compute $I(t, c)$ according to Eq. 1

13:      Append(V, $< P_i, MI >$)

14: Descending order V by MI

15: $Q \leftarrow$ first k elements of V

16: **return** Q

---

where the N_s are counts of images that have the values of e_t and e_c that are indicated by the two subscripts. For example, N_{10} is the number of images that contain t ($e_t = 1$) and are not in c ($e_c = 0$). $N_{1.} = N_{10} + N_{11}$ is the number of images that contain t ($e_t = 1$). $N = N_{00} + N_{01} + N_{10} + N_{11}$ is the total number of images.

In this paper, category c is logo type and term t is visual word. For each logo in the training set, the images containing the logo are positive samples and the others are negative samples. For each positive sample, we compute the MI of each visual word according to Eq. 1. The top k visual words whose MI are greater than the others are retained finally, since visual words with greater MI have more relevance to the target logo. In practice, the top k visual words are mainly within the target logo region, as shown in Fig. 1(b), thus most irrelevant keypoints are filtered out. The detailed algorithm for feature selection is shown in Algorithm 1.

3.2   Recognition

The second step of our approach is recognition. Denote an image I_q and a matched image I_t are found to share N pairs of matched keypoints[1]. Then the corresponding k-Nearest-Neighbors (kNN) of these matched keypoints for both

---

[1] two keypoints are matched if they are quantized into the same visual word.

I_q and I_t can be generated and denoted as kNN_q and kNN_t. As the Improved Topological Constraint (ITC) described above, we add an extra symmetric point for each point in kNN_q and kNN_t, as shown in Fig. 2, and the points of kNN_q and kNN_t are doubled, denoted as kNN'_q and kNN'_t. Then kNN'_q and kNN'_t are represented as strings according to cyclic order and their cyclic Longest Common Subsequence (LCS) is computed using the algorithm proposed by Gregor et al. [10], denote as $LCS(kNN'_q, kNN'_t)$. Ideally, if all N matched pairs are true, the length of $LCS(kNN'_q, kNN'_t)$ is equivalent to the size of kNN'_q or kNN'_t, but if some false matches exist, the former will be smaller than the latter. The similarity between two matched keypoints is defined as Eq. 2:

$$r = \frac{Length\ of\ LCS(kNN'_q, kNN'_t)}{min\{\#(kNN'_q),\ \#(kNN'_t)\}} \tag{2}$$

Two keypoints are truly matched if r is bigger than a predefined threshold α, which controls the strictness of topological constraint and impacts the verification performance.

We formulate the logo recognition as a voting problem. Each matched keypoint in the test image votes on its matched image. Intuitively, the MI weight of feature selection can be used to distinguish different matched keypoints. However, from our experiments, we find that simply counting the number of matched keypoints which are quantized to different visual words yields similar or better results. To recognize if the test image contains any logo and which logo does it contain, it need to be matched with every image in the training set, which are regarded as reference images. Since those reference images have been preprocessed by feature selection in the last step and only a few irrelevant keypoints are kept, the matching process is quite fast. Denote test image is I_q, reference images are $S = \{I_i\}$ and corresponding numbers of matched features are $M = \{m_i\}$. Then the recognition result is define as Eq. 3:

$$c_q = \begin{cases} \{c_t \mid m_t = max(M)\}, & \text{if } max(M) \geq \beta \\ \text{no-logo}, & \text{if } max(M) < \beta \end{cases} \tag{3}$$

where c_q is the recognition result of I_q, c_t is the logo class of reference image I_t, and β is a predefined threshold which determines whether an image contains any logo.

The detailed algorithm for recognition with improved topological constraint is shown in Algorithm 2.

4   Experiments

We evaluate the proposed approach on two challenging and commonly used logo recognition datasets, FlickrLogos-32 [3] and FlickrLogos-27 [1]. They are specially designed for real-world logo recognition. FlickrLogos-32 contains photos showing brand logos and is used for the evaluation of logo retrieval and multi-class logo detection/recognition systems on real-world images. It contains 8240

Algorithm 2. Recognition for test image I_q

Input:

$I_q = \{P_i\} :=$ BoW features of I_q

$S' = \{I_j\} :=$ Images pre-processed by feature selection in the training set

$C = \{c_j\} :=$ Corresponding logo class of images in S'

$\alpha :=$ Similarity threshold of matched keypoints

$\beta :=$ Similarity threshold of images

Output:

$c_q :=$ The recognition result of I_q

1: $M \leftarrow []$
2: **for each** $I_j \in S'$
3: **do** $V \leftarrow []$
4:     **for each** $P_i \in I_q$
5:     **do for each** $P_k \in I_j$
6:         **do if** $P_i == P_k$ **then** Append($V, < P_i, P_k >$)
7:     $T \leftarrow []$
8:     **for each** $< P_i, P_k > \in V$
9:     **do** $kNN_i \leftarrow$ k-Nearest-Neighbors of P_i
10:         $kNN_k \leftarrow$ k-Nearest-Neighbors of P_k
11:         $kNN_i' \leftarrow$ adding extra symmetry points to kNN_i as Fig. 2
12:         $kNN_k' \leftarrow$ adding extra symmetry points to kNN_k as Fig. 2
13:         $s_i \leftarrow$ cyclic order of kNN_i'
14:         $s_k \leftarrow$ cyclic order of kNN_k'
15:         $l \leftarrow$ cyclic Longest Common Subsequence of s_i and s_k
16:         $r \leftarrow Length(lcs) \ / \ min\{\#(kNN_i'), \ \#(kNN_k')\}$
17:         **if** r $\geq \alpha$ **then** Append(T, P_i)
18:     $m_j \leftarrow$ number of distinct visual words in T
19:     Append(M, m_j)
20: $c_q \leftarrow$ *no-logo*
21: **if** $max(M) \geq \beta$
22: **then** $c_q \leftarrow \{c_t | m_t = max(M)\}$
23: **return** c_q

---

images and is split into three disjoint subsets, each containing images of 32 logo classes. FlickrLogos-27 dataset is an annotated logo dataset downloaded from Flickr. Different from FlickrLogos-32, it contains more than four thousand classes in total.

4.1 Impact of Parameters

The performance of the proposed approach is affected by three main parameters: the similarity threshold of matched keypoints α, the similarity threshold of images β and the number of selected features. We evaluate the impact of these parameters on FlickrLogos-32 dataset, since the large number of images and distractors makes this dataset more analogous to natural scenario.

The similarity threshold of matched keypoints α controls the strictness of the topological constraint. The performance of precision, recall and F1 score for

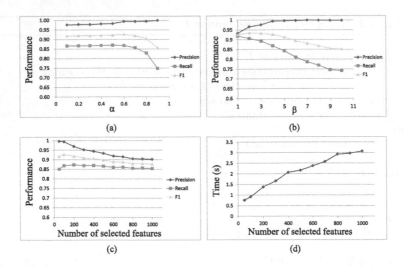

Fig. 4. (a) Performance with different similarity threshold of matched keypoints α; (b) Performance with different similarity threshold of images β; (c) Performance with different number of selected features; (d) Time cost for the reconition of an image with different number of selected features.

different values of α is shown in Fig. 4(a) (with $\beta = 4$ and the number of selected features equal to 100). Note that $\alpha = 0$ means that no topological constraint is performed. When α increases, the key performance indicator F1 score, which considers both precision and recall, first increases slowly and then decreases sharply. Since the F1 score reaches the maximum when $\alpha = 0.6$, we fix it as 0.6 in the following experiments.

The similarity threshold of images β determines whether a test image contains any logo or not. The experimental results for different values of β are shown in Fig. 4(b) with $\alpha = 0.6$ and the number of selected features equal to 100. As we can see, with the increasing value of β, the F1 score first increases and then decreases. When β reaches the value of 4, F1 score and the precision are both comparatively high, so we fix it as 4 in the following experiments.

The third important parameter is the number of selected features. Figure 4(c) and (d) show how it affects the performance of F1 score, as well as the speed of recognition, respectively. With the increasing number of selected features, the F1 score first increases and then decreases sharply, and the time cost for recognition keeps increasing. We found that the selected features with number 100 give the best tradeoff between F1 score and the time cost. So in the following experiments, we fix the number of selected features as 100.

4.2   FlickrLogos-32 Dataset

Romberg et al. [4] proposed a Bundle min-Hashing (BmH) approach for logo recognition which achieves the best result of FlickrLogos-32 so far. They used

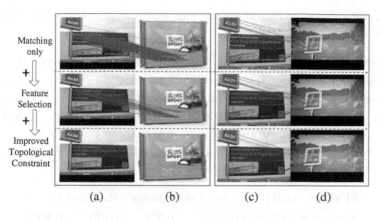

Fig. 5. An illustration of the process how the proposed approach filters out the mismatched keypoints between (a) and (b) and keeps the truly matched keypoints between (c) and (d), resulting in getting more accurate recognition results.

the DoG detector and the SIFT descriptor, and employed Approximate K-Means (AKM) to quantize the descriptor vectors to visual words. For fair comparisons, we also use the 1 M dimension BoW features they provided in our experiment.

Besides Bundle min-Hashing (BmH), we report the comparison results of the proposed approach with several common methods including Scalable Logo Recognition (SLR) [3], Tree-based Shape Descriptor (TSD) [1] and Correlation-Based Burstiness (CBB) [2] in terms of F1 score. In order to verify the necessity and effectiveness of each step in our approach, we design three baselines: Baseline1 does not use feature selection or any constraint, Baseline2 only uses feature selection and Baseline3 uses feature selection and the constraint proposed by Tell et al. [12]. The difference between ITC and Baseline3 is that every three keypoints of the cyclic LCS obtained by ITC satisfy Lemma 1. As shown in Table 1, the result of ITC outperforms the state-of-the-art result of BmH. Moreover, each step of our approach does make certain contribution to the final result.

In Fig. 5, we show how the proposed approach filters out the mismatched keypoints between Fig. 5(a) and (b), and keeps the truly matched keypoints between Fig. 5(c) and (d). Since there are many similar keypoints around the letters in Fig. 5(a) and (b), the matched keypoints between them are more than the matched keypoints between Fig. 5(c) and (d). After the process of feature selection, about half of the matched keypoints between Fig. 5(a) and (b) are filtered out since they are irrelevant to the target logo "Ritter SPORT" of Fig. 5(b). Moreover, by further applying the topological verification with the improved topological constraint, the matched keypoints between Fig. 5(a) and (b) are all removed because they have different spatial distribution. Finally, the number of matched keypoints between Fig. 5(c) and (d) becomes larger than the number of matched keypoints between Fig. 5(a) and (b), thus the images in Fig. 5(a) and (c) which contain the logo "ALDI" will be recognized as "ALDI" rather than "Ritter SPORT".

Table 1. Performance of ITC (the proposed approach) against SLR, TSD, CBB, BmH and three baselines on FlickrLogos-32.

Method	SLR	TSD	CBB	BmH	Baseline1	Baseline2	Baseline3	ITC
Precision	0.980	0.980	0.980	0.999	0.957	0.992	0.994	0.994
Recall	0.610	0.680	0.730	0.832	0.023	0.782	0.821	0.868
F1	0.752	0.803	0.837	0.908	0.045	0.875	0.899	**0.927**

Table 2. Performance of ITC (the proposed approach) against msDT, TSD, Baseline1, Baseline2 and Baseline3 on FlickrLogos-27.

Method	msDT	TSD	Baseline1	Baseline2	Baseline3	ITC
Accuracy	0.520	0.580	0.500	0.542	0.619	**0.637**

4.3   FlickrLogos-27 Dataset

Following the experimental settings in Kalantidis et al. [1], we apply the SURF descriptors as local features. Then a vocabulary of 5 K visual words is built to quantize all these descriptors. We compare the proposed approach with msDT [1], TSD [5] and the three baselines. TSD achieves the state-of-the-art result on FlickrLogos-27 so far when the distractor set is used. The comparison results are shown in Table 2, in terms of accuracy as in Kalantidis et al. [1].

We can see that, our best result is about 6 % higher than the state-of-the-art result of TSD. As same as the result on FlickrLogos-32, each step of our approach does make certain contribution to the final result, suggesting the proposed approach is steady when logo classes increase to several thousands.

5   Conclusion

In this paper, we have proposed an Improved Topological Constraint (ITC) and a new logo recognition approach which combines both feature selection and ITC. Firstly, feature selection is used to filter out most keypoints which are irrelevant to the target logo. Then ITC is used to reduce the number of mismatching by considering relative position between a keypoint and its neighboring points. The experimental results on challenging logo datasets have shown the effectiveness of our approach.

Acknowledgements. This work was supported by National Natural Science Foundation of China under Grants 61371128 and 61532005, and National Hi-Tech Research and Development Program of China (863 Program) under Grants 2014AA015102 and 2012AA012503.

References

1. Kalantidis, Y., Pueyo, L. G., Trevisiol, M., Van Zwol, R., Avrithis, Y.: Scalable triangulation-based logo recognition. In: ACM International Conference on Multimedia Retrieval (ICMR), p. 20. ACM (2011)
2. Revaud, J., Douze, M., Schmid, C.: Correlation-based burstiness for logo retrieval. In: ACM International Conference on Multimedia (ACM-MM), pp. 965–968. ACM (2012)
3. Romberg, S., Pueyo, L.G., Lienhart, R., Van Zwol, R.: Scalable logo recognition in real-world images. In: ACM International Conference on Multimedia Retrieval (ICMR), p. 25. ACM (2011)
4. Romberg, S., Lienhart, R.: Bundle min-hashing for logo recognition. In: ACM International Conference on Multimedia Retrieval (ICMR), pp. 113–120. ACM (2013)
5. Wan, C., Zhao, Z., Guo, X., Cai, A.: Tree-based shape descriptor for scalable logo detection. In: Visual Communications and Image Processing (VCIP), pp. 1–6. IEEE (2013)
6. Romberg, S.: From local features to local regions. In: ACM International Conference on Multimedia (ACM-MM), pp. 841–844. ACM (2011)
7. Wu, X., Kashino, K.: Image retrieval based on spatial context with relaxed gabriel graph pyramid. In: IEEE International Conference on Acoustics, Speech and Signal Processing (ICASSP), pp. 6879–6883. IEEE (2014)
8. Zhou, W., Lu, Y., et al.: Spatial coding for large scale partial-duplicate web image search. In: ACM International Conference on Multimedia (ACM-MM), pp. 511–520. ACM (2010)
9. Lowe, D.G.: Distinctive image features from scale-invariant keypoints. Int. J. Comput. Vis. (IJCV) $60(2)$, 91–110 (2004)
10. Gregor, J., Thomason, M., et al.: Dynamic programming alignment of sequences representing cyclic patterns. IEEE Trans. Pattern Anal. Mach. Intell. (TPAMI) $15(2)$, 129–135 (1993)
11. Manning, C.D., Raghavan, P., Schütze, H.: Introduction to Information Retrieval, vol. 1. Cambridge University Press, Cambridge (2008)
12. Tell, D., Carlsson, S.: Combining appearance and topology for wide baseline matching. In: Heyden, A., Sparr, G., Nielsen, M., Johansen, P. (eds.) ECCV 2002, Part I. LNCS, vol. 2350, pp. 68–81. Springer, Heidelberg (2002)

Compound Figure Separation Combining Edge and Band Separator Detection

Mario Taschwer[1]([⊠]) and Oge Marques[2]

[1] ITEC, Klagenfurt University (AAU), Klagenfurt, Austria
`mario.taschwer@aau.at`
[2] Florida Atlantic University (FAU), Boca Raton, FL, USA
`omarques@fau.edu`

Abstract. We propose an image processing algorithm to automatically separate compound figures appearing in scientific articles. We classify compound images into two classes and apply different algorithms for detecting vertical and horizontal separators to each class: the edge-based algorithm aims at detecting visible edges between subfigures, whereas the band-based algorithm tries to detect whitespace separating subfigures (separator bands). The proposed algorithm has been evaluated on two datasets for compound figure separation (CFS) in the biomedical domain and compares well to semi-automatic or more comprehensive state-of-the-art approaches. Additional experiments investigate CFS effectiveness and classification accuracy of various classifier implementations.

1 Introduction

Due to a substantial amount of compound figures in the biomedical literature[1], the automatic separation of these figures into subfigures has been recently identified as a relevant research problem for content-based analysis and image-based information retrieval in collections of biomedical articles [1,4,9,11]. From the few approaches known from literature [1,3,6–8,11], all but one [7] focus on the detection of homogeneous image regions separating subfigures, which we call *separator bands*, as illustrated by Fig. 1(a). These approaches fail for compound images where subimages are stitched together without separator bands, as shown in Fig. 1(b).

We therefore propose a method that provides separate algorithms for detecting separator bands and separator edges and selects the appropriate algorithm for a given compound image using the prediction of an image classifier. The classifier is trained to distinguish between graphical illustrations and other images in biomedical articles, based on the observation that compound images containing graphical illustrations almost always contain separator bands between subfigures, whereas most subfigures in other compound images show rectangular border edges.

---

[1] In recently published datasets drawn from open access biomedical literature, between 40 % and 60 % of figures occurring in articles are compound figures [1,3,4].

© Springer International Publishing Switzerland 2016
Q. Tian et al. (Eds.): MMM 2016, Part I, LNCS 9516, pp. 162–173, 2016.
DOI: 10.1007/978-3-319-27671-7_14

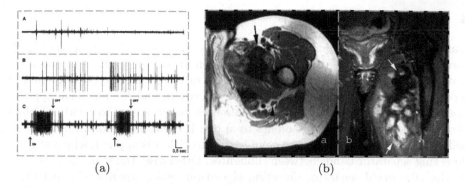

Fig. 1. Sample compound images (of the ImageCLEF 2015 CFS dataset [5]) suitable for two different separator detection algorithms. Subfigures are separated by (a) whitespace, (b) a vertical edge. Dashed lines represent the expected output of CFS.

The proposed approach builds upon previous work [10] and adds the following new research contributions: (1) the proposed algorithm outperforms state-of-the-art automatic and semi-automatic CFS approaches on two recently published biomedical datasets; and (2) several implementation options for the illustration classifier have been evaluated with respect to effectiveness for CFS and classification accuracy.

2   Related Work and Context

Until now there has been little research on the CFS problem in the literature, probably because its relevance had not been recognized by the research community until ImageCLEF 2013, where a CFS task was introduced as one of the challenges in the biomedical domain [4]. Task organizers provided training and test datasets, and evaluated CFS results submitted by participants for the test dataset. The few approaches resulting from participation [1,3,6] as well as other proposed approaches [8,11] focus on the detection of separator bands, and hence fail for compound images where subimages are stitched together without separator bands. Yuan and Ang [11] additionally used an edge-based approach involving Hough transform to separate overlayed zoom-in views from background figures, a case that is not considered in this work.

The next CFS task at ImageCLEF in 2015 [5] stimulated further work on the CFS problem. The approach of NLM (U.S. National Library of Medicine) [7] and our previous approach [10] independently proposed to address compound images without separator bands by processing edge detection results. Besides algorithmic differences in edge-based separator detection, our approach incorporates a classifier to automatically select edge- or band-based separator detection, whereas NLM's approach uses manual image classification for evaluation[2].

---

[2] We therefore call NLM's approach [7] *semi-automatic*, although an automatic classifier could be easily integrated.

3   Proposed Algorithm

Our approach to compound figure separation is a recursive algorithm (see Fig. 2) comprising the following steps: (1) classification of the compound image as illustration or non-illustration image, (2) removal of border bands, (3) detection of separator lines, (4) vertical or horizontal separation, and (5) recursive application to each subfigure image. The *illustration classifier* is used to decide which of two separator line detection modules to apply: if the compound image is classified as an illustration image, the *band-based* algorithm is applied, which aims at detecting separator bands between subfigures. Otherwise, the image is processed by the *edge-based* separator detection algorithm, which applies edge detection and Hough transform to locate candidate separator edges. The algorithm selection is based on the assumption that edge-based separator detection is better suited for non-illustration compound images due to visible vertical or horizontal edges separating subfigures. Note that this assumption is not violated by non-illustration compound images with separator bands where subfigures have a visible rectangular border. The following four sections describe the illustration classifier, the main recursive algorithm, and the two separator detection modules in more detail.

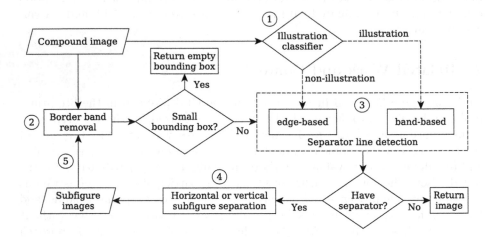

Fig. 2. Recursive algorithm for compound figure separation. Numbers denote the main algorithmic steps described in the beginning of Sect. 3.

3.1   Illustration Classifier

The illustration classifier is used to decide which separator detection algorithm to apply to a given compound image. If the image is predicted to be a graphical illustration with probability greater than `decision_threshold`, the band-based separator detection is applied, otherwise the edge-based separator module is used. This decision is made only once for each compound image, so all recursive invocations use the same separator detection algorithm.

For training the classifier, we use the dataset of the ImageCLEF 2015 multi-label image classification task [5]. The training dataset consists of 1071 images annotated with one or more labels of 29 classes (organized in a class hierarchy), which have been mapped to two meta classes: the *illustration* meta class comprises all "general biomedical illustration" classes except for chromatography images, screenshots, and non-clinical photos. These classes and all classes of diagnostic images have been assigned to the *non-illustration* meta class. About 36 % of the images in the training set are labeled with multiple classes; most of them represent compound images.

Classifier training requires mapping the set of labels of a given image to a single meta class. We implemented four mapping strategies and selected the most effective one during parameter optimization (Sect. 4). All mapping strategies first assign each image label to the *illustration* or *non-illustration* meta class as described above, and then operate differently on the list L of meta labels associated with a given image: (1) the *first* strategy simply assigns the first meta label of L to the image; (2) the *majority* strategy selects the meta label occurring most often in L, dropping the image from the training set if both meta labels occur equally often; (3) the *unanimous* strategy only assigns a meta label to the image if all meta labels in L are equal, otherwise the image is dropped from the training set; and (4) the *greedy* strategy maps an image to the *illustration* label if L contains at least one such meta label, otherwise the image is assigned the *non-illustration* label. The *greedy* strategy is inspired by the assumption that a compound image containing an illustration subfigure can be processed more effectively by the band-based separator detection algorithm than by the edge-based algorithm. Whereas mapping strategies *first* and *greedy* could use all 1071 images in the original training set, *majority* and *unanimous* strategies resulted in reduced training sets of 895 and 867 images, respectively.

Due to promising effectiveness for CFS in early experiments, we use four sets of global image features as classifier input, computed after gray-level conversion: (1) *simple2* is a two-dimensional feature consisting of image entropy, estimated using a 256-bin histogram, and mean intensity; (2) *simple11* extends *simple2* by 9 quantiles of the intensity distribution; (3) *CEDD* is the well-known color and edge directivity descriptor [2] (144-dimensional); and (4) *CEDD_simple11* is the concatenation of *CEDD* and *simple11* features (155-dimensional).

As machine learning algorithms we consider support vector machines (SVM) with radial basis function kernel (RBF) and logistic regression. Although logistic regression is generally inferior to kernel SVM due to its linear decision boundary, it has the advantage of providing prediction probabilities, which allow us to tune the selection of separator detection algorithms using the `decision_threshold` parameter. Classifier performance and its effectiveness for CFS are evaluated in Sect. 5.

3.2 Recursive Algorithm

Before applying the main algorithm (Fig. 2) to a given compound figure image, it is converted to 8-bit gray-scale. *Border band removal* detects a rectangular bounding box surrounded by a maximal homogeneous image region adjacent to image borders (border band). If the resulting bounding box is empty or smaller than elim_area or if maximal recursion depth has been reached, an empty bounding box is returned, terminating recursion. The *separator line detection* modules are invoked separately for vertical and horizontal directions, so they deal with a single direction θ and return a list of corresponding separator lines. An empty list is returned if the respective image dimension (width or height) is smaller than mindim or if no separator lines are found. If the returned lists for both directions are empty, recursion is terminated and the current image (without border bands) is returned. The *decision about vertical or horizontal separation* is trivial if one of both lists of separator lines is empty. Otherwise the decision is made based on the regularity of separator distances: locations of separator lines and borders are normalized to the range [0,1], and the direction (vertical or horizontal) yielding the lower variance of adjacent distances is chosen. Finally, the current figure image is divided into subimages along the chosen separation lines, and the algorithm is applied recursively to each subimage.

3.3   Edge-Based Separator Detection

The edge-based separator line detection algorithm aims at detecting full-length edges of a certain direction θ (vertical or horizontal) in a given gray-scale image. It comprises the following processing steps: (1) unidirectional edge detection, (2) peak selection in one-dimensional Hough transform, and (3) consolidation and filtering of candidate edges.

Edge detection is implemented by a one-dimensional Sobel filter und subsequent thresholding (edge_sobelthresh) to produce a binary edge map. The one-dimensional Hough transform counts the number of edge points aligned on each line in direction θ. So the peaks correspond to the longest edges, and their locations identify candidate separator edges. To make borders appear as strong Hough peaks, we add an artificial high-contrast border to the image prior to edge detection. Peaks are identified by an adaptive threshold t that depends on the recursion depth k (zero-based), the maximal value m of the current Hough transform, and the fill ratio f of the binary edge map (fraction of non-zero pixels, $0 \leq f \leq 1$), see (1). α and β are internal parameters (edge_houghratio_min and edge_houghratio_base).

$$h = \alpha * \beta^k, \qquad t = m * \left(h + (1 - h) * \sqrt{f} \right). \tag{1}$$

The rationale behind these formulas is to cope with noise in the Hough transform. Hough peaks were observed to become less pronounced as image size decreases (implied by increasing recursion depth) and as the fill ratio f increases (more edge points increase the probability that they are aligned by chance). Equation (1) ensures a higher threshold in these cases. Additionally, as recursion depth increases, the algorithm should detect only more pronounced separator edges, because further figure subdivisions become less likely.

Hough peak selection also includes a similar regularity criterion as used for deciding about vertical or horizontal separation (see Sect. 3.2): the list of candidate peaks is sorted by their Hough values in descending order, and candidates are removed from the end of the list until the variance of normalized edge distances of remaining candidates falls below a threshold (`edge_maxdistvar`). Candidate edges resulting from Hough peak selection are then consolidated by filling small gaps (of maximal length given by `edge_gapratio`) between edge line segments (of minimal length given by `edge_lenratio`). Finally, edges that are too short in comparison to image height or width (threshold `edge_minseplength`), or too close to borders (threshold `edge_minborderdist`) are discarded.

3.4 Band-Based Separator Detection

The band-based separator detection algorithm aims at locating homogeneous rectangular areas covering the full width or height of the image, which we call *separator bands*. Since this algorithm is intended primarily for gray-scale illustration images with light background, we assume that separator bands are white or light gray. The algorithm consists of four steps: (1) image binarization, (2) computation of mean projections, (3) identification and (4) filtering of candidate separator bands.

Initially, we binarize the image using the mean intensity value as a threshold. We then compute mean projections along direction θ (vertical or horizontal), that is, the mean value of each line of pixels in this direction. A resulting mean value will be 1 (white) if and only if the corresponding line contains only white pixels. Candidate separator bands are then determined by identifying maximal runs of ones in the vector of mean values that respect a minimal width threshold (`band_minsepwidth`). They are subsequently filtered using a regularity criterion similar to Hough peak selection (see Sect. 3.3), this time using distance variance threshold `band_maxdistvar`. Finally, selected bands that are close to the image border (threshold `band_minborderdist`) are discarded, and the center lines of remaining bands are returned as separator lines.

4 Parameter Optimization

The proposed CFS algorithm takes a number of internal parameters[3], whose initial values were chosen manually by looking at the results produced for a few training images. They were used during participation in ImageCLEF 2015 [10]. For parameter optimization, the CFS algorithm was evaluated for various parameter combinations on the ImageCLEF 2015 CFS training dataset (3,403 compound images, 14,531 ground-truth subfigures) using the evaluation tool provided by ImageCLEF organizers. Due to the number of parameters and the run time of a single evaluation run (about 17 min), a grid-like optimization evaluating all possible parameter

---

[3] Section 3 describes 17 internal parameters. A table with initial and optimized parameter values could not be included due to space constraints, but will be provided by authors upon request.

combinations in a certain range was not feasible. Instead, we applied a hill-climbing optimization strategy to locate the region of a local maximum and then used grid optimization in the neighborhood of this maximum.

More precisely, we defined up to five different values per parameter, including the initial values, on a linear or logarithmic scale, depending on the parameter. Then a set of parameter combinations was generated where only one parameter was varied at a time and all other parameters were kept at their initial values, resulting in a feasible number of parameter combinations to evaluate (linear in the number of parameters). After measuring accuracy on the training set, the most effective value of each parameter was chosen as its new *optimal* value. For parameters whose optimal values differed from the initial ones, the range was centered around the optimal value. Other parameters were fixed at their latest value. The procedure was repeated until accuracy improved by no more than 5 %, which happened after three iterations. Finally, after sorting parameter combinations by achieved accuracy, the five most effective parameters were chosen for grid optimization, where only two "nearly optimal" values (including the latest optimal value) per parameter were selected.

The effect of parameter optimization was surprisingly strong: whereas the initial parameter configuration achieved an accuracy of 43.5 % on the training set, performance increased to 84.5 % after hill-climbing optimization, and finished at 85.5 % after grid optimization.

5   Evaluation

The proposed CFS algorithm is evaluated on two datasets and compared to state-of-the-art approaches. To shed light on the effectiveness of the illustration classifier, we evaluate the CFS algorithm without a classifier and with different classifier implementations on the ImageCLEF 2015 dataset and compare results with the best algorithm submitted to ImageCLEF 2015 (Sect. 5.1). To evaluate the generalization capability of our algorithm, we apply it to the dataset of Apostolova et al. [1] (U.S. National Library of Medicine, NLM) using the same parameters optimized on the ImageCLEF dataset, and compare it to the image panel segmentation approach of the authors (Sect. 5.2). Finally, we compare classification accuracy of the classifier implementations used before on a third dataset and confirm that classification performance is not a critical factor for CFS effectiveness (Sect. 5.3).

5.1   Evaluation on ImageCLEF Dataset

The ImageCLEF 2015 CFS test dataset [5] contains 3,381 compound images with 12,789 ground-truth subfigures. The evaluation tool provided by ImageCLEF organizers computes the accuracy of detected subfigures for a given compound figure as follows. A detected subfigure is associated with at most one ground-truth subfigure G of maximal overlap if the overlap ratio is greater than 2/3 and if G is not already associated with a different detected subfigure. The number C of such associations for a given compound figure represents the number of correctly detected subfigures (true positives), and accuracy is defined by $C/\max(N_G, N_D)$, where N_G and N_D are the numbers of ground-truth and detected subfigures, respectively. Accuracy on the test set is the average of accuracy values computed for each compound figure.

Experimental results are shown in Table 1. For comparison, we also included a previous version of our approach [10] that did not use optimized parameters, and the best approach submitted to ImageCLEF 2015 (by NLM). We evaluated the proposed algorithm with optimized parameters (see Sect. 4) and with different implementations and feature sets for the illustration classifier, as described in Sect. 3.1. Because logistic regression using *simple2* features was found to be most effective by parameter optimization when trained on the *greedy* set, we focused on this training set when evaluating other classifier implementations. Internal SVM parameters were optimized on the entire ImageCLEF 2015 multi-label classification test dataset (see Sect. 5.3) to maximize classification accuracy. The optimized `decision_threshold` parameter for deciding between edge-based and band-based separator detection is effective only for logistic regression classifiers, because SVM predictions do not provide class probabilities. To confirm the effectiveness of the illustration classifier, we also included results for algorithm variants where the classifier has been replaced by a random decision selecting band-based separator detection with probability p. The value $p = 0.741$ corresponds to the decision rate of the most effective classifier (LogReg,simple11,greedy). $p = 0$ and $p = 1$ represent algorithms that always use edge-based or band-based separator detection, respectively. Finally, we considered an algorithm variant (*SubfigureClassifier*) that applied the illustration classifier not only once per compound image, but also to each subimage that is to be further divided by recursive figure separation.

When comparing our results to NLM's approach, we note that the authors of [7] manually classified the test set into stitched (4.3 %) and non-stitched (95.7 %) images, whereas our approach uses automatic classification. Using band-based separator detection for all test images (no classifier, $p = 1$) works surprisingly well (82.2 % accuracy), which can be explained by the low number of stitched compound images in the test set. On the other hand, using edge-based separator detection for all test images (no classifier, $p = 0$) results in modest performance (58 % accuracy), which we attribute to a significant number of subfigures without rectangular borders (illustrations) in the test set. Selecting edge-based or band-based separator detection using the illustration classifier improved accuracy for all tested classifier implementations. In fact, it turned out to be effective to bias the illustration classifier towards band-based separator detection and apply edge-based separator detection only to high-confidence non-illustration images. This happened in two ways: by using the *greedy* training set, and by optimizing the `decision_threshold` parameter for the logistic regression classifier. This explains why best results were obtained by logistic regression classifiers trained on the *greedy* training set.

Table 1. Experimental results on the ImageCLEF 2015 CFS test set. Illustration classifiers are described in Sect. 3.1 (LogReg = logistic regression). BB denotes the percentage of images (or decisions*) where band-based separator detection was applied.

Algorithm	Classifier	BB %	CFS Accuracy %
Previous [10]	LogReg,simple2,first		49.4
NLM [7]	manual	95.7	84.6
Proposed	LogReg,simple2,first	61.6	84.2
Proposed	LogReg,simple2,majority	61.1	84.1
Proposed	LogReg,simple2,unanimous	61.8	84.2
Proposed	LogReg,simple2,greedy	75.8	84.8
Proposed	LogReg,simple11,greedy	74.1	**84.9**
Proposed	SVM,simple2,greedy	58.6	83.5
Proposed	SVM,simple11,greedy	60.3	83.5
Proposed	SVM,CEDD,greedy	59.2	82.8
Proposed	SVM,CEDD_simple11,greedy	59.6	83.2
Proposed	random,p=0.741	74.7	75.4
Proposed	no classifier,p=0	0	58.0
Proposed	no classifier,p=1	100	82.2
SubfigureClassifier	LogReg,simple11,greedy	60.1*	84.0

To further analyze the effectiveness of separator detection selection, we partitioned the CFS test dataset into two classes according to decisions of the most effective CFS algorithm variant (LogReg,simple11,greedy) and evaluated detection results of this algorithm separately on the two partitions. Resulting accuracy values of 85.7 % on the edge-based partition and 84.6 % on the band-based partition show that the classifier was successful in jointly optimizing detection performance for both separator detection algorithms.

5.2 Evaluation on NLM Dataset

Apostolova et al. [1] used a different criterion to evaluate the accuracy of detected subfigures. Each detected subfigure F is decided to be *true positive* or *false positive* according to the following rule. Let $\{G_i \mid i \in I\}$ be the set of ground-truth subfigures overlapping with F, let A_i be the area size of G_i, and O_i the size of the overlapping area between F and G_i. The subfigure F is considered true positive if and only if there is an index $j \in I$ with $O_j/A_j > 0.75$ and $O_i/A_i < 0.05$ for all $i \in I, i \neq j$. That is, subfigure F has a notable overlap with one ground-truth subfigure only.

Given the total number N of ground-truth subfigures in the dataset, the total number D of detected subfigures, and the number T of detected true positive subfigures, the usual definitions for classifier evaluation measures can be applied to obtain precision P, recall R, and F_1 measure, see (2). Note that accuracy is not well-defined in this situation, because the number of negative results (not detected arbitrary bounding boxes) is theoretically unlimited.

$$P = \frac{T}{D}, \quad R = \frac{T}{N}, \quad F_1 = \frac{2 * P * R}{P + R}. \tag{2}$$

The dataset created by Apostolova et al. [1] contains 398 images with 1754 ground-truth subfigures. Table 2 shows the results of evaluating our proposed algorithm on this dataset using the measures described above. Note that we used the same parameter settings as in Sect. 5.1 to demonstrate the generalization capability of our algorithm. We selected the most effective illustration classifiers using logistic regression and SVM, respectively. They both use *simple11* features and the *greedy* training set. For convenience, we also included the results reported in [1] for a direct comparison with our approach[4].

Table 2. Evaluation results on the NLM CFS dataset [1]. Precision, recall, and F_1 score are computed from the total number of detected (D) and true positive (T) subfigures.

Algorithm	D	T	Precision %	Recall %	F_1 %
Proposed (LogReg)	1646	1407	85.5	**80.2**	**82.8**
Proposed (SVM)	1681	1392	82.8	79.4	81.1
Apostolova et al. [1]	1482	1276	**86.1**	72.3	78.6

Results show that the relative performance of the proposed algorithm using different classifiers is consistent with evaluation results in Sect. 5.1. The proposed algorithm could detect 10 % more true positive subfigures than the image panel segmentation algorithm of Apostolova et al. [1], leading to a higher recall rate. On the other hand, precision is only slightly lower. Note that algorithm [1] has been used as a component in NLM's CFS approach [7] referenced in Sect. 5.1.

5.3   Illustration Classifier Accuracy

To investigate the correlation of illustration classifier performance and effectiveness for CFS, we evaluated classification accuracy for the various classifier implementations considered in Sect. 5.1 on the test dataset of the ImageCLEF 2015 multi-label image classification task [5]. Labels of test images were mapped to binary meta classes using the same procedure as described in Sect. 3.1, resulting in 497 images for *first* and *greedy* test sets, 428 images for *majority*, and 398 images for *unanimous* test set. Evaluation results are shown in Table 3. The decision threshold for logistic regression was set to 0.5 to provide a fair comparison with SVM. Internal parameters of SVM (box constraint C and standard

---

[4] The dataset reported in [1] contains 400 images with 1764 ground-truth subfigures, so reported recall may be up to 0.4 % higher if evaluated on the 398 images of the dataset available to us.

Table 3. Classification accuracy on ImageCLEF 2015 multi-label image classification test dataset (497 images) for different implementation options of illustration classifier. Features and training sets are described in Sect. 3.1, LogReg = logistic regression.

Classifier	Features	Training Set	Accuracy %
LogReg	simple2	first	82.5
LogReg	simple2	majority	86.5
LogReg	simple2	unanimous	**88.2**
LogReg	simple2	greedy	84.7
LogReg	simple2	greedy	84.7
LogReg	simple11	greedy	83.7
SVM	simple2	greedy	84.3
SVM	simple11	greedy	84.3
SVM	CEDD	greedy	**87.1**
SVM	CEDD_simple11	greedy	86.7

deviation σ of RBF kernel) were optimized using two-fold cross-validation on the test set.

The upper part of Table 3 tells us that *majority* and *unanimous* training sets improve classification performance, although we know from Sect. 5.1 that this does not help CFS effectiveness. From the lower part of Table 3 we note that, interestingly, SVM does not perform better on *simple2* features than logistic regression and causes only a modest improvement (around 3 %) on CEDD features (144-dimensional). This may indicate the need to select more discriminative features for this classification task in future work, although results of Sect. 5.1 suggest that accuracy of the illustration classifier is not a critical factor of the proposed CFS algorithm.

6   Conclusion and Further Work

We proposed a recursive image processing algorithm for automatic separation of compound figures appearing in scientific articles. The algorithm has been evaluated on two recently published CFS datasets and achieved a detection performance slightly better than state-of-the-art approaches, even though it was compared to a semi-automatic approach [7] and not all known useful techniques (e.g. image markup removal and subfigure label recognition [1]) have been incorporated. Future work may therefore include the integration of such techniques into the proposed algorithm.

The use of the illustration classifier to select either edge-based or band-based separator detection for a given compound figure proved to be effective to improve CFS detection accuracy in conducted experiments. From evaluation of two classifier implementations, four image features, and four training sets, we conclude that classification accuracy is not a critical factor for CFS effectiveness, but biasing the classifier towards the illustration class improves CFS detection accuracy. We explain these results by the observation that band-based separator detection works well for almost all compound figures in the dataset except for "high-confidence" non-illustration images, where edge-based separator detection is the more effective choice. Finding more discriminative image features and better training sets for the classifier to improve CFS effectiveness of the proposed algorithm may be an additional subject of future work.

Acknowledgements. We thank Sameer Antani (NLM) and the authors of [1] for providing their compound figure separation dataset for evaluation purposes, and Laszlo Böszörmenyi (ITEC, AAU) for valuable discussions and comments on this work.

References

1. Apostolova, E., You, D., Xue, Z., Antani, S., Demner-Fushman, D., Thoma, G.R.: Image retrieval from scientific publications: text and image content processing to separate multipanel figures. J. Assoc. Inf. Sci. Technol. **64**(5), 893–908 (2013)
2. Chatzichristofis, S.A., Boutalis, Y.S.: CEDD: color and edge directivity descriptor: a compact descriptor for image indexing and retrieval. In: Gasteratos, A., Vincze, M., Tsotsos, J.K. (eds.) ICVS 2008. LNCS, vol. 5008, pp. 312–322. Springer, Heidelberg (2008)
3. Chhatkuli, A., Foncubierta-Rodríguez, A., Markonis, D., Meriaudeau, F., Müller, H.: Separating compound figures in journal articles to allow for subfigure classification. In: Proceedings of the SPIE, vol. 8674, pp. 86740J–86740J-12 (2013)
4. García Seco de Herrera, A., Kalpathy-Cramer, J., Demner-Fushman, D., Antani, S., Müller, H.: Overview of the ImageCLEF 2013 medical tasks. CLEF 2013 Working Notes. CEUR Proc., vol. 1179 (2013). http://ceur-ws.org/Vol-1179/
5. García Seco de Herrera, A., Müller, H., Bromuri, S.: Overview of the ImageCLEF 2015 medical classification task. CLEF 2015 Working Notes. CEUR Proc., vol. 1391 (2015). http://ceur-ws.org/Vol-1391/
6. Kitanovski, I., Dimitrovski, I., Loskovska, S.: FCSE at medical tasks of ImageCLEF 2013. CLEF 2013 Working Notes. CEUR Proc., vol. 1179 (2013). http://ceur-ws.org/Vol-1179/
7. Santosh, K., Xue, Z., Antani, S., Thoma, G.: NLM at ImageCLEF 2015: biomedical multipanel figure separation. CLEF 2015 Working Notes. CEUR Proc., vol. 1391 (2015). http://ceur-ws.org/Vol-1391/
8. Shatkay, H., Chen, N., Blostein, D.: Integrating image data into biomedical text categorization. Bioinformatics **22**(14), 446–453 (2006). http://dx.doi.org/10.1093/bioinformatics/btl235
9. Simpson, M.S., You, D., Rahman, M.M., Xue, Z., Demner-Fushman, D., Antani, S., Thoma, G.: Literature-based biomedical image classification and retrieval. Comput. Med. Imag. Graph. **39**, 3–13 (2015)
10. Taschwer, M., Marques, O.: AAUITEC at ImageCLEF 2015: Compound figure separation. CLEF 2015 Working Notes. CEUR Proc., vol. 1391 (2015). http://ceur-ws.org/Vol-1391/
11. Yuan, X., Ang, D.: A novel figure panel classification and extraction method for document image understanding. Int. J. Data Min. Bioinform. **9**(1), 22–36 (2014). http://dx.doi.org/10.1504/IJDMB.2014.057779

Camera Network Based Person Re-identification by Leveraging Spatial-Temporal Constraint and Multiple Cameras Relations

Wenxin Huang[1], Ruimin Hu[1,2]([✉]), Chao Liang[1,2], Yi Yu[3], Zheng Wang[1], Xian Zhong[4], and Chunjie Zhang[5]

[1] National Engineering Research Center for Multimedia Software, School of Computer, Wuhan University, Wuhan, China
wenxin.huang@whu.edu.cn
[2] Collaborative Innovation Center of Geospatial Technology, Wuhan, China
hrm1964@163.com
[3] National Institute of Informatics, Tokyo, Japan
[4] School of Computer, Wuhan University of Technology, Wuhan, China
[5] University of Chinese Academy of Sciences, Beijing, China

Abstract. With the rapid development of multimedia technology and vast demand on video investigation, long-term cross-camera object tracking is increasingly important in the practical surveillance scene. Because the conventional Paired Cameras based Person Re-identification (PCPR) cannot fully satisfy the above requirement, a new framework named Camera Network based Person Re-identification (CNPR) is introduced. Two phenomena have been investigated and explored in this paper. First, the same person cannot simultaneously appear in two non-overlapping cameras. Second, the closer two cameras, the more relevant they are, in the sense that persons can transit between them with a high probability. Based on these two phenomena, a probabilistic method is proposed with reference to both visual difference and spatial-temporal constraint, to address the novel CNPR problem. (i) Spatial-temporal constraint is utilized as a filter to narrow the search space for the specific query object, and then the Weibull Distribution is exploited to formulate the spatial-temporal probability indicating the possibility of pedestrians walking to a certain camera at a certain time. (ii) Spatial-temporal probability and visual feature probability are collaborated to generate the ranking list. (iii) The multiple camera relations related to the transitions are explored to further optimize the obtained ranking list. Quantitative experiments conducted on TMin and CamNeT datasets have shown that the proposed method achieves a better performance to the novel CNPR problem.

Keywords: Person re-identification · Spatial-temporal constraint · Camera relation · Ranking optimization · Camera network

1 Introduction

Person re-identification is a task of visually matching person images, obtained from different cameras deployed in non-overlapping surveillance scenes [1–3] with

© Springer International Publishing Switzerland 2016
Q. Tian et al. (Eds.): MMM 2016, Part I, LNCS 9516, pp. 174–186, 2016.
DOI: 10.1007/978-3-319-27671-7_15

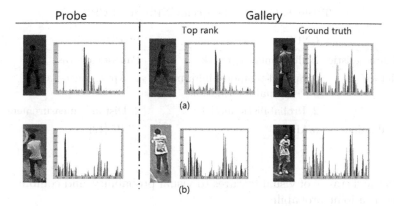

Fig. 1. Examples of the difficulties in exploiting visual features. The first column stands for the probe images, and the other two columns represent the candidates in the gallery. All the images have their color histograms in right boxes. The person in the second column is not the one in the first column but has the similar color histogram profile based on appearance information; the person in the third column is the ground truth but might be overlooked due to the large difference in the color histogram. (a) illustrates the difficulty caused by the different illuminations and (b) illustrates the difficulty caused by similar wearing.

place and time difference. The existing person re-identification problem is mainly regarded as a retrieval task on a pair of cameras [4,6,7,14,15,18,20]. With the rapid development of multimedia technology and vast demand for video investigation, camera networks have become increasingly deployed in public spaces such as airports, road intersections and campuses [14–18], which calls for the long-term cross-camera object tracking and human behavior analysis. However, the conventional Paired Camera based Person Re-identification (PCPR) methods cannot well solve this problem [11–13]. To achieve the retrieval task on multiple cameras, we introduce a new framework named Camera Network based Person Re-identification (CNPR).

As we know, some of the existing PCPR approaches focus on constructing visual features which are both distinctive and stable under various conditions [4–7], while others focus on learning an optimal metric [8–10], in which images of the same person are closer than those of different persons. All these methods mainly depend on the visual information, which faces intrinsic challenges caused by various changes in viewpoints, poses and illumination conditions (see Fig. 1(a)) in the practical surveillance environment [11,12]. To be worse, for special conditions where different persons wear highly similar clothes (see Fig. 1(b)), it is difficult to provide sufficient identity discrimination power when purely relying on visual information. According to the characteristic of CNPR, the spatial-temporal information exists among the images of persons, which motivates us to combine original visual features with it together to assist person re-identification. The distance of these two types of information cannot be simply fused and measured. Then, the proposed method introduces a spatial-temporal probability,

Table 1. Contrast between CNPR and PCPR.

	CNPR	PCPR
Characteristic	Multi-camera network	Pair of cameras
Method	1. Spatial-temporal information + Visual information	1. Appearance-based
	2. Probabilistic model	2. Distance measurement
Evaluation	mAP [21]	CMC [19]

converts the distance of visual features to visual probability, and combines them together as a joint probability.

Meanwhile, we find two phenomena in the CNPR (the details will be illustrated in Sect. 2), that can be utilized to optimize the re-identification performance in the camera network. (i) **Spatial-Temporal Constraint.** A person of interest who appears in one camera cannot appear in another non-overlapping camera at the same time or in a certain time period. Based on this phenomenon, Hinge loss function is exploited to construct a filter model to reduce the query scope, and the Weibull Distribution is utilized to formulate the spatial-temporal probability model. Probabilities of probe-to-gallery images are all calculated, then the initial ranking lists are generated. (ii) **Multiple Cameras Relations.** Inspired by the idea of image retrieval re-ranking methods [18], the probe-to-gallery ranking list is influenced by the relationships of different gallery-to-gallery images from different cameras. However, the relationships are different originated from different relations of cameras. We consider that if two cameras are more relevant according to deployed locations and transition time, it reveals that the probability of the corresponding person should be elevated. Exploiting the gallery-to-gallery probabilities and camera relations, the initial ranking list is optimized.

We summarize the differences of CNPR and PCPR in Table 1, and further explain in the following: (i) PCPR only focuses on a pair of cameras in the network, neglecting the relationships among different cameras, while CNPR involves a camera network which includes many cameras deployed in different places. (ii) It is distinctly different to solve CNPR and PCPR problems. PCPR methods [8,10,22,23] regard PCPR as a ranking problem and solve it by measuring the differences among visual features, while the proposed CNPR exploits spatial-temporal information besides visual features. In order to solve the CNPR problem, a probabilistic model, instead of the distance metric, is used to compute the similarity of two images. (iii) The Cumulated Matching Characteristics (CMC) curve [20] is typically used in PCPR to evaluate the performance. This evaluation measurement is valid only if there is only one ground truth for a probe. For the CNPR problem, there may be more than one cross-camera ground truth for each query. Therefore, we adopt the mean average precision (mAP) [21] as the metric to evaluate the overall performance.

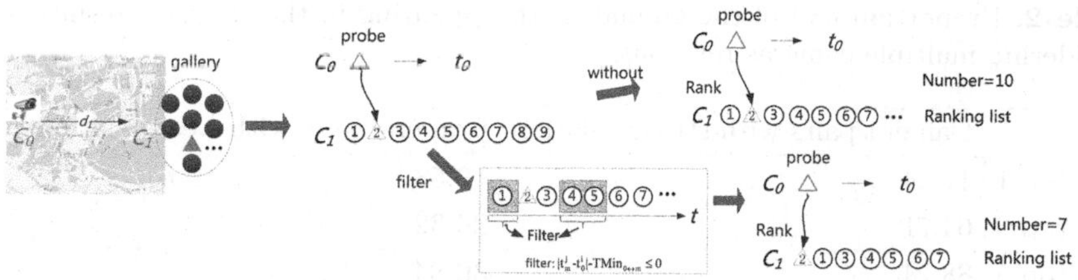

Fig. 2. Spatial-temporal constraint. There are always noises in the candidates in the gallery, especially in the case of ever increasing surveillance videos. Spatial-temporal constraint can effectively remove a number of noises in the non-overlapping camera network. That is to say, if the probe appears in camera C_0 at a certain time t_0, the images of candidates observed in camera C_1 at time t with the interval of $TMin_{0\leftrightarrow1}$ can be filtered out. Here, $TMin_{0\leftrightarrow1}$ is the minimized walking time which is related to the distance between C_0 and C_1.

Technical contributions of this paper are three-fold, as follows: (1) This paper puts forward CNPR as a new approach for the person re-identification problem. (2) Two phenomena, spatial-temporal constraint and camera relations, are exploited by a probabilistic model. (3) This paper adopts mAP as a new performance criterion particularly designed for CNPR. It considers both precision and recall, thus providing a more comprehensive evaluation than CMC widely used for PCPR.

2   Observations

In order to acquire additional information such as the time stamps and make full use of the captured videos, the association of the information of each camera can be explored since there are relations among a large number of videos. In this work, spatial-temporal constraint and relations of cameras in the network are investigated.

Spatial-Temporal Constraint. Spatial-temporal information represents more strict constraint to limit the query scope. It could help to improve the query efficiency and matching rate. There is a common sense that a person of interest appeared in a camera cannot be in another non-overlapping camera at the same time. In addition, significant spatial-temporal gaps exists since there is a distance interval between two cameras in a non-overlapping camera network. Based on the statement, it can describe when and where the target had stayed in the entire camera network. Therefore, we can reduce the confusing images which are in different places at the same time of interests. Figure 2 illustrates the phenomenon.

Multiple Cameras Relations. We assume that the walk of pedestrian is continuous, then a person will have a high appearance probability in two adjacent cameras. In CNPR problem, a person may appear in multiple cameras in the

Table 2. Proportion (%) of the ground truth appearing in the returned results with considering multiple cameras relations.

	Camera pairs with strong relation	Camera pairs with weak relation
top 1	17.64	10.53
top 3	64.71	26.32
top 5	88.24	36.84

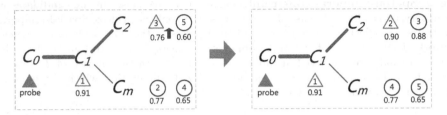

Fig. 3. Multiple cameras relations. There are relations of C_0, C_1, C_2 and C_m, where the degree of thickness of lines represents how the relations are. The numbers in the triangle and circle reveal the ranking number of the candidates. The numbers under the triangle and circle indicate the matching probabilities of the candidates. Left box indicates the situation of regular ranking and right box indicates the situation of the new ranking list after considering relations of cameras. Since the relation of C_1 and C_2 is closer than that of C_1 and C_m, the matching probabilities of C_2 rises.

network. In this assumption, the matching probability of the person is affected by the distance relations of cameras. A preliminary experiment conducted to explain this phenomenon. Two adjacent pairs of cameras and two furthest pairs of cameras in TMin [13] were selected respectively as the pairs with strong relation and the pairs with weak relation. For each pair, we choose one as the probe, and the other as the retrieval candidate. We performed statistical analysis on these data and the statistical result is illustrated in Table 2. We can find that a person will have a high appearance probability in cameras with strong relations. In the initial ranking list, the persons in the top results have high matching probabilities to be the probe person. For each top result, he can also obtain the corresponding pedestrians with high probabilities in other gallery cameras. These corresponding pedestrians may be highly possible to be the probe person. Therefore, the ranking number of these pedestrians in the initial ranking list should be elevated. The degree of elevation depends on the relation between the gallery camera of the top result and that of the selected pedestrian. Figure 3 reveals that if the relation of camera is closer, the evaluation will be higher. However, the selected pedestrians may have multiple relations with different cameras, then the elevation is related to multiple cameras relations.

In brief, the core idea of solving the CNPR problem is that we construct a probabilistic framework based on the whole camera network, further optimizing with spatial-temporal constraint and multiple cameras relations.

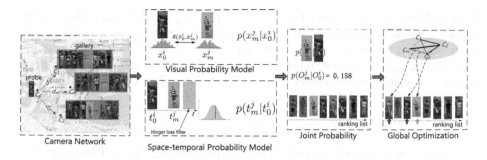

Fig. 4. Framework of the proposed method in CNPR

3   Our Approach

3.1   Problem Definition

This subsection gives a brief definition of the CNPR problem. We consider a camera network $C = \{C_0, C_1, C_2, ..., C_M\}$, which is composed of $M + 1$ cameras with non-overlapping field of views. Here, C_0 denotes the probe camera, and M is the number of gallery cameras.

We denote the representative image of person j, captured by camera C_m as x_m^j, $x_m^j \in R^n$, and $0 < m \le M$. Here, n represents the dimension of the visual feature of the image. The time of each observation is recorded as well. Then, an observation O_m^j can be described as a combination $O_m^j = (x_m^j, t_m^j)$, where t_m^j denotes the time person j walks into the view of camera C_m.

In PCPR, merely the visual feature is exploited. That is to say, the distance $d(x_0^i, x_m^j)$ is calculated between the probe image i and an image j in the gallery, where the gallery images are all from camera C_m. Then, a ranking list is obtained depending on the calculated distances. In comparison, for the CNPR problem, we exploit the probability theory instead of a distance metric to represent not only the similarity of visual features but also the similarity of spatial-temporal relationship. If the observation O_m^j gets a high conditional probability $p(O_m^j|O_0^i)$ based on the probe O_0^i, it will obtain a high ranking number. Besides, the probability is also related to the relations of cameras, as a result, we can exploit the relations of cameras to refine the observations under the entire camera network. The framework is shown in Fig. 4.

$$p(x_m^j|x_0^i) = e^{-\alpha \cdot d(x_0^i, x_m^j)} \tag{1}$$

Equation 1 converts the visual distance into a probability. Here, the distance can be obtained by any existing algorithm for PCPR.

3.2   Probabilistic Model with Spatial-Temporal Constraint

In a fixed camera network, the minimum walking time between each pair of cameras is given. It is assumed that the minimum walking time between C_0 and

C_m is $TMin_{0 \leftrightarrow m}$. As discussed in Sect. 2, if $|t_m^j - t_0^i| < TMin_{0 \leftrightarrow m}$, the spatial-temporal probability $p(t_m^j|t_0^i)$ will be zero for the person i appearing in C_m. In other words, if the person i appears in camera C_0, within the minimum walking time $TMin_{0 \leftrightarrow m}$, he cannot appear in camera C_m. In this situation, we do not need to calculate the probability $p(O_m^j|O_0^i)$ any more. Here, we introduce hinge loss acting as a filter described in Eq. 2.

$$h(t_0^i, t_m^j) = \max(0, |t_m^j - t_0^i| - TMin_{0 \leftrightarrow m}) \tag{2}$$

Considering the process of pedestrian's walk, when $|t_m^j - t_0^i| = TMin_{0 \leftrightarrow m}$, the matching probability equals to zero; when $|t_m^j - t_0^i| > TMin_{0 \leftrightarrow m}$, the matching probability increases at first and then reaches the peak value. As the time interval gets too long, the matching probability will get down and tend to zero with the assumption that the person is continuously walking. We assume the time for the transition between cameras follows a Weibull distribution [26], which has been successfully applied to nearly all scientific disciplines, such as biological, environmental, health, physical and social sciences. By fitting time data to Weibull distributions, Weibull analysis enables risk assessment and planning of corrective actions. Then, if we have two observations between the probe camera and another one in the fixed camera network, the conditional probability of the transition from C_0 to C_m is described as following:

$$p(t_m^j|t_0^i) = \frac{k}{\lambda}\left(\frac{h(t_0^i, t_m^j)}{\lambda}\right)^{k-1} e^{-(h(t_0^i, t_m^j)/\lambda)^k}, \tag{3}$$

where $k > 0$ is the shape parameter and $\lambda > 0$ is the scale parameter of the distribution.

Joint Probability Model. The visual distance between the image from a camera in the gallery and the probe image can be formulated as a conditional probability in Eq. 1. It is assumed spatial-temporal relations, independent of visual features, could assist to increase the ranking number. Then the conditional probability is obtained as shown in Eq. 4. The ranking list could be obtained after repeating the process on all the images in the gallery from the cameras in the network. Through this way, the ranking is related to both visual features and spatial-temporal constraint.

$$p(O_m^j|O_0^i) = p(x_m^j|x_0^i)p(t_m^j|t_0^i) \tag{4}$$

3.3   Optimization with Multiple Camera Relations

Optimization includes two aspects: (i) We exploit the relations between two cameras which are related to the spatial-temporal interval to construct spatial-temporal probabilities, and combine them with visual features probabilities to get the probabilistic model. (ii) We utilize the relations of entire camera network to improve the matching probability of candidates in the gallery. The optimization of the ranking can be concluded as follows: (i) By constructing the joint probability model, the initial ranking lists are obtained. (ii) With each image

Fig. 5. Example of maP. This example indicates the value of AP. For all three persons, the CMC curve remains 1, which is not proper for CNPR but AP shows the differences. mAP is the mean average precision of the three persons which is $(1 + 1 + 0.41)/3 = 0.803$.

in top-k of the initial ranking list as a probe, we find its top-k similar images from the gallery. (iii) The relevance of two cameras which is related to spatial-temporal interval in the network can be obtained by the spatial-temporal difference as $r(m, m')$ $(r(m, m') > 1)$, and the closer the cameras are, the larger the value of $r(m, m')$ is. The matching probability of the candidates in the initial ranking list can be adjusted to $\sum_{m' \in KNN(m)} p(O_m^j | O_0^i) * r(m, m')$. Here, for an image in the initial ranking list, m is its camera ID, and m' is the camera ID of an image in its KNN. Finally, the new ranking list is achieved, which involves the relationship between cameras.

4 Experiments

4.1 Baselines

mAP. The CMC curve is exploited by most papers on the person re-identification problem [4,8–10,19]. The value of CMC which tells the rate of the correct match indicates the identification results for every pair of camera views [7]. As each probe image may correspond to multiple ground truths in the gallery, precision and recall over entire camera network should be considered as a metric while evaluating the CNPR problem. In this case, we adopt a metric named mean average precision (mAP) to indicate the percentage of the real matches over the camera network. There is an example of mAP shown in Fig. 5.

4.2 TMin Data Set

Our task requires a number of cameras, walking time between pair of cameras and the topology of cameras. As far as we know, there is no appropriate public database in non-overlapping multi-camera person re-identification field. Thus, we utilize two public databases, TMin [13] and CamNeT [24], often used in the multi-camera tracking field which can satisfy our requirements. The version 1 of the TMin database contains 1680 images from 30 subjects. All the images are extracted from 6 cameras and the video acquisition time starts at twenty

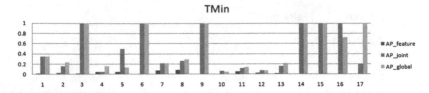

Fig. 6. Performance of different methods on the TMin dataset AP_ feature, AP_ joint and AP_ global represent the matching accuracy of each probe which are calculated by visual features, probabilistic model of combined spatial-temporal and visual features and optimization approach. Most of the results reveals that there is evident improvement after introducing spatial-temporal probabilistic model and relations of multiple cameras.

to twelve in the morning and ends at a quarter to two in the afternoon. Each pedestrian of TMin appears in at least two different cameras.

We employed camera 1# of TMin database which contains 17 of 30 pedestrians as our probe set in this paper. First, the noisy images were filtered out based on a filter introduced above. Then, we exploited Local Maximal Occurrence (LOMO) [25] as the feature representation which is effective and robust to illumination and viewpoint changes. We calculated the visual appearance probability of C_0 and C_m respectively, and unified them in a matrix. After that, spatial-temporal probability which was calculated according to hinge loss was joined with the appearance probability to construct the overall probability of an observation in the entire camera network compared with probe camera 1#. At last, we used the relations between cameras to refine the ranking. The results of visual features, joint probability and optimization are shown in Fig. 6. Moreover, in Table 3 we compare the performance of visual features, joint probability and global optimization in the range of the first 10, 20, 30, 40, 50, 75 ranks respectively. As can be seen, our approach greatly improves mAP compared with conventional methods, and the improvements are evident after introducing spatial-temporal probabilistic model and global optimization respectively[1].

4.3   CamNeT

CamNeT [24] is the database of non-overlapping camera network tracking data set for multi-target tracking, it consists of 5 to 8 cameras which cover both indoor and outdoor scenes at a university. The paths of around 10–25 people are predefined while several unknown persons move through the scene. There are 6 scenarios in which every scenario lasts at least 5 min with 5 to 8 cameras. We employed camera 1# of CamNeT database which contains 11 pedestrians as our probe set in this paper. To evaluate our method, the experimental process mainly followed the way for the TMin data set. The result is presented in Fig. 7.

---

[1] Here, we set $\alpha = 0.5$, $\lambda = 75$, $\beta = 2.5$, and $K = 5$, when evaluating on TMin data set.

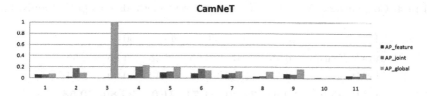

Fig. 7. Performance of different methods on the CamNeT dataset AP_ feature, AP_ joint and AP_ global represent the matching accuracy of each probe which are calculated by visual features, probabilistic model of combined spatial-temporal and visual features and optimization approach. Most of the results reveals that there is evident improvement after introducing spatial-temporal probabilistic model and relations of multiple cameras.

Table 3. Comparing mAP value (%) with different methods on top K (TMin).

K	10	20	30	40	50	75
mAP_ feature	2.39	2.28	2.39	2.36	2.46	2.57
mAP_ joint	50.08	49.10	48.44	47.97	47.83	47.73
mAP_ global	**53.96**	**51.92**	**51.06**	**50.50**	**50.33**	**50.20**

Our approach improves evidently. And we compared the performance of visual features, joint probability and optimization in the range of the first 10, 20, 30, 40, 50, 75 ranks respectively in Table 4[2].

However, we found that the joint probability of person 9 and person 11 which combined spatial-temporal and features had been decreased in Fig. 7. The reason is that there is excess layover time during the pedestrians' walk among cameras, which is beyond our assumption that the walk is continuous in this paper. For another unexpected situation, we also found that the optimization results of person 5 and person 16 had been decreased in Fig. 6. These two persons appeared only twice in other cameras, in other words, negative samples appears more which influenced the optimization effect. Although there are few disharmonious members, the overall results are still evident improved by the proposed method.

4.4 Running Time

In this subsection, we test the running time of each procedure. The additional running time, when compared with the running time of the baseline method using the distance of visual features, is no more than 1 ms even though we add the processes of calculating spatial-temporal probability and optimization. In fact, we use spatial-temporal constraint to filter 2.6 % and 27.35 % images of gallery of TMin and CamNeT, it somehow increases the efficiency. The experimental

---

[2] Here, we set $\alpha = 0.5$, $\lambda = 50$, $\beta = 1.5$, and $K = 10$, when evaluating on CamNeT data set.

Table 4. Comparing mAP value (%) with different methods on top K (CamNeT).

K	10	20	30	40	50	75
mAP_ feature	6.35	5.13	4.99	4.98	4.85	4.59
mAP_ joint	13.66	11.02	9.72	9.14	9.00	8.94
mAP_ global	**24.41**	**23.10**	**21.71**	**21.01**	**20.80**	**20.58**

environment is as follows [18]: our computer includes a dual core 2.80 GHz CPU and 2 GB RAM.

5   Conclusion

This paper puts forward a new framework CNPR as a fresh person re-identification solution. While solving the CNPR problem, two phenomena of spatial-temporal constraint and the relations among cameras in the network are investigated and leveraged in this work. On this basis, this paper proposes an approach of optimization, based on a probabilistic model taking into account spatial-temporal constraint and visual probability. The experiments, conducted on two public databases TMin and CamNeT, show significant improvement in efficiency and accuracy compared with conventional methods.

Acknowledgement. The research was supported by National Nature Science Foundation of China (61303114, 61231015, 61170023), the Specialized Research Fund for the Doctoral Program of Higher Education (20130141120024), the Technology Research Project of Ministry of Public Security (2014JSYJA016), the China Postdoctoral Science Foundation funded project (2013M530350), the major Science and Technology Innovation Plan of Hubei Province (2013AAA020), the Guangdong-Hongkong Key Domain Break-through Project of China (2012A090200007), and the Special Project on the Integration of Industry, Education and Research of Guangdong Province (2011B090400601). Nature Science Foundation of Hubei Province (2014CFB712). Jiangxi Youth Science Foundation of China(20151BAB217013).

References

1. Gong, S., Cristami, M., Yan, S., Loy, C.: Person Re-Identification. Advances in Computer Vision and Pattern Recognition. Springer, London (2014)
2. Wang, Z., Hu, R., Liang, C., Leng, Q., Sun, K.: Region-based interactive ranking optimization for person re-identification. In: Ooi, W.T., Snoek, C.G.M., Tan, H.K., Ho, C.-K., Huet, B., Ngo, C.-W. (eds.) PCM 2014. LNCS, vol. 8879, pp. 1–10. Springer, Heidelberg (2014)
3. Wang, Z., Hu, R., Liang, C., Jiang, J., Sun, K., Leng, Q., Huang, B.: Person re-identification using data-driven metric adaptation. In: He, X., Luo, S., Tao, D., Xu, C., Yang, J., Hasan, M.A. (eds.) MMM 2015, Part II. LNCS, vol. 8936, pp. 195–207. Springer, Heidelberg (2015)

4. Farenzena, M., Bazzani, L., Perina, A., Murino, V., Cristani, M.: Person re-identification by symmetry-driven accumulation of local features. In: Computer Vision and Pattern Recognition (CVPR) (2010)
5. Hu, Y., Liao, S., Lei, Z., Yi, D., Li, S.Z.: Exploring structural information and fusing multiple features for person re-identification. In: IEEE Workshop on Camera Networks and Wide Area Scene Analysis (in conjunction with CVPR 2013) (2013)
6. Gheissari, N., Sebastian, T.B., Hartley, R.: Person re-identification using spatiotemporal appearance. In: Computer Vision and Pattern Recognition (CVPR) (2006)
7. Wang, X., Doretto, G., Sebastian, T., Rittscher, J., Tu, P.H.: Shape and appearance context modeling. In: International Conference on Computer Vision (ICCV) (2007)
8. Prosser, B., Zheng, W.-S., Gong, S., Xiang, T.: Person re-identification by support vector ranking. In: British Machine Vision Conference (BMVC) (2010)
9. Zheng, W.S., Gong, S., Xiang, T.: Person re-identification by probabilistic relative distance comparison. In: Computer Vision and Pattern Recognition (CVPR) (2011)
10. Kostinger, M., Hirzer, M., Wohlhart, P., Roth, P.M., Bischof, H.: Large scale metric learning from equivalence constraints. In: Computer Vision and Pattern Recognition (CVPR) (2012)
11. Ma, L., Yang, X., Tao, D.: Person re-identification over camera networks using multi-task distance metric learning. IEEE Trans. Image Process. (TIP) 23(8), 3656–3670 (2014)
12. Das, A., Chakraborty, A., Roy-Chowdhury, A.K.: Consistent re-identification in a camera network. In: Fleet, D., Pajdla, T., Schiele, B., Tuytelaars, T. (eds.) ECCV 2014, Part II. LNCS, vol. 8690, pp. 330–345. Springer, Heidelberg (2014)
13. Hu, Y., Liao, S., Yi, D., et al.: Multi-camera trajectory mining: database and evaluation. In: International Conference on Pattern Recognition (ICPR) (2014)
14. Loy, C.C., Xiang, T., Gong, S.: Multi-camera activity correlation analysis. In: Computer Vision and Pattern Recognition (CVPR) (2009)
15. Javed, O., Shafique, K., Rasheed, Z., Shah, M.: Modeling intercamera spacetime and appearance relationships for tracking across non-overlapping views. In: Computer Vision and Image Understand (CVIU) (2008)
16. Wang, Y., Velipasalar, S., Casares, M.: Cooperative object tracking and composite event detection with wireless embedded smart cameras. IEEE Trans. Image Process. (TIP) 19(10), 2614–2613 (2010)
17. Ding, C., Song, B., Morye, A., Farrell, J.A., Roy-Chowdhury, A.K.: Collaborative sensing in a distributed PTZ camera network. IEEE Trans. Image Process. (TIP) 21(7), 3282–3295 (2012)
18. Leng, Q., Hu, R., Liang, C., et al.: Bidirectional ranking for person re-identification. In: International Conference on Multimedia and Expo (ICME) (2013)
19. Li, X., Tao, D., Jin, L., Wang, Y., Yuan, Y.: Person re-identification by regularized smoothing kiss metric learning. IEEE Trans. Circuits Syst. Video Technol. (TCSVT) 23(10), 1675–1685 (2013)
20. Wang, Y., Hu, R., Liang, C., Zhang, C., Leng, Q.: Camera compensation using a feature projection matrix for person reidentification. IEEE Trans. Circ. Syst. Video Technol. 24(8), 1350–1361 (2014)
21. Zheng, L., Shen, L., Tian, L., et al.: Person re-identification meets image search. arXiv (2015)
22. Zheng, W.-S., Gong, S., Xiang, T.: Re-identification by relative distance comparison. IEEE Trans. Pattern Anal. Mach. Intell. (TPAMI) 35(3), 653–668 (2013)
23. Mignon, A., Jurie, F.: PCCA: a new approach for distance learning from sparse pairwise constraints. In: Computer Vision and Pattern Recognition (CVPR) (2012)

24. Zhang, S., Staudt, E., Faltemier, T., et al.: A camera network tracking (CamNeT) dataset and performance baseline. In: IEEE Winter Conference on Applications of Computer Vision (WACV) (2015)
25. Liao, S., Hu, Y., Zhu, X., et al.: Person re-identification by local maximal occurrence representation and metric learning. In: Proceedings of the IEEE Conference on Computer Vision and Pattern Recognition (CVPR) (2015)
26. Liu, C., Ryen, W., Susan, D.: Understanding web browsing behaviors through Weibull analysis of dwell time. In: Proceedings of the 33rd International ACM SIGIR Conference on Research and Development in Information Retrieval (2010)

Global Contrast Based Salient Region Boundary Sampling for Action Recognition

Zengmin Xu[1,3], Ruimin Hu[1,2]([✉]), Jun Chen[1,2], Huafeng Chen[1],
and Hongyang Li[1]

[1] National Engineering Research Center for Multimedia Software,
School of Computer, Wuhan University, Wuhan, China
`hrm@whu.edu.cn`
[2] Collaborative Innovation Center of Geospatial Technology, Wuhan, China
[3] School of Mathematics and Computing Science,
Guangxi Colleges and Universities Key Laboratory of Data Analysis
and Computation, Guilin University of Electronic Technology, Guilin, China

Abstract. Although the excellent representation ability of improved
Dense Trajectory (iDT) based features for action video had been proved
on several action datasets, the performance of action recognition still suffers from large camera motion of videos. In this paper, we improve the
iDT method by advancing a novel salient region boundary based dense
sampling strategy, which reduces the number of trajectories while pre-
serves the discriminative power. We first implement the iDT sampling
based on motion boundary image, then introduce a global contrast based
salient object segmentation method in interest points sampling step of
action recognition. To overcome the flaws of global color contrast-based
salient region sampling, we apply morphological gradient to generate a
more robust mask for sampling dense points, as motion boundaries are
much clearer. To evaluate the proposed method, we conduct extensive
experiments on two benchmarks including HMDB51 and UCF50. The
results show that our sampling strategy can improve the performance of
action recognition with minor computational cost of mask production. In
particular, on the HMDB51 dataset, the improvement over the original
iDT result is 3 %. Meanwhile, any other dense features of action recognition can achieve more competitive performance by utilizing our sampling
strategy and Fisher vector encoding method simply.

Keywords: Salient region boundary · Sampling strategy · Action representation · Improved dense trajectories

1 Introduction

Human action recognition, as an important biometric technology, has become
an active research topic for decades due to their wide applications in video
surveillance, video understanding, *etc.* Researchers used to focus on simple
datasets [1,2] collected from controlled experimental settings. As the increasing

© Springer International Publishing Switzerland 2016
Q. Tian et al. (Eds.): MMM 2016, Part I, LNCS 9516, pp. 187–198, 2016.
DOI: 10.1007/978-3-319-27671-7_16

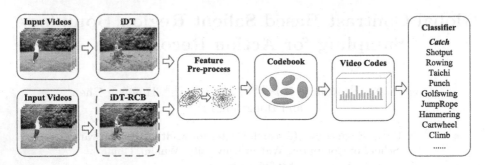

Fig. 1. Comparison of proposed approach (iDT-RCB) to traditional approach (iDT) for action recognition. Points sampled by iDT-RCB are more effective than iDT. Green trajectories indicate the sampled points have been tracked for fixed length of frames (Color figure online).

demand for understanding the content of real world video, action recognition still remains a challenging task on realistic data sets, which were collected from movies [3], web videos [4–6], *etc.* The diversity of realistic video data has resulted in significant challenges due to camera motion, viewpoint changes, occlusion, intra-class variations, complex background, *etc.*

How to represent human action in these realistic videos has been a fundamental problem in action recognition. By far, local space-time features [7–13] were shown to be successful on these datasets. Laptev [7] extracted space-time interest points (STIP) by extending the Harris detector from image to video. Dollar *et al.* [8] used 2D spatial Gaussian and 1D temporal Gabor filters to develop salient interest points in video. Willems *et al.* [9] applied the scale-space theory and Hessian matrix to detect interest points. Kläser *et al.* [10] proposed a HOG3D descriptor based on 3D-gradients. Sun *et al.* [11] modeled hierarchical spatio-temporal context via SIFT-based Trajectory. Wang *et al.* [12] sampled interest points on dense grid in each frame, and tracked them based on dense optical flow.

Among the state-of-the-art local space-time features, improved dense trajectories [13] have been shown to perform best on various datasets. The main idea is to remove camera motion from optical flow by homographic matrix, and explore the Fisher vector as a feature encoding approach. A large set of evaluations was presented to demonstrate the excellent performance of this feature. However, the iDT based representation is expensive in memory storage and computation due to the large number of densely sampled points.

In this paper, we develop a salient Region-based Contrast Boundary sampling strategy named iDT-RCB to refine improved dense trajectory approach. We start from densely sampled points on grid in each frame, and separate a large-scale region from its surroundings by a global contrast based method. Then we perform Morphological Gradient to construct a spatial saliency map with region boundary, and remove those sampled points which have no overlaps with foreground in the mask. The iDT-RCB is motivated by the fact that the trajectories on motion boundary are the most meaningful ones. This is also inspired by DT-MB based sampling strategy [14] and the high performance of the MBH descriptor [23]. Under the control of our sampling method, the action representation can be focused on the most effective IDTs.

2   Methodology

In this section, we first briefly review the iDT features [13] including three steps: dense sampling, point tracking and trajectory estimating. We also illustrate the principle of motion boundary sampling method, then implement the improved dense trajectories sampling based on motion boundary.

2.1   Improved Dense Trajectories

Dense trajectories approach [12] densely sample feature points on a grid in each frame spaced by W pixels. If the eigenvalues of the auto-correlation matrix are very small, it is impossible to track any point in homogeneous image areas. Hence the DT approach sets a threshold T on the eigenvalues for each frame I as:

$$T = 0.001 \times \max_{i \in I} \min(\lambda_i^1, \lambda_i^2), \tag{1}$$

where $(\lambda_i^1, \lambda_i^2)$ are the eigenvalues of point i in the image I. Experiments showed that a value of 0.001 represents a good compromise between saliency and density of the sampled points. The sampled points are tracked through the video for L =15 frames. Then they are removed and replaced by new interest points.

For each frame I_t, its dense optical flow field $\omega_t = (u_t, v_t)$ is computed w.r.t. the next frame I_{t+1}, where u_t and v_t are the horizontal and vertical components of the optical flow. Given a point $P_t = (x_t, y_t)$ in frame I_t, its tracked position in frame I_{t+1} is smoothed by applying a median filter on ω_t:

$$P_{t+1} = (x_{t+1}, y_{t+1}) = (x_t, y_t) + (M * \omega_t)|_{(x_t, y_t)}, \tag{2}$$

For each point, there are no feature points matching between every two frames, as the trajectory is only predicted by the points position in consecutive frames and the computed optical flow.

The improved dense trajectories approach samples and tracks feature points the same way as dense trajectories, but improve the dense trajectory by explicit camera motion estimation. The iDT approach utilizes human detector as a mask to remove feature matches on humans, the rest matches extracted from consecutive frames are applied to estimate the homography. Then iDT warps the second frame with the estimated homography and re-computes dense optical flow. HOF and MBH descriptors are computed on the warped optical flow. The homography and warped optical flow are estimated for every two frames.

For each trajectory, its shape is described by a sequence $(\Delta P_t, ..., \Delta P_{t+L-1})$ of displacement vectors $\Delta P_t = P_{t+1} - P_t = (x_{t+1} - x_t, y_{t+1} - y_t)$. If the maximal displacement of the trajectory vectors is less than a threshold, the trajectory is considered to camera motion and will be removed. Due to the spatial neighbor-hood information and temporal motion properties of dense sampled points, the iDT method matches the visual fixation of video representation very well. Hence, iDT can always outperform DT, STIP and dense cuboids. The video frame and iDT sampling trajectories are illustrated in the 1st column of Fig. 2 respectively.

2.2 Motion Boundary Based Sampling

Although the iDT is benefited from the camera motion compensation, the per-formance of action recognition still suffers from the large camera movements. The truth is that most of challenging action datasets contains lots of camera motion, for example, HMDB51 has 59.9 % videos including camera motion [5]. Hence, we should study how to improve dense trajectories approaches.

Among the approaches improving dense trajectories, Vig *et al.* [15] uses saliency-mapping algorithms to prune background features. Wang *et al.* [16] extracts video patches only from human body regions instead of the whole videos, it only works well on those action videos including simple background. Shi *et al.* [17] explores sampling over high density with local spatio-temporal features extracted from a *Local Part Model*, but its sampling process cost a lot of time. Jain *et al.* [18] decomposes visual motion into dominant and residual motions, and designs a new descriptor to capture additional information on the local motion patterns. Jiang *et al.* [19] clusters dense trajectories, and use the cluster centers as reference points so that the relationship between them can be mod-eled. Ballas *et al.* [20] does not use saliency information to sample features but to pool them. Simonyan *et al.* [21] propose a two-stream ConvNet architecture which incorporates spatial and temporal networks.

All approaches mentioned above cannot solve the problem of reducing irrel-evant trajectories caused by large camera movements. Therefore, we focus on discovering points need to be tracked. Unlike the Dense Trajectories based on Motion Boundary (DT-MB) sampling method [14], we implement improved Dense Trajectories sampling based on Motion Boundary (iDT-MB) to save mean-ingful points. We follow [14] to create the mask named Motion Boundary Image (MBI) by Otsus algorithm, and retain the regions with motion boundary fore-grounds. The magnitude of each position in MBI is calculated as

$$MBI(i,j) = Otsu(\max(\sqrt[2]{\mathcal{I}_u^u * \mathcal{I}_u^u + \mathcal{I}_v^u * \mathcal{I}_v^u}, \sqrt[2]{\mathcal{I}_u^v * \mathcal{I}_u^v + \mathcal{I}_v^v * \mathcal{I}_v^v})), \quad (3)$$

where $\mathcal{I}^u, \mathcal{I}^v$ denote images containing the u (horizontal) and v (vertical) com-ponents of optical flow, $\mathcal{I}^\omega = (\mathcal{I}^u, \mathcal{I}^v)$ denote the 2D flow image ($\omega = (u, v)$), e.g., $\mathcal{I}_v^u = \frac{d}{dv}\mathcal{I}^u$ is the v-derivative of the u component of optical flow.

The MBI is a middle result of iDT, so we do not need to add complexity. Note that iDT-MB can save fewer trajectories than DT-MB because of the dif-ference between iDT and DT. The 4th column of Fig. 2 exhibits an MBI and the trajectories from historical points by iDT-MB. The detailed comparisons of complexity and performance are given in Sect. 4.

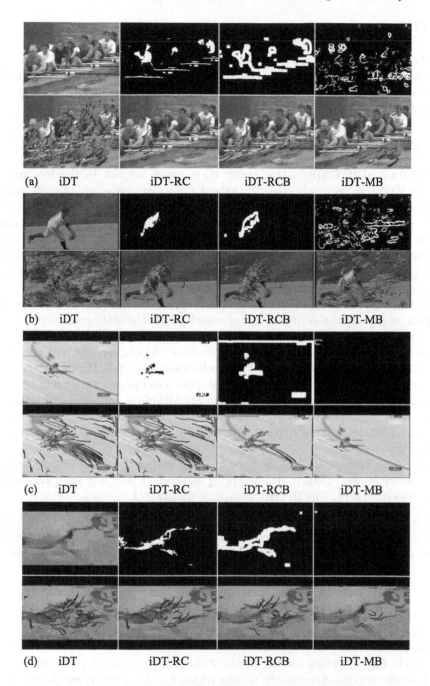

Fig. 2. Visualization of iDT, iDT-RC, iDT-RCB and iDT-MB sampling strategies for 4 actions. Compared to iDT, iDT-RCB is more robust to salient regions, in particular at shot boundaries (see 3rd column). iDT-RC can also handle salient regions, but it cannot capture the salient boundaries accurately. iDT-MB can reduce irregular motions, but it is not stable.

3   Our Approach

In this section, we introduce a Global Contrast based Salient Region Detection Algorithm in interest points sampling step of action recognition. We also explain why this method does not perform well in points sampling, then we present our new sampling strategy based on Salient Region Boundary in details.

3.1   Global Contrast Based Salient Region Sampling

A limitation of iDT-MB is that many trajectories are not in the foreground area, as iDT approach only considers salient points on dense grid of each frame, not the whole salient region of image. Meanwhile, the iDT-MB method is not stable since the motion boundaries are significantly influenced by the threading on gradient variation of optical flow. See the 4th column of Fig. 2(a), it shows the effective sampling example, but fail to capture the meaningful ones in Fig. 2(b) and Fig. 2(d) due to the unstable performance of MBI threading. Another worse result is given in the 4th column of Fig. 2(c), nothing is left in some cases.

To highlight the salient regions for action representation, we take into account the human detection algorithm. Unfortunately, even the state-of-the-art human detector cannot work well on action video datasets [13]. Furthermore, the salient region may be not in human body area but other objects, like the oars are more attractive in rowing action, see the 2nd column of Fig. 2(a). Hence, in order to find out the attractive salient regions, we propose a sampling method for applying the state-of-the-art salient region detection algorithm on improved dense trajectories. It is implemented by the iDT sampling based on salient Region-based Contrast (iDT-RC). This is partly inspired by Global Contrast based Salient Region Detection [22].

The main idea of iDT-RC is to automatic estimate salient object regions across every frame, enhances iDT sampling method without any prior knowledge of the video content. The iDT-RC sampling includes three steps:

(a) We first use a graph-based image segmentation method [22] to cut every frame into regions, and build the color histogram for each region. For a region r_k, we assign its saliency value by measuring its color contrast to other regions:

$$S(r_k) = \sum_{r_k \neq r_i} w(r_i) D_r(r_k, r_i), \tag{4}$$

where $w(r_i)$ is the weight of region defined by the number of pixels in r_i, and $D_r(r_k, r_i)$ is the color distance metric between regions r_k and r_i.

(b) We further incorporate spatial information by introducing a spatial weighting term in Eq. (4) to increase the effects of closer regions and decrease the effects of farther regions. Specifically, for any region r_k, the spatially weighted region contrast based saliency is

$$S(r_k) = \sum_{r_k \neq r_i} \exp(-\frac{D_s(r_k, r_i)}{\sigma_s^2}) w(r_i) D_r(r_k, r_i), \tag{5}$$

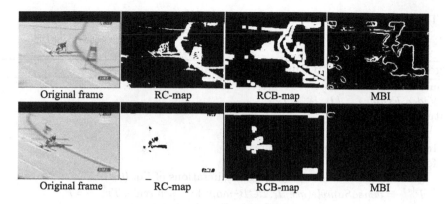

Fig. 3. RCB-map using Morphological Gradient is more robust than RC-map and MBI for salient region segmentation in action videos.

where $D_s(r_k, r_i)$ is the spatial distance between the two regions, σ_s controls the strength of spatial distance weighting.

(c) To save the useful interest points in every frame, we follow the RC-map [22] approach to get a segmentation mask, and apply the estimated salient mask to iDT sampling method. Those interest points sampled by the iDT-RC but not in global contrast based salient regions will be deleted.

3.2 Optimization with Salient Region Boundary

However, the iDT-RC combined iDT with salient regions straightly does not perform well in points sampling. Several reasons may account for this issue: Firstly, the Global Contrast based Salient Region Detection, which uses image contrast under the assumption that a salient object exists in an image, aim to model saliency for image pixels using color statistics of the input image. Hence, the RC-map approach does not always work perfectly, it will get some unexpected masks due to its global color contrast, see the 2nd column of Fig. 2(c). Secondly, sometimes the salient region generated by RC-map is too limited to track enough interest points for representing an action, the discriminative ones may be not saved, see the 2nd column of Fig. 2(a). Last but not the least, not all trajectories from salient regions may lead to valid trajectories, the performance of the codebook is influenced by the noise trajectory samples. Therefore, in order to handle the issue mentioned above, we propose another iDT sampling method based on salient Region Boundary named iDT-RCB. Unlike [15–18], our iDT-RCB sampling strategy constrains the sampled points on salient region boundaries in the sampling step. We use two iterations of the Morphological Gradient on RC-map to generate a robust RCB-map. The Morphological Gradient can be expressed as

$$RCBmap = morph_{grad}\,(RCmap) = dilate\,(RCmap) - erode\,(RCmap), \quad (6)$$

The proposed iDT-RCB sampling process is described below in detail.

Algorithm 1. iDT-RCB Sampling Procedure

Input:
  $Video\ Frames = \{I_1, I_2, ..., I_N\}$;
Output:
  $Valid\ Trajectories = Tr_1, Tr_2, ..., Tr_M$;
  1: Initialize the sampling parameters
  2: **for** $i = 1$ to N **do**
  3:    generate the $RCB\text{-}map$ by using two iterations of Eq. (6)
  4:    $P_j^{(1)} \Leftarrow denseSample(greyI_i, RCB\text{-}map)$ for each scale. $Tr_j^{(1)} \Leftarrow P_j^{(1)}$
  5:    $\omega_i \Leftarrow$ compute dense optical flow by Farnebäck algorithm
  6:    $matches_i \Leftarrow matchFromSurfandFlow(greyI_i, \omega_i, RCB\text{-}map)$
  7:    $H_i \Leftarrow findHomography(matches_{i-1}, matches_i, RANSAC)$
  8:    warp the second frame with H_i
  9:    $\omega_i' \Leftarrow$ re-compute dense optical flow by warped second frame
 10:    predict the motion of $P_j^{(t+1)}$ by using ω_i' and Eq. (2)
 11:    $Tr_j \Leftarrow \{P_j^{(1)}, P_j^{(2)}, ..., P_j^{(t)}, P_j^{(t+1)}, ..., P_j^{(L)}\}$
 12:    **if** Tr_j is valid $\&\&Tr_j$ is not camera motion **then**
 13:       $Valid\ Trajectories \Leftarrow Tr_j$
 14:    **end if**
 15: **end for**

---

where $P_j^{(1)}$ denote the first position of the j-th sampled point. Points from $P_j^{(1)}$ to $P_j^{(L)}$ of subsequent L frames are concatenated to form the j-th trajectory Tr_j.

We hold that those points on the salient region boundary are the most discriminative ones. This is indeed partly implied by MBH descriptor [23], Dmask including narrow strip surrounding the persons contour [24], and motion boundary contour system in neural dynamics of motion perception [25].

Although many action recognition approaches have been developed and inspiring progresses can achieve advanced levels, our iDT-RCB sampling method is more effective for large camera motion. It is very suitable for feature extraction in action videos, see the 3rd column of Fig. 3.

4   Experiments

In this section, we describe the details of extensive experiments to evaluate the usefulness of the proposed method in action recognition.

4.1   Datasets

We conduct experiments on two action datasets, namely HMDB51 [5] and UCF50 [6]. Some example frames are illustrated in Fig. 4. We summarize them and the experimental protocols as follows.

Fig. 4. Sample frames from HMDB51 (top) and UCF50 (bottom) datasets.

The HMDB51 dataset is collected from a variety of sources ranging from digitized movies to YouTube videos. There are 51 action categories and 6,766 video sequences in HMDB51. We follow the original protocol using three train-test splits and perform experiments on the original videos not the stabilized ones. We report average accuracy over the three splits as performance measure.

The UCF50 dataset has 50 action categories, consisting of real-world videos taken from YouTube. The actions range from general sports to daily life exercises. For all 50 categories, the videos are split into 25 groups. For each group, there are at least 4 action clips. In total, there are 6,618 video clips in UCF50. We apply the Leave-One-Group-Out Cross Validation for UCF50 dataset and report average accuracy over the twenty five splits.

4.2   Experimental Setup

In all the following experiments, we densely extract improved trajectories based on the code from Wang [13]. The iDT-MB is implemented by using the code from Peng [14]. The iDT-RC and iDT-RCB is partly implemented by using the code from Wang [13] and Cheng [22].

To recognize actions, we run five feature sampling methods at the same server cluster with multithreading, follow [13,26] to train a GMM codebook with $K = 256$ Gaussians based on 256,000 randomly sampled trajectories. The default parameters of descriptor in the spatio-temporal grid, the size of the volume and the tracked frames length are the same as [13]. Each trajectory is described by concatenating HOG, HOF, and MBH descriptors, which is a 396-dimensional vector. We reduce the descriptors dimension to 200 by performing PCA and Whitening. Then, each video is represented by a $2DK$ dimensional Fisher vector for each descriptor type. Finally, we apply Power L2-normalization to the Fisher vector. To combine different descriptor types, we concatenate their normalized Fisher vectors. In our experiments, we choose linear SVM as our classifier with the implementation of LIBSVM [27]. For multi-class classification, we use the one-vs-rest approach and select the class with the highest score.

We compare our approach to recent methods [13–21]. The mean run-time of sampling process and mean number of sampled trajectories are compared to iDT. The processing speed is reported in frames per second (fps), run at a single-core Intel Xeon X3430 (2.4 GHz) without multithreading.

4.3   Results and Analysis

Since the SaliencyCut [22] is an iterative process of using graphcut and GMM appearance mode, there may be a slight difference in generalized results. However, its performance still improves, as salient region boundaries are much clearer, see the 3rd column of Fig. 2. Compared to the baseline (iDT without HD [13]), we have 3 % improvement on HMDB51 and 1.5 % improvement on UCF50.

Table 1. Comparison of our results (HOG+HOF+MBH) to the state of art. We present our results for FV encoding without automatic human detection (HD).

HMDB51		UCF50	
Peng et al. [14]	49.2 %	Reddy et al. [6]	76.9 %
Jain et al. [18]	52.1 %	Shi et al. [17]	83.3 %
Simonyan et al. [21]	**59.4 %**	Wang et al. [28]	85.7 %
iDT without HD [13]	55.9 %	iDT without HD [13]	90.5 %
iDT with HD [13]	57.2 %	iDT with HD [13]	91.2 %
iDT-MB	53.3 %	iDT-MB	88.4 %
iDT-RC	55.7 %	iDT-RC	90.8 %
iDT-RCB	58.9 %	iDT-RCB	**92.0 %**

We use the bounding box provided from [13] in iDT sampling with HD. As the human detector does not always work perfectly, it will miss humans due to pose or viewpoint changes. Table 1 reports action recognition average accuracy compared to other dense trajectories approaches. Our iDT-RCB sampling method achieves the best result on UCF50. The result on HMDB51 is slightly decreased than [21], which have used the trained deep Convolutional Networks.

We evaluate the average number of trajectories per video clip and fps within 10 videos randomly selected from each dataset, and the run-time is obtained. Table 2 illustrates the minor computational cost of iDT-RCB. Fewer trajectories also can lead to faster speed in the feature encoding process.

Table 2. Comparison of sampled trajectories number and features extraction speed to iDT [13]. Note that we only randomly select 10 videos from each dataset.

Sampling strategy	HMDB51		UCF50	
	Trajectories/clip	fps	Trajectories/clip	fps
iDT without HD [13]	489,865	3.59	2,383,147	3.90
iDT with HD [13]	492,456	3.60	2,452,230	3.97
iDT-MB	164,643	3.75	911,280	4.36
iDT-RC	263,670	2.81	1,329,635	3.25
iDT-RCB	380,511	2.75	1,737,075	3.16

5 Conclusion

This paper proposes a novel dense sampling approach named iDT-RC for improved dense trajectories. We first implement iDT sampling based on MBI, and applies a salient region contrast based segmentation method in interest points sampling step. To overcome the flaws of salient region contrast based method in action recognition, we apply morphological gradient to RC-map for generating more robust salient mask named RCB-map. The improved sampling method named iDT-RCB constrains sampled points on the salient region boundary which can improve the performance with minor computational cost. The comparisons of the sampling strategies demonstrate that salient region boundary information is more effective. Finally, our method improves the performance of current action recognition systems on two challenging datasets which represents a good compromise between speed and accuracy.

Acknowledgement. The research was supported by the National Nature Science Foundation of China (61231015, 61170023, 61367002), the National High Technology Research and Development Program of China (863 Program) (2015AA016306, 2013AA014602), the Internet of Things Development Funding Project of Ministry of industry in 2013(25), the Technology Research Program of Ministry of Public Security (2014JSYJA016), the Major Science and Technology Innovation Plan of Hubei Province (2013AAA020), the Nature Science Foundation of Hubei Province (2014CFB712).

References

1. Schuldt, C., Laptev, I., Caputo, B.: Recognizing human actions: a local SVM approach. In: ICPR 2004, vol. 3, pp. 32–36 (2004)
2. Gorelick, L., Blank, M., Shechtman, E., Irani, M., Basri, R.: Actions as space-time shapes. Pattern Anal. Mach. Intell. **29**(12), 2247–2253 (2007)
3. Laptev, I., Marszałek, M., Schmid, C., Rozenfeld, B.: Learning realistic human actions from movies. In: CVPR 2008, pp. 1–8 (2008)
4. Liu, J.G., Luo, J.B., Shah, M.: Recognizing realistic actions from videos in the wild. In: CVPR 2009, pp. 1996–2003 (2009)
5. Kuehne, H., Jhuang, H., Garrote, E., Poggio, T., Serre, T.: HMDB: a large video data-base for human motion recognition. In: ICCV 2011, pp. 2556–2563 (2011)
6. Reddy, K., Shah, M.: Recognizing 50 human action categories of web videos. Mach. Vis. Appl. **24**(5), 971–981 (2013)
7. Laptev, I.: On space-time interest points. Int. J. Comput. Vis. **64**(2), 107–203 (2005)
8. Dollar, P., Rabaud, V., Cottrell, G., Belongie, S.: Behavior recognition via sparse spatio-temporal features. In: PETS 2005, pp. 65–72 (2005)
9. Willems, G., Tuytelaars, T., Van Gool, L.: An efficient dense and scale-invariant spatio-temporal interest point detector. In: Forsyth, D., Torr, P., Zisserman, A. (eds.) ECCV 2008, Part II. LNCS, vol. 5303, pp. 650–663. Springer, Heidelberg (2008)
10. Kläser, A., Marszałek, M., Schmid, C.: A spatio-temporal descriptor based on 3D-gradients. In: BMVC 2008 (2008)

11. Sun, J., Wu, X., Yan, S.C., Cheong, L.F., Chua, T.S., Li., J.T.: Hierarchical spatio-temporal context modeling for action recognition. In: CVPR 2009, pp. 2004–2011 (2009)
12. Wang, H., Kläser, A., Schmid, C., Liu, C.L.: Dense trajectories and motion boundary descriptors for action recognition. Int. J. Comput. Vis. **103**(1), 60–79 (2013)
13. Wang, H., Schmid, C.: Action recognition with improved trajectories. In: ICCV 2013, pp. 3551–3558 (2013)
14. Peng, X.J., Qiao, Y., Peng, Q.: Motion boundary based sampling and 3D co-occurrence descriptors for action recognition. Image Vis. Comput. **32**(9), 616–628 (2014)
15. Vig, E., Dorr, M., Cox, D.: Space-variant descriptor sampling for action recognition based on saliency and eye movements. In: Fitzgibbon, A., Lazebnik, S., Perona, P., Sato, Y., Schmid, C. (eds.) ECCV 2012, Part VII. LNCS, vol. 7578, pp. 84–97. Springer, Heidelberg (2012)
16. Wang, B., Liu, Y., Xiao, W.H., Xiong, Z.H., Wang, W., Zhang, M.J.: Human action recognition with optimized video densely sampling. In: ICME 2013, pp. 1–6 (2013)
17. Shi, F., Petriu, E., Laganiere, R.: Sampling strategies for real-time action recognition. In: CVPR 2013, pp. 2595–2602 (2013)
18. Jain, M., Jegou, H., Bouthemy, P.: Better exploiting motion for better action recognition. In: CVPR 2013, pp. 2555–2562 (2013)
19. Jiang, Y.-G., Dai, Q., Xue, X., Liu, W., Ngo, C.-W.: Trajectory-based modeling of human actions with motion reference points. In: Fitzgibbon, A., Lazebnik, S., Perona, P., Sato, Y., Schmid, C. (eds.) ECCV 2012, Part V. LNCS, vol. 7576, pp. 425–438. Springer, Heidelberg (2012)
20. Ballas, N., Yang, Y., Lan, Z.Z., Delezoide, B., Preteux, F., Hauptmann, A.: Space-time robust video representation for action recognition. In: ICCV 2013, pp. 2704–2711 (2013)
21. Simonyan, K., Zisserman, A.: Two-stream convolutional networks for action recognition in videos. In: NIPS 2014 (2014)
22. Cheng, M., Mitra, N.J., Huang, X., Torr, P.H.S., Hu, S.: Global contrast based salient region detection. IEEE Trans. Pattern Anal. Mach. Intell. **37**(3), 569–582 (2015)
23. Dalal, N., Triggs, B., Schmid, C.: Human detection using oriented histograms of flow and appearance. In: Leonardis, A., Bischof, H., Pinz, A. (eds.) ECCV 2006. LNCS, vol. 3952, pp. 428–441. Springer, Heidelberg (2006)
24. Jhuang, H., Gall, J., Zuffi, S., Schmid, C., Black, M.J.: Towards understanding action recognition. In: ICCV 2013, pp. 3192–3199 (2013)
25. Grossberg, S., Mingolla, E.: Neural dynamics of motion perception: direction fields, apertures, and resonant grouping. Percept. Psychophysics **53**(3), 243–278 (1993)
26. Peng, X., Zou, C., Qiao, Y., Peng, Q.: Action recognition with stacked fisher vectors. In: Fleet, D., Pajdla, T., Schiele, B., Tuytelaars, T. (eds.) ECCV 2014, Part V. LNCS, vol. 8693, pp. 581–595. Springer, Heidelberg (2014)
27. Chang, C.C., Lin, C.J.: Libsvm: a library for support vector machines. ACM Trans. Intell. Syst. Tech. **2**(3), 27 (2011)
28. Wang, L.M., Qiao, Y., Tang, X.O.: Mining motion atoms and phrases for complex action recognition. In: ICCV 2013, pp. 2680–2687 (2013)

Elastic Edge Boxes for Object Proposal on RGB-D Images

Jing Liu[1,2], Tongwei Ren[1,2(✉)], and Jia Bei[1,2]

[1] State Key Laboratory for Novel Software Technology, Nanjing University,
Nanjing, China
rentw@nju.edu.cn, {ljing12,beijia}@software.nju.edu.cn
[2] Software Institute, Nanjing University, Nanjing, China

Abstract. Object proposal is utilized as a fundamental preprocessing of various multimedia applications by detecting the candidate regions of objects in images. In this paper, we propose a novel object proposal method, named *elastic edge boxes*, integrating window scoring and grouping strategies and utilizing both color and depth cues in RGB-D images. We first efficiently generate the initial bounding boxes by edge boxes, and then adjust them by grouping the super-pixels within elastic range. In bounding boxes adjustment, the effectiveness of depth cue is explored as well as color cue to handle complex scenes and provide accurate box boundaries. To validate the performance, we construct a new RGB-D image dataset for object proposal with the largest size and balanced object number distribution. The experimental results show that our method can effectively and efficiently generate the bounding boxes with accurate locations and it outperforms the state-of-the-art methods considering both accuracy and efficiency.

Keywords: Elastic edge boxes · Object proposal · RGB-D image

1 Introduction

Object proposal aims to detect candidate regions possibly containing class-independent objects in an image [1], which is widely used as a fundamental of various multimedia applications, such as scene analysis [2], image annotation [3] and retrieval [4], object recognition [5] and matching [6], visual tracking [7], and social media mining [8].

Typically served as a preprocessing procedure, object proposal is usually needed to satisfy the following requirements: First, object proposal should cover all or most objects in images with a limited number of candidate regions. In this way, the candidate regions can provide a majority of image content and reduce the further processing cost. Second, the candidate regions, usually bounding boxes, should accurately cover the objects. As shown in [9], improving intersection over union (IoU) of candidate regions and objects is as important as increasing recall of objects in several applications, such as object detection. Finally, the

© Springer International Publishing Switzerland 2016
Q. Tian et al. (Eds.): MMM 2016, Part I, LNCS 9516, pp. 199–211, 2016.
DOI: 10.1007/978-3-319-27671-7_17

processing of object proposal should be efficient, which will benefit its usage in realtime or large-scale applications.

The existing methods address object proposal problem with two typical strategies: window scoring and grouping [9]. The methods using window scoring strategy typically score quantities of candidate bounding boxes according to some features which measure the likelihood of a box containing an object. They usually have high efficiency but fail to detect most of objects under high IoU. And the methods using grouping strategy generally initialize a number of segments and then merge the similar segments to produce final results. They can obtain the accurate bounding boxes especially under high IoU, but they are usually time consuming.

Generally speaking, the current object proposal methods suffer two problems. In one aspect, both window scoring strategy and grouping strategy have their drawbacks either in accuracy or efficiency, which limits their usage in many applications. An interesting idea is to combine these two strategies together to obtain both high accuracy and efficiency, but the related research is still in embryonic stage [10]. In the other aspect, depth information has been proved to be effective in discriminating objects from complex scenes [11], but most current methods merely focus on color cue and ignore depth in object proposal.

In this paper, we propose a novel object proposal method, named *elastic edge boxes*, by integrating window scoring and grouping strategies and exploring both color and depth cues in RGB-D images. Figure 1 shows an overview of the proposed method. To each RGB-D image (Fig. 1(a)), we first utilize window scoring strategy to identify the potential object locations with boxes according to edge cue (Fig. 1(b)). Then, we represent the RGB-D image with super-pixels and select the undetermined super-pixels for each box (Fig. 1(c)). Finally, we adjust the boundary of each box by applying grouping strategy on the undetermined super-pixels

Fig. 1. An overview of the proposed method. (a) RGB-D image. (b) Initial bounding boxes (orange boxes) by window scoring strategy and ground truths (red boxes). (c) Super-pixel representation. (d) Box boundary adjustment by grouping strategy. (e) Final bounding boxes (green and orange boxes) and ground truths (red boxes) (Color figure online).

(Fig. 1(d)) and generate the final object proposal result (Fig. 1(e)). To the best of our knowledge, it is the first object proposal method integrating window scoring and grouping strategies for RGB-D images. To validate the performance of the proposed method, we construct an RGB-D image dataset, named *NJU1500*, on the base of stereo objectness dataset. The experiments show that our method can generate the bounding boxes with accurate locations under both low and high accuracy, and it outperforms the existing methods considering both accuracy and efficiency.

Our major contribution can be summarized as:

- We propose a novel object proposal method for RGB-D images, which can obviously improve the recall of objects with high IoU boxes.
- We provide a new RGB-D image dataset for object proposal, which can be used as a benchmark for the future research.

The rest of the paper is organized as follows. Section 2 provides a brief review of the related work. Section 3 describes the details of the proposed method. Section 4 shows the performance evaluation of the proposed method. Finally, the paper is concluded in Sect. 5.

2 Related Work

The strategies of the existing object proposal methods can be roughly classified into three categories: window scoring, grouping and integration of them.

Window Scoring. Window scoring based methods generate a pool of candidate windows and score the windows by their probabilities of containing an object with objectness measurements. Alexe *et al.* [1] first propose an objectness measurement based on a variety of appearance and geometry properties. Cheng *et al.* [12] utilize binarized normed gradient by training a linear classifier over edge features. Zitnick *et al.* [13] use edge cue to guide window refinement, which can be specially optimized for different IoU thresholds. Xu *et al.* [11] explore the effectiveness of depth cue in handling complex scenes. Overall, window scoring based methods can efficiently generate the bounding boxes as proposal results, but their performance under high IoU is usually limited.

Grouping. Grouping based methods over-segment the images into tiny parts, such as super-pixels, and merge the segments to generate the candidates of objects. Carreira *et al.* [14] use constrained parametric mincuts in merging by several different seeds and multiple features, and Humayun *et al.* [15] improve it by applying multiple graph cut segmentations and using edge detectors. Uijlings *et al.* [16] propose selective search method to merge super-pixels greedily. Rantalankila *et al.* [17] propose a similar merging strategy with different features in similarity measurement. Xiao *et al.* [18] extend selective search by specializing merging in high-complexity scenarios, and Wang *et al.* [19] improve it with multi-branch hierarchical segmentation. Manen *et al.* [20] use randomised super-pixel connectivity graph during merging. Long *et al.* [21] utilize bottom-up merging to generate initial object candidates, and train a supervised descent model to greedily adjust the boxes. Arbelaez *et al.* [22] perform hierarchical segmentation and multiscale combinatorial grouping. Krähenbühl *et al.* [23] judiciously place object-like seeds and identify critical level sets in geodesic distance transforms as object proposal results. Overall, grouping based methods can generate accurate bounding boxes as well as object boundaries, especially under high IoU, but they are usually inefficient due to bottom-up merging.

Integration of Window Scoring and Grouping. It is interesting to integrate window scoring and grouping strategies together, for example, utilizing the object proposal result by window scoring strategy as the input of further grouping. Chen *et al.* [10] first apply this strategy in object proposal to achieve accurate bounding boxes while retaining high efficiency, but their method only focuses on RGB images and completely ignores depth cue.

3   Elastic Edge Boxes

3.1   Initial Bounding Boxes Generation

We first generate the initial bounding boxes by window scoring strategy. In the proposed approach, we utilize edge boxes method [13], which can efficiently detect the approximate locations of most objects by exploiting edge cue.

Edge boxes first generate candidate objects utilizing sliding window approach, and then scoring the boxes according to the number of contours completely inside each box which is highly indicative of the possibility of the box including an object. Score of box b_k is defined using:

$$score(b_k) = \frac{\sum_i \rho_k(e_i)\hat{m}_i}{2(w_k + h_k)^\eta} - \frac{\sum_{p \in b^{ct}} m_p}{2(w_k^{ct} + h_k^{ct})^\eta}, \tag{1}$$

where w_k and h_k are width and height of the box b_k; b_k^{ct} is a box centered in b_k with size $w_k^{ct} \times h_k^{ct}$ which equal $w_k/2$ and $h_k/2$ respectively; $\eta = 1.5$ is a parameter to offset the bias of larger windows generally containing more edges; m_p represents the edge magnitude of each pixel and $\hat{m}_i$ is obtained by summing up edge magnitude of each pixel in the ith edge group e_i enclosed by box b_k; ρ_k equals zero if e_i overlaps b_k's boundaries. Finally, non-maximal suppression is performed for the boxes to decrease the candidate number.

Though its performance under high IoU is not satisfactory, edge boxes can achieve high recall under low IoU. It means that edge boxes can detect the approximate location of objects but cannot provide bounding boxes with high accuracy. Hence, we adjust the initial bounding boxes to provide more accurate object proposal results.

3.2   Elastic Range Determination

Based on the initial object proposal result, we further determine the elastic range for each bounding box. Obviously, too small elastic range will limit the adjustment and prevent from providing accurate bounding boxes, while too large elastic range may cause high computational cost and reduce the effect of initial proposal results. We represent images with super-pixels and utilize super-pixel as the basic operation unit in bounding boxes adjustment because super-pixel can well describe object boundaries, increase the robustness to depth map inaccuracy, and reduce the computational cost.

We represent an image as a set of super-pixels $S = \{s_1, \ldots, s_N\}$ which are generated by [24]. Given an initial bounding box b_k, we define $S_{in}^{b_k}$ as a set of super-pixels which are completely inside b_k (cyan ones in Fig. 2(c)), $S_{out}^{b_k}$ as a set of super-pixels which are completely outside b_k, and $S_e^{b_k}$ as a set of the rest super-pixels which are crossed by b_k (yellow ones in Fig. 2(c)). In our method, $S_e^{b_k}$ is used as elastic range.

For the number of super-pixels in $S_{in}^{b_k}$ and $S_{out}^{b_k}$ are usually unbalanced, we select a subset $\hat{S}_{out}^{b_k}$ of $S_{out}^{b_k}$ (blue ones in Fig. 2(c)) to avoid bias in bounding box adjustment. We sort the super-pixels in $S_{out}^{b_k}$ in ascending order according to their weights of minimum center distances to the super-pixels $S_{in}^{b_k}$. To each super-pixel s_i in $S_{out}^{b_k}$, its weight is calculated as:

$$\omega(s_i) = \arg \min_{s_j \in S_{in}^{b_k}} \left(dis(s_i, s_j) \right), \tag{2}$$

where $dis(,)$ denotes the distance between the centers of two super-pixels.

Then we select the super-pixels in $\hat{S}_{out}^{b_k}$ according to their weights, and the number of selected super-pixels is required similar to the number of super-pixels in $S_{in}^{b_k}$:

$$\frac{1}{\lambda} \mid S_{in}^{b_i} \mid \leq \mid \hat{S}_{out}^{b_i} \mid \leq \lambda \mid S_{in}^{b_i} \mid \tag{3}$$

where λ is the parameter which equals to 1.25 in our experiments.

3.3 Bounding Box Adjustment

As shown in Fig. 2(d), to each super-pixel in elastic range $S_e^{b_k}$, we calculate its similarities to all the super-pixels in $S_{in}^{b_k}$ and $\hat{S}_{out}^{b_k}$, and determine whether it should be included in the bounding box. Here, we utilize both color channel and depth channel of an RGB-D image, and define four decision parameters φ_{in}^c, φ_{in}^d, φ_{out}^c and φ_{out}^d as follows:

$$\varphi_{in}^c = \sum_{s_j \in S_{in}^{b_k}} sim^c(s_i, s_j), \quad \varphi_{in}^d = \sum_{s_j \in S_{in}^{b_k}} sim^d(s_i, s_j), \tag{4}$$

$$\varphi_{out}^c = \sum_{s_l \in \hat{S}_{out}^{b_k}} sim^c(s_i, s_l), \quad \varphi_{out}^d = \sum_{s_l \in \hat{S}_{out}^{b_k}} sim^d(s_i, s_l), \tag{5}$$

where $sim^c(,)$ denotes the average color similarity of two super-pixels in HSV space, and $sim^d(,)$ denotes the depth similarity of two super-pixels.

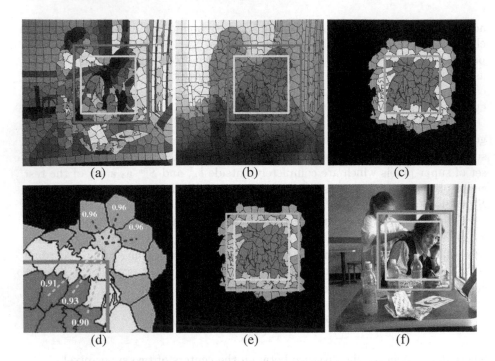

Fig. 2. Bounding box adjustment. (a) and (b) Color and depth channels of RGB-D image with ground truth (red box) and initial bounding box (orange box). (c) Elastic range (yellow super-pixels). (d) Details of decision in bounding box adjustment. (e) Adjusted bounding box (green box). (f) Final bounding boxes (green and orange boxes) and ground truth (red box) (Color figure online).

Based on the four parameters, we extend $S_{in}^{b_k}$ by adding the super-pixels satisfying the following requirements:

$$S_{in}^{b_k}{}^* = S_{in}^{b_k} \cup \left\{ s_i \in S_e^{b_i} \mid \varphi_{in}^c > \varphi_{out}^c \text{ and } \varphi_{in}^d > \varphi_{out}^d \right\}.$$

Based on the extended super-pixel set $S_{in}^{b_k}{}^*$, we generate a new bounding box $\tilde{b}_i$ (green box in Fig. 2(e)). By adjusting each bounding box, we can obtain the final object proposal result as follows:

$$B^* = B \cup \left\{ \tilde{b}_i \mid \forall b_i \in B \text{ and } b_i \neq \tilde{b}_i \right\}, \tag{6}$$

where B is the initial object proposal result.

4   Experiments

4.1   Dataset Construction

To validate the performance of our method, we construct a new RGB-D image dataset, named *NJU1500*, by extending the stereo objectness dataset [11].

Stereo objectness dataset provided in [11] is an RGB-D image dataset including 1,032 stereo images. As far as we know, it is the only RGB-D image dataset for object proposal. However, with the analysis of stereo objectness dataset, we find that it has some obvious drawbacks. We divide the images in stereo objectness dataset into six groups according to their object numbers, including 1, 2, 3, 4, 5, and 5+ (more than five), and the image numbers of the groups are 90, 417, 251, 133, 66, 75, respectively. It is easy to find that the distribution of object number among images is not balanced, which may lead to the bias in evaluation. Moreover, the numbers of objects contained in nearly half of the images are no more than 2, and the average number of objects per image is only 2.98, which makes object proposal task on it less challenging.

To overcome the drawbacks of stereo objectness dataset, we construct a new RGB-D image dataset named *NJU1500* based on it. To keep the balance of object number distribution among images and increase the average number of objects per image, we remove all the images with one object and a part of images with two objects, and supplement the images containing more than two objects. The selection of images with two objects only depends on their identifier in stereo objectness dataset, and the images of large identifiers are removed. 825 images are retained from the 1,032 images of stereo objectness dataset, and 675 images are supplemented. The supplemented images are collected from several 3D movies and videos, and the depth maps are calculated with Sun's optical flow method [25]. Similar to stereo objectness dataset construction, we annotate the ground truths of object locations according to PASCAL VOC2007 annotation guidelines. Five participants, including three males and two females, are invited to annotate the object bounding boxes for each supplementary image. The final constructed dataset includes five groups with 300 images per group, and the average number of objects per image increases from 2.98 to 4.22.

4.2   Performance Evaluation

We validate the performance of our method on *NJU1500* dataset. All the experiments are carried out on a computer with Intel i5 2.8 GHz CPU and 8 GB memory.

Figure 3 shows some examples of object proposal results generated by our method. The best bounding boxes to each ground truth within top 1,500 of each image are marked with green bounding boxes, and the IoU values of the bounding boxes to their corresponding ground truths are indicated with yellow numbers. We can find that almost all the objects are detected by our method with high IoU values, including obscure objects (such as the sword in the fourth image of the first row), small objects (such as the blue and red dustbins in the fifth image of the second row) and occluded objects (such as Papa Smurf in the third image of the third row).

We compare our method with the state-of-the-art methods, including binarized normed gradients (BING) [12], edge boxes (EB) [13], objectness (OBJ) [1], geodesic object proposal (GOP) [23], multiscale combinatorial grouping (MCG)

Fig. 3. Examples of object proposal results generated by our method (Color figure online).

[22], and selective search (SS) [25]. These methods of window scoring strategy or grouping strategy have excellent performance in object proposal [9]. In addition, we compare our method with two latest and somewhat similar methods, adaptive integration of depth and color (AIDC) [11] and multi-thresholding straddling expansion of edge boxes (M-EB) [10]. The former is also proposed for object proposal on RGB-D images, and the latter also utilizes integration proposal strategy for RGB images.

Accuracy. We validate the proposal accuracy with three criteria. Figure 4(a) and (b) show the recall vs. proposal number curves of all the methods when IoU equals 0.5 and 0.8, respectively. We can find that our method has similar proposal accuracy to the existing best methods when IoU equals 0.5, and it outperforms all the methods when IoU equals 0.8. It means that our method can handle object proposal requirements under both low and high IoU. Figure 4(c) shows the average recall (AR) vs. proposal number curve [9]. It is found that our method outperforms all the other methods except MCG, which is more than 10 times slower than our method as shown in Table 2. Figure 4(d) shows the recall vs. IoU curve. We can find that our method outperform the other methods when IoU is in range of [0.5, 0.8] and it is only worse than GOP, MCG and SS when IoU is larger than 0.8. However, all these three methods use grouping strategy and have low efficiency.

Table 1 provides more details of comparison results, in which "prop" denotes the proposal number, "0.5-DR" and "0.8-DR" denote the recall when IoU equals 0.5 and 0.8, and "AR" denotes average recall. It shows that our method is only slightly worse than EB in average recall when proposal number equals 500 but it has the best recall and average recall with different proposal numbers under all the other conditions.

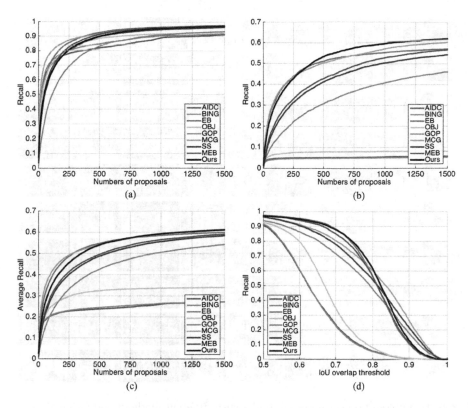

Fig. 4. Comparison of our method with the state-of-the-art methods. (a) and (b) Recall vs. proposal number curves when IoU equals 0.5 and 0.8. (c) Average recall vs. proposal number curve. (d) Recall vs. IoU curve with 1,500 bounding boxes.

Table 1. Comparison of our method and the state-of-the-art methods with different proposal numbers.

Method	Type	#prop=500			#prop=1000			#prop=1500		
		0.5-DR	0.8-DR	AR	0.5-DR	0.8-DR	AR	0.5-DR	0.8-DR	AR
AIDC	scoring	0.82	0.05	0.24	0.89	0.05	0.26	0.91	0.06	0.27
BING	scoring	0.85	0.05	0.24	0.90	0.05	0.26	0.92	0.05	0.27
EB	scoring	**0.92**	0.51	**0.55**	**0.96**	0.55	**0.59**	**0.98**	0.57	0.60
OBJ	scoring	0.90	0.08	0.32	0.92	0.08	0.34	0.93	0.08	0.34
GOP	grouping	0.83	0.30	0.43	0.91	0.41	0.51	0.93	0.46	0.54
MCG	grouping	**0.92**	0.50	0.54	0.95	0.57	**0.59**	0.97	0.60	**0.61**
SS	grouping	0.90	0.41	0.49	0.95	0.50	0.56	0.96	0.54	0.59
M-EB	integration	0.91	0.43	0.50	**0.96**	0.53	0.57	0.97	0.57	0.60
Ours	integration	**0.92**	**0.53**	0.54	**0.96**	**0.60**	**0.59**	**0.98**	**0.62**	**0.61**

Fig. 5. Examples of object proposal results with different methods. (a) Original image with ground truth. (b)-(j) Object proposal results of AIDC [11], BING [12], EB [13], OBJ [1], GOP [23], MCG [22], SS, M-EB [10] and our method (Color figure online).

Figure 5 shows some examples of object proposal results generated by different methods. In the examples, red boxes indicate the ground truths, green boxes indicate the bounding boxes generated by different methods, and blue boxes indicate the missed ground truths when IoU equals 0.8. Though the structure of some images is complex and some objects are inconspicuous, our method can detect almost all the objects with high IoU.

Speed. We also compare the efficiency of all the methods. Table 2 presents the running time of all methods. Though some methods require much less time than other methods processing an image, such as BING and AIDC, they are obviously worse than our method in proposal accuracy (Fig. 4). The three methods, which have better proposal accuracy than our method when IoU is larger than 0.8, are obviously worse than our method in efficiency. The most efficient method among them, that is SS, is about 10 % slower than our method, and MCG is even 10 times slower than our method.

Table 2. Comparison of our method and the state-of-the-art methods in running time.

Method	Type	Language	Time (s)
AIDC	window	C++	0.07
BING	window	C++	0.06
EB	window	C++ & Matlab	0.69
OBJ	window	C++ & Matlab	4.13
GOP	grouping	C++ & Matlab	7.25
MCG	grouping	C++ & Matlab	60.12
SS	grouping	C++ & Matlab	6.39
M-EB	integration	C++ & Matlab	0.99
Ours	integration	C++ & Matlab	5.78

5 Conclusions

In this paper, we propose an object proposal method for RGB-D images by integrating window scoring and grouping strategies. The method generates the initial bounding boxes by an efficient edge-based window scoring method, and adjusts the bounding boxes by grouping the super-pixels in elastic range, which improves proposal accuracy while retaining high efficiency. Moreover, the effectiveness of depth cue is explored as well as color cue, which benefits to handle the images with complex scenes. The experiments show that our method can effectively and efficiently generate the bounding boxes with high IoU, and it outperforms state-of-the-art object proposal methods considering both accuracy and efficiency.

Acknowledgments. This work is supported by the National Science Foundation of China (61321491, 61202320), Research Project of Excellent State Key Laboratory (61223003), National Undergraduate Innovation Project (G1410284074) and Collaborative Innovation Center of Novel Software Technology and Industrialization.

References

1. Alexe, B., Deselaers, T., Ferrari, V.: Measuring the objectness of image windows. IEEE Trans. Pattern Anal. Mach. Intell. **34**(11), 2189–2202 (2012)
2. Li, T., Chang, H., Wang, M., Ni, B., Hong, R., Yan, S.: Crowded scene analysis: a survey. IEEE Trans. Circ. Syst. Video Technol. **25**(3), 367–386 (2015)
3. Sang, J., Changsheng, X., Liu, J.: User-aware image tag refinement via ternary semantic analysis. IEEE Trans. Multimedia **14**(3–2), 883–895 (2012)
4. Xu, X., Geng, W., Ju, R., Yang, Y., Ren, T., Wu, G.: OBSIR: object-based stereo image retrieval. In: IEEE International Conference on Multimedia and Expo, pp. 1–6 (2014)
5. Bao, B.-K., Liu, G., Hong, R., Yan, S., Changsheng, X.: General subspace learning with corrupted training data via graph embedding. IEEE Trans. Image Process. **22**(11), 4380–4393 (2013)

6. Hong, C., Zhu, J., Yu, J., Cheng, J., Chen, X.: Realtime and robust object matching with a large number of templates. Multimedia Tools Appl., 1–22 (2014)

7. Ren, T., Qiu, Z., Liu, Y., Tong, Y., Bei, J.: Soft-assigned bag of features for object tracking. Multimedia Syst. **21**(2), 189–205 (2015)

8. Tang, J., Tao, D., Qi, G.-J., Huet, B.: Social media mining and knowledge discovery. Multimedia Syst. **20**(6), 633–634 (2014)

9. Hosang, J., Benenson, R., Dollár, P., Schiele, B.: What makes for effective detection proposals? arXiv preprint. arXiv:1502.05082 (2015)

10. Chen, X., Ma, H., Wang, X., Zhao, Z.: Improving object proposals with multi-thresholding straddling expansion. In: IEEE Conference on Computer Vision and Pattern Recognition (2015)

11. Xu, X., Ge, L., Ren, T., Wu, G.: Adaptive integration of depth and color for objectness estimation. In: IEEE International Conference on Multimedia and Expo (2015)

12. Cheng, M.-M., Zhang, Z., Lin, W.-Y., Torr, P.: BING: Binarized normed gradients for objectness estimation at 300fps. In: IEEE Conference on Computer Vision and Pattern Recognition, pp. 3286–3293 (2014)

13. Zitnick, C.L., Dollár, P.: Edge boxes: locating object proposals from edges. In: Fleet, D., Pajdla, T., Schiele, B., Tuytelaars, T. (eds.) ECCV 2014, Part V. LNCS, vol. 8693, pp. 391–405. Springer, Heidelberg (2014)

14. Carreira, J., Sminchisescu, C.: Constrained parametric min-cuts for automatic object segmentation. In: IEEE Conference on Computer Vision and Pattern Recognition, pp. 3241–3248 (2010)

15. Humayun, A., Li, F., Rehg, J.M.: Rigor: reusing inference in graph cuts for generating object regions. In: IEEE Conference on Computer Vision and Pattern Recognition, pp. 336–343 (2014)

16. Uijlings, J.R.R., van de Sande, K.E.A., Gevers, T., Smeulders, A.W.M.: Selective search for object recognition. Int. J. Comput. Vis. **104**(2), 154–171 (2013)

17. Rantalankila, P., Kannala, J., Rahtu, E.: Generating object segmentation proposals using global and local search. In: IEEE Conference on Computer Vision and Pattern Recognition, pp. 2417–2424 (2014)

18. Xiao, Y., Lu, C., Tsougenis, E., Lu, Y., Tang, C.-K.: Complexity-adaptive distance metric for object proposals generation. In: IEEE Conference on Computer Vision and Pattern Recognition, pp. 778–786 (2015)

19. Wang, C., Zhao, L., Liang, S., Zhang, L., Jia, J., Wei, Y.: Object proposal by multi-branch hierarchical segmentation. In: IEEE Conference on Computer Vision and Pattern Recognition (2015)

20. Manen, S., Guillaumin, M., Van Gool, L.: Prime object proposals with randomized prims algorithm. In: IEEE International Conference on Computer Vision, pp. 2536–2543 (2013)

21. Long, C., Wang, X., Hua, G., Yang, M., Lin, Y.: Accurate object detection with location relaxation and regionlets re-localization. In: Cremers, D., Reid, I., Saito, H., Yang, M.-H. (eds.) ACCV 2014. LNCS, vol. 9003, pp. 260–275. Springer, Heidelberg (2015)

22. Arbelaez, P., Pont-Tuset, J., Barron, J., Marques, F., Malik, J.: Multiscale combinatorial grouping. In: IEEE Conference on Computer Vision and Pattern Recognition, pp. 328–335 (2014)

23. Krähenbühl, P., Koltun, V.: Geodesic object proposals. In: Fleet, D., Pajdla, T., Schiele, B., Tuytelaars, T. (eds.) ECCV 2014, Part V. LNCS, vol. 8693, pp. 725–739. Springer, Heidelberg (2014)

24. Achanta, R., Shaji, A., Smith, K., Lucchi, A., Fua, P., Susstrunk, S.: Slic superpixels compared to state-of-the-art superpixel methods. IEEE Trans. Pattern Anal. Mach. Intell. **34**(11), 2274–2282 (2012)
25. Sun, D., Roth, S., Black, M.J.: Secrets of optical flow estimation and their principles. In: IEEE Conference on Computer Vision and Pattern Recognition, pp. 2432–2439 (2010)

Pairing Contour Fragments
for Object Recognition

Wei Zheng$^{(\boxtimes)}$, Qian Zhang, Zhixuan Li, and Junjun Xiong

Beijing Samsung Telecom R&D Center, Beijing 100028, China
{w0209.zheng, q0712.zhang,
zxl228.li, jjun.xiong}@samsung.com

Abstract. Contour fragments are adept to interpret the characteristics of object boundaries, but difficult to encode the information of object interior region. In this paper, inspired by the Gestalt principles that people can perceive object interior region by grouping similar and proximate fragments, we propose to pair contour fragments to encode more information of object interior region. To this end, we propose a pairing algorithm to generate Contour Fragment Pairs (CFPs). According to the proposed algorithm, the fragments of a valid CFP are required to be: co-occurrent over the training images, similar in shape, and proximate with each other. With a valid CFP, we can represent object shape using its fragments and object interior region using the region between its fragments. Finally, we design a boosting algorithm to select and assemble many CFPs into a classifier. The proposed classifier is competent for localizing objects with bounding boxes, delineating boundary and segmenting foreground. Moreover, the method possesses another merit that it only requires annotated bounding boxes as training data. Experiments on the public datasets show that the proposed approach achieves very promising performance.

1 Introduction

Contour fragments can accurately describe local shapes of objects, thus widely-used for object recognition [2, 9]. Comparing with region-based features, contour fragments can not only localize objects with bounding boxes but also delineate the object boundaries [3]. However, a weakness of contour fragments is that they have difficulty in encoding the information of object interior region. As shown in Fig. 1(b), we can answer where the horseback boundary is, but cannot figure out where the interior regions of the horse by a single contour fragment. Such weakness limits the application of contour fragments in segmentation.

To overcome the weakness, this paper proposes to pair contour fragments to capture more information of object interior region. The proposed approach is inspired by some psychological evidence from Gestalt studies [6]. The similarity and proximity principles show that people tend to group a pair of similar and proximate fragments together and perceive the region between them as object interior region (i.e., foreground). Figure 1(a) shows that two fragments that are near-by and similar in shape indicate the object foreground. People can easily infer the object foreground according

© Springer International Publishing Switzerland 2016
Q. Tian et al. (Eds.): MMM 2016, Part I, LNCS 9516, pp. 212–225, 2016.
DOI: 10.1007/978-3-319-27671-7_18

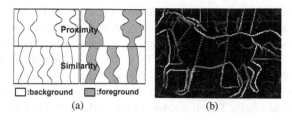

Fig. 1. (a) Illustration of Gestalt principles. (b) CFPs help people perceive interior region.

to such fragment pairs. Figure 1(b) shows a horse image marked with a few CFPs, from which we can recognize the interior regions of the legs, back and neck.

Guided by such psychology evidence, we propose to pair contour fragments together and use such pairs to encode the shape information of object boundary and the appearance information of object interior region. The two fragments of a valid CFP should conform to the following three requirements. Firstly, they should be co-occurrent over the training images. To measure the co-occurrence, we define an energy function for matching two fragments to images. Secondly, the two fragments should be similar in shape according to the similarity principle. Thirdly, the two fragments should be close to each other according to the proximity principle. The above requirements play important roles in filtering out vast amount of meaningless pairs and preserving the perceptually salient pairs. For a valid CFP, we treat its two fragments as object boundaries and the region between the two fragments as object interior region. Thus, we can extract the shape and appearance information of the object from the CFPs. Based on such information, we train the classifiers using the boosting algorithm. The proposed object classifiers are capable of delineating and segmenting object foreground besides localizing objects with bounding boxes. Extensive experiments show that CFPs convey both shape and appearance information of objects and the proposed classifiers achieve promising performance.

2 Related Works

Object recognition has been extensively studied and lots of novel approaches are proposed in recent years. Among these approaches, Convolutional Neural Networks (CNNs) [18, 19] show extraordinary performance on detection and segmentation. But this paper will not compare with CNNs. On one side, the CNNs need lots of segmented data for training while our approach does not need any. On the other side, the CNN does not utilize shape prior and our work may be complementary for CNNs. In this section, we review the related works from three aspects, namely, local features based on contour fragments, grouping algorithms and region descriptors.

Researchers have proposed lots of features based on the contour fragments. These approaches generate geometric curves [5] or directly use contour fragments [2, 9] as shape templates and match these templates to images. Lots of researchers propose to improve these shape templates by deformation or perturbation schemes [5]. To design reasonable deformation schemes, a recent work [16] proposes to learn a shape subspace from training data before matching and deform the contour fragments in the shape subspace during matching. Such deformable contour fragments, namely Active Contour Fragments (ACFs), show the effectiveness in the object detection and boundary delineation tasks. Therefore, we adopt ACFs as the shape features and contour fragments in this paper refer to ACFs. We will briefly review ACFs in Sect. 3.

A problem of contour fragment features is that they frequently mismatch to the cluttered background edges. Therefore, lots of researchers propose to bundle multiple contour fragments together to suppress the false matches. One way of bundling the contour fragments is to perceptually group them together. The basic assumption of perceptual grouping is that the contour fragments that belong to the same object should conform to some perceptual properties (e.g., continuation [16] and closure [7]). Guided by these properties, some works concatenate k Adjacent contour Segments (kAS) together to encode more complex shapes [2, 3] and another recent work chains multiple ACFs into a group, namely Active Contour Group (ACG) [16], to describe long and continuous boundaries. Different from kAS and ACG, our work bundles the contour fragments with different perceptual properties (i.e., similarity and proximity principles) and the CFPs describe not only the information of object boundaries but also the information of object interior region. The other way of bundling contour fragments is to assemble multiple contour fragments into a global shape model, such as particle filter grouping [7], active skeleton [1], Boundary Structure Segmentation (BoSS) [11]. Most of these approaches require one or more segmented shape masks to learn the global properties of the object shapes. Differently, our approach does not require any segmented shape mask in the training process.

CFPs can be also treated as region descriptors. We extract the appearance information from the region between two fragments of a CFP and utilize such information in a similar manner with that of region descriptors. But CFPs are different from the existing region descriptors [20] in two aspects: (1) CFPs are not predefined rectangular regions and they are automatically generated from pairs of contour fragments; (2) CFPs explicitly reflect the shapes of object structures and the deformation of object shapes can be leant from training data without manual design.

3   Contour Fragment Pairs

In this section, we firstly define a matching energy to measure the co-occurrence of two ACFs. Then, we present the details of the pairing algorithm. Finally, we give an approach to accelerate the CFPs' matching process.

3.1   Matching Energy

An ACF is a deformable extension of contour fragments. Its deformation characteristic is learnt from the training data and represented by a K-dimensional subspace:

$$\psi_i : \mathbf{t}_i = \bar{\mathbf{s}}_i + \sum_{k=1}^{K} \alpha_{i,k} \mathbf{v}_{i,k} \quad s.t. \ |\alpha_{i,k}| \leq 2\sqrt{\lambda_{i,k}}, \tag{1}$$

where ψ_i represent the ith ACF in the feature pool, $\bar{s}_i$ is the averaged fragment, $v_{i,k}$ and $\lambda_{i,k}$ are respectively the eigenvectors and eigenvalues returned by the Principle Components Analysis (PCA). As shown in Fig. 2(a), the ACFs encode the deformation in local regions. We can match ACFs to images over the 2-dimensional subspace spanned by the top two eigenvectors according to the matching energy:

$$E_{\psi_i}(\mathbf{E}) = \sum_{p=1}^{|\mathbf{t}_i^{\alpha_{i,1},\alpha_{i,2}}|} \min_{\mathbf{e}_p \in \mathbf{E}} \left\| \mathbf{t}_i^{\alpha_{i,1},\alpha_{i,2}}(p) - \mathbf{e}_p \right\|_D, \tag{2}$$

where $\mathbf{E}$ is the edge map of the image and $\|\bullet\|_D$ represents a certain distance measurement between two points. Similar to [16], we define $\|\bullet\|_D$ as

$$\left\| \mathbf{e}_i - \mathbf{e}_j \right\|_D = \beta \left((x_i - x_j)^2 + (y_i - y_j)^2 \right) + (\theta_i - \theta_j)^2, \tag{3}$$

where $\mathbf{e}_i$ represents a point of ACFs or image edges, $(x_i; y_i; \theta_i)$ represents the coordinates and normal orientation of $\mathbf{e}_i$. We let β equal to $25/A$ and A is the area of the object window. More details of ACFs can be found in [16].

Supposing that we have generated CFPs from ACFs, we design the energy function for matching CFPs to images. The proposed energy function measures not only the matching energy of two ACFs but also the relationship between two ACFs. To this end, we formulate the energy function of CFPs as

$$E_{\{\psi_i,\psi_j\}}(\mathbf{E}) = E_{\psi_i}(\mathbf{E}) + E_{\psi_j}(\mathbf{E}) + E_p(\psi_i, \psi_j), \tag{4}$$

where $E_{\psi_i}(\bullet)$ (or $E_{\psi_j}(\bullet)$) is the matching energy for the ACF ψ_i (or ψ_j) and $E_p(\psi_i, \psi_j)$ measures the relationship between two ACFs. We give an intuitive definition for the pairwise energy term

$$E_p(\psi_i, \psi_j) = \left| A(\mathbf{t}_i^{\alpha_{i,1},\alpha_{i,2}}, \mathbf{t}_j^{\alpha_{j,1},\alpha_{j,2}}) - A(\bar{s}_i, \bar{s}_j) \right|, \tag{5}$$

where $A(\bullet, \bullet)$ represents the area of the region between the two fragments that can be obtained by connecting the corresponding end points with line fragments (see Fig. 2 (c)). The pairwise energy requires that the area between the two fragments should keep stable during deformation. The reason is that all the training horses are normalized into the same scale and their torsos should be of similar size. Therefore, we constrain that the region between two ACFs should be of similar size.

Supposing the two fragments are of the same length, we can approximate the area difference as follows

$$\left| A(\mathbf{t}_i^{\alpha_{i,1},\alpha_{i,2}}, \mathbf{t}_j^{\alpha_{j,1},\alpha_{j,2}}) - A(\bar{s}_i, \bar{s}_j) \right| \approx \sum_{p=1}^{|\mathbf{t}_i^{\alpha_{i,1},\alpha_{i,2}}|} \left\| \left(\mathbf{t}_i^{\alpha_{i,1},\alpha_{i,2}}(p) - \mathbf{t}_j^{\alpha_{j,1},\alpha_{j,2}}(p) \right) - (\bar{s}_i(p) - \bar{s}_j(p)) \right\|_2, \tag{6}$$

where $\|\bullet\|_2$ represents the Euclidean distance. To make the pairwise term and unary terms equidimensional, we finally modify the pairwise terms as

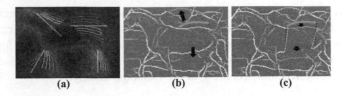

Fig. 2. (a) ACF examples. (b) Independently matched results. (c) Matched results by pair.

$$E(\psi_i, \psi_j) = \sum_{p=1}^{\left|\mathbf{t}_i^{\alpha_{i,1}, \alpha_{i,2}}\right|} \left\| \left(\mathbf{t}_i^{\alpha_{i,1}, \alpha_{i,2}}(p) - \mathbf{t}_j^{\alpha_{j,1}, \alpha_{j,2}}(p) \right) - \left(\bar{\mathbf{s}}_i(p) - \bar{\mathbf{s}}_j(p) \right) \right\|_S, \qquad (7)$$

where $\|\bullet\|_S$ measures the spatial distance in the same manner with Eq. (3)

$$\left\| \mathbf{e}_i - \mathbf{e}_j \right\|_S = \beta \left((x_i - x_j)^2 + (y_i - y_j)^2 \right). \qquad (8)$$

To be mentioned, Eq. (7) plays the same role as Eq. (6) and can be used to keep the area of the region between two ACFs stable. To match the CFPs to images, we should minimize the energy function of Eq. (4) over the parameters $\{\alpha_{i,1}, \alpha_{i,2}, \alpha_{j,1}, \alpha_{j,2}\}$. If we quantize the top two dimensions into N_1 bins and N_2 bins for each ACF, the complexity of minimizing Eq. (4) is $(N_1 N_2)^2$.

3.2   Pairing Algorithm

We generate a feature pool of ACFs for each object class as listed in Table 1. If we pair the ACFs with each other, the number of ACF pairs will be extremely large. Hereby, we propose a pairing algorithm to solve this problem. The proposed algorithm plays two important roles. On one hand, it greatly reduces the total number of CFPs and saves lots of time and storage. On the other hand, it helps to select the perceptually salient features among millions of candidates.

Table 1. Sizes of detection windows and feature pools for different object classes.

	Horses	Applelogos	Bottles	Giraffes	Mugs	Swans
Sizes	150 × 100	80 × 80	50 × 120	150 × 150	90 × 70	120 × 60
ACFs	11352	3107	3595	8234	4745	3320
CFPs	6221	409	1915	5757	1822	1287

For a valid CFP, both of the two ACFs in it should be capable of describing the object shapes. Meanwhile, they should be perceptually salient according to the Gestalt properties. We require that the two fragments of a CFP should have three properties:

(1) *Co-occurrence*: the two ACFs should frequently appear on the training samples at the same time. As shown in Fig. 3(a), the horseback (i.e., fragment 4) and the horse

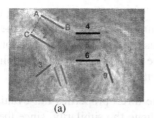

CFPs		Coocurr.	Similarity	Proximity
2	1	✓	✓	✓
2	3	✗	✗	✗
4	5	✗	✓	✓
4	6	✓	✓	✓
7	8	✓	✓	✓
8	9	✓	✓	✗

(a)   (b)

Fig. 3. (a) Some shape bases of the ACFs. (b) Properties of the pairs of ACFs.

belly (i.e., fragment 6) should appear simultaneously for all the horses. Thus, fragment 4 and 6 satisfy the co-occurrence property. On the contrary, the fragment 5 may correspond to an inner edge and it may only appear on some horses. Therefore, fragment 4 and 5 does not satisfy the co-occurrence property. We use the matching energy in Eq. (4) to measure the co-occurrence property for two ACFs as

$$coo(\psi_i, \psi_j) = \mathbf{I}\left(\sum_k E_{\{\psi_i, \psi_j\}}(\mathbf{E}_k) < T_c\right) \tag{9}$$

where $\mathbf{I}(\bullet)$ is the indicator function, k is the index of a training image, T_c is a threshold for judging the co-occurrence. The threshold T_c can be learnt from training data without manual design. To this end, we collect many background samples. We randomly sample some pairs of ACFs and match them to the objects and backgrounds. Our assumption is that the fragments of a valid CFP are likely to be co-occurrent on objects. Then, we select the T_c by minimizing the classification error using the naïve Bayesian probabilistic model.

(2) *Similarity*: this property shows that people tend to group two curves together if they have similar shapes (see the upper figure in Fig. 1(a)) and the regions between these two curves are perceived as foreground. Intuitively, we constrain that the two similar ACFs should have parallel shape bases and should be of the same length. The similarity property is measured by

$$sim(\psi_i, \psi_j) = \mathbf{I}\left(\mathbf{b}_{\psi_i} \parallel \mathbf{b}_{\psi_j}\right)\mathbf{I}\left(|\mathbf{b}_{\psi_i}| == |\mathbf{b}_{\psi_j}|\right) \tag{10}$$

where $\mathbf{b}_{\psi_i}$ is the shape basis (i.e., edgelet [14]) that generates ψ_i and | represents two fragments are parallel. Due to the constraint of similarity property, we suppose the two ACFs are of the same length in Eqs. (6) and (7). According to Eq. (10), Fragments 7 and 8 are similar while fragments 7 and 3 are not (see Fig. 3(a)).

(3) *Proximity*: this property states another psychological evidence that people tend to pair curves together when they are close with each other (see the lower figure in Fig. 1 (a)) and perceive the region between them as foreground. We formulate this property as

$$pro(\psi_i) = \arg\min_{\psi_j} A(\bar{\mathbf{s}}_{\psi_i}, \bar{\mathbf{s}}_{\psi_j}), \quad s.t., \vec{\mathbf{b}}_{\psi_i}(0)\mathbf{b}_{\psi_i}(1)\perp\vec{\mathbf{b}}_{\psi_i}(0)\mathbf{b}_{\psi_j}(0), \tag{11}$$

where $A(\bullet, \bullet)$ represent the area between the two fragments. According to Eq. (11), ψ_j is proximate to ψ_i if ψ_j can generate the minimum area with ψ_i over the feature pool. We hope that ψ_i and ψ_i are not collinear, thus we constrain the direction of the shape bases is perpendicular with the direction of the connections of the end points. As shown in Fig. 3(a), the line AB is perpendicular to the line AC.

For efficiency, we do not evaluate all the three properties but evaluate them in a cascaded manner. For each pair, we first evaluate the similarity, since the computation for this property is simplest. Only when the pair satisfies similarity property, we evaluate the co-occurrence property. At last, we evaluate the proximity property. If two ACFs conform all the above three properties, we treat them as a CFP.

3.3   Learning Subspace for CFPs

As discussed above, the computational complexity of matching CFPs is $(N_1 N_2)^2$. For accelerating, we propose to use a shape subspace to estimate the deformation of CFPs. We match a CFP to the kth training image and get a pair of best matched fragments, i.e., $\mathbf{t}_{i,k}^{\alpha_{i,1}*,\alpha_{i,2}*}$ and $\mathbf{t}_{j,k}^{\alpha_{j,1}*,\alpha_{j,2}*}$. We represent the two matched fragments in a $2\left(\left|\mathbf{t}_{i,k}^{\alpha_{i,1}*,\alpha_{i,2}*}\right| + \left|\mathbf{t}_{j,k}^{\alpha_{j,1}*,\alpha_{j,2}*}\right|\right)$ dimensional space by concatenating their coordinates, i.e.,

$$\{x_{i,k}^{\alpha_{i,1}*,\alpha_{i,2}*}(1), y_{i,k}^{\alpha_{i,1}*,\alpha_{i,2}*}(1), \ldots, x_{j,k}^{\alpha_{j,1}*,\alpha_{j,2}*}(1), y_{j,k}^{\alpha_{j,1}*,\alpha_{j,2}*}(1), \ldots\}. \tag{12}$$

Then, we learn the subspace using PCA over the concatenated shape space and represent the CFP as

$$\Omega_i^{\gamma_{i,1},\ldots,\gamma_{i,K}} = \overline{\mathbf{d}}_i + \sum_{k=1}^{K} \gamma_{i,k}\mathbf{u}_{i,k} \quad s.t. |\gamma_{i,k}| \leq 2\sqrt{\eta_{i,k}}, \tag{13}$$

where $\overline{\mathbf{d}}_i$ is the averaged fragments of the CFP, $\mathbf{u}_{i,k}$ and $\eta_{i,k}$ are respectively the eigenvectors and eigenvalues returned by the PCA. To be mentioned, the subspace does not only encode the deformation of the two ACFs but also encode the relationship between two ACFs. Fixing the coefficients $\gamma_{i,1}$ and $\gamma_{i,2}$ in Eq. (13), we can estimate the two ACFs and obtain a matching energy from Eq. (4). Thus, we can match the CFP over the 2-dimensional subspace instead of directly minimizing Eq. (4). The computation complexity of matching CFPs becomes $N_1 N_2$. In Fig. 4, we show the top two components of three CFPs. Obviously, the CFPs cannot only encode the local shapes of object boundaries but also reflect the information of object interior regions.

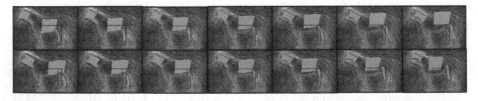

Fig. 4. Top two components of three CFPs. Blue curves are shape bases (i.e., edgelet), red curves are deformed fragment pairs and green regions indicate foreground regions (Color figure online).

4 Recognition Algorithm

In this section, we firstly utilize the CFPs as local features and learn weak classifiers based on the CFPs. Then, we give a naïve voting scheme to delineate boundaries and segment foreground based on the detected object windows.

4.1 Learning Weak Classifiers Based on CFPs

To extract appearance information of objects, we firstly deform the CFPs in the subspace generated by Eq. (13) and find a best match according to Eq. (4). The region corresponds to the best match may describe the object interior region. Therefore, we extract the appearance information in this region. We represent the feature vector of the ith CFP on the pth sample as follows

$$\mathbf{x}_{i,p}^{\gamma_{i,1}*,\gamma_{i,2}*}, s.t., \{\gamma_{i,1}*, \gamma_{i,2}*\} = \arg\min_{\gamma_{i,1},\gamma_{i,2}} E_{\{\psi_i,\psi_j\}}(\mathbf{E}_p), \tag{14}$$

where $\gamma_{i,1}*, \gamma_{i,2}*$ is the shape coefficients correspond to the best match on the sample and $\mathbf{x}_{i,p}^{\gamma_{i,1}*,\gamma_{i,2}*}$ is the feature vector based on the best match. Since the region between two ACFs is considered as object interior region, we can extract object appearance information from this region (e.g., red region in Fig. 5). For extracting the texture information, we build a 59-dimensional LBP histogram using the uniform patterns [13]. For extracting color information, we build a 20-dimensional histogram by quantizing H channel into 8 bins, S channel into 8 bins and V channel into 4 bins. For describing the statistical shape information, we extract 6-dimensional HOG features [16] in the bounding box of the CFPs (e.g., brown dashed bounding box in Fig. 5). For describing the object shape, we use the matching energy of Eq. (4) as feature response. Finally, $\mathbf{x}_{i,p}^{\gamma_{i,1}*,\gamma_{i,2}*}$ becomes an 85-dimensional feature vector by concatenating all the above histograms and matching energy.

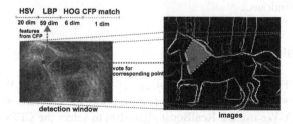

Fig. 5. Extracting features from CFP and voting for object boundary and foreground (Color figure online).

Based on multiple dimensional features, lots of linear classifiers can be used as the weak classifier. In this paper, we choose the logistic function as weak classifier since it provides non-linear prediction and is fast in training. The logistic function is:

$$f_\omega(\mathbf{x}_{i,p}^{\gamma_{i,1}*,\gamma_{i,2}*}) = \frac{1}{1 + e^{\omega \mathbf{x}_{i,p}^{\gamma_{i,1}*,\gamma_{i,2}*}}}, \tag{15}$$

where ω is the coefficients. For learning the parameter ω, we let $f_\omega(\bullet)$ equal to 1 for objects and $f_\omega(\bullet)$ equal to 0 for backgrounds. We can obtain ω by minimizing the cross-entropy error function over the training set.

4.2   Voting Boundaries and Foregrounds

We treat the CFPs as local features and use the probability returned by Eq. (15) as the CFP's response. Then, we use the RealBoost to select the discriminative CFPs from the feature pool and assemble these CFPs in a rejection-based cascaded classifier that is identical to [16].

The boundary delineation and foreground segmentation are only conducted on the object windows. An object is identified when a detection window is classified as positive at a false positive rate of 10^{-5}. We propose a naïve algorithm for voting the boundaries and foreground in the detected windows. Each CFP will cast a vote for each detected window. Supposing the point $\mathbf{m}$ is a point on the two ACFs or in the region between two ACFs, it votes for the corresponding point according to

$$P(\mathbf{m}') = W_{f_\mathbf{m}}^+ / (W_{f_\mathbf{m}}^+ + W_{f_\mathbf{m}}^-), \tag{16}$$

where $\mathbf{m}'$ is a point in the image that is the corresponding to $\mathbf{m}$, $f_\mathbf{m}$ is the feature response of a CFP (returned by Eq. (15)) that is corresponding to the matching point $\mathbf{m}$, $W_{f_\mathbf{m}}^+$ (or $W_{f_\mathbf{m}}^+$) is the distribution of positive (or negative) samples when the feature response equals to $f_\mathbf{m}$. As shown in Fig. 5, the point $\mathbf{m}$ will vote for the object boundaries if it is on the green boundary. Likewise, the point $\mathbf{m}$ will vote for the object foreground if it is in the red region. To be mentioned, the green curves and red region can be automatically generated during off-line training. The final results of boundary delineation and foreground segmentation are obtained by accumulating the votes from all the detected windows.

5   Experiments

We utilize the CFPs for recognizing objects of different classes. We use different detection windows and generate different feature pools for different object classes (as shown in Table 1). We use the RealBoost algorithm to select the CFPs as discussed in Sect. 4.1. Figure 6 shows the top two selected CFPs for each object class. Intuitively, these CFPs do not only encode the object shapes but also indicate the interior regions of objects.

In the following, we give both qualitative and quantitative evaluation for object detection, boundary delineation and foreground segmentation on two widely-used public datasets, namely Weizmann Horses dataset [8] and ETHZ shape dataset [2].

Fig. 6. Top two selected CFPs for each object class. Blue curves are averaged fragments. Green and red curves show the top component of each CFP (Color figure online).

5.1 Experiments on Weizmann Horses Dataset

Weizmann Horses dataset contains horses of near-side views in real scenes under varying scales and poses. We use the training-testing split as used in [4, 9].

We evaluate the object detection performance using the Average Precision (AP) under PASCAL IoU 50 % criterion. In Fig. 7(a), we plot the recall rate versus the False Positive rate Per Image (FPPI). We compare the proposed approach with six other shape-only features, namely, *contour fragment & Berkeley* [8], *HOG* [16], *edgelet* [14], *strip* [15], *ACF* [16] and *ACG* [16]. For a fair comparison, we implement all these features by ourselves and train the classifiers based on these features in the same boosting framework as proposed in Sect. 4. Apparently, *CFP* outperforms all the other approaches. Comparing to these features, *CFP* describes not only the object shapes but also encodes the object appearance. Such appearance information (e.g., color and texture) of the horses may be very useful for discriminating the horses from the grasses or sands. For verifying our inference, we only use the minimum energy of Eq. (4) as the feature response for a CFP and train the classifiers using such feature responses (namely, *CFP match only*). It can be seen that HSV, LBP and HOG histograms improve the AP rate about 2.4 %. The proposed approach is comparable to the state-of-the-art classifier [4] whose AP rate is 95.2 % on this dataset.

To evaluate the boundary localization performance, we adopt the coverage-against-precision curve suggested by [2]. In Fig. 7(b), we compare the proposed approach with other approaches, namely, *edgelet* [14], *strip* [15], *ACF* [16], *ACG* [16], *bounding box*, *GrabCut* [8] and *particle filter* [7]. All these methods are evaluated in

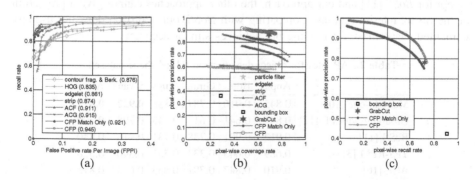

Fig. 7. Performance evaluation on Weizmann Horses dataset. (a) Detection. (b) Boundary delineation. (c) Foreground segmentation.

the correctly detected images. For *edgelet, strip, ACF and ACG*, we can vote for the boundary according to the same voting algorithm in Sect. 4.1. For *bounding box*, we directly use the boundary of the detected bounding boxes to estimate the object boundary. For *GraphCut*, we use the detected bounding boxes as input of the GraphCut algorithm [8] and use the boundary of the foreground returned by GrabCut to estimate the object boundary. For *particle filter*, we refer to the reported result [17]. We can see that *CFP match only* and *ACG* both perform better than *ACF*. Comparing *CFP match only* and *CFP,* we can see the appearance information improves the accuracy by about 8 % for boundary delineation. Obviously, *CFP* can give a much more precise interpretation for object boundary than *bounding box*. Moreover, *CFP* achieves slightly better accuracy than *GrabCut*. One possible reason is that *CFP* learns the class-specific information from training data, while *GrabCut* is a generic segmentation algorithm without the class-specific information.

We treat the foreground segmentation as a pixel-wise classification problem, and plot the pixel-wise precision versus recall curves for the correctly detected images. Since *ACF* and *ACG* cannot capture object boundaries, we compare the propose approach with three approaches, namely *bounding box, GrabCut* and *CFP match only*. Comparing with *CFP match only,* we can see that the appearance information improves the segmentation accuracy by 6 %. The reason may be that the appearance information is discriminative for classifying foreground and background, thus it may suppress many false matches on background. As a result, *CFP* returns more accurate foreground and boundary.

5.2   Experiments on ETHZ Shape Dataset

For comprehensively evaluating on the effectiveness of the proposed approach, we also give quantitative evaluations on the widely-used ETHZ shape dataset. We follow the same evaluation method suggested by [2].

We compare the proposed approach with other recent works and list the AP rates in Table 2 for evaluating the detection performance. Comparing with the shape-only features (e.g., *ACF* and *ACG*), *CFP* improves about 5 %. Table 2 shows that the proposed approach achieves comparable performance with the state-of-the-art methods. Except for *BoSS* [11] and our approach, the other approaches cannot give a prediction for the object interior regions. Comparing with *BoSS*, our approach does not use any segmented shape mask for training but achieves similar accuracy.

Table 2. AP rates for object detection on ETHZ shape dataset.

	Apple.	Bott.	Giraf.	Mugs	Swans	Avg.
Many-to-one [10]	0.845	0.916	0.787	0.888	**0.922**	**0.872**
Grouping with PF [17]	0.844	0.641	0.617	0.643	0.798	0.709
BoSS [11]	0.912	0.901	0.738	0.731	0.918	0.840
TPS-RPM [3]	0.689	0.643	0.333	0.585	0.390	0.528
ACF [16]	0.910	0.847	0.791	0.853	0.674	0.815
ACG [16]	**0.920**	0.864	0.782	0.856	0.653	0.815
CFP	0.909	0.845	0.815	**0.894**	0.869	0.866

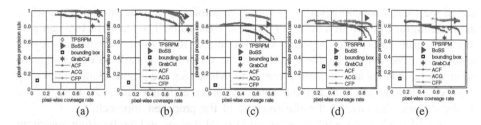

Fig. 8. Boundary delineation performance on ETHZ shape dataset. (a) Applelogos. (b) Bottles. (c) Giraffes. (d) Mugs. (e) Swans.

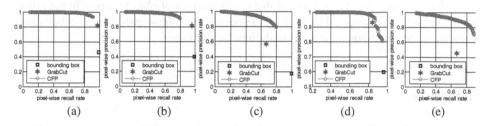

Fig. 9. Foreground segmentation performance on ETHZ shape dataset. (a) Applelogos. (b) Bottles. (c) Giraffes. (d) Mugs. (e) Swans.

Fig. 10. Some recognition results returned by our approach. Red points indicate object foregrounds. Bright color means higher confidence and vice versa (Color figure online).

We evaluate the pixel-wise performance for boundary delineation and foreground segmentation in Figs. 8 and 9 respectively. As discussed in Sect. 5.1, we only evaluate the correctly detected images. Comparing to other shape-features, *CFP* gives improvements for boundary delineation especially on the natural object classes. The boundary delineation is generally comparable to the state-of-the-art methods (e.g., *TPSRPM* [3] and *BoSS* [11]). Comparing to *bounding box* and *GrabCut* [8], *CFP* gives substantially better accuracy for both boundary delineation and foreground segmentation. We show some detection results in Fig. 10.

6   Conclusion and Future Works

This paper proposes a novel approach to pair contour fragments (i.e., ACFs) to encode both shape and appearance information of objects. To this end, we design a new matching energy to measure the co-occurrence of ACF pairs. Furthermore, we make use of the Gestalt psychological properties (i.e., similarity and the proximity) to select the meaningful CFPs from millions of possible ACF pairs. The learned CFPs can give more precise shape and appearance information than those rectangular region descriptors. The object classifiers based on the CFPs are versatile to detect objects, delineate boundaries and segment foreground. Moreover, the proposed algorithm only requires bounding boxes as training data. Extensive experimental evaluations on the public datasets demonstrate the effectiveness of the proposed approach.

The CFPs only encode the information in local regions and the boosting algorithm uses them independently. Therefore, the proposed algorithm cannot make use of the global properties of object shape. We may handle this problem from two aspects. On one hand, we may group more contour fragments according to some global constraints (e.g., shape closure) to encode more global characteristics of object shape. On the other hand, we may model the CFPs with graph-based algorithms instead of the boosting framework. We will pursue these possibilities in the future.

References

1. Bai, X., Wang, X., Latecki, L.J., Liu, W., Tu, Z.: Active skeleton for non-rigid object detection. In: ICCV (2009)
2. Ferrari, V., Fevrier, L., Jurie, F., Schmid, C.: Groups of adjacent contour segments for object detection. In: PAMI (2008)
3. Ferrari, V., Jurie, F., Schmid, C.: From images to shape models for object detection. In: IJCV (2010)
4. Ferrari, V., Tuytelaars, T., Van Gool, L.: Object detection by contour segment networks. In: Leonardis, A., Bischof, H., Pinz, A. (eds.) ECCV 2006. LNCS, vol. 3953, pp. 14–28. Springer, Heidelberg (2006)
5. Hu, W., Wu, Y.N., Zhu, S.C.: Image representation by active curves. In: ICCV (2011)
6. Koffka, K.: Principles of Gestalt Psychology (1935)
7. Levinshtein, A., Sminchisescu, C., Dickinson, S.: Optimal contour closure by superpixel grouping. In: ECCV (2012)
8. Rother, C., Kolmogorov, V., Blake, A.: GrabCut: interactive foreground extraction using iterated graph cuts. In: ACM Transactions on Graphics (2004)
9. Shotton, J., Blake, A., Cipolla, R.: Multi-scale categorical object recognition using contour fragments. In: PAMI (2008)
10. Srinivasan, P., Zhu, Q., Shi, J.: Many-to-one contour matching for describing and discriminating object shape. In: CVPR (2010)
11. Toshev, A., Taskar, B., Daniilidis, K.: Shape-based object detection via boundary structure segmentation. In: IJCV (2012)
12. Tuzel, O., Porikli, F., Meer, P.: Pedestrian detection via classification on riemannian manifolds. In: PAMI (2008)
13. Ojala, T., Pietikainen, M., Maenpaa, T.: Multiresolution gray-scale and rotation invariant texture classification with local binary patterns. In: PAMI (2002)
14. Wu, B., Nevatia, R.: Detection and tracking of multiple, partially occluded humans by bayesian combination of edgelet based part detector. In: IJCV (2007)
15. Zheng, W., Liang, L.: Fast car detection using image strip features. In: CVPR (2009)

16. Zheng, W., Song, S., Chang, H., Chen, X.: Grouping active contour fragments for object recognition. In: Lee, K.M., Matsushita, Y., Rehg, J.M., Hu, Z. (eds.) ACCV 2012, Part I. LNCS, vol. 7724, pp. 289–301. Springer, Heidelberg (2013)
17. Yang, X., Latecki, L.J.: Weakly supervised shape based object detection with particle filter. In: Daniilidis, K., Maragos, P., Paragios, N. (eds.) ECCV 2010, Part V. LNCS, vol. 6315, pp. 757–770. Springer, Heidelberg (2010)
18. Krizhevsky, A., Sutskever, I., Hinton, G.E.: Imagenet classification with deep convolutional neural networks. In: NIPS (2012)
19. Sermanet, P., Eigen, D., Zhang, X., Mathieu, M., Fergus, R., LeCun, Y.: OverFeat: integrated recognition, localization and detection using convolutional networks. In: ICLR (2015)
20. Zhu, Q., Avidan, S., Yeh, M.C., Cheng, K.T.: Fast human detection using a cascade of histograms of oriented gradients. In: CVPR (2006)

Instance Search with Weak Geometric Correlation Consistency

Zhenxing Zhang$^{(\boxtimes)}$, Rami Albatal, Cathal Gurrin, and Alan F. Smeaton

School of Computing, Insight Centre for Data Analytics, Dublin City University,
Glasnevin, Co. Dublin, Ireland
{zzhang,cgurrin,asmeaton}@computing.dcu.ie, rami.albatal@dcu.ie
https://www.insight-centre.org

Abstract. Finding object instances from large image collections is a challenging problem with many practical applications. Recent methods inspired by text retrieval achieved good results; however a re-ranking stage based on spatial verification is still required to boost performance. To improve the effectiveness of such instance retrieval systems while avoiding the computational complexity of a re-ranking stage, we explored the geometric correlations among local features and incorporate these correlations with each individual match to form a transformation consistency in rotation and scale space. This weak geometric correlation consistency can be used to effectively eliminate inconsistent feature matches and can be applied to all candidate images at a low computational cost. Experimental results on three standard evaluation benchmarks show that the proposed approach results in a substantial performance improvement compared with recent proposed methods.

Keywords: Multimedia indexing · Information retrieval

1 Introduction

Given a query image of an object, the objective of this work is to find images which contain a recognisable instance of the object from a large image collection, henceforth referred to as "instance search". A successful application requires efficient retrieval of instance images with high accuracy, possibly under various imaging conditions, such as rotation, viewpoint, zoom level, occlusion and so on.

Instance search is an interesting yet challenging problem and attracts significant research attention in recent years. Most of the state-of-the-art approaches [11,16,19] have been developed based on the Bag-of-Visual-Words (BoVW) framework firstly introduced by Sivic and Zisserman [3]. This framework successfully made use of the discriminative power of the local feature descriptors (e.g. SIFT [1], SURF [2]) which are generally robust to changes in image condition and are applied to build a statistical representation for each image in the database. At query time, the BoVW representation may take advantage of indexing techniques such as inverted files [4] to provide fast retrieval speed, even over large

© Springer International Publishing Switzerland 2016
Q. Tian et al. (Eds.): MMM 2016, Part I, LNCS 9516, pp. 226–237, 2016.
DOI: 10.1007/978-3-319-27671-7_19

collections. However this representation leads to a loss of the ability to encode spatial information between local features, so spatial verification [16] was introduced to improve retrieval accuracy. Based on the observation that there can be only one local feature correspondence to any given feature from query object, the geometric layout of objects was adopted to verify the spatial consistency between matched local feature points. Generally, the spatial verification algorithms were applied to refine the ranked results by iteratively optimizing the transformation models and fitting them to the initial correspondences to eliminate inconsistent matches. However those techniques such as RANSAC are normally computationally expensive; they can be applied only as a post-processing step to the top ranked images in the initial result set.

In this work, we address the challenges of improving the efficiency and robustness of examining the consistency between local feature matches to enhance the retrieval performance of instance search systems. Recent work of Jégou et al. [5] proposed a novel approach to efficiently apply spatial verification and made it suitable for very large datasets. They used the weak geometric constraints, specifically in the rotational and scale spaces, to examine each individual feature match and filter out those inconsistent feature matches at a very low computational cost. Although it improved retrieval performance for instance search, we observed that their approach considered feature matches independently and ignored the geometric correlation between local features, thus performed less effectively when searching more challenging datasets like FlickrLogos-27 [18]. In this work, we believe that the geometric correlation between reliable feature matches should also be consistent to the weak geometric constraints, just like each individual feature match. Based on that, we propose a scheme to incorporate the geometric correlations between matched feature correspondences to form a weak geometric correlation consistency to improve the effectiveness of spatial verification.

The experimental results on three standard evaluation benchmarks, in Sect. 5, illustrate that the proposed method is more reliable, and also more tractable for large image collections, which leads to an overall significant improvement of instance search performance compared to state-of-the-art methods.

2 Related Work

In this section, we briefly review the development of visual instance retrieval systems and discuss existing approaches to improve retrieval performance with the geometric information.

Sivic and Zisserman [3] were the first to address instance search using a BoVW representation combined with scalable textual retrieval techniques. Subsequently, a number of techniques have been proposed to improve the performance. The work in [11] suggested using very high dimensional vocabulary (1 million visual words) during the quantization process. This method has improved the retrieval precision with more discriminative visual words, and also increased the retrieval efficiency with more sparse image representations, especially for large scale database. Chum et al. [13] brought query expansion techniques to visual search domain and improved instance recall by expanding the query information. For further improvement on the retrieval performance, both approaches added the spatial verification stage to re-rank the results in order to remove noisy or ambiguous visual words. Recent works in [7–10] extended the BoVW approach by encoding the geometric information around the local features into the representation and refine the matching based on visual words. Those methods were very sensitive to the change in imaging condition and made them only suitable for partial-duplicate image search.

Recently, alternative approaches have been developed to implicitly verify the feature matches with respect to the consistency of their geometric relations, i.e., scaling, orientation, and location, in the Hough transformation space. Avrithis and Tolias [12] developed a linear algorithm to effectively compute pairwise affinities of correspondences in 4-dimensional transformation space by applying a pyramid matching model constructed from each single feature correspondence. Jégou et al. [5], increased the reliability of feature matches against imagining condition changes by applying weak constraints to verify the scaling and orientation relations consistency according to the dominant transformation found in the transformation space. Similarly, Zhang and Ngo [15] proposed to represent the feature points geometric information using topology-base graphs and verified the spatial consistency by performing a graph matching.

Our proposed method follows the direction of implicitly verifying the feature matches to reduce the computational cost. However compared to existing work, which focused on individual correspondences, our proposed method also considers the spatial consistency for the geometric correlations between matched feature correspondences, while maintaining the efficiency and increasing the effectiveness of the instance search systems.

3   Weak Geometric Correlation Consistency

In BoVW architecture, the local features are firstly extracted from each image to encode the invariant visual information into feature vectors. Generally, a feature vector is defined as $v(x, y, \theta, \sigma, q)$, where variables $\{x, y, \theta, \sigma\}$ stand for the local salient point's 2-D spatial location, dominant orientation, and most stable scale respectively. While q represents a 128-D feature vector to describe the local region. For a query image I_q and candidate image I_c, a set of initial matching features $C_{initial}$ could be established by examining feature vector q. The task of spatial verification is to eliminate the unreliable feature matches and only retain the matches set C_{stable} that linked the patches of the same object. The following equation formatted this process:

$$C_{stable} = \{m_i \in C_{initial} \quad \text{and} \quad f_{sp}(m_i) = 1\} \tag{1}$$

where m_i stands for the ith feature match in the initial match set. f_{sp} stands for the spatial verification function for assessing its geometric consistency. Take the weak geometric consistency [5] for example, the verification function in their work could be expressed as follows:

$$f_{sp} = \begin{cases} 1 & \text{if } \Delta\theta \in D_\theta \text{ and } \Delta\sigma \in D_\sigma \\ 0 & \text{if otherwise} \end{cases} \tag{2}$$

where $\Delta\theta$ and $\Delta\sigma$ is the geometric transformation for individual feature match and D_θ and D_σ is the dominated transformation in orientation and scale space.

3.1 Motivation

We take the geometric correlation among local features into consideration and hypothesize that the pairwise geometric correlation between consistent matches should also be consistent and follow the same spatial transformation between objects. So instead of verifying the geometric consistency for each match individually, we proposed a novel approach to verify the consistency between pairwise geometric correlations along with their corresponding feature points. So for a given pair of feature matches m_l and m_n, we then define the proposed spatial verification function as following:

$$f_{sp} = \begin{cases} 1 & \text{if } \Delta\theta, \Delta\theta_{l\rightarrow n} \in D_\theta \text{ and } \Delta\sigma, \Delta\sigma_{l\rightarrow n} \in D_\sigma \\ 0 & \text{if otherwise} \end{cases} \tag{3}$$

where $\Delta\theta_{l\rightarrow n}$ and $\Delta\sigma_{l\rightarrow n}$ represents spatial transformation of the geometric correlation from feature match m_l to m_n.

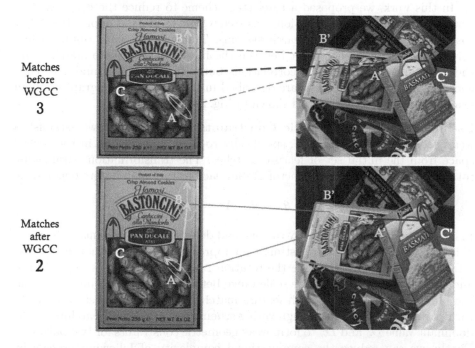

Fig. 1. An illustration of verifying consistency of feature matches using geometric correlations. The green(red) line indicates the consistent(inconsistent) feature matches (Color figure online).

We named our proposed approach to be Weak Geometric Correlation Consistency (WGCC) and Fig. 1 demonstrates our idea of using geometric correlations to assess reliability of feature matches. The object of interest (front cover of a box) is highlighted with dark yellow box. To begin with, we have three initial feature matches for spatial validation. Matches (A, A'), (B, B') are considered to be consistent because the spatial transformation is consistent between match (A, A'), (B, B') and their correlation $(AB, A'B')$. On the other hand, match (C, C') is filtered out due to the fact that geometric correlation between $(AC, A'C')$ is not consistent with the spatial transformation. Hence, we can successfully eliminate the inconsistent feature matches despite the fact that they may obey weak spatial constraints individually.

3.2   Implementation

To explicitly examine all the correlations between the initial feature matches is a non-trivial problem. If we take a total number of N initial matches as example, the potential pairwise correlation could be modeled as $O(C_N^2)$. The initial feature matching number N is usually large in practical systems, and this will cause a high computational cost to verify all the correlations and makes this solution less attractive for large image collections.

In this work, we proposed a three-step scheme to reduce the complexity of verifying the geometric correlation consistency, and to make it applicable at low cost for large-scale instance search systems. The key idea is to obtain a feature match as a reference point between the initial set of feature matches and then examine only the $O(N)$ correlations between each match and the reference match. These three steps are described in the following paragraphs and an example output for each step is shown in Fig. 2.

Estimating Weak Geometric Constraints. To begin with, we establish a weak geometric transformation, specifically rotation and scaling, in the spatial space from the initial set of feature matches. The transformation parameters, rotation angle $\Delta\theta$ and scaling factor $\Delta\sigma$ for each feature match were denoted as:

$$\Delta\theta = \theta_m - \theta_i, \ \Delta\sigma = \sigma_m / \sigma_i \tag{4}$$

In order to reduce the sensitivity to non-rigid deformation, we quantize the value of the parameters into bins to estimate an approximated transformation. We use a factor of 30 degrees to divide the rotation range of 360 degrees into 12 bins, and a factor of 0.5 to divide the scale range between 0 to 4 into 8 bins. To avoid the bin quantization error, each feature match votes to the closest two bins in each parameter space. The Hough voting scheme was applied in searching of the dominant value D_θ and D_σ to form weak geometric constraints for two purposes. Firstly, we can reduce the computational complexity of following process by eliminating the matches who are not obey the constraints. Secondly, these weak constraints will be used to assess the transformation consistency for geometric correlation to obtain the reliable matches.

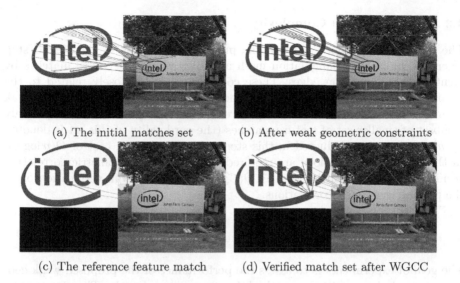

(a) The initial matches set   (b) After weak geometric constraints

(c) The reference feature match   (d) Verified match set after WGCC

Fig. 2. An illustration of applying WGCC on the initial set of feature matches to obtain the consistent feature matches.

Identifying the Reference Matching Correspondence. In this step, we aim to determine the strongest feature matches which will be served as a reference match in the step of verifying geometric correlations. We follow the approach of Zhang and Ngo [15] and adopt a topology-based graph match for this purpose. To represent the topology structure for objects, we created Delaunay Triangulation mesh from the geometric layout among the feature points in object plane. Then we could find the strongest feature matches which corresponding to the common edges between topology graphs by performing a graph matching.

Verifying Weak Consistency for Geometric Correlations. The last step focused on identify the reliable feature matches by verifying the consistency of the geometric correlations from each feature match to the reference match. Suppose we have a feature match m_l and a reference match m_n between image Q and D, the geometric correlation from m_l to m_n in image Q could be expressed as vector $v_{l->n} = (x_l, y_l) - (x_n, y_n)$ where x, y represent the 2D location of corresponding feature points in image Q for match m_l and m_n respectively. Similarly we can express the geometric correlation between m_l and m_n in image D as vector $v'_{l->n} = (x'_l, y'_l) - (x'_n, y'_n)$. Then the transformation parameters in orientation $\Delta\theta_{i->n}$ and the scale $\Delta\sigma_{l->n}$ between geometric correlations can be defined as:

$$\Delta\theta_{i->n} = \arccos \frac{\|v_{l->n}\| \|v'_{l->n}\|}{v_{l->n} \cdot v'_{l->n}}, \qquad \Delta\sigma_{i->n} = \frac{\|v'_{l->n}\|}{\|v_{l->n}\|} \qquad (5)$$

It is now possible to assess the spatial consistency by verifying the transformation parameters values with the weak constraints according to Eq. 3 and further filter out the inconsistent matches to obtain the final set of reliable feature matches.

3.3   Computational Complexity

The major computational cost in the proposed scheme is in the second step where we build the triangulation mesh and discover the reference matches by identifying the common edges. These computations are closely related to the total number of feature matches. The good news is that we already build weak geometric constraints in the first step to verify the initial feature matches, so only a subset of smaller set of feature matches (the cardinality of this set is denoted by n) needs to be conducted in this step, which leads to a cost of $O(n \log n)$. In the end, $O(n)$ operations are required to perform the geometric correlation verification which is much less than $O(C_n^2)$ required for a full verification of all the possible geometric correlations.

4   Experiments

The goal of experiments is to assess the performance of the proposed weak geometric correlation consistency methods in instance search tasks. Therefore a complete instance search system was developed and comparative experiments were designed to evaluate retrieval performance against state-of-the-art approaches on three standard and publicly available benchmark datasets. The datasets chosen were the Oxford, Pairs6K and FlickrLogis-32 datasets. Each of these datasets includes a set of queries and relevance judgements.

In the rest of this section, we introduce the three chosen benchmark datasets, describe the evaluation protocol and analyse the experimental results by comparing them to the three state-of-the-art approaches.

4.1   Datasets

The Oxford Dataset. This dataset [11] contains 5,062 high resolution images crawled from Flickr using texture queries for famous Oxford landmarks. 11 building topics with 55 images queries was provided with manually annotated ground truth for users to evaluate the retrieval performance. The images are considered to be positive if more than 25 % of the instance is clearly visible.

Pairs6K. This collection [16] consists of 6,412 images collected by searching for particular Paris landmarks from Flickr. In total, 11 Landmarks with 55 images queries was provided with manually annotated ground truth for users to evaluate the retrieval performance. The images are considered to be positive if more than 25 % of the instance is clearly visible.

FlickrLogos-27. This dataset [18] consists of 5,107 images including 810 annotated positive images corresponding to 27 classes of commercial brand logos and 4,207 distraction images that depict its own logo class. It is a very challenging dataset because the positive images share much more visually similar regions with the distraction images and have more noisy background. For each logo, 5 query example images are given for evaluation purposes.

4.2   Evaluation Protocol

The standard evaluation protocol based on the classic BoVW scheme was adopted to assess the improvements of our proposed method for instance search. The Hessian detector and SURF descriptor implemented in the OpenCV Library [20] were used to extract the local features from database images. Subsequently, a visual vocabulary was generated using the approximate K-means algorithm [11] to quantize each feature into visual words for indexing. After that, the represented visual words (along with auxiliary information, e.g. the geometric information) are indexed in an inverted structure for the retrieval process. When performing the search tasks, the candidate images sharing same visual words are retrieved from the database collections. Auxiliary information is used to perform the spatial verification to improve retrieval performance.

We measured the mean Average Precision (mAP) score of the top 1,000 results to evaluate the retrieval accuracy. mAP is defined as the mean of the average precision (AP) over all the queries. To evaluate the retrieval efficiency, we also record the response time accurate to one hundredths of a second. Each approach was implemented and evaluated on the same computing hardware.

4.3   Experiment Settings

The Weak Geometric Correlation Consistency (WGCC) was compared against the standard BoVW approach as the baseline, but it was also compared against two other advanced approaches; Weak Geometric Consistency (WGC) and Delaunay Triangulation (DT), both of which also adopt the geometric information to enhance the baseline method.

Baseline. The baseline approach was based on [11] with a vocabulary of 1M words which had been shown to give the best performance. The $tf-idf$ weighting scheme and hard assignment was used to keep a consistent setting for all the systems.

WGC [5]. We chose this approach for evaluation because this method assessed each feature match by verifying its transformation against a weak geometric consistency to increase the robustness in changing of rotation and scale space. The constraints for geometric consistency was obtained by converting the parameter values into Hough transformation space.

DT [15]. This approach make use of relations between matched points in a 2-dimensional translation space to improve the matching reliability between two sets of features. It used the Delaunay Triangulation (DT) based graph representation to model and match the layout topology of initial matched feature points. A Hamming embedding signature was used to enforce an point-to-point matches and ensure the number of nodes in each graph is identical.

WGCC. This is our proposed method and the contribution of this work, as described in Sect. 3. We follow the recent work in feature search in high-dimensional space and use the product-quantization based algorithm [6] to build up search components for initial feature matching. Then we applied the proposed weak geometric correlation consistency (WGCC) for spatial verification.

5   Results and Discussion

Table 1 presents the experimental results of comparing our proposed approach WGCC with the baseline and two enhanced approaches, on the three benchmarks. To study the impact of adopting geometric information for enhancing the retrieval performance in instance search system, we compared the advanced systems against baseline system on the three described datasets. We observed that the advanced approaches for spatial verification consistently improves performance in mAP compared to the baseline. Compared to the other two advanced systems, our proposed approach achieved the best results. Especially on the FlickrLogos dataset, our approach have a 59 % relative improvement in the mAP performance from the baseline's 0.145 to 0.231 in our method. This proved that our approach are strong enough to reject inconsistent feature matches, while also flexible for keep the evidence from locally consistent patches.

Table 1. mAP comparison between our proposed WGCC and the baseline and two other state-of-the-art advanced approaches, on Oxford, Pairs6K and FlickrLogo-27 datasets.

Methods	Oxford		Pairs6K		FlickrLogos	
	mAP	Time[1]	mAP	Time[1]	mAP	Time[1]
BoF	0.489	0.46	0.526	0.62	0.145	0.26
WGC	0.530	1.06	0.576	1.12	0.193	0.41
DT	0.542	0.86	0.546	0.89	0.201	0.31
WGCC	**0.693**	1.07	**0.607**	1.23	**0.231**	1.06

Time[1] measures the average response time per query in second, excluding feature extraction.

Figure 3 shows some examples of the improvement in mAP obtained by the proposed WGCC approach compared to the baseline system. The interested object delimited in the yellow box from the query image on left side of each subfigure. The Precision-Recall curve is displayed on the right side with baseline results shown in blue line and WGCC method shown in red. The gap area between two lines indicates the performance improvement for our methods. The high precision value in low recall range indicates that our proposed approaches improved retrieval performance by ranking the good images higher in the ranked list.

Figure 4 displays some good queries results retrieved from the experiment benchmark collections. The results demonstrates the robustness of the proposed methods to the considerable variations in viewpoint, scale, lighting and partial occlusion from practical environment.

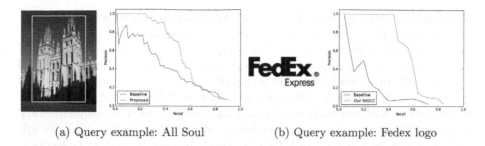

(a) Query example: All Soul   (b) Query example: Fedex logo

Fig. 3. Examples of the improvement in the precision-recall curve obtained by using proposed approaches compared to the baseline system.

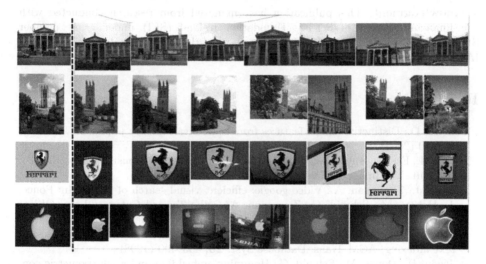

Fig. 4. Some good examples of searching experiment benchmark: First row: "Magdalen" in Oxford, Second row: "Ashmolean" in Oxford, Third row: "Ferrari" in FlickrLogo-27, and Fourth row: "Apple" in FlickrLogo-27

The experiments were carried out on a desktop computer with 4-core 2.3 GHz CPU and 8 G RAM. Only one core was used when performing the query task. At run-time, our proposed method achieved comparable retrieval efficiency with the two advanced approaches while providing better accuracy. Although the retrieval efficiency of WGCC was slightly less than the other approaches, retrieval efficiency could be optimised by adopting parallel computing approaches.

6 Conclusion

This paper proposed novel approach to improve retrieval performance of instance search systems by combining the pairwise geometric correlations with individual feature match transformation to form a weak geometric correlation consistency. This model is strong enough to eliminate inconsistent feature matches while keeping reliable matches using locally spatial correlation. Our experiments shows that our approach consistently outperformed the baseline system in three standard benchmark evaluations and achieved improved results when compared with two advanced systems. This indicates the effectiveness of our method for spatial verification. Another positive aspect is that other advanced technologies, such as automatic query expansion [13], re-ranking base on full spatial verification [11] are compatible with our proposed method and could be used as complementary components to further improve the retrieval performance. In future work, we will investigate how to incorporate WGCC methods in vary large data collections, e.g. collection with millions of images.

Acknowledgment. This publication has emanated from research conducted with the financial support of Science Foundation Ireland (SFI) under grant number SFI/12/RC/2289, as well as the financial support of the Norwegian Research Council's iAD project under grant number 174867.

References

1. Lowe, D.: Distinctive image features from scale-invariant key points. IJCV **60**(2), 91–110 (2004)
2. Bay, H., Ess, A., Tuytelaars, T., Van Gool, L.: Speeded-up robust features (SURF). Comput. Vis. Image Underst. **110**(3), 346–359 (2008)
3. Sivic, J., Zisserman, A.: Video google: efficient visual search of videos. In: Ponce, J., Hebert, M., Schmid, C., Zisserman, A. (eds.) Toward Category-Level Object Recognition. LNCS, vol. 4170, pp. 127–144. Springer, Heidelberg (2006)
4. Zobel, J., Moffat, A., Ramamohanarao, K.: Inverted files versus signature files for text indexing. ACM Trans. Database Syst. **23**, 453–490 (1998)
5. Jégou, H., Douze, M., Schmid, C.: Hamming embedding and weak geometric consistency for large scale image search. In: Forsyth, D., Torr, P., Zisserman, A. (eds.) ECCV 2008, Part I. LNCS, vol. 5302, pp. 304–317. Springer, Heidelberg (2008)
6. Jégou, H., Douze, M., Schmid, C.: Product quantization for nearest neighbor search. IEEE Trans. Pattern Anal. Mach. Intell. **33**, 117–128 (2011)
7. Wu, Z., Ke, Q., Isard, M., Sun, J.: Bundling features for large scale partial-duplicate web image search. In: Proceeding of the IEEE Conference on Computer Vision and Pattern Recognition (2009)
8. Albatal, R., Mulhem, P., Chiaramella, Y.: Visual phrases for automatic images annotation. In: Proceedings of CBMI (2010)
9. Romberg, S., Lienhart, R.: Bundle min-hashing for logo recognition. In: Proceedings of International Conference of Multimedia Retrieval (2013)
10. Zhou, W., Lu, Y., Li, H., Song, Y., Tian, Q.: Spatial coding for large scale partial-duplicate web image search. In: Proceedings of the ACM Conference in Multimedia (2010)
11. Philbin, J., Chum, O., Isard, M., Sivic, J., Zisserman, A.: Object retrieval with large vocabularies and fast spatial matching. In: Proceedings of the IEEE Conference on Computer Vision and Pattern Recognition (2007)
12. Avrithis, Y., Tolias, G.: Hough pyramid matching: speeded-up geometry re-ranking for large scale image retrieval. IJCV **107**(1), 1–19 (2014)

13. Chum, O., Philbin, J., Sivic, J., Isard, M., Zisserman, A.: Total recall: automatic query expansion with a generative feature model for object retrieval. In: IEEE International Conference on Computer Vision (2007)
14. Zhang, Y., Jia, Z., Chen, T.: Image retrieval with geometry-preserving visual phrases. In: Proceedings of the IEEE Conference on Computer Vision and Pattern Recognition (2011)
15. Zhang, W., Ngo, C.-W.: Searching visual instances with topology checking and context modeling. In: Proceedings of ICMR (2013)
16. Philbin, J., Chum, O., Isard, M., Sivic, J., Zisserman, A.: Lost in quantization: improving particular object retrieval in large scale image databases. In: Proceedings of the IEEE Conference on Computer Vision and Pattern Recognition (2008)
17. Revaud, J., Douze, M., Schmid, C.: Correlation-based burstiness for logo retrieval. In: Preceeding of the ACM International Conference on Multimedia, October 2012
18. Romberg, S., Pueyo, L.G., Lienhart, R., van Zwol, R.: Scalable logo recognition in real-world images. In: ACM International Conference on Multimedia Retrieval 2011 (ICMR 2011), Trento, April 2011
19. Kalantidis, Y., Pueyo, L.G., Trevisiol, M., van Zwol, R., Avrithis, Y.: Scalable triangulation-based logo recognition. In: Proceedings of ACM International Conference on Multimedia Retrieval (ICMR 2011), Trento, Italy, April 2011
20. Bradski, G.: The OpenCV library. Dr. Dobbs J. Softw. Tools 25, 120–126 (2000)

Videopedia: Lecture Video Recommendation for Educational Blogs Using Topic Modeling

Subhasree Basu[1]([⊠]), Yi Yu[2], Vivek K. Singh[3], and Roger Zimmermann[1]

[1] National University of Singapore, Singapore 117417, Singapore
{sbasu,rogerz}@comp.nus.edu.sg
[2] National Institute of Informatics, Tokyo 101-8430, Japan
yiyu@nii.ac.jp
[3] University of South Florida, Tampa, FL 33620, USA
vivek4@mail.usf.edu

Abstract. Two main sources of educational material for online learning are e-learning blogs like Wikipedia, Edublogs, etc., and online videos hosted on various sites like YouTube, Videolectures.net, etc. Students would benefit if both the text and videos are presented to them in an integrated platform. As the two types of systems are separately designed, the major challenge in leveraging both sources is how to obtain video materials, which are relevant to an e-learning blog. We aim to build a system that seamlessly integrates both the text-based blogs and online videos and recommends relevant videos for explaining the concepts given in a blog. Our algorithm uses content extracted from video transcripts generated by closed captions. We use topic modeling to map videos and blogs in the common semantic space of topics. After matching videos and blogs in the space of topics, videos with high similarity values are recommended for the blogs. The initial results are plausible and confirm the effectiveness of the proposed scheme.

1 Introduction

The lecture videos captured in classrooms contain a substantial portion of the instructional content [12]. These videos have several advantages over graphic and textual media, e.g., portrayal of concepts involving motion, the alteration of space and time, the observation of dangerous processes in a safe environment, etc. [14]. They also provide the viewer with the benefit of pausing and reviewing at leisure, which is not possible in traditional classrooms. These videos are often hosted on the intranet of the universities and also on numerous online sites like Coursera[1], Khan Academy[2], etc. Leading universities like MIT and Stanford have made their lectures available online for distance learning. Besides, there are numerous lecture videos uploaded to video sharing sites like YouTube[3],

---

[1] https://www.coursera.org/.
[2] https://www.khanacademy.org/.
[3] https://www.youtube.com/.

© Springer International Publishing Switzerland 2016
Q. Tian et al. (Eds.): MMM 2016, Part I, LNCS 9516, pp. 238–250, 2016.
DOI: 10.1007/978-3-319-27671-7_20

Videolectures.net[4], etc., to facilitate further learning. With the increase in Internet speed globally, the demand for multimedia content has been increasing over the past couple of years, with a projected Internet video consumption of 56,800 PB/month in 2018 [1]. E-learning and smart learning videos constitute a huge fraction of this consumption. Hence there is an increase in the demand and consumption of multimedia content related to e-learning.

The entire e-learning process is not limited to videos alone. Text based blogs and websites like Wikipedia[5] and Edublogs[6] are also important sources of learning material. Wikipedia is a user generated knowledge base available on the Internet that is frequently visited for learning and reference by users from both academia and industry. In essence, wikis offer an online space for collaborative authorship and writing. A survey demonstrates that students reported that they frequently, if not always, consult Wikipedia at some point during their course-related research [8]. Wikipedia serves as a convenient go-to source when students are stuck on some concepts and can not move forward [8]. However, Wikipedia usually only provides a brief explanation of concepts without the detailed descriptions. To provide comprehensive learning for students over the Internet, the contents from multiple sources, e.g., text and videos are required and should be provided in an integrated platform [15]. For example, it will be more effective to enhance the Wikipedia contents with multimedia information such as videos that cover some online lectures about the topic described in the wikipage.

In this paper, we design a system for integrating the blogs and the videos containing educational multimedia data and propose a novel recommendation system called *Videopedia*. We provide the necessary matching between the texts in the webpages and the lecture videos based on their contents. Retrieving suitable videos for a particular webpage is challenging mainly because webpages and videos are represented in different forms and finding a correlation between them is not straightforward. Also e-learning lecture videos mostly consist of lecture slides, whiteboard/blackboard content and sometimes the instructors delivering the lecture. Hence extracting the content of these videos based on visual words is not a plausible solution. Video transcripts have been generated using automatic speech recognition (ASR) of the audio as an attempt for content detection [16]. Most of the extracted texts by ASR contain errors due to poor audio recording quality, which degrades the spoken text's usability for video retrieval [6]. Recently, the majority of video sharing platforms like Ted[7], YouTube, etc., provide videos with automatic video transcripts. These are manually created closed captions (CC) and contain less error than ASR. These transcripts provide the content of the lecture videos which we can then mine to find the correlation with the text-based blogs.

---

[4] http://videolectures.net/.
[5] http://www.wikipedia.org/.
[6] https://edublogs.org/.
[7] http://www.ted.com/.

Our novel algorithm implemented as a part of *Videopedia* thus employs techniques of topic modeling in extracting the contents of the webpages and the videos. We use Latent Dirichlet Allocation (LDA) [3] to find the topics in a particular website and those in a particular video. These topics provide better representation of the content of the video compared to video metadata. The topics for the videos are extracted from video transcripts, stored on the server and indexed by their topic distributions. The topics representing the contents of a blog are generated at run-time. *Videopedia* determines the similarity between the topics of a webpage and those of a video and recommends the relevant videos based on the similarity. In summary, our specific contributions in this paper are as follows:

- For a webpage recommend the relevant videos that would elaborate the concepts introduced in the webpage.
- Provide a comprehensive framework which combines multimodal data, i.e., text and video for an enhanced e-learning experience.
- Effectively use video transcripts for lecture video recommendation.

The rest of the paper is organized as follows. In Sect. 2 we provide a brief summary of the related work. In Sect. 3 we give a description of the model of our system and Sect. 4 summarizes our experiments and the results. We conclude with the direction of our future work in Sect. 5.

2   Related Work

Chen et al. attempted to link web-videos to wikipages by leveraging Wikipedia categories (WikiCs) and content-duplicated open resources (CDORs) [5]. Later they adopted a multiple tag property exploration (mTagPE) approach to hyperlink videos to webpages [6]. Okuoka et al. have used Wikipedia entries to label news videos [15]. This labeling was primarily done by using date information and labels that are extended along the topic thread structure following a time-series semantic structure. Roy et al. have used Wikipedia articles to model the topic of a news story [17]. Topic modeling has found its use in a lot of research areas to find the latent topics in documents, many of which being in the e-learning scenario. Wang and Blei combined probabilistic topic modeling with collaborative filtering to recommend scientific papers [19]. Chen, Cooper, Joshi, and Girod have proposed a Multi-modal Language Model (MLM) that uses latent variable techniques to explore the co-occurrence relation between multi-modal data [4]. They mainly use MLMs to index text from slides and speech in lecture videos. One of the major contributions to indexing and searching lecture videos has been done by Adcock et al. [2]. They have designed a system that provides a keyword-based search on a database of 62,406 video lectures and talks. Zhu, Shyu and Wang used topic modeling techniques in video recommendation systems and designed VideoTopic [20]. The recommendation system generates a personalized video recommendation list to fit the user's interests. The bag-of-words model used in VideoTopic contains both visual as well as textual features.

In contrast, *Videopedia* is designed to automatically recommend relevant videos based on the content of the webpages and not for a particular user. The aim of our system is to integrate e-learning materials from different online sources and not to develop a latent variable model as MLM [4]. One of the main problems in extracting the content of the lecture videos is the absence of meaningful visual words. Hence we need to extract the content covered in the speech of the lecture videos. Compared to Chen et al. [5], our system extracts the educational material covered in the videos and not just metadata. Also the videos in *Videopedia* are indexed by the latent topics present in them and not timing information as is done by Okuoka et al. [15]. We are recommending lecture videos for Wikipedia entries and not using the content of wikipages for mining topics in our videos like Roy et al. [17]. In the domain of online education, topic modeling has been used only for mining the text and not utilized on lecture video content analysis. Hence ours is a first attempt in that direction. Also by indexing the videos by topics, we propose a content based search and not just keyword based search as proposed by Adcock et al. [2].

3   System Model

Videopedia proposes a novel video search technique based on topic extraction and automatically recommends videos based on the content of a particular webpage or educational blog. The existing e-learning systems provide keyword-based video search like Videotopic [2] and hence they can not be used to automatically recommend videos for wikipages and other blogs. The webpage and the video represent two different media formats and finding a correlation between them is the main challenge in designing such a system. An important factor in our recommendation system is that we are recommending lecture videos and hence mining the visual content of the videos extracted through visual words do not provide too much semantic information. This is because the lecture content is delivered either through slides or blackboard/whiteboards and the visual content does not change much throughout the entire video. However, the instructor explains most of the points written in the slides and/or on boards, and the speech of the videos will represent the content of these videos. Therefore, we use the transcripts of these videos and use them as the representative of the content of these lecture videos.

Figure 1 provides a pictorial description of the proposed model. A user views the webpage (e.g., a Wikipedia page) via a browser. The request for videos comes from the webpage on the browser by sending the URL of this webpage to the server. The content of the webpage is extracted and then the latent topics are generated from the content. The video repository is stored on the server and the topic models of these videos are generated beforehand. The matching between the topic models of the webpage and those of the video transcripts in the repository is also done on the server side in real time. The recommended videos are then linked to the webpage on the browser.

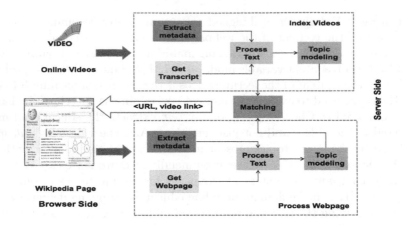

Fig. 1. System model for Videopedia

3.1   Dataset Used for Recommendation

For this research, we collected the videos from the YouTube channel for National Programme on Technology Enhanced Learning (NPTEL) and also from Videolectures.net. Our video repository now consists of 3,000 videos covering the following subjects – Humanities and Social Sciences, Metallurgical and Material Sciences, Civil Engineering, Electrical Engineering, Chemistry, Mathematics, Electronics Engineering and Management. All the videos have English as the medium of instruction. Hence the transcripts generated consist of words in English.

3.2   Extracting Video Content

The contents of the videos in *Videopedia* are represented by their transcripts. The transcripts are generated from the YouTube platform using the CC feature of YouTube. We also extract the metadata for each of the videos in the form of the title, description and keywords of the videos. The subtitle files available along with the videos uploaded on the Videolectures.net are used. The text available after removing the timing information from the subtitle files provide the text for the transcripts of these videos. These text files are then processed as illustrated in Sect. 3.4.

3.3   Extracting Webpage Content

When the browser sends a request for relevant videos, the content of the Wikipedia page is extracted at the server. For the metadata of the webpages, we consider the topic of the Wikipedia page and the summary at the top of the page. The content of the remaining part of the wikipage along with all the subsections are extracted for topic modeling. Processing according to Sect. 3.4 provides us the bag of words to be used as representative for the webpage.

3.4   Processing the Extracted Text

Table 1. Comparison of precision before and after dictionary matching.

Methods	Precision	
	Before dict Matching	After dict Matching
VSM	0.024	0.27
PLSA	0.244	0.27
LDA	0.146	0.475
Videopedia	0.38	0.642

The transcripts extracted from the videos are not 100 % accurate, which necessitates correcting the transcripts by checking with a dictionary. We match the words against a dictionary provided by Wordnet and used the nearest word as a substitute for the incorrect words. Next we remove the stop-words from the remaining words in the transcripts. This leaves us with the main bag of words of the documents. The videos are educational in nature and the topics mined from the transcripts should also be educational. Hence we generate the representative words in these transcripts by removing the non-academic words from the bag of words. This is achieved by matching them against a set of 20,000 academic words [7]. To find the best matching among the wikipages and the lecture videos, we extend the bag-of-words to include synonyms of the academic words already present in it. This will minimize the errors in case the web pages and the videos are illustrating the same topic but use synonymous words. We apply the deep learning library Word2vec [13] to extract words similar to the words in the transcript of the videos and use the newly generated vector of similar words as the representative bag-of-words for the videos. A similar approach is taken while processing the content of the Wikipedia webpages also. The improvement in the precision of the recommendation by matching against a dictionary is shown in Table 1. We define the methods VSM and PLSA in Sect. 4.1.

As explained in Sects. 3.2 and 3.3, we extract the metadata for both the videos and the Wikipedia pages. The metadata in both cases are also matched to the academic dictionary to remove the non-academic words. Similar to the case of processing the content of the videos and webpages, the bag-of-words for metadata are also extended by Word2vec. For each of the words in the metadata for the videos and wikipages, we find the vector of words similar to it by using Word2vec with the set of academic words as the dictionary. The union of the vectors returned by the Word2vec forms the bag of words for metadata.

3.5   Topic Modeling

After the transcripts and the bag-of-words are processed for each video, the topics of the videos are generated using LDA. LDA posits that each word in a document is generated by a topic and each document is a mixture of a finite number of topics. Each topic is represented as a multinomial distribution over words. There are a number of outputs from the LDA. We will consider $Z = \{P(z_i)\}$ – the probability distribution of the topics inside a document. These topics along with their probability distribution are then stored on the server indexed by the corresponding video-ids.

3.6   Definition of Similarity Matching

In probability theory and information theory, the Kullback-Leibler divergence (KL Div) is a measure of the difference between two probability distributions P and Q [10]. KL Div being a non-symmetric metric, we calculate the KL Div of both P from Q and Q from P and then take the average of the two. The lower the value of the KL Div between two distributions, the closer they are semantically. Hence the similarity between the videos and the webpages are calculated as the inverse of the KL Div(D_{KL}). Let $Z_{LDA}(Blog_{wiki}) = \{P(z_i)\}$, where z_i are the latent topics present in $Blog_{wiki}$ of blog pages which are returned by LDA and $Z_{LDA}(Trans_{v^m}) = \{P(z_k)\}$, where z_k are the latent topics present in $Trans_{v^m}$ of videos. We define the similarity measure in $Videopedia$ as:

$$Sim_{Videopedia}(Blog_{wiki}, Trans_{v^m}) = [\frac{1}{2}(D_{KL}(Z_{LDA}(Blog_{wiki}) \| Z_{LDA}(Trans_{v^m}))$$
$$+ D_{KL}(Z_{LDA}(Trans_{v^m}) \| Z_{LDA}(Blog_{wiki})))]^{-1} \quad (1)$$

This matching between the topic models of the webpage and those of the video transcripts in the repository is also done on the server side. The recommended videos are then linked to the webpage on the browser.

3.7   Algorithm for Video Recommendation

The algorithm used to recommend the videos is given in Algorithm 1. As described in the procedure PREPAREVIDEOTOPICMODELS, our system extracts the transcripts of the videos, removes the stop-words from these transcripts and checks the remaining words against an academic dictionary to eliminate spelling mistakes and extract the academic words from the transcripts. This facilitates the extraction of the topics related to academia. Then, the topic distribution of these videos is obtained by using LDA on these transcripts. The metadata of the videos are extracted as described in Sect. 3.2.

When a request is received from a website, the text of the webpage is retrieved by using the provided URL and then relevant videos are recommended as described in the procedure RECOMMENDVIDEO. The actual matching between the webpage and videos is divided into two stages. In the filtering stage, the meta data is compared in the vector space model by the cosine similarity, which removes most non relevant videos. These bag-of-words for metadata of videos are compared against the bag-of-words for metadata for the incoming website as is illustrated in the procedure MATCHMETA. The top 10 % of the videos with the maximum similarity are then selected and used for matching based on topics. Let the content of the webpage W_j after checking against a dictionary be C_j'. LDA will return a set of topics $l_{C_j'}$ along with the probability distribution of the topics. Similarly, the transcript t_i' of a video v_i, has a set of topics $l_{t_i'}$ with their distribution. To find the video closest to C_j' we calculate the KL Div between $l_{C_j'}$ and $l_{t_i'}$s for the videos v_i in the set V', returned by the procedure MATCHMETA. Equation 1 is used to calculate the similarity and it is described in the procedure MATCHTOPICS.

Algorithm 1. *Videopedia*: Video recommendation based on topic modeling.

Online Processing
procedure RECOMMENDVIDEO(W_j)
   v_k = PROCESSWIKIPAGE(W_j)
   Recommend v_k for webpage W_j
end procedure

Offline Processing
procedure PREPAREVIDEOTOPICMODELS
   V = Set of videos in the Video Repository
   for $v_i \in V$ **do**
      Extract the transcript t_i of v_i
      Remove stop-words from t_i and get t_i'
      Check the words in t_i' against an academic dictionary to correct spelling mistakes and extract academic
words
      Use LDA to find the latent topics in the t_i'
      Index the v_is based on the topics
   end for
end procedure
procedure PROCESSWIKIPAGE(W_j)
   Extract the content of W_j as C_j
   Remove stop-words from C_j and get C_j'
   Check the words in C_j' against an academic dictionary to correct spelling mistakes and extract academic words
   Select V' = MATCHMETA (W_j, V)
   Use LDA to find the latent topics in the C_j'
   for $v_i \in V'$ **do**
      t_i' contains the transcript of v_i
      r_i = MATCHTOPICS (C_j', t_i')
   end for
   $k = argmax_i r_i$
   Return v_k for webpage W_j
end procedure
procedure MATCHMETA(W_j, V)
   Extract the metadata of W_j as M_j
   Remove stop-words from M_j and get M_j'
   Check the words in M_j' against an academic dictionary to correct spelling mistakes and extract academic words
   for $v_i \in V$ **do**
      v_i' contains the metadata of v_i
      Cosine Sim(M_j', v_i') = $\dfrac{M_j' \cdot v_i'}{\|M_j'\| \|v_i'\|}$
   end for
   V' = Top 10 % of $argmax_i CosineSim(M_j', v_i')$
   Return V'
end procedure
procedure MATCHTOPICS(C_j', t_i')
   $l_{t_i'}$ = Set of latent topics in t_i' for video v_i
   $l_{C_j'}$ = Set of latent topics in C_j' of webpage W_j
   for each pair (l_t, l_C) in the combination of $(l_{t_i'}, l_{C_j'})$ **do**
      KLDiv(l_C, l_t) = $\dfrac{D_{KL}(l_C \| l_t) + D_{KL}(l_t \| l_C)}{2}$
   end for
   KLDiv(C_j', t_i') = $\dfrac{\sum KLDiv(l_C, l_t)}{|l_{C_j'}||l_{t_i'}|}$
   Return $[\text{KLDiv}(C_j', t_i')]^{-1}$
end procedure

---

4 Experimental Results

The videos used in this research paper were extracted from the repository of
publicly available videos on YouTube and Videolectures.net. The webpages used
as input to *Videopedia* were Wikipedia pages. In this section we provide the

evaluation for our recommendation system against three baselines. The baselines are defined in the Sect. 4.1 and the results are presented in Sect. 4.2.

4.1 Baselines Used for Comparison

Topic modeling methods have been used to find latent topics from a collection of documents using probabilistic analysis of documents. However to find the semantic similarity between the documents, we need to find the similarity between the topics of these documents. There are broadly two categories of similarity measures in existing literature: (1) count similarity measures [11], (2) probabilistic similarity models [10].

In this paper, we choose one widely accepted technique from each category of models. Cosine similarity represents the count similarity measures paradigm and KL divergence represents the probabilistic similarity models. To compare the efficacy of our system, we use the VSM model [18] which uses cosine similarity and PLSA [9] which is another probabilistic topic modeling. After generating the topics through PLSA we use the KL Divergence method to find the similarity. Since in *Videopedia*, we first prune the set of videos based on metadata before applying LDA, we compare the efficiency of our algorithm with calculating the similarity between the topics generated by LDA without matching the metadata. The following subsections give the definitions necessary for the similarity measures and the topic models.

Vector Space Model. Vector Space Models represent text documents in an algebraic structure of vectors of identifiers [18]. The identifiers are often the term frequency-inverse document frequency (tf-idf) of the words in the document. Cosine similarity, which calculates the cosine of the angle between two vectors, provides a measure of the deviation between the identifiers of the document. Let the Wikipedia page be denoted by $Blog_{wiki}$ and the transcripts for the mth video, v^m, as $Trans_{v^m}$ In our case, while constructing the baseline, we thus represented $Blog_{wiki}$ by its tf-idf vector $v_{wiki} = (v_{wiki_1}, v_{wiki_2}, \ldots, v_{wiki_k})$. $Trans_{v^m}$ is also represented by the tf-idf vector for its transcript $v_{vid}^m = (v_{vid_1}^m, v_{vid_2}^m, \ldots, v_{vid_l}^m)$, where $m = 1, \ldots n$, n represents the number of videos in the repository and l represents the number of terms in the vid^i. Thus the similarity between the Wikipedia page and the video is calculated in the VSM model by the following formula:

$$Sim_{VSM}(Blog_{wiki}, Trans_{Vid}^m) = \frac{v_{wiki} \cdot v_{vid}^m}{\|v_{wiki}\| \|v_{vid}^m\|} = \frac{\sum_{i=1}^{n} v_{wiki_i} \times v_{vid_i}^m}{\sqrt{\sum_{i=1}^{n} (v_{wiki_i})^2} \times \sqrt{\sum_{i=1}^{n} (v_{vid_i}^m)^2}} \quad (2)$$

PLSA. Probabilistic Latent Semantic Analysis (PLSA) is a technique to statistically analyze the co-occurrence of words and documents [9]. Let $Z_{PLSA}(Blog_{wiki}) = \{P(z_i)\}$, where z_i are the latent topics present in $Blog_{wiki}$ and $Z_{PLSA}(Trans_{v^m}) = \{P(z_k)\}$, where z_k are the latent topics present in $Trans_{v^m}$. Using the probability distribution of the topics in a Wikipedia page and that in the transcripts of the videos, we find the KL Div between them and subsequently the similarity with the following formula:

$$Sim_{PLSA}(Blog_{wiki}, Trans_{v^m}) = [\frac{1}{2}(D_{KL}(Z_{PLSA}(Blog_{wiki})\|Z_{PLSA}(Trans_{v^m})) \\ + D_{KL}(Z_{PLSA}(Trans_{v^m})\|Z_{PLSA}(Blog_{wiki})))]^{-1} \tag{3}$$

We calculate the Sim_{PLSA} of $Blog_{wiki}$ with each of the n videos in the repository using the formula given above and the one having the highest value is recommended as the relevant video for the wikipage.

4.2   Evaluation of Recommendation

We use Precision, Recall and F-measure to evaluate the recommendation by our system. The definitions for them are provided below:

$$Precision = \frac{|RelevantVideos \cap RetrievedVideos|}{|RetrievedVideos|}$$

$$Recall = \frac{|RelevantVideos \cap RetrievedVideos|}{|RelevantVideos|}$$

$$\text{F-measure} = 2 \times \frac{Precision \times Recall}{Precision + Recall}$$

We tested the system by recommending videos for 1,000 Wikipedia pages categorized under the same eight subjects as the e-learning videos. The distribution of the videos across the subjects is given in Fig. 2a and the distribution of the subjects among the wikipages used for testing the system are shown in Fig. 2b. We have created the ground truth of whether a video is relevant or not by checking if the wiki-category of the Wikipedia page, which is an input to *Videopedia*, matches the subject in the metadata of the relevant videos. Hence, the number of Relevant Videos for a particular webpage is often more than 1 and thus $|RelevantVideos|$ may be greater than the $|RetrievedVideos|$. The Precision, Recall and F-measure values for *Videopedia* and the three baselines VSM, PLSA and LDA are presented in Table 2. We proceed to measure the effectiveness of our method by recommending more than 1 video for a webpage. The columns @n are the results of Precision (Table 2a), Recall (Table 2b) and F-measure (Table 2c) when we recommend n videos for a webpage, where $n = 1,5,10$. When *Videopedia* is recommending more than one video for a particular webpage, we use the mean average precision defined below to find the precision.

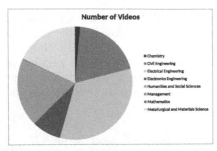

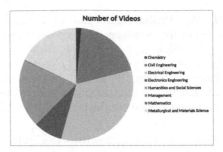

(a) Distribution of videos according to subjects

(b) Distribution of webpages according to subjects

Fig. 2. Distribution of videos and webpages according to subjects

Table 2. Videopedia compared with VSM, PLSA and LDA

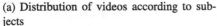

Methods	Precision		
	@1	@5	@10
VSM	0.198	0.179	0.185
PLSA	0.0.213	0.239	0.203
LDA	0.475	0.323	0.309
Videopedia	0.642	0.658	0.655

(a) Precision

Methods	Recall		
	@1	@5	@10
VSM	0.022	0.099	0.206
PLSA	0.021	0.133	0.226
LDA	0.053	0.180	0.343
Videopedia	0.256	0.244	0.2368

(b) Recall

Methods	F-measure		
	@1	@5	@10
VSM	0.040	0.128	0.195
PLSA	0.039	0.171	0.214
LDA	0.095	0.23	0.325
Videopedia	0.366	0.356	0.348

(c) F-Measure

$$mAP = \frac{\sum_{i=1}^{n} AveragePrecision_i}{n}$$

where n represents the total number of webpages for which the videos are retrieved and $AveragePrecision_i$ is the Average Precision for retrieving video for the ith webpage.

As is evident from the results, *Videopedia* performs much better than previous methods, by using a refining stage based on metadata. To decide the actual percentage of videos that should be selected after matching the metadata between the incoming webpage and the videos in the repository, we calculate the precision of the recommendations made by *Videopedia* for different percentages of videos selected. This is reported in Fig. 3b. From the precision values for each of the three recommendations, we find that *Videopedia* performs best when we select the top 10 % of the videos in the repository are selected after matching the metadata. We also perform a comparative study to select the optimal number of topics mined from the transcripts by LDA and report the findings in Fig. 3a.

From the results reported in Figs. 3b and a, we find that our algorithm in *Videopedia*, performs best when it selects 10 % of the videos after matching the metadata of the incoming wikipage with the metadata of the videos. The results also show that the best precision is obtained when LDA returns 10 topics from the content of the webpages and videos. This also provides a more comprehensive topic modeling of the documents.

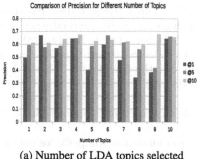

(a) Number of LDA topics selected

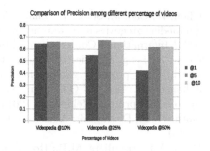

(b) Percentage of video selected for filtering

Fig. 3. Comparison of various methods

5 Conclusions

We designed and demonstrated a novel technique in *VIDEOPEDIA* which integrates multiple media formats – namely making the educational blogs and Wikipedia pages more illustrative with multimedia contents like videos. The main contribution of this work is to automatically recommend relevant educational videos for blogs like Wikipedia. We effectively used topic modeling on automatically generated video transcripts from various video sharing platforms and using them as a representation of the video content. Our promising results show that topic modeling is a good way of video recommendations. This integrated framework reduces the users efforts in searching relevant e-learning videos. Moreover, it provides a mechanism for content-based search. We aim to scale our system for global use in academia and industry in near future. Although we focused on the educational blogs in an e-learning setup, this technique can be extended to other blogs as well.

Acknowledgments. This research has been partially supported by the Singapore National Research Foundation under its International Research Centre @ Singapore Funding Initiative and administered by the IDM Programme Office through the Centre of Social Media Innovations for Communities (COSMIC). This work was also partially supported by JSPS KAKENHI Grant Number 15H06829.

References

1. Cisco visual networking index: Forecast and methodology, 2013–2018 (2014)
2. Adcock, J., Cooper, M., Denoue, L., Pirsiavash, H., Rowe, L.A.: Talkminer: a lecture webcast search engine. In: Proceedings of the ACM International Conference on Multimedia, MM 2010, pp. 241–250 (2010)
3. Blei, D.M., Ng, A.Y., Jordan, M.I.: Latent dirichlet allocation. J. Mach. Learn. Res. **3**, 993–1022 (2003)
4. Chen, H., Cooper, M., Joshi, D., Girod, B.: Multi-modal language models for lecture video retrieval. In: Proceedings of the ACM International Conference on Multimedia, MM 2014, pp. 1081–1084 (2014)

5. Chen, Z., Cao, J., Song, Y., Zhang, Y., Li, J.: Web video categorization based on wikipedia categories and content-duplicated open resources. In: Proceedings of the ACM International Conference on Multimedia, MM 2010, pp. 1107–1110 (2010)
6. Chen, Z., Feng, B., Xie, H., Zheng, R., Xu, B.: Video to article hyperlinking by multiple tag property exploration. In: Gurrin, C., Hopfgartner, F., Hurst, W., Johansen, H., Lee, H., O'Connor, N. (eds.) MMM 2014, Part I. LNCS, vol. 8325, pp. 62–73. Springer, Heidelberg (2014)
7. Gardner, D., Davies, M.: A new academic vocabulary list. Appl. Linguist. **35**, 1–24 (2013)
8. Head, A.J., Eisenberg, M.B.: How today's college students use wikipedia for course-related research. First Monday **15**(3) (2010)
9. Hofmann, T.: Probabilistic latent semantic indexing. In: Proceedings of the 22nd International ACM SIGIR Conference on Research and Development in Information Retrieval, SIGIR 1999, pp. 50–57 (1999)
10. Huang, A.: Similarity measures for text document clustering. In: Proceedings of the Sixth New Zealand Computer Science Research Student Conference, NZCSRSC 2008 (2008)
11. Lee, M.D., Welsh, M.: An empirical evaluation of models of text document similarity. In: Proceedings of XXVIIth Annual Conference of Cognitive Science Society, CogSci 2005, pp. 1254–1259 (2005)
12. Liu, T., Kender, J.R.: Lecture videos for e-learning: current research and challenges. In: Proceedings of the 6th IEEE International Symposium on Multimedia Software Engineering, ISM 2004, pp. 574–578 (2004)
13. Mikolov, T., Sutskever, I., Chen, K., Corrado, G.S., Dean, J.: Distributed representations of words and phrases and their compositionality. In: Advances in Neural Information Processing Systems, pp. 3111–3119 (2013)
14. Misanchuk, E.R., Schwier, R.A., Boling, E.: Visual design for instructional multimedia. In: Proceedings of World Conference on Educational Multimedia, Hypermedia and Telecommunications, pp. 1621–1621 (1999)
15. Okuoka, T., Takahashi, T., Deguchi, D., Ide, I., Murase, H.: Labeling news topic threads with wikipedia entries. In: Proceedings of the 11th IEEE International Symposium on Multimedia, ISM 2009, pp. 501–504 (2009)
16. Repp, S., Linckels, S., Meinel, C.: Question answering from lecture videos based on an automatic semantic annotation. ACM SIGCSE Bull. **40**(3), 17–21 (2008)
17. Roy, S., Mak, M.-T., Wan, K.W.: Wikipedia based news video topic modeling for information extraction. In: Lee, K.-T., Tsai, W.-H., Liao, H.-Y.M., Chen, T., Hsieh, J.-W., Tseng, C.-C. (eds.) MMM 2011 Part II. LNCS, vol. 6524, pp. 411–420. Springer, Heidelberg (2011)
18. Salton, G., McGill, M.J.: Introduction to Modern Information Retrieval. McGraw-Hill Inc., New York (1986)
19. Wang, C., Blei, D.M.: Collaborative topic modeling for recommending scientific articles. In: Proceedings of the 17th ACM SIGKDD International Conference on Knowledge Discovery and Data Mining, pp. 448–456 (2011)
20. Zhu, Q., Shyu, M.-L., Wang, H.: Videotopic: content-based video recommendation using a topic model. In: Proceedings of IEEE International Symposium on Multimedia, ISM 2013, pp. 219–222 (2013)

Towards Training-Free Refinement for Semantic Indexing of Visual Media

Peng Wang[1]([✉]), Lifeng Sun[1], Shiqang Yang[1], and Alan F. Smeaton[2]

[1] National Laboratory for Information Science and Technology, Department of
Computer Science and Technology, Tsinghua University, Beijing, China
{pwang,sunlf,yangshq}@tsinghua.edu.cn
[2] Insight Centre for Data Analytics, Dublin City University, Glasnevin,
Dublin 9, Ireland
alan.smeaton@dcu.ie

Abstract. Indexing of visual media based on content analysis has now
moved beyond using individual concept detectors and there is now a
focus on combining concepts or post-processing the outputs of individual
concept detection. Due to the limitations and availability of training corpora which are usually sparsely and imprecisely labeled, training-based
refinement methods for semantic indexing of visual media suffer in correctly capturing relationships between concepts, including co-occurrence
and ontological relationships. In contrast to training-dependent methods
which dominate this field, this paper presents a training-free refinement
(TFR) algorithm for enhancing semantic indexing of visual media based
purely on concept detection results, making the refinement of initial concept detections based on semantic enhancement, practical and flexible.
This is achieved using global and temporal neighbourhood information
inferred from the original concept detections in terms of weighted non-
negative matrix factorization and neighbourhood-based graph propagation, respectively. Any available ontological concept relationships can
also be integrated into this model as an additional source of external *a
priori* knowledge. Experiments on two datasets demonstrate the efficacy
of the proposed TFR solution.

Keywords: Semantic indexing · Refinement · Concept detection
enhancement · Context fusion · Factorization · Propagation

1 Introduction

Video in digital format is now widespread in everyday scenarios. While mainstream consumer-based use of video such as YouTube and Vine are based on user

P. Wang—This work was part-funded by 973 Program under Grant No.
2011CB302206, National Natural Science Foundation of China under Grant No.
61272231, 61472204, 61502264, Beijing Key Laboratory of Networked Multimedia
and by Science Foundation Ireland under grant SFI/12/RC/2289. We also thank
Prof. Philip S. Yu for helpful discussions.

Q. Tian et al. (Eds.): MMM 2016, Part I, LNCS 9516, pp. 251–263, 2016.
DOI: 10.1007/978-3-319-27671-7_21

tags and metadata, prevailing methods to indexing based on *content* detect the presence or absence of semantic concepts which might be general (e.g., *indoor*, *face*) or abstract (e.g., *violence*, *meeting*). The conventional approach to content-based indexing, as taken in the annual TRECVid benchmarking [12,13], is to annotate a collection covering both positive and negative examples, for the presence of each concept and then to train a machine learning classifier to recognise the presence of the concept. This typically requires a classifier for each concept without considering inter-concept relationships or dependencies yet in reality, many concept pairs and triples will co-occur rather than occur independently. It is widely accepted and it is intuitive that detection accuracy for concepts can be improved if concept correlation can be exploited.

Context-Based Concept Fusion (CBCF) is an approach to refining the detection results for independent concepts by modeling relationships between them [2]. Concept correlations are either learned from annotation sets [3,4,7,15,16] or inferred from pre-constructed knowledge bases [6,18] such as WordNet. However, annotation sets are almost always inadequate for learning correlations due to their limited sizes and the annotation having being done with independent concepts rather than correlations in mind. In addition, training sets may not be fully labeled or may be noisy. The use of external knowledge networks also limits the flexibility of CBCF because it uses a static lexicon which is costly to create. When concepts do not exist in an ontology, these methods cannot adapt to such situations.

In this paper we propose a training-free refinement (TFR) method to exploit inherent co-occurrence patterns for concepts which exist in testing sets, exempt from the restrictions of training corpus and external knowledge structures. TFR can fully exploit global patterns of multi-concept appearance and an ontology (if available), as well as sampling the distribution of concept occurrences in the neighbourhood to enhance the original one-per-class concept detectors, all within a unified framework. Although this reduces the learning/training process, we set out here to see if TFR can still obtain better or comparable performance than the state-of-the-art as such an investigation into refinement of semantic indexing has not been done before.

2   Related Work

The task of automatically determining the presence or absence of a semantic concept in an image or a video shot (or a keyframe) has been the subject of at least a decade of intensive research. The earliest approaches treated the detection of each semantic concept as a process independent of the detection of other concepts, but it was quickly realised that such an approach is not scalable to large numbers of concepts, and does not take advantage of inter-concept relationships. Based on this realisation, there have been efforts within the multimedia retrieval community focusing on utilization of inter-concept relationships to enhance detection performances, which can be categorized into two paradigms: multi-label training and detection refinement or adjustment.

In contrast to isolated concept detectors, *multi-label training* tries to classify concepts and to model correlations between them, simultaneously. A typical multi-label training method is presented in [11], in which concept correlations are modeled in the classification model using Gibbs random fields. Similar multi-label training methods can be found in [20]. Since all concepts are learned from one integrated model, one shortcoming is the lack of flexibility, which means that the learning stage needs to be repeated when the concept lexicon is changed. Another disadvantage is the high complexity when modeling pairwise correlations in the learning stage. This also hampers the ability to scale up to large-scale sets of concepts and to complex concept inter-relationships.

As an alternative, *detection refinement or adjustment* methods post-process detection scores obtained from individual detectors, allowing independent and specialized classification techniques to be leveraged for each concept. Detection refinement has attracted interest based on exploiting concept correlations inferred from annotation sets [2,7,15,16] or from pre-constructed knowledge bases [6,9,18]. However, these depend on training data or external knowledge. When concepts do not exist in the lexicon ontology or when extra annotation sets are insufficient for correlation learning as a result of the limited size of the corpus or of sparse annotations, these methods cannot adapt to such situations. Another difficulty is the matter of determining how to quantify the adjustment when applying the correlation. Though concept similarity [6], sigmoid function [18], mutual information [7], random walk [15,16], random field [2], etc. have all been explored, this is still a challenge in the refinement of concept detections. In a state-of-the-art refinement method for indexing TV news video [3,4], the concept graph is learned from the training set. The migration of concept alinement to testing sets, is based on the assumption of the homogeneity of two data sets, which is not always the case and can reduce the performance of indexing user-generated media, for example. The proposed TRF method in this paper is indeed a refinement methods but tries to tackle the above challenges.

3 Motivation and Proposed Solution

Fusing the results of concept detection to provide better quality semantic analysis and indexing is a challenge. Current research is focused on learning inter-concept relationships explicitly from training corpora and then applying these to test sets. Since the initial results of semantic concept detection will always be noisy because of the accuracy level at which they operate, little work has investigated a refinement approach which directly uses the original detection results to exploit correlations. However, according to the TRECVid benchmark, acceptable detection results can now be achieved, particularly for concepts for which there exists enough annotated training data [12,14]. These detections with high accuracies should be used as cues to enhance overall multi-concept detections since the concepts are highly correlated, though the bottleneck is in the correlation itself which is difficult to precisely model.

For much of the visual media we use in our everyday lives there is a temporal aspect. For example video is inherently temporal as it captures imagery over time and thus video shots or keyframes from shots may have related content because they are taken from the same scene or have the same characters of related activities. Likewise still images of a social event captured in sequence will have semantic relationships based on shared locations, activities or people. For such "connected" visual media it makes sense to try to exploit any temporal relationships when post-processing initial concept detection, and to use the "neighbourhood" aspect of visual media.

Our TFR method is thus motivated based on the following:

- **Reliability:** Detection results for at least some concepts should be accurate enough to be exploited as reliable cues for a refinement process.
- **Correlation:** Instead of occurring in isolation, concepts usually co-occur or occur mutually exclusively among the same samples.
- **Compactness:** Since concept occurrences are not fully independent, detection results can be projected to a compact semantic space.
- **Re-Occurrence:** Concepts will frequently occur across semantically similar samples so where the visual media has temporal relationships such as video keyframes, neighbourhood relationships can be exploited.

Based on the above motivations, the TFR method is proposed which will combine the correlation of individual concepts with various detection accuracies, to improve the performance of overall semantic indexing. The overview of this proposed solution is illustrated in Fig. 1. In Fig. 1(a), initial concept detection is first applied to a set of visual media inputs, returning results denoted as matrix C where each row $c_i (1 \leq i \leq N)$ represents a sample media element such as an image or video shot, while each column corresponds to a concept $v_j (1 \leq j \leq M)$ in the vocabulary. We use different gray levels to represent matrix elements in C, namely the confidences of concept detections.

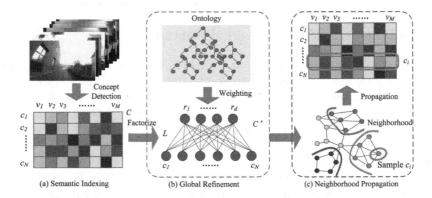

(a) Semantic Indexing   (b) Global Refinement   (c) Neighborhood Propagation

Fig. 1. Illustration of the TFR framework. (a) Semantic Indexing: Media samples indexed through concept detections, returning C. (b) Global Refinement (GR): Refining C as C' using global contextual patterns. (c) Neighbourhood propagation (NP): Refining C' by similarity propagation between nearest neighbours (Color figure online).

As shown in Fig. 1, the refinement procedure involves two stages of global refinement (GR) and neighborhood propagation (NP). The intuition behind GR is that, the high-probable correct detection results are selected to construct an incomplete but more reliable matrix which is then completed by a factorization method. GR in Fig. 1(b) is a weighted matrix factorization process and performs an estimation of concept detection results which were less accurate in the original matrix C. Ontological relationships among concepts if they exist may also be employed to appropriately choose the entry value in the weighted matrix in correspondence to C. In Fig. 1(c), reconstructed concept detection results C' are used to calculate the sample-wise similarity in order to identify a number of nearest neighbours of the target sample c_i. The propagation algorithm is then applied to infer labels iteratively based on neighbours connected to each sample.

4   Training-Free Refinement (TFR)

As illustrated in Fig. 1, GR and NP in the TFR framework are implemented by ontological factorization and graph propagation, which exploit global patterns and local similarities respectively.

4.1   Factorizing Detection Results

In GR, the task of detection factorization is to modify the $N \times M$ matrix C to overlay a consistency on the underlying contextual pattern of concept occurrences. Non-negative matrix factorization (NMF) has shown advantages in scalably detecting the essential features of input data with sparsity, which is more suitable to the semantic indexing refinement task where the annotations are sparse and the confidences in C are non-negative.

The application of NMF here is to represent C as $\tilde{C} = LR$, where vectors in $L_{N \times d}$ and $R_{d \times M}$ can be referred to as d-dimensional sample-related and concept-related latent factors. By applying rules of customized optimization, each confidence value in C can be refined as $\tilde{c}_{ij} = \sum_{k=1}^{d} l_{ik} r_{kj}$. In GR, we optimize the factorization problem in weighted low rank to reflect different accuracies of concept detections. Because each value c_{ij} in C denotes the probability of the occurrence of concept v_j in sample c_i, the estimation of the existence of v_j is more likely to be correct when c_{ij} is high, which is also adopted by [7,17] under the same assumption that the initial detectors are reasonably reliable if the returned confidences are larger than a threshold. To distinguish contributions of different concept detectors to the cost function, we employ a weight matrix $W = (w_{ij})_{N \times M}$ whose elements are larger for reliable and lower for less reliable detections, and optimizing the weighted least square form:

$$F = \frac{1}{2} \sum_{ij} w_{ij} (c_{ij} - L_{i.} R_{.j})^2 + \frac{\lambda}{2} (\|L\|_F^2 + \|R\|_F^2) \tag{1}$$

such that $L \geq 0, R \geq 0$ where $\|\cdot\|_F^2$ denotes the Frobenius norm and the quadratic regularization term $\lambda(\|L\|_F^2 + \|R\|_F^2)$ is applied to prevent over-fitting. After factorization, refinement can be expressed as a fusion of confidence matrices:

$$C' = \alpha C + (1 - \alpha)\tilde{C} = \alpha C + (1 - \alpha)LR \tag{2}$$

To solve the factorization problem, we use a multiplicative method [8] which has the advantage of re-scaling the learning rate instead of optimization with a fixed and sufficient small rate. Without loss of generality, we focus on the update of R in the following derivation and the update rule for L can be obtained in a similar manner. Inspired by [8], we construct an auxiliary function $G(r, r^k)$ of $F(r)$ for fixed L and each corresponding column r, c, w in R, C and W respectively. $G(r, r^k)$ should satisfy the conditions $G(r, r^k) \geq F(r)$ and $G(r, r) = F(r)$. Therefore, $F(r)$ is non-increasing under the update rule [8]:

$$r^{t+1} = argmin_r G(r, r^t) \tag{3}$$

where r^t and r^{t+1} stand for r values in two successive iterations. For function F defined in Eq. (1), we construct G as

$$G(r, r^t) = F(r^t) + (r - r^t)^T \nabla F(r^t) + \frac{1}{2}(r - r^t)^T K(r^t)(r - r^t) \tag{4}$$

where r^t is the current update of optimization for Eq. (1). Denoting $D(\cdot)$ as a diagonal matrix with elements from a vector on the diagonal, $K(r^t)$ in Eq. (4) is defined as $K(r^t) = D(\frac{(L^T D_w L + \lambda I)r^t}{r^t})$, where $D_w = D(w)$ and the division is performed in an element-wise manner.

According to Eq. (3), r can be updated by optimizing $G(r, r^t)$. By solving $\frac{\partial G(r, r^t)}{\partial r} = 0$, we obtain

$$\nabla F(r^t) + K(r^t)r - K(r^t)r^t = 0 \tag{5}$$

where

$$\nabla F(r^t) = L^T D_w(Lr^t - c) + \lambda r^t \tag{6}$$

The combination of Eqs. (5) and (6) achieves the update rule

$$R_{kj}^{t+1} \leftarrow R_{kj}^t \frac{[L^T(C \circ W)]_{kj}}{[L^T(LR \circ W)]_{kj} + \lambda R_{kj}} \tag{7}$$

where $\circ$ denotes Hadamard (element-wise) multiplication and each element in L can be updated similarly. Note that it is not hard to prove convergence under the update rule of Eq. (7) by proving $G(r, r^t)$ as an auxiliary function of F.

4.2   Integration with Ontologies

In Sect. 4.1, we applied weighted NMF (WNMF) to perform low-accuracy concept estimation based on the assumption that the credibility of concepts in C is high enough if their detection confidence is larger than a predefined threshold. If we assign uniform weights for low-confidence concepts, WNMF will adjust confidences in terms of equal chance over these concepts. However, this is not the case in real world applications, where we often have biased estimations. To reflect concept semantics in W we introduce an ontological weighting scheme for WNMF-based global refinement.

To model concept semantics, an ontology is employed to choose appropriate weights for different concepts based on their semantics. The goal is to correctly construct the matrix W which can reflect the interaction between concepts and their detection accuracy. The confidence of sample x belonging to concept v being returned by a detector is represented as $Conf(v|x)$. We introduce the multi-class margin factor [9] as $Conf(v|x) - max_{v_i \in D}Conf(v_i|x)$, where D is the universal set of disjoint concepts of v which contains all concepts exclusively occurring with v.

By employing an ontology we assign each element in W as

$$w_{ij} \propto 1 - [c_{ij} - max_{v_k \in D}c_{ik}] \tag{8}$$

The interpretation of the weighting scheme is that if the disjoint concepts of v_j have higher detection confidences, it is less likely that v_j exists in sample x_i. In this case, the weight for concept v_j needs to be larger, otherwise the weight is lowered by ontology relationships using the multi-class margin.

4.3   Temporal Neighbourhood-Based Propagation

As shown in Fig. 1(c), temporal neighbourhood-based propagation further refines C' to achieve better indexing by exploiting local information between samples which are semantically similar.

Following GR, detection results will have been adjusted in a way consistent with the latent sample/concept factors modeled in WNMF. While this procedure exploits general contextual patterns which are modeled globally by matrix factorization, the similarity propagation method can further refine the result by exploiting any local relationships between samples as demonstrated in Fig. 1(c). In this, it is important to localize highly related temporal neighbours for similarity-based propagation, for which the results C' after GR can provide better measures.

To derive the similarity between samples c_i and c_j, we calculate based on the refined results C' formulated in Eq. (2) by Pearson Correlation, defined as:

$$P_{i,j} = \frac{\sum_{k=1}^{M}(c'_{ik} - \bar{c}'_i)(c'_{jk} - \bar{c}'_j)}{\sqrt{\sum_{k=1}^{M}(c'_{ik} - \bar{c}'_i)^2}\sqrt{\sum_{k=1}^{M}(c'_{jk} - \bar{c}'_j)^2}}$$

where $c'_i = (c'_{ik})_{1 \leq k \leq M}$ is the i-th row of C', and $\bar{c}'_i$ is the average weight for c'_i. To normalize the similarity, we employ the Gaussian formula and denote the similarity as

$$P'_{i,j} = e^{-\frac{(1-P_{i,j})^2}{2\delta^2}} \tag{9}$$

where δ is a scaling parameter for sample-wise distance. Based on this we can localize the k nearest neighbours of any target sample c_i which is highlighted with an orange circle in Fig. 1(c). Neighbours of c_i are indicated with green dots connected with edges quantified by Eq. (9).

For implementing graph propagation, the NP procedure localizes k nearest neighbours for further propagation which are connected with the target sample in an undirected graph. The label propagation algorithm [19] is derived to predict more accurate concept detection results based on this fully connected graph whose edge weights are calculated by the similarity metric in Eq. (9). Mathematically, this graph can be represented with a sample-wise similarity matrix as $G = (P'_{i,j})_{(k+1)\times(k+1)}$, where the first k rows and columns stand for the k nearest neighbours of a target sample to be refined which is denoted as the last row and column in the matrix. The propagation probability matrix T is then constructed by normalizing G at each column as $t_{i,j} = P'_{i,j}/\sum_{l=1}^{k+1} P'_{l,j}$, which guarantees the probability interpretation at columns of T. By denoting the row index of k nearest neighbours of a sample c'_i to be refined as $n_i (1 \leq i \leq k)$ in C' and stacking the corresponding rows one below another, the neighbourhood confidence matrix can be constructed as $C_n = (c'_{n_1}; c'_{n_2}; ...; c'_{n_k}; c'_i)$. The propagation algorithm is carried out iteratively by updating $C_n^t \leftarrow TC_n^{t-1}$, where the first k rows in C_n stand for the k neighbourhood samples in C' indexed by subscript n_i and the last row corresponds to the confidence vector of the target sample c'_i.

Since C_n is a subset of C', the graph G constructed on C_n is indeed a subgraph of the global graph constructed on C' as shown in Fig. 1(c). During each iteration, the neighbourhood concept vector c'_{n_i} needs to be clamped to avoid fading away. After a number of iterations, the algorithm converges to a solution in which the last row of C_n is a prediction based on similarity propagation. In this way, the local temporal relationships between neighbours can be used for a more comprehensive refinement.

5   Experiments and Discussion

We assessed the performance of the TFR approach on two heterogenous datasets, a dataset of still images collected from wearable cameras (Dataset1) and the videos used in the TRECVid 2006 evaluation (Dataset2). We adopted per-concept average precision (AP) for evaluation based on manual groundtruth as well as mean AP (MAP) for all concepts.

5.1   Evaluation on Wearable Camera Images (Dataset1)

For this evaluation, we assess TFR method on the same dataset as in [17], indexed by a set of 85 everyday concepts with 12,248 images collected from 4 users with wearable cameras. To test the performance on different levels of concept detection accuracy, detectors were simulated using the *Monte Carlo* method following the work in [1]. By varying the controlling parameter μ_1 in the range [1.0...5.0], the original detection accuracy results for individual concepts

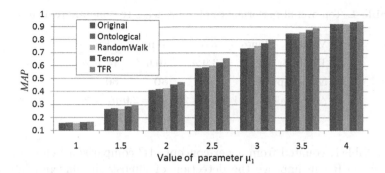

Fig. 2. MAP of TFR refinement, Ontological, Random Walk, Tensor and Original on the wearable sensing dataset (mean over 20 runs)

are simulated and MAP is shown in Fig. 2 (denoted as Original) as semantic indexing results before refinement.

In Fig. 2, the TFR method is compared with a variety of concept detection refinement methods including ontological refinement [18], a Random Walk-based method [15], as well as the state-of-the-art Tensor-based refinement for wearable sensing [17]. An ontology is constructed on 85 concepts with *subsumption* and *disjointness* concept relationships and applied to TFR. Note that the ontology is not a pre-requisite to TFR as shown in Sect. 5.2 in which TFR can still achieve a comparable result to the state-of-the-art without an ontology and training step. To be fair, the Random Walk is performed in the same training-free manner, which means the concept co-occurrence is also inferred from thresholded pseudo-positive samples. We empirically choose the number of latent features as $d = 10$ and we threshold the detection results with 0.3. The fusion parameter in Eq. (2) is simply set to $\alpha = 0.5$, assigning equal importance to the two matrices. We also use 30 nearest neighbours in the propagation step.

As we can see, TFR out-performs all the other methods at all levels of original detection MAP from $0.15 @ \mu_1 = 1.0$ to $0.92 @ \mu_1 = 4.0$. At $\mu_1 = 1.0$, the less significant performance of all refinement approaches makes sense as initial detection accuracy is low. In this case, very few correctly detected concepts are selected for further enhancement which is impractical in real world applications and counter to our assumption of reliability (Sect. 3). When original detection performance is good, as shown in Fig. 2 if $\mu_1 \geq 4.0$, there is no space to improve detection accuracy. Therefore, the improvement is not that significant at $\mu_1 \geq 4.0$ for all refinements. However, TFR still achieves the best refinement in both extreme cases.

The best of the overall improvements of different approaches are shown in Table 1, in which the corresponding accuracy levels are depicted with μ_1 values. As shown, TFR out-performs other approaches significantly and obtains the highest overall MAP improvement of 14.6 %. Recall that Tensor-based refinement uses the temporal neighbourhood patterns within image sequences but is still out-performed by the TFR method. The number of improved concepts is

Table 1. Top overall performance of approaches to semantic refinement.

Method	Onto	RW	Tens	TFR
Top Impr	3.2 %	3.9 %	10.6 %	**14.6 %**
Num Impr	30	56	**80**	**80**
Accu level	$\mu_1 = 1.5$	$\mu_1 = 2.5$	$\mu_1 = 2.0$	$\mu_1 = 2.0$

shown in Table 1, counted from a per-concept AP comparison before and after refinement. TFR can improve the detection of almost all concepts (80 out of 85). Due to the constraints of the ontology model with its fixed lexicon, only a limited number of concepts can be refined in the ontological method (only 30 concepts are improved). However, this does not limit the TFR methods which exploit various semantics.

5.2   Evaluation on TRECVid Video (Dataset2)

Experiments were also conducted in the domain of broadcast TV news to assess the generality of TFR using the TRECVid 2006 video dataset [3,4]. Dataset2 contains 80 h broadcast TV news video segmented into 79,484 shots in total. As a multi-concept detection task, in TRECVid 2006 the dataset is indexed by a lexicon of 374 LSCOM concepts [10] and 20 concepts are selected for performance evaluation with their groundtruth provided.

We employed the reported performance of the official evaluated concepts by VIREO-374 as a baseline[1], which is based on building SVM models of 374 LSCOM concepts [5]. The performance of TFR is also compared to the state-of-the-art domain adaptive semantic diffusion (DASD) [3] technique on the same 20 evaluated concepts by TRECVid using the official metric of $AP@2000$, as shown in Fig. 3.

In our evaluation, TFR is implemented without using a concept ontology. The same parameters are applied directly as were used in Dataset1 without further optimization. As demonstrated, the results on Dataset2 are also promising using the same parameter values of d, α, etc., showing these parameters to be dataset independent. Similar as DASD, TFR achieves consistent enhancement gain against the baseline except for the concept of "Corporate_Leader", which is degraded in terms of performance. This is because "Corporate_Leader" only has 22 positive samples within the 79,484 samples in Dataset2, which makes accurately exploiting contextual patterns from such few samples quite difficult. Over all other 19 concepts, the performance of TFR is comparable with DASD. Interestingly, according to our evaluation TFR does not require many positive samples in order to achieve satisfactory refinement. In Dataset2, the number of positive samples ranges from 150 to 1,556 and there are 10 of the 20 concepts which have less than 300 positive samples but still achieve satisfactory refinement by TFR. Note that DASD is still a training-based refinement method which

---

[1] http://vireo.cs.cityu.edu.hk/research/vireo374/.

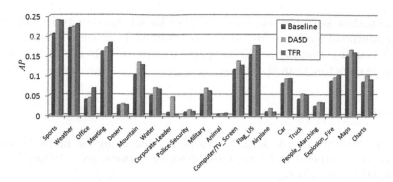

Fig. 3. Per-concept $AP@2000$ comparison on the TRECVid 2006 dataset.

needs to construct an initial concept semantic graph through learning from the TRECVid 2005 dataset whereas training data or *a priori* knowledge are not a pre-requisite for TFR.

5.3 Efficiency Analysis of TFR

In each iteration using Eq. (7), the computational complexity is only relevant to the dimensionality of the matrix C and the selection of low rank d. For a total of *iter* iterations to converge, the running time is thus $O(iter \cdot NMd^2)$.

Recall that $d \leq min\{N, M\}$ and the number of concepts M in the lexicon is usually much smaller than the number of instances in the corpus N. Hence the computational complexity can be simplified as $O(iter \cdot N)$. In our experiments, the updating step of the approximation of L and R only takes several hundred iterations to obtain satisfactory approximation. Thus we empirically fix *iter* = 1, 000 and for Dataset1, it takes approximately 30 s to execute the factorization on a conventional desktop computer.

Similarly, the computational complexity for graph propagation on one target sample can be represented as $O(iter \cdot kM * k^2)$. Since a small fixed value for k is enough in the implementation, the total complexity for neighbourhood-based refinement is also $O(iter \cdot N)$ which indicates the TFR method can be easily scaled up to much larger corpora.

6 Conclusions

Heterogenous multimedia content generated for various purposes usually have high visual and semantic diversities, thus presenting a barrier to the current approaches usually taken to refinement for concept-based semantic indexing, which highly depend on the quality of a training corpus. To ease these challenges, we presented the motivation for a training-free semantic refinement (TFR) of visual concepts, aimed at maximizing indexing accuracy by exploiting trustworthy annotations. TFR can take advantage of various semantics including global contextual patterns, ontologies or other knowledge structures and temporal neighbourhood relationships, all within a unified framework. Though exempt from the training/learning steps, the performance of TFR is still found to be comparable or better than the state-of-the-art.

References

1. Aly, R., Hiemstra, D., de Jong, F., Apers, P.: Simulating the future of concept-based video retrieval under improved detector performance. Multimedia Tools Appl. **60**(1), 203–231 (2012)
2. Jiang, W., Chang, S.-F., Loui, A.: Context-based concept fusion with boosted conditional random fields. In: ICASSP, p. I-949 (2007)
3. Jiang, Y.-G., Dai, Q., Wang, J., Ngo, C.-W., Xue, X., Chang, S.-F.: Fast semantic diffusion for large-scale context-based image and video annotation. IEEE Trans. Image Proc. **21**(6), 3080–3091 (2012)
4. Jiang, Y.-G., Wang, J., Chang, S.-F., Ngo, C.-W.: Domain adaptive semantic diffusion for large scale context-based video annotation. In: ICCV, pp. 1420–1427 (2009)
5. Jiang, Y.-G., Ngo, C.-W., Yang, J.: Towards optimal bag-of-features for object categorization and semantic video retrieval. In: Proceedings of the 6th ACM International Conference on Image and Video Retrieval (CIVR), pp. 494–501. ACM (2007)
6. Jin, Y., Khan, L., Wang, L., Awad, M.: Image annotations by combining multiple evidence & WordNet. In: ACM Multimedia, pp. 706–715 (2005)
7. Kennedy, L.S., Chang, S.-F.: A reranking approach for context-based concept fusion in video indexing and retrieval. In: CIVR, pp. 333–340. ACM (2007)
8. Lee, D.D., Seung, H.S.: Algorithms for non-negative matrix factorization. In: NIPS, pp. 556–562. MIT Press, April 2001
9. Li, B., Goh, K., Chang, E.Y.: Confidence-based dynamic ensemble for image annotation and semantics discovery. In: ACM Multimedia, pp. 195–206 (2003)
10. Naphade, M., Smith, J.R., Tesic, J., Chang, S.-F., Hsu, W., Kennedy, L., Hauptmann, A., Curtis, J.: Large-scale concept ontology for multimedia. IEEE Multimedia **13**(3), 86–91 (2006)
11. Qi, G.-J., Hua, X.-S., Rui, Y., Tang, J., Mei, T., Zhang, H.-J.: Correlative multi-label video annotation. In: ACM Multimedia, pp. 17–26 (2007)
12. Smeaton, A., Over, P., Kraaij, W.: High level feature detection from video in TRECVid: a 5-year retrospective of achievements. In: Divakaran, A. (ed.) Multimedia Content Analysis, Theory and Applications, pp. 151–174 (2008)
13. Smeaton, A.F., Over, P., Kraaij, W.: Evaluation campaigns and TRECVid. In: Proceedings of the ACM International Workshop on Multimedia Information Retrieval, pp. 321–330. ACM (2006)
14. Snoek, C.G.M., Worring, M.: Concept-based video retrieval. Found. Trends Inf. Retrieval **2**(4), 215–322 (2008)
15. Wang, C., Jing, F., Zhang, L., Zhang, H.-J.: Image annotation refinement using random walk with restarts. In: ACM Multimedia, pp. 647–650 (2006)
16. Wang, C., Jing, F., Zhang, L., Zhang, H.-J.: Content-based image annotation refinement. In: CVPR, pp. 1–8 (2007)

17. Wang, P., Smeaton, A.F., Gurrin, C.: Factorizing time-aware multi-way tensors for enhancing semantic wearable sensing. In: He, X., Luo, S., Tao, D., Xu, C., Yang, J., Hasan, M.A. (eds.) MMM 2015, Part I. LNCS, vol. 8935, pp. 571–582. Springer, Heidelberg (2015)
18. Wu, Y., Tseng, B., Smith, J.: Ontology-based multi-classification learning for video concept detection. In: ICME, vol. 2, pp. 1003–1006 (2004)
19. Xu, D., Cui, P., Zhu, W., Yang, S.: Find you from your friends: graph-based residence location prediction for users in social media. In: ICME, pp. 1–6 (2014)
20. Xue, X., Zhang, W., Zhang, J., Wu, B., Fan, J., Lu, Y.: Correlative multi-label multi-instance image annotation. In: ICCV, pp. 651–658. IEEE (2011)

Deep Learning Generic Features
for Cross-Media Retrieval

Xindi Shang[⊠], Hanwang Zhang, and Tat-Seng Chua

School of Computing, National University of Singapore, Singapore, Singapore
{xindi.shang,hanwang,chuats}@comp.nus.edu.sg

Abstract. Cross-media retrieval is an imperative approach to handle the explosive growth of multimodal data on the web. However, how to effectively uncover the correlations between multimodal data has been a barrier to successful retrieval of cross-media data. The traditional approaches learn the connection between multiple modalities by direct utilization of hand-crafted low-level heterogeneous features and the learned correlation are merely constructed in terms of high-level feature representation. To well exploit the intrinsic structures of multimodal data, it is essential to build up an interpretable correlation between multimodal data. In this paper, we propose a deep model to learn the high-level feature representation shared by multiple modalities for cross-media retrieval. We learn the discriminative high-level feature representation in a data-driven manner before faithfully encoding the multimodal correlations. We use the large-scale multimodal data crawled from Internet to train our deep model and evaluate its effectiveness on cross-media retrieval based on NUS-WIDE dataset. The experimental results show that the proposed model outperforms other state-of-the-arts approaches.

Keywords: Multimodal analysis · Retrieval · Deep learning

1 Introduction

The web contents have evolved from text-based to increasingly multimedia with the emergence of huge amount of multimodal data such as text, image, video and audio. For example, users upload 15 million images per hour on Facebook while a high proportion of these images are without proper text annotations. This brings up a huge challenge in multimedia information retrieval, because the traditional approaches that fundamentally rely on text-based search will inevitably fail when there is little or no surrounding text for the multimedia data. In order to resolve this challenge, we need to find the correlations between different modalities. Without loss of generality, in this paper, we focus on the typical cross-modality correlation paradigm between text and image. In particular, there are two retrieval scenarios we need to consider: text-to-image and image-to-text. For the first case, given a text query, search through the database to return a small set of images that are most relevant to the textual description. Likewise, the second image-to-text task retrieves a list of relevant words

© Springer International Publishing Switzerland 2016
Q. Tian et al. (Eds.): MMM 2016, Part I, LNCS 9516, pp. 264–275, 2016.
DOI: 10.1007/978-3-319-27671-7_22

or matching paragraphs according to some image query, which is also known as automatic image annotation [4].

We propose to deeply learn the generic features for multiple modalities. Similar to Canonical Correlation Analysis (CCA), which finds a linear projection that maximizes the correlation of the two projected sets, our model maps image modality and text modality into some shared space where the visual and textual representations are connected in the sense of maximizing correlation. We adopt deep architecture to make the mapping non-linear and generate proper generic features through its top shared layer. Because practical retrieval system is supposed to fit the constantly evolving incoming data, we also propose to train and test the model separately on different datasets to demonstrate the generality of the learned features. We want to highlight three key properties of the generic features that make our model more effective and different from other state-of-the-arts models.

Accurate Reconstruction. The generic features produced by our model are trained to reconstruct both image and text modalities when given only one modality as input. This training setting enables the features to contain useful cross-media information. Besides, it is widely known that a deep model can result in high-level discriminative feature at its top layer. Thus, our generic features fully exploit the robustness and semantic nature of large-scale multimodal data.

Sparse Representation. We regularize the generic features to be sparse. This regularization makes only a few variables selected in the shared layer to generate the representation for multimodal data, while keeping others zero. So the units can be more independent from each other, which leads to rich representations in the shared layer.

High Correlation. We embed Canonical Correlation Analysis in the proposed model to correlate the variables generated from different modalities. In particular, we can produce a representation in the shared layer for each modality by only providing one modality while the other is absent. The two representations are then tuned for maximizing the correlation between them, so they are more closely interrelated.

The overview of the proposed approach is illustrated in Fig. 1. In order to learn the generic features for cross-media retrieval, we propose a novel deep model, named Correlation Multimodal Auto-Encoder (Corr-MAE). Corr-MAE has two pathways for image and text modalities to learn high-level features in a layer-wise manner(i.e., from 'Layer0' to 'Layer2'). The two pathways are then merged into a shared layer, where the generic features of both modalities are correlated at high-order. Based on Corr-MAE, the generic feature of image (text) query is generated through image (text) pathway and finally gets its representation in the shared layer. The generic feature of query is then compared with those of the other modality using some metric. The closest one is regarded as the most similar retrieved result in semantic according to the query. In order to demonstrate the effectiveness of the learned features, we train the proposed model on User-Generated Content (UGC) and test it on a separate standard dataset.

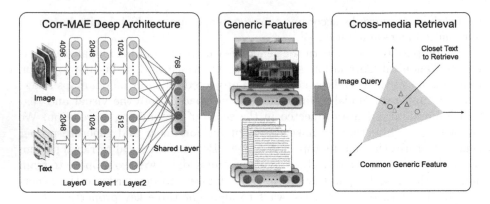

Fig. 1. The overview of the proposed Corr-MAE with the number of units in each layer marked. The generic features produced are then used for cross-media retrieval. The circles in the right box represent the generic feature of image query, and the triangles represent the generic feature of texts.

The experimental results show that the proposed generic feature is effective in supporting real-world cross-media retrieval.

The rest of the paper is organized as follows. Section 2 reviews related works. Section 3 describes the machine learning framework of the proposed Corr-MAE. The experimental settings and results are reported in Sect. 4, followed by conclusions in Sect. 5.

2   Related Work

To perform cross-media retrieval, most existing methods bridge the heterogeneity-gap by mapping different feature spaces of multimodal data into a common (or shared) space. In the common space, some shared high-level features of different modalities should be extracted for enhancing the retrieval system to return more semantic-based results with respect to user's query. This goal has encouraged researchers to propose many criteria to find such a mapping. Rasiwasia et al. [13] introduced CCA into cross-media retrieval, so the mapping is a pair of linear transformations that maximize the correlation of two set of features in some subspace. Wu et al. [19] proposed a supervised approach by minimizing an empirical ranking risk, within which two linear mappings project similar samples as neighbours in the common space. They formulated the optimization procedure as a Structural-SVM problem. Dictionary learning is also studied and adapted to capture the intrinsic heterogeneous features by learning different dictionaries for multimodal data [8].

However, all the above methods fall into the category of linear model, which can hardly adapt to real-world application because of the high non-linearity and inherent correlations between modalities in such data. Even though some methods have kernelized extensions, it is still difficult to choose a proper kernel for real-world data and the learned feature space is limited to manually pre-defined kernels. In order to address these problems, Andrew et al. [1] developed

Deep-CCA (DCCA) using deep networks. DCCA constructs a deep network for each modality and pre-trains them separately using denoising auto-encoder [18]. In the fine-tuning phase, it correlates the two networks by updating their weights simultaneously while maximizing the correlation of two top layers as CCA does. Due to the empirical success of deep model on various tasks, DCCA is able to capture more intrinsic high-level features than CCA. But the DCCA features will be gradually distorted as the fine-tuning proceeds, because its objective function only considers the correlation of two top levels. Therefore, the final learned features might poorly reflect the inherent properties of actual data.

Another multimodal deep architecture is Multimodal Auto-Encoder (MAE) [12]. It is pre-trained using sparse Restricted Boltzmann Machines and fine-tuned by minimizing reconstruction error. So its top levels can produce proper representations from the original data. To correlate multiple modalities, MAE reconstructs one modality given the others missing, which can indirectly generate shared representations. Our deep model follows this reconstruction strategy. Besides, we propose to incorporate the criterion of CCA into the fine-tuning phase, since maximizing the correlation is a more straightforward way to correlate two modalities. Deep Belief Nets are also well studied in modelling heterogeneous data [16,20]. Srivastava and Salakhutdinov developed a multimodal Deep Boltzmann Machine [16] that models joint probability of image and text samples. This model also learns some high-level shared features from different modalities but it may suffer from the early-stopping issue because there is no fine-tuning phase.

3 Deep Learning Generic Features

The basic architecture of the proposed Corr-MAE is similar to the deep network described in [12]. As shown in Fig. 1, it has two parallel pathways separately constructed for image and text modalities. Each pathway consists of stacked encoders and transforms the input into lower dimension layer by layer. In this paper, we set the number of layers to 3 and the dimensions of the three layers to 4096, 2048, 1024 (2048, 1024, 512) for image (text) pathway. We also map the top layers of two pathways into a 768-D shared layer, denoted as $\mathbf{z}$, in order to model the relationship between the two modalities. The architecture is formulated as follows:

$$\mathbf{p}^k = \sigma(\mathbf{W}_p^k \mathbf{p}^{k-1} + \mathbf{c}_p^k), \quad \mathbf{t}^k = \sigma(\mathbf{W}_t^k \mathbf{t}^{k-1} + \mathbf{c}_t^k), \quad k = 1, 2,$$

$$\mathbf{z} = \sigma([\mathbf{W}_p^3 \ \mathbf{W}_t^3] \begin{bmatrix} \mathbf{p}^2 \\ \mathbf{t}^2 \end{bmatrix} + \mathbf{c}^3), \tag{1}$$

where $\sigma(\cdot)$ is the element-wise sigmoid function, $\mathbf{p}^k$ and $\mathbf{t}^k$ are denoted as the kth layers of the image pathway and text pathway, $\mathbf{z}_p$ and $\mathbf{z}_t$ are the representations in the shared layer produced from the two pathways. Additionally, $\mathbf{W}_p^k$ and $\mathbf{c}_p^k$ are trainable weight and encoding bias in the image pathway; while $\mathbf{W}_t^k$ and $\mathbf{c}_t^k$ are the parameters for the textual data.

3.1 Layer-Wise Pre-training

Pre-training is an important part of training deep network, since it initializes the parameters to a good solution [2]. We use the Restricted Boltzmann Machine (RBM) [15] to pre-train the proposed model in a greedy layer-wise manner. RBM is a generative stochastic neural network that can learn a joint distribution over its visible layer and hidden layer. We assign any two connected layers in our deep network to the RBM, so that $\mathbf{W}_p^k$, $\mathbf{W}_t^k$, $\mathbf{c}_p^k$ and $\mathbf{c}_t^k$ mentioned above can be properly initialized. In particular, the proposed deep model can be viewed as stacked RMBs. We train these RBMs one by one from the bottom to the top. As the inputs of the whole model are real-valued, we use the continuous version of RBM, Gaussian-Binary RBM [2], for the bottom layers in the image and text pathways; while the other layers are pre-trained by classical RBM. To sum up, there are four RBMs in the two pathways and one RBM for the parameters between the joint layer and the shared layer. We use Contrastive Divergence [17] to estimate the gradients in the RBMs.

3.2 Overall Fine-Tuning

In order to learn generic multimodal features, the objective function of the fine-tuning should not only encourage the network to preserve the intrinsic information in each modality but also correlate the two modalities. We minimize the reconstruction error as deep auto-encoder [6] does. We define the decoding functions of image pathway as:

$$\hat{\mathbf{p}}^2 = \sigma(\mathbf{W}_p^{3^T}\mathbf{z} + \mathbf{b}_p^3), \quad \hat{\mathbf{p}}^1 = \sigma(\mathbf{W}_p^{2^T}\hat{\mathbf{p}}^2 + \mathbf{b}_p^2), \quad \hat{\mathbf{p}}^0 = \mathbf{W}_p^{1^T}\hat{\mathbf{p}}^1 + \mathbf{b}_p^1, \quad (2)$$

where $\mathbf{b}_p^k$ is the decoding bias and $\mathbf{W}_p^k$ is the transformation matrix in the encoding function. Note that the decoding function from layer 1 to layer 0 is not sigmoidal; this is because we need to retain the intrinsic Gaussian of the input data. The decoding function of text pathway can be defined in a similar way.

Based on the encoding/decoding functions in Eqs. (1) and (2), we write the reconstructed image as a function of all the input modalities, i.e. $\hat{\mathbf{p}}^0(\mathbf{p}^0, \mathbf{t}^0)$. We also denote $\hat{\mathbf{p}}^0(\mathbf{p}^0, \mathbf{0})$ and $\hat{\mathbf{p}}^0(\mathbf{0}, \mathbf{t}^0)$ as the reconstructed result when one input modality is absent, and minimize these two reconstruction errors. This is to prevent different part of the representation in the shared layer from overfitting to different single modality. The three terms for the reconstructed text can be written out likewise. Therefore, we have three reconstruction error terms to minimize: $R_{p\bar{t}}$, $R_{\bar{p}t}$ and R_{pt}. For example,

$$R_{p\bar{t}} = \sum_i \left(\left\| \mathbf{p}_i^0 - \hat{\mathbf{p}}^0(\mathbf{p}_i^0, \mathbf{0}) \right\|_2^2 + \left\| \mathbf{t}_i^0 - \hat{\mathbf{t}}^0(\mathbf{p}_i^0, \mathbf{0}) \right\|_2^2 \right) \quad (3)$$

represents the reconstruction error when we only provide the image input but fill the text input with zeros.

Furthermore, we regularize the representations $\mathbf{z}(\mathbf{p}^0, \mathbf{0})$ and $\mathbf{z}(\mathbf{0}, \mathbf{t}^0)$ in the shared layer, so that the units in each one are as independent as possible while the correlation between the two is maximized. We apply l_1-penalty on the representations, which is related to a relaxed version of Independent Component Analysis [10], so the resulted representations can be more robust and abundant. To correlate the two representations, we apply Canonical Correlation Analysis on them, which is inspired by DCCA [1]. Denoting $\mathbf{Z}_1, \mathbf{Z}_2$ as matrices whose columns are $\mathbf{z}(\mathbf{p}^0, \mathbf{0})$ and $\mathbf{z}(\mathbf{0}, \mathbf{t}^0)$ produced from n training samples, let $\bar{\mathbf{Z}}_1 = \mathbf{Z}_1 - \frac{1}{n}\mathbf{Z}_1\mathbf{1}$ be the centered matrix (resp. $\bar{\mathbf{Z}}_2$) and define cross-covariance matrix $\boldsymbol{\Sigma}_{12} = \frac{1}{n-1}\bar{\mathbf{Z}}_1\bar{\mathbf{Z}}_2^T$ as well as covariance matrix $\boldsymbol{\Sigma}_{11} = \frac{1}{n-1}\bar{\mathbf{Z}}_1\bar{\mathbf{Z}}_1^T + r_1\mathbf{I}$ (resp. $\boldsymbol{\Sigma}_{22}$). Herein, $r_1 > 0$ is a regularization parameter guaranteeing the positive definiteness of $\boldsymbol{\Sigma}_{11}$. The total correlation of the top k components of $\mathbf{Z}_1$ and $\mathbf{Z}_2$ is the sum of the top k singular values of the matrix $\mathbf{T} = \boldsymbol{\Sigma}_{11}^{-1/2}\boldsymbol{\Sigma}_{12}\boldsymbol{\Sigma}_{22}^{-1/2}$ [5]. In this paper, we set k equal to the dimension of the shared layer, so that the correlation is exactly the matrix trace norm of $\mathbf{T}$:

$$Corr\left(\mathbf{Z}_1, \mathbf{Z}_2\right) = \|\mathbf{T}\|_{tr} = tr(\mathbf{T}^T\mathbf{T})^{1/2}. \tag{4}$$

In summary, the objective function of the overall fine-tuning is showed in Eq. (5), where θ denotes the set of all trainable parameters of the model and λ, γ are positive trade-off hyperparameters. In order to maximize the correlation and guarantee a lower bound, we take the inverse of $Corr\left(\mathbf{Z}_1, \mathbf{Z}_2\right)$, which also assures that non-correlated (i.e. close to zero) representations are heavily penalized. We adopt stochastic gradient descent to update θ and apply weight decay to regularize the learned parameters.

$$L(\mathbf{p}^0, \mathbf{t}^0; \theta) = R_{pt} + R_{p\bar{t}} + R_{\bar{p}t} + \lambda(\|\mathbf{Z}_1\|_1 + \|\mathbf{Z}_2\|_1) + \gamma Corr^{-1}\left(\mathbf{Z}_1, \mathbf{Z}_2\right) \tag{5}$$

3.3   Cross-Media Retrieval

We use the proposed Corr-MAE to produce generic features for cross-media retrieval task. We consider the following two tasks.

Text-to-image Retrieval: Given a text query, we first extract the 2048-D low-level feature of the text, then pass the feature into text pathway to obtain the representation in the shared layer as the generic feature. This generic feature is then compared to the generic features of the other modality data (i.e. image) according to some metric, and retrieve the K nearest samples as the retrieval results. Generally, l_1, l_2-distance, normalized correlation (NC) and Kullback-Leibler divergence (KL) can be used to measure the similarity of features.

Image-to-text Retrieval: Given an image query, we transform its visual feature into the shared layer and use the generic feature to retrieve a set of tags that accurately describe it. This task is more challenging than the above one, since going from a generic feature in the shared layer to a coherent set of tags requires a sophisticated reconstruction or decoding function [7,21]. Designing such a algorithm is beyond the scope of our work. In this paper, we search the database that consists of pre-defined tag-lists to find the most similar tag-list whose generic feature is closest to the query's.

4   Experiments and Results

4.1   Experimental Setup

Training Dataset. We constructed the training dataset by crawling image-text multimodal data from Flickr. The crawling protocol is as follows. First, we found 500 seed users who contribute the most number of interesting photos using Flickr API. Then, for each of the seed user, we selected up to 100 users from its contact, resulting in around 50,000 users in total. Finally, we crawled all their uploaded photos together with the corresponding tags. We obtained 1,321,496 image-text pairs crawled during 10 continuous days in Jan, 2014.

We extracted image features through a deep CNN, AlexNet [9], trained on ILSVRC-2012 dataset [14]. Given an image, we first resized its shorter edge to 256 pixels and passed it into AlexNet. Then, we used the output of the second fully connection layer as its visual feature, which is 4096-D. For textual features, we developed a novel feature extraction pipeline as follows. First, we transformed each word in a tag list into a 300-D vector according to a pre-trained Word2Vec [11] model trained on Google News data of 100 billion words. In our training dataset, we excluded those words that do not appear in the pre-trained model, resulting in about 100,000 words. Second, we learned a 2048-D dictionary from the overlapped 100,000 words by applying K-means. Third, we obtained 2048-D sparse codes for all the words using a prevailing sparse coding toolbox, and max-pooled the sparse codes of the words in a tag list to form a 2048-D textual feature for the tag list. Different from traditional word-frequency methods, the third step of our pipeline sufficiently utilizes the linguistic property of the 300-D Word2Vec vectors, which is known as a closure under linear combination, so as to produce features having more semantic meanings and non-zero dimensions (30–50 non-zero). Finally, we normalized both image features and textual features into a zero-mean unit-variance Gaussian distribution.

Test Dataset and Evaluation. We performed the comparative experiments on a well-labelled dataset: NUS-WIDE [3]. The dataset consists of 269,648 images with 5,018 unique tags and ground-truth for 81 concepts. We adopted the visual feature as mentioned above. For textual feature, we conducted the same feature extraction procedure but used the clustering centres obtained on training dataset, because we found that almost all the words in NUS-WIDE dataset are contained in the training dataset if we ignore those words not in the Word2Vec model.

For evaluation, we randomly selected 100 images (texts) for each concept to form the query set in image-query-text (text-query-image) task. Denoting both image and text as document, we define two documents that share at least one concept as relevant. Then, given a query and the top K retrieved documents, the precision of the retrieval is the percentage of relevant documents. We used mean Average Precision (mAP) as our performance measures of text-query-image and image-query-text tasks. Note that larger mAP score corresponds to better performance.

Comparing Methods. To verify the effectiveness of our proposed model on real-world cross domain multimodal dataset, we compared Corr-MAE with three state-of-the-arts models: (a) Bi-CMSRM [19], (b) DCCA [1] and (c) MAE [12]. The first chosen model is a linear one; this is because our model employs the high non-linearity of deep architecture, so we want to demonstrate its ability to model the complex real-world data by comparing it with the best linear model. The other two chosen models are deep model, but both of them fail to jointly considerate the reconstruction error and correlation of multimodalities; we want to show the effectiveness of our model in jointly considering both features.

Note that all the abovementioned models, except Bi-CMSRM, accept training examples that are presented as pairs of text and image. Bi-CMSRM on the other hand trains the model in a supervised manner, so it needs one training example to be the query and the corresponding ranking results in other modality as the answer, as stated in the original paper [19]. In particular, for each text (image) query, we randomly selected 40 images (texts) in the other modality in our training dataset as candidates, and then formed a ranking example according to whether the candidate is relevant to the query.

Implementation Details. We started pre-training RBMs with initial weights uniformly sampled from the range of $[-0.05, +0.05]$ and initialized the biases as -2. The mini-batch size and weight decay are set as 500 and 0.0001 respectively. The learning rate of the pre-training phase was initialized to 0.1 and decayed by dividing 1.000015 at each epoch. We stopped the training of each RBM after 1000 epochs. For the fine-tuning phase, we initialized the parameters according to the pre-trained RBMs. We reduced the starting learning rate to 0.001 and dynamically adjusted it by monitoring the objective function. As cross-media retrieval, we used l_2-distance to compute the similarity.

Table 1. Mean Average Precision for top 10 retrieved results (mAP@10).

Methods	mAP@10	
	Text query	Image query
Bi-CMSRM	0.72 %	1.86 %
DCCA	1.50 %	1.23 %
MAE	1.10 %	2.75 %
Corr-MAE	**1.52 %**	**5.15 %**

4.2 Experimental Results

Table 1 reports the performance of the proposed model, Corr-MAE, and the other comparative models on the NUS-WIDE testing dataset, showing that Corr-MAE outperforms all the comparative approaches on both image-to-text and text-to-image retrieval tasks. It is worth noting that the performance of Corr-MAE on

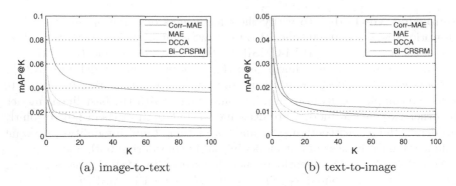

Fig. 2. Mean Average Precision for top $K(K = 1 \ldots 100)$ retrieved results (mAP@K) over 8,100 image (text) queries (100 queries for each of 81 concepts).

image-to-text retrieval task is about twice higher than the second best one. These results demonstrate that the proposed model can learn effective generic features for multimodal data, leading to a better performance on cross-media retrieval.

The improvement is firstly due to the deep architecture's ability to capture the complex and semantic information from the real-world data. By comparing the deep models (i.e. DCCA, MAE, Corr-MAE) with the linear one (Bi-CMSRM), we can see that the deep architecture usually archives better performance than the shallow model. Secondly, the improvement also comes from the novel method we used to fine-tune the overall deep model. Since both MAE and DCCA consider only one of reconstruction and correlation, they fail to learn features that not only well reconstruct the original data but also correlate multi-modalities. Corr-MAE takes both of these two critical factors into consideration, thus we find that Corr-MAE outperforms MAE and DCCA from Table 1. Figure 3 shows the detailed performance of retrieval on each of 81 concepts in NUS-WIDE.

Figure 2(a) and (b) report the mAP@K curves on image-to-text and text-to-image retrieval respectively. These curves illustrate the trends of the retrieval performance as the number of retrieved documents increases. We can see that Corr-MAE still outperforms the other comparative methods for $K = 1 \ldots 100$. In addition, we emphasize again that the training dataset and testing dataset come from different sources. Because the training set cannot indicate the distribution of testing set in this experimental setting, it is hard for a model to perform very well simply using the learned distribution, but requires the model to learn robust invariant features of data. The experimental results also show that the proposed model can learn more robust generic features than the other models.

Figures 4 and 5 present actual retrieval results by different comparing methods. Shown in Fig. 4 is the text-to-image retrieval results based on the list of keywords given in the left box. We can see that Corr-MAE could respond better to those words containing more abstractive content, such as 'abandoned' (corresponds to the 2nd result), while other methods only retrieved material items

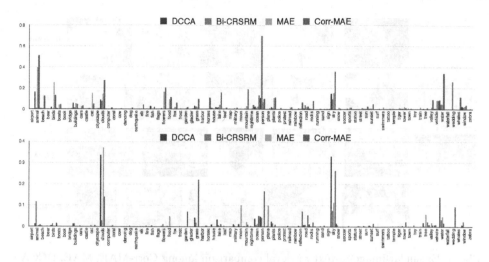

Fig. 3. Detailed performance of retrieval (mAP@10) over 81 concepts of NUS-WIDE. The upper one is the result of image-to-text retrieval; the lower one is the result of text-to-image retrieval.

Fig. 4. Exemplar text-to-image retrieval comparison among Corr-MAE, MAE, DCCA and Bi-CMSRM on the NUS-WIDE dataset, where a list of keywords is used as query.

such as 'sea'. Figure 5 shows the results of image-to-text retrieval. We manually replaced the retrieved textual samples with the corresponding images for better visualization. From the results, we can see that Corr-MAE and MAE returned almost all the correct terms (i.e. 'cat'), while the other two methods only returned irrelevant content, which may due to the noisy data in the training set.

Fig. 5. Exemplar image-to-text retrieval comparison among Corr-MAE, MAE, DCCA and Bi-CMSRM on the NUS-WIDE dataset. The retrieved texts are replaced by the corresponding images.

5   Conclusions

In this paper, we presented a novel approach to learn generic features for cross-media retrieval. We adopted deep architectures to model multiple modalities and related them using a shared layer. In particular, we trained the proposed model, named Correlation Multimodal Auto-Encoder (Corr-MAE), to generate generic features in the shared layer that well reconstructs multimodal data and correlates cross-modalities. We also used the sparsity penalty to enrich the features in the shared layer. By training and evaluating on different datasets, we have demonstrated that the proposed model can learn robust and effective generic features.

Acknowledgments. This research is supported by the Singapore National Research Foundation under its IRC@Singapore Funding Initiative and administered by IDMPO.

References

1. Andrew, G., Arora, R., Bilmes, J., Livescu, K.: Deep canonical correlation analysis. In: ICML (2013)
2. Bengio, Y., Lamblin, P., Popovici, D., Larochelle, H., et al.: Greedy layer-wise training of deep networks. In: NIPS (2007)
3. Chua, T.-S., Tang, J., Hong, R., Li, H., Luo, Z., Zheng, Y. Nus-wide: a real-world web image database from national university of singapore. In: CIVR (2009)
4. Dong, J., Cheng, B., Chen, X., Chua, T.-S., Yan, S., Zhou, X.: Robust image annotation via simultaneous feature and sample outlier pursuit. In: TOMCCAP (2013)

5. Hardoon, D.R., Szedmak, S., Shawe-Taylor, J.: Canonical correlation analysis: an overview with application to learning methods. Neural Comput. **16**, 2639–2664 (2004)
6. Hinton, G.E., Salakhutdinov, R.R.: Reducing the dimensionality of data with neural networks. Science **313**, 504–507 (2006)
7. Hsu, D., Kakade, S., Langford, J., Zhang, T.: Multi-label prediction via compressed sensing. In: NIPS (2009)
8. Jia, Y., Salzmann, M., Darrell, T.: Factorized latent spaces with structured sparsity. In: NIPS (2010)
9. Krizhevsky, A., Sutskever, I., Hinton, G.E.: Imagenet classification with deep convolutional neural networks. In: NIPS (2012)
10. Le, Q.V., Karpenko, A., Ngiam, J., Ng, A.Y.: Ica with reconstruction cost for efficient overcomplete feature learning. In: NIPS (2011)
11. Mikolov, T., Chen, K., Corrado, G., Dean, J.: Efficient estimation of word representations in vector space. arXiv preprint arXiv:1301.3781 (2013)
12. Ngiam, J., Khosla, A., Kim, M., Nam, J., Lee, H., Ng, A.Y.: Multimodal deep learning. In: ICML (2011)
13. Rasiwasia, N., Costa Pereira, J., Coviello, E., Doyle, G., Lanckriet, G.R., Levy, R., Vasconcelos, N.: A new approach to cross-modal multimedia retrieval. In: ACMMM (2010)
14. Russakovsky, O., Deng, J., Su, H., Krause, J., Satheesh, S., Ma, S., Huang, Z., Karpathy, A., Khosla, A., Bernstein, M., Berg, A.C., Fei-Fei, L.: ImageNet large scale visual recognition challenge. Int. J. Comput. Vis. (IJCV) 1–42 (2014)
15. Smolensky, P.: Information processing in dynamical systems: Foundations of harmony theory (1986)
16. Srivastava, N., Salakhutdinov, R.: Multimodal learning with deep boltzmann machines. In: NIPS (2012)
17. Tieleman, T.: Training restricted boltzmann machines using approximations to the likelihood gradient. In: ICML (2008)
18. Vincent, P., Larochelle, H., Bengio, Y., Manzagol,P.-A.: Extracting and composing robust features with denoising autoencoders. In: ICML (2008)
19. Wu, F., Lu, X., Zhang, Z., Yan, S., Rui, Y., Zhuang, Y.: Cross-media semantic representation via bi-directional learning to rank. In: ACMMM (2013)
20. Yuan, Z., Sang, J., Liu, Y., Xu, C.: Latent feature learning in social media network. In: Proceedings of the 21st ACM International Conference on Multimedia, pp. 253–262. ACM (2013)
21. Zhang, Y., Schneider, J.G.: Multi-label output codes using canonical correlation analysis. In: AI Statistics (2011)

Cross-Media Retrieval via Semantic Entity Projection

Lei Huang and Yuxin Peng$^{(\boxtimes)}$

Institute of Computer Science and Technology, Peking University,
Beijing 100871, China
{lei.huang,pengyuxin}@pku.edu.cn

Abstract. Cross-media retrieval is becoming increasingly important nowadays. To address this challenging problem, most existing approaches project heterogeneous features into a unified feature space to facilitate their similarity computation. However, this unified feature space usually has no explicit semantic meanings, which might ignore the hints contained in the original media content, and thus is not able to fully measure the similarities among different media types. By considering the above issues, we propose a new approach to cross-media retrieval via semantic entity projection (SEP) in this paper. Our approach consists of three main steps. Firstly, an entity level with fine-grained semantics between low-level features and high-level concepts are constructed, so as to help bridge the semantic gap to a certain extent. Then, an entity projection is learned by minimizing both cross-media correlation error and single-media reconstruction error from low-level features to the entity level, with which a unified feature space with explicit semantic meanings can be obtained from low-level features. Finally, the semantic abstraction of high-level concepts is generated by using logistic regression to conduct cross-media retrieval. Experimental results on the Wikipedia dataset show the effectiveness of the proposed approach.

Keywords: Cross-media retrieval · Entity · Entity projection · Semantic abstraction

1 Introduction

Nowadays there are a large amount of multimedia resources on the Internet. In order to better utilize these multimedia resources, great efforts have been devoted to content-based multimedia retrieval. In the beginning, many researchers have concentrated on single-media retrieval, where the queries and the retrieved results are from the same modality, such as text retrieval [10], image retrieval [5] and video retrieval [11]. Subsequently, in order to exploit the complementation among different modal data, multi-modal retrieval has attracted much research interest. In this case, the queries and the retrieved results usually share the same multiple modalities, and the approaches to conducting multi-modal fusion in feature level, decision level or hybrid level have been prevalently adopted [1]. Moreover, other approaches, such as image annotation [8] and video annotation [19], have also been developed rapidly to support keyword-based retrieval.

© Springer International Publishing Switzerland 2016
Q. Tian et al. (Eds.): MMM 2016, Part I, LNCS 9516, pp. 276–288, 2016.
DOI: 10.1007/978-3-319-27671-7_23

Till now, user demands have been developing towards diversification. For instance, a user provides an image of the Great Wall in hand, and wants to retrieve the corresponding articles, images or videos, which can present a full description for the query. However, the above mentioned techniques are not suitable in this situation. Hence, in recent years, cross-media retrieval has become an important topic in the literature. The goal is to uniformly manage multimedia objects of different media types, where the queries and the retrieved results are not necessary to be of the same media type. Through cross-media retrieval, users can retrieve whatever they want by providing what they have.

The most challenging problem in cross-media retrieval is how to exploit the semantic correlation between low-level features of multimedia objects and high-level concepts. Generally speaking, the existing approaches to cross-media retrieval can be divided into two main categories: the unified feature representation based approaches and the unified graph model based approaches. The unified feature representation based approaches try to obtain a unified feature representation by projecting heterogeneous features of different modalities into a unified feature space, and then compute the similarities between two arbitrary multimedia objects based on the unified feature representation. The representative approaches of this kind include the unsupervised ones, such as Canonical Correlation Analysis (CCA) [9] and Cross-modal Factor Analysis (CFA) [12], as well as the supervised ones [7, 14–18, 21]. However, the unified feature space obtained by these approaches usually has no explicit semantic meanings, which might ignore the hints contained in the original media content and thus is not able to fully measure the similarities among different media types. The unified graph model based approaches [13, 20, 22] utilize manifold learning techniques and construct a unified graph model in which media objects of different modalities are denoted as nodes, and the similarity measures are learned by the unified graph model. However, this kind of approaches lack out-of-sample generalization capability and they cannot effectively project the new coming queries into the manifold. In these cases, relevance feedback is usually required to guarantee the retrieval performance.

As we all know, an entity can be distinguished from other entities. Generally, entities can appear among different modalities, and the related entities often appear in the related high-level concepts. In addition, entities are scalable to different domain due to their objective existence. Therefore, in consideration of distinguishability, correlation and scalability of entities, the entity level can be a natural bridge between low-level features and high-level concepts. The entities appearing in a certain document can be regarded as fine-grained entity-level semantic labels, in contrast to coarse-grained high-level concepts, and then the unified representation of different modalities can be obtained at the entity level.

By considering the above issues, a new approach to cross-media retrieval via semantic entity projection (SEP) is proposed in this paper. Our approach consists of three main steps. Firstly, the fine-grained entities with explicit semantic meanings are automatically extracted by applying the mature techniques in text processing domain. Thus, we build an entity-level connection between low-level features and high-level concepts, which can help bridge the semantic gap to a certain extent. Then, an entity projection from low-level features to the entity level is learned, and finally the semantic abstraction of high-level concepts is generated by using logistic regression to conduct cross-media retrieval.

The contribution of the proposed approach is two-fold. Firstly, the proposed approach constructs an entity level with fine-grained semantics between low-level features and high-level concepts, which can help bridge the semantic gap to a certain extent. Secondly, by the entity projection and the semantic abstraction, we build a concise and effective framework for cross-media retrieval. The proposed approach is experimentally evaluated on a publicly available cross-media dataset (the Wikipedia dataset [17]), and the experimental results demonstrate the effectiveness of this approach compared with other prevailing approaches.

The rest of the paper is organized as follows. Section 2 reviews the related works on cross-media retrieval. The details of the proposed approach to cross-media retrieval are presented in Sect. 3. Then the experimental results are shown in Sect. 4. Finally, Sect. 5 draws the conclusions.

2   Related Works

Cross-media retrieval is a retrieval technique which does not impose restrictions on media types of the queries and the retrieved results. In other words, users can retrieve multimedia objects of different media types by providing the query of any media type they can access, e.g., to retrieve the related images by providing an audio query. As we discussed in the previous section, the existing approaches to cross-media retrieval can be divided into two main categories: the unified feature representation based approaches and the unified graph model based approaches. In this section, we will briefly review these related works in the literature.

The unified feature representation based approaches project heterogeneous features into a unified feature space so as to compute the similarities among different modalities. Pereira et al. [15] propose correlation matching (CM), semantic matching (SM) and semantic correlation matching (SCM) which combines both CM and SM. They utilize unsupervised methods like CCA to obtain a unified subspace, but there exists no explicit semantic meanings in this unified subspace. In [18], Sharma et al. present the generalized multiview analysis (GMA), which is a supervised extension of CCA and is useful for cross-media retrieval. Zhuang et al. [21] propose the supervised coupled dictionary learning with group structures for multi-modal retrieval $(SliM^2)$, which extends uni-modal dictionary learning to multi-modal dictionary learning and jointly learns a set of mapping functions across different modalities for cross-media retrieval. In order to make full use of both the media instances and their patches in one graph, Peng et al. [14] integrate the sparse and unified patch graph regularization to learn a unified feature representation. Furthermore, different modalities are mapped into the unified representation space by the deep architecture in [7].

As for the unified graph model based approaches, Zhuang et al. [22] analyze the features of media objects and their co-existence information, and construct a unified cross-media correlation graph. In this graph, media objects of different modalities are presented uniformly, and thus the semantic correlations among them can be mined. In [20], Yang et al. construct a Laplacian media object space for each modality respectively and a multimedia document semantic graph to learn the semantic correlations. Then, the characteristics of media objects propagate along the semantic graph of multimedia document to preform cross-media retrieval. In [13], Vasconcelos et al. propose the maximum covariance unfolding (MCU). With manifold techniques, MCU implicitly constructs a unified graph from different input modalities, and computes a common low dimensional embedding that maximizes the cross-modal correlations while preserving the local distances.

3   Our Approach

In this section, we firstly present the formal definition of the cross-media retrieval problem, and introduce the framework of the proposed approach. Then, the details of the proposed approach are depicted.

Given a dataset of multimedia documents, denoted as $D = \{D_1, D_2, \cdots, D_N\}$, where $D_i(i = 1, \cdots, N)$ denotes the i-th multimedia document. For simplicity, here we consider the case where each document consists of an image and its corresponding text, which belong to the same concept. Thus, the i-th multimedia document can be represented as $D_i = (x_T^i, x_I^i)$, where $x_T^i \in R^{d(T)}$ is the low-level textual feature vector of $d(T)$ dimension to represent the i-th text, and $x_I^i \in R^{d(I)}$ is the low-level visual feature vector of $d(I)$ dimension to represent the i-th image. Given a text or image query $x_T^q \in R^{d(T)}$ ($x_I^q \in R^{d(I)}$), the goal of cross-media retrieval is to retrieve the most related images and texts with the same semantic concept.

As shown in Fig. 1, the framework of the proposed approach mainly consists of three steps as follows.

- Entity level Construction. We utilize an entity extraction tool to extract entities from the texts of multimedia documents, and select the entities which can help to generate accurate semantic abstraction, for the sake of effectiveness and efficiency. Finally, the entity level with fine-grained semantics is constructed, which can help bridge the semantic gap to a certain extent.
- Entity projection learning. In this step, by minimizing cross-media correlation error and single-media reconstruction error, the entity projection from low-level features to the entity level can be learned, and a unified entity-level representation can be obtained.
- Semantic abstraction generation. In order to retrieve the most related images and texts belonging to the same high-level concepts, we utilize logistic regression to generate the unified semantic abstraction of these concepts.

Finally, based on the unified semantic abstraction, we can compute the similarities between media objects of different media types and conduct the cross-media retrieval. The details of each step are presented as follows.

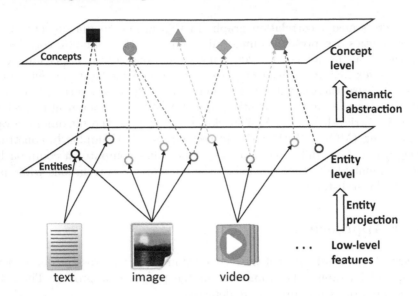

Fig. 1. The framework of the proposed approach.

3.1 Entity Level Construction

In the text processing domain, people have conducted a plenty of research and achieved outstanding performance on entity extraction. In order to extract entities, we adopt the tool of the Illinois Wikifier [3], which can identify "the important entities" in texts and disambiguate them into the unique and unambiguous ones. Due to the large quantity of the entities extracted by the tool, it will bring heavy computational burden to the subsequent steps. In addition, some entities may have no benefit for the cross-media retrieval, but cause confusion instead. So it is necessary to pick out those entities that can help to generate accurate semantic abstraction.

Therefore, we firstly make some heuristic rules to filter the entities.

(1) The top linked score of the entity generated by the tool should be greater than a predefined threshold, which means that we only select the entities with high confidence.
(2) The entity title should not contain numbers. Though these entities correspond to the specific Wikipedia pages, they do not contain visual information and usually have no benefit for the cross-media retrieval, for example, "2012", "January 9", "16th century".

Furthermore, we also select the more helpful entities by mutual information [4], which is a measure of the variables' mutual dependence. We utilize the mutual information to measure the correlation coefficient between the entities and the concepts, and pick out the entities which carry more information. Formally, we denote the entity variable as $V_E \in \{e_i | i = 1, 2, \ldots, n_E\}$, and denote the concept variable as $V_C \in \{c_i | i = 1, 2, \ldots, n_C\}$, where n_E is the number of the entities and n_C is the number of the concepts. The mutual information can be defined as follows.

$$
\begin{aligned}
P(e_i = 1, c_i = 1) &= n_{(e_i=1,c_i=1)}/N \\
P(e_i = 1) &= n_{(e_i=1)}/N \\
P(c_i = 1) &= n_{(c_i=1)}/N
\end{aligned}
\tag{1}
$$

where $n_{(e_i=1,c_i=1)}$ represents the number of the documents containing the entity e_i and the concept c_i, $n_{(e_i=1)}$ represents the number of the documents containing the entity e_i, and $n_{(c_i=1)}$ represents the number of the documents containing the concept c_i.

$$
I(V_E, V_C) = \sum_{e_i \in \{0,1\}} \sum_{c_i \in \{0,1\}} P(V_E = e_i, V_C = c_i) \log \frac{P(V_E = e_i, V_C = c_i)}{P(V_E = e_i) P(V_C = c_i)}
\tag{2}
$$

When $e_i = 1$ and $c_i = 1$, we can compute the above probabilities by Eq. 1. Similarly, the probabilities can be computed in other situations ($e_i = 1 \& c_i = 0$, $e_i = 0 \& c_i = 1$ and $e_i = 0 \& c_i = 0$). Then, the mutual information of e_i and c_i ($I(e_i, c_i)$) is obtained by Eq. 2, and the mutual information of the entity e_i ($I(e_i)$) is obtained by computing the average of $I(e_i, c_i)$ on these concepts. According to the mutual information of each entity, we can pick out the entities which carry more information related to these concepts.

From the Wikipedia dataset, we can obtain over ten thousand entities by the Illinois Wikifier, and then we pick out the entities which can help to generate accurate semantic abstraction by the above approach. Consequently, we can finally construct an entity level with fine-grained semantics, which can help bridge the semantic gap to a certain extent.

3.2 Entity Projection Learning

Given a training dataset of m multimedia documents, it can be denoted as $D_{tr} = (X_T, X_I, Y_E, Y_C)$, where $X_T \in R^{m \times d(T)}$, $X_I \in R^{m \times d(I)}$, $Y_E \in R^{m \times n_E}$, $Y_C \in R^{m \times n_C}$. Here, X_T is the textual feature matrix of the labeled texts, X_I is the visual feature matrix of the labeled images, Y_E is the label matrix of the entity level, and Y_C is the label matrix of the high-level concepts. It should be noted that Y_E can be obtained by entity extraction in texts. We respectively denote $P_T \in R^{d(T) \times n_E}$ and $P_I \in R^{d(I) \times n_E}$ as the projection matrices to project the two modalities into the entity-level space. Then, we can exploit the correlation between the low-level features and the entity level by the projection. Due to the sparsity of the entity-level label matrix, we adopt linear projection here, and the objective function for the entity projection is defined as follows.

$$
\min_{P_T, P_I} \|X_T P_T - X_I P_I\|_F^2 + \mu(\|X_T P_T - Y_E\|_F^2 + \|X_I P_I - Y_E\|_F^2) + \lambda(\|P_T\|_F^2 + \|P_I\|_F^2)
\tag{3}
$$

Here, $\| \cdot \|_F^2$ is the Frobenius norm. $\|X_T P_T - X_I P_I\|_F^2$ is the cross-media correlation error item between texts and images, which can exploit the correlation among different media objects sharing the same entities. $\|X_T P_T - Y_E\|_F^2$ and $\|X_I P_I - Y_E\|_F^2$ are the single-media reconstruction error items to ensure that multimedia objects are respectively accordant with their initial entity labels as close as possible. $\|P_T\|_F^2$ and $\|P_I\|_F^2$ are used to avoid the overfitting problem, and μ and λ are the parameters for balancing the weights of different items.

This objective function in Eq. 3 can be efficiently optimized by an iterative method. When P_I is fixed, the projection P_T reduces to a ridge-regression problem and can be solved in the following closed form.

$$P_T = ((1 + \mu)X_T^\top X_T + \lambda I)^{-1}(X_T^\top X_I P_I + X_T^\top Y_E) \tag{4}$$

Similarly, when P_T is fixed, the solution to Eq. 3 can be expressed as follows.

$$P_I = ((1 + \mu)X_I^\top X_I + \lambda I)^{-1}(X_I^\top X_T P_T + X_I^\top Y_E) \tag{5}$$

Finally, we have learned the entity projection.

3.3 Semantic Abstraction Generation

So far, we have learned the entity projection (P_T and P_I). Given a test dataset ($D^* = \{X_T^*, X_I^*\}$), we can obtain the unified feature representation of each multimedia document at the entity level as follows.

$$Y_{E_T}^* = X_T^* P_T \tag{6}$$

$$Y_{E_I}^* = X_I^* P_I \tag{7}$$

However, we have no knowledge of the concepts which the users want to retrieve. Hence, we should also exploit the semantic correlation between the entity level and the high-level concepts. In order to make the retrieval framework concise and effective, we utilize the logistic regression to compute the posterior probability of a particular concept from the unified feature representation at the entity level. We call this step as the semantic abstraction, and the objective function is presented as follows.

$$\min_w \frac{1}{2} w'w + C \sum_i log(1 + exp(-y_{c_i} w' x_i)) \tag{8}$$

In Eq. 8, x_i is the unified feature vector at the entity level which can be obtained by Eqs. 6 and 7, y_{c_i} is the concept label of x_i, and w is the vector of parameters. Following the same settings as in [15], we adopt the Liblinear software package [6] to implement the multiclass logistic regression.

Eventually, we have obtained the probabilistic vector as the unified semantic abstraction of the high-level concepts through the entity-level bridge. We can directly compute the similarities between any multimedia objects to conduct the cross-media retrieval.

4   Experimental Evaluation

In this section, we evaluate the performance of our approach by comparing with several existing approaches in the literature for cross-media retrieval.

4.1   Dataset Description

We evaluate the proposed approach on the Wikipedia dataset [17]. The document corpus is collected from Wikipedia's "featured articles". Each article is split into sections and each image is assigned to the section according to the position in the article. Then, the sections and the images are assigned the same concept with the article. Through pruning, the final corpus contains 2866 documents, and there is only one text-image pair in a certain document labeled as one of the 10 concepts. For fair comparison, we adopt the same split as in [15], and the dataset is randomly split into a training set of 2173 documents and a test set of 693 documents.

Regarding to the low-level features, we take the same strategy as in [15] to generate both text and image representations. Each text is represented as a 200-topic histogram by the latent Dirichlet allocation model [2], and each image is represented as a SIFT histogram with a 4096-codeword codebook.

4.2   Evaluation Metrics

According to the experiments on distance measures in [15], the centered normalized correlation has the best average performance in almost all the experiments. Therefore, we adopt the same distance measure, i.e., the centered normalized correlation, for computing similarities to retrieve the multimedia objects.

$$d(p, q) = \frac{(p - \mu_p)^{\top}(q - \mu_q)}{\|p - \mu_p\| \cdot \|q - \mu_q\|} \tag{9}$$

where μ_p and μ_q are the sample averages of p and q respectively.

The final performance is evaluated using the mean average precision (MAP) since it is widely adopted to evaluate the performance of ranked retrieval results.

4.3   Experimental Results

Firstly, in order to determine the appropriate number of entities needed in the Wikipedia dataset, we conduct the experiments with various numbers of the entities obtained by mutual information selection. The results are shown in Fig. 2, and we can find that when the entity number surpasses five thousand, the results gradually stay stable. Therefore, we fix the entity number as five thousand in the following experiments.

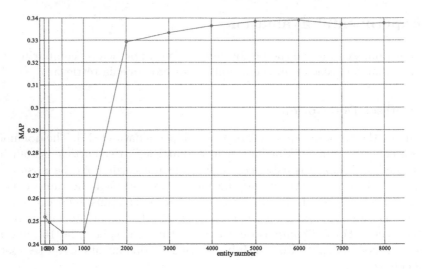

Fig. 2. Experimental results on various number of entities.

In order to validate the effectiveness of the entity level, we design three baseline approaches as follows.

(1) Correlation projection (CP). This approach directly learns the correlation projection from low-level features to high-level concepts, and then measures the similarities of multimedia objects. Its formulation is similar to Eq. 3, and can be obtained by replacing Y_C with Y_E in Eq. 3. Note that CP does not have the intermediate feature space and directly gets the unified representation on high-level concepts from low-level features to compute the similarities.

(2) Semantic correlation projection (SCP). This approach is the same as CP to generate the unified representation by the correlation projection, and then utilizes the logistic regression to obtain the probabilistic vector as the unified semantic abstraction of the high-level concepts. Finally, with the semantic abstraction, we can compute the similarities among the multimedia objects.

(3) Semantic random entity projection (SREP). This approach is similar to the proposed approach where an entity level of five thousand entities is constructed. The difference is that the entity level is constructed randomly here, thus contains no semantic meanings.

In Table 1, the first two baseline approaches (CP and SCP) do not have the fine-grained intermediate representation, and directly generate the unified representation on the high-level concepts. The proposed approach outperforms these two baseline approaches (CP and SCP), demonstrating the effectiveness of the entity level. Furthermore, the SREP approach achieves the similar results with CP and SCP which is inferior to the results of SEP, since the random entity level used in SREP has no semantic meanings and cannot help to generate accurate

Table 1. MAP scores of the three baseline approaches and the proposed approach.

Approaches	img. query	txt. query	avg.
CP	0.331	0.269	0.300
SCP	0.355	0.268	0.312
SREP	0.344	0.273	0.309
SEP(ours)	**0.387**	**0.290**	**0.339**

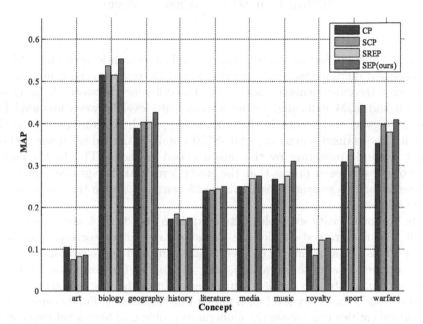

Fig. 3. MAP scores in each concept of the three baseline approaches and the proposed approach.

semantic abstraction. It further validates the effectiveness of the entity level construction with fine-grained semantics. Moreover, Fig. 3 shows the MAP scores in each concept of the three baseline approaches and the proposed approach. We can see that in most concepts, SEP outperforms the other approaches. Additionally, the selected entities make different contributions to the related concepts, which results in the difference between the MAP scores in each concept.

Then, we compare the proposed approach with several existing approaches in the literature including TTI [16], GMA [18], $SliM^2$ [21] and SCM [15]. The results of TTI and SCM are reported in [15]. GMA and $SliM^2$ are implemented with the codes provided by their authors respectively. The experimental results of these approaches are shown in Table 2. It can be seen that the proposed approach achieves the best results among these approaches, demonstrating the effectiveness of the proposed approach. TTI projects heterogeneous features into a unified latent topic space without the exact semantic meanings. GMA also

Table 2. Experimental results on Wikipedia dataset.

Approaches	img. query	txt. query	avg.
TTI [16]	0.237	0.137	0.187
GMA [18]	0.283	0.214	0.249
$SliM^2$ [21]	0.230	0.191	0.211
SCM [15]	0.362	0.273	0.318
SEP(ours)	**0.387**	**0.290**	**0.339**

generates a unified feature space without explicit semantic meanings. $SliM^2$ learns coupled dictionaries across multi-modal data, but these dictionaries cannot indicate the exact semantic meanings. The difference between the proposed approach and SCM is located in the intermediate level between low-level features and high-level concepts. In the proposed approach, we construct the entity level with fine-grained semantics, while SCM obtains a unified subspace without explicit semantic meanings by the unsupervised methods. The better performance of our approach proves that the entity level with fine-grained semantics has benefit for the semantic abstraction and can help bridge the semantic gap to a certain extent.

Finally, some entity examples extracted from the dataset are presented in Table 3. The fine-grained entities are almost related to the corresponding concepts, such as "Death Valley" (a desert valley, related to geography), "Ovid" (a Roman poet, related to literature), and "Simpsons" (an American adult animated sitcom, related to media), which can help bridge the semantic gap by the transition of the entity level with fine-grained semantics. Furthermore, these fine-grained entities can relieve the ambiguous problem of high-level concepts to some degree.

Table 3. The entity examples of the 10 concepts.

Concepts	Entity examples		
art	frescoes	putti	effigy
biology	genome	nest	breed
geography	Manchester	colliery	Death Valley
history	Ming Dynasty	earthenware	Blackbeard
literature	Romeo and Juliet	Ovid	Hamlet
media	Movie Award	Simpsons	Paramount
music	aria	Mozart	blues
royalty	Queen	Prince William	Caesar
sport	Olympics	coach	sportsmanship
warfare	General Staff	minefields	missile

5   Conclusions

In this paper, we have proposed the semantic entity projection for cross-media retrieval. Experimental results have validated that our approach achieves better performance than the existing approaches, and the improvement mainly lies in the entity level with fine-grained semantics that is automatically constructed by our approach. Moreover, it is worth noting that our approach has good scalability due to the objective existence of entities. For the future work, we intend to fully exploit the correlation among the entities to construct a more powerful entity level.

Acknowledgments. This work was supported by National Natural Science Foundation of China under Grants 61371128 and 61532005, and National Hi-Tech Research and Development Program of China (863 Program) under Grants 2014AA015102 and 2012AA012503.

References

1. Atrey, P.K., Hossain, M.A., El Saddik, A., Kankanhalli, M.S.: Multimodal fusion for multimedia analysis: a survey. Multimedia Syst. **16**(6), 345–379 (2010)
2. Blei, D.M., Ng, A.Y., Jordan, M.I.: Latent dirichlet allocation. J. Mach. Learn. Res. **3**, 993–1022 (2003)
3. Cheng, X., Roth, D.: Relational Inference for Wikification. In: EMNLP (2013)
4. Church, K.W., Hanks, P.: Word association norms, mutual information, and lexicography. Comput. Linguist. **16**(1), 22–29 (1990)
5. Deng, J., Dong, W., Socher, R., Li, L.J., Li, K., Fei-Fei, L.: Imagenet: a large-scale hierarchical image database. In: IEEE Conference on Computer Vision and Pattern Recognition, pp. 248–255 (2009)
6. Fan, R.E., Chang, K.W., Hsieh, C.J., Wang, X.R., Lin, C.J.: LIBLINEAR: a library for large linear classification. J. Mach. Learn. Res. **9**, 1871–1874 (2008)
7. Feng, F., Wang, X., Li, R.: Cross-modal retrieval with correspondence autoencoder. In: Proceedings of the ACM International Conference on Multimedia, pp. 7–16 (2014)
8. Guillaumin, M., Mensink, T., Verbeek, J., Schmid, C.: Tagprop: discriminative metric learning in nearest neighbor models for image auto-annotation. In: IEEE 12th International Conference on Computer Vision, pp. 309–316 (2009)
9. Hotelling, H.: Relations between two sets of variates. Biometrika **42**(1), 321–377 (1936)
10. Jacobs, P.S.: Text-Based Intelligent Systems: Current Research and Practice in Information Extraction and Retrieval. Psychology Press, New York (2014)
11. Jiang, Y., Ngo, C.W., Yang, J.: Towards optimal bag-of-features for object categorization and semantic video retrieval. In: Proceedings of the 6th ACM International Conference on Image and Video Retrieval, pp. 494–501 (2007)
12. Li, D., Dimitrova, N., Li, M., Sethi, I.K.: Multimedia content processing through cross-modal association. In: Proceedings of the 11th ACM International Conference on Multimedia, pp. 604–611 (2003)
13. Mahadevan, V., Wong, C.W., Pereira, J.C., Liu, T., Vasconcelos, N., Saul, L.K.: Maximum covariance unfolding: manifold learning for bimodal data. In: Advances in Neural Information Processing Systems, pp. 918–926 (2011)

14. Peng, Y., Zhai, X., Zhao, Y., Huang, X.: Semi-supervised cross-media feature learning with unified patch graph regularization. IEEE Trans. Circ. Syst. Video Technol. (2015). http://ieeexplore.ieee.org/xpl/articleDetails.jsp?arnumber=7036070&tag=1
15. Pereira, J.C., Coviello, E., Doyle, G., Rasiwasia, N., Lanckriet, G., Levy, R., Vasconcelos, N.: On the role of correlation and abstraction in cross-modal multimedia retrieval. IEEE Trans. Pattern Anal. Mach. Intell. **36**(3), 521–535 (2014)
16. Qi, G.J., Aggarwal, C., Huang, T.: Towards semantic knowledge propagation from text corpus to web images. In: Proceedings of the 20th International Conference on World Wide Web, pp. 297–306 (2011)
17. Rasiwasia, N., Pereira, J.C., Coviello, E., Doyle, G., Lanckriet, G., Levy, R., Vasconcelos, N.: A new approach to cross-modal multimedia retrieval. In: Proceedings of the ACM International Conference on Multimedia, pp. 251–260 (2010)
18. Sharma, A., Kumar, A., Daume III, H., Jacobs, D.W.: Generalized multiview analysis: a discriminative latent space. In: IEEE Conference on Computer Vision and Pattern Recognition, pp. 2160–2167 (2012)
19. Wang, H., Kläser, A., Schmid, C., Liu, C.L.: Dense trajectories and motion boundary descriptors for action recognition. Int J. Comput. Vis. **103**(1), 60–79 (2013)
20. Yang, Y., Zhuang, Y., Wu, F., Pan, Y.: Harmonizing hierarchical manifolds for multimedia document semantics understanding and cross-media retrieval. IEEE Trans. Multimedia **10**(3), 437–446 (2008)
21. Zhuang, Y., Wang, Y., Wu, F., Zhang, Y., Lu, W.: Supervised coupled dictionary learning with group structures for multi-modal retrieval. In: Proceedings of the 27th AAAI Conference on Artificial Intelligence, pp. 1070–1076 (2013)
22. Zhuang, Y., Yang, Y., Wu, F.: Mining semantic correlation of heterogeneous multimedia data for cross-media retrieval. IEEE Trans. Multimedia **10**(2), 221–229 (2008)

Visual Re-ranking Through Greedy Selection and Rank Fusion

Bin Lin, Ai Wei, and Xinmei Tian$^{(\boxtimes)}$

Department of Electronic Engineering and Information Science,
University of Science and Technology of China, Hefei, Anhui, China
{lb1991,clytieai}@mail.ustc.edu.cn, xinmei@ustc.edu.cn

Abstract. Image search re-ranking has proven its effectiveness in the text-based image search system. However, traditional re-ranking algorithm heavily relies on the relevance of the top-ranked images. Due to the huge semantic gap between query and the image, the text-based retrieval result is unsatisfactory. Besides, single re-ranking model has large variance and is easy to over-fit. Instead, multiple re-ranking models can better balance the biased and the variance. In this paper, we first conduct label de-noising to filter false-positive images. Then a simple greedy graph-based re-ranking algorithm is proposed to derive the resulting list. Afterwards, different images are chosen as the seed images to perform re-ranking multiple times. Using the rank fusion, the results from different graphs are combined to form a better result. Extensive experiments are conducted on the INRIA web353 dataset and demonstrate that our method achieves significant improvement over state-of-the-art methods.

Keywords: Image search re-ranking · Rank fusion · Greedy graph-based re-ranking

1 Introduction

The importance of image searching has gained more and more attention to the society due to the explosive growth of the social media and multimedia information. Most web image search engines rely on textual information to determine the relevance between images and search keywords. Due to the lack of visual information and the lack of context of images, the searching results are unsatisfying at most of the time. Therefore, image re-ranking, which incorporates visual features of images to improve text-based image-searching, is introduced as the post-process of core search. It is defined as re-ordering the visual documents based on the initial text-based search results and their visual patterns [1]. With the help of the image re-ranking techniques, the quality of image search engine can be improved to a certain extent.

Image search re-ranking is mainly based on two assumptions. (1) The top ranked images are expected to possess the same semantic meaning with the query. (2) Images relevant to the query are expected to share similar visual

© Springer International Publishing Switzerland 2016
Q. Tian et al. (Eds.): MMM 2016, Part I, LNCS 9516, pp. 289–300, 2016.
DOI: 10.1007/978-3-319-27671-7_24

patterns more often than the irrelevant ones [2]. The key aspect of image re-ranking is the descriptive ability of the image feature and the robustness and strength of the image re-ranking model.

Researchers have proposed many informative local and holistic features to dig out the image information. Krupac et al. extracted SIFT features [4] on the dense grid and applied the BOW (bag of visual words) model [5] to present the image visual information [3]. Wang and Hua proposed an intuitive way to analyze the spatial distribution of color for desired images, and color map was applied for their interactive system to enhance the text-based image search [6]. Besides, other non-visual features can also help us determine the relevance between different images. Click information, as an example, can describe the relevance between images and queries accurately. Yu et al. employed the click data combined with the multi-model sparse coding method to build the image retrieval system [7]. Apart from that, multi-view features are also powerful features which can be combined with click features in the re-ranking model. Yu et al. proposed a re-ranking method using both click constraints and multi-view features to improve the retrieval performance [8]. Besides, different models are applied to conduct the re-ranking procedure. Yan et al. adopted pseudo relevance feedback which assumed that the top-ranked images were the few relevant ones. Those pseudo-relevant samples were further used in SVM to classify the remaining images into different classes [9]. Motivated by the well-known PageRank technique, Jing and Baluja proposed the Visual-Rank algorithm to treat images as the visual pages and analyze the visual link structures among images [10]. Tian et al. treated the re-ranking as a global optimization problem and proposed a Bayesian framework to derive the re-ranking model [11,12]. Yang and Hanjalic were inspired by the learning-to-rank paradigm and derived a re-ranking function in a supervised way from the human-labeled training data [13]. Luo and Tao proposed a manifold regularized multi-task method to learn a discriminative subspace to deal with multiple labels, thus images with different labels are divided [14]. Data mining techniques are also applied in image re-ranking model. Deng et al. proposed a weakly supervised multi-graph learning based on the mining of the intrinsic attributes among the instances [15]. Liu et al. proposed a noise-resistant graph and performed a graph ranking scheme to improve web image search results [16].

There are deficiencies existing in these re-ranking systems. First, among most of the re-ranking models, the credibility of images ranked on the top of the list contributes a lot to the system. However, the image set is usually filled with noisy samples, thus the performance of the retrieval system is often degraded. Therefore, specific de-noising method should be applied to avoid this circumstance. Second, we adopt graph-based learning, which can capture the intrinsic manifold structure underlying the images in the query, to replace traditional feature learning method as our re-ranking model. However, using just one graph for re-ranking is easy to over-fit for its low bias and high variance. Therefore, rank fusion method should be introduced to solve this problem.

We propose a framework where simple label de-noising is performed before re-ranking. Then a simple yet effective greedy graph-based re-ranking method is proposed for each query. Finally, multiple graphs are combined to reduce the high variance at the expense of a small increase in the bias and some loss of interpretability. It can significantly boost the performance. Extensive experiments are conducted on a web image data set to prove the necessity of our re-ranking architecture.

The remainder of this paper is organized as follows: Sect. 2 introduces our architecture of algorithm, including feature extraction, label de-noising, greedy selection, and rank fusion. Section 3 describes the experiments on the benchmark dataset which prove the effectiveness of our methods. Section 4 concludes the paper and raises a suggestion for future work.

2   Effective Visual Re-ranking

The key of image re-ranking is to find images relevant to the query and re-rank those images to the top of the list. With the large semantic gap between image and textual query, it is extremely hard to rank images only basing on their textual data. A large number of false positive samples are ranked on the top of the list as in Fig. 1, which makes it harder to re-rank images.

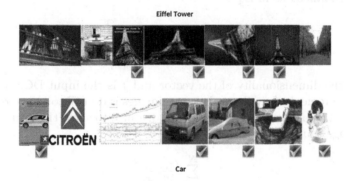

Fig. 1. Top-7 ranked images returned by a text-based image search engine for two queries: "Eiffel tower" and "car", ordered left to right. Query-relevant images are marked by the ticking sign. It illustrates that there are many irrelevant images lying on the top of the list.

Our architecture mainly consists of four parts. (1) Descriptive feature extraction and similarity calculation. We extract the informative holistic features from images and calculate the similarity matrix for each query. (2) Label de-noising. We define a simple confident criterion to evaluate the relevance between image and query. Then images with high confidence scores will replace those top-ranked images. We re-order the initial list by their confidence scores and create the "re-arranged list". (3) Graph-based re-ranking. After the pre-filtering, the "re-arranged list" is more relevant to the query. We select the image ranked 1st from the "re-arranged list" as the seed image and a simple graph-based greedy algorithm is proposed to incorporate the image which shares the highest similarity with the "seed image" as the relevant image. (4) Multiple graph fusion. Different image is selected as the "seed image" and we perform the graph-based re-ranking several times to generate multiple graphs. Rank fusion method is applied to combine different results into a more reasonable one.

2.1 Informative Feature Extraction

In order to capture the similarities between images, we first extract a powerful feature, which is the DCNN (deep convolutional neural network) feature, to present the holistic information of the image. Convolutional neural network was proposed by Le-Cun et al. [17] to solve the handwritten digit recognition problem. The convolutional neural network shows its superior ability in imitating the human biological vision system. The image can be well described using this network. Krizhevsky et al. [18] proposed a deep convolutional neural network with millions of parameters and applied it on the ImageNet dataset.

In this paper, DCNN feature is extracted on the network trained on the ILSVRC-2012 (ImageNet Large-Scale Visual Recognition Challenge) dataset. We select the 4096-dimensional vector which is induced from the fc7 level. The feature is normalized as in Eq. 1

$$\mathbf{x} = [x_1, x_2, .., x_N], x_k = \frac{\sqrt{f_k}}{\sum\limits_{m=1}^{N} f_m} \tag{1}$$

where N is the dimensionality of the vector and f is the input DCNN vector.

The measurement of similarity between images is crucial to image re-ranking system as well. We calculate the chi-square distance to show the visual distance between image i and image j as in Eq. 2

$$d_{ij} = \frac{\sum\limits_{k=1}^{N} \frac{(x_k^i - x_k^j)^2}{x_k^i + x_k^j}}{2}. \tag{2}$$

where x_k^i is the k-th feature of image i.

An inverse proportional function is applied as in Eq. 3 to transfer the distance into similarity score. s_{ij} denotes the similarity between image i and j and d_{ij} denotes the chi-square distance between two feature vector of images i and j. $\lambda = 0.5$ is chosen to avoid the situation where $d_{ij} = 0$.

$$s_{ij} = \frac{1}{d_{ij} + \lambda} \tag{3}$$

2.2 Label De-noising

In this part, our main goal is to filter out those false positive samples which are ranked on the front of the list. The procedure of filtering those images is based on a simple confidence score counting. After label de-noising, those outliers will be ranked lower, which significantly increase the reliability of the list.

Our label de-noising method is based on an intuition that relevant images share similar visual patterns with each other more often than irrelevant images [19]. We assume that the relevant images take up the majority of the query set. For each image, we sum up their similarities with every other image in the list. Thus we calculate the confidence score $C(I_j)$ for image i as in Eq. 4

$$C(I_j) = \sum_{i=1, i \neq j}^{N} \mu_i s_{ij}, \tag{4}$$

where s_{ij} denotes the similarity score between image i and image j and μ_i is a damping value which demonstrates the importance of the similarity with image i. When i is small, μ_i should be larger to increase the weight that come from the top-ranked images of the initial list because these images are more likely to share the same visual information with the query.

Images with high confidence scores are similar to most of the images in the list, which show their high probability to be relevant to the query. We simply move images with high confidence scores to the front of the list and thus the "re-arranged list" is created. The false positive samples will be removed from the front of the list due to their low confidence scores.

2.3 Graph-Based Re-ranking

The similarity matrix can be treated as the graph where each node represents a single image and the connectivity of a node reflects its visual similarity to others. Thus graph-based re-ranking is applied to derive the result image list.

After label de-noising, the high-ranked images are more credible. We treat the top-ranked image as the "seed image". Instead of finding images relevant to the query, we can simplify the task into finding images relevant to the "seed image" [20]. We apply a simple yet effective greedy algorithm to deal with the graph-based re-ranking problem.

Our goal is to find images which share more similarity with the "seed image set". At first, the "seed image set" only includes one image which is ranked 1st on the list. Then the image sharing the largest similarity value with images in the "seed image set" is included to the set. This process is iteratively conducted until all images in the query are included to the set. The algorithm is listed as in Algorithm 1.

The advantage of this method is that we only focus on images which are similar to the certificated relevant images. As long as those "certificated images" are truly relevant to the query, this method is fully reliable. Our experiment shows that the more credible the candidate images are, the more accurate the average precision will be.

Algorithm 1. Image Greedy Selection.

Input:

The initial image list, $L = [I_i]_{i=1}^N$;

The seed image set, $V = \emptyset$;

A given query, Q;

Output:

The image sequence set, Seq;

1: We start by choosing a seed image I_{seed} from L and add this image to the seed image set V.

2: Greedy Selection. We find the candidate image by calculating $I_{candidate} = \max\limits_{i \in L} \sum\limits_{j \in V} s_{ij}$ where s_{ij} denotes the similarity value between two images i and j.

3: $V = V \bigcup I_{candidate}$, $L = L - I_{candidate}$.

4: We iteratively conduct Step 2 and Step 3 until all images in the query set are included into set V.

5: We order images by their sequence of being added to the image set V and get the sequence list Seq.

6: **return** Seq;

2.4 Multiple Graph Fusion

Instead of using only one graph to derive the result lists, multiple graphs are generated to raise the credibility of the re-ranking results. In our greedy algorithm aforementioned, the result is closely related to the relevance of the seed image. In order to reduce the high variance, other high-ranked images on the lists are selected to be the "seed image". Therefore multiple graphs are generated using these different seed images. Different ranks are derived from these graphs and simple rank fusion method is applied to derive the final result. Three classic rank fusion methods are introduced.

Borda Fusion. The Borda fusion is a simple fusion method which turns the ranking information into the score [21]. For those top-ranked visual documents, higher scores are allocated to them. The Borda count function is defined as in Eq. 5.

$$s = \frac{num - R}{num - 1} \tag{5}$$

where num is the number of images in the list and R is the rank of this image. We sum up the ranking score for each image and order the list by their scores.

Condorcet Fusion. The Condorcet voting algorithm is a majoritarian method which specifies that the winner of the election is the candidate(s) that beats or ties with every other candidate in a pair-wise comparison [22]. For each iteration, we find a single image which beat every other image in a pair-wise comparison, then we exclude this image on the list and repeat the iteration until there is no image on the list. We order these images by their sequence of being excluded from the list.

RRF Fusion. Reciprocal Rank Fusion simply sorts the documents according to a naive scoring formula as in Eq. 6 [23].

$$RRFscore(d \in D) = \sum_{r \in R} \frac{1}{k + r(d)} \tag{6}$$

where k is a fixed number of 60. This formula derived from facts that while high-ranked documents are more important, the importance of lower-ranked documents still exists. This method is straightforward and effective. We order these images by their RRF scores.

3 Experiments

In this section, several experiments are conducted on a web image dataset called Web353 to prove the effectiveness of our re-ranking strategy.

3.1 Dataset

Our experiment is conducted on a diversified dataset - INRIA web353 dataset, which was collected by Krapac et al. [3]. This dataset includes 353 queries, where the original textual query is also included. For 80 % of queries, there are more than 200 images. Each image is resized to 150×150 pixels square. The ground-truth relevance label for every image is divided to two levels, which are "relevant" and "irrelevant". The 353 queries are diverse in topics, covering "object" items (e.g., "flag" and "car"), "celebrity" (e.g., "Justin Timberlake" and "will smith"), and abstract terms (e.g., "tennis court"), as shown in Fig. 2. Queries are diverse in ratio of relevance as well. In all, there are about 43.86 % images in this dataset labeled as relevant samples.

Fig. 2. Example pictures in INRIA web353 dataset.

3.2 Performance Metrics

Average precision (AP), which can reflect the occurrence of the relevant images, is adopted as our criterion to measure the effectiveness of the algorithm [18]. Average precisions are calculated at several truncation levels of T, i.e. $AP@T$; $T = \{5, 10, 20, 40, 60, 80, 100\}$, which reflect the precision for the top-T-ranked images, and they are defined in Eq. 7.

$$AP@T = \frac{1}{Z_T} \sum_{i=1}^{T} [precision(i) \times rel(i)] \qquad (7)$$

$rel(i)$ is a binary function which reflects the relevance of the ith-ranked image. The precision value is the precision of top i-ranked images.

3.3 Experiments for Label De-noising

The "Search Engine" result is conducted based on the textual information on the meta-data file. It is listed as the baseline result for re-ranking.

To evaluate the power of label de-nosing, we compare this method with the Visual-Rank method [10]. From the Table 1, it is clear that after label de-noising, the MAP result is higher than Visual-Rank when T is larger than 20. Furthermore, we conduct Visual-Rank algorithm on the "re-arranged list", the result is even better.

Table 1. MAP for evaluating label de-noising.

T	Search engine	Visual-Rank	"Re-arranged list"	"Re-arranged list" +Visual-Rank
5	0.611	0.799	0.743	**0.771**
10	0.553	0.743	0.715	**0.742**
20	0.503	0.656	0.676	**0.699**
40	0.452	0.552	0.633	**0.650**
60	0.431	0.557	0.612	**0.628**
80	0.426	0.567	0.605	**0.620**
100	0.431	0.581	0.610	**0.624**
ALL	0.569	0.679	0.699	**0.710**

3.4 Experiments for Greedy Selection

To evaluate the power of greedy selection, we propose two different schemes to test its performance.

Scheme 1: We treat the image on the top of the initial list as the "seed image" and then perform the greedy selection.

Table 2. The MAP result for greedy selection using different schemes.

T	Search engine	Scheme 1	Scheme 2	Scheme 3
5	0.611	0.670	0.773	**0.794**
10	0.553	0.664	0.754	**0.767**
20	0.503	0.655	0.728	**0.739**
40	0.452	0.639	0.689	**0.694**
60	0.431	0.629	0.667	**0.671**
80	0.426	0.627	0.656	**0.660**
100	0.431	0.634	0.657	**0.662**
ALL	0.570	0.715	0.736	**0.740**

Table 3. The number of graphs for graph fusion.

T	1	2	3	4	5	6	7	8	9	10
5	0.794	**0.818**	0.811	0.814	0.817	0.810	0.809	0.804	0.809	0.806
10	0.767	0.786	0.784	0.786	**0.789**	0.781	0.783	0.783	0.780	0.780
20	0.739	0.751	0.752	0.753	**0.757**	0.755	0.754	0.752	0.751	0.751
40	0.694	0.700	0.703	0.704	**0.708**	0.707	0.704	0.703	0.702	0.702
60	0.671	0.676	0.677	0.679	**0.683**	0.681	0.679	0.679	0.678	0.677
80	0.660	0.663	0.665	0.666	**0.670**	0.668	0.667	0.667	0.666	0.666
100	0.662	0.664	0.667	0.667	**0.670**	0.669	0.667	0.667	0.667	0.666
ALL	0.740	0.742	0.744	0.745	**0.748**	0.746	0.745	0.745	0.744	0.743

Scheme 2: We treat the image on the top of list after label de-noising as the "seed image" and then perform the greedy selection.

After simply conducting the greedy selection, the MAP for $AP@T$ gains a lot for every T. Adding label de-nosing can significantly improve the performance. $MAP@ALL$ achieves as high as 73.60 %. In comparison to Visual-Rank method, greedy selection beats it when T is larger than 20. But when T is small, the performance of greedy selection is quite poor. The reason for that is this method hugely depends on the relevance of images on the top. Visual-Rank performs poorly when T is large. But when we only focus on the small Ts, Visual-Rank achieves the best result. In order to get a better performance, we can sacrifice the time to perform Visual-Rank first and conduct the greedy selection based on the "Visual-Rank list". The result is listed as Scheme 3 in Table 2. And the performance is elevated as expected.

3.5 Experiments for Rank Fusion

Using one single graph to perform the re-ranking is of huge variance. Thus multiple graphs are introduced to boost the performance.

The Number of Graphs for Rank Fusion. First the number of graphs for each query is needed to confirm. Since employing Visual-Rank first and then conduct greedy selection can achieve a better result, we select the seed image from the "Visual-Rank" list. And Borda fusion is conducted first to evaluate the parameter. We evaluate the results while choosing the graph number ranging from 1 to 10. Table 3 shows that five is an appropriate number to choose since it beats other numbers when T is larger than 10.

Different Rank Fusion Methods. We perform Borda count method, Condorcet Fusion and RRF scoring method to test their performance. Five is selected as the number of graphs.

Table 4 shows that the RRF method is the best among these three methods. And the MAP@ALL achieves 74.94 %, which is one percent point higher than the situation when we don't conduct rank fusion.

3.6 Experiments to Compare with the State-of-Art Methods

In this part, we compare our results with state-of-the-art re-ranking methods, including pseudo relevance feedback [9], Bayesian re-ranking [11], query relative

Table 4. The experiment for testing different rank fusion methods.

T	Borda	Condorcet	RRF
5	0.817	0.817	**0.824**
10	0.789	0.792	**0.796**
20	0.757	0.759	**0.762**
40	0.708	0.709	**0.710**
60	0.683	0.683	**0.685**
80	0.670	0.670	**0.671**
100	0.670	0.671	**0.672**
ALL	0.748	0.749	**0.750**

Table 5. The experiment for comparing the state-of-the-art methods.

Methods	MAP
Search engine	0.569
PRF [9]	0.658
Bayesian [11]	0.665
Query relative [3]	0.666
Two-stage learning [24]	0.705
Noise-resistant [16]	0.736
Our method	**0.750**

re-ranking [3], two stage learning [24], and noise-resistant graph ranking [16]. Since most of them only reported the results on MAP@ALL, we compare the results under this metric. The results are listed in Table 5. Our method demonstrates better performance than the state-of-the-art re-ranking methods.

4   Conclusion

In this paper, we propose a simple graph-based re-ranking model. The contributions of our proposed method lie in three aspects. (1) A simple yet effective label de-noising method is conducted to filter out those false-positive images for each query. (2) A simple greedy selecting scheme is performed to fastly and accurately find the images relevant to the query. The experiment results show its superior performance over the Visual-Rank algorithm. (3) Different "seed images" are selected to conduct the greedy selection to avoid the huge biased. Then rank fusion is conducted to further improve the re-ranking performance. Extensive experiments are conducted on a web image dataset, which show the effectiveness of our method. There are many avenues for future explorations. To begin with, our greedy selection method is quite raw and straightforward. More delicate and carefully designed methods can be applied to further improve the selection performance. In addition, only the holistic information of the image is employed. Other useful feature can be applied to enrich the re-ranking model.

Acknowledgment. This work was supported by the 973 project under Contract 2015CB351803, by the NSFC under Contracts 61390514 and 61201413, by the Youth Innovation Promotion Association CAS No. CX2100060016, by the Fundamental Research Funds for the Central Universities No. WK2100060011 and No. WK2100100021, and by the Specialized Research Fund for the Doctoral Program of Higher Education No. WJ2100060003.

References

1. Mei, T., Rui, Y., Li, S., Tian, Q.: Multimedia search re-ranking: a literature survey. ACM Comput. Surv. **46**(3), 1–38 (2014)
2. Wang, X., Qiu, S., Liu, K., Tang, X.: Web image re-ranking using query-specific semantic signatures. IEEE Trans. Pattern Anal. Mach. Intell. **36**(4), 810–823 (2014)
3. Krapac, J., Allan, M., Verbeek, J., Jurie, F.: Improving web image search results using query-relative classifiers. In: Computer Vision and Pattern Recognition, pp. 1094–1101 (2010)
4. Lowe, D.G.: Distinctive image features from scale-invariant keypoints. Int. J. Comput. Vision **60**(2), 91–110 (2004)
5. Jun, Y., Jiang, Y.G., Hauptmann, A.G., Ngo, C.-W.: Evaluating bag of-visual-words representations in scene classification. In: Proceedings of the International Workshop on Multimedia Information Retrieval, pp. 197–206 (2007)
6. Wang, J., Hua, X.-S.: Interactive image search by color map. ACM Trans. Intell. Syst. Technol. **3**(1), 12–37 (2011)

7. Yu, J., Rui, Y., Tao, D.: Click prediction for web image re-ranking using multimodal sparse coding. IEEE Trans. Image Process. **23**(5), 2019–2032 (2014)
8. Yu, J., Rui, Y., Chen, B.: Exploiting click constraints and multiview features for image re-ranking. IEEE Trans. Multimedia **16**(1), 159–168 (2014)
9. Yan, R., Hauptmann, A., Jin, R.: Multimedia search with pseudo relevance feedback. In: Image and Video Retrieval, pp. 238–247 (2003)
10. Jing, Y., Baluja, S.: Visualrank: applying pagerank to large-scale image search. IEEE Trans. Pattern Anal. Mach. Intell. **30**(1), 1877–1890 (2008)
11. Tian, X., Yang, L., Wang, J., Yang, Y., Wu, X., Hua, X.-S.: Bayesian video search re-ranking. In: Proceedings of the 16th ACM International Conference on Multimedia, pp. 131–140 (2008)
12. Tian, X., Yang, L., Wang, J., Wu, X., Hua, X.-S.: Bayesian visual re-ranking. IEEE Trans. Multimedia **13**(4), 639–652 (2011)
13. Yang, L., Hanjalic, A.: Supervised re-ranking for web image search. In: Proceedings of the International Conference on Multimedia, pp. 183–192 (2010)
14. Luo, Y., Tao, D., Geng, B., Xu, C., Maybank, S.J.: Manifold regularized multitask learning for semi-supervised multilabel image classification. IEEE Trans. Image Process. **22**, 523–536 (2013)
15. Deng, C., Ji, R., Liu, W., Tao, D., Gao, X.-B.: Visual re-ranking through weakly supervised multi-graph learning. In: International Conference on Computer Vision, pp. 2600–2607 (2013)
16. Liu, W., Jiang, Y.-G., Luo, J., Chang, S.-F.: Noise resistant graph ranking for improved web image search. In: IEEE Conference on Computer Vision and Pattern Recognition, pp. 849–856 (2011)
17. Le Cun, Y., Boser, B., Denker, J.S., Henderson, D., Howard, R.E., Hubbard, W., Jackel, L.D.: Handwritten digit recognition with a backpropagation network. In: Advances in Neural Information Processing Systems, pp. 396–404 (1990)
18. Krizhevsky, A., Sutskever, I., Hinton, G.E.: Imagenet classification with deep convolutional neural networks. In: Advances in Neural Information Processing Systems, pp. 1097–1105 (2012)
19. Morioka, N., Wang, J.: Robust visual re-ranking via sparsity and ranking constraints. In: Proceedings of the 16th ACM International Conference on Multimedia, pp. 533–542 (2011)
20. Zhang, S., Yang, M., Cour, T., Yu, K., Metaxas, D.N.: Query specific rank fusion for image retrieval. IEEE Trans. Pattern Anal. Mach. Intell. **37**(4), 803–815 (2015)
21. Aslam, J.A., Montague, M.: Models for metasearch. In: Proceedings of the 24th Annual International ACM SIGIR Conference on Research and Development in Information Retrieval, pp. 276–284 (2001)
22. Montague, M., Aslam, J.A.: Condorcet fusion for improved retrieval. In: Proceedings of Acmcikm, pp. 538–548 (2002)
23. Cormack, G.V., Clarke, C., Buettcher, S.: Reciprocal rank fusion outperforms condorcet and individual rank learning methods. In: Proceedings of SIGIR, pp. 758–759 (2009)
24. Yang, L., Hanjalic, A.: Learning to re-rank web images. IEEE Multimedia **20**, 13–21 (2013)

No-reference Image Quality Assessment Based on Structural and Luminance Information

Qiaohong Li[1]([✉]), Weisi Lin[1], Jingtao Xu[2], Yuming Fang[3], and Daniel Thalmann[1]

[1] School of Computer Engineering, Nanyang Technological University,
Singapore 639798, Singapore
qli013@e.ntu.edu.sg
[2] School of Information and Communication Engineering,
Beijing University of Posts and Telecommunications, Beijing, China
[3] School of Information Technology,
Jiangxi University of Finance and Economics, Nanchang, China

Abstract. Research on no-reference image quality assessment (IQA) aims to develop a computational model simulating the human perception of image quality accurately and automatically without any prior information about the reference clean image signals. In this paper, we introduce a novel no-reference IQA metric, based on the analysis of structural degradation and luminance changes. Since the human visual system (HVS) is highly sensitive to structural distortion, we encode the image structural information as the local binary pattern (LBP) distribution. Besides, image quality is also affected by luminance changes, which cannot be captured properly by LBP threshold mechanism. Hence, the distribution of normalized luminance magnitudes is also included in the proposed IQA metric. Extensive experiments conducted on two large public image databases have demonstrated the effectiveness and robustness of the proposed metric in comparison with the relevant state-of-the-art metrics.

Keywords: Image quality assessment · Blind image quality assessment · No-reference · Human vision system (HVS) · Local binary pattern (LBP) · Structural distortion

1 Introduction

With the increasing proliferation of information and communication technologies, accurate estimation of the perceived quality of visual signals is becoming a crucial issue. Quality of visual signals can be degraded by a variety of distortions during the process of acquisition, compression, transmission and reproduction. Subjective viewing test is considered to be the most accurate and reliable way for quality evaluation. However, it is time consuming and expensive, and cannot be embedded into real-time quality monitoring and prediction applications. Objective image quality assessment (IQA) metrics which aim to emulate the human judgment of perceived quality have attracted much research effort [1].

© Springer International Publishing Switzerland 2016
Q. Tian et al. (Eds.): MMM 2016, Part I, LNCS 9516, pp. 301–312, 2016.
DOI: 10.1007/978-3-319-27671-7_25

The investigations into objective image quality assessment have led to considerable progress in developing IQA models based on the human perceptual system. Generally, objective IQA metrics can be categorized into full-reference (FR), reduced-reference (RR) and no-reference (NR) algorithms according to the availability of original image signals [1]. In this work, we will focus on the development of no-reference (a.k.a blind) IQA models. A priori knowledge of the image distortion types (such as white Gaussian noise, JPEG compression) permits the use of distortion specific NR IQA models. Early research works on NR IQA mainly focus on the development of distortion specific models, as the knowledge about specialized distortion characteristics can simplify such NR IQA model design. Wang et al. [17] develop a NR IQA metric for JPEG compressed images based on analysis of in-block activities and between-block discontinuities. Sheikh et al. [16] propose a novel way to evaluate JPEG2000 compressed images by the use of natural scene statistics (NSS) models in wavelet domain. In [19], the concept of just noticeable blur (JNB) is introduced to quantify the sharpness or blurriness of examined images using local contrast features. Recently, the study [18] makes an attempt to measure image contrast distortion and achieves some promising results. Although distortion specific NR IQA models can perform very well in a given application scope, their practical usage is limited by the targeted distortion type. By contrast, general purpose NR IQA models which require no priori information about both reference image and distortion types are versatile and can be used in many different scenarios.

The development of general purpose NR IQA models is approached in two broad ways. One is using the natural scene statistics (NSS) models with the assumption that natural images share similar statistical regularities while various distortions may change these statistics and render the images unnatural. Several state-of-the-art NR models fall under this category. NR image quality metrics based on spatial domain NSS [2,3], DCT domain NSS [4] and wavelet domain NSS [5,6] have been proposed and achieve promising results. Another approach is feature based models which first extract quality-aware features from visual signals and then learn a machine learning model to map these extracted features to the final quality score. Representative metrics include [7,8] where neural network and support vector regression (SVR) are used to learn the mapping function between extracted features and quality score, respectively.

In this paper, we propose a novel NR IQA model by accounting for both structural and luminance degradation. First, the spatial local divisive normalization is applied to the image to mimic early visual system and remove redundancy in the visual input. Second, we employ the LBP descriptor to extract the structural information and calculate the LBP histogram. And then, the normalized luminance magnitude distribution in the form of histogram is extracted from the same processed image. Finally, the luminance and LBP histograms are concatenated together as the input to the SVR to learn the nonlinear mapping from extracted quality-aware features to subjective quality scores. Figure 1 shows the flow chart of the proposed NR IQA model.

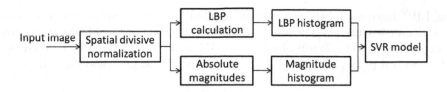

Fig. 1. The block diagram of proposed NR IQA model.

The rest of this paper is organized as follows. The proposed IQA model NRSL (No-Reference image quality assessment based on Structural and Luminance information) is detailed in Sect. 2. Experimental results of the proposed method are presented in Sect. 3. Finally, we draw the conclusions in Sect. 4.

2   The Proposed NR IQA Model

2.1   Spatial Divisive Normalization

Divisive normalization has been widely adopted as a nonlinear preprocessing stage in the research areas of image processing and computer vision to emulate the nonlinear masking phenomenon of visual perception and reduce statistical dependencies of input visual signals [9]. Generally, each coefficient is divided by the square root of a linear combination of the squared amplitudes of its neighbors. In this work, we employ the same preprocessing model as [2,3].

$$\tilde{I}(i,j) = \frac{I(i,j) - \mu}{\sigma + C} \tag{1}$$

where i and j are the spatial indices of the image, and

$$\mu(i,j) = \sum_{k=-K}^{K} \sum_{l=-L}^{L} \omega_{k,l} I(i+k, j+l) \tag{2}$$

$$\sigma(i,j) = \sqrt{\sum_{k=-K}^{K} \sum_{l=-L}^{L} \omega_{k,l} \left[I(i+k, j+l) - \mu(i,j) \right]^2} \tag{3}$$

are the local mean and standard deviation of the surrounding local patch, where $\omega = \{\omega_{k,l} | k = -K, ..., K, l = -L, ..., L\}$ defines a unit-volume Gaussian window. And C is a small constant to guarantee numerical stability when the denominators are close to zero.

2.2   The LBP Histogram

Since HVS is extremely sensitive to structural degradation, we extract image structural information in the form of LBP histogram for quality evaluation.

The LBP histogram has been widely applied as image feature for, e.g., face recognition, texture classification, background subtraction, and interest region description, *etc.* [11]. Traditionally, the LBP value is calculated by comparison of the central pixel p_c with its circularly symmetric neighborhood p_i [11]:

$$LBP_{P,R} = \sum_{i=0}^{P-1} s(p_i - p_c)2^i \tag{4}$$

$$s(p_i - p_c) = \begin{cases} 1, & p_i - p_c > 0 \\ 0, & p_i - p_c < 0 \end{cases} \tag{5}$$

where P is the number of neighbors and R is the radius of the neighborhood. The rotation invariant LBP operator can be defined as

$$LBP_{P,R}^{riu2} = \begin{cases} \sum_{i=0}^{P-1} s(p_i - p_c) & if \ \mathcal{U}(LBP_{P,R}) \leq 2 \\ P+1 & else \end{cases} \tag{6}$$

where superscript $riu2$ refers as rotation invariant "uniform" patterns with $\mathcal{U}$ value less than 2, and $\mathcal{U}$ is the uniformity measure which is calculated as:

$$\mathcal{U}(LBP_{P,R}) = \|s(p_{P-1} - p_c) - s(p_0 - p_c)\|$$
$$+ \sum_{i=0}^{P-1} \|s(p_i - p_c) - s(p_{i-1} - p_c)\| \tag{7}$$

Then, the rotation invariant uniform LBP would have $P + 2$ distinct patterns ($P+1$ for the different uniform patterns and 1 for the non-uniform pattern) and can be mapped to a histogram of $P + 2$ bins. In this work, we set $P = 8$, thus there would be 10 bins for the LBP histogram. Figure 2 shows the pristine and three distorted versions of parrots image (white Gaussian noise (WN), Gaussian blur (GB) and JPEG compression), and their corresponding LBP histograms are shown in Fig. 3. As can be seen in Fig. 3, different distortions alter the LBP histograms in their own characteristic way.

2.3   The Normalized Luminance Histogram

After spatial divisive normalization, the resultant normalized luminance coefficients of natural images are observed to follow a Gaussian-like distribution [10]. Previous works [2,3] have fitted a generalized Gaussian distribution (GGD) to these coefficients and extracted the GGD parameters as quality-aware features. In this work, we directly calculate the histogram of resultant coefficient magnitudes to represent the image luminance information.

$$\bar{I}(i,j) = \left| \tilde{I}(i,j) \right| \tag{8}$$

Figure 4 shows the normalized luminance magnitude histograms corresponding to images in Fig. 2. As shown in Fig. 4, different distortion types result in

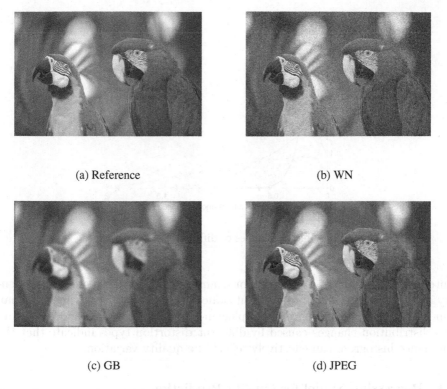

(a) Reference

(b) WN

(c) GB

(d) JPEG

Fig. 2. Image samples of different distortion types. All images come from LIVE database.

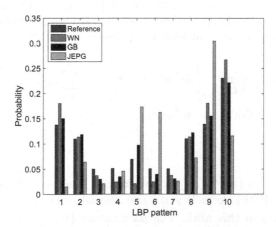

Fig. 3. LBP histograms of different distortion types.

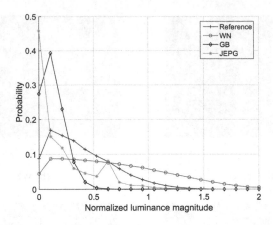

Fig. 4. Luminance changes of different distortion types.

quite different luminance changes. For example, white noise renders the intensity distribution more uniform as it introduces random disturbs; while Gaussian blur makes the distribution more Laplacian-like with high peak and small tail. The distribution changes caused by different distortion types indicate that the luminance histogram can effectively reflect the quality variation.

2.4   Regression Model for Quality Prediction

Generally, a mapping is required for feature pooling from feature space to quality measure. In this work, we adopt the SVR to learn the metric for image quality estimation. Considering a set of training data $\{(\boldsymbol{x}_1, y_1), ..., (\boldsymbol{x}_l, y_l)\}$, where $\boldsymbol{x}_i \in R^n$ is a feature vector and y_i is the target output. Under given parameters $C > 0$ and $\epsilon > 0$, the standard form of SVR is represented as [12]:

$$\min_{\boldsymbol{w}, b, \boldsymbol{\xi}, \boldsymbol{\xi}^*} \frac{1}{2} \boldsymbol{w}^T \boldsymbol{w} + C \{ \sum_{i=1}^{l} \xi_i + \sum_{i=1}^{l} \xi_i^* \} \tag{9}$$

$$\text{subject to} \quad \boldsymbol{w}^T \phi(\boldsymbol{x}_i) + b - y_i \leq \epsilon + \xi_i, \tag{10}$$

$$y_i - \boldsymbol{w}^T \phi(\boldsymbol{x}_i) - b \leq \epsilon + \xi_i^*, \tag{11}$$

$$\xi_i, \xi_i^* \geq 0, i = 1, ..., l. \tag{12}$$

where $K(\boldsymbol{x}_i, \boldsymbol{x}_j) = \phi(\boldsymbol{x}_i)^T \phi(\boldsymbol{x}_j)$ is the kernel function. We use the Radial Basis Function (RBF) kernel with the kernel function of $K(\boldsymbol{x}_i, \boldsymbol{x}_j) = \exp(-\gamma \|\boldsymbol{x}_i - \boldsymbol{x}_j\|^2)$ in this work. The parameters $\{C, \gamma\}$ are selected through cross validation on the training data.

3   Experimental Results and Analysis

3.1   Implementation Details

During the implementation, the constant C in Eq. 1 is set to $(KL)^2$, where L is the dynamic range of pixel gray scale levels (255 for 8-bit grayscale image), and $K \ll 1$ is a small constant set to be 0.005. The bin number for luminance histogram is 10. For LBP calculation, the number of neighbors P is 8 and the radius of the neighborhood R is 1. Both luminance and LBP histograms are calculated at two scales to account for the multi-scale property of natural images. Thus, the extracted features for SVR learning are 40 dimensions in total.

3.2   Database Description and Evaluation Methodology

We have tested the proposed method NRSL on LIVE [13] and TID2008 [14] databases. The LIVE IQA database includes 29 reference images and 779 distorted images corrupted by five types of distortions: JP2K compression, JPEG compression, WN, GB, and FF (transmission errors in the JP2K using Fast-fading Rayleigh channel model). Subjective quality scores are in the form of DMOS ranging from 0–100. The TID2008 database includes 25 reference images distorted by 17 types of distortions. We only test 24 natural images with the four distortion versions that also appear in LIVE database.

It is advised in [20] that a monotonic logistic function can be used to provide a nonlinear mapping between objective scores and subjective scores:

$$f(x) = \beta_1 \left(\frac{1}{2} - \frac{1}{\exp(\beta_2(x - \beta_3))} \right) + \beta_4 x + \beta_5 \qquad (13)$$

where x is the original IQA score, $f(x)$ is the fitted IQA score, $\beta_j(j = 1, 2, ...5)$ are regression parameters trained per database.

The performance of IQA metrics can be evaluated by three different criteria: Pearson linear correlation coefficient (LCC) for prediction accuracy, Spearman rank order correlation coefficient (SROCC) for prediction monotonicity and root mean squared error (RMSE) between the subjective MOS and predicted scores.

3.3   Experimental Results

We compare the proposed algorithm NRSL with several state-of-the-art NR IQA methods, including BIQI [5], DIIVINE [6], BLIINDS2 [4], CORNIA [21] and BRISQUE [2]. We also include the performance of FR IQA methods PSNR and SSIM [15] for reference.

Performance on Whole Database. First, we evaluate the overall performance of competing NR IQA models on LIVE and TID2008 databases. Since the proposed method adopts SVR learning for quality estimation, we need to divide the database into training and testing subsets. In our experiments, 80 % of the reference images and their associated distorted versions are used for training, and the rest are used for testing. This training-testing split is repeated 1000 times and the median performance is reported as the final result. Although FR-IQA metrics PSNR and SSIM do not require training on the database, we also report the median performance across 1000 trials on the testing subset for consistent comparison. The results are listed in Table 1.

Table 1. Overall performance on LIVE and TID2008 databases. The best two NR IQA models are shown in boldface.

IQA model	LIVE			TID2008		
	SROCC	LCC	RMSE	SROCC	LCC	RMSE
PSNR	0.885	0.883	12.793	0.879	0.861	0.806
SSIM	0.940	0.935	9.690	0.910	0.940	0.538
BIQI	0.813	0.831	15.072	0.801	0.849	0.834
DIIVINE	0.913	0.913	11.119	0.882	0.902	0.684
BLIINDS2	0.930	0.936	9.498	0.892	0.918	0.630
CORNIA	**0.943**	**0.946**	**8.810**	0.897	**0.930**	0.582
BRISQUE	0.942	**0.946**	8.826	**0.913**	**0.930**	**0.579**
NRSL	**0.944**	**0.948**	**8.657**	**0.948**	**0.959**	**0.448**

From Table 1, we can see that several NR IQA models can achieve satisfactory performance on LIVE database, among them the best three performing IQA models are NRSL, BRISQUE and CORNIA. The quality prediction performance of competing IQA models generally decrease when tested on TID2008 database. Among them, NRSL and BRISQUE can still provide promising performance. In summary, our proposed NRSL model can always achieve consistently better performance than competing NR IQA models and the two widely used FR IQA models PSNR and SSIM.

Performance on Individual Distortion Types. In this section, we evaluate the performance of competing IQA models on individual distortion types. For NR IQA models, we train on the 80 % of images of various distortion types and then test on the left 20 % of images with the specific distortion type. The SROCC comparison on both LIVE and TID2008 are tabulated in Table 2. The best two IQA models for each distortion type are shown in boldface. It should be noted that similar results can be obtained for LCC and RMSE criteria, and we only list SROCC here for brevity.

From the results presented in Table 2, we can see that for the 9 distortion groups in two databases, our proposed model NRSL are always in the best two positions. For the cases when NRSL is the second best model, the performance is quite similar to the best performing one. However, for the four distortion groups on TID2008 database, NRSL is always the best performing model and advance the second best one at a significant margin.

Table 2. SROCC comparison on individual distortion types. The best two NR IQA models are shown in boldface.

DB	D-Type	BIQI	DIIVINE	BLIINDS2	CORNIA	BRISQUE	NRSL
LIVE	JP2K	0.777	0.902	**0.929**	0.922	0.915	**0.935**
	JPEG	0.800	0.897	0.949	0.941	**0.964**	**0.952**
	WN	0.958	**0.981**	0.945	0.963	0.979	**0.980**
	GB	0.859	0.935	0.914	**0.955**	0.947	**0.950**
	FF	0.728	0.855	0.873	**0.907**	0.881	**0.880**
TID2008	JP2K	0.857	0.902	**0.925**	0.920	0.901	**0.950**
	JPEG	0.859	0.869	0.857	**0.909**	0.905	**0.938**
	WN	0.798	0.764	0.704	0.565	**0.860**	**0.908**
	GB	0.901	**0.905**	0.902	0.901	0.888	**0.929**

Performance on Cross-Database Validation. In the experiments of pervious two sections, the training and testing samples are from the same database. In this section, we'd like to test the generality of machine-learning based IQA models through cross-database validation. Specifically, we develop the NR IQA models by training on all the images from LIVE database and use it to test on the images from TID2008 database. The SROCC results of cross-database validation on TID2008 database are tabulated in Table 3.

Table 3. SROCC comparison on TID2008 database with models trained on LIVE. The best two NR IQA models are shown in boldface.

IQA model	JP2K	JPEG	WN	GB	ALL
BIQI	0.825	0.775	0.742	0.857	0.795
DIIVINE	0.845	0.845	**0.821**	0.860	0.872
BLIINDS2	0.882	0.902	0.715	0.786	0.861
CORNIA	**0.915**	**0.915**	0.683	**0.901**	0.890
BRISQUE	0.904	0.909	**0.827**	0.879	**0.897**
NRSL	**0.955**	**0.921**	0.817	**0.904**	**0.918**

As can be observed in Table 3, our proposed model NRSL achieves consistently better performance on the overall database and individual distortion types. The cross-database validation has demonstrated the database independency and robustness of the proposed NRSL method.

In Fig. 5, we show the relationship of subjective MOS and the objective scores predicted by IQA models on TID2008 database when all the models are trained on LIVE database. One can see that the distribution of NRSL is more compact and deviates less across different distortion types, which explains the good overall performance.

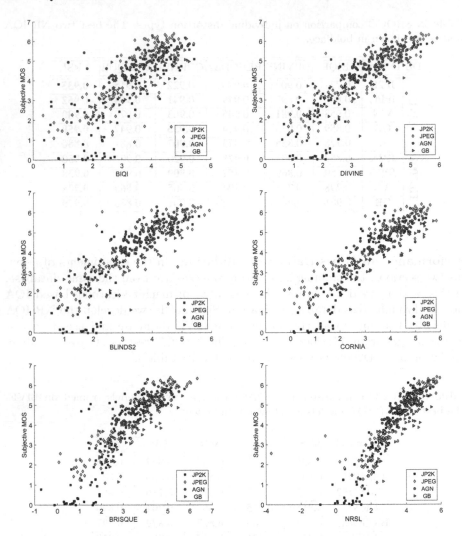

Fig. 5. Scatter plots of predicted quality scores against subjective MOS by representative IQA models on TID2008 database. The four distortion types are represented by different shaped colors (Color figure online).

4   Conclusions

In this paper, we propose a novel no-reference image quality assessment algorithm based on structural and luminance information. After spatial divisive normalization, the LBP histogram and luminance histogram are extracted to construct image quality-aware feature. Support Vector Regression (SVR) is employed to estimate human subjective ratings from structural and luminance features as the input. Extensive experimental results on two public image databases have demonstrated that the proposed method NRSL is quite competitive to state-of-the-art IQA methods in terms of prediction accuracy, distortion consistency, and database independency.

Acknowledgments. This work was funded by the Ph.D. Grant from the Institute for Media Innovation, Nanyang Technological University, Singapore.

References

1. Lin, W., Kuo, C.-C.J.: Perceptual visual quality metrics: a survey. J. Vis. Commun. Image Represent. **22**(4), 297–312 (2011)
2. Mittal, A., Moorthy, A.K., Bovik, A.C.: No-reference image quality assessment in the spatial domain. IEEE Trans. Image Process. **21**(12), 4695–4708 (2012)
3. Mittal, A., Soundararajan, R., Bovik, A.C.: Making a completely blind image quality analyzer. IEEE Signal Process. Lett. **20**(3), 209–212 (2013)
4. Saad, M.A., Bovik, A.C., Charrier, C.: Blind image quality assessment: a natural scene statistics approach in the DCT domain. IEEE Trans. Image Process. **21**(8), 3339–3352 (2012)
5. Moorthy, A.K., Bovik, A.C.: A two-step framework for constructing blind image quality indices. IEEE Signal Process. Lett. **17**(5), 513–516 (2010)
6. Moorthy, A.K., Bovik, A.C.: Blind image quality assessment: from natural scene statistics to perceptual quality. IEEE Trans. Image Process. **20**(12), 3350–3364 (2011)
7. Li, C., Bovik, A.C., Wu, X.: Blind image quality assessment using a general regression neural network. IEEE Trans. Neural Networks **22**(5), 793–799 (2011)
8. Xue, W., Mou, X., Zhang, L., Bovik, A.C., Feng, X.: Blind image quality assessment using joint statistics of gradient magnitude and laplacian features. IEEE Trans. Image Process. **23**(11), 4850–4862 (2014)
9. Lyu, S., Simoncelli, E.P.: Nonlinear image representation using divisive normalization. In: IEEE Conference on Computer Vision and Pattern Recognition, CVPR 2008, pp. 1–8. IEEE (2008)
10. Ruderman, D.L.: The statistics of natural images. Network: Comput. Neural Syst. **5**(4), 517–548 (1994)
11. Ojala, T., Pietikäinen, M., Mäenpää, T.: Multiresolution gray-scale and rotation invariant texture classification with local binary patterns. IEEE Trans. Pattern Anal. Mach. Intell. **24**(7), 971–987 (2002)
12. Schlkopf, B., Smola, A.J.: Learning with Kernels: Support Vector Machines, Regularization, Optimization, and Beyond. MIT Press, Cambridge (2002)
13. Sheikh, H.R., Wang, Z., Cormack, L., Bovik, A.C.: LIVE image quality assessment database release 2 (2005)
14. Ponomarenko, N., Lukin, V., Zelensky, A., Egiazarian, K., Carli, M., Battisti, F.: TID2008-a database for evaluation of full-reference visual quality assessment metrics. Adv. Mod. Radioelectronics **10**(4), 30–45 (2009)
15. Zhou, W., Bovik, A.C., Sheikh, H.R., Simoncelli, E.P.: Image quality assessment: from error visibility to structural similarity. IEEE Trans. Image Process. **13**(4), 600–612 (2004)
16. Sheikh, H.R., Bovik, A.C., Cormack, L.: No-reference quality assessment using natural scene statistics: JPEG2000. IEEE Trans. Image Process. **14**(11), 1918–1927 (2005)

17. Wang, Z., Sheikh, H.R., Bovik, A.C.: No-reference perceptual quality assessment of JPEG compressed images. In: Proceedings of the International Conference on Image Processing, vol. 1, pp. I–477. IEEE (2002)
18. Fang, Y., Ma, K., Wang, Z., Lin, W., Fang, Z., Zhai, G.: No-reference quality assessment of contrast-distorted images based on natural scene statistics. IEEE Signal Process. Lett. **22**(7), 838–842 (2015)
19. Ferzli, R., Karam, L.J.: A no-reference objective image sharpness metric based on the notion of just noticeable blur (JNB). IEEE Trans. Image Process. **18**(4), 717–728 (2009)
20. V.Q.E. Group, Final report from the video quality experts group on the validation of objective models of video quality assessment, VQEG, March 2000
21. Ye, P., Kumar, J., Kang, L., Doermann, D.: Unsupervised feature learning framework for no-reference image quality assessment. In: 2012 IEEE Conference on Computer Vision and Pattern Recognition (CVPR), pp. 1098–1105. IEEE (2012)

Learning Multiple Views with Orthogonal Denoising Autoencoders

TengQi Ye[1](✉), Tianchun Wang[2], Kevin McGuinness[1], Yu Guo[3],
and Cathal Gurrin[1]

[1] Insight Centre for Data Analytics, Dublin City University, Dublin, Ireland
{yetengqi,kevin.mcguinness}@gmail.com
[2] School of Software, TNList, Tsinghua University, Beijing, China
wtc13@mails.tsinghua.edu.cn
[3] Department of Computer Science, City University of Hong Kong,
Hong Kong, China

Abstract. Multi-view learning techniques are necessary when data is described by multiple distinct feature sets because single-view learning algorithms tend to overfit on these high-dimensional data. Prior successful approaches followed either consensus or complementary principles. Recent work has focused on learning both the shared and private latent spaces of views in order to take advantage of both principles. However, these methods can not ensure that the latent spaces are strictly independent through encouraging the orthogonality in their objective functions. Also little work has explored representation learning techniques for multi-view learning. In this paper, we use the denoising autoencoder to learn shared and private latent spaces, with orthogonal constraints — disconnecting every private latent space from the remaining views. Instead of computationally expensive optimization, we adapt the backpropagation algorithm to train our model.

Keywords: Denoising autoencoder · Autoencoder · Representation learning · Multi-view learning · Multimedia fusion

1 Introduction

In many machine learning problems, data samples are collected from diverse sensors or described by various features and inherently have multiple disjoint feature sets (conventionally referred as views) [1,2]. For example, in video classification, the videos can be characterized with respect to vision, audio and even attached comments; most article search engines take title, keywords, author, publisher, date and content into consideration; images have different forms of descriptors: color descriptors, local binary patterns, local shape descriptors, etc. The last example reveals the noteworthy case of views obtained from manual descriptors instead of natural splits. With multiple descriptors, important information concerning the task, which may be discarded by single descriptor, is hopefully retained by others [3].

© Springer International Publishing Switzerland 2016
Q. Tian et al. (Eds.): MMM 2016, Part I, LNCS 9516, pp. 313–324, 2016.
DOI: 10.1007/978-3-319-27671-7_26

Traditional machine learning methods may fail when concatenating all views into one for learning because the concatenation can cause over-fitting due to the high-dimensionality of the features (the curse of dimensionality [4]) and also ignores the specific properties of each view [2]. Compared with traditional single-view data, the multi-view data contains significant redundancy shared by its views. Previous successful multi-view learning algorithms follow two principles: consensus and complementary principles [2]. The consensus principle aims to maximize the agreement or shared knowledge between views. The complementary principle asserts each view may contain useful knowledge that the rest do not have, and errors can be corrected by this private knowledge.

To generalize over previous multi-view learning approaches, recent works focused on explicitly accounting for the dependencies and independencies of views, i.e., decompose the latent space into a shared common one (of all views) and several private spaces (of each view) [5,6]. The intuition is that each view is generated from the combination of the shared latent space and a corresponding private latent space. Although these methods benefit from the idea, they embed the orthogonality requirement into the objective function with a weight to set its relative influence, i.e., they encourage rather than restrict the orthogonality. The main reason is that it is hard to optimize an objective function with complex constraints. However, in this case, orthogonality may not be satisfied and extra costly computation effort is needed.

Representation learning, which seeks good representations or features that facilitate learning functions of interest, has promising performance in single-view learning [7]. From the perspective of representation learning, multi-view learning can be viewed as learning several latent spaces or features with orthogonality constraints. Only very few works have discussed the similarity between representation learning, in the form of sparse learning, and multi-view learning [5,8].

In this paper, we propose using the denoising autoencoder to learn the shared and private latent spaces with orthogonality constraints. In our approach, the constraints are satisfied by disconnecting every private latent space from the remaining views. The advantages of our proposed method are: (i) By disconnecting the irrelevant latent spaces and views, the orthogonality constraints are enforced. (ii) Such constraints keep the chain rule almost the same, thus simplify training the model, i.e., no extra effort is needed for tuning weights or complex optimization. (iii) No preprocessing is required for denoising because the denoising autoencoder learns robust features.

2   Related Work

Existing multi-view learning algorithms can be classified into three groups: co-training, multiple kernel learning, and subspace learning [2]. Co-training [9] was the first formalized learning method in the multi-view framework. It trains repeatedly on labeled and unlabeled data until the mutual agreement on two distinct views is maximized. Further extensions include: an improved version with the Bayesian undirected graphical model [10], and application in active multi-view learning [11]. Multiple Kernel Learning naturally corresponds to different modalities (views) and combining kernels either linearly or non-linearly improves learning performance [12]. Multiple Kernel Learning always comes with diverse linear or nonlinear constraints, which makes the objective function complex [13–15]. Subspace learning-based approaches aim to obtain latent subspaces that have lower dimensions than input views, thus effective information is learned and redundancy is discarded from views. As those latent spaces can be regarded as effective features, subspace learning-based algorithms allow single-view learning algorithms to be capable for learning on multi-view data.

Subspace learning-based algorithms initially targeted on conducting meaningful dimensional reduction for multi-view data [2]. Canonical correlation analysis based approaches, following the consensus principle, linearly or non-linearly project two different views into the same space where the correlation between views is maximized [16,17]. Because the dimension of the space to be projected on equals to the smaller one of the two views, the dimension is reduced by at least half. Other similar approaches include Multi-view Fisher Discriminant Analysis, Multi-view Embedding and Multi-view Metric Learning [2].

Unlike other types of subspace learning-based algorithms, latent subspace learning models resort to explicitly building a shared latent space and several private latent spaces (a private latent space corresponds to a view). The Factorized Orthogonal Latent Space model proposes to factorize the latent space into shared and private latent spaces by encouraging these spaces to be orthogonal [6]. In addition, it penalized the dimensions of latent spaces to reduce redundancy. Similarly, a more advanced version is employed with sparse coding of structured sparsity for the same purpose [5]. Nevertheless, in their approaches, orthogonality is encouraged by penalizing inner products of latent spaces or encouraging the structured sparsity in the objective function. These approaches to not guarantee orthogonality.

Representation learning focuses on learning good representations or extracting useful information from data that simplifies further learning tasks [18]. A well-known successful example of representation learning is deep learning. By stacking autoencoders to learn better representations for each layer, deep learning drastically improves the performance of neural networks in tasks such as image classification [19], speech recognition [20], and natural language processing [21]. An autoencoder is simply a neural network that tries to copy its input to its output [7].

Multi-view feature learning was first analyized from the perspective of representation learning through sparse coding in [8], but little work has employed representation learning for studying multi-view data so far, thus far only sparse coding has been used [5,22]. Since the autoencoder is the most prevalent representation learning algorithm, our model enables various existing work on autoencoder to inherently extend it to multi-view settings.

3   Approach

3.1   Problem Formulation

Let $X = \{X^{(1)}, X^{(2)}, \cdots, X^{(V)}\}$ be a data set of N observations from V views and $X^{(v)}$ be the vth view of data, where $X^{(v)} \in \mathbb{R}^{N \times P(X^{(v)})}$. $P(\cdot)$ is the number of columns of the matrix; and $D(\cdot)$ is the number of features that the space or matrix has. Additionally, $Y \in \mathbb{R}^{N \times P(Y)}$ is the shared latent space across all views; $Z = \{Z^{(1)}, Z^{(2)}, \cdots, Z^{(V)}\}$ is the set of private latent spaces of each individual view, where $Z^{(v)} \in \mathbb{R}^{N \times P(Z^{(V)})}$. Because the latent spaces are required to have no redundancy, then $P(Y) = D(Y)$ and $P(Z^{(v)}) = D(Z^{(v)})$. Moreover, $X^{(v)}$ is expected to be linearly represented by Y and $Z^{(v)}$, $D(X^{(v)}) = D(Y) + D(Z^{(v)})$. Our goal is to learn Y and the Z from X where Y is independent of Z and the arbitrary two private latent spaces $Z^{(v_i)}, Z^{(v_j)}$ are orthogonal.

3.2   Basic Autoencoder

A standard autoencoder takes an input vector x and initially transforms it to a hidden representation vector $y = s(Wx + b)$ through an activation function s and weights W. Note the activation function can be linear or nonlinear. The latent representation y is subsequently mapped back to a reconstructed vector $z = s(W'y + b')$. The objective is that the output z is as close as possible to x, i.e., the parameters are optimized to minimize the average reconstruction error:

$$W^\star, b^\star, W'^\star, b'^\star = \underset{W,b,W',b'}{\arg\min} L(x, z) \tag{1}$$

where L is a loss function to measure how good the reconstruction is, and often least squares $L = \|(x - z)\|^2$ is used. The expectation is that the hidden representation y could capture the main factors of data x [23].

3.3   Orthogonal Autoencoder for Multi-view Learning

In the scenarios of multi-view learning, the hidden or latent representation (neuron) is expected to be consist of shared and private spaces. To this end, we modify the autoencoder such that every private latent space only connects to its own view. Figure 1 depicts the graphical model of such improved autoencoder given that the first view has two original features while the second one has three.

Because the first view is disconnected from the second private latent space (the third hidden neuron), the second private latent space is strictly independent of the first view. Similarly, the first private latent space is independent of the second view. In order to maintain orthogonality of private latent spaces, the bias is disconnected from private latent spaces (proof is given below).

In addition, in order to retain that views are linearly dependent of latent spaces, the hidden representation before the nonlinear mapping $(Wx + b)$ is regarded as latent spaces. If the activation function is nonlinear, then y and

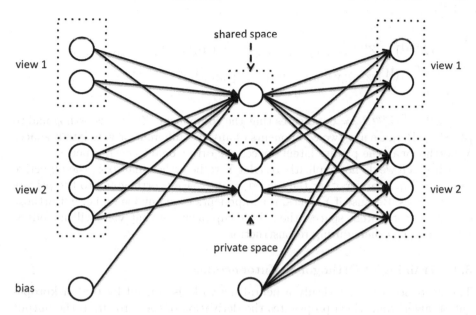

Fig. 1. Graphical model of an orthogonal autoencoder for multi-view learning with two views.

$W'y + b'$ can be considered as mappings of latent representation and reconstruction of input in a nonlinear space. And the last activation function maps the $W'y + b'$ back to original space.

Also note that the number of neurons representing the shared latent space equals the dimensions of its features $D(Y)$ and it is the same for private latent space, i.e., a neuron represents a feature in the space. Our model is inherently able to fit in any number of views and arbitrary numbers of features for each view and latent space.

Following the aforementioned notations, we further define $I(A|B)$ as the indices of columns of A in terms of B if matrix A is a submatrix of matrix B and they have same row numbers. The orthogonality constraints on weights can be formulated as:

$$W_{I(Z^{(v_2)}|[Y,Z]),I(X^{(v_1)}|X)} = 0 \qquad (v_1 \neq v_2) \tag{2}$$

$$W'_{I(X^{(v_1)}|X),I(Z^{(v_2)}|[Y,Z])} = 0 \qquad (v_1 \neq v_2) \tag{3}$$

$$b_{I(Z|[Y,Z])} = 0 \tag{4}$$

For a matrix A, the symbol $A_{I,\cdot}$ denotes a sub-matrix consisting of row vectors indexing by I (of A); similarly, $A_{\cdot,I}$ is a sub-matrix from such column vectors. We provide the rigorous proof that arbitrary two private latent spaces ($Z^{(v_1)}$ and $Z^{(v_2)}$) are orthogonal:

$$(Z^{(v_1)})^T \cdot Z^{(v_2)} = (W_{I(Z^{(v_1)}|[Y,Z]),\cdot} \cdot x + \mathbf{0})^T \cdot (W_{I(Z^{(v_2)}|[Y,Z]),\cdot} \cdot x + \mathbf{0})$$
$$= x^T \cdot ((W_{I(Z^{(v_1)}|[Y,Z]),\cdot})^T \cdot W_{I(Z^{(v_2)}|[Y,Z]),\cdot}) \cdot x = x^T \cdot \mathbf{0} \cdot x$$
$$= \mathbf{0}$$

$[(Z^{(v_1)})^T \cdot Z^{(v_2)}]_{ij} = 0$ indicates that the component $[Z^{(v_1)}]_{.,i}$ is orthogonal to $[Z^{(v_2)}]_{.,j}$. Because any two components of any two different private latent spaces are orthogonal, the private latent spaces are orthogonal to each other.

Although we do not explicitly restrict Y to be orthogonal to Z, the objective function (Eq. 1) prefers that the shared latent space is orthogonal to the private latent spaces. Because if Y contains components from any views, i.e. not orthogonal to private latent spaces, then the components of that view will introduce noise into the others during reconstruction.

3.4 Training of Orthogonal Autoencoder

The autoencoder, intrinsically a neural network, is trained by the backpropagation algorithm, which propagates the derivation of the error from the output to the input, layer-by-layer [4]. As the 0 values automatically break gradients passing through that connection, we only need to set Eqs. 5, 6 and 7 after basic backpropagation as follows:

$$\frac{\partial L}{W_{I(Z^{(v_2)}|[Y,Z]),I(X^{(v_1)}|X)}} = 0 \qquad (v_1 \neq v_2) \tag{5}$$

$$\frac{\partial L}{W'_{I(X^{(v_1)}|X),I(Z^{(v_2)}|[Y,Z])}} = 0 \qquad (v_1 \neq v_2) \tag{6}$$

$$\frac{\partial L}{b_{I(Z|[Y,Z])}} = 0 \tag{7}$$

3.5 Orthogonal Denoising Autoencoder for Robust Latent Spaces

The denoising autoencoder was formally proposed to enforce the autoencoder to learn robust features or latent spaces in our case [24]. The idea is based on the assumption that robust features can reconstruct or repair input which is partially destroyed or corrupted. A typical way of producing $\tilde{x}$, the corrupted version of initial input x, is randomly choosing a fixed number of components and forcing their values to be 0; while other components stays the same. As a consequence, the autoencdoer is encouraged to learn robust features which are most likely to recover the wiped information.

Afterwards the $\tilde{x}$ is fed as input to the autoencoder, then mapped to the hidden representation $y = s(W\tilde{x} + b)$ and finally transformed to output $z = s(W'y + b')$. In the process, the aforementioned orthogonality constraints (Eqs. (2) and (3)) remain the same. In the end, the objective function L enforces the output z to be as close as possible to the original, uncorrupted x, instead of input $\tilde{x}$.

4   Experiments

In this section, we introduce two datasets to evaluate our method: the synthetic dataset is straightforward to demonstrate the ability of our method to learn shared and private latent spaces from data with noise; and the real-world dataset is employed to compare our approach with other state-of-the-art algorithms and display optimization on the number of neurons for vieww using random search.

4.1   Synthetic Dataset

We evaluated our approach with a toy dataset similar to [6], which can be generated in 10 lines of MATLAB code (listed in Algorithm 1).

Algorithm 1. Toy data generation in MATLAB notation

```
t = -1:0.02:1;
x = sin(2*pi*t);
z1 = cos(pi*pi*t);
z2 = cos(5*pi*t);
v1 = (rand(20, size(x, 1)+1)) * [x;z1];
v1 = v1 + randn(size(v1))*0.01;
v2 = (rand(20, size(x, 1)+1)) * [x;z2];
v2 = v2 + randn(size(v2))*0.01;
v1 = [v1; 0.02*sin(3.6*pi*t)];
v2 = [v2; 0.02*sin(3.6*pi*t)];
```

In words, v1 and v2 (first and second views in Fig. 2b respectively) are two views generated by randomly projecting two ground truths, [x;z1] and [x;z2], to 20 dimensions spaces. The first component (blue curve in Fig. 2a) of the two ground truths are shared latent space and their second components (green curves in first and second ground truth of Fig. 2a respectively) are individual private latent spaces. The two views are then added to Gaussian noise with standard deviation 0.01 and correlated noise on their last dimension (both views have 21 dimensions now).

In the experiment, we adopt the Hyperbolic Tangent ($tanh(x) = \frac{1-e^{-2x}}{1+e^{-2x}}$) as the activation function. Three hidden neurons are used as latent spaces, one represents the shared space and the other two represent the private ones. Features of the original data are evenly corrupted for denoising.

The result of our method is depicted in Fig. 2d, where the first graph employs modified autoencoder while the second one is modified denoising autoencoder. The shared latent factor is in blue, the private latent factor of first view is in green and private latent factor of second view is in red. The denoising autoencoder generates more robust latent spaces than those from the autoencoder.

As expected, Canonical Correlation Analysis (Fig. 2c) extracts the true shared signal (in blue) and the correlated noise (in red), while fails to discover the two private signals. Notice that both true recovered signals are scaled and the correlated noise is even inverted. The reason is that we can multiply a number

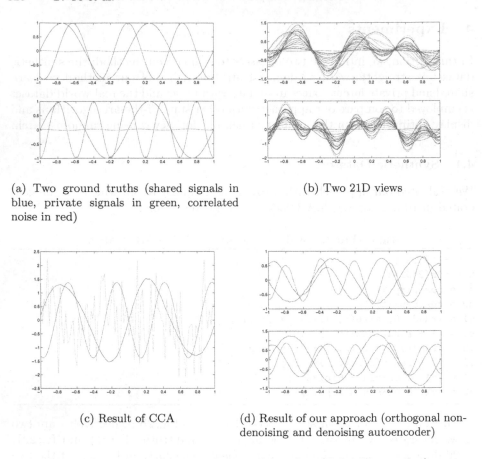

(a) Two ground truths (shared signals in blue, private signals in green, correlated noise in red)

(b) Two 21D views

(c) Result of CCA

(d) Result of our approach (orthogonal non-denoising and denoising autoencoder)

Fig. 2. Latent spaces recovered on synthetic data (Color figure online).

and its multiplicative inverse respectively to the latent spaces and corresponding coefficient to attain same product.

The experiment confirms our approach is able to effectively learn robust shared and private latent spaces, while CCA is sensitive to noise. Moreover, we do not need to use PCA like previous work [5][1]. Because the local minimum differs due to the random initialization of weights, methods of optimization, etc., Fig. 2d may vary slightly in repeat experiments [25].

4.2   Real-World Dataset

We applied our model to the PASCAL VOC'07 data set [26] for multi-label object classification, i.e., an image may contain more than one object label. Through the experiment, we compare the results from using only images, tags, and their combinations to demonstrate multi-view learning algorithms are capable of using

---

[1] With PCA, CCA can also perform well.

Table 1. Mean and variance of AP of different approaches (best results of classification for each class in bold).

	aeroplane	bicycle	bird	boat	bottle	bus	car
Images	0.596 ± 0.012	0.367 ± 0.028	0.277 ± 0.044	0.568 ± 0.018	0.127 ± 0.021	0.319 ± 0.023	0.613 ± 0.011
Tags	0.734 ± 0.016	0.539 ± 0.012	0.688 ± 0.003	0.495 ± 0.016	$\mathbf{0.237 \pm 0.017}$	0.430 ± 0.016	0.591 ± 0.003
MKL	$\mathbf{0.847 \pm 0.031}$	0.532 ± 0.030	0.711 ± 0.027	0.596 ± 0.044	0.210 ± 0.032	0.569 ± 0.026	0.694 ± 0.025
SVM2K(0)	0.584 ± 0.157	0.360 ± 0.271	0.576 ± 0.356	0.290 ± 0.290	0.117 ± 0.098	0.116 ± 0.033	0.431 ± 0.115
SVM2K(1)	0.682 ± 0.132	0.385 ± 0.377	0.113 ± 0.356	$\mathbf{0.796 \pm 0.122}$	0.080 ± 0.038	0.133 ± 0.098	0.579 ± 0.292
ODAE	0.837 ± 0.009	$\mathbf{0.548 \pm 0.008}$	$\mathbf{0.725 \pm 0.005}$	0.682 ± 0.020	0.205 ± 0.024	$\mathbf{0.588 \pm 0.029}$	$\mathbf{0.714 \pm 0.013}$

	cat	chair	cow	diningtable	dog	horse	motorbike
Images	0.369 ± 0.010	$\mathbf{0.364 \pm 0.006}$	0.225 ± 0.079	$\mathbf{0.291 \pm 0.102}$	0.220 ± 0.019	0.607 ± 0.040	0.394 ± 0.039
Tags	0.715 ± 0.009	0.174 ± 0.004	0.471 ± 0.014	0.106 ± 0.012	0.678 ± 0.006	0.775 ± 0.004	0.607 ± 0.006
MKL	0.722 ± 0.018	0.296 ± 0.018	0.513 ± 0.037	0.193 ± 0.017	$\mathbf{0.689 \pm 0.022}$	$\mathbf{0.815 \pm 0.014}$	$\mathbf{0.660 \pm 0.032}$
SVM2K(0)	0.231 ± 0.307	0.095 ± 0.013	0.089 ± 0.011	0.074 ± 0.034	0.137 ± 0.056	0.068 ± 0.013	0.521 ± 0.305
SVM2K(1)	0.366 ± 0.199	0.311 ± 0.164	0.187 ± 0.170	0.070 ± 0.028	0.486 ± 0.293	0.516 ± 0.353	0.258 ± 0.222
ODAE	$\mathbf{0.725 \pm 0.012}$	0.347 ± 0.011	$\mathbf{0.521 \pm 0.015}$	0.219 ± 0.014	0.665 ± 0.009	0.760 ± 0.015	0.624 ± 0.021

	person	pottedplant	sheep	sofa	train	tvmonitor
Images	0.689 ± 0.016	0.115 ± 0.018	0.163 ± 0.026	0.220 ± 0.016	0.589 ± 0.035	0.230 ± 0.014
Tags	0.686 ± 0.002	0.329 ± 0.012	0.592 ± 0.031	0.180 ± 0.006	0.811 ± 0.005	0.378 ± 0.013
MKL	$\mathbf{0.782 \pm 0.028}$	$\mathbf{0.388 \pm 0.047}$	0.584 ± 0.043	0.225 ± 0.026	$\mathbf{0.845 \pm 0.017}$	0.395 ± 0.022
SVM2K(0)	0.643 ± 0.227	0.053 ± 0.003	0.599 ± 0.329	$\mathbf{0.510 \pm 0.404}$	0.188 ± 0.156	$\mathbf{0.711 \pm 0.299}$
SVM2K(1)	0.739 ± 0.038	0.256 ± 0.308	0.126 ± 0.115	0.151 ± 0.064	0.290 ± 0.214	0.197 ± 0.115
ODAE	0.749 ± 0.015	0.350 ± 0.012	$\mathbf{0.644 \pm 0.008}$	0.244 ± 0.026	0.838 ± 0.002	0.435 ± 0.006

information from different views. We also provide a comparison of our method to other methods, and test the sensitivity of our model.

The data set contains around 10,000 images of 20 different categories of object with standard train/test sets provided. The dataset also provides tags for each image: textual descriptions of the images. A total of 804 tags, which appear at least 8 times, form the bag-of-words vectors (bit-based) of a view. 15 image features are chosen from [15] as visual descriptors: local SIFT features [27]; local hue histograms [28] (both were computed on a dense multi-scale grid and on regions found with a Harris interest-point detector); global color histograms over RGB, HSV, and LAB color spaces; the above histogram image representations computed over a 3×1 horizontal decomposition of the image; and GIST descriptor [29], which roughly encodes the image layout. This produces two views: a visual modality with 37,152 features; and a textual modality with 804 features.

The default experiment setup used 6 SVMs (sigmoid kernel) with different C parameters[2] for classification. Table 1 reports the mean and variance of the average precision (AP) values from these SVMs. For the single modality image classification ("Images" in Table 1), we report AP using the single visual feature that performed best for each class.

In Multiple Kernel Learning (MKL), we computed textual kernel $k_t(\cdot, \cdot)$ and visual kernels $k_v(\cdot, \cdot)$ following [15]. Because the final kernel $k_f(\cdot, \cdot) = d_v k_v(\cdot, \cdot) + d_t k_t(\cdot, \cdot)$, where $d_v, d_t > 0$ and $d_v + d_t = 1$, is a convex combination, the d_v and d_t

---

[2] C from $\{10^{-3}, 10^{-2}, 10^{-1}, 10^0, 10^1, 10^2\}$.

were chosen by grid search with step 0.1 and $C = 1$. Instead of original features, kernels $k_f(\cdot, \cdot)$ were then used in place of the original features.

We tested two variants of SVM2K, both containing two views, since SVM2K [17] only supports two-view learning. SVM2K(0) concatenates all visual features into one view, with the other view comprising the textual features; SVM2K(1) uses only the 2 SIFT-based features. SVM2K contains 4 penalty parameters: we set the penalty value for the first SVM to 0.2; for the second SVM to 0.1; tolerance for the synthesis to 0.01; and varied the penalty value for synthesis in the same way as previously described for linear SVM parameter C.

Our orthogonal denoising autoencoder for multi-view learning (ODAE) uses the sigmoid function ($y = \frac{1}{1+e^{-x}}$) as the activation function. We reduplicate the data 10 times and randomly corrupted 10 % of total features each time for denoising. Random search was used to find the optimal numbers of neurons for each hidden views [30].

Generally speaking, multi-view learning approaches for object classification outperform those using only the visual or textual view. SVM2K methods have the worst performance and are highly unstable among all the multi-view learning approaches, since they can only accept 2 views. Specifically, the visual view of SVM2K(0) tends to overfit while that of SVM2K(1) tends to underfit. ODAE performs best for 7 classes and second best for another 10 classes (slightly worse).

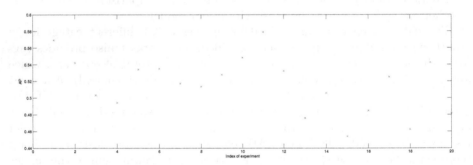

Fig. 3. AP values for 20 experiments with random hyperparameters (max 300 SGD iterations). The number of hidden nodes in each experiment are uniformly drawn from: $[3, 40]$ for the shared view, $[500, 700]$ for the textual view, and $[50, 200]$ for all visual views with the exception of Harris hue ranges, which were drawn from $[50, 100]$.

We used random search [30] to select reasonable values for the number of hidden neurons for each view[3]. Figure 3 plots AP values for different parameter configurations of an orthogonal autoencoder, demonstrating the relative robustness of the method with respect to the hyperparameters. The best configuration found was: 29 nodes of hidden layer for the shared view, 672 for the textual view, and $[99, 148, 149, 143, 74, 51, 113, 121, 96, 164, 113, 91, 148, 68, 56]$ for the visual views.

---

[3] Because of pages limit, we can not provide more details.

5 Conclusions

In this paper, we modify the basic autoencoder for multi-view learning to enable private latent spaces to be absolutely orthogonal to each other while simultaneously encouraging the shared latent space to be orthogonal to private latent spaces as well. Inheriting from the denoising autoencoder, our model is able to learn robust features, i.e., features that are strongly resistant to noise. We also extend back-propagation algorithm elegantly to train the model, which means our model is exempt from extra complex optimization tricks.

Acknowledgement. The research was supported by the Irish Research Council (IRC-SET) under Grant Number GOIPG/2013/330. The authors wish to acknowledge the DJEI/DES/SFI/HEA Irish Centre for High-End Computing (ICHEC) for the provision of computational facilities and support. Amen.

References

1. Sun, S.: A survey of multi-view machine learning. Neural Comput. Appl. **23**(7–8), 2031–2038 (2013)
2. Xu, C., Tao, D., Xu, C.: A survey on multi-view learning. arXiv preprint arXiv:1304.5634 (2013)
3. Dasgupta, S., Littman, M.L., McAllester, D.: Pac generalization bounds for co-training. Adv. Neural Inf. Process. Syst. **1**, 375–382 (2002)
4. Bishop, C.M.: Pattern Recognition and Machine Learning. Springer, New York (2006)
5. Jia, Y., Salzmann, M., Darrell, T.: Factorized latent spaces with structured sparsity. In: Advances in Neural Information Processing Systems, pp. 982–990 (2010)
6. Salzmann, M., Ek, C.H., Urtasun, R., Darrell, T.: Factorized orthogonal latent spaces. In: International Conference on Artificial Intelligence and Statistics, pp. 701–708 (2010)
7. Bengio, Y., Goodfellow, I.J., Courville, A.: Deep learning. Book in preparation for MIT Press (2015)
8. Memisevic, R.: On multi-view feature learning. arXiv preprint arXiv:1206.4609 (2012)
9. Blum, A., Mitchell, T.: Combining labeled and unlabeled data with co-training. In: Proceedings of the Eleventh Annual Conference On Computational Learning Theory, pp. 92–100. ACM (1998)
10. Yu, S., Krishnapuram, B., Rosales, R., Rao, R.B.: Bayesian co-training. J. Mach. Learn. Res. **12**, 2649–2680 (2011)
11. Muslea, I., Minton, S., Knoblock, C.A.: Active learning with multiple views. J. Artif. Intell. Res. **27**, 203–233 (2006)
12. Gönen, M., Alpaydın, E.: Multiple kernel learning algorithms. J. Mach. Learn. Res. **12**, 2211–2268 (2011)
13. Rakotomamonjy, A., Bach, F., Canu, S., Grandvalet, Y.: More efficiency in multiple kernel learning. In: Proceedings of the 24th International Conference On Machine Learning, pp. 775–782. ACM (2007)
14. Akaho, S.: A kernel method for canonical correlation analysis. arXiv preprint cs/0609071 (2006)

15. Guillaumin, M., Verbeek, J., Schmid, C.: Multimodal semi-supervised learning for image classification. In: 2010 IEEE Conference on Computer Vision and Pattern Recognition (CVPR), pp. 902–909. IEEE (2010)

16. Sun, S., Hardoon, D.R.: Active learning with extremely sparse labeled examples. Neurocomputing **73**(16), 2980–2988 (2010)

17. Farquhar, J., Hardoon, D., Meng, H., Shawe-taylor, J.S., Szedmak, S.: Two view learning: Svm-2k, theory and practice. In: Advances in Neural Information Processing Systems, pp. 355–362 (2005)

18. Bengio, Y., Courville, A., Vincent, P.: Representation learning: a review and new perspectives. IEEE Trans. Pattern Anal. Mach. Intell. **35**(8), 1798–1828 (2013)

19. Krizhevsky, A., Sutskever, I., Hinton, G.E.: Imagenet classification with deep convolutional neural networks. In: Advances in Neural Information Processing Systems, pp. 1097–1105 (2012)

20. Hinton, G., Deng, L., Yu, D., Dahl, G.E., Mohamed, A., Jaitly, N., Senior, A., Vanhoucke, V., Nguyen, P., Sainath, T.N., et al.: Deep neural networks for acoustic modeling in speech recognition: the shared views of four research groups. IEEE Signal Process. Mag. **29**(6), 82–97 (2012)

21. Collobert, R., Weston, J.: A unified architecture for natural language processing: deep neural networks with multitask learning. In: Proceedings of the 25th International Conference On Machine Learning, pp. 160–167. ACM (2008)

22. Liu, W., Tao, D., Cheng, J., Tang, Y.: Multiview hessian discriminative sparse coding for image annotation. Comput. Vis. Image Underst. **118**, 50–60 (2014)

23. Bengio, Y., Lamblin, P., Popovici, D., Larochelle, H., et al.: Greedy layer-wise training of deep networks. Adv. Neural Inf. Process. Syst. **19**, 153 (2007)

24. Vincent, P., Larochelle, H., Bengio, Y., Manzagol, P.-A.: Extracting and composing robust features with denoising autoencoders. In: Proceedings of the 25th International Conference on Machine Learning, pp. 1096–1103. ACM (2008)

25. LeCun, Y.A., Bottou, L., Orr, G.B., Müller, K.-R.: Efficient BackProp. In: Montavon, G., Orr, G.B., Müller, K.-R. (eds.) Neural Networks: Tricks of the Trade, 2nd edn. LNCS, vol. 7700, pp. 9–48. Springer, Heidelberg (2012)

26. Everingham, M., Van Gool, L., Williams, C.K.I., Winn, J., Zisserman, A.: The PASCAL Visual Object Classes Challenge 2007 (VOC2007) Results. http://www.pascal-network.org/challenges/VOC/voc2007/workshop/index.html

27. Lowe, D.G.: Object recognition from local scale-invariant features. In: The proceedings of the Seventh IEEE International Conference on Computer Vision, vol. 2, pp. 1150–1157. IEEE (1999)

28. van de Weijer, J., Schmid, C.: Coloring local feature extraction. In: Leonardis, A., Bischof, H., Pinz, A. (eds.) ECCV 2006. LNCS, vol. 3952, pp. 334–348. Springer, Heidelberg (2006)

29. Oliva, A., Torralba, A.: Modeling the shape of the scene: a holistic representation of the spatial envelope. Int. J. Comput. Vis. **42**(3), 145–175 (2001)

30. Bergstra, J., Bengio, Y.: Random search for hyper-parameter optimization. J. Mach. Learn. Res. **13**(1), 281–305 (2012)

Fast Nearest Neighbor Search
in the Hamming Space

Zhansheng Jiang[1]([✉]), Lingxi Xie[2], Xiaotie Deng[1], Weiwei Xu[3],
and Jingdong Wang[4]

[1] Shanghai Jiao Tong University, Shanghai, People's Republic of China
jzhsh1735@sjtu.edu.cn, deng-xt@cs.sjtu.edu.cn
[2] Tsinghua University, Beijing, People's Republic of China
198808xc@gmail.com
[3] Hangzhou Normal University, Hangzhou, People's Republic of China
weiwei.xu.g@gmail.com
[4] Microsoft Research, Beijing, People's Republic of China
jingdw@microsoft.com

Abstract. Recent years have witnessed growing interests in computing compact binary codes and binary visual descriptors to alleviate the heavy computational costs in large-scale visual research. However, it is still computationally expensive to linearly scan the large-scale databases for nearest neighbor (NN) search. In [15], a new approximate NN search algorithm is presented. With the concept of *bridge vectors* which correspond to the cluster centers in Product Quantization [10] and the *augmented neighborhood graph*, it is possible to adopt an *extract-on-demand* strategy on the online querying stage to search with priority. This paper generalizes the algorithm to the Hamming space with an alternative version of *k*-means clustering. Despite the simplicity, our approach achieves competitive performance compared to the state-of-the-art methods, *i.e.*, MIH and FLANN, in the aspects of search precision, accessed data volume and average querying time.

Keywords: Approximate nearest neighbor search · Hamming space · Bridge vectors · Augmented neighborhood graph

1 Introduction

Compact binary codes [7,8] and binary visual descriptors [4–6] play a significant role in computer vision applications. Binary vectors have a lot of advantages, including the cheap storage cost and the low consumption in computing the distance between binary vectors through a bitwise XOR operation. Despite the efficiency of binary vector operations, it is often time-consuming to linearly scan a large-scale database, seeking for the nearest neighbor.

This work was done when Zhansheng Jiang was an intern at Microsoft Research, P.R. China.

© Springer International Publishing Switzerland 2016
Q. Tian et al. (Eds.): MMM 2016, Part I, LNCS 9516, pp. 325–336, 2016.
DOI: 10.1007/978-3-319-27671-7_27

A lot of efforts are made in accelerating nearest neighbor search. MIH [2] is an exact nearest neighbor search algorithm. However, the index construction and the query process for large-scale search databases are very time-consuming and therefore impractical. FLANN [3] is an approximate nearest neighbor search method, the precision of which is often low in the scenarios of long binary codes or large-scale search databases, even with a large number of accessed data points. In [15], the authors present the concept of *bridge vectors*, which are similar to the cluster centers in Product Quantization [10], and construct a *bridge graph* by connecting each bridge vector to its nearest vectors in the database. In the online querying procedure, it is possible to organize data in a priority queue and design an *extract-on-demand* strategy for efficient search.

Neighborhood graph search has attracted a lot of interests [12,15] because of the low cost in extracting neighbor vectors and the good search performance. The index structure is a directed graph connecting each vector to its nearest neighbors in a search database. The graph search procedure involves measuring the priority of each candidate. Accessed vectors are organized in a priority queue ranked by their distance to the query vector. The top vector in the priority queue is popped out one by one and its neighborhood vectors are pushed into the queue. This process continues until a fixed number of vectors are accessed.

In this paper, we generalize the algorithm in [15] to the Hamming space. The major novelty of this work lies in that we generalize the previous algorithm to the Hamming space with an alternative version of k-means. Experiments reveal that, despite the simplicity, our approach achieves superior performance to the state-of-the-art methods, *i.e.*, MIH and FLANN, in the aspects of search precision, accessed data volume and average querying time.

The rest of this paper is organized as follows. In Sect. 2, we review previous works of nearest neighbor search in the Hamming space. We introduce our approach in Sect. 3. Experiment results are presented in the next section. In the end, we summarize this paper and state the conclusions.

2   Related Works

In this section, we will introduce two previous works which are popular for nearest neighbor search in the Hamming space.

2.1   Multi-index Hashing

Multi-index hashing (MIH) algorithm is presented in [2] to achieve a fast NN search in the Hamming space. The algorithm indexes the database m times into m different hash tables, then during the search for each query, it collect NN candidates through looking up the hash tables by taking advantage of the Pigeonhole Principle.

In detail, each b-bit binary code $\mathbf{h}$ in the database is split into m disjoint substrings $\mathbf{h}^{(1)}, \ldots, \mathbf{h}^{(m)}$, each of which is b/m in length, assuming that b is divisible by m. Then the binary code is indexed in m hash tables according to its m substrings. The key idea rests on the following proposition: Given $\mathbf{h}, \mathbf{g}$, and $\|\mathbf{h} - \mathbf{g}\|_H \leq r$, where $\|\cdot\|_H$ denotes the Hamming norm, there exists k, $1 \leq k \leq m$, such that

$$\|\mathbf{h}^{(k)} - \mathbf{g}^{(k)}\| \leq \lfloor \frac{r}{m} \rfloor.$$

The proposition can be proved by the Pigeonhole Principle. During the search for query $\mathbf{q}$ with substrings $\{\mathbf{q}^{(i)}\}_{i=1}^m$, we collect from i^{th} hash table those entries which are within the Hamming distance $\lfloor \frac{r}{m} \rfloor$ to $\mathbf{q}^{(i)}$, denoted as $\mathcal{N}_i(\mathbf{q})$. Then the set $\mathcal{N} = \bigcup_{i=1}^m \mathcal{N}_i(\mathbf{q})$ will contain all binary codes within the Hamming distance r to $\mathbf{q}$.

The key idea stems from the fact that, in the case of n binary b-bit code and $2^b \gg n$, when we build a full Hash table, most of the buckets are empty. When we retrieve the r-neighbors of a query q, we need to traverse 2^r buckets, which is very costly when r is large. However, in this approach, many buckets are merged together by marginalizing over different dimensions of the Hamming space. As a result, the number of visited buckets is greatly reduced from 2^r to $m \cdot 2^{\lfloor \frac{r}{m} \rfloor}$. The downside is that not all the candidates in these merged buckets are the r-neighbors of the query, so we need to examine them one by one.

In this approach, the search cost depends on the number of visited buckets and the number of accessed candidates. As a result, a trade-off has to be made between them by choosing a proper m. When m is small, $\lfloor \frac{r}{m} \rfloor$ is large so we have to traverse many buckets. When m is large, many buckets are merged together so we have to check a large number of candidates one by one.

The major disadvantage of MIH algorithm lies in the heavy computational costs. When the code length b is large, $e.g.$, $b = 512$, either m or $\lfloor \frac{r}{m} \rfloor$ is too large so that the algorithm becomes less efficient, as shown in later experiments.

2.2   FLANN

FLANN [3] has been a well-know library for approximate nearest neighbor (ANN) search in the Euclidean space. In [3], the authors of FLANN introduce a method for ANN search in the Hamming space.

First, the algorithm performs a hierarchical decomposition of the Hamming space to build a tree structure. It starts with clustering all the data points into K clusters, where K is a parameter. The cluster centers are K points which are randomly selected and each data point is assigned to its nearest center in the Hamming distance. The decomposition is repeated recursively for each cluster until the number of points in a cluster is less than a threshold, in which case this cluster becomes a leaf node. The hierarchical decomposition of the database is repeated for several times and multiple hierarchical trees are constructed. The search performance significantly improves as the number of trees increases. The search process is performed by traversing multiple trees in parallel, which is presented in Algorithm 1 [3].

The approach of constructing multiple hierarchical trees is similar to k-d trees [13] or hierarchical k-means trees [14]. However, in these approaches, the strategy of building trees is well-optimized so the decomposition results are always

---

Algorithm 1. Searching Parallel Hierarchical Clustering Trees

---

Input hierarchical trees $\mathcal{T} = \{T_i\}$, query point **q**
Output K nearest approximate neighbors of **q**
Parameters the max number of examined points $L_{\max}$
Variables $\mathcal{PQ}$: the priority queue storing the unvisited branches; $\mathcal{R}$: the priority queue storing the examined data points; L: the number of examined data points.
  1: $L \leftarrow 0$
  2: $\mathcal{PQ} \leftarrow \emptyset$
  3: $\mathcal{R} \leftarrow \emptyset$
  4: **for** each tree T_i **do**
  5:     call TraverseTree($T_i, \mathcal{PQ}, \mathcal{R}$)
  6: **end for**
  7: **while** $\mathcal{PQ} \neq \emptyset$ **and** $L < L_{\max}$ **do**
  8:     $N \leftarrow$ top of $\mathcal{PQ}$
  9:     call TraverseTree($N, \mathcal{PQ}, \mathcal{R}$)
10: **end while**
11: **return** K nearest points to **q** from $\mathcal{R}$
procedure TraverseTree($T, \mathcal{PQ}, \mathcal{R}$)
  1: **if** T is a leaf node **then**
  2:     examine all data points in T and add them in $\mathcal{R}$
  3:     $L \leftarrow L + |T|$
  4: **else**
  5:     $\mathcal{C} \leftarrow$ child nodes of T
  6:     $N_q \leftarrow$ nearest node to **q** in $\mathcal{C}$
  7:     $\bar{\mathcal{C}} \leftarrow \mathcal{C} \setminus \{N_q\}$
  8:     add all nodes in $\bar{\mathcal{C}}$ to $\mathcal{PQ}$
  9:     call TraverseTree($N_q, \mathcal{PQ}, \mathcal{R}$)
10: **end if**

---

the same and constructing multiple trees is unnecessary. In this approach, since the K cluster centers are randomly selected and no further optimization is performed, multiple different trees can be constructed. The benefit of constructing multiple hierarchical trees is much more significant due to the fact that when explored in parallel, each tree will retrieve different candidates, thus the probability of finding the exact NNs is increased.

3 Our Approach

3.1 Data Structure

The major part of our data structure is nearly the same as in [15], but we will later generalize it to the Hamming space. The index structure consists of two components: the *bridge vectors* and the *augmented neighborhood graph*.

Bridge Vectors. Inspired by Product Quantization [10] and inverted multi-index [9], we propose to construct a set of bridge vectors, which are similar to the cluster centers in Product Quantization. We split the vectors in the database into m dimensional chunks and cluster i-th dimensional chunk into n_i centers. The bridge vectors are defined as follows:

$$\mathcal{Y} = \times_{i=1}^{m} \mathcal{S}_i \triangleq \{\mathbf{y}_j = [\mathbf{y}_{j_1}^T \ \mathbf{y}_{j_2}^T \ \cdots \ \mathbf{y}_{j_m}^T]^T | \mathbf{y}_{j_i} \in \mathcal{S}_i\}.$$

where $\mathcal{S}_i$ is the center set for the i-th dimensional chunk. To be simplified and without loss of generality, we assume that $n_1 = n_2 = \cdots = n_m = n$. There is a nice property that finding the nearest neighbor over the bridge vectors is really efficient, which takes only $O(nd)$ in spite of the size of the set is n^m. Moreover, by adopting the Multi-sequence algorithm in [9], we can identify the second, the third and more neighbors each in $O(m^2 \log n)$ after sorting n_i centers in each dimensional chunk $\mathcal{S}_i$ in $O(mn \log n)$.

The major difference between this work and [15] lies in that we need to deal with the Hamming space, i.e., **the conventional Euclidean distance shall be replaced by the Hamming distance**. In this respect, we adopt an alternative version of k-means clustering for binary codes and the Hamming distance to obtain the center set for each dimensional chunk. We first initialize k cluster centers by randomly drawing k different data points from the database. Then we run two steps, the assignment step and the update step, iteratively. In the assignment step, we find the nearest center for each data point and assign the point to the corresponding cluster. In the update step, for each cluster, we find a Hamming vector which has a minimum average Hamming distance to all cluster members as the new cluster center. This is simply done by voting for each bit individually, i.e., each bit of each cluster center takes the dominant case of this bit (0 or 1) assigned to this center.

Although our approach seems straightforward on the basis of k-means, it provides an opportunity to transplant other algorithms based on the Euclidean distance to the Hamming space. One of the major costs in k-means clustering lies in the distance computation, and the current version which computes the Hamming distance is much faster than the previous one which computes the Euclidean distance (5× faster). In the large-scale database, this property helps a lot in retrieving the desired results in reasonable time.

Augmented Neighborhood Graph. The augmented neighborhood graph is a combination of the neighborhood graph **G** over the search database vectors $\mathcal{X}$ and the bridge graph **B** between the bridge vectors $\mathcal{Y}$ and the search database vectors $\mathcal{X}$. The neighborhood graph **G** is a directed graph. Each node corresponds to a vector $\mathbf{x}_i$, and each node $\mathbf{x}_i$ is connected with a list of nodes that correspond to its neighbors.

In the bridge graph **B**, each bridge vector $\mathbf{y}_j$ in $\mathcal{Y}$ is connected to its nearest vectors in $\mathcal{X}$. To avoid expensive computation cost, we build the bridge graph approximately by finding the top t (typically 1000 in our experiments) nearest bridge vectors for each search database vector through the Multi-sequence

---

Algorithm 2. Fast Nearest Neighbor Search in the Hamming Space (FNNS)

Input query point q, bridge graph $\mathbf{B}$, neighborhood graph $\mathbf{G}$
Output K nearest approximate neighbors of q
Parameters the max number of accessed points $L_{\max}$
Variables $\mathcal{PQ}$: the priority queue storing the accessed points; L: the number of accessed data points; $\mathbf{b}$: the current bridge vector.

 1: $\mathbf{b} \leftarrow$ nearest bridge vector to q through the *Multi-sequence algorithm* in [9]
 2: $\mathcal{PQ} \leftarrow \{\mathbf{b}\}$
 3: $L \leftarrow 0$
 4: **while** $L < L_{\max}$ **do**
 5:     **if** $\mathbf{b}$ is top of $\mathcal{PQ}$ **then**
 6:         call ExtractOnDemand($\mathbf{b}, \mathcal{PQ}$)
 7:     **else**
 8:         $\mathbf{t} \leftarrow$ top of $\mathcal{PQ}$
 9:         examine and push neighbor points of $\mathbf{t}$ in neighborhood graph $\mathbf{G}$ to $\mathcal{PQ}$
10:         $L \leftarrow L +$ number of new points pushed to $\mathcal{PQ}$
11:         pop $\mathbf{t}$ from $\mathcal{PQ}$
12:     **end if**
13: **end while**
14: **return** K nearest points to q popped from $\mathcal{PQ}$

procedure ExtractOnDemand($\mathbf{b}, \mathcal{PQ}$)
 1: examine and push neighbor points of $\mathbf{b}$ in bridge graph $\mathbf{B}$ to $\mathcal{PQ}$
 2: $L \leftarrow L +$ number of new points pushed to $\mathcal{PQ}$
 3: pop $\mathbf{b}$ from $\mathcal{PQ}$
 4: $\mathbf{b} \leftarrow$ next nearest bridge vector to q through the *Multi-sequence algorithm* in [9]
 5: push $\mathbf{b}$ to $\mathcal{PQ}$

---

algorithm in [9] and then keeping the top b (typically 50 in our experiments) nearest search database vectors for each bridge vector.

3.2   Search over the Augmented Neighborhood Graph

We first give a brief view of the search procedure in a neighborhood graph. In this procedure, we first select one or more seed points, and then run a priority search starting from these seed points. The priority search is very similar to the breadth-first search. In the breadth-first search, we organize all visited points in a queue, in which the points are sorted by the order that they are pushed into the queue. However, in the priority search, we organize the visited points in a priority queue, in which the points are sorted by the distance to the query point. In detail, we first push all seed points into the priority queue. During each iteration, we pop out the top element, which is the nearest to the query point, from the priority queue and push its unvisited neighbors in the neighborhood graph into the priority queue.

To exploit bridge vectors and the augmented neighborhood graph, we adopt an extract-on-demand strategy. At the beginning of the search, we push the bridge vector nearest to the query into the priority queue, and during the entire search procedure, we maintain the priority queue such that it consists exactly one bridge vector. In each iteration, if the top element is a data point, the algorithm proceeds as usual; if the top element is a bridge vector, we extract its neighbors in the bridge graph and push the unvisited ones into the priority queue, and in addition we extract the next nearest bridge vector using the Multi-sequence algorithm [9] and push it into the priority queue. The algorithm ends when a fixed number of data points are accessed.

4 Experiments

4.1 Datasets and Settings

Datasets. We evaluate our algorithm on three datasets: 1 million BRIEF dataset, 1 million BRISK dataset and the 80 million tiny images dataset [1]. The BRIEF and BRISK datasets are composed of the BRIEF and BRISK features of 1 million Flickr images crawled from the Internet. BRIEF and BRISK features are 128-bit and 512-bit binary codes, respectively. For the 80 million tiny images dataset, we learn Hash codes based on GIST features [11] through LSH [8] and MLH [7] resulting in 512-bit binary codes. We carry on experiments in three different scales of search databases, *i.e.*, 1 M, 10 M and 79 M.

Evaluation. We precompute the exact k-NNs as ground truth based on the Hamming distance to each query in the test set. We use the precision to evaluate the search quality. For k-NN search, the precision is computed as the ratio of retrieved points that are the exact k nearest neighbors. We conduct experiments on multiple settings with the number of neighbors $k = 1$, $k = 10$ and $k = 50$.

We construct the test set by randomly choosing 10 K images excluded from the search database. We download the source codes of FLANN and MIH method and compare our approach with these two algorithms.

Settings. All the experiments are conducted on a single CPU core of server with 128 G memory. In our approach, binary codes are split into 4 dimensional chunks and each chunk is clustered into 50 centers. In FLANN, hierarchical clustering uses the following parameters: tree number 4, branching factor 32 and maximum leaf size 100.

4.2 Results

For FLANN and our approach, we report the search time and the precision with respect to the number of accessed data points. Since MIH is an exact search algorithm, the precision is always 1. We also report the best search time and the number of accessed data points by tuning the parameter m.

According to Table 1, FLANN takes about 50 % more search time than our approach to achieve the same precision. Meanwhile, our approach only accesses half of data points that FLANN accesses to achieve the same precision. For a larger k, the advantage of our approach is more significant. MIH has similar

Table 1. BRIEF 1 M Data

Algorithm	Volume	Time(ms)	Precision
k = 1			
FNNS	3000	0.350	0.974
	5000	0.569	0.984
	10000	1.248	0.993
FLANN	5000	0.508	0.947
	10000	0.939	0.975
	20000	1.856	0.990
MIH	21064	1.5	1.000
k = 10			
FNNS	3000	0.395	0.978
	5000	0.599	0.989
	10000	1.342	0.996
FLANN	5000	0.509	0.831
	10000	0.945	0.914
	20000	1.831	0.964
MIH	49655	3.6	1.000
k = 50			
FNNS	3000	0.517	0.971
	5000	0.838	0.985
	10000	1.726	0.995
FLANN	5000	0.522	0.693
	10000	0.963	0.840
	20000	1.855	0.932
MIH	72888	5.5	1.000

Table 2. BRISK 1 M Data

Algorithm	Volume	Time(ms)	Precision
k = 1			
FNNS	1000	0.122	0.755
	6000	0.865	0.971
	20000	3.296	0.997
FLANN	10000	1.566	0.859
	30000	4.578	0.959
	60000	9.263	0.988
MIH	776410	87.4	1.000
k = 10			
FNNS	1000	0.142	0.698
	6000	0.899	0.957
	20000	3.517	0.995
FLANN	10000	1.620	0.540
	30000	4.691	0.836
	60000	9.508	0.948
MIH	851400	97.6	1.000
k = 50			
FNNS	1000	0.231	0.612
	6000	1.281	0.932
	20000	4.655	0.991
FLANN	10000	1.641	0.234
	30000	4.758	0.633
	60000	9.691	0.870
MIH	891010	103.4	1.000

Table 3. LSH 1 M Data

Algorithm	Volume	Time(ms)	Precision
k = 1			
FNNS	6000	0.953	0.913
	20000	3.976	0.983
	50000	11.504	0.996
FLANN	30000	4.917	0.909
	50000	8.299	0.955
	100000	17.083	0.990
MIH	516118	84.1	1.000
k = 10			
FNNS	6000	0.989	0.894
	20000	4.159	0.978
	50000	12.101	0.995
FLANN	30000	5.000	0.724
	50000	8.331	0.850
	100000	17.011	0.960
MIH	596586	97.5	1.000
k = 50			
FNNS	6000	1.307	0.859
	20000	5.197	0.968
	50000	14.547	0.992
FLANN	30000	4.875	0.521
	50000	8.209	0.715
	100000	16.906	0.917
MIH	645739	105.3	1.000

Table 4. MLH 1 M Data

Algorithm	Volume	Time(ms)	Precision
k = 1			
FNNS	3000	0.551	0.971
	5000	0.981	0.989
	10000	1.905	0.997
FLANN	10000	1.506	0.905
	30000	4.413	0.980
	50000	7.415	0.992
MIH	302475	52.0	1.000
k = 10			
FNNS	3000	0.563	0.952
	5000	1.006	0.981
	10000	1.947	0.995
FLANN	10000	1.507	0.679
	30000	4.476	0.902
	50000	7.479	0.960
MIH	399496	69.9	1.000
k = 50			
FNNS	3000	0.747	0.920
	5000	1.295	0.964
	10000	2.448	0.988
FLANN	10000	1.534	0.415
	30000	4.382	0.774
	50000	7.375	0.904
MIH	472219	79.7	1.000

Table 5. LSH 10 M Data

Algorithm	Volume	Time(ms)	Precision
$k = 1$			
FNNS	20000	4.931	0.913
	50000	13.835	0.962
	100000	**31.880**	**0.980**
FLANN	80000	16.525	0.847
	100000	19.900	0.873
	200000	**57.020**	**0.937**
MIH	4302972	719.1	1.000
$k = 10$			
FNNS	20000	5.028	0.901
	50000	14.115	0.959
	100000	**32.456**	**0.981**
FLANN	80000	16.585	0.588
	100000	19.867	0.639
	200000	**59.723**	**0.799**
MIH	5320692	890.6	1.000
$k = 50$			
FNNS	20000	6.361	0.873
	50000	16.929	0.945
	100000	**37.501**	**0.974**
FLANN	80000	16.544	0.377
	100000	19.936	0.432
	200000	**65.875**	**0.648**
MIH	5774213	965.6	1.000

Table 6. MLH 10 M Data

Algorithm	Volume	Time(ms)	Precision
$k = 1$			
FNNS	10000	2.640	0.978
	20000	5.407	0.992
	50000	**14.830**	**0.998**
FLANN	80000	15.809	0.939
	100000	19.447	0.954
	200000	**63.566**	**0.983**
MIH	2289901	404.3	1.000
$k = 10$			
FNNS	10000	2.603	0.968
	20000	5.337	0.988
	50000	**14.569**	**0.998**
FLANN	80000	15.647	0.802
	100000	19.583	0.841
	200000	**67.699**	**0.934**
MIH	3173628	558.5	1.000
$k = 50$			
FNNS	10000	3.262	0.947
	20000	6.308	0.979
	50000	**16.175**	**0.995**
FLANN	80000	15.908	0.632
	100000	19.939	0.696
	200000	**67.609**	**0.868**
MIH	3735299	665.3	1.000

Table 7. LSH 79 M Data

Algorithm	Volume	Time(ms)	Precision
$k = 1$			
FNNS	20000	9.682	0.774
	50000	25.215	0.862
	100000	**53.279**	**0.909**
FLANN	80000	36.267	0.734
	100000	44.545	0.756
	200000	**91.143**	**0.824**
MIH	23172087	7042.2	1.000
$k = 10$			
FNNS	20000	9.011	0.784
	50000	23.288	0.881
	100000	**49.710**	**0.930**
FLANN	80000	40.065	0.383
	100000	49.659	0.413
	200000	**92.129**	**0.527**
MIH	37399803	11513.3	1.000
$k = 50$			
FNNS	20000	11.113	0.750
	50000	28.187	0.859
	100000	**58.036**	**0.916**
FLANN	80000	40.467	0.194
	100000	47.267	0.219
	200000	**91.466**	**0.318**
MIH	41137255	12583.8	1.000

Table 8. MLH 79 M Data

Algorithm	Volume	Time(ms)	Precision
$k = 1$			
FNNS	20000	8.150	0.969
	50000	21.911	0.989
	100000	**47.077**	**0.995**
FLANN	80000	36.120	0.869
	100000	48.690	0.884
	200000	**75.446**	**0.929**
MIH	11380955	3396.6	1.000
$k = 10$			
FNNS	20000	8.227	0.961
	50000	22.026	0.987
	100000	**47.036**	**0.995**
FLANN	80000	40.873	0.600
	100000	49.348	0.637
	200000	**76.483**	**0.755**
MIH	20314128	5994.4	1.000
$k = 50$			
FNNS	20000	10.522	0.940
	50000	26.882	0.980
	100000	**55.765**	**0.992**
FLANN	80000	39.798	0.354
	100000	46.901	0.398
	200000	**73.510**	**0.554**
MIH	24028084	6697.4	1.000

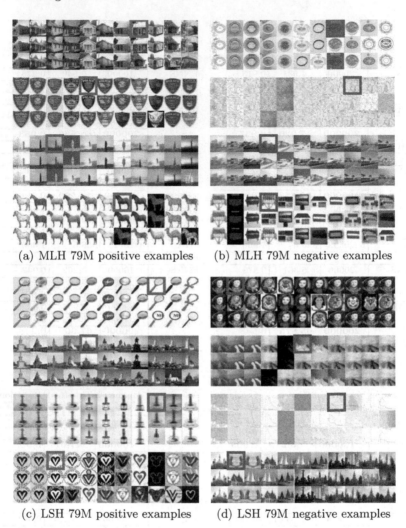

(a) MLH 79M positive examples   (b) MLH 79M negative examples

(c) LSH 79M positive examples   (d) LSH 79M negative examples

Fig. 1. The three lines of each group represent the ground truth, the result of our approach and the result of FLANN, respectively. The leftmost image in each case is the query. For positive examples (the left-hand side), the red box indicates the sample that our approach hits, but FLANN misses. For negative examples (the right-hand side), the blue box indicates the sample that our approach misses, but FLANN hits (Color figure online).

performance to our approach when k is small while our approach, with a larger k, only takes about half search time cost to achieve a 0.99 precision.

Table 2 shows the performance on 512-bit binary codes. For a large binary code length, our approach significantly outperforms the other two methods. Compared to FLANN, our approach achieves a 0.99 precision in merely 1/3 search time cost. MIH accesses most of the data points in the search database to get the exact nearest neighbors, which makes it too time-consuming.

From Tables 3, 4, 5, 6, 7 and 8, experiments on the 80 million tiny images dataset show that our approach is scalable to large-scale search databases. Among 79M candidates, FLANN only has a 0.318 precision for LSH and a 0.554 precision for MLH when $k = 50$ while our approach can reaches a 0.916 precision for LSH and a 0.992 precision for MLH with less search time cost. However, MIH takes extremely high search time cost to retrieve the exact nearest neighbors.

Besides, Fig. 1(a) and (c) show positive examples that our approach outperforms FLANN while Fig. 1(b) and (d) show negative examples that FLANN outperforms our approach.

4.3   Analysis

Graph Construction Costs. The construction of the neighborhood graph and the bridge graph is the major part of extra computation compared to FLANN, but it is an offline task which is performed only once, thus acceptable. Exact neighborhood search for each data point in the search database may be impractical especially for large-scale search databases, therefore in experiments, we adopt FLANN as an approximate algorithm to build the neighborhood graph among the search database.

Graph Storage Costs. Both the neighborhood graph and the bridge graph are organized by attaching an adjacent list to each data point or bridge vector. By analyzing the total number of bridge vectors and adjacent lists, we can easily derive that the additional storage is $O(nd + Nk + n^m b)$, with N the number of data points, k the length of data point adjacent lists, n^m the number of bridge vectors and b the length of bridge vector adjacent lists.

Advantages over FLANN and MIH. Compared to FLANN and MIH, our approach enjoys two-fold advantages. On the one hand, both the neighborhood graph and the bridge graph structure provide an efficient way to retrieve high-quality NN candidates thanks to the close relationship built on the graph. On the other hand, candidates retrieved in such a manner are usually better than those retrieved by FLANN because the neighborhood graph provides a better order of data access. FLANN does not produce as good performance as our approach especially in large-scale search databases and the precision is very low even when a large number of data points are accessed. For MIH, the index construction and the query process are very time-consuming and therefore impractical for large-scale search databases.

5   Conclusions

In this paper, we generalize the algorithm in [15] to the Hamming space with an alternative version of k-means clustering based on the Hamming distance. The simple approach also inspires later research related to the Euclidean space and the Hamming space. Experiments show that our algorithm outperforms state-of-the-art approaches and has significant improvement especially in the scenarios of longer binary codes or larger databases.

Acknowledgments. Weiwei Xu is partially supported by NSFC 61322204.

References

1. Torralba, A., Fergus, R., Freeman, W.T.: 80 million tiny images: a large data set for nonparametric object and scene recognition. IEEE Trans. Pattern Anal. Mach. Intell. **30**(11), 1958–1970 (2008)
2. Norouzi, M., Punjani, A., Fleet, D.J.: Fast search in hamming space with multi-index hashing. In: IEEE Conference on Computer Vision and Pattern Recognition, pp. 3108–3115. IEEE (2012)
3. Muja, M., Lowe, D.G.: Fast Matching of Binary Features. In: 9th Conference on Computer and Robot Vision, pp. 404–410. IEEE (2012)
4. Calonder, M., Lepetit, V., Strecha, C., Fua, P.: BRIEF: binary robust independent elementary features. In: Daniilidis, K., Maragos, P., Paragios, N. (eds.) ECCV 2010, Part IV. LNCS, vol. 6314, pp. 778–792. Springer, Heidelberg (2010)
5. Leutenegger, S., Chli, M., Siegwart, R.Y.: BRISK: binary robust invariant scalable keypoints. In: IEEE International Conference on Computer Vision, pp. 2548–2555. IEEE (2011)
6. Rublee, E., Rabaud, V., Konolige, K., Bradski, G.: ORB: an efficient alternative to SIFT or SURF. In: IEEE International Conference on Computer Vision, pp. 2564–2571. IEEE (2011)
7. Norouzi, M., Blei, D.M.: Minimal loss hashing for compact binary codes. In: Proceedings of the 28th International Conference on Machine Learning, pp. 353–360 (2011)
8. Charikar, M.S.: Similarity estimation techniques from rounding algorithms. In: Proceedings of the Thirty-Fourth Annual ACM Symposium on Theory of Computing, pp. 380–388. ACM (2002)
9. Babenko, A., Lempitsky, V.: The inverted multi-index. In: IEEE Conference on Computer Vision and Pattern Recognition, pp. 3069–3076. IEEE (2012)
10. Jegou, H., Douze, M., Schmid, C.: Product quantization for nearest neighbor search. IEEE Trans. Pattern Anal. Mach. Intell. **33**(1), 117–128 (2011)
11. Oliva, A., Torralba, A.: Modeling the shape of the scene: a holistic representation of the spatial envelope. Int. J. Comput. Vis. **42**, 145–175 (2001)
12. Wang, J., Wang, J., Zeng, G., Tu, Z., Gan, R., Li, S.: Scalable k-NN graph construction for visual descriptors. In: IEEE Conference on Computer Vision and Pattern Recognition, pp. 1106–1113. IEEE (2012)
13. Bentley, J.L.: Multidimensional binary search trees used for associative searching. Commun. ACM **18**, 509–517 (1975)
14. Nister, D., Stewenius, H.: Scalable recognition with a vocabulary tree. In: IEEE Conference on Computer Vision and Pattern Recognition, pp. 2161–2168. IEEE (2006)
15. Wang, J., Wang, J., Zeng, G., Gan, R., Li, S., Guo, B.: Fast neighborhood graph search using cartesian concatenation. In: IEEE International Conference on Computer Vision, pp. 2128–2135. IEEE (2013)

SOMH: A Self-Organizing Map Based Topology Preserving Hashing Method

Xiao-Long Liang, Xin-Shun Xu$^{(\boxtimes)}$, Lizhen Cui, Shanqing Guo, and Xiao-Lin Wang

School of Computer Science and Technology, Shandong University, Jinan, China
rembern@126.com, {xuxinshun,clz,guoshanqing,xlwang}@sdu.edu.cn

Abstract. Hashing based approximate nearest neighbor search techniques have attracted considerable attention in media search community. An essential problem of hashing is to keep the neighborhood relationship while doing hashing map. In this paper, we propose a self-organizing map based hashing method–SOMH, which cannot only keep similarity relationship, but also preserve topology of data. Specifically, in SOMH, self-organizing map is introduced to map data points into hamming space. In this framework, in order to make it work well on short and long binary codes, we propose a relaxed version of SOMH and a product space SOMH, respectively. For the optimization problem of relaxed SOMH, we also present an iterative solution. To test the performance of SOMH, we conduct experiments on two benchmark datasets–SIFT1M and GIST1M. Experimental results show that SOMH can outperform or is comparable to several state-of-the-arts.

Keywords: Hashing · Self-organizing map · Media research · Approximate nearest neighbor search

1 Introduction

With the development of the Internet and electronic devices, data, e.g., texts, images and videos are growing rapidly. For many scenarios, people need to retrieve relevant content from such large scale data. However, for large scale data, it is computationally infeasible to find the closest points of a query in a given database. To tackle this problem, recently, a lot of efforts have been devoted to approximate nearest neighbor (ANN) search, which returns approximate nearest neighbors instead of exact nearest neighbor, but is fast [1–4].

Hashing based ANN search is a well known technique for approximate nearest neighbor search [5], which first maps data points into hamming space and makes the comparison. They have efficient constant query time and can also reduce storage by storing compact codes of data points. Thus, in these years, hashing based methods have attracted more attention; many methods have been proposed [6–9]. Usually, such methods can be divided into two classes: data independent hashing and data dependent hashing. In data independent hashing, a hash function family

© Springer International Publishing Switzerland 2016
Q. Tian et al. (Eds.): MMM 2016, Part I, LNCS 9516, pp. 337–348, 2016.
DOI: 10.1007/978-3-319-27671-7_28

is defined independently of training data set, and then the hash function family is directly used to generate hashing codes of data points. A typical data independent hashing approach is Locality-Sensitive Hashing (LSH) proposed by Gionis [10]. In LSH, the hashing functions are constructed by random projections. Shift Invariant Kernel Hashing [11] is another kind of data independent hashing, which is an extension of LSH. It is a distribution free method based on random features mapping for shift invariant kernels, and the expected hamming distance between the binary codes is related to the distance between data points in a kernel space.

A problem of data independent hashing is that the hashing code should be very long to get a good result. However, it would be desirable to learn a compact code for large scale data retrieval task. In order to solve this problem, data dependent hashing is proposed, in which, hash functions are learned from a given training data set. In these years, many data dependent approaches have been proposed. For instance, Spectral Hashing (SH) [12] is a representation of data dependent hashing technique. In SH, a similar matrix is constructed to reflect the similar relationship of data points; then, the matrix is used to guide the learning of hashing function [13]. PCAH and ITQ [14] both make use of Principal Component Analysis (PCA) to project the input data space into a low dimension space. Locally Linear Hashing [15] and Inductive Hashing [16] learn hashing function by constructing a manifold structure. K-means hashing (KMH) [17] adopts a k-means algorithm to cluster data points and assign hashing codes to every cluster while minimizing the difference between hashing distance and euler distance of data points.

Many hashing based approximate nearest neighbors have been proposed; but, most them just consider to keep the similarity relationship of data points during mapping. However, for some time, we hope that a model could also preserve topology of data. To this, in this paper, we propose a new hashing method based on self-organizing map–SOMH, in which, SOM is introduced for mapping. Moveover, to make it perform well on both short and long code, we also propose a relaxed version of SOMH, and a product space SOMH.

The rest of this paper is organized as follows. We briefly review related work, e.g., Vector Quantization (VQ) and SOM, in Sect. 2. Section 3 presents the details of our proposed methods. Section 4 provides experimental results on several benchmark datasets. The conclusions are given in Sect. 5.

2   Background

In this section, we introduce the background of the proposed method, including Vector Quantization (VQ) and Self-Organizing map (SOM).

2.1   Vector Quantization

VQ is classical quantization technique, which models the probability density of input data by the distribution of prototype vectors. It works by dividing the data space into a set of regions, and uses a single vector to represent all the data points in a region. The vectors used to represent data points are called codes. Once the "codebook" is chosen, every input data point x can be represented by its nearest code. One kind of optimal codebook M can be got by minimizing a loss function E.

$$E = \int ||x - m_c||p(x)\,dx. \tag{1}$$

where $p(x)$ is the probability density of x. The index c is the nearest index of codes in codebook M, that is

$$c = \operatorname*{argmin}_i(||x - m_i||) \tag{2}$$

In general, no closed-form solution is available for the optimal codebook M. But an iterative gradient-descent algorithm could be used.

2.2 Self-Organizing Map

SOM is a kind of vector quantization; but, in SOM, the codes are chosen spatially, globally orderly. This means that the topological structure of the input data could be preserved during the mapping [18].

As that in VQ, we could construct an energy function for an SOM:

$$E = \int ||x - m_i||N(i,c)p(x)\,dx. \tag{3}$$

where $N(i,c)$ is a neighborhood function used to control the topology of the SOM. Note that there are many methods to choose neighborhood; here, we use the neighborhood function as follows:

$$N(x,c) = exp(\frac{dist(i,c)}{2\theta}) \tag{4}$$

where the $dist(i,c)$ is the regular distance of the two nodes m_i and m_c, m_c is the best matching nodes for x.

For discrete data, the energy function could be expressed as follows:

$$E = \frac{1}{n}\sum x||x - m_i||N(i,c) \tag{5}$$

For this problem, we can use a gradient-descent approach to get a solution,

$$m(t+1) = m(t) + \alpha(t)N(i,c)[x(t) - m(t)] \tag{6}$$

where $\alpha(t)$ is a scalar-valued "adaption gain", $0 < \alpha(t) < 1$.

3   SOMH

This section describes the proposed SOMH. In the first subsection, we introduce the basic idea of SOMH, which is called Naive SOMH; then, we introduce a relaxed version of SOMH which works better when hashing code is short. In addition, we also give an iterative solution for the optimization problem in relaxed SOMH. Finally, we present a product space SOMH for the scenarios of long hashing codes.

3.1 Naive SOMH

The first problem to construct the map from real space into a hamming space using SOM is how to define a corresponding relation between SOM nodes and hashing codes. The desired map should capture the input data structure and also could convert input data into binary code. To do this, a direct approach is to assign a hashing code to every node in the SOM. Then, to compute the neighborhood function, we can use the hamming distance of hashing codes as the distance of two nodes.

Thus, the neighborhood function using hamming distance can be written as:

$$N_{ham}(x, c) = exp(\frac{ham(i, c)}{2\theta})$$ (7)

where i, c is node indexes.

Based on this, then we can define a loss function for projection from the input data space to the hamming space.

$$E = \frac{1}{n} \sum_{x} ||x - m_c|| N_{ham}(i, c)$$ (8)

The above approach, which is called naive SOMH, is summarized in Algorithm 1. From Algorithm 1, we can find that one advantage of SOMH is that there is no quantization step. However, that does not mean the quantization error disappears in naive SOMH. Actually, it is merged in the projection error.

Algorithm 1. Naive SOMH

Input: Training data X.
  Initialize hashing code H, Node weights M;
  while not convergent **do**
    Sample a data point x.
    $m(t + 1) = m(t) + \alpha(t) N_{ham}(c, i)[x(t) - m(t)]$
  end while

3.2 Relaxed SOMH

In SOMH, we actually try to find a corresponding relation between original data and binary codes in hamming space. When the hashing code is long enough, it is no problem to construct a complex topological structure in hamming space as that in original data space. However, if hashing code is too short, it will be much hard to map the original data points into hamming space directly while preserving topology. So, we introduce a relaxed version of SOMH. Specifically, we relax each binary code to a vector with real values, e.g., R_i. Let $d(i, j)$ represent distance between two relaxed vectors R_i and R_j, $d_h(i, j)$ the distance between hashing codes h_i and h_j. The desired regular should be topologically consistent with hamming codes, which means if $d(i, j) > d(j, k)$ then $d_h(i, j) > d_h(j, k)$.

Then, we could define the energy function as follows.

$$E_{relax} = -\sum_{i,j,k} sign((d(i,j) - d(j,k))sign(d_h(i,j) - d_h(j,k)))$$
(9)

By minimizing this relaxed error E_{relax}, we can minimize the topological difference between relaxed vector matrix R and hamming codes matrix. On the other hand, we want the regular could fit input data well, which means the following project error should be minimized.

$$E_{project} = \frac{1}{n} \sum_x ||x - m_c||N_{reg}(i,c)$$
(10)

where

$$N_{reg}(x,c) = exp(\frac{dist(i,c)}{2\theta})$$
(11)

as defined previously, $dist(i,c)$ is the distance between R_i and R_c, that is

$$dist(i,c) = ||R_i - R_c||^2$$
(12)

We can find that it is not hard to get a solution of $E_{project}$; however, it will get a minimum when all vectors R_i take the same value. In order to avoid this, we add a regularization term in $E_{project}$. Thus, it becomes

$$E_{project} = \frac{1}{n} \sum_x ||x - m_i||N_{reg}(i,c) + ||R||$$
(13)

Then, we can combine Eqs. (9) and (13), and get the following error function:

$$E = E_{project} + \alpha E_{relax}$$
(14)

where α is a tradeoff parameter.

3.3   An Iterative Solution

Directly minimizing the error function in Eq. 14 is intractable. However, we can divide the error function into two sub-problems; then, use an iterative method to optimize it. The detailed steps are listed below.

Step(i): Fix R, update M.
When the regular R is fixed, the second item of Eq. 14 does not influence the value of M; then, the problem degenerates into the following problem.

$$\min \frac{1}{n} \sum x||x - m_i||N(i,c) + constant$$
(15)

which can be easily solved according to methods in standard SOM algorithm.

Step(ii): Fix M, update R

If we fix the codebook M, the sub problem becomes

$$min\frac{1}{n}\sum_x ||x - m_i||N(i,c) + \alpha \sum_{i,j,k} W \tag{16}$$

where

$$W = sign((d(i,j) - d(j,k))sign(d_h(i,j) - d_h(j,k))) \tag{17}$$

It is still a hard problem because the sign function is not differentiable. To tackle this, we relax the $sign(\cdot)$ function to $(d(i,j) - d(j,k))(d_h(i,j) - d_h(j,k))$. Then, the problem becomes

$$min\frac{1}{n}\sum_x ||x - c||N(i,c) + \alpha \sum_{i,j,k}(d(i,j) - d(j,k))(d_h(i,j) - d_h(j,k)) \tag{18}$$

Apparently, this problem can be solved by the Quasi-Newton method. This relaxed SOMH is summarized in Algorithm 2. Note that the above iterative method needs to initialize the codebook M, regular R. In SOM, the initialization of node weights could greatly influence the iterative times before convergence. In practice, we use PCA to initialize the node weights because the PCA is a kind of dimensionality reduction preserving most information. Specifically, we choose the K largest eigen vectors as the initial codebook M.

Algorithm 2. Relaxed SOMH Algorithm

Input: Training data X.

    Initialize the relaxed matrix R, node weights M, hashing codes H;

    while not convergent **do**

        Update M by minimizing Eq. 15.

        Update R by minimizing Eq. 16.

    end while

3.4   Product Space SOMH

One problem of SOMH is that it will become unpracticable when the length of hashing code is very long. For example, when the code length is 32 bits, it will need 2^{32} nodes to represent every possible hash codes, which will make it very inefficient. To tackle this problem, we further propose the product space SOMH, which divides the input data space into a set of subspaces, and trains an SOMH in each subspace.

The method of dividing feature space has been used in cartesian K-means and PQ [19, 20]. Similar with those methods, our proposed product space SOMH decomposes the D-dimensional vector space into M subspaces. For instance, a vector $x \in R^D$ is represented as a concatenation of M subvectors: $x = [x^1, x^2, x^3, ..., x^M]$, where the superscription m in x^m denotes the mth subspace. Based on this, the destination hashing code is also decomposed into M subspaces, that is

$$H(x) = [h^1(x^1), ..., h^m(x^m), ..., h^M(x^M)]$$

Note that every subspace is computed independently. Then, for every subspace, hamming distance can be used to approximate the actual distance between two vectors, i.e.,

$$d(x^i, y^i) \simeq d(h(x)^i, h(y)^j) \qquad (19)$$

Then, we can represent the distance between two vectors as the sum distance of every subspace.

$$d(x, y) = \sqrt{\sum_{i=1}^{n} (d(x^i, y^i))^2} \simeq \sqrt{\sum_{i=1}^{n} (d(h(x)^i, h(y)^i))^2} \qquad (20)$$

From this equation, we can also find that the computing of every subspace is independent. Once we have SOMH's for every subspace; then, we can use the results of every subspace to construct the final result. The product space SOMH is summarized in Algorithm 3.

Algorithm 3. Product Space SOMH

Input: Training data X.
output: A set of SOMH.
    Divide the data X into m subspace $x = [x^1, x^2, x^3, ..., x^m]$
    for $i = 1 \rightarrow m$ **do**
        Train an SOMH SOM^i using x^i by Algorithm 1 or Algorithm 2.
    end for

---

4   Experiments

4.1   Dataset

In order to evaluate our method, we choose two widely used databases, i.e., SIFT1M and GIST1M. In every database, there are three vector subsets: learning, database, and query. SIFT1M contains 1 million 128-D SIFT features in learning set and 10000 independent queries in query set. The learning set is extracted form Flickr images and the query set is from the INRIA holidays images. GIST1M contains 1 million 384-D GIST features in learning set and 500 queries in query set. The learning set is extracted from a tiny image set.

The search quality is measured by the recall in the first N retrieved data points. We follow the search strategy of hamming ranking commonly used in many hashing methods. The data points are sorted according their hamming distances to the query, and the first N samples will be retrieved. Hamming ranking exhausts the whole database to get the result because it is very fast to search in hamming space. For example, it takes about 1.5 ms to scan 1 million 64-bit codes in a single core c++ implementation (Intel Core i7 2.93 GHz CPU and 8 GB RAM).

4.2   Baselines

We compare our SOMH method with five state-of-the-art unsupervised hashing methods: Spectral Hashing (SH), Principal Component Analysis Hashing (PCAH), Iterative Quantization (ITQ) and K-means Hashing (KMH).

4.3   Performance Evaluation on Short Binary Code

In this experiment, we test the proposed SOMH and relaxed SOMH when the length of binary code is short, e.g., 8 bits and 16 bits. The tradeoff parameter α is set to 0.3 and the neighborhood parameter θ is set to 1.5. The top 10 Euclidean nearest neighbors are considers as the ground truth. In addition, in KMH and SOMH, the dimension of each subspace is set to 2.

The results on SIFT1M and GIST1M are shown in Figs. 1 and 2, respectively. From these figures, we can observe that the relaxed SOMH works better than SOMH and other state-of-the-arts. This confirms that the relaxed SOMH can work well when the length of hashing code is short.

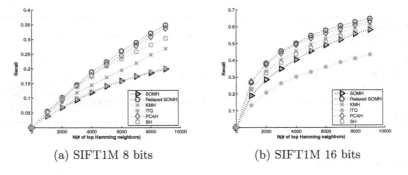

(a) SIFT1M 8 bits   (b) SIFT1M 16 bits

Fig. 1. Performance of six hashing methods on SIFT1M with 8 and 16-bit codes.

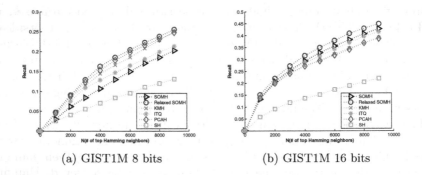

(a) GIST1M 8 bits   (b) GIST1M 16 bits

Fig. 2. Performance of six hashing methods on GIST1M with 8 and 16-bit codes.

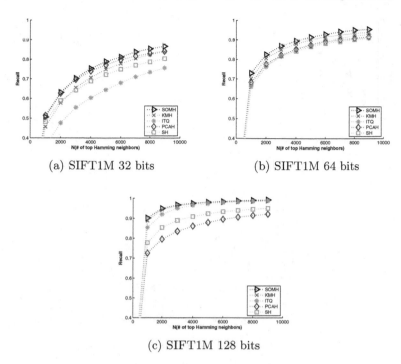

(a) SIFT1M 32 bits

(b) SIFT1M 64 bits

(c) SIFT1M 128 bits

Fig. 3. Performance of six hashing methods on SIFT1M with 32, 64 and 128-bit codes. In this experiment, the top 10 Euclidean nearest neighbors are considered as the ground truth. In KMH, the dimension of subspace is set to 2 for 32-bit case, and 4 for 64 and 128-bit cases. In our method, the dimension of subspace is set to 4 for all cases.

4.4 Performance Evaluation on Long Hashing Code

Figures 3 and 4 show the results of our method and other state-of-the-arts on datasets with 32, 64 and 128 bits hashing code length. Note that, in this experiment, SOMH is the product space SOMH. From these figures, we have the following observations.

- SOMH achieves better results than other state-of-the-arts on SIFT1M, and comparable results on GIST1M.
- PCAH performs very well on SIFT1M with 32 bits; however, degrades quickly when the code length becomes longer, e.g., 64 bits and 128 bits.
- ITQ performs better than SH and PCA on SIFT1M with 64 bits and 128 bits and on GIST1M with all cases. This further confirms that reducing the quantization error is effective to improve the performance.

4.5 Training Time Evaluation

In this experiment, we evaluate the efficiency of our method and other state-of-the-arts in training stage. All algorithms run on the same machine (CPU: Intel

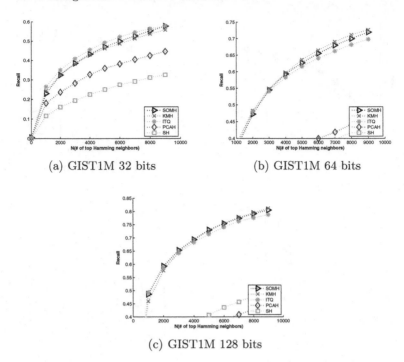

(a) GIST1M 32 bits   (b) GIST1M 64 bits

(c) GIST1M 128 bits

Fig. 4. Performance of six hashing methods on GIST1M with 32, 64 and 128-bit codes. In this experiment, the top 10 Euclidean nearest neighbors are considered as the ground truth. In KMH, the dimension of subspace is set to 8 for all cases; In our method, it is set to 8 for all cases.

Xeon E5650 2.67 GHz, RAM 16 GB). The results on SIFT1M and GIST1M are listed in Table 1. From this table, we can find that

- PCAH uses the shortest time on both SIFT1M and GIST1M.
- For SOMH, there is no significant increase of training time on SIFT1M with code length increasing.
- For SOMH, there is some increase on GIST1M with code length increasing. However, its training time is still much shorter than KMH.
- For KMH, there is significant increase when the code length increases. This means that, although KMH can obtain good results on some cases, it uses much time in training stage.

4.6   Parameter Evaluation

It is well known that the neighborhood function is very important in SOM as it controls the topology of SOM. Especially, parameter θ in the neighborhood function, e.g. Eq. 11, directly controls its width. In this section, we conduct experiments to show the influence of θ on the performance of SOMH. The results are

Table 1. Training time of different methods on SIFT1M and GIST1M with different code lengths (in second).

(a) Training time on SIFT1M

Method	32 bits	64 bits	128 bits
SOMH	16.313	16.638	19.021
SH	7.057	10.550	21.064
PCAH	2.522	2.705	2.561
ITQ	70.490	155.364	381.193
KMH	212.216	725.955	9733.670

(b) Training time on GIFT1M

Method	32 bits	64 bits	128 bits
SOMH	64.353	91.954	1184.172
SH	56.152	73.862	95.036
PCAH	45.344	57.585	58.891
ITQ	170.325	378.242	796.018
KMH	9087.233	20472.437	40482.338

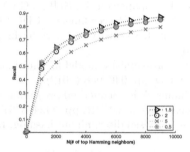

Fig. 5. Performance comparison of SOMH with different values of θ.

shown in Fig. 5. From this figure, we can find that SOMH indeed have different performance with different values of θ, and obtains the best results when θ is set to 1.5.

5   Conclusion

In this paper, we propose a new hashing method–SOMH, which is based on self-organizing map. It cannot only keep the similarity relationship, but also preserving the topology of data. To make it practicable on short code length, we propose a relaxed version of SOMH. In addition, to make it efficient on long code length, we further propose a product space SOMH. All proposed methods are tested on two benchmark data sets–SIFT1M and GIST1M. Generally, SOMH outperforms or is comparable to other state-of-the-arts. Especially, on some cases, SOMH obtains similar results compared with some methods, e.g., KMH; however, it is much faster that those methods.

In this paper, we just show how to make the proposed method work on single-modal data; however, it is an interesting work on how to generalize it to work on multimodal data.

Acknowledgments. This work is partially supported by National Natural Science Foundation of China (61173068, 61573212, 61572295), Program for New Century Excellent Talents in University of the Ministry of Education, the Key Science Technology

Project of Shandong Province (2014GGD01063), the Independent Innovation Foundation of Shandong Province (2014CGZH1106) and the Shandong Provincial Natural Science Foundation (ZR2014FM020, ZR2014FM031).

References

1. Wang, J., Shen, H.T., Song, J., Ji, J.: Hashing for similarity search: a survey. arXiv (2014)
2. Andoni, A., Indyk, P.: Near-optimal hashing algorithms for approximate nearest neighbor in high dimensions. In: Proceedings of FOCS, pp. 459–468 (2006)
3. Athitsos, V., Potamias, M., Papapetrou, P., Kollios, G.: Nearest neighbor retrieval using distance-based hashing. In: Proceedings of ICDE, pp. 327–336 (2008)
4. Jin, Z., Li, C., Lin, Y., Cai, D.: Density sensitive hashing. IEEE Trans. Cybern. **44**(8), 1362–1371 (2013)
5. Zhen, Y., Yeung, D.Y.: A probabilistic model for multimodal hash function learning. In: Proceedings of KDD, pp. 940–948 (2012)
6. Wang, J., Kumar, S., Chang, S.-F.: Sequential projection learning for hashing with compact codes. In: Proceedings of CVPR, pp. 3424–3431 (2010)
7. Kim, S., Choi, S.: Semi-supervised discriminant hashing. In: Proceedings of ICDM, pp. 1122–1127 (2011)
8. Wang, S.S., Huang, Z., Xu, X.-S.: A multi-label least-squares hashing for scalable image search. In: Proceedings of SDM 2015, pp. 954–962 (2015)
9. Xu, B., Bu, J., Lin, Y., Chen, C., He, X., Cai, D.: Harmonious hashing. In: Proceedings of IJCAI, pp. 1820–1826 (2013)
10. Gionis, A., Indyk, P., Motwani, R.: Similarity search in high dimensions via hashing. In: Proceedings of VLDB, pp. 518–529 (1999)
11. Raginsky, M., Lazebnik, S.: Locality-sensitive binary codes from shift-invariant kernels. In: NIPS, vol. 22, pp. 1509–1517 (2009)
12. Weiss, Y., Torralba, A., Fergus, R.: Spectral hashing. In: NIPS, vol. 21, pp. 1753–1760 (2008)
13. Belkin, M., Niyogi, P.: Laplacian eigenmaps for dimensionality reduction and data representation. Neural Comput. **15**, 1373–1396 (2003)
14. Gong, Y., Lazebnik, S.: Iterative quantization: a procrustean approach to learning binary codes. In: Proceedings of CVPR, pp. 817–824 (2011)
15. Irie, G., Li, Z., Wu, X.M., Chang, S.F.: Locally linear hashing for extracting non-linear manifolds. In: Proceedings of CVPR, pp. 2123–2130 (2014)
16. Shen, F., Shen, C., Shi, Q., van den Hengel, A.: Inductive hashing on manifolds. In: Proceedings of CVPR, pp. 1562–1569 (2013)
17. He, K., Wen, H., Sun, J.: K-means hashing: an affinity-preserving quantization method for learning binary compact codes. In: Proceedings of CVPR, pp. 2938–2945 (2013)
18. Johnson, T.: Networks. In: Klouche, T., Noll, T. (eds.) MCM 2007. CCIS, vol. 37, pp. 311–317. Springer, Heidelberg (2009)
19. Norouzi, M., Fleet, D.J.: Cartesian K-means. In: Proceedings of CVPR, pp. 3017–3024 (2013)
20. Ge, T., He, K., Ke, Q., Sun, J.: Optimized product quantization for approximate nearest neighbor search. In: Proceedings of CVPR, pp. 2946–2953 (2013)

Describing Images with Ontology-Aware Dictionary Learning

Chengyue Zhang[1,2] and Yahong Han[1,2(✉)]

[1] School of Computer Science and Technology, Tianjin University, Tianjin, China
[2] Tianjin Key Laboratory of Cognitive Computing and Application, Tianjin, China
yahong@tji.edu.cn

Abstract. In this paper, we focus on the generation of contextual descriptions for images by learning an ontology-aware dictionary. Ontology deals with questions concerning what entities exist and how such entities can be related with a hierarchy. Thus, if we incorporate the semantic hierarchies of visual concepts into a learned visual dictionary, which consists of visual atoms, we can generate contextual descriptions of testing images through the reconstruction. This paper proposes to learn the ontology-aware dictionary by integrating hierarchical dictionary learning and multi-task regression into a joint framework. By utilizing a hierarchical regularization term defined on the multiple semantic categories, the hierarchical structures are introduced into the multi-task regression. The joint optimization of the sparse coding and multi-task regression makes the semantic hierarchies embedded into the learned dictionary. Experiments on two benchmark datasets show the better performance of the proposed algorithm. Examples of the ontology-aware dictionary and generated image descriptions successfully demonstrate the effectiveness of the proposed framework.

Keywords: Ontology · Dictionary learning · Image describing · Semantic hierarchy

1 Introduction

Ontology deals with questions concerning what entities exist and how such entities can be related with a hierarchy. If we incorporate the semantic hierarchies of visual concepts into a learned visual dictionary, which consists of visual atoms, we can generate contextual descriptions of testing images through the reconstruction. Much existing work has shown that using concept hierarchy to guide the model learning for classification can bring in improvements in both efficiency and accuracy [1,5,13]. The categories are usually organized in the form of tree-structured hierarchy [11] and treated as the leaf nodes in the bottom of the tree.

Y. Han was partly supported by the NSFC (under Grant 61202166 and 61472276) and the Major Project of National Social Science Fund of China (under Grant 14ZDB153).

Q. Tian et al. (Eds.): MMM 2016, Part I, LNCS 9516, pp. 349–358, 2016.
DOI: 10.1007/978-3-319-27671-7_29

Each internode corresponds to one hyper-category that is composed of a group of categories with semantic relevance or visual similarity, so that the structure reflects the hierarchical correlation among categories. Although much work focus on how to exploit and introduce semantic hierarchy to tasks like image classification, region tagging, less work has been done to incorporate semantic hierarchy into dictionary learning. Learning a compact, discriminative dictionary is crucial for generating effective image representations. If semantic hierarchy can be introduced into dictionary training, image representations would be more discriminative and semantic meaningful.

It has been shown in [9,14] that dictionaries via supervised learning are beneficial for generating image representations which are discriminative and rich in semantics. Therefore, recent studies on dictionary learning focused on training dictionary in supervised or semi-supervised way [11,12]. In [12], authors proposed a semi-supervised dictionary learning method by developing a new structured sparse regularization to incorporate the supervision information into the dictionary learning process. In [17], authors proposed a supervised dictionary learning method that jointly learns multi-layer hierarchical dictionaries and corresponding linear classifiers for region tagging. This method first generates a node-specific dictionary for each tag node in the taxonomy, and then concatenates the node-specific dictionaries from each level to construct a level-specific dictionary. However, as this method mainly focus on two-layer level dictionaries, some hierarchical correlations may be lost among categories.

In this paper, to introduce the hierarchical relations among the concepts into dictionary learning process, we propose an ontology-aware dictionary learning method. We combine hierarchical dictionary learning with multi-task regression in the proposed method. Semantic hierarchy induced from WordNet is introduced by a novel devised structured regularization and hierarchical image semantic categories serve as the supervised information. Figure 1 shows the framework of our method. We first extract low-level features for images and hierarchical structure among semantic categories from WordNet. Then we optimize the ontology-aware dictionary learning by a well-devised random gradient decent algorithm. Our method can learn the hierarchical dictionary D and classifier W simultaneously, which are used to generate sparse codes for testing images and describing test images, respectively. Since ontology-based semantic information has been integrated in dictionary, it would provide a flexible semantic cue for general image semantics understanding. In our method, each column of the dictionary D is associated with a specific class, and the atoms in each column can be considered as the basic atoms of specific class.

In the experiments, we evaluate our method on two benchmark datasets, i.e., Caltech256 and ILSVRC2013. The improved classification accuracy demonstrates better performance of dictionary learning. Examples of the learned dictionaries and generated image descriptions successfully demonstrate the effectiveness of the proposed framework.

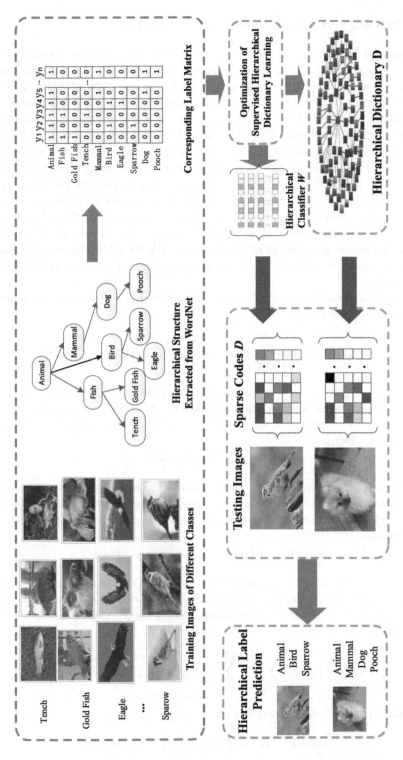

Fig. 1. The flowchart of the proposed framework.

2   Ontology-Aware Dictionary Learning

Assume that we have a training set of N labeled images from J semantic categories. Matrix $X = \{x_1, x_2....x_N\}$ represents the image data matrix, where $x_i \in \mathbb{R}^p$ denotes the ith training image, and p is the dimensionality of image features. Denote a semantic hierarchical graph $H = (C, \varepsilon_C)$, consisting of a set of semantic categories $C = \{c_1, c_2, \ldots, c_J\}$ and a set of edges ε_C, where each edge is an ordered pair in $C \times C$. Denote the ith edge as a vector $e^{(i)} \in \mathbb{R}^J$ defined as follows: $e_{c_p}^{(i)}$ and $e_{c_q}^{(i)}$ are set to 1 and -1 respectively if the two nodes p and q are connected in the hierarchical graph. The graph is encoded in the matrix $R = [e^{(1)}, e^{(2)}, ..., e^{(\|\varepsilon\|)}]$. $Y = \{y_1, y_2....y_N\}$ represents the label indicator matrix of image data, where $y_i \in \{0, 1\}^J$ denotes the label indicator vector of ith training image, $Y_{(j,i)} = 1$ if the ith image belongs to the jth categories and $Y_{(j,i)} = 0$, otherwise. Let $D \in \mathbb{R}^{P \times K}$ denote the dictionary, $W = \mathbb{R}^{K \times J}$ denote the model parameters of J categories. We let $||M||_F$ denote the matrix Frobenius norm, x^T and M^T represent the transpose of vector x and Matrix M, respectively.

In unsupervised dictionary learning, the dictionary D and sparse representation α are obtained by:

$$(\alpha, D) = arg \min_{\alpha, D} \sum_{i=1}^{N} \frac{1}{2}||x_i - D\alpha_i||_2^2 + \lambda_1||\alpha_i||_1 + \frac{\lambda_2}{2}||\alpha_i||_2^2 \qquad (1)$$

Where $\lambda_1 \geq 0$ and $\lambda_2 \geq 0$ are regularization parameters. We use α to denote the sparse representations for training samples. After obtained sparse representations α, we can predict semantic to testing images by solving the following regression problem:

$$\min_{W} ||Y - W^T\alpha||_F^2 + \frac{\nu}{2}||W||_F^2 \qquad (2)$$

Where $\alpha = \{\alpha_1, \alpha_2, ...\alpha_N\}$, $W_{(:,i)}$ is the parameter vector for ith concept and ν is a regularization parameter. Here we consider the case where concepts are organized in a hierarchical tree. Therefore, we use a hierarchical multi-task regression as following:

$$\min_{W} ||Y - W^T\alpha||_F^2 + \frac{\nu}{2}||W||_F^2 + \rho_1 \Omega_H(W) \qquad (3)$$

Where ρ_1 is a regularization parameter, and $\Omega_H(W)$ is a structure sparsity-inducing norm penalty with a structure H defined over the output direction of W. H encodes the hierarchical structure among the concepts, which is defined as following:

$$\Omega_H(W) = \sum_{e=(c_p, c_q) \in \varepsilon_C} \sum_{i=1}^{P} (W_{(i,p)} - W_{(i,q)})^2 = ||WR||_F^2 \qquad (4)$$

Therefore, Eq. (3) can be rewritten as:

$$\min_{W} ||Y - W^T\alpha||_F^2 + \frac{\nu}{2}||W||_F^2 + \rho_1||WR||_F^2 \qquad (5)$$

where ρ_1 is a regularization parameter and $R \in \{1, -1\}^{J \times ||\epsilon||}$ is a matrix that encodes the hierarchical relations among the concepts.

So far, the dictionary D is learned in an unsupervised way, which means there are no supervised or even semantic hierarchy information encoded in D. As designing a supervised hierarchical dictionary method could allow the semantic-related concepts share some common basic visual atoms, we propose to learn a supervised hierarchical dictionary and obtain the model parameters W simultaneously by optimizing the following equation:

$$\sum_{i=1}^{N} \min_{D,W} [\frac{1}{2}||y_i - W^T \alpha_i^*(x_i, D)||_F^2] + \frac{\nu}{2}||W||_F^2 + \rho_1||WR||_F^2 \qquad (6)$$

$$\alpha_i^*(x_i, D) = arg \min_{\alpha \in \mathbb{R}^K} \frac{1}{2}||x_i - D\alpha_i||_2^2 + \lambda_1||\alpha_i||_1 + \frac{\lambda_2}{2}||\alpha_i||_2^2 \qquad (7)$$

Each column of the dictionary D is associated with a specific semantic concept and each column can be considered as the basic atoms of specific semantic. The hierarchical structure has been incorporated by R, i.e., through the term $||WR||_F^2$, which also induces hierarchical structure to the sparse representation α^*.

3   Solution and Algorithm

Motivated by [8], we propose a stochastic gradient descent algorithm to optimize the objective function. We rewrite Eq. (6) as:

$$\sum_{i=1}^{N} \min_{D,W} f^i(D, W) + \frac{\nu}{2}||W||_F^2 + \rho_1||WR||_F^2, \qquad (8)$$

where $f^i(D, W) = [\frac{1}{2}||y_i - W^T \alpha_i^*(x_i, D)||_F^2]$. We use notation $f^i(D, W)$ to emphasize that the loss function associated with the ith training sample. We compute the gradient of $f^i(D, W)$ with respect to D and W as following:

$$\frac{\partial f^i}{\partial W} = \alpha_i^* \alpha_i^{*T} W - \alpha_i y_i^T \text{ and } \frac{\partial f^i}{\partial D} = \frac{\partial f^i}{\partial \alpha_i^*} \frac{\partial \alpha_i^*}{\partial D}$$

where α_i^* is short for $\alpha_i^*(X, D)$. However, $\alpha_i^*(X, D)$ is not differentiable. Note that $\alpha_i^*(X, D)$ is uniformly Lipschitz continuous and differentiable almost everywhere. The only points where $\alpha_i^*(X, D)$ is not differentiable are points where the set of nonzero coefficients of α_i change. In order to compute the gradient $\alpha^* i(X, D)$ with respect to D, we use implicit differentiation on the fixed point equation, which is the same as [8]. We Denote $\Lambda \triangleq \{j \in \{1, ...K\}, s.t. \alpha_j^* \neq 0\}$ the active set. Therefore, we can obtain gradient of $f^i(D, W)$ with respect to D:

$$\frac{\partial f}{\partial D} = -D\beta^* \alpha^{*T} + (x - D\alpha^*)\beta^{*T}$$

where β^* is a vector in $\mathbb{R}^K$ that depends on Y, X, W, D with: $\beta_{\Lambda^c}^* = 0$ and $\beta_\Lambda^* = (D_\Lambda^T D_\Lambda + \lambda_2 I)^{-1}(W(W^T \alpha^* - Y))_\Lambda$. We summarize our optimization algorithm in Algorithm 1.

Algorithm 1. Stochastic gradient descent algorithm for supervised hierarchical dictionary learning

---

Require: $X, Y, \lambda_1, \lambda_2, \rho_1, \nu \in \mathbb{R}$ regularization parameters, $R, D \in \mathcal{D}$ (initial dictionary), $W \in \mathcal{W}$ (initial parameters), T(number of interations), t_0, ρ(learning rate parameters).

1. **for** $i \leftarrow 1$ to T **do**
2.   Sample randomly (x_i, y_i) from X, Y
3.   Compute α^K by sloving Eq. (7) using LARS [3]
4.   Compute the active set: $\Lambda \leftarrow j \in \{1, \ldots, p\} : \alpha^*[j] \neq 0$
5.   Compute β^*: Set $\beta^*_{\Lambda_C} = 0$ and

$$\beta^*_\Lambda = (D^T_\Lambda D_\Lambda + \lambda_2 I)^{-1}(W(W^T \alpha^* - y_t))_\Lambda$$

6.   Choose the learning rate $\rho_t \leftarrow \min\{\rho, \rho\frac{t_0}{t}\}$
7.   Update the parameters by a projected gradient step
    $W \leftarrow \Pi_\mathcal{W}[W - \rho_t((W^T(\alpha^* \alpha^{*T})))^T - (y_t \alpha^{*T})^T + \nu W + \rho_1 W R R^T)],$
    $D \leftarrow \Pi_\mathcal{D}[D - \rho_t(-D\beta^* \alpha^{*T} + (x_t - D\alpha^*)\beta^{*T})]$
    where $\Pi_\mathcal{W}$ and $\Pi_\mathcal{D}$ are respectively orthogonal projections on the sets $\mathcal{W}$ and $\mathcal{D}$
8. **end for**
9. **return** D(learned dictionary), W(learned classifiers)

---

4 Experiments

4.1 Datasets and Parameters

We evaluate the proposed method on two datasets: the Caltech256 [4] and the ILSVRC2013[1]. For each dataset, we use the sift features. In Caltech256, a concept hierarchy is pre-defined using the 256 concepts as leaf nodes. ILSVRC2013 contains 1000 object categories that contains both internal nodes and leaf nodes of ImageNet, which do not overlap with each other. Starting from the 1,000 object categories, a hierarchy is extracted from the WordNet hierarchy in our experiment. Since the ground truth labels of test data are not provided, we build a test dataset by randomly extracting 40 % images from the training set of ILSVRC2013. We summarize the hierarchical structure and training dataset size as well as testing dataset size in Table 1.

Table 1. Dataset statistics: number of leaf nodes (#Leaf) and internal nodes (#Int), depth of the hierarchy (#Dep), average number of instances of each concept for training (#Trn), and testing (#Tst) respectively.

Dataset	#Leaf	#Int	#Dep	#Trn	#Tst
Caltech256	256	62	6	58	29
ILSVRC2013	1000	797	10	1261	150

We have five parameters to set: regularization parameters ν, λ_1, λ_2, ρ_1 and learning rate ρ. We tune all the five parameters by a grid search in the scope of $\{0.1, 0.2, \ldots, 0.9, 1\}$ for both datasets. Parameter values with the best performance are used in the following experiments. In addition, according to Algorithm 1, we also need to initialize dictionary $D \in \mathcal{D}$, classifiers $W \in \mathcal{W}$, T

---

[1] http://www.image-net.org/challenges/LSVRC/2013/.

(number of iterations), t_0 (numbers of iterations at constant learning rate). In the experiment, T is set to twice the number of training samples, and $t_0 = T/10$. We initialize D by the unsupervised formulation of Eq. (1) using the SPAMS toolbox[2]. With this initial dictionary D, we optimize Eq. (6) with respect to W.

4.2 Comparison Methods

We compare our method with two state-of-the-art multi-label learning methods ML-KNN [16] and LIFT [15], one graph-guided structured sparse multi-task learning method SPG [2], one unsupervised dictionary learning method UDL (Eq. (1)), one hierarchical unsupervised dictionary learning method PSH [6], and one discriminative supervised dictionary leaning method DSDL [7].

4.3 Evaluation Metric

We evaluate the performance of compared methods with Average Precision (AP) and two commonly used multi-label criteria: One Error (1-Err) and Coverage (Cov.), which measure the performance from different aspects and the definitions can be found in [10].

4.4 Experimental Results

Example Dictionary Visualization. To verify that our dictionary contains semantic hierarchy, we visualize our dictionary obtained from ILSVRC2013. Note that each column of the dictionary D is associated with a specific semantic class. Thus the atoms in each column can be considered as the basic atoms of specific semantic class. Since we have deleted some meaningless nodes (such as entity) in WordNet hierarchy, the hierarchical structure used in our experiments is a forest instead of a tree. This forest contains 53 trees in total with each corresponding to a semantic entity. We assume each column of D is associated with a specific semantic class. Figure 2(a) shows an example of the learned dictionary which corresponds to a semantic hierarchy whose root node represents "bird". Furthermore, we compute the nearest neighbor (in Euclidean distance) of each column of the dictionary D to see whether its the nearest neighbor is just the semantic class it corresponds to. Figure 2(b) shows an example with each node in the hierarchy annotated with a specific semantic class. Moreover, the nearest neighbor of each column is just the class it should corresponds to, which demonstrates that the learned dictionary has been encoded with the semantic hierarchy.

Comparison Results. Firstly, we evaluate the performance of predicting multiple semantic concepts to images. The results are reported in Tables 2 and 3. From the results we can see: (1) Compared to the state-of-the-art methods of multi-label learning, graph-guided multi-task learning, unsupervised (hierarchical) dictionary

---

[2] http://www.di.ens.fr/willow/SPAMS/.

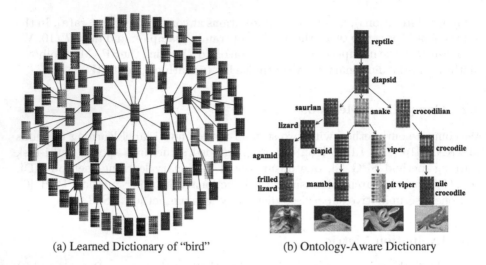

(a) Learned Dictionary of "bird"   (b) Ontology-Aware Dictionary

Fig. 2. Examples of the learned dictionary. The hierachy is built according to the WordNet. Each node that corresponds to a specific semantic class is one of the columns (reshaped to a matrix) of the learned dictionary.

Table 2. Comparison results on Caltech256 in terms of AP (↑), 1-Err (↓) and Cov. (↓). ↑ (↓) implies the larger (smaller), the better.

	ML-KNN	LIFT	SPG	UDL	PSH	DSDL	Ours
AP	0.4379	0.5105	0.5408	0.5934	0.5945	0.6290	**0.6482**
1-Err	0.2786	0.2469	0.3542	0.2543	0.2423	0.2458	**0.2368**
Cov	130.68	65.65	151.57	45.99	39.58	37.87	**36.13**

Table 3. Comparison results on ILSVRC2013 in terms of AP (↑), 1-Err (↓) and Cov. (↓). ↑ (↓) implies the larger (smaller), the better.

	ML-KNN	LIFT	SPG	UDL	PSH	DSDL	Ours
AP	0.4652	0.4158	0.4644	0.4516	0.4226	0.4661	**0.4891**
1-Err	0.2960	0.3943	0.3689	0.3202	0.3546	0.3843	**0.2840**
Cov	289.51	297.39	312.88	232.48	235.84	243.12	**226.17**

learning, our method obtains the better performance in terms of all the three criteria. (2) Compared with DSDL [7], which is also a supervised dictionary learning method, the performance of our method is better, which demonstrates the strength of incorporating the ontology-aware supervised information into the learned dictionary.

Secondly, in Fig. 3, we show sample images with predicted semantic hierarchies. Because there are no hierarchical structure considered in ML-KNN, no structured concepts are associated with images in the results and there are more

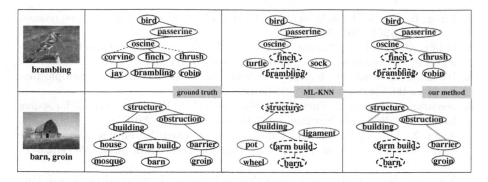

Fig. 3. Sample images with annotations and predicted semantic hierarchies. For ground truth, dashed lines denote the brother nodes of ground truth extracted from Word-Net. For ML-KNN and our method, circled concepts are predicted by corresponding methods. Dashed circles denote the concepts that are not recalled. Red circles denote the false positives. Lines in green indicate true positive paths in the hierarchy (Color figure online).

false positives. For our method, owing to the strength of ontology-aware dictionary learning, hierarchical concepts are predicted. For example, for the image in the second row, we have two structured paths "structure" - "buildings" and "structure" - "obstruction". For the first image, though "brambling" was not recalled by our method, we successfully describe the images with a structured positive path, i.e., "bird" - "passerine"- "oscine" - "thrush"- "robin", which includes brother nodes of ground truth from WordNet.

5   Conclusions

We have presented an ontology-aware dictionary learning for image understanding, which utilizes the semantic relationship among concept nodes in a hierarchy for better describing of images. Hierarchical dictionary learning is combined with multi-task regression framework with annotated hierarchies. We proposed a stochastic gradient descent algorithm to solve our optimization problem. Experiments on two benchmark datasets demonstrate the effectiveness and better performance of the proposed method. Future works include multi-feature dictionary learning and generating the human readable descriptions for images under the help of semantic hierarchies.

References

1. Balcan, N., Blum, A., Mansour, Y.: Exploiting ontology structures and unlabeled data for learning. In: ICML, pp. 1112–1120 (2013)
2. Chen, X., Lin, Q., Kim, S., Carbonell, J.G., Xing, E.P., et al.: Smoothing proximal gradient method for general structured sparse regression. Ann. Appl. Stat. **6**(2), 719–752 (2012)

3. Efron, B., Hastie, T., Johnstone, I., Tibshirani, R., et al.: Least angle regression. Ann. Stat. **32**(2), 407–499 (2004)
4. Griffin, G., Holub, A., Perona, P.: Caltech-256 object category dataset. CNS-TR-2007-001, Technical report, California Institute of Technology (2007)
5. Hwang, S.J., Grauman, K., Sha, F.: Semantic kernel forests from multiple taxonomies. In: NIPS, pp. 1718–1726 (2012)
6. Jenatton, R., Mairal, J., Bach, F.R., Obozinski, G.R.: Proximal methods for sparse hierarchical dictionary learning. In: ICML, pp. 487–494 (2010)
7. Jiang, Z., Lin, Z., Davis, L.S.: Label consistent k-svd: learning a discriminative dictionary for recognition. IEEE Trans. Pattern Anal. Mach. Intell. **35**(11), 2651–2664 (2013)
8. Mairal, J., Bach, F., Ponce, J.: Task-driven dictionary learning. IEEE Trans. Pattern Anal. Mach. Intell. **34**(4), 791–804 (2012)
9. Mairal, J., Ponce, J., Sapiro, G., Zisserman, A., Bach, F.R.: Supervised dictionary learning. In: NIPS, pp. 1033–1040 (2009)
10. Schapire, R.E., Singer, Y.: Boostexter: a boosting-based system for text categorization. Mach. Learn. **39**(2–3), 135–168 (2000)
11. Shen, L., Wang, S., Sun, G., Jiang, S., Huang, Q.: Multi-level discriminative dictionary learning towards hierarchical visual categorization. In: CVPR, pp. 383–390. IEEE (2013)
12. Wang, H., Nie, F., Cai, W., Huang, H.: Semi-supervised robust dictionary learning via efficient l-norms minimization. In: ICCV, pp. 1145–1152. IEEE (2013)
13. Wang, M., Gao, Y., Lu, K., Rui, Y.: View-based discriminative probabilistic modeling for 3d object retrieval and recognition. IEEE Trans. Image Process. **22**(4), 1395–1407 (2013)
14. Zhang, L., Wang, M., Hong, R., Yin, B.C., Li, X.: Large-scale aerial image categorization using a multitask topological codebook. IEEE Trans. Cybern. (2015). doi:10.1109/TCYB.2015.2408592
15. Zhang, M.L., Wu, L.: Lift: multi-label learning with label-specific features. IEEE Trans. Pattern Anal. Mach. Intell. **37**(1), 107–120 (2015)
16. Zhang, M.L., Zhou, Z.H.: Ml-knn: a lazy learning approach to multi-label learning. Pattern Recogn. **40**(7), 2038–2048 (2007)
17. Zheng, J., Jiang, Z.: Tag taxonomy aware dictionary learning for region tagging. In: CVPR, pp. 369–376. IEEE (2013)

Quality Analysis on Mobile Devices for Real-Time Feedback

Stefanie Wechtitsch, Hannes Fassold, Marcus Thaler, Krzysztof Kozłowski,
and Werner Bailer[✉]

JOANNEUM RESEARCH Forschungsgesellschaft mbH,
DIGITAL - Institute for Information and Communication Technologies,
Steyrergasse 17, 8010 Graz, Austria
{stefanie.wechtitsch,hannes.fassold,marcus.thaler,
krzysztof.kozlowski,werner.bailer}@joanneum.at

Abstract. Media capture of live events such as concerts can be improved by including user generated content, adding more perspectives and possibly covering scenes outside the scope of professional coverage. In this paper we propose methods for visual quality analysis on mobile devices, in order to provide direct feedback to the contributing user about the quality of the captured content. Thus, wasting bandwidth and battery for uploading/streaming low-quality content can be avoided. We focus on real-time quality analysis that complements information that can be obtained from other sensors (e.g., stability). The proposed methods include real-time capable algorithms for sharpness, noise and over-/ underexposure which are integrated in a capture app for Android. Objective evaluation results show that our algorithms are competitive to state-of-the art quality algorithms while enabling real-time quality feedback on mobile devices.

1 Introduction

Many cultural and sports live events do not take place at only a single spot, but are spread out over different stages, halls, cities or even regions, with different actions happening in parallel in each of these places. Examples are music festivals with several stages or tents, city festivals, parades, marathons or bike races. Except for some few high-profile events, it is not possible to fully cover such events with professional capture equipment.

The video capabilities of mobile devices have improved tremendously in recent years, reaching a level that only used to be found in mid-range consumer cameras. Thus, users are encouraged to contribute their videos and images captured on their mobile devices for use in professional productions. However, despite the improvements in capture technology, many user generated videos suffer from quality issues. These may occur due to inappropriate actions of the user, e.g. (partly) covering lenses, unintentional capture while moving or fast and shaky movements of the device. Others are caused by the limitations of the optical and electronic components under certain capture conditions, such as underexposure or clipping, noise (in particular under low light condition), failure of

© Springer International Publishing Switzerland 2016
Q. Tian et al. (Eds.): MMM 2016, Part I, LNCS 9516, pp. 359–369, 2016.
DOI: 10.1007/978-3-319-27671-7_30

automatic focus adjustment or coding artifacts (loss of details, macro-blocking). Some of these problems may be indicated by other sensors of the mobile device (e.g., accelerometer readings may indicate shakiness), while others can only be identified by inspecting the image content. The integrated processing in modern consumer devices tries to detect and conceal some of these issues, however, the resulting quality is often not sufficient for including user generated content in professional productions. For example, reacting to exposure conditions will often cause strong sudden changes of the luminance level, which is not acceptable.

In this paper we focus on visual quality analysis in real-time on mobile devices in order to enable the utilization of user generated content of live events. We have developed technologies for analyzing and filtering streams based on quality and content properties. An Android app provides immediate feedback about quality problems while recording in order to avoid wasting bandwidth and battery for uploading/streaming low-quality content. Users are encouraged to contribute their content by receiving immediate feedback, rather than frustrating them by discarding their contribution later. The obtained quality metadata can also serve as a filter or ranking criterion for selecting the best quality out of multiple contributions. We focus on quality analysis that complements information we can obtain from other sensors, and we do not consider quality problems that may occur due to encoding and transmission later in the processing chain. In this setup, only no-reference quality metrics are of interest. Thus, we have developed a reliable sharpness measure, a noise estimator which may also indicate problematic lighting conditions and an over/underexposure detector.

This paper is organized as follows. The remainder of this section reviews the application scenario and related work. Section 2 presents the three quality analysis algorithms in detail. The implementation within an Android app is discussed in Sect. 3. Section 4 contains evaluation results for the algorithms and Sect. 5 concludes the paper.

1.1 Application Scenario

Many cultural and sports live events do not take place at only a single spot, but are spread out over different stages, halls, cities or even regions, with different actions happening in parallel in each of these places. Examples are music festivals with several stages or tents, city festivals, parades, marathons or bike races. Except for some few high-profile events, it is not possible to fully cover such events with professional capture equipment. In order to enable the inclusion of user generated content we are developing technologies for live capture and streaming from professional and consumer devices, and for fusion of audio and video content from heterogeneous devices into a format agnostic representation. Bandwidth and processing resources are scarce in such a live scenario, so methods for analyzing and filtering streams based on quality and content properties are provided. Users are encouraged to contribute their content by providing immediate feedback on the quality of their recordings, rather than frustrating them by discarding their contribution later. Quality metadata can also serve as a filter or ranking criterion for selecting the best quality out of multiple contributions.

1.2   Related Work

In our scenario, we only have the incoming live stream. Thus, only no-reference methods are applicable, and the large body of literature for reference-based quality assessment (mostly from video coding) is not directly relevant for this work. In addition, the real-time setting and the limited computing capabilities on mobile devices make a number of state-of-the art algorithms not applicable.

No-reference sharpness measures typically can be grouped into two categories: Spatial domain methods utilize the gradient, edge or slope information and evaluate specific statistical properties, e.g. [4]. In contrast, transform based methods are usually more complex and consider the fact that sharp edges increase the high-frequency components. A spatial method presented in [3] measures the edge spread for obtaining a sharpness value and shows encouraging results that are well correlated with the human perception. However, due to its computational complexity this approach is not applicable on mobile devices. A fast approach for blur estimation on mobile devices is presented in [7], utilizing the fact that blurred images have a lower amount of gradients and decreased mean and standard deviation values of the gradient magnitude compared to sharper images. Based on the statistics of the image gradients the approach makes a binary decision about sharpness, ignoring local sharpness variations. Thus, false estimates are to be expected for images containing regions with different levels of sharpness.

Noise level estimation under the real-time requirement is only addressed by the work presented in [1]. Many of the robust and high-quality approaches for noise level estimation, e.g. [13] which analyzes the distribution of spatial and temporal wavelet coefficients or [2] which does a statistical analysis of the motion-compensated difference image, are not suitable due to the low computational resources available on mobile devices.

Related work on over- or underexposure is mainly geared towards still images and usually analyzes the intensity distribution (upper and lower bound) in order to apply a threshold in an appropriate color space (e.g., [6]). The algorithm described in [12] calculates the joint probability density function (PDF) of a specific pixel and its right neighbor, followed by an analysis of certain regions (corresponding to the dark and bright areas of the image) of the joint PDF matrix. The method proposed in [11] takes into account the properties of the human visual system (HVS) for a more precise detection of the perceived overexposed image regions. However, all these approaches are tailored to still images without considering the specifics of video.

2   Quality Analysis Algorithms

2.1   Sharpness

Obtaining a reliable sharpness value for images and videos is computationally complex. For providing immediate feedback about the quality of recordings via a mobile device, real-time capability of the algorithms is essential. Thus, we use the Laplace operator for the sharpness estimation instead of measuring the edge width or operating in the frequency domain, which is commonly done in related work. The Laplace operator, defined as

$$I_{Lap} = \frac{\partial^2 I}{\partial x^2} + \frac{\partial^2 I}{\partial y^2}, \tag{1}$$

is performed as the sum of the second derivatives of the input image using the 3×3 kernel

$$k_{Lap} = \begin{vmatrix} 0 & 1 & 0 \\ 1 & -4 & 1 \\ 0 & 1 & 0 \end{vmatrix}. \tag{2}$$

Basically, the Laplace operator is used for edge detection, where the response values indicate the slope of the corresponding edge. The response image is sub-sampled into equally sized blocks, e.g., of size 32×32 pixels. For each block a representative value is statistically computed, e.g., by using the maximum response within the block.

The HVS tends to judge sharpness based on the sharpest regions within an image [5], which usually correspond with the regions in focus. According to this empirical finding, we finally determine a sharpness score from a subset of blocks yielding the highest values. Typically, using 5–25 % of the blocks yields good results.

The proposed approach is computationally less complex than the methods presented in [3,4], as measuring the edge width is avoided. This step involves edge detection followed by identifying local minimum and maximum at each edge point, which is clearly not a trivial problem and needs several pixel operations in contrast to a simple convolution.

2.2 Noise

The workflow of the proposed lightweight noise level estimation algorithm is described in this section. First, the luminance component of the input image is calculated. All further processing is done using only the luminance component. A Laplacian operator is applied and the filtered image is then divided into equally sized blocks. Typically, block sizes of 16×16 or 32×32 are used. In the next steps, blocks which are not homogeneous or which correspond to nearly black (i.e., average luminance lower than 40 for an 8 bit intensity range) image areas (as not much noise is present in nearly black areas) are discarded. A block is discarded as being not homogeneous (too much texture present) if the maximum response within the block is higher than 30 % of the maximum response in the whole Laplacian image. For all remaining blocks, a noise level is calculated separately for each block. The per-block noise level value is calculated by applying a median filter of size 3×3 to the luminance component, subtracting the median-filtered block from the original block and calculating the average (over the block) of the absolute values of the differences. Finally, an overall noise level value is obtained as the median of all per-block noise level values.

2.3 Over-/Underexposure

The detection of over- or underexposure would be straightforward if one could base it solely on the luminance values of a captured image. But clearly, this is not enough as most cameras of mobile devices do an automatic exposure and brightness adjustment when they detect a possible over- or underexposure situation, which usually cannot be turned off. The principle of our proposed algorithm is to detect whether it is likely that the automatic brightness adjustment was applied by analyzing the overall brightness progression of the camera images within a certain temporal analysis window. We start the analysis phase whenever the intensity falls beyond a user-specified lower bound (e.g. 50) or rises above an upper bound (e.g. 180). Within the analysis phase, we estimate for each pair of consecutive camera images I_t and I_{t+1} a brightness correction function $C_t(g)$ which models the brightness change between the two images for each gray level g. The estimation of the brightness correction function is based on the algorithm from [10] which is used there for the correction of flicker in archival sequences. This algorithm sets up for each gray level a histogram of the intensity differences between the two images and compacts the histogram into one difference value by determining the bin with the highest number of hits. From this, one gets a raw correction function $R(g)$ into which a smooth polynomial is fitted into, yielding the final correction function $C_t(g)$ for time point t. After calculating $C_t(g)$, we obtain the overall amount of positive $E_+(g)$ and negative brightness correction $E_-(g)$ via

$$E_+(g) = \sum_{\substack{1 \leq g \leq 255 \\ C_t(g) > 0}} C_t(g)\, w_g, \quad E_-(g) = \sum_{\substack{1 \leq g \leq 255 \\ C_t(g) < 0}} C_t(g)\, w_g \qquad (3)$$

where w_g is a weight for the intensity g used for over-weighting intensity values which are occurring more often in the image. Within the temporal analysis window (usually in the range of 0.5–1.5 s) we sum up the positive brightness correction values $E_+(g)$ and the negative brightness correction values $E_-(g)$ individually. If either of these sums exceeds a certain threshold, the detector returns either an overexposure or underexposure event.

3 Implementation

To provide an integrated capture app enabling real-time quality analysis on mobile devices the proposed sharpness, noise and over-/underexposure detection algorithms have been implemented in an Android app. The capture app uses basic image processing functionality of the OpenCV[1] library version for Android. The implementation makes use of the fact that the sharpness and noise estimation both consider blocks of Laplacian filtered input images. As a result, the computation effort can be reduced if both quality measures are estimated. In order to support quality analysis the application captures additional data

---

[1] http://opencv.org/.

using the available sensors of the mobile device. For example, the accelerometer readings are used to detect fast and shaky movements of the mobile device and thus unstable image sequences. Detailed sensor information of the following sensors is captured: location, accelerometer, gyroscope, magnetic field, orientation, rotation vector, ambient light, proximity and pressure. To run the capture application at least Android version 4.3 with API level 18 is required to benefit from hardware encoding utilized by methods from the grafika project[2] to capture audiovisual data. The methods used by the proposed implementation enable encoding of the captured image sequence and image quality analysis of data from image data buffers in parallel.

The application provides a live preview to the user, as shown in Fig. 1. The application continuously measures stability, sharpness, noise, exposure and detects the use of brightness compensation by analyzing the captured images in real-time. If one of the quality measurements is outside the target range, an overlay with an icon and a message is displayed in order to inform the user about the quality problem. The user has the chance to react and avoid the quality impairment. If the problem persists for some time, recording will be terminated automatically. The user is presented with feedback of each quality measure that violates the threshold. Together with additional quality analysis performed on the server side after uploading, an integrated measure is determined as a weighted sum of the individual measures. The captured video is uploaded to a server in segments as a background tasks. Each segment is posted along with the related sensor an quality metadata.

Using a Samsung Galaxy S5 the runtime of the sharpness and noise detection algorithm is about 140 ms for one HD image. The exposure and brightness adjustment detection needs about 60 ms. That implies that the proposed quality analysis algorithms need about 200 ms to provide feedback. For online image quality analysis it is sufficient to analyze every sixth frame of the captured image sequence since image quality problems (e.g. noise) are continuous and do not disappear instantly.

4   Evaluation

4.1   Sharpness

To our knowledge, no evaluation database exists which provides sharpness scores obtained on blurred videos. Consequently, we were forced to evaluate our algorithm by using an image database. A well established and frequently used image database in this research area is the LIVE Quality Assessment Database [9]. The database consists of 174 Gaussian blurred images with corresponding Mean Opinion Scores (MOS), obtained from a user experiment. 29 reference images were artificially blurred by using a circular-symmetric 2-D Gaussian kernel of five varying standard deviations. 24 participants were shown both the original and the distorted image. Subsequently, they rated the quality of the distorted image on a five

---

[2] https://github.com/google/grafika.

Fig. 1. Screenshot of the Android capture app, including quality feedback (bottom)

point scale: excellent, good, fair, poor and bad. The user ratings for each image were averaged and linearly transformed in order to fit them into an interval of 0 to 1. The results of the algorithm are transformed into values from 0 to 1 as well. An image with sharpness 1 would have the same nominal resolution as the original, not blurred image. A non-linear scaling of the algorithm output has been applied in order to approximate the nominal resolution of the processed image or video to the best effect. As shown in Fig. 2 the obtained sharpness scores are well correlated with the subjective user scores. As parameter setting for this test we used a block size of 32×32 pixels and $10\,\%$ of the sharpest blocks are contributing to the final sharpness score. The evaluation results yield a Pearson correlation of 0.96, and a Spearman rank correlation coefficient of also 0.96.

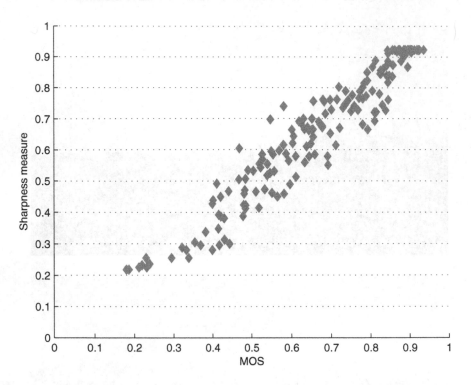

Fig. 2. Result of sharpness estimation algorithm (non-linearly scaled) compared to objective MOS

4.2 Noise

We evaluate the proposed lightweight noise level estimation algorithm on a noise data set collected and created within the PrestoPRIME project[3]. The content in the reference dataset is chosen so that it contains different motion types (static scene / object motion / global camera motion), different magnitudes of motion (no motion / low motion / fast motion) and different degrees of texturedness (scene mostly homogenous / scene partly textured / scene fully textured). Half of the clips have been created at JOANNEUM RESEARCH and the other ones have been taken from publicly available videos (e.g. the Vimeo Group Public Domain HD Videos and the NASA HD video gallery) and from a dataset from the 2020 3D Media project[4]. The clips contain no visible noise and are encoded with a high-quality encoder. Each clip is exactly 10 seconds, with 25 frames per second. After setup of the reference video clips, artificial noise/grain of different magnitudes has been added to the clean clips. In order to make the addition of noise/grain as natural as possible, we do not go the usual way of adding synthetic Gaussian-distributed noise with a specified standard deviation. Instead, we take

---

[3] http://www.prestoprime.org.
[4] http://www.20203dmedia.eu/.

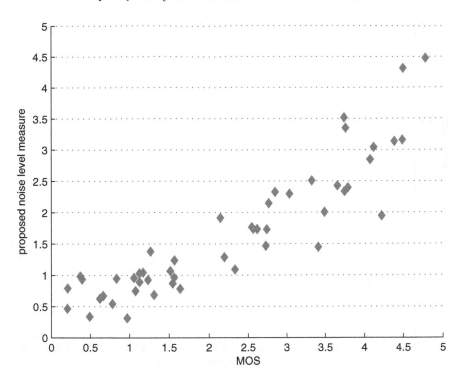

Fig. 3. Result of proposed lightweight noise level estimation algorithm compared to objective MOS

a noise/grain template, captured from homogenous regions of a noisy image, and use the texture synthesis method from [8] to create artificial grain with the same appearance as the noise/grain template. We use two different noise templates, the first one for simulating the appearance of fine electronic noise and the second one for simulating the appearance of coarse film grain. For each noise template and for each clip, we add noise in three different magnitudes (low/medium/high). So finally we have 7 variants from each video clip: The clean reference clip, 3 variants with fine noise added and 3 variants with coarse noise added. The data has been subjectively judged by 20 persons. Each person assessed the perceived noise level of each video of the noise level dataset. The judgments are represented as mean opinion scores (MOS) on a five point scale.

The lightweight noise estimation algorithm has been compared to the MOS obtained from the experiment (see Fig. 3) as well as to the results of a highly sophisticated, but computationally expensive (even employing GPU-acceleration) noise level estimation algorithm from [2] (see Fig. 4). In both cases the correlation is excellent, yielding a correlation coefficient of 0.902 when comparing the lightweight algorithm with the MOS and 0.948 when comparing it with the algorithm from [2]. The lightweight noise level estimation algorithm,

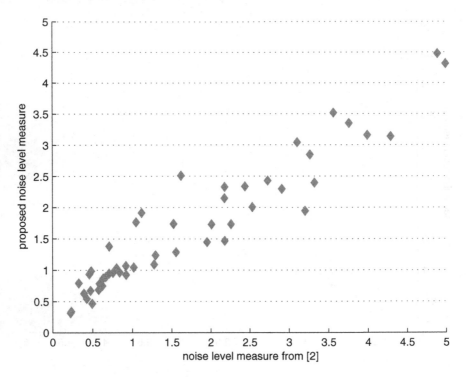

Fig. 4. Result of proposed lightweight noise level estimation algorithm (vertical axis) versus result of noise level estimation algorithm from [2] (horizontal axis)

which runs in real-time on mobile devices, is able to give nearly the same performance as the computationally much more expensive GPU-accelerated algorithm.

5   Conclusion

In this paper we have presented real-time quality algorithms tailored for mobile devices integrated in an Android app. These algorithms address quality parameters (sharpness, noise, over/underexposure) that may be influenced by the user (in contrast to transmission problems), thus they are providing immediate feedback. User generated content captured at concerts or another live events can be checked constantly while capturing before wasting bandwidth when uploading it to a server. The proposed algorithms have low computational complexity and are thus suited for real-time implementations on mobile devices with limited processing capabilities. Nonetheless, objective evaluation shows that their performance is competitive to more complex algorithms for many practical problems for video quality assessments.

Acknowledgements. The authors would like to thank their project partners bitmovin for providing the video streaming implementation and VRT for the graphics design. The research leading to these results has received funding from the European Union's Seventh Framework Programme (FP7/2007-2013) under grant agreement n° 610370, ICoSOLE ("Immersive Coverage of Spatially Outspread Live Events", http://www.icosole.eu).

References

1. Bosco, A., Bruna, A., Messina, G., Spampinato, G.: Fast method for noise level estimation and integrated noise reduction. IEEE Trans. Consum. Electron. **51**(3), 1028–1033 (2005)
2. Fassold, H., Wechtitsch, S., Hofmann, A., Bailer, W., Schallauer, P., Borgotallo, R., Messina, A., Liu, M., Ndjiki-Nya, P., Altendorf, P.: Automated visual quality analysis for media production. In: IEEE International Symposium on Multimedia (2012)
3. Feichtenhofer, C., Fassold, H., Schallauer, P.: A perceptual image sharpness metric based on local edge gradient analysis. IEEE Sig. Proc. Letters **20**(4), 379–382 (2013)
4. Ferzli, R., Karam, L.J.: A no-reference objective image sharpness metric based on the notion of just noticeable blur (JNB). IEEE Trans. Image Process. **18**(4), 717–728 (2009)
5. Hassen, R., Wang, Z., Salama, M.: No-reference image sharpness assessment based on local phase coherence measurement. In: IEEE ICASSP (2010)
6. Hou, L., Ji, H., Shen, Z.: Recovering over-/underexposed regions in photographs. SIAM J. Imaging Sci. **6**(4), 2213–2235 (2013)
7. Ko, J., Kim, C.: Low cost blur image detection and estimation for mobile devices. In: 11th ICACT, vol. 3 (2009)
8. Schallauer, P., Mörzinger, R.: Film grain synthesis and its application to regraining. In: SPIE Electronic Imaging (2006)
9. Sheikh, H., Wang, Z., Cormack, L., Bovik, A.: LIVE image quality assessment database, release 2 [Online]. http://live.ece.utexas.edu/research/quality
10. Vlachos, T.: Flicker correction for archived film sequences using a nonlinear model. IEEE Trans. Circuits Syst. Video Tech. **14**(4), 508–516 (2004)
11. Yoon, Y.-J., Byun, K.-Y., Lee, D.-H., Jung, S.-W., Ko, S.-J.: A new human perception-based over-exposure detection method for color images. Sensors **14**(9), 17159–17173 (2014)
12. Yousefi, S., Rabiee, H.R., Mianjy, P.: Optimal exposure detection function for digital and smart-phone camera applications. In: IEEE ICCE (2012)
13. Zlokolica, V., Pizurica, A., Philips, W.: Noise estimation for video processing based on spatio-temporal gradients. IEEE Sig. Proc. Lett. **13**(6), 337–340 (2006)

Interactive Search in Video: Navigation With Flick Gestures vs. Seeker-Bars

Klaus Schoeffmann, Marco A. Hudelist$^{(\boxtimes)}$, Bonifaz Kaufmann,
and Kevin Chromik

Klagenfurt University, Universitaetsstr. 65-67, 9020 Klagenfurt, Austria
{ks,marco}@itec.aau.at, bonifaz.kaufmann@gmail.com, kchromik@edu.aau.at

Abstract. On touch-based devices such as smartphones and tablets users are accustomed to browse through lists and collections by using flick gestures. For video navigation, however, mobile touch devices still use the seeker-bar interaction concept. In this paper, we evaluate the performance of a flick gesture-based video player in direct comparison to a default video player with seeker-bar navigation for the purpose of interactive search in video. We have developed a special video player on a tablet device and performed a user study with 16 users with two different types of interactive search: target/known-item search and scene counting. Our results show that the flick-based video player is less performant than the default video player in terms of search time, but more efficient in finding target scenes and the preferred interface by the vast majority of tested users.

1 Introduction

When users want to navigate in a video they typically use a seeker-bar and scrub to the corresponding temporal position in the video sequence. Almost all video players and video interaction tools use this navigational interaction concept for random access in video [13,14]. This works fine as long as the video duration is rather small and, hence, the required scrubbing accuracy is rather low. However, if the video has a length of a few hours, scrubbing a seeker-bar becomes inconvenient, especially on small displays, since even small movements result in large temporal jumps (e.g., several minutes instead of seconds). Many researchers have focused on how to improve navigation behavior and performance for such situations and proposed improved navigation tools; an overview is given in Sect. 2. However, in practice these sophisticated navigation means are unfortunately not available to the majority of users, since their devices typically only feature a default video player. Therefore, many users often simply switch to fast-forward and reverse mode when they want to find a specific scene in a long video (e.g., for *known-item search* tasks [10,11]), or use the available seeker-bar for navigation.

With mobile touch devices such as smartphones and tablets, a very natural and intuitive way of enabling fast-forward and reverse functionality would be a *flick* gesture: The user touches the screen with his/her finger and pushes the video forward or backward. Once pushed to either direction the playback rate

© Springer International Publishing Switzerland 2016
Q. Tian et al. (Eds.): MMM 2016, Part I, LNCS 9516, pp. 370–381, 2016.
DOI: 10.1007/978-3-319-27671-7_31

suddenly changes (to fast-forward or reverse, depending on the pushed direction) and then linearly returns to normal, except the user performs another flick gesture. Otherwise, if the user flicks again, the fast-forward or reverse speed increases even more. Flicking is a very well-known gesture that is available on most (if not all) smartphones and tablets for browsing lists and collections, and we can assume that users are very accustomed to this type of interaction. However, currently it is not used for controlling the video playback rate, i.e., for interactive search and navigation in video, which is exactly the focus of this work.

The idea of navigating in video through flicking was already investigated in [8]. Unfortunately, from their evaluation we cannot directly deduce whether flick-based navigation in video works worse or better for interactive search tasks than seeker-bar navigation. Their interaction models – called *dynamic/static flicking* and *dynamic/static panning* – were designed for stylus-based interaction on a PDA and did not consider the actual speed of the flick gesture. Instead, they used the distance of the flick gesture as granularity for navigation in video. Moreover, in their evaluation the authors did not compare flick-based navigation to seeker-bar navigation but evaluated flicking against panning and could not find any statistical differences, neither for task performance nor for subjective ratings.

In this paper we investigate the performance of flick gesture-based video navigation for interactive search tasks in videos. For that purpose we have implemented a special video player on an Apple iPad device and performed a user study with 16 users and two different types of tasks: (i) known-item/target search and (ii) specific scene counting. We evaluate the achievable performance of a flick gesture-based player in direct comparison to the one achievable with a common video player using seeker-bar interaction. Our evaluation considers several aspects: (i) achievable search time, (ii) retrieval performance, and (iii) user experience and preference. Our results show that for the majority of tested tasks/videos, users are faster with the common video player but less efficient in finding target scenes than with the flick gesture-based player. At the same time they find the flick gesture-based player more convenient and the better interface than the default video player.

2 Related Work

In the last two decades, several proposals for improving video content navigation were made. An overview of recent work in this area can be found in [13,14]. Here, we only summarize works proposed for mobile devices with touch screen interaction, a topic addressed by only a few authors.

Hürst et al. proposed the *Mobile ZoomSlider* interface [7] for stylus-based navigation on handheld devices. The basic idea is a virtual seeker-bar that can be used at any screen position. The vertical click position is utilized as a parameter for navigation resolution. Moving the stylus left or right results in backward or forward navigation and the vertical position of the drag operation defines how fast the navigation action is performed. The same interaction concept has been proposed for video content navigation on PDAs and smartphones in a follow-up work [8]. The authors further suggested circular navigation gestures for stylus-based navigation in video and presented a few different interaction concepts in [6]. However, unfortunately in the evaluation no direct comparison to seeker-bar navigation was performed.

Karrer et al. [9] proposed the *PocketDRAGON* interface for touch-based video content navigation on mobile devices. Instead of an overlaid seeker-bar, they suggest direct manipulation of objects in the scene. A similar concept was already proposed by Dragicevic for video navigation on desktop PCs in an earlier work [2]. In their work, motion tracking is performed and object motion is used as basis for the dragging operation along a motion trajectory. Additionally to this object-based navigation mode, which rather improves the navigation accuracy and typically does not allow quicker navigation over longer segments in the video, they also support two-finger gestures. A horizontal wipe gesture with two fingers allows to jump to the previous or next scene in the video. However, as no evaluation has been performed by the authors, it remains unclear how well this navigation concept supports navigation tasks in videos.

Huber et al. [4] focused on improving navigation in e-learning videos rather than entertaining videos and proposed the *Wipe'n'Watch* interface. Instead of a seeker-bar, they suggest to use wipe gestures for touch-based navigation on smartphones. Their interface that operates in portrait mode is subdivided into two areas: (i) the upper area shows the actual video content and (ii) the lower area shows an overview of all available keyframes, i.e., available slides that act as direct access points. Their work also targets inter-video navigation, similarly to the idea of the *RotorBrowser*, proposed by De Rooij et al. for desktop use [1]. Hence, vertical wiping allows to jump between semantically similar segments among videos; e.g., topically related segments. The availability of such related segments is indicated with an arrow in the upper right corner of the interface. However, in their evaluation they did not directly compare their approach to seeker-bar navigation. Therefore, it is not clear if this interaction concept is superior than a seeker-bar.

In a recent work [12], we did already investigate navigation performance in video with a multi-touch navigation model that uses different navigation granularity according to the vertical position the wipe gesture is performed. Our results showed that uses like this kind of navigation and – for short videos – were even faster than with the default video player.

In [5], we proposed the *Keyframe Navigation Tree* interface for touch-based video content navigation on tablets. The main component of its navigation concept are three scrollable horizontal stripes with very compact keyframes, which can significantly outperform seeker-bar navigation at known-item search tasks.

To best of our knowledge, no one has however investigated the navigation performance of a video player using flick gestures, in direct comparison to seeker-bar navigation.

3   Video Navigation with Flick Gestures

To evaluate the performance of video browsing/navigation with flick gestures, we developed a new video navigation tool, which we call *flick player*.

3.1   Interaction Concept

The interaction concept of the flick player is based on the same idea proposed in [8]: instead of providing a seeker-bar for navigation, we allow the user to flick over the screen with his/her finger (see Fig. 1(a)). More precisely, when the user performs a flick gesture to the right the video player switches to fast-forward mode with a playback speed that corresponds to the velocity of the flick gesture (up to a maximum of 32x). If the user performs only one flick gesture, the playback speed will automatically return to normal playback rate (1x) after a while. We use a linear *easing function* for that purpose, which has been selected after several experiments with different easing functions and early test users. If, however, another flick gesture is performed by the user, the momentum of the gesture will be added to the current value of the playback speed. Hence, the fast-forward or reverse rate will be increased even more, until the maximum of 32x is reached. We use the same interaction concept for reverse mode, which can be enabled via a flick gesture to the left.

3.2   Implementation Details and Issues

We implemented the flick gesture-based interaction concept on an Apple iPad device with a 9.7-inch screen. Our implementation does always show the current playback rate with a colored line in a dedicated visualization area at the top of the screen. At normal playback rate (1x) a purple bubble is shown in the

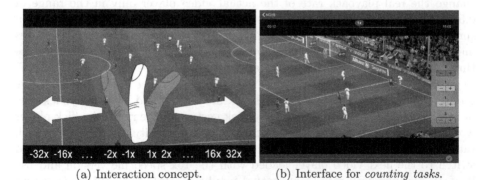

(a) Interaction concept.   (b) Interface for *counting tasks*.

Fig. 1. The *flick player* allows users to push the video forward and backward, i.e., switch to fast-forward and reverse mode by a simple flick gesture. Playback speed can vary between -32x and 32x. Without any further interaction the player returns to single playback speed (1x), according to a linear easing function (Color figure online).

(a) Interfac for *target search* tasks (reverse mode).   (b) Interface for *target search* tasks (fast-forward mode).

Fig. 2. Our implementation of the flick player also shows the current playback speed (top) as well as the current position in the video (bottom) (Color figure online).

middle of the line (see top of Fig. 1(b)). If the user switches to fast-forward, the bubble becomes blue and moves to the right. Similarly, in reverse mode the bubble becomes red and moves to the left (see Fig. 2). Additionally, our implementation shows the current playback time (top left), the duration of the video (top right), and the current position in the video with a non-interactive time-line visualization (bottom).

Even though we used a recent edition of a tablet device (an Apple iPad Air), its performance was not good enough to decode and play HD video at a speed of 32 times, even not with several optimizations and special video encodings with very small GOP (group-of-pictures) sizes. The only working solution we found was to encode 10 different versions of each video file, each with a different "encoded playback speed" (-16x, -8x, -4x, -2x, -1x, 1x, 2x, 4x, 8x, and 16x) and change the real playback rate of the actual video player only in the range of 0.5x to 2x. For example, with the video file version encoded at 8x speed, we can simulate any actual playback rate from 4x to 16x, although the video player uses only a real playback rate from 0.5x to 2x. Hence, in our implementation we use three simultaneous video players that are loaded with three "adjacent instances" (e.g., 4x, 8x, and 16x), where only the video player instance in the middle of this group is currently visible (e.g., with the 8x file version and a real playback rate of 1x). When the user would perform a flick gesture to the right, the flick player would increase the playback rate accordingly to the velocity. If the real playback rate of the video player reaches 2x (actual playback rate of 16x), our tool would switch to the next video player loaded with the 16x speed file version and use it at 1x real playback rate (same actual playback rate). With this instance we can further increase the real playback speed from 1x up to 2x, and simulate an actual playback rate of up to 32x.

This idea is consequently used for both directions (faster and slower playback) and implemented in a way that at any time, all three video player instances are loaded with the correct file version. Our implementation allows a very smooth transition from one video player instance to another, where the current video player is brought to the background and the other one is brought to the foreground, such that the user will not notice this "implementation trick".

4   Evaluation

To evaluate the performance of the flick player in direct comparison to a default video player, we conducted a user study with 16 participants (10 men, 6 women; age: 25.38 ± 3.2 years) that had to solve interactive search tasks in videos. All participants were experienced smartphone users, i.e., had been actively using a smartphone for at least one year. Each participant performed search tasks in four different videos for two different kind of search tasks (see Table 1). For *target search* tasks (also known as known-item search tasks) we presented the target scene of interest to the user and requested him/her to find it as fast as possible with the corresponding tool (flick player or default video player). For *scene counting* tasks we requested the user to find several instances of specific scenes as fast as possible, and increase counter values by using available stepper buttons in the right part of the interface, as shown in Fig. 1(b). Such specific scenes were goals, corners, throw-ins, and free-kicks in the soccer videos, and the appearance of specific answers (A, B, C, and D) in the "Who Wants to Be A Millionaire" videos. We used a maximum of four different kinds of such scenes for scene counting tasks, since our interface provides only four counter buttons.

Each participant tested both interfaces (flick player and default video player) with different instances of test videos (however with the same genre). For the selection of tasks and interfaces we followed a latin-square principle to avoid any familiarization effects. For example, if for target search with the flick player the video file *1_news* was used, *2_news* was used for target search with the default video player, and vice versa.

In the following we perform statistical analysis on the collected log data and evaluate the performance of both interfaces in terms of run-time, retrieval efficiency, and user ratings.

4.1   Target Search Tasks

Search-Time. A paired-samples t-test was used to determine whether there was a statistically significant mean difference between the interfaces regarding search time for documentary videos. Data are mean seconds $\pm$ standard deviation, unless otherwise stated. One outlier was detected that was more than 1.5 box-lengths away from the edge of the box in a boxplot. Inspection of its value did not reveal it to be extreme and it was kept in the analysis. The assumption of normality was not violated, as assessed by Shapiro-Wilk's test ($p = .08$). The default player (67.813 ± 30.259 s) was statistically significant faster than the flick player (128.813 ± 93.293 s) with $t(15) = 3.026, p = .05, d = .756$.

Table 1. Videos used for the evaluations.

Video	Genre	Duration (hh:mm:ss)	Search task
1_documentary	Documentary about	00:50:04	Target
2_documentary	Nature/animals	00:44:32	Search
1_news	News	00:53:14	Target
2_news	Show	00:53:30	Search
1_football	Excerpt of a	00:15:02	Scene
2_football	Soccer match	00:15:00	Counting
1_gameshow	Excerpt of "Who Wants	00:13:38	Scene
2_gameshow	To Be A Millionaire"	00:15:01	Counting

The same test was performed for news videos. No outlier was detected in this case. The assumption of normality was not violated, as assessed by Shapiro-Wilk's test ($p = .835$). The test revealed that the default player (81.313 ± 31.33 s) was again statistically significant faster than the flick player (142.313 ± 60.643 s) with $t(15) = 4.608, p = .0005, d = 1.152$.

4.2   Scene Counting Tasks

In terms of scene counting we analyzed the data in different aspects.

Search-Time. In a first step, we investigated whether the required search times to complete a search task for a video type ("Who Wants To Be A Millionaire", soccer) differed between the interfaces. We concentrated first on the "Who Wants To Be A Millionaire" videos. Four outliers were detected, which were more than

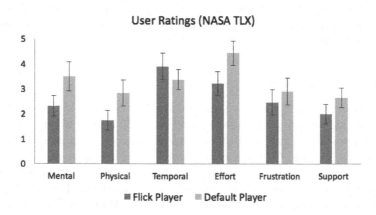

Fig. 3. Perceived workload ratings (according to NASA Task-Load-Index [3], with Likert-scale 0-10) for both interfaces (error bars: $\pm$ s.e. of the mean)

1.5 box-lengths from the edge of the box in a boxplot. Inspection of their values did not reveal them to be extreme and it they were kept in the analysis. As normality was violated in Shapiro-Wilk's test ($p = .05$) we used Wilcoxon signed-rank test in this case. A statistically significant median difference was detected between the default video player (median $= 105$ s) and the flick player (median $= 148$ s), $z = -2.638, p = .05$.

In case of the soccer videos, two outliers were detected, which were more than 1.5 box-lengths from the edge of the box in a boxplot. Inspection of their values did not reveal them to be extreme and it they were kept in the analysis. Normality was violated in Shapiro-Wilk's test ($p = .05$). Therefore, we again used the Wilcoxon signed-rank test to detect differences. This time, however, no statistically significant median difference could be detect between the default video player (median $= 219$ s) and the flick player (median $= 261$ s).

Retrieval Efficiency. As a next step, we investigated if there are differences in the number of found scenes between the interfaces. In case of the "Who Wants To Be A Millionaire" videos no outliers were detected, but normality was violated in Shapiro-Wilk's test. Therefore, a Wilcoxon signed-rank test was performed.

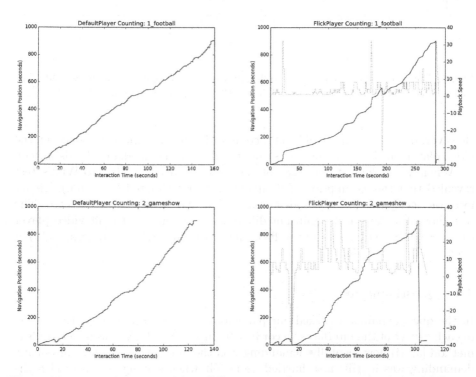

Fig. 4. Navigation behavior of four different users for scene counting in two different videos with the default video player (left) and the flick player (right).

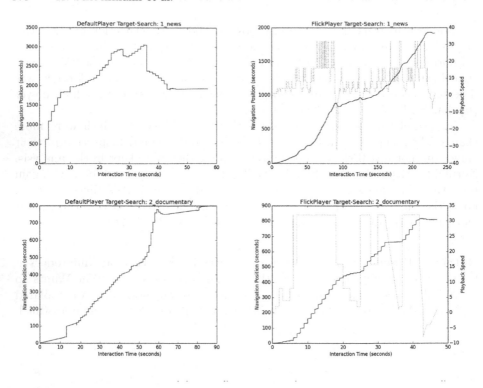

Fig. 5. Navigation behavior of four different users for target search in two different videos with the default video player (left) and the flick player (right).

However, no statistically significant median difference was measured between the video player (median = 13 instances) and the flick player (median = 14 instances).

When we looked at the data of the soccer videos, a Shapiro-Wilk's test revealed that the assumption of normality was not violated ($p = .406$). Therefore, we decided to use a paired-samples t-test. No outliers were detected. The test showed a statistical significant difference between the default video player (mean = 6.938 instances) and the flick player (mean = 8.188 instances), $t(15) = 2.825, p = .05, d = .706$.

4.3   Questionnaires

In the questionnaires we asked our participants to perform a subjective rating of each aspect of the interfaces, according to the NASA-TLX (Task-Load-Index) method [3]: (i) how mentally demanding was the interaction, (ii) how physically demanding was it, (iii) how hurried users felt when solving the task with the interface, (iv) how hard they had to work to accomplish their level of performance, (v) how insecure, stressed or annoyed they felt when using the interface, and (vi) how successful they felt in accomplishing the tasks. They could choose a value between 0 and 10, where 0 in all cases but for question (iv) represented "very low" and 10 represented "very high". In contrast, for question (vi) 0 indicated "perfect" and 10 indicated "failure".

To determine whether there are statistical significant differences between the interfaces we performed a Wilcoxon signed-rank test for each question. The flick player scored statistically significantly better for questions (i) $z = 2.145, p = .05$, (ii) $z = 2.419, p = .05$, and (iv) $z = 2.621, p = .05$. In contrast, the default video player performed significantly better for question (vi) $z = 2.224, p = .05$. In all other cases the difference was not statistically significant.

4.4   Preferred Interface

We further asked the users to rate both interfaces on a range between 1 and 6, where 1 is best and 6 is worst. The average rating of all 16 users was 2.5 ± 0.18 for the default video player, and 1.75 ± 0.17 for the flick player, which is statistically significant better. Out of 16 users, 11 users (68.75 %) gave a better rating to the flick player, three gave the same rating as to the default video player, and only two gave a worse rating. When asked about the reasons for their ratings, many users mentioned that with the default video player they did not like that they always had to keep their finger on the seeker-bar and move it while watching the content. They much more preferred to just push the video and then concentrate on the playback. Similarly, most users mentioned that they like that less physical interaction is required for the flick player.

5   Discussion and Conclusions

We found that users perceived the flick player as less mentally demanding and less physically demanding than the default video player, and felt that the flick player requires less effort for interactive video search tasks in general. More than 68 % of the participants rated the flick player as the better interface for interactive video search. At the same time, however, they think that the default video player with a common seeker-bar provides better support for such navigational search tasks in video content.

The evaluations of the run-time and the retrieval efficiency revealed that for three out of four video files users were significantly faster with the default video player for both kinds of tested search tasks (target search and scene counting). However, in terms of retrieval efficiency the flick player performed significantly better, at least for one of the two tested videos (cf. Table 1 and Sect. 4.2).

Figure 4 shows navigation plots of four different users, which reveal the navigation behavior in the same video file for the two different interfaces over time. In the right-top part of the figure we can see the reason why many users were slower with the flick player: many users did not make use of available high-speed playback rates (e.g., 16x and 32x), instead they mainly used playback rates of 1x to 4x and, hence, required a lot of time for the scene counting task.

With the default video player (left two plots) users could quite quickly navigate over the video and easily count the target scenes by constantly scrubbing the seeker-bar at moderately high speed. Only those users that made use of the available high playback rates of the flick player, such as the one in the bottom right of Fig. 4, could achieve a similarly high – or even better – run-time performance.

The same is true for the navigation behavior at target search tasks (Fig. 5), where only users with a high amount of 32x playback rate could outperform users of the default video player (bottom-right in the figure). The top-left part in the figure shows another interesting observation: this user obviously had a rough clue about the temporal position of the target scene in the video, for any reason whatsoever, and navigated very quickly to the last quarter of the video where he/she looked around more carefully. Such a navigation behavior is currently unfortunately not supported by the interaction concept of the flick player.

From these findings we can conclude that there is high potential for video navigation with flick gestures and users seem to welcome such an alternative navigation model (see Fig. 3). Also, a flick-based video player can enable higher interactive retrieval efficiency, but with the current interaction model cannot outperform the default video player with seeker-bar navigation in terms of search time. This means that additional studies should be performed, where seeker-bar navigation is used in combination with flick-gestures and/or the flick gesture flexibility is further improved.

Acknowledgments. The work was funded by the Federal Ministry for Transport, Innovation and Technology (bmvit) and Austrian Science Fund (FWF): TRP 273-N15, supported by Lakeside Labs GmbH, Klagenfurt, Austria and funded by the European Regional Development Fund and the Carinthian Economic Promotion Fund (KWF) under grant 20214/26336/38165.

References

1. de Rooij, O., Snoek, C.G.M., Worring, M.: Mediamill: semantic video search using the rotorbrowser. In: Proceedings of the ACM International Conference on Image and Video Retrieval, pp. 649–649. ACM Press (2007)
2. Dragicevic, P., Ramos, G., Bibliowitcz, J., Nowrouzezahrai, D., Balakrishnan, R., Singh, K.: Video browsing by direct manipulation. In: Proceedings of the SIGCHI Conference on Human Factors in Computing Systems, CHI 2008, pp. 237–246. ACM, New York, NY, USA (2008)
3. Hart, S., Staveland, L.: Development of NASA-TLX (Task Load Index): results of empirical and theoretical research. In: Hancock, P.A., Meshkati, N. (eds.) Human Mental Workload, pp. 139–183. Elsevier, Amsterdam (1988)
4. Huber, J., Steimle, J., Lissermann, R., Olberding, S., Mühlhäuser, M.: Wipe'n'watch: spatial interaction techniques for interrelated video collections on mobile devices. In: Proceedings of the 24th BCS Interaction Specialist Group Conference, BCS 2010, pp. 423–427. British Computer Society, Swinton, UK (2010)

5. Hudelist, M.A., Schoeffmann, K., Xu, Q.: Improving interactive known-item search in video with the keyframe navigation tree. In: He, X., Luo, S., Tao, D., Xu, C., Yang, J., Hasan, M.A. (eds.) MMM 2015, Part I. LNCS, vol. 8935, pp. 306–317. Springer, Heidelberg (2015)
6. Hürst, W., Götz, G.: Interface designs for pen-based mobile video browsing. In: Proceedings of the 7th ACM Conference on Designing Interactive Systems, DIS 2008, pp. 395–404. ACM, New York, NY, USA (2008)
7. Hürst, W., Götz, G., Welte, M.: Interactive video browsing on mobile devices. In: Proceedings of the 15th international Conference on Multimedia, MULTIMEDIA 2007, pp. 247–256. ACM, New York, NY, USA (2007)
8. Hürst, W., Meier, K.: Interfaces for timeline-based mobile video browsing. In: Proceedings of the 16th ACM International Conference on Multimedia, pp. 469–478. ACM (2008)
9. Karrer, T., Wittenhagen, M., Borchers, J.: Pocketdragon: a direct manipulation video navigation interface for mobile devices. In: Proceedings of the 11th International Conference on Human-Computer Interaction with Mobile Devices and Services, MobileHCI 2009, pp. 47:1–47:3. ACM, New York, NY, USA (2009)
10. Over, P., Awad, G., Michel, M., Fiscus, J., Sanders, G., Kraaij, W., Smeaton, A.F., Quéenot, G.: Trecvid 2013 - an overview of the goals, tasks, data, evaluation mechanisms and metrics. In: Proceedings of TRECVID 2013. NIST, USA (2013)
11. Schoeffmann, K.: A user-centric media retrieval competition: the video browser showdown 2012–2014. IEEE MultiMedia 21(4), 8–13 (2014)
12. Schoeffmann, K., Chromik, K., Boeszoermenyi, L.: Video navigation on tablets with multi-touch gestures. In: 2014 IEEE International Conference on Multimedia and Expo Workshops (ICMEW), pp. 1–6 (July 2014)
13. Schoeffmann, K., Hopfgartner, F., Marques, O., Boeszoermenyi, L., Jose, J.M.: Video browsing interfaces and applications: a review. SPIE Reviews 1(1), 018004 (2010)
14. Schoeffmann, K., Hudelist, M.A., Huber, J.: Video interaction tools: A survey of recent work. ACM Computing Surveys, 1–36 (2015) accepted for publication

Second-Layer Navigation in Mobile Hypervideo for Medical Training

Britta Meixner[1,2]($\boxtimes$) and Matthias Gold[1]

[1] University of Passau, 94032 Passau, Germany
meixner@fim.uni-passau.de
[2] FX Palo Alto Laboratory, Inc., Palo Alto, CA, USA
meixner@fxpal.com

Abstract. Hypervideos yield to different challenges in the area of navigation due to their underlying graph structure. Especially when used on tablets or by older people, a lack of clarity may lead to confusion and rejection of this type of medium. To avoid confusion, the hypervideo can be extended with a well known table of contents, which needs to be created separately by the authors due to an underlying graph structure. In this work, we present an extended presentation of a table of contents for hypervideos on mobile devices. The design was tested in a real world medical training scenario with the target group of people older than 45 which is the main target group of these applications. This user group is a particular challenge since they sometimes have limited experience in the use of mobile devices and physical deficiencies with growing age. Our user interface was designed in three steps. The findings of an expert group and a survey were used to create two different prototypical versions of the display, which were then tested against each other in a user test. This test revealed that a divided view is desired. The table of contents in an easy-to-touch version should be on the left side and previews of scenes should be on the right side of the view. These findings were implemented in the existing SIVA HTML5 open source player (https://code.google.com/p/siva-producer/ (accessed February 06, 2015)) and tested with a second group of users. This test only lead to minor changes in the GUI.

Keywords: Hypervideo · User interface · Navigation · Preview · Table of contents

1 Introduction

With new technologies like HTML5 and high internet bandwidths, the embedding of high quality video into homepages is nothing special anymore. Easily portable end user devices like tablets or smart-phones can be used to watch online videos. Using these technologies, appealing hypervideos can be created while providing a good quality of experience to the viewers even on mobile devices. Hypervideos in general consist of different types of media (like video, audio, images, and text) which are linked with each other. They provide interactive elements (links and other clickable elements). Using these interactive

© Springer International Publishing Switzerland 2016
Q. Tian et al. (Eds.): MMM 2016, Part I, LNCS 9516, pp. 382–394, 2016.
DOI: 10.1007/978-3-319-27671-7_32

elements, a viewer can chose her/his own path through the hypervideo. Accordingly, the shown elements of one whole hypervideo vary from viewer to viewer. A detailed explanation and definition of the term hypervideo is given in [13]. The hypervideos described in this work have video as a main medium, which are cut into scenes and linked to a scene graph. Additional information in form of text, images, audio files, or other videos are linked to these main video scenes. Hypervideos are very suitable for the transfer of knowledge due to their structure and the creative potential they offer [3, 6, 19]. They can be used to support learning in an every day work live with increasing mobility. Benefits of a video based learning portal like reduced overall costs, better learning results, improved employee satisfaction, and a higher reachability of the employees[1] can also be applied to training videos in healthcare or fitness scenarios. With further savings in the healthcare sector, the rehabilitation treatments in clinics become shorter and the patients have to do their training at home after leaving the clinic. Thereby a hypervideo with different trainings may be helpful for a proper execution of the exercises.

In this paper, we propose interfaces for medical training scenarios. The target group were people above 45. This group is more likely to need medical treatment which often results in longer recovery times than for people of younger age. Their computer skills are ranging from nonexistent to expert. A more precise description of the scenario and the target group can be found in Tonndorf et al. [18].

1.1 Problem Statement

This works focuses on navigation in a whole hypervideo scene graph, not in one scene or a very limited part of the scene graph, like for example by clicking on a tracked object (which may invoke the display of an annotation) or different levels of detail for a scene and different camera angles (which are usually implemented as parallel strands of scenes with time synchronization between scenes).

Three types of navigation may exist in a hypervideo (see also Fig. 1):

- *Graph-based navigation*: navigation in the underlying graph from scene to scene; the follow-up scene at the end of an already watched scene depends on the user interaction; for example, the user clicks on a button in a button panel
- *Graph-independent navigation*: jumps in the hypervideo from one scene to another which are not necessarily connected by an edge; for example, by the selection of a result from a search on the whole graph
- *Second-layer navigation*: navigation in a second navigation layer that leads to specific scenes in the graph structure; for example, a table of contents (tree structure) which provides entry points to certain nodes in the scene graph

The graph-based and the graph-independent navigation can be easily and intuitively implemented as lists of buttons or clickable areas, because the lists are usually very short (3–8 items). The more complex tree structure of a table of

---

[1] http://site.kaltura.com/rs/kaltura/images/TheStateofEnterpriseVideo2014-Kaltura Report-Final.pdf (accessed August 04, 2015).

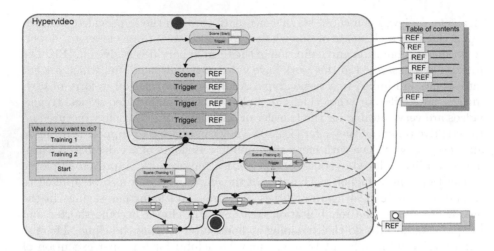

Fig. 1. Graph-based navigation (purple color), graph-independent navigation (orange color), and second-layer navigation (red color) in hypervideos (Color figure online).

contents contains some more challenges in contrast, because they contain many entries which afford a structure. Traditionally, tables of contents are presented as indented lists of formatted links. The indentation indicates the hierarchy between the links. These formatted links are comparatively small but easy to handle with a mouse on a PC. Using the same traditional representation on a tablet with touch input appears to be much more difficult due to the fat finger problem. Accordingly, another appearance is needed. It should be easy to interact with, especially on smart-phones and tablets. Furthermore, it should provide an overview of all levels with simple interactions (like scrolling). Design patterns for menus/link structures representing different levels of contents exist for web pages, like the "Accordion Menu"[2], but they are unusable on smart-phones or tablets. The iOS design guidelines provide a similar working pattern called "Table View"[3] which also could be used for a table of contents, but only provides an overview of one level at a time. The underlying level has to be folded out to see its contents. Navigation from the top level of the structure to lower levels is very laboriously because the lower levels have to be expanded one by one and no quick overview is provided which can be accessed with simple gestures. To the best of the author's knowledge, no design pattern suitable for smart-phones or tablets could be found which represents an indented list with different levels and gives an overview of all contents. Used with the underlying graph structure of the hypervideo, an intuitive linking between the hypervideo structure and the table of contents is necessary. Furthermore, a traditionally designed table of

---

[2] http://ui-patterns.com/patterns/AccordionMenu (accessed August 04, 2015).

[3] https://developer.apple.com/library/ios/documentation/UserExperience/
Conceptual/MobileHIG/ContentViews.html#//apple_ref/doc/uid/
TP40006556-CH13-SW1 (accessed August 04, 2015).

contents does not provide any preview of the linked video contents which further complicates making the right selection in a hypervideo. When a video is paused or stopped and the player is closed and reopened, the viewer usually has to start from the beginning which can be annoying in larger trainings.

1.2   Research Contributions

Dealing with the problems previously described, this work proposes the following solutions:

– A design for a table of contents which is usable on smart-phones or tablets and provides a preview from the table of contents into the hypervideo structure.
– Additional orientation in the table of contents to find entry points after a break.

In this paper we firstly present related work. Then, our usability evaluation consists of two parts. A concept is created and evaluated, then the GUI is implemented, and afterwards the implementation is tested and refined.

2   Related Work

A study on mobile multimedia summarized "that usability aspects, like an intuitive UI, are strongly related to the users? desire for being effective and ambitious" [10]. Keeping that finding in mind, we now try to give an overview of related work for navigation in hypervideos on mobile devices. It has to be noted, that by the best of the authors knowledge, no work exists which unites the terms "hypervideo", "navigation", and "mobile device". Related work can mainly be found in two areas: Navigation in hypervideos and video navigation on mobile devices. We will present and discuss papers from both areas.

Implementations for navigation in linear videos are examined by Cunha et al. [2] and Hurst et al. [7]. Hurst et al. evaluate methods of "pen-based navigation of videos on PDAs" [7]. They implement gestures to "skim a video along the timeline on different granularity levels" [7] and to manipulate the replay speed. However, this navigation is very basic and provides no additional information to the viewer to help her/him to find the desired contents. Cunha et al. explore "the generation of textual annotations on [linear] videos played on mobile device" [2]. Their "approach is to offer an application that allows associating annotations to a navigation line decorated with frames that are representative of the points of interest" [2]. Users should be able to find interesting points in the video intuitively using a combination of thumbnails and text as points of reference.

Related work regarding the navigation in hypervideos can be found in two forms. Either the GUI is evaluated or waiting times should be reduced. Shipman et al. [17] and Girgensohn et al. [4] introduce the Hyper-Hitchcock detail-on-demand video player. Thereby, detail-on-demand videos are a restricted form of hypervideo consisting nothing but videos and providing only one link at a time. This type of video is used to "watch short video segments and to follow hyperlinks to see additional detail" [4]. This desktop player provides no table of contents. The proposed interface, which should "provide users with the appropriate affordances to understand the hypervideo structure and to navigate it effectively" [4] was evaluated for its usability and refined in this process. A successful navigation in this type of video was tested afterwards. Girgensohn et al. summarize that "the user interface needs to present users with an intuitive view of the hypervideo structure. Such a view should be suitable for different tasks and guide the users towards the most appropriate interaction" [4]. Grigoras et al. [5] use a form of hypervideo which is composed of video scenes that are linked in a graph structure. Their work focuses on the streaming and reduction of latency through prefetching after analyzing the user behavior, but not on the creation of the user interface. Their findings are not especially tailored for mobile devices, but can be applied nevertheless.

Online course platforms usually provide linear videos or sequences of videos which can be navigated by lists of entry points. Furthermore, it is possible to jump from one video to its successor or predecessor. Watched videos are marked with symbols. The Coursera[4] and Udacity[5] apps provide lists of video parts which can be selected for viewing. A more complex structure of the video can only be provided by structuring the labels, for example with numbers or a repetition of the main headline followed by a separator and the sub-headline. The Khan Academy app[6] provides lists with categories, which results in a menu-like structure, but does not provide a good overview for navigation to certain contents or explanations. The Lynda app[7] has a contents list with two layers. Thereby, headlines are formatted differently. None of these apps provides a table of contents-like representation or has a scene graph as an underlying structure.

Other related work can be found in the areas of multi-view video on mobile devices and on direct manipulation video navigation. Apostu et al. [1] study navigation in linear multi-view videos on mobile devices based on spatial information. Miller et al. [14] explore multi-view video with hyperlinks on mobile devices. They "offer several mechanisms for viewing hypermedia and perspective selection" [14]. No user study can be found about the prototypes in this work. In both works, the only used medium is video, none provides a table of contents or any other navigational structures. The research is more focused on the switching between the different view instead of navigation in the whole video. Karrer et al. [8] and Nguyen et al. [15] examined direct manipulation video navigation with gestures in linear videos on mobile devices. These works reveal important findings on the fat finger problem, which is also an issue in our work.

3   Usability Evaluation in the Design Phase

The design phase for the extended table of contents was separated into tree major parts. An expert group was conducted to get first hints on how the table

---

[4] https://www.coursera.org/ (accessed August 05, 2015).
[5] https://www.udacity.com/ (accessed August 05, 2015).
[6] https://www.khanacademy.org/ (accessed August 05, 2015).
[7] http://www.lynda.com/ (accessed August 05, 2015).

of contents should be presented to the users and which information should be displayed. In a second step, a survey was composed to find out how people above 45 use technologies and media, as well as to find out about habits while searching for information. The third step was the implementation of a prototype, which was used to make a first usability test.

3.1 Expert Group

The expert group was conducted to find answers to the following questions: Which information should be presented to users considering the technical constraints of hypervideos? How should the table of contents be presented?

Study Method and Participants. Five experts ($N = 5$) from different disciplines were invited: a media and communication scientist, a web developer, a programmer, a multimedia researcher, and an expert for media philology and media analysis. All of them had experience with hypervideos and/or the usability of web and mobile GUIs. At least three of them had experience in experiments with the target age group. A discussion guideline as well as first paper prototypes from a brain storming session were presented to the group.

Result and Discussion. As a result it can be noted that a history is considered necessary but should be implemented separately from the table of contents. According to the experts, most hypervideos that will be used for learning or training will have a table of contents and not a structure which is created from the graph structure automatically. A result of an automated creation has the disadvantage that it is dependent from the graph structure and the used algorithms. Unfitting combinations thereof may result in an unusable table of contents which may need a lot of manual restructuring. This lead to the conclusion that no algorithms are needed to analyze the whole scene graph and bring it into a tree-like structure. In the remainder of this work, we focus on the table of contents created by the author of the video. The expert group agreed that the view of the table of contents should be divided. There should be a structure inspired by a table of contents on the left side and a preview on the selected scene and follow-up scenes on the right side. The upright table of contents should be on the left side due to the reading direction and common practices for websites. A video preview or representative thumbnail image is preferred to a simple textual description of the scene.

3.2 Survey

Our survey consisted of several sections which tried to give hints for the answers of the following questions: Which knowledge about tablets does the target group have? Which attitude towards learning and training videos does the target group have? Which devices are used for learning and training videos by the target group? What do users expect from a table of contents in a hypervideo?

Study Method and Participants. The process of creating the survey consisted of several steps. A first version of the survey was created and then tested for its functionality by two people with knowledge about empirical social research. After the proposed changes were made, the survey underwent a pre-test with twenty people with knowledge about surveys. A second revision of the survey with the comments of the pre-test followed and lead to the final survey which was available in two versions, on paper and online.

The survey was propagated in three different ways: in a forum where it was seen 200 times, in a social network where it was seen 113 times, and on paper where it was printed 125 times. The survey had 193 participants, whereby 130 people used the online version and 63 people filled out the paper form. 104 of the 193 participants were over 45 years old. In the remainder of this section, we focus on the over 45 year olds ($N = 104$). In this group, 47 were male and 55 female (2 did not answer the question). 38 participants had a degree from a university or university of applied sciences, 29 have completed an apprenticeship.

Result and Discussion. The following statements can be made about the results of the survey: Most of the devices are used daily/several times a week (like TV, computer, notebook, and smart-phone) or rarely/not regularly (like TV with internet access, tablets, mobile phone without internet access, MP3-player). Only 5.8 % of the participants above 45 do not use the internet at all. Regarding devices with touch screens, the most used device is an ATM with a touchscreen. Smart phones or mobile phones are the mainly used devices in the private area followed by tablets and notebooks with touch screen. 74 participants claimed that they have watched learning and training videos at some time. With multiple answers possible, 39 participants had watched tutorials, 38 participants had watched repair instructions, and 26 participants had watched assembly instructions. Besides those types of materials, videos about language courses, sport exercises, or styling tips were watched. Especially sport exercises were watched repeatedly. The question about the devices used to watch the learning and training videos revealed the following findings (multiple answers possible): 44 participants used the PC, 41 used a laptop, 25 used the TV, 25 used a tablet, and 25 used a smart-phone.

According to the fact that currently neither exists a commonly recognized structure for hypervideos, nor are hypervideos widespread and thus not known to many people, we used a more analog scenario to find out how the participants would navigate in contents. The scenario of searching the name of an unknown bird in an encyclopedia was introduced. This type of book is usually not read in a linear way and provides links between different entries, which is very similar to hyper-linked structures on the Web. The participants were asked how they would search for the bird. The most preferred structure was searching in an index, secondly they would just browse the book and thirdly they would use the table of contents. Using the table of contents for this task, the participants were asked how it should look like. They considered images of the bird with a short description as most useful. The grouping of the birds in their categories as well as images of the birds without text were also contemplated as very advantageous. Textual descriptions were acknowledged as less helpful. In another scenario, namely the trip to a foreign city, the participants were asked how they would note down the sightseeing route. Thereby the options of marking the sights and marking the taken path were preferred to numbering the sights consecutively, marking the taken path at crossroads, and noting down the street names.

Summarizing the results of the survey, it can be noted that 90 % of the participants know touch screens and smart-phones. Tablets are widespread but not always used regularly. Learning and training videos are well known among the participants, about 1/3 of them has watched these videos on a tablet. Regarding the table of contents, a structure of the entries is important, a combination of image and text are favored, and already visited contents should be marked somehow. These results match the findings of the expert group and will be implemented in the prototypes for the user test.

3.3 User Test with Prototype

The user test with the prototypes was conducted to find answers to the following questions:

- Is the table of contents used for navigation by the users?
- Is a structured or a graphical (menu-like) approach more clear for the users?
- Is the separation between a structured table of contents and the follow-up scenes comprehensible for every user?
- Are the used markers for already visited scenes understood correctly?
- Are the symbols large enough?
- How comfortable is the user handling the device (tablet)?

Study Method and Participants. The results from the expert group and the survey were implemented in software supported paper-like prototypes created with Balsamiq[8]. Two versions, the structured version and the graphical (menu-like) version were designed. An example of the structured prototype can be found in Fig. 2, an example of the graphical (menu-like) version is illustrated in Fig. 3.

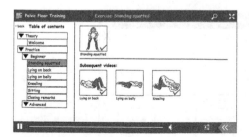
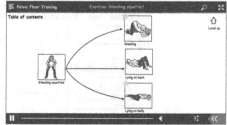

Fig. 2. Structured version of the prototype.

Fig. 3. Graphical (menu-like) version of the prototype.

---

[8] https://balsamiq.com/ (accessed January 25, 2015).

To answer the previously described questions, the test users got different tasks in a scenario with a medical training. The tasks included:

- Use of the table of contents as a help for navigation.
- Navigate through the video to the end scene.
- Recognize already visited video scenes.
- Recognize follow-up scenes in the representation.

The user test was conducted with a group of eight people ($N = 8$) all aged above 45. To make sure every tester had the same preconditions for the test, it was ensured that the participants did not meet between the tests. An introductory and explanatory text was read to the participants from a paper, then some basic questions about age and educational background were asked. After that, the tasks enlisted above were presented to the test users one after the other after a brief introduction to the GUI of the prototype. The users had to perform each task for both versions. While completing their tasks they were asked to comment what they were doing and why they were doing it with the think-aloud technique [16, p.256].

Result and Discussion. The user test revealed the following results: Both, volume and play/pause button were recognized correctly by all of the test users. The buttons for the table of contents, the search, and the full screen were recognized by half of the participants. The settings button was not recognized by anyone. The presentation of the follow-up scenes was problematic, connections were assumed that did not exist. A more clear separation may solve this problem. The currently selected scene was not recognized and should be highlighted. The video controls (play/pause, timeline, and volume control) were visible when the table of contents was open. This irritated the test users, because no connection between the table of contents and the video control existed. Accordingly, the video control panel should be hidden when the table of contents is displayed. Furthermore, it was suggested that the "<back" button should be removed from the top left side and an "X" should be added to the right upper corner of the view, which is well known from other applications. Already watched scenes should be

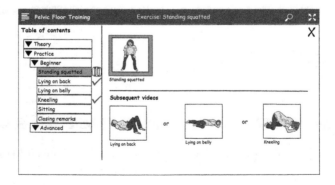

Fig. 4. Refined version of the paper prototype.

marked in the structure at the left side, instead on the scene thumbnails in the preview area. The structured version was preferred to the graphic version by the test users. This leads to the refined prototype in Fig. 4. It has less elements than the structured version in Fig. 2 which leads to a better overview.

4   Implementation

The implementation in the player software afforded extensions of the meta data format and the authoring tool. The existing meta data format described by Meixner and Kosch in [11] had to be extended. Besides a video for a scene, a thumbnail image for that video has to be referenced in the scene element in the XML file. Furthermore the authoring tool described by Meixner et al. in [12] needs to export a thumbnail for each scene. This was accomplished with an editor that allows the selection of the thumbnail.

The implementation of the table of contents in the player GUI was realized according to the results of the user test as described in the previous section and illustrated in Fig. 4. Thereby, the design was matched with the design of the existing player. A screen-shot of the implementation of the table of contents can be found in Fig. 5. The drop down fields of the prototype are implemented with the commonly known triangles. The current scene is highlighted in reversed colors. Instead of marking the thumbnail of the current video with a colored frame, it is displayed enlarged.

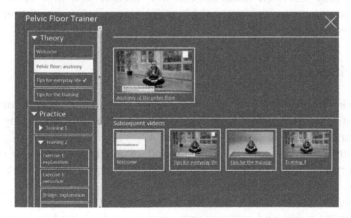

Fig. 5. Implemented version of the table of contents in the hypervideo player.

5   Final Evaluation

The implemented software was tested in a second user test.

Study Method and Participants. The test users had to perform similar tasks like those from the first test, but with the player and videos from a real scenario. The tasks included:

- Use of the table of contents as a help for navigation.
- Navigate trough the video to the end scene.
- Recognize already visited video scenes.

The second user test was also conducted with a group of eight people ($N = 8$) all aged above 45. None of the participants was a member of the first test user group. The same arrangements were made as in the first test (participants did not meet, introductory and explanatory text read from paper, basic questions). After that, the tasks enlisted above were presented to the test users one after the other after a brief introduction to the GUI of the prototype. The users had to perform each task on the tablet using the implemented software. In combination with that, the think-aloud technique [16, p.256] was used. Furthermore, the test users were asked to fill out the UEQ [9].

Result and Discussion. With the small number of answers, only a qualitative statement can be made, but the results indicate, that the implementation was evaluated positive in all categories of the UEQ [16, p.256], namely attractiveness, perspicuity, efficiency, dependability, stimulation, and novelty.

Only few adaptations were made in the software after the user test. The basic structure was not changed. The representation of some elements was further improved: the button areas were enlarged and the distance between the thumbnails of the follow-up scenes was increased. The test users furthermore preferred a fully unfolded table of contents when it was first opened, which speaks against a menu structure which unfolds step-by-step and confirms the result of the survey.

6   Conclusion

In this work, we presented an extended presentation of a table of contents for hypervideos. The design was tested in a real world medical training scenario with the target group of people older than 45. This user group is a particular challenge since they sometimes have limited experience in the use of mobile devices. Furthermore, age-related physical deficiencies have to be taken into consideration for this user group.

We designed the user interface for an extended table of contents in three steps. An expert group and a survey were used to get hints on how a prototypical implementation should look like. The findings thereof were implemented in two different paper prototyped versions of the display. These were then tested in a user test with an in-group design, where each user had to test and evaluate both versions. As a result it can be noted that a divided view is desired. The table of contents should be presented in an easy-to-touch version on the left side of the view. A preview on the selected scene and follow-up scenes extracted from the underlying scene graph should be presented on the right side of the view. Unnecessary buttons from the underlying video player should be hidden to limit confusion. These findings were implemented in the existing open source SIVA

HTML5 player and tested with a second group of users. This test only revealed minor problems which were implemented in the final GUI.

However, this work has some limitations. The GUI was only tested on 7" and 10" tablets in landscape mode. Accordingly, the design should also be usable on monitors, but the usage in portrait mode needs a revision of the design. This is part of our future work.

References

1. Apostu, S., Al-Nuaimi, A., Steinbach, E., Fahrmair, M., Song, X., Möller, A.: Towards the design of an intuitive multi-view video navigation interface based on spatial information. In: Proceedings of the 15th International Conference on Human-Computer Interaction with Mobile Devices and Services, MobileHCI 2013, pp. 103–112. ACM, New York, NY, USA (2013)
2. Cunha, B., Pedrosa, D., Goularte, R., Pimentel, M.: Video annotation and navigation on mobile devices. In: Proceedings of the 18th Brazilian Symposium on Multimedia and the Web, WebMedia 2012, pp. 261–264. ACM, New York, NY, USA (2012)
3. Gerjets, P., Kirschner, P.: Learning from multimedia and hypermedia. Technology-Enhanced Learning, pp. 251–272. Springer, The Netherlands (2009)
4. Girgensohn, A., Wilcox, L., Shipman, F., Bly, S.: Designing affordances for the navigation of detail-on-demand hypervideo. In: Proceedings of the Working Conference on Advanced Visual Interfaces, AVI 2004, pp. 290–297. ACM, New York, NY, USA (2004)
5. Grigoras, R., Charvillat, V., Douze, M.: Optimizing hypervideo navigation using a Markov decision process approach. In: Proceedings of the Tenth ACM International Conference on Multimedia, MULTIMEDIA 2002, pp. 39–48. ACM, New York, NY, USA (2002)
6. Höffler, T., Leutner, D.: Instructional animation versus static pictures: a meta-analysis. Learn. Instr. **17**(6), 722–738 (2007)
7. Hürst, W., Götz, G., Welte, M.: Interactive video browsing on mobile devices. In: Proceedings of the 15th International Conference on Multimedia, MULTIMEDIA 2007, pp. 247–256. ACM, New York, NY, USA (2007)
8. Karrer, T., Wittenhagen, M., Borchers, J.: Pocketdragon: a direct manipulation video navigation interface for mobile devices. In: Proceedings of the 11th International Conference on Human-Computer Interaction with Mobile Devices and Services, MobileHCI 2009, pp. 47:1–47:3. ACM, New York, NY, USA (2009)
9. Laugwitz, B., Held, T., Schrepp, M.: Construction and evaluation of a user experience questionnaire. In: Holzinger, A. (ed.) USAB 2008. LNCS, vol. 5298, pp. 63–76. Springer, Heidelberg (2008)
10. Leitner, M., Wolkerstorfer, P., Sefelin, R., Tscheligi, M.: Mobile multimedia: identifying user values using the means-end theory. In: Proceedings of the 10th International Conference on Human Computer Interaction with Mobile Devices and Services, MobileHCI 2008, pp. 167–175. ACM, New York, NY, USA (2008)
11. Meixner, B., Kosch, H.: Interactive non-linear video: definition and XML structure. In: Proceedings of the 2012 ACM Symposium on Document Engineering, DocEng 2012, pp. 49–58. ACM, New York, NY, USA (2012)
12. Meixner, B., Matusik, K., Grill, C., Kosch, H.: Towards an easy to Use authoring tool for interactive non-linear video. Multimedia Tools and Applications, pp. 1–26. Springer, US (2012)

13. Meixner, B.: Annotated interactive non-linear video - software suite, download and cache management. PhD thesis, Universität Passau (2014)
14. Miller, G., Fels, S., Ilich, M., Finke, M.M., Bauer, T., Wong, K., Mueller, S.: An end-to-end framework for multi-view video content: creating multiple-perspective hypervideo to view on mobile platforms. In: Anacleto, J.C., Fels, S., Graham, N., Kapralos, B., Saif El-Nasr, M., Stanley, K. (eds.) ICEC 2011. LNCS, vol. 6972, pp. 337–342. Springer, Heidelberg (2011)
15. Nguyen, C., Niu, Y., Liu, F.: Direct manipulation video navigation on touch screens. In: Proceedings of the 16th International Conference on Human-Computer Interaction with Mobile Devices and Services, MobileHCI 2014, pp. 273–282. ACM, New York, NY, USA (2014)
16. Rogers, Y., Sharp, H., Preece, J.: Interaction Design: Beyond Human - Computer Interaction. Wiley, New York (2011)
17. Shipman, F., Girgensohn, A., Wilcox, L.: Combining spatial and navigational structure in the hyper-hitchcock hypervideo editor. In: Proceedings of the Fourteenth ACM Conference on Hypertext and Hypermedia, HYPERTEXT 2003, pp. 124–125. ACM, New York, NY, USA (2003)
18. Tonndorf, K., Handschigl, C., Windscheid, J., Kosch, H., Granitzer, M.: The effect of non-linear structures on the usage of hypervideo for physical training. In: Proceedings of the 2015 IEEE International Conference on Multimedia and Expo (ICME 2015), IEEE, pp. 1–6 (2015)
19. Zahn, C., Barquero, B., Schwan, S.: Learning with hyperlinked videosdesign criteria and efficient strategies for using audiovisual hypermedia. Learn. Instr. **14**(3), 275–291 (2004)

Poster Papers

Reverse Testing Image Set Model Based Multi-view Human Action Recognition

Z. Gao[1,2(✉)], Y. Zhang[1,2], H. Zhang[1,2], G.P. Xu[1,2], and Y.B. Xue[1,2]

[1] Key Laboratory of Computer Vision and System, Tianjin University of Technology, Ministry of Education, Tianjin 300384, China
zangaonsh4522@gmail.com, zhangyantjut@126.com, {hzhang62,yanbingxue}@163.com, xugp2008@aliyun.com
[2] Tianjin Key Laboratory of Intelligence Computing and Novel Software Technology, Tianjin University of Technology, Tianjin 300384, China

Abstract. Recognizing human activities from videos becomes a hot research topic in computer vision, but many studies show that action recognition based on single view cannot obtain satisfying performance, thus, many researchers put their attentions on multi-view action recognition, but how to mine the relationships among different views still is a challenge problem. Since video face recognition algorithm based on image set has proved that image set algorithm can effectively mine the complementary properties of different views image, and achieves satisfying performance. Thus, Inspired by these, image set is utilized to mine the relationships among multi-view action recognition. However, the studies show that the sample number of gallery and query set in video face recognition based on image set will affect the algorithm performance, and several ten to several hundred samples is supplied, but, in multi-view action recognition, we only have 3–5 views (samples) in each query set, which will limit the effect of image set.

In order to solve the issues, reverse testing image set model (**called RTISM**) based multi-view human action recognition is proposed. We firstly extract dense trajectory feature for each camera, and then construct the shared codebook by k-means for all cameras, after that, Bag-of-Word (**BoW**) weight scheme is employed to code these features for each camera; Secondly, for each query set, we will compute the compound distance with each image subset in gallery set, after that, the scheme of the nearest image subset (**called RTIS**) is chosen to add into the query set; Finally, **RTISM** is optimized where the query set and **RTIS** are whole reconstructed by the gallery set, thus, the relationship of different actions among gallery set and the complementary property of different samples among query set are meanwhile excavated. Large scale experimental results on two public multi-view action3D datasets - **Northwestern UCLA** and **CVS-MV-RGBD-Single**, show that the reconstruction of query set over gallery set is very effectively, and **RTIS** added into query set is very helpful for classification, what is more, the performance of **RTISM** is comparable to the state-of-the-art methods.

Keywords: Action recognition · Multi-view · Image set · Reverse testing

© Springer International Publishing Switzerland 2016
Q. Tian et al. (Eds.): MMM 2016, Part I, LNCS 9516, pp. 397–408, 2016.
DOI: 10.1007/978-3-319-27671-7_33

1 Introduction

Recognizing human activities from videos becomes a hot research topic in computer vision field, whose application includes moving object detection, moving object classification, human motion tracking and behavior recognition and understanding. So far, researchers [1–10] have proposed a lot of action recognition algorithms, for example, Turaga et al. [1] and Ke et al. [2] have discussed how to discover an efficient representation for human actions, how to extract the discriminant features from real-time video and how to construct the action recognition models. These papers [1, 2] demonstrate that single view action dataset is utilized in most the-state-of-art methods, but in real-world applications, we often observe the same object by different view spaces, which are highly related but sometimes look different from each other. Action samples from multi-view observation spaces are shown in Fig. 1.

From it, we can observe that in a multi-view setting, each view of the data not only contain some common knowledge, but also different knowledge is included where other views do not have. Therefore, multiple views can be employed to comprehensively and accurately describe the data. Thus, many multi-view learning methods have been recently proposed [3–15] to dig the complementary and correlative information of different views. For example, In Song et al. [3], multi-view latent variable discriminative models was presented to jointly learn both view-shared and view-specific sub-structures to capture the interaction between views for human action recognition; Cai et al. [4] proposed a new global representation, Multi-view Super Vector, which is composed of relatively independent components derived from a pair of descriptors; a multi-view discriminant analysis method [5] was proposed to seek for a discriminant common space by jointly learning multiple view-specific linear transforms for robust object recognition from multiple views; Multi-task learning and model building [6–8] were proposed to dig the relationships among different views; Multiple sequential or views were jointly learned in [9–11] to mine the relationships among different views. In fact, multi-view learning algorithms are also concerned in other research domains, such as, multi-target tracking [12], multimedia analysis [13], 3D model retrieval [14] and Multi-modality fusion [15]. Although many multi-view action recognition algorithms have been proposed, the current methods focus on the single-view learning, which was the lack of discovering correlations among multiple views, and the latent relationship is always heuristically defined with prior knowledge, but seldom literatures work on adaptive latent correlation discovery by model learning.

Recently, classification based on image sets has attracted increasingly interest in the computer vision and pattern recognition community, especially in video face recognition [16, 17], whose performance has shown superior performance to single image based face recognition, since many sample images in query set and gallery set belong to the same class and cover large variation in the object's appearance due to camera pose change, different lighting conditions, or different facial expressions [16].

In fact, many parametric and non-parametric approaches for image set modeling are proposed [18], however, no matter how the set is modeled, in almost all the previous works [16–18], several ten to several hundred samples are supplied, but in the multi-view action recognition, we often have 3–5 views (samples) for each query set, which will limit the effect of image set. What is worse, the query set is compared to each of the gallery sets separately, and then classified to the class closest to it. Such a classification scheme does not consider the correlation between gallery sets, like the nearest neighbor or nearest subspace classifier in single image based face recognition.

To solve above issues, and we are inspired by SRC (Sparse Representation Classification) [19] and image set based face recognition [16–18], reverse testing image set model based multi-view human action recognition model is proposed. To add the number of the query set, the shared codebook for each camera is constructed, which can made samples from different views have the same feature structure, and then for each query set, we will compute the compound distance with each image subset in gallery set, and the nearest image subset is chosen to add into the query set. And then, in order to excavate the hidden relationship of different samples and different actions, the original query set and reverse testing image set are whole reconstructed by the gallery set, thus, the relationship of different actions among the gallery set and the complementary property of different samples among the reverse query set are meanwhile excavated by model learning.

The main contributions lie in the following aspects: (1) The consistency properties of different views is explored by the shared codebook, which can made that the features from different views have the same meaning, and the relationship among them can be simply excavated; (2) The reverse testing image set approach is proposed to remedy the lacking of the samples in the query set. (3) The problem of exploring the complementary properties among different views and different actions is turned into reverse testing image set model learning. The rest of this paper is organized as follows: In Sect. 2, the proposed approach is introduced, in Sect. 3, experiment and analysis is discussed. Finally, the conclusions are shown in Sect. 4.

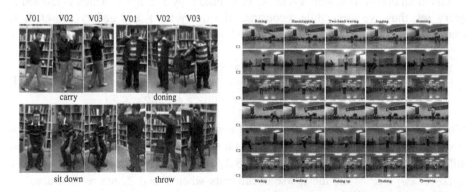

Fig. 1. Action samples from different views and different action datasets, and the left is from Northwestern **UCLA** dataset where 'V0#' means that the samples are from view #, and 'carry', 'doning', 'sit down' and 'throw' means that different actions are performed and each action has three samples. And the right is from **CVS-MV-RGBD-Single** dataset, where 'C#' means that the samples are from view #, and each object and each action are observed by three cameras.

2   Reverse Testing Image Set Model Based Multi-view Human Action Recognition

2.1   The Scheme of Adding Samples in Query Set

In the task of video face recognition, there are many frames in each query, which belong to the same class and cover large variation in the object's appearance, and can supply much more complementary information. In fact, we have evaluated a lot of existing image set based face recognition on Honda face dataset[1], and experimental results show that the accuracy will improve with the increase of the sample number of the query and gallery set. However, in action recognition, each video be considered as a sample, thus, even in multi-view setting, we only can obtain three $\sim$ five samples which will limit the effect of image set. Lucky, we have a lot of samples in the gallery set, which can be added into the query set, but what kinds of image set should be joined. In fact, if the same action samples are added, the performance of image set algorithms can be improved, but if different action samples are chosen, the accuracy of most of image set algorithms will be decreased.

Supposed that we can observe K action classes from V views, thus, they can simultaneously obtain V samples for each observation and each action, and then the feature is extracted for each sample, whose dimension is d_n, thus, for each observation, the image set is naturally constructed. If there are N observations for all actions in the training dataset, where each action is taken by different people and different times, we can denote each observation by $Z_n \in R^{d_n \times V}(n = 1, 2, \ldots N)$, and denote the training set by $Z = \{Z_1, Z_2, \ldots Z_n \ldots Z_N\} \in R^{d_n \times N}$ which is considered as the gallery set. To represent the gallery set into much more compact set, dictionary learning methods such as KSVD are employed for each action, thus, $D = \{D_1, D_2, \ldots D_k \ldots, D_K\} \in R^{d_n \times M}$ is constructed where D_k indicates the dictionary for action k, and M is much smaller than N. Given an observation with V-views, we denote it by $Y \in R^{d_n \times V}$ which is labeled as query set, thus, we meanwhile reconstruct the query set Y and gallery subset Z_n to measure their similarities. Therefore, we define the energy function as follows:

$$<\hat{b}, \hat{c}> = \arg\min_{(b,c)}\{||Y\alpha - Db||_2^2 + ||Z_n\alpha - Dc||_2^2$$
$$+ \lambda_1||b||_1 + \lambda_2||c||_1 + \lambda_3 w_n||b - c||_2^2\} \tag{1}$$

Where $\lambda_i (i = 1, 2)$ is a scale constant and control the weights of each regularization term, and $\alpha = [\alpha_1, \alpha_2, \ldots \alpha_V] \in R^{V \times 1}$ also is a scale constant, which is utilized to fuse different views, and its weight is equal to $1/V$. $b = [b_1, b_2, \ldots b_K] \in R^{M \times 1}$ and $c = [c_1, c_2, \ldots c_K] \in R^{M \times 1}$ are the sparse coefficients when Y and Z_n are reconstructed by dictionary D, $b_i \in R^{N_i \times 1}$ and $c_i \in R^{N_i \times 1}$ are the sparse sub-coefficients corresponding to sub-dictionary D_i. In this object function, there are three kinds of regularization terms:

---

[1] http://vision.ucsd.edu/~leekc/HondaUCSDVideoDatabase/HondaUCSD.html.

(1) **Fidelity term based on image set** $-||Y\alpha - Db||_2^2$ and $||Z_n\alpha - Dc||_2^2$, the query set and one image subset in gallery set are meanwhile reconstructed, and we will measure the fitting error of the query set and reverse image subset.

(2) **Sparsity term** - $||b||_1$ and $||c||_1$ these terms are utilized to regularize the sparse coding coefficients and control the sparsity of them. It is well known that the ridge penalty with strict convexity can preserve consistence for the decomposed coefficient.

(3) **Interconnection term** $-w_n||b - c||_2^2$, the term is utilized to control the difference of reconstruction coefficients, and the more similar between query set Y and sub-gallery set Z_n, the less difference between reconstruction coefficient b and c should has. As for w_n, it is defined as follows:

$$w_n = {}^1\!/\!_V \sum_{i=1}^{V} \min_{z \in Z_n} ||y_i - z||_2^2 \tag{2}$$

Where y_i is the sample feature in query set Y, and V is the number of samples in this observation. In addition, z is the sample feature of sub-gallery set Z_n.

After obtaining the spare coefficients, we will compare the differences of them, and if Y and Z_n are belong to the same action, the difference of b and c will be small, or it will be very different. Then we define the following function to evaluate the difference between the query set and image subset in gallery set.

$$error(k) = ||Y\alpha - D_k\hat{b}_k||_2^2 + ||Z_n\alpha - D_k\hat{c}_k||_2^2 \tag{3}$$

Thus, the minimum of $error(k)$ is adopted as the similarity evaluation between Y and Z_n, and we repeat the above processing for each image subset in gallery set, then, the minimum of them is chosen as the reverse testing image set whose image set will be added into the query set.

2.2 Reverse Testing Image Set Model Based Multi-view Action Recognition Model

On the basis of Sect. 2.1, the reverse testing image set which comes from the gallery set, is put into the query set, thus, we can obtain enough samples in query set, which can supply much more complementary information to the original query set. However, we cannot make sure that we always choose the same class image subset to add into the query set. If done, the updating query set will contains much noise, thus, we need reduce or even remove the influence of these noise samples. Therefore, the following object function is defined:

$$<\hat{a}, \hat{b}, \hat{c}> = \arg \min_{(\alpha,b,c)} \{||Y\alpha - Db||_2^2 + \lambda||Z_n\beta - Dc||_2^2$$

$$+ \lambda_0||\alpha||_1 + \lambda_1||b||_1 + \lambda_2||c||_1 + \lambda_3||\beta||_1\} \tag{4}$$

$$s.t. \quad \sum a_i = 1, \quad \sum \beta_i = 1$$

Y indicates the updating query set where **RTIS** is added into the original query set, and the number of samples in it is larger than the original query set, $\alpha = \{\alpha_1, \alpha_2, \ldots \alpha_{V'}\}' \in R^{V' \times 1}$ and $\beta = \{\beta_1, \beta_2, \ldots \beta_{V'}\}' \in R^{(V'-V) \times 1}$ are the hull coefficients where we construct the convex hull for the updating query set and **RTIS** respectively, and the meaning of other parameters in Eq. (4) is the same to Eq. (1). In Eq. (4), there are three kinds of regularization terms:

(1) **Fidelity term based on image set** $-||Y\alpha - Db||_2^2$ and $||Z_n\beta - Dc||_2^2$, the fidelity term also be used to measure the reconstruction error of the query set and reverse testing image set, at the same time, we also need compare the difference of b and c by $||Y - D_k\hat{b}_k||_2^2$ and $||Z_n - D_k\hat{c}_k||_2^2$. Since we hope the class of Y and Z_n will be same, thus, the difference of their representation coefficients will be small, thus, λ will be set to one, $\alpha = \{\alpha_1, \ldots \alpha_{V'}\}$ will be obtained by optimization. However, if it is too large, λ and the corresponding weight $\alpha = \{\alpha_4, \ldots \alpha_{V'}\}$ will be zero, $\alpha = \{\alpha_1, \ldots \alpha_V\}$ will be obtained by optimization.

(2) **Convex hull term** - $||\alpha||_1$ and $||\beta||_1$ in order to measure the relationship between query set and gallery set, we construct the convex hull $Y\alpha$ and $Z_n\beta$ where the sum of each sub-term in α or β is equal to 1. In Eq. (4), the hull $Y\alpha$ of the query set Y is collaboratively represented over the gallery sets. However, the coefficients in α will make the samples in Y be treated differently in the representation and the subsequent classification process. The parameter β has the same means with α.

(3) **Sparsity term** - $||b||_1$ and $||c||_1$, the meanings of sparse coefficients are same to Eq. (1).

Suppose that the coefficient vectors $\hat{\alpha}$ $\hat{b}$ and $\hat{c}$ are obtained by solving Eq. (4), then we can write $\hat{b} = [\hat{b}_1, \hat{b}_2, \ldots \hat{b}_K]$ where $\hat{b}_k$ is the sub-vector of coefficients associated with gallery set D_k, and then we also adopt the representation residual of hull $Y\alpha$ by each set D_k to determine the class label of $Y\alpha$. The classifier is defined as follows:

$$error(k) = \{||Y\alpha - D_k\hat{b}_k||_2^2 + \lambda||Z_n\beta - D_k\hat{c}_k||_2^2\} \tag{5}$$

The class of the test samples can be inferred by choosing the action class k with the minimum $error(k)$.

2.3   Solution and Inference

In the solving of the energy minimization of the object function, there are three variables -α, b, c and β, thus, we employ the alternating minimization method, which is very efficient to solve multiple variable optimization problems [20]. Where at each iteration, we first fix the one variable, and then optimize the other variables [20], at the same time, we update these variables by repeated iterative.

3 Experimental and Discussion

In order to evaluate the proposed algorithms, we extensively perform the experiments on two benchmark 3D activity datasets including **CVS-MV-RGBD-Single 3D** [11] and **Northwestern UCLA Dataset** [21].

3.1 Experimental Setting

(1) Feature Extraction For each sample, the popular feature- Dense trajectories, is extracted, and then for different views, the shared codebook is constructed by K-Means clustering, after that, Bag-of-Words scheme is employed to normalize each video. The shared codebook sizes of **UCLA** and **CVS-MV-RGBD-Single** datasets are 1000 and 500 respectively.

(2) **Parameter Setting** We will strictly follow the experimental setup for the evaluation scheme in the relative reference [11, 21]. As for the parameters in **SRC**, **CRC**, **and RTISM** models are selected by cross validation within the range of [1, 0.5, 0.1, 0.05, 0.01, 0.005, 0.001, 0.0005, 0.0001]. In addition, average accuracy is utilized as evaluation criteria.

3.2 Evaluation the Relationships of Different Views

In order to discuss the relationships among multi-view samples, we first assess the performance of view to view setting, where the training and testing samples are from the same camera, and then evaluate the effectiveness of cross view, and finally, we will add the training samples step by step to discuss the effectiveness of the number of training samples. The experimental results on **UCLA** and **CVS-MV-RGBD-Single** are shown in Table 1. In the Table, C1 + C2 + C3 indicate that samples from different views are put together in the training processing to construct new training dataset.

From Table 1, we can observe that the accuracy cannot obtain stable improvement with the increase of samples in the training dataset, for example, in **UCLA** dataset, when the training dataset comes from C1, C1 + C2, and C1 + C2 + C3 respectively, and then models are evaluated on C1 camera, their accuracies are 76.4 %, 72.2 % and 75.7 % respectively. However, in **CVS-MV-RGBD-Single** dataset, the same experimental method is utilized, and they can obtain 90.3 %, 92.5 % and 92.5 % respectively. Thus, that is to say, different multi-view datasets have different relationship, sometimes, the adjacent view has is a strong correlation, but sometimes, they also have a lot of difference. We also can know that samples from different views have some relationships, but their relationships will vary with the change of different cameras setting

Table 1. Evaluation the relationship among different views on different datasets by SVM model, and left column is **UCLA** dataset, but right column is **CVS-MV-RGBD-Single** dataset. In both tables, the leftmost column denotes the training samples are from corresponding view where the model is built on this dataset, and the top row indicates that the test samples are from corresponding view.

SVM	C1	C2	C3	C1+C2	C1+C3	C2+C3	C1+C2+C3	Aver	SVM	C1	C2	C3	C1+C2	C1+C3	C2+C3	C1+C2+C3	Aver
C1	76.4	16.9	11.7	43.4	44.1	11.3	32.8	33.8	C1	90.3	84.1	81.9	87.2	86.1	83	85.4	85.4
C2	17	62.7	10.8	39.6	13.9	36.8	30.2	30.1	C2	88	89.4	85.6	88.7	86.8	87.5	87.7	87.7
C3	10.1	47.3	69.1	28.5	39.2	58.1	42	42.0	C3	84.5	83.8	91.6	84.2	88.1	87.7	86.6	86.6
C1+C2	72.2	62	58.9	66.9	65.4	60.4	64.4	64.3	C1+C2	92.5	91.7	84.7	92.1	88.6	88.2	89.6	89.6
C1+C3	72.6	45	68.6	58.8	70.5	56.8	62.1	62.1	C1+C3	91.7	86.1	92.8	88.9	92.3	89.5	90.2	90.2
C2+C3	12.5	65.8	69.5	39.2	41	67.7	49.3	49.3	C2+C3	90.2	91.1	92.3	90.7	91.3	91.7	91.2	91.2
C1+C2+C3	75.7	68.5	70.9	72.1	73.3	69.7	71.7	71.7	C1+C2+C3	92.5	92.2	92.7	92.4	92.6	92.6	92.5	92.5

and the number of cameras. Thus, the sample stacking scheme is not robust and efficient, thus, image set is proposed to mine the relationships among different views, which will be introduced in the following sections.

3.3   Evaluation the Effect of the Number of Samples in Query Set - RTIS

In order to assess the effect of the number of samples in query set, we perform the experiments on **CVS-MV-RGBD-Single** dataset. Thus, firstly, we randomly choose a query set, and then employ the **RTIS** scheme to select reverse testing set in the gallery set, finally, the original query set and the reverse testing image set are reconstructed respectively, and the sparse representation coefficients, the reconstruction errors and the hull coefficients α are given in Fig. 2, Tables 2 and 3 respectively.

In our experiment, the class of query set is belong to action 4, and the non-zero coefficients should be nearby 200, but from left figure in Fig. 2, we can observe that the reconstruction sparse coefficients distribute at different locations, and the coefficients not only locate in nearby 200, but also they distribute at dictionary 450 and 600, what is worse, the size of them is very similar, which will be confused for classification. However, in the right figure in Fig. 2, although the coefficient distribution is similar with left figure, but the reconstruction sparse coefficients in nearby 200 are much bigger than others, which will be very helpful for classification. From Table 2, we also can see that the minimum reconstruction error in left figure is 0.377, which is belong to action 8, but its minimum reconstruction error in right figure achieves 0.259, and it is assigned to action 4 whose label is same to ground truth. Thus, the reverse testing image set has much more distinguishing ability In addition, we can find from Table 3 that with the increase of samples in query set, its hull coefficients changes which will change the roles in the reconstruction.

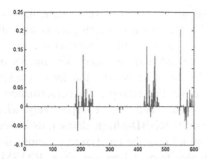 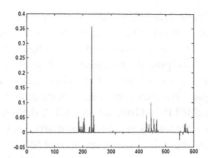

Fig. 2. Different sparse representation coefficients on **CVS-MV-RGBD-Single** dataset when the reverse testing image set is added into the query set or not. In left figure, the original query set is adopted, but the updating query set is utilized in right figure. In this Figure, horizontal axis is the index of dictionaries, and the vertical axis is the reconstruction coefficients.

Table 2. The reconstruction errors when different query sets are utilized on **CVS-MV-RGBD-Single** dataset, and A_k means that the reconstruct error where only the sub-vector of coefficients associated with gallery set D_k is adopted to reconstruct the query set.

Action	A1	A2	A3	A4	A5	A6	A7	A8	A9	A10
Error-IS	0.695	0.695	0.696	**0.527**	0.695	0.702	0.69	**0.377**	0.696	0.588
Error-RTIS	0.648	0.649	0.649	**0.259**	0.649	0.651	0.649	**0.474**	0.649	0.634

Table 3. The change of hull coefficients α when different query set is utilized on **CVS-MV-RGBD-Single** dataset, and S1 $\sim$ S6 means the sample index in the query set, **a-IS** and **a-RTIS** indicate that the distribution of hull coefficients α when the original query set and the updating query set are employed respectively.

Para a	S1	S2	S3	S4	S5	S6
a-IS	0.220	0.168	0.611	--	--	--
a-RTIS	0.111	0.144	0.052	0.082	0.200	0.410

3.4 Performance Evaluation of the Proposed Algorithm

In order to assess our proposed algorithms, in our experiments, we will perform experiments by three steps as follows: (1) **IS** -Image set, the original query set is employed, and then the query set is reconstructed by gallery set; (2) **RTISM** – Reverse testing image set model, in this approach, the sub image set in gallery set are added into the original query set by the scheme introduced in Sect. 2.1 to construct new query set, and then reconstruct the new query set by gallery set; (3) **RTISM-Optimal**, in this algorithm, we also construct new query set, and then reconstruct the query set by gallery set, but in the processing of choosing samples, we will select the same label samples with the test sample from the gallery set. In addition, **SRC** and **CRC** classification schemes are also assessed, and their results are shown in Table 4.

From Table 4, we can observe that image set approach is very efficient, whose accuracy is better than that of **SRC, CRC** and **SVM**, what is more, its performance also is much better than that of sample stacking and features connection. As for **RTISM-Optimal**, the same label samples are added into the query set, and its performance can obtain a big enhancement to **IS**. In other words, the idea of adding samples into query set is very efficient and effective. For example, the performances of **IS** and **RTISM-Optimal** on UCLA dataset are 82.6 % and 92.3 % respectively, whose improvement reaches 10 %, similarly, on **CVS-MV-RGBD-Single** dataset, its increase obtain about 5 %. However, in real life, we cannot know the label of test sample, thus, we cannot always choose the same label samples from gallery set, thus, **RTISM** is proposed to select samples from gallery set, whose performance is given in Table 4. From it, we can observe that although **RTISM** cannot obtain the same performance with **RTISM-Optimal**, but it also is much better than **IS, SRC, CRC** and **SVM**. Thus, we can conclude that the proposed approach is effective and efficient.

Table 4. Recognition accuracies of our method compared to the state-of-the-art method, and left column is **UCLA** Dataset, but right column is **CVS-MV-RGBD-Single** dataset

Method	Accuracy (%)	Method	Accuracy (%)
RTISM-(optimal)	92.3	RTISM- (optimal)	99.5
RTISM	85.7	RTISM	96.3
IS	82.6	IS	94
CRC (C1_C2_C3)	79.1	CRC (C1_C2_C3)	93.2
CRC (C1+C2+C3)	72.6	CRC (C1+C2+C3)	91.6
SRC (C1_C2_C3)	77.5	SRC (C1_C2_C3)	93.7
SRC (C1+C2+C3)	72.3	SRC (C1+C2+C3)	91.2
SVM (C1+C2+C3)	71.7	SVM (C1_C2_C3)	93.5
SVM (C1_C2_C3)	78.4	SVM (C1+C2+C3)	92.5
Poselet [22]	54.9	K-SVD [11]	80.47
Mixture of DPM [21]	74.8	DK-SVD [11]	81.25
MST-AOG w/o Low-S [21]	78.9	GM-GS-DSDL [11]	92.19
MST-AOG w Low-S [21]	81.6		

4   Conclusions

In this paper, reverse testing image set model based multi-view human action recognition is proposed. From large scale experimental results on two public multi-view action3D datasets - **Northwestern UCLA** and **CVS-MV-RGBD-Single**, we can know that different views have highly related, but the scheme of samples stacking and the approach of features connection are not effective and stable. However, image set approach is very helpful for mining the relationships among different views, and the reconstruction of query set over the gallery set is very effectively. What is more, the samples number of query set will affect its performance, and the idea of adding samples into query set is very efficient and effective, and then the proposed **RTIS** is very helpful for classification, what is more, the performance of **RTISM** is comparable to the state-of-the-art methods.

Acknowledgments. This work was supported in part by the National Natural Science Foundation of China (No. 61572357, No. 61502337, No. 61472275, No. 61201234, No. 61202168), Tianjin Municipal Natural Science Foundation (No. 14JCZDJC31700, No. 13JCQNJC0040), Tianjin Education Committee science and technology development Foundation (No. 20120802).

References

1. Turaga, P., Chellappa, R., Subrahmanian, V.S., Udrea, O.: Machine recognition of human activities: a survey. IEEE Trans, Circ. Syst. Video Technol. **18**(11), 1473–1488 (2008)
2. Ke, S.-R., Thuc, H.L.U., Lee, Y.-J., Hwang, J.-N., Yoo, J.-H., Choi, K.-H.: A review on video-based human activity recognition. Computers **2**, 88–131 (2013)
3. Song, Y., Davis, R.: Multi-view latent variable discriminative models for action recognition. In: CVPR 2012, pp. 1–8 (2012)
4. Cai, Z., Wang, L., Peng, X.: Multi-view super vector for action recognition. In: CVPR 2014, pp. 1–8 (2014)
5. Kan, M., Shan, S., Zhang, H., Lao, S., Chen, X.: Multi-view discriminant analysis. In: Fitzgibbon, A., Lazebnik, S., Perona, P., Sato, Y., Schmid, C. (eds.) ECCV 2012, Part I. LNCS, vol. 7572, pp. 808–821. Springer, Heidelberg (2012)
6. Liu, A., Su, Y., Jia, P., Gao, Z., Hao, T., Yang, Z.: Multipe/single-view human action recognition via part-induced multi-task structural learning. IEEE Trans. Cybern. **45**(6), 1194–1208 (2015)
7. Gao, Z., Zhang, H., Liu, A., Xue, Y., Xu, G.: Human action recognition using pyramid histograms of oriented gradients and collaborative multi-task learning. KSII Trans. Internet Inf. Syst. **8**(2), 483–503 (2014)
8. Liu, A., Xu, N., Su, Y., Lin, H., Hao, T., Yang, Z.: Single/multi-view human action recognition via regularized multi-task learning. Neurocomputing **151**(2), 544–553 (2015)
9. Gao, Z., Zhang, H., Xu, G.P., Xue, Y.B.: Multi-perspective and multi-modality joint representation and recognition model for 3D action recognition. Neurocomputing **151**(2), 554–564 (2015). doi:10.1016/j.neucom.2014.06.085
10. Liu, A., Su, Y., Nie, W., Yang, Z.: Jointly learning multiple sequential dynamics for human action recognition. PLoS ONE **10**(7), e0130884. doi:10.1371/journal.pone.0130884
11. Gao, Z., Zhang, H., Xu, G-P., Xue, Y.-B., Hauptmann, A.G.: Multi-view discriminative and structure dictionary learning with group sparsity for human action recognition. Sig. Process. (2014). doi:10.1016/j.sigpro.2014.08.034
12. Nie, W., Liu, A., Su, Y., et al.: Single/cross-camera multiple-person tracking by graph matching. Neurocomputing **139**, 220–232 (2014)
13. Gao, Z., Zhang, L., Chen, M., Hauptmann, A., Zhang, H., Cai, A.: Enhanced and hierarchical structure algorithm for data imbalance problem in semantic extraction under massive video dataset. Multimedia Tools Appl. **68**(3), 641–657 (2014)
14. Liu, A., Wang, Z., Nie, W., Su, Y.: Graph-based characteristic view set extraction and matching for 3D model retrieval. Inf. Sci. (2015). doi:10.1016/j.ins.2015.04.042
15. Gao, Z., Song, J., Zhang, H., Liu, A., Xu, G., Xue, Y.: Human action recognition via multi-modality information. J. Electr. Eng. Technol. **9**(2), 739–748 (2014)

16. Hu, Y., Mian, A.S., Owens, R.: Sparse approximated nearest points for image set classification. In: IEEE Conference on Computer Vision and Pattern Recognition, pp. 121–128. IEEE (2011)
17. Cui, Z., Shan, S., Zhang, H., Lao, S., Chen, X.: Image sets alignment for video-based face recognition. In: IEEE Conference on Computer Vision and Pattern Recognition, pp. 2626–2633. IEEE (2012)
18. Chen, Y.-C., Patel, V.M., Phillips, P.J., Chellappa, R.: Dictionary-based face recognition from video. In: Fitzgibbon, A., Lazebnik, S., Perona, P., Sato, Y., Schmid, C. (eds.) ECCV 2012, Part VI. LNCS, vol. 7577, pp. 766–779. Springer, Heidelberg (2012)
19. Wright, J., Yang, A., Ganesh, A., Sastry, S., Ma, Y.: Robust face recognition via sparse representation. IEEE Trans. Pattern Anal. Mach. Intell. 31(2), 210–227 (2009)
20. Gunawardana, A., Byrne, W.: Convergence theorems for generalized alternating minimization procedures. J. Mach. Learn. Res. 6, 2049–2073 (2005)
21. Wang, J., Nie, X., Xia, Y., Wu, Y., Zhu, S.-C.: Cross-view action modeling, learning, and recognition. In: Proceedings of IEEE Conference on Computer Vision and Pattern Recognition (CVPR) (2014)
22. Maji, S., Bourdev, L., Malik, J.: Action recognition from a distributed representation of pose and appearance. In: CVPR, IEEE, June 2011 (2, 6, 7, 8)

Face Image Super-Resolution Through Improved Neighbor Embedding

Kebin Huang[1,5], Ruimin Hu[1,2,3]($\boxtimes$), Junjun Jiang[4], Zhen Han[1,2,3],
and Feng Wang[5]

[1] National Engineering Research Center for Multimedia Software,
Computer School of Wuhan University, Wuhan, China
hrm1964@163.com
[2] Collaborative Innovation Center of Geospatial Technology,
Wuhan University, Wuhan, China
[3] Research Institute of Wuhan University, Shenzhen, China
[4] School of Computer Science, China University of Geosciences, Wuhan, China
[5] Department of Digital Media Technology, Huanggang Normal University,
Huangzhou, China

Abstract. In the process of investigating a case, face image is the most interesting clue. However, due to the limitations of the imaging conditions and the low-cost camera, the captured face images are often Low-Resolution (LR), which cannot be used for criminal investigation. Face image super-resolution is the technology of inducing a High-Resolution (HR) face image from the observed LR face image. It has been a topic of wide concern recently. In this paper, we propose a novel face image super-resolution method based on Tikhonov Regularized Neighbor Representation, which is called TRNR for short. It can overcome the technological bottlenecks (e.g., instable solution) of the patch representation problem in traditional neighbor embedding based image super-resolution method. Specially, we introduce the Tikhonov regularization term to regularize the representation of the observation LR patch, which can give rise to a unique and stable solution for the least squares problem and produce detailed and discriminant HR faces. Extensive experiments on face image super-resolution are carried out to validate the generality, effectiveness, and robustness of the proposed algorithm. Experimental results on the public FEI face database show that the proposed method plays a better subjective and objective performance, which can recover more fine structures and details from an input low-resolution image, when compared to previously reported methods.

Keywords: Super-resolution · Neighbor embedding · Tikhonov regularized · Low-resolution

1 Introduction

In real-world scenarios, the low-resolution (LR) images are generally captured in many electronic imaging applications, such as surveillance video, consumer photographs, remote sensing, magnetic resonance (MR) imaging and video standard

© Springer International Publishing Switzerland 2016
Q. Tian et al. (Eds.): MMM 2016, Part I, LNCS 9516, pp. 409–420, 2016.
DOI: 10.1007/978-3-319-27671-7_34

conversion [1]. The resolution of captured images is limited by the image charge-coupled device (CCD) and CMOS image sensors, the hardware storage and other constraints in electronic imaging systems. However, high-resolution (HR) images or videos are usually desired and often required for subsequent image processing and analysis in most real applications. Image super-resolution reconstruction, as an effective way to solve this problem, aims to reconstruct HR images from the observed LR images. It increases high-frequency components and removes the undesirable effects, e.g., the resolution degradation, blur and noise. In this paper, we mainly focus on the face image super-resolution (sometimes also called face hallucination) problem. In the following, we will review some representative works in this field. For a more complete and comprehensive survey of image super-resolution technology, please refer to the work in [2].

In 2000, Baker and Kanade [3] developed a learning-base method named "face hallucination", and this is the pioneering work on face image super-resolution. Given an input LR image, this approach infers the high frequency components from a parent structure with the assistance of training samples. Abandoning the MRF (Markov random field) assumption as in [4], the super-resolution of this method is established based on training images (pixel by pixel) using Gaussian, Laplacian and feature pyramids. Since the introduction of this work, a number of different methods and models have been introduced for estimating the image information lost in the down-sampling process. These systems differ in how they model the HR image. A successful face image super-resolution algorithm must meet the global constraints, which means that the results must have common human characteristics, and the local constraints, which means that the results must have specific characteristics of a particular face image. To fulfill these two constraints, Liu et al. [5] described a two-step approach integrating a principal component analysis (PCA) based global parametric model and a patch-based non-parametric Markov network. In the first step, the relationship between the HR face images and their smoothed, down-sampled LR ones is learned to obtain the global HR image. In the second step, the residual between an original HR image and a reconstructed one is compensated to learn the residual image. Both the original work in [3] and later work by Liu et al. in [5] have shown that knowing that the image contains a face makes it possible for a super-resolution system to perform much better because the system can leverage regularities in face appearance to recover more detail than could be created from a general image model.

However the above methods use probabilistic models and are based on an explicit resolution reduction function, which is sometimes difficult to acquire in practice. Instead of using a probabilistic model, Wang and Tang [6] proposed a face image super-resolution approach using eigentransformation algorithm, which can well capture the structural similarity of face images. This method treats the image super-resolution problem as the transformation between LR images and their HR counterparts. An input LR image can be expressed as a linear combination of the LR images in the training set by using principal component analysis (PCA). Then, the corresponding target HR image can be reconstructed by linear combination of those corresponding HR images in the training set while keeping the mixture coefficients. Following [6], many PCA based face image super-resolution methods have been proposed recently [7–9]. However, they are highly dependent on the training set, as its core concept is to use a linear combination of the HR images in the training set to reconstruct the target HR face image. To achieve a good performance, the algorithm requires that the training set size is very large, so that the test images have great potential to be similar to the images in the training set. Since the poor representative ability, decomposing a complete face image into smaller face patches has attracted much attention recently [10–19].

For example, Chang et al. [10] introduced locally linear embedding [20] to estimate HR patches by linearly combining K candidate HR patches in the training dataset. They assumed that the LR image patch and its HR counterparts (or their feature representations) are locally isometric (share the same neighborhoods have the same representation coefficients). To incorporate much more prior information about the human face, which is a highly structured object, Ma et al. [12] assumed that similar textures are shared by the patches located at the same position of face image, and proposed a position-patch based face image super-resolution method. In their proposed method, the HR patches are reconstructed as a linear combination of training position-patches by the same representation coefficients generated on the LR training samples by least squares representation. When the number of the training position-patches is much larger than the dimension of the patch, which is a common occurrence in image super-resolution, the least squares solution will be not unique. This will lead to unstable face image super-resolution results. To address this problem, Jung and Yang et al. [14,15] regarded the training position-patches as an over-complete dictionary and obtained the reconstruction weights by solving an L_1-norm minimization problem. By decomposing a complete face image into smaller face patch and incorporating the position-patch as well as the sparsity prior, they can well capture the face structure and produce good HR results. However, when the input LR face is disturbed by noise, the sparse representation coefficients may be very unstable. Most recently, our previous works [18,19] further improved these position-patch based face image super-resolution methods by incorporating the local geometrical constraint of manifold instead of incorporating the sparsity constraint into the patch reconstruction objective function as in [14,15], thus reaching sparsity and locality simultaneously. Because of the introducing of locality constraint, it can well extract the maximum amount of facial information from the LR face image and is robust to the noise. As far as we know, this Locality-constraint Representation (LcR) based face image super-resolution method [18,19] obtained the best performance reported in the literatures, especially with noise input. In [21–23], they further expanded LcR to multilayer LcR to improve the reconstruction results. However, when they encounters with noise in the iteration, the SR performance is very poor.

Inspired by [18,19], in this paper we propose a novel noise robust position-patch based representation method called Tikhonov Regularized Neighbor Representation (TRNR). Our proposed inherits the merits of neighbor embedding strategy, but there are essential differences between the two. Traditional neighbor embedding based image super-resolution method uses a fixed number K neighbors for reconstruction often results in blurring effects. We attributed this disadvantage to the over- or under-fitting of the linear combination problem of the selected K neighbors. To this end, we introduce the Tikhonov regularization [24] to the ordinary least squares in neighbor embedding to give preference to a particular solution with desirable properties. Experimental results demonstrated that our proposed TRNR method that can preserve mainly face structure suppress most of the noise is promising and competitive to the state-of-the-art methods, and outperforms other leading super-resolution methods both visually and quantitatively in most cases.

2   Notations

Given an LR observation image I_t^L (subscript "t" distinguishes the test patch from the training patches), our goal is to construct its HR version I_t^H by learning the relationship between the LR and HR training sets, $I^L = \{I_1^L, I_2^L, \cdots, I_N^L\}$ and $I^H = \{I_1^H, I_2^H, \cdots, I_N^H\}$, where N is the number of samples in the training set. As for a local patch based method, we divide the LR observation image into M patches, $\{x_t(p,q)|1 \leq p \leq U, 1 \leq q \leq V\}$, according to the predefined patch_size and overlap pixels. $x_t(p,q)$ denotes a small patch at the position (p,q) of the LR observation image, U represents the patch number in every column, V represents the patch number in every row, thus we have $M = UV$. Specially, the values of U and V can be calculated by $U = \text{ceil}\left\{\frac{imrow - overlap}{patch\_size - overlap}\right\}$ and $V = \text{ceil}\left\{\frac{imcol - overlap}{patch\_size - overlap}\right\}$, where $imrow$ and $imcol$ denote the rows and columns of one face image, respectively, $\text{ceil}(x)$ is the function that rounds the elements of x to the nearest integers towards infinity. In the same way we divide the LR and HR training face image pairs into M patches respectively, $\{x_i(p,q)|1 \leq p \leq U, 1 \leq q \leq V\}_{i=1}^N$ and $\{y_i(p,q)|1 \leq p \leq U, 1 \leq q \leq V\}_{i=1}^N$. $x_i(p,q)$ denotes a small patch at the position (p,q) of the i-th training sample in the LR training set, while $y_i(p,q)$ denotes a small patch at the position (p,q) of the i-th training sample in the HR training set. For more details about the dividing strategy, please refer to [18]. For the LR observation patch $x_t(p,q)$ located at position (p,q), its HR patch $y_t(p,q)$ is estimated using the LR and HR training image patch pairs at the same position. Acquiring all the super-resolved HR patches $\{y_t(p,q)|1 \leq p \leq U, 1 \leq q \leq V\}$, the final HR image I_t^H can be generated by averaging pixel values in the overlapping regions according to the original position. When it does not lead to a misunderstanding, we drop the term (p,q) for convenient from now on.

3   Position-Patch Based Face Image Super-Resolution

For those position-patch based methods, the common idea is firstly to exploit the relationship in the LR patch space formed by all the LR training image patches, and then preserve this relationship for the HR patch space formed by all the HR training image patches. They first represent the given LR patch x_t with the LR training image patch set $\{x_i\}_{i=1}^{N}$, and then transform the representation coefficients w (i.e., the outcome of different representation methods) to faithfully represent each corresponding (unknown) HR patch y_t by replacing the LR training image patch set $\{x_i\}_{i=1}^{N}$ with its HR counterpart $\{y_i\}_{i=1}^{N}$, $y_t = Yw$. To handle the compatibility problem between adjacent patches, simple averaging in the overlapping regions is performed. From the process presented above, we learn that the key issue of these position-patch based methods is to obtain the optimal representation coefficients w, and this section reviews some existing representation schemes.

3.1   Least Square Representation

Given an LR observation patch x_t on the input LR face image, LSR uses patches from all training samples at the same position to represent it:

$$x_t = \sum_{i=1}^{N} w_i x_i + e, \tag{1}$$

where e is the reconstruction error. The optimal weight can be solved by the following constrained least square fitting problem:

$$\min \|x_t - \sum_{i=1}^{N} w_i x_i\|_2^2, \quad \text{s.t.} \quad \sum_{i=1}^{N} w_i = 1, \tag{2}$$

where $w = [w_1, w_2, ..., w_N]^T$ is the optimal N-dimensional weight vector for LR observation patch x_t. Above least square estimation can produce biased solutions when the dimension of the patch is smaller than the size of the training image set.

3.2   Sparse Representation

To solve the biased problem (2), a possible way is to impose several regularization terms onto it. For example, [14,15,25] introduce the sparse representation (SR) theory and use a small subset of patches to represent LR observation patch x_t instead of performing collaboratively over the whole training samples,

$$\min \|w\|_1, \quad \text{s.t.} \quad \|x_t - \sum_{i=1}^{N} w_i x_i\|_2^2 \leq \varepsilon, \tag{3}$$

where $\| \bullet \|_1$ denotes the ℓ_1-norm. This sparsity constraint can ensure that the under-determined equation have an exact solution and allow the learned representation to capture salient properties.

3.3   Locality-Constrained Representation

Recent theoretical results in machine learning have shown that the learning performance can be significantly enhanced if the local geometrical structure is exploited [20]. Motivated by these facts, in our previous work [19], we impose a locality constraint onto the least squares problem and proposed the LcR algorithm to represent an observation LR patch,

$$\min ||dist \odot w||_2^2 \quad \text{s.t.} \quad ||\mathbf{x_t} - \sum_{i=1}^{N} \mathbf{w_i x_i}||_2^2 \leq \varepsilon, \tag{4}$$

where $\odot$ denotes a point wise vector product, and $dist$ is a N-dimensional locality adaptor that gives different freedom for each training patch x_i to its similarity to the input LR patch x_t. It is simply determined by the squared Euclidean distance. By assigning different freedom to the training samples, e.g., these patches near to the input LR patch will be preferable selected while these patches distant from the target point should be penalized severely, thus LcR [19] will achieve sparsity and locality simultaneously.

4   Face Image Super-Resolution Through Tikhonov Regularized Neighbor Representation (TRNR)

In this paper, we introduce the Tikhonov regularization [24] to the ordinary least squares in neighbor embedding to give preference to a particular solution with desirable properties. Specially, for each LR patch x_t in the input LR image, the optimal reconstruction weights are obtained by minimizing the Tikhonov Regularized reconstruction error:

$$J_{\text{TRNR}}(w) = ||x_t - \sum_{k \in C(x_t)} w_k x_k||_2^2 + ||\Gamma w||_2^2 \quad \text{s.t.} \sum_k w_k = 1, \tag{5}$$

where Γ is the Tikhonov matrix, $C(x_t)$ is the index set of the K-NN of x_t in the LR training patches $\{x_m\}_{m=1}^M$. Minimizing Eq. (5) is a Tikhonov regularized least squares problem. In this paper, this Tikhonov matrix is chosen as a multiple of the identity matrix ($\Gamma = \lambda I$). Denote X_K as a matrix with its columns being the K-NN of x_t in the LR space. Thus, Eq. (5) can be written by the following matrix form,

$$J_{\text{TRNR}}(w) = ||x_t - X_K w||_2^2 + \lambda ||w||_2^2, \quad \text{s.t.} \sum_k w_k = 1. \tag{6}$$

Denote $Q = (x_t \mathbf{1}^{\text{T}} - X_K)^{\text{T}} (x_t \mathbf{1}^{\text{T}} - X_K) + \lambda I$, where $\mathbf{1}$ is a column vector of ones, and "T" denotes the Matrix transpose. Thus, Eq. (6) can be optimized by solving a linear system equation $Qw = 1$, thus

$$w = Q \backslash 1. \tag{7}$$

and then normalize the weights so that $\sum_{k=1}^{K} w_k = 1$.

The target HR image patch can be generated by applying the same reconstruction weights to corresponding neighbor HR patches in the HR space:

$$y_t = \sum_{k \in C(x_t)} w_k y_k. \tag{8}$$

The steps of TRNR are summarized in Algorithm 1.

Algorithm 1. Face Image Super-Resolution through TRNR.

1: **Input**: The training sets of LR and HR face images, $I^L = \{I_1^L, I_2^L, \cdots, I_N^L\}$ and $I^H = \{I_1^H, I_2^H, \cdots, I_N^H\}$, an input LR face image I_t^L, *patch_size, overlap, K* and, λ.

2: **Output**: HR super-resolved face image I_t^H .

3: Compute U and V:
   $U = ceil((imrow - overlap)/(patch\_size - overlap))$
   $V = ceil((imcol - overlap)/(patch\_size - overlap))$

4: Divide each of the LR and HR training images and the input LR image into N small patches according to the same location of face, $\{x_i(p,q)|1 \le p \le U, 1 \le q \le V\}_{i=1}^{N}$, $\{y_i(p,q)|1 \le p \le U, 1 \le q \le V\}_{i=1}^{N}$ and $\{y_t(p,q)|1 \le p \le U, 1 \le q \le V\}$, respectively.

5: **for** $p = 1$ to U **do**

6:   **for** $q = 1$ to V **do**

7:      Compute the distance between $x_t(p,q)$ and $x_i(p,q)$, $i = 1, 2, ..., N$.

8:      Obtain the index set $C(x_t)$ of the K-NN of x_t in the LR training patches.

9:      Obtain the K-NN set X_K.

10:     Calculate the matrix Q.

11:     Compute the representation coefficients w vis solving $Qw = 1, \mathbf{s.t.} \sum_k w_k = 1$.

12:     Reconstruct the target HR image patch $y_t = \sum_{k \in C(x_t)} w_k y_k$.

13:   **end for**

14: **end for**

15: Integrating all the obtained HR patches above according to the position. The final HR image I_t^H can be generated by averaging pixel values in the overlapping regions.

---

5   Experiments and Result Analysis

In this section, we describe the details of extensive experiments performed to evaluate the generality, effectiveness, and robustness of the proposed method for face image super-resolution. To evaluate this, we make the comparative analysis of our proposed method with some recent state-of-the-art methods. The experiments are performed on the public FEI face database[1] [26]. Two quality measures, PSNR and SSIM [27] are adopted to evaluate the objective quality of the reconstructed results.

---

[1] http://fei.edu.br/~cet/facedatabase.html.

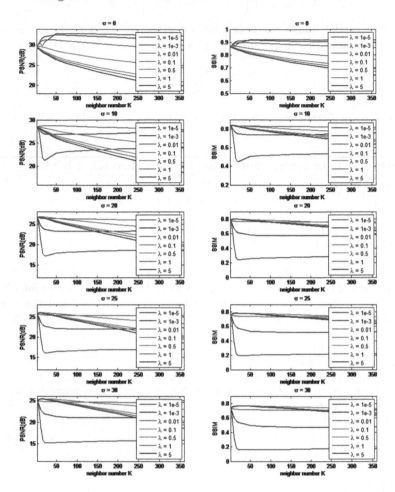

Fig. 1. The PSNR (dB) and SSIM results according to neighbor numbers K and Tikhonov regularization parameters λ. For different noise levels, the best performance is reached at $K = 75$ and $\lambda = 1e - 5$ ($\sigma = 0$), $K = 20$ and $\lambda = 1e - 2$ ($\sigma = 10$), $K = 10$ and $\lambda = 1e - 1$ ($\sigma = 20$), $K = 15$ and $\lambda = 1e - 1$ ($\sigma = 25$), $K = 20$ and $\lambda = 1e - 1$ ($\sigma = 30$), respectively.

Database and Parameter Settings. All experiments described in this paper are conducted on the FEI face database, which consists of 400 facial images. All the images are cropped to 120×100 pixels to form the HR training faces. Human faces in the database are mainly from 19 to 40 years old with distinct appearances, e.g., hairstyles and adornments. The LR images are formed by smoothing (by a 4×4 mean filter) and down-sampling (by a factor of 4 resulting the size of LR face images to be 30×25 pixels) the corresponding HR images, and then the Additive White Gaussian Noise (AWGN) with different noise levels (denoted as σ) is added. In our experiments, we use 360 images to train the

Table 1. PSNR (dB) and SSIM comparisons of different super-resolution methods

Noise levels	Wang	NE	LSR	SR	LcR	TRNR
0	27.51	32.89	32.36	32.77	**32.96**	32.90
	0.7653	0.9197	0.9132	0.9186	**0.9207**	0.9198
10	26.62	28.32	23.60	23.20	28.56	**28.71**
	0.7467	0.8288	0.5206	0.4955	0.8343	**0.8406**
20	25.24	26.14	18.32	17.89	26.41	**26.73**
	0.7386	0.7728	0.2657	0.2224	0.7840	**0.7984**
25	24.89	25.31	16.62	16.18	25.69	**26.01**
	0.7309	0.7438	0.2002	0.1821	0.7662	**0.7810**
30	24.49	24.57	15.26	14.82	25.08	**25.37**
	0.7217	0.7180	0.1557	0.1405	0.7515	**0.7653**

proposed algorithm, leaving the rest 40 images for testing. As in reported in [19], we recommend to set the size as 16×16 pixels for HR patch and the overlap between neighbor patches as 12 pixels, while corresponding LR patch size is set to 4×4 pixels with an overlap of three pixels.

Influence of K and λ. In the following, we evaluate the performance of different noise levels (e.g., $\sigma = 0, 10, 20, 25, 30$) according to different parameter settings, i.e., neighbor number K and Tikhonov regularization parameter λ. We find in this experiment that (i) in addition to K, the regularization parameter λ also has a great influence on the face image super-resolution performance. This implies the effectiveness of Tikhonov regularization in neighbor embedding; (ii) when the value of λ is set to zero, our proposed TRNR method reduce to the traditional NE method. We can see from Fig. 1 that when the input is noiseless, the performance of NE and TRNR are substantially equivalent. However, when the noise level becomes large, by setting a large value of λ, the performance gain of TRNR over NE is remarkable; (iii) when the input image is with strong noise corruption, it needs a relatively large regularization parameter λ to balance the data term and the Tikhonov regularization term, which further indicates that the effectiveness of introducing Tikhonov regularization especially when the noise level is very high.

Comparisons with State-of-the-art Approaches. We evaluate our proposed TRNR method and some representative work, such as Wang et al. [6]'s eigentransformation method, Chang et al.'s Neighbor Embedding (NE) method [10], Ma et al.'s Least Squares Representation (LSR) method [28], Yang et al.'s Sparse Representation (SR) [29] method, and our previously proposed Localityconstrained Representation (LcR) method [19].

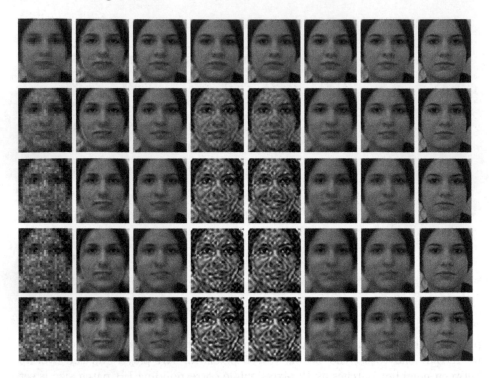

Fig. 2. Super-resolved faces with different noise levels (from top row to bottom row, the noise levels are $\sigma = 0$, $\sigma = 10$, $\sigma = 20$, $\sigma = 25$, and $\sigma = 30$ respectively) of different methods(from left column to right column, they are the input LR image, Wang et al.'s [6] results, NE [10] method's results, LSR [28] method's results, SR [29] method's results, LcR [19] method's results, our results and the ground truth HR face images.)

Table 1 tabulates the PSNR (dB) and SSIM comparisons of different super-resolution methods with different noise levels. We find that the proposed TRNR method is the best among the competitors with noise input (in the noiseless case, TRNR is slightly worse than LcR [19]). With the increase of the noise level, the gain of TRNR over NE [10], LSR [28] and SR [29] becomes increasingly apparent.

Some visual results are shown in Fig. 2. *In the absence of noise,* the performance of these position-patch based methods are equally, and Wang et al.'s global face method [6] has severe "ghosting" effects, especially around face contours. *As the noise becomes large,* LSR [28] and SR [29] can hardly reconstruct the reasonable faces. The main reason is the unstable solution of solving the least squares problem in LSR [28] and SR [29]. Wang et al.'s method can well preserve the characteristics of human face and remove most of the noise, but the reconstructed faces are dirty and different from the ground truth especially when the noise increases. NE [10] method has obvious blocking effect. The super-resolved HR faces by LcR [19] are smooth and lack of details when compared with our proposed TRNR methods. We attribute this to that LcR [19] unitizes the entire training patch samples to represent the observation, thus lacking the discriminant facial features. In contrast, we select several similar samples and exclude the impact of *"unrelated"* samples, thus enhancing the discriminant of the reconstructed HR faces.

6 Conclusion

In this paper, we propose a novel position-patch based face super-resolution method through Tikhonov Regularized Neighbor Representation (TRNR). By incorporating the Tikhonov regularization term to the neighbor embedding, TRNR can select several representative patch samples to reconstruct the HR version of the observed LR image patch, thus giving rise to discriminant HR face image with detailed features. Experimental results demonstrate the generality, effectiveness, and robustness of the proposed algorithm.

Acknowledgement. The research was supported by the National Natural Science Foundation of China (61501413, 61231015, 61172173, 61303114, and 61170023), the EU FP7 QUICK project under Grant Agreement No. PIRSES-GA-2013-612652, the technology Research Program of Ministry of Public Security (2014JSYJA016), the major Science and Technology Innovation Plan of Hubei Province (2013AAA020), the Guangdong-Hongkong Key Domain Breakthrough Project of China (2012A090200007), the China Postdoctoral Science Foundation funded project (2013M530350), and the Specialized Research Fund for the Doctoral Program of Higher Education (20130141120024).

References

1. Park, S.C., Park, M.K., Kang, M.G.: Super-resolution image reconstruction: a technical overview. IEEE Signal Process. Mag. **20**(3), 21–36 (2003)
2. Wang, N., Tao, D., Gao, X., Li, X., Li, J.: A comprehensive survey to face hallucination. Int. J. Comput. Vision **106**, 1–22 (2013)
3. Baker, S., Kanade, T.: Hallucinating faces. In: FG, pp. 83–88 (2000)
4. Freeman, W., Pasztor, E., Carmichael, O.: Learning low-level vision. Int. J. Comput. Vision **40**, 25–47 (2000)
5. Liu, C., Shum, H.-Y., Zhang, C.-S.: A two-step approach to hallucinating faces: global parametric model and local nonparametric model. In: CVPR, vol. 1, pp. 192–198 (2001)
6. Wang, X., Tang, X.: Hallucinating face by eigentransformation. IEEE Trans. Syst. Man Cybern. Part C-Appl. Rev. **35**(3), 425–434 (2005)
7. Chakrabarti, A., Rajagopalan, A., Chellappa, R.: Super-resolution of face images using kernel pca-based prior. IEEE Trans. Multimedia **9**(4), 888–892 (2007)
8. Park, J.-S., Lee, S.-W.: An example-based face hallucination method for single-frame, low-resolution facial images. IEEE Trans. Image Process. **17**(10), 1806–1816 (2008)
9. Jia, K., Gong, S.: Generalized face super-resolution. IEEE Trans. Image Process. **17**(6), 873–886 (2008)
10. Chang, H., Yeung, D., Xiong, Y.: Super-resolution through neighbor embedding. In: CVPR, vol. 1, pp. 275–282 (2004)

11. Zhuang, Y., Zhang, J., Wu, F.: Hallucinating faces: Lph super-resolution and neighbor reconstruction for residue compensation. Pattern Recogn. **40**(11), 3178–3194 (2007)
12. Ma, X., Zhang, J., Qi, C.: Position-based face hallucination method. In: ICME, pp. 290–293 (2009)
13. Li, B., Chang, H., Shan, S., Chen, X.: Aligning coupled manifolds for face hallucination. IEEE Signal Proc. Let. **16**(11), 957–960 (2009)
14. Jung, C., Jiao, L., Liu, B., Gong, M.: Position-patch based face hallucination using convex optimization. IEEE Signal Proc. Let. **18**(6), 367–370 (2011)
15. Yang, J., Wright, J., Huang, T., Ma, Y.: Image super-resolution via sparse representation. IEEE Trans. Image Process. **19**(11), 2861–2873 (2010)
16. Hu, Y., Lam, K.-M., Qiu, G., Shen, T.: From local pixel structure to global image super-resolution: a new face hallucination framework. IEEE Trans. Image Process. **20**(2), 433–445 (2011)
17. Liang, Y., Xie, X., Lai, J.-H.: Face hallucination based on morphological component analysis. Signal Process. **93**(2), 445–458 (2013)
18. Jiang, J., Hu, R., Han, Z., Lu, T., Huang, K.: Position-patch based face hallucination via locality-constrained representation. In: ICME, pp. 212–217 (2012)
19. Jiang, J., Hu, R., Wang, Z., Han, Z.: Noise robust face hallucination via locality-constrained representation. IEEE Trans. Multimedia **16**(5), 1268–1281 (2014)
20. Roweis, S.T., Saul, L.K.: Nonlinear dimensionality reduction by locally linear embedding. Science **290**(5500), 2323–2326 (2000)
21. Jiang, J., Hu, R., Han, Z., Wang, Z., Lu, T., Chen, J.: Locality-constraint iterative neighbor embedding for face hallucination. In: ICME, pp. 1–6 (2013)
22. Jiang, J., Hu, R., Wang, Z., Han, Z.: Face super-resolution via multilayer locality-constrained iterative neighbor embedding and intermediate dictionary learning. IEEE Trans. Image Process. **23**(10), 4220–4231 (2014)
23. Jiang, J., Hu, R., Wang, Z., Han, Z., Ma, J.: Facial image hallucination through coupled-layer neighbor embedding. IEEE Trans. Circuits Syst. Video Technol. (2015)
24. Golub, G.H., Hansen, P.C., O'Leary, D.P.: Tikhonov regularization and total least squares. SIAM J. Matrix Anal. Appl. **21**(1), 185–194 (1999)
25. Ma, X., Luong, H.Q., Philips, W., Song, H., Cui, H.: Sparse representation and position prior based face hallucination upon classified over-complete dictionaries. Signal Process. **92**(9), 2066–2074 (2012)
26. Thomaz, C.E., Giraldi, G.A.: A new ranking method for principal components analysis and its application to face image analysis. Image Vision Comput. **28**(6), 902–913 (2010)
27. Wang, Z., Bovik, A., Sheikh, H., Simoncelli, E.: Image quality assessment: from error visibility to structural similarity. IEEE Trans. Image Process. **13**(4), 600–612 (2004)
28. Ma, X., Zhang, J., Qi, C.: Hallucinating face by position-patch. Pattern Recogn. **43**(6), 2224–2236 (2010)
29. Yang, J., Tang, H., Ma, Y., Huang, T.: Face hallucination via sparse coding. In: Proceedings of IEEE Conference on Image Processing (ICIP), pp. 1264–1267 (2008)

Adaptive Multichannel Reduction Using Convex Polyhedral Loudspeaker Array

Lingkun Zhang[1,3], Ruimin Hu[1,2(✉)], Dengshi Li[1,2], Xiaochen Wang[1,3], and Weiping Tu[1,3]

[1] National Engineering Research Center for Multimedia Software, School of Computer, Wuhan University, Wuhan, China
{zhanglingkunwhu,reallds}@126.com, {hrm1964,clowang,echo_tuwp}@163.com
[2] Collaborative Innovation Center of Geospatial Technology, Wuhan University, Wuhan, China
[3] Research Institute of Wuhan University in Shenzhen, Shenzhen, China

Abstract. Multichannel audio systems always need large numbers of loudspeakers and special placement, or sometimes need many subject evaluations to find an optimal arrangement. The application of multichannel systems is difficult in a typical home environment, where few number of loudspeakers, few subject evaluations and the placement flexibility are highly desirable.In this paper, a design of convex polyhedral loudspeaker arrays surround the target region is proposed for reducing multichannel system adaptively, and then derive an error metric to narrow down the subjective evaluation list. In such a way, a well-performed loudspeaker arrangement can be chosen by few subjective evaluations and placed in a restricted environment. The reproduction accuracy of the method is verified through numerical simulations, and the subjective evaluations indicate the effectiveness of the method.

Keywords: Polyhedral loudspeaker array · Three-dimensional sound · Multichannel audio

1 Introduction

With the popularization of three dimension films, people want to experience the spatial impressions at home. Although such advanced systems can be arranged in a cinema, they are difficults to set up in a typical home environment. MPEG proposed a call (N12610) for 3D audio technology in the home cinema and personal TV environment in 2012. To reach this goal, a technology for reproducing or converting sound fields is necessary.

Wave field synthesis(WFS) [1,7] and higher order Ambisonics(HOA) [5,6] are two common reproduction methods. WFS can reproduce the wave generated

R. Hu—This work is supported by National Nature Science Foundation of China (No. 61231015,61201340, 61201169); National High Technology Research and Development Program of China (863 Program) No. 2015AA016306; Science and Technology Plan Projects of Shenzhen(ZDSYS2014050916575763).

© Springer International Publishing Switzerland 2016
Q. Tian et al. (Eds.): MMM 2016, Part I, LNCS 9516, pp. 421–431, 2016.
DOI: 10.1007/978-3-319-27671-7_35

by a virtual source accurately, but it requires large numbers of loudspeakers equally spaced to implement 3D systems for high-frequency reproduction over large regions. HOA uses an array of loudspeakers on a sphere and reproduce sound field accurately in a target region, however, it still needs many loudspeakers and there is a critical problem in the placement of loudspeakers.

On the other hand, to convert between two systems, a method proposed by Ando [2] can keep the physical properties (sound pressure and direction of particle velocity) at a receiving point, the subjective evaluation results showed that it is effective in a nearby area too. It controls the placement of loudspeakers by a rule of thumb and needs many subjective evaluations to find a well-performed one.

The methods mentioned above attempt to reproduce sound field in a different room, however, they become invalid when the environment is restricted, such as a typical home environment. In such environment, fewer loudspeakers and limited space, even fewer subjective evaluations are allowed. The reproduction methods need special placement of loudspeakers to satisfying their theory, it's not adapt to the restricted environment. On the other hand, the conversion method has fewer limitations with the placement of loudspeakers and the effect in the nearby area can be used to narrow down the list of candidate placements, but only a few attempts have so far been made.

In this paper, a new reduction method is presented, first we propose a construction method of convex polyhedron loudspeaker array which satisfy the limitation of the conversion method, and then derive an error metric on the basis of the effect in the nearby area. Here, a convex polyhedron is a convex set of points in space. In such a way, a well-performed loudspeaker arrangement can be chosen by few subjective evaluations and placed in a restricted environment.

2   Related Work

2.1   Sound Fields Reproduction Model

All reproduction methods try to reproduce sound fields $S(x)$, the sound field produced by loudspeakers are $T(x)$, where $x \in \chi$ is the interested region. Because the physical property of sound is a linear function

$$T(x) = \sum_l w_l T_l(x) \tag{1}$$

where $T_l(x)$ represents the sound fields due to the lth loudspeaker, for a given loudspeaker array, $T_l(x)$ is a function of x, to a general situation, $T_l(x)$ is also a function of the position of loudspeaker $(\sigma_l\, \theta_l\, \varphi_l)$, $w_l\ (l = 1, ..., n)$ is a complex value to be solved.

Reproduction of sound fields is to find w_l $(l = 1, ..., n)$ to satisfy

$$\min_{\chi} \| S(x) - T(x) \| \tag{2}$$

where $\|\cdot\|$ is the error metric, it always uses the mean square error.

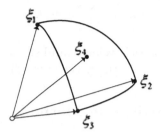

Fig. 1. The Conversion Method

2.2   The Conversion Method

We will review the main idea of the conversion method mentioned above. It is about four loudspeakers ξ_1, ξ_2, ξ_3 and ξ_4. The basic concept of the method is to replace ξ_4 whose direction is included by a spherical triangle formed by ξ_1, ξ_2 and ξ_3 as Fig. 1. The interested region is the receiving point.

   The following assumptions are made: (1) each loudspeaker can be modeled as a point source; (2) the sound pressure at a unit distance from a loudspeaker is proportional to the input to the loudspeaker (the proportionality coefficient is denoted as G); (3) only the outgoing wave from the loudspeaker is considered; and (4) the reflected sound can be neglected in the original and converted fields. (5) $k\sigma_{\min} \gg 1$ is assumed, where k is the wave number and $\sigma_{\min}$ is minimum distance between the loudspeakers and the receiving point. Assumption (5) is valid except for the low-frequency sound that does not contribute to the perception of sound localization.

   If a loudspeaker whose input signal is s(t) is located at $(\sigma\,\theta\,\varphi)$, receive point is located at origin. The Fourier transforms of sound pressure and particle velocity at the receiving point are represented as

$$p(\omega) = G\frac{e^{-ik\sigma}}{\sigma} s(\omega) \tag{3}$$

$$u(\omega) = -\frac{G}{\rho c} h s(\omega) \tag{4}$$

where i is the imaginary unit, ρ is the density of air, c is the speed of sound, and

$$h \triangleq \frac{e^{-ik\sigma}}{\sigma} \begin{pmatrix} \cos\theta\cos\varphi \\ \sin\theta\cos\varphi \\ \sin\varphi \end{pmatrix} \tag{5}$$

Its basic concept is convert four signals into three signals through equations:

$$p_4(\omega) = \sum_{j=1}^{3} w_j p_j (\omega) \tag{6}$$

$$u_4(\omega) = \sum_{j=1}^{3} w_j u_j (\omega) \tag{7}$$

where $u_j (r,\omega)$, $p_j (r,\omega)$ is particle velocity and sound pressure due to the loud-speakers at $(\sigma_l, \theta_l, \varphi_l), (l = 1, \ldots, 4)$, w_j is the weight to be solved.

3   The Proposed Reduction Method

3.1   The Reduction Scheme

The method we proposed is to find a well-performed loudspeaker array convert from an original system in a restricted environment by few subjective evaluations. Figure 2 shows the block diagram of the proposed method, where the conversion method is from [2], $R(x)$ is the limitation of space, satisfy

$$R(x) = \begin{cases} 1, \ x \notin restricted \\ 0, \ x \in restricted \end{cases} \tag{8}$$

the error metricε_v and the placement method will be presented next.

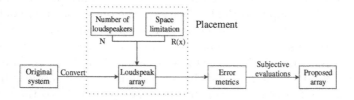

Fig. 2. Block diagram of proposed method

3.2   Convex Polyhedral Loudspeaker Array

First we will derive the convex polyhedral from the conversion method on a sphere. According to assumption (1), loudspeaker is modeled as a point. To convert a loudspeaker from any direction, the sphere must be divided into spherical triangles.

Definition 1. *A convex polyhedron is that for an arbitrary plane of the polyhedron, all the vertexes of the polyhedron are on the same side of the plane.*

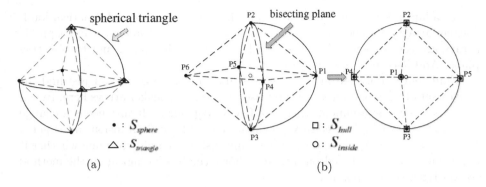

Fig. 3. Sphere triangulation

Proposition 1. *A set of points divide a sphere into spherical triangles is also a convex set of points surround the origin, as Fig. 3.*

Proof. We assume that the set of points on the sphere as S_{sphere}, arbitrary set of three points which can form a spherical triangle as $S_{triangle}$, and the spherical triangle can not divide by other points in S_{sphere} more, it just needs to prove that $\forall p \in S_{sphere} \rightarrow$ p is on the same side of the triangle formed by the points of $S_{triangle}$ (according to the definition of convex polyhedron).

Because of the left-right symmetry of ears, we assume that loudspeaker arrays are set up left-right symmetrical.

In a case that S_{sphere} divide a sphere into spherical triangles and surround the receiving point (origin). $\tilde{S}_{sphere}$ is the projection of S_{sphere} on bisecting plane. $\tilde{S}_{sphere}$ can be classified into two types S_{hull} and S_{inside}. S_{hull} is the set of points which belong to the convex hull (two dimension convex set) of $\tilde{S}_{sphere}$, and S_{inside} is the set of other points.

Proposition 2. *If the convex hull S_{hull} surrounds the receiving point, the convex polyhedron S_{sphere} surrounds the receiving point o.*

Proof. If S_{sphere} surround the receiving point o, but S_{hull} do not surround o, $\exists d \cap S_{hull} = \emptyset$, where d is a diameter of the projected circle. Then S_{sphere} is on the half sphere, do not surround o, that in turn implies that S_{hull} surround o.

If S_{hull} surround o, but S_{sphere} do not surround o, $\exists$ plane P of the convex polyhedron $\rightarrow$ S_{sphere} and o are at different sides of P. Because of the left-right symmetry, $P \perp$ the bisecting plane. Then S_{hull} is on the half circle of the projection, do not surround o, that in turn implies that the convex polyhedron S_{sphere} surround o.

If we want to find all symmetrical convex polyhedrons on a sphere, we can find all the convex hulls on bisecting plane and then project them to the sphere. There are classical method to find convex hull [3], and then just need to find points of S_{inside}.

To adapt a typical home environment, it is needed to extend spherical loud-speaker arrays to aspheric. The differences between spherical arrays and aspheric arrays are the distances from loudspeakers to receiving point and to bisecting plane. And it is restricted by the room size.

Find convex polyhedron in a general situation is a hard work, it just find the convex set of a set of points and can't use to find loudspeaker arrays here. But in a typical home environment, there are few loudspeakers. It just needs a change from the step of finding convex polyhedron in spherical situation: after get a $\tilde{S}_{sphere}$ (S_{hull} and S_{inside}), project it to any distances and determine whether it can form a convex polyhedron. Figure 4 shows the block diagram of the method to construct convex polyhedrons.

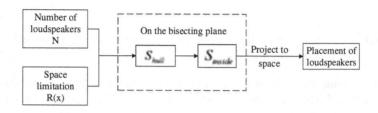

Fig. 4. Block diagram of construct convex polyhedron in a restricted environment

3.3   Error Metric

Because the conversion method is effective in a nearby area. There is the relationship below

$$\begin{cases} \varepsilon_{v_0} = 0 \\ \varepsilon_{v-v_0} < C \end{cases} \Rightarrow \varepsilon_v < C \qquad (9)$$

where v is the predefined region, v_0 is the neighbourhood of the receiving point. If the reproduced sound fields are the same with the original one, the impression in the region must be the same. So error metric in a region can narrow down the list of candidate placements when the sound fields are the same at receiving point.

The relative mean square error (MSE):

$$\varepsilon(r,\omega) \triangleq \frac{\int_0^r \int_0^{2\pi} \int_{-\frac{\pi}{2}}^{\frac{\pi}{2}} |P_r(x,\omega) - P_o(x,\omega)|^2 dx}{\int_0^r \int_0^{2\pi} \int_{-\frac{\pi}{2}}^{\frac{\pi}{2}} |P_o(x,\omega)|^2 dx} \qquad (10)$$

where the integration is over a spherical ball of radius r which is predefined, $dx = r_x^2 \sin\theta_x d\theta_x d\varphi_x dr_x$. $P_0(x,\omega)$ and $P_r(x,\omega)$ are the original sound pressure and reproduced sound pressure by three loudspeakers, respectively.

This error is not depend on the input signal but on the position of loudspeakers and receiving point. It represents the approximation of four loudspeakers convert into three loudspeakers in the predefined region.

We choose the average relative mean square error (AMSE) as (11) of the converted system to represent the approximation of two loudspeaker arrays.

$$\tilde{\varepsilon}(r,\omega) \triangleq \frac{1}{n} \sum_{j=1}^{n} \varepsilon_j(r,\omega) \tag{11}$$

where n is the number of loudspeakers converting original system, $\varepsilon_j(r,\omega)$ is the MSE of jth group of four loudspeakers convert into three.

4 Simulation and Subjective Evaluation Results

4.1 Simulation

To evaluate the performance of the proposed method of reduction, we used 22.2 multichannel system without two low-frequency effect channels in Fig. 5(a) as the original system, and reduced it to 8 and 6 channels, and compared with the system proposed in [2] in Fig. 5(b) where set up on a sphere with a radius of 2m. The signal of 22 loudspeakers are all the same to offset the difference of loudspeakers at different position. The loudspeaker weights were found from [2]. Free-field source conditions are assumed; The operation frequency f = 1000HZ and radius of target region r = 0.085m which is the distance from each ear to the center of head. 22 multichannel system is set up on a sphere with a radius of $\sigma = 2m$. Receiving point with a height of $h = \sqrt{2}m$ to the floor and loudspeakers have a minimum distance $\sigma_1 = 1m$ from receiving point. The step length of azimuthal and elevation angle is 45 degree, the step length from the bisecting plane is 0.2m. The AMSE over the target region is used as the error metric.

4.2 Example Loudspeaker Arrays and Sound Fields

The reduced loudspeaker arrays of 8, 6 are shown in Fig. 6. The AMSE in target region of those two cases are 3.25 % and 4.5 %, and that of the compared system

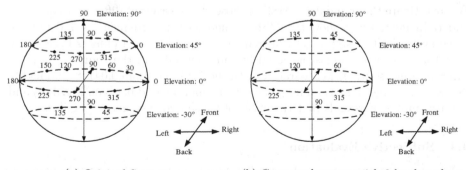

(a) Original System   (b) Compared system with 8 loudspeakers

Fig. 5. 22 Channel system and compared system. Black points are the position of loudspeakers.

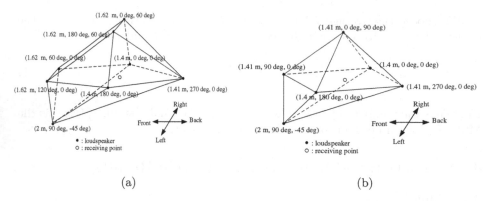

Fig. 6. Proposed System in the restricted environment. (a) Proposed system with 8 loudspeakers. (b) Proposed system with 6 loudspeakers.

are 5.04 %. Because the 22.2 multichannel system has more loudspeakers in front than in back, the conversion system with low AMSE has more loudspeakers in front and few in back, and loudspeakers at elevation of 0 are more than loudspeakers above it and below it.

The real part of the sound fields in the $x - y$ plane are shown in Fig. 7 as an example. These figures are displayed as density plots, where the display is limited the maximum sound pressure of the field, i.e., sound pressure greater than 0.5 are red and less than -0.5 are blue. The circle indicates the target region at the particular height shown. In both cases, conversion system can reproduce sound field well in the target region, the area is even larger than it at given frequency. The main difference between Fig. 7 (b) and (c) are the region with radius more than 8.5 cm, our 8 loudspeakers can reproduce the sound field better than the compared one.

4.3   Conversion Error

We investigate the sound fields MSE of those three arrays at different frequencies and radii next, Fig. 8 show the MSE change with the frequencies and radii of three systems. The step length of frequency is 50HZ and radius is 0.1 cm. We note that the two systems proposed by our method and the compared system can reproduce sound field accurately(with the MSE less than 0.04) [8] and highly distorted at large radius and frequency. The general trend between Fig. 8 (a) and (b) are similar, it reveals that the effect of those two systems are similar.

4.4   Subjective Evaluation

Subjective evaluations used the "double-blind triple-stimulus with hidden reference" method [4], which can be used for the subjective assessment of small impairments in a multichannel sound system. Figure 9 shows the method. In Fig. 9, stimulus "R" indicates the reference sound and stimuli "A" and "B" are

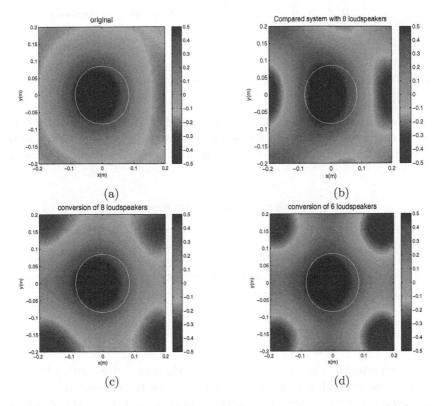

Fig. 7. Example Sound fields. (a) Sound Fields of 22 channel system. (b) Fields of compared system with 8 loudspeakers, 0.24 % MSE with original fields in circle. (c) and (d) Fields of proposed system with 8 and 6 loudspeakers, 0.03 % and 0.08 % MSE with original fields in circle

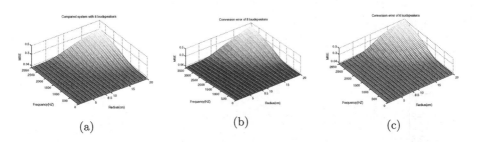

Fig. 8. Conversion Error in Different Frequencies and Radii. (a) Conversion Error of compared system with 8 loudspeakers, the reproduction accuracy with a upper bound of 2850HZ at radius of 8.5 cm. (b) and (c) Proposed systems with 8 and 6 loudspeakers, the reproduction accuracy with a upper bound of 3550HZ and 2850HZ at radius of 8.5 cm.

Fig. 9. Double-blind triple-stimulus with hidden reference method

the sounds for evaluation. The subject was asked to assess the impairment on A and B compared with R. Table 1 shows the scale, according to a continuous five-grade impairment scale. The impairment was assessed from the sound localization.

Table 1. Scales used for subjective evaluation.

Impairment	Grade
Imperceptible	5.0
Perceptible, but not annoying	4.0
Slightly annoying	3.0
Annoying	2.0
Very annoying	1.0

In the experiment, the reference was the 22-channel system with white noise to evaluate the method in all band. "A" and "B" are the compared system and the candidate system, respectively, with sounds convert from the 22-channel system. There were 3 layouts with 8 loudspeakers and 3 layouts with 6 loudspeakers in the candidate list. Subjects were 20 people in their twenties, thirties and forties. They evaluated each system twice. After the evaluation, calculated the "difference grade" for each system by subtracting the grade given to the hidden reference.

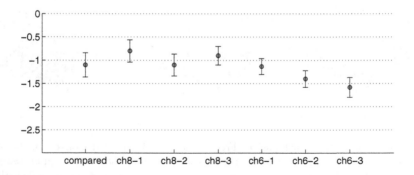

Fig. 10. Results of subjective evaluation. Mean scores and 95 % limits. ch8-1 and ch6-1 are the proposed system mentioned in Sect. 4.2. ch8-2, ch8-3, ch6-2 and ch6-3 are the systems in candidate list

The results are shown in Fig. 10. The systems in candidate list have a much different score. It reveals that the error metric proposed in this paper do not narrow down the candidate list accurate, but it is effective, as the proposed systems have a score similar to that of the compared one. This again in agreement with the derivation of the error metric mentioned in Sect. 3.3.

5   Conclusion

This paper proposed a new reduction method to adapt the restricted environment and with few subjective evaluations. Our method introduce a construction of convex polyhedron loudspeaker array to adapt the restricted environment, and derive an error metrics to narrow down the candidate list of placement of arrays. Finally, compared to the conversion method, a well-performed loudspeaker array placement can be found by few subjective evaluations and set up adaptively. Simulation and subjective evaluation obtained by converting 22-channel system into 8 and 6 channels in a restricted environment indicate the effectiveness of the method.

References

1. Ahrens, J.: Analytic Methods of Sound Field Synthesis. Springer, Heidelberg (2012)
2. Ando, A.: Conversion of multichannel sound signal maintaining physical properties of sound in reproduced sound field. IEEE Trans. Audio Speech Lang. Process. **6**(19), 1467–1475 (2011)
3. Andrew, A.M.: Another efficient algorithm for convex huls in two dimensions. Inform. Process. Lett. **9**, 216–219 (1979)
4. ITU-R Rec. BS.1116-1, Geneva, Switzerland: Methods for the subjective assessment of small impairments in audio systems including multichannel sound systems (1997)
5. Sena, E.D., Hacihabiboglu, H., Cvetkovic, Z.: On the design and implementation of higher order differential microphones. IEEE Trans. Audio, Speech, Lang. Process. **20**(1), 162–174 (2012)
6. Sena, E.D., Hacihabiboglu, H., Cvetkovic, Z.: Analysis and design of multichannel systems for perceptual sound field reconstruction. IEEE Trans. Audio, Speech, Lang. Process. **21**(8), 1653–1665 (2013)
7. Spors, S., Wierstorf, H., Geier, M.: Comparison of modal versus delay-and-sum beamforming in the context of data-based binaural synthesis. In: Audio Engineering Society Convention 132 Audio Engineering Society (2012)
8. Ward, D.B., Abhayapala, T.D.: Reproduction of a plane-wave sound field using an array of loudspeakers. IEEE Trans. Speech Audio Process. **9**(6), 697–707 (2001)

Dominant Set Based Data Clustering and Image Segmentation

Jian Hou[1]([✉]), Chunshi Sha[1], Hongxia Cui[2], and Lei Chi[1]

[1] College of Engineering, Bohai University, Jinzhou 121013, China
dr.houjian@gmail.com
[2] School of Information Science, Bohai University, Jinzhou 121013, China

Abstract. Clustering is an important approach in image segmentation. While various clustering algorithms have been proposed, the majority of them require one or more parameters as input, making them a little inflexible in practical applications. In order to solve the parameter dependent problem, in this paper we present a parameter-free clustering algorithm based on the dominant sets. We firstly study the influence of regularization parameters on the dominant sets clustering results. As a result, we select an appropriate regularization parameter to generate over-segmentation in clustering results. In the next step we merge clusters based on the relationship between intra-cluster and inter-cluster similarities. While being simple, our algorithm is shown to improve the clustering quality significantly in comparison with the dominant sets algorithm in data clustering and image segmentation experiments. It also performs comparably to or better than some other clustering algorithms with manually selected parameters input.

Keywords: Dominant set · Clustering · Over-segmentation

1 Introduction

Clustering is an important unsupervised learning tool and widely applied in computer vision and image processing tasks, e.g., image segmentation. In past decades, a large amount of clustering algorithms have been proposed from different perspectives. Based on the strategies used in clustering, clustering algorithms can often be categorized into five types, i.e., partitioning, hierarchical, density-based, model-based and grid-based algorithms. Some of the typical clustering algorithms falling into these types include k-means, BIRCH (Balanced Iterative Reducing and Clustering using Hierarchies), DBSCAN (Density-Based Spatial Clustering of Applications with Noise) [1], EM (Expectation Maximization) and CLIQUE (CLustering In QUEst) [2]. In recent progresses, spectral clustering, e.g., normalized cuts (NCuts) [3], makes use of the eigen-structure of the pairwise data similarity matrix in partitioning data into clusters. Another popular clustering method is the so-called affinity propagation (AP) [4], which is also based on the pairwise similarity matrix of input data and identifies cluster members

© Springer International Publishing Switzerland 2016
Q. Tian et al. (Eds.): MMM 2016, Part I, LNCS 9516, pp. 432–443, 2016.
DOI: 10.1007/978-3-319-27671-7_36

by means of passing affinity messages among data. In [5] the authors propose to identify cluster centers based on local density and minimum distance to higher density, and then assign the remaining data to different clusters in the decreasing order of local density. Some other progresses in this field include [6,7].

In various clustering approaches, graph-based clustering has attracted increasing interest in recent years. With the pairwise data similarity matrix as input, this kind of clustering algorithms exploit the rich data structure information encoded in the similarity matrix and generate impressive clustering clustering. One typical example of graph-based clustering is the so-call spectral clustering, which depends on the spectrum of the similarity matrix to perform dimension reduction and then clusters data in fewer dimensions. The well-known NCuts and some other algorithms [8–10] belongs to this field. Another popular graph-based clustering algorithm is the AP method, which identifies cluster centers and cluster members gradually by means of passing affinity messages among data. More recently, [6] presented an unsupervised approach to generate robust similarity matrices by identifying and exploiting discriminative features, which are shown to improve the performance of spectral clustering.

On the other hand, the majority of existing clustering algorithms, including the above-mentioned graph-based clustering algorithms, require as input user-specified parameters, and their clustering performance is heavily influenced by the parameters. Typically, many algorithms, e.g., k-means and normalized cuts, require as input the number of clusters, which is not easy to determine in many cases. Although some algorithms determine the number of clusters automatically, they require other parameters as input. Examples of this kind include DBSCAN where a radius and a minimum cluster size are required as input, and AP where the preference values of all data need to be specified. In either cases the clustering results are influenced by the parameters selection, and a careful tuning process is required to generate satisfactory clustering results. In the algorithm proposed in [5], while the cutoff distance d_c influences the clustering results evidently, it seems that the optimal d_c can only be determined empirically at present.

Some efforts have been made to determine the required input parameters. For example, a set of methods were presented to determine the appropriate number of clusters [11–13], and the authors of [1,14] provided methods to estimate the range of required parameters. In addition, some parameter-independent algorithms, e.g., correlation clustering [15,16], have been proposed. However, the parameter-tuning problem in clustering algorithms is still open in general and the appropriate parameters are mainly determined empirically in existing works.

In order to reduce the dependence on user-specified parameters, we study existing clustering algorithms and notice the dominant sets algorithm as a promising approach. As a graph-based clustering approach, dominant sets (DSets) clustering [17,18] uses the pairwise data similarity matrix as input and does not involve any parameters explicitly. By defining a dominant set as a graph-theoretic concept of a cluster, DSets clustering extracts clusters sequentially and determines the number of clusters automatically. DSets clustering has been successfully used in such various domains as image segmentation [17], object detection [19], object classification [20] and human activity analysis [21], etc. However, we find that DSets clustering is not parameter-independent implicitly. This algorithm requires the pairwise data similarity matrix as input, and the similarity between two data x and y is usually represented by $s(x, y) = exp(-d(x, y)/\sigma)$, where $d(x, y)$ is the Euclidean distance of two data items and σ is a regularization parameter. With the same set of data to be clustered, different σ's result in different similarity matrices, which then lead DSets clustering to generate clustering results. This means that an appropriate σ is necessary to obtain satisfactory clustering results. Unfortunately, there is still no methods to determine the best value of this parameter.

In this paper we firstly study the influence of σ on DSets clustering results. As a result, we find that a large σ tends to generate under-segmentation and a small one tends to result in over-segmentation. Since it is hard to find the σ corresponding to the switch between over-segmentation and under-segmentation, we propose to generate over-segmentation to some extent and then merge clusters to overcome over-segmentation. Specifically, we select σ as the mean of all pairwise distances in DSets clustering, and then merge clusters based on the relationship between intra-cluster and inter-cluster similarities. While being simple, our algorithm is shown to be quite effective in improving the DSets clustering results. In experiments on data clustering and image segmentation, our algorithm performs comparably to or better than other algorithms with user-specified parameters.

The remainder of this paper is organized as follows. In Sect. 2 we briefly introduce the concept and properties of the dominant sets clustering algorithm. We then study the influence of σ's on DSets clustering results and present our clustering algorithm in Sect. 3. Data clustering and image segmentation experiments are conducted to validate the proposed algorithm in Sect. 4. Finally, Sect. 5 concludes this paper.

2   Dominant Sets Clustering

Existing clustering algorithms usually choose to partition the input data and the clusters are obtained as a by-product of the partitioning process. In contrast, the DSets clustering algorithm extracts clusters in a sequential manner. We present a brief introduction of the DSets clustering algorithm in the following, and refer interested readers to [17,18,22] for more details.

The DSets clustering algorithm defines dominant set as a graph-theoretic concept of a cluster. Informally, a dominant set is a maximally subset of input data with internal coherency. In other words, the data in a dominant sets are similar to each other, and they are dissimilar to those outside the dominant set. Evidently, the dominant sets satisfy the constraint of high intra-cluster similarity and low inter-cluster similarity, and can be regarded as clusters.

The formal definition of dominant set is presented as follows. Given n data to be clustered, we represent them in an undirected, edge-weighted graph $G = (V, E, w)$ with no self-loops, where V, E and w denotes the node set, the edge set and the weight function, respectively. As customary, we represent the graph G

by the pairwise $n \times n$ similarity matrix $A = (a_{ij})$, where $a_{ij} = w(i, j)$ if $(i, j) \in E$ and $a_{i,j} = 0$ otherwise. Since G has no self loops, all the elements on the main diagonal of A will be set to zero.

In the first step we define a criterion to measure the internal coherency of a data subset. With a non-empty subset $D \subseteq V$, the average similarity between one data $i \in D$ with the whole subset D is

$$aw_D(i) = \frac{1}{|D|} \sum_{j \in D} a_{ij}. \tag{1}$$

Then for one data $j \notin D$ we define

$$\phi_D(i, j) = a_{ij} - aw_D(i, j). \tag{2}$$

Obviously, $\phi_D(i, j)$ reflects the relationship of the similarity between i and j, with respect to the average similarity between i and the data in D. With $D \subseteq V$ and $i \in D$, the weight of i with respect to D is defined as

$$w_D(i) = \begin{cases} 1, & \text{if } |D| = 1, \\ \sum_{j \in D \setminus \{i\}} \phi_{D \setminus \{i\}}(j, i) w_{D \setminus \{i\}}(j), & \text{otherwise.} \end{cases} \tag{3}$$

The weight $w_D(i)$ can be understood approximately as follows. A positive $w_D(i)$ means that the average similarity between i and $D \setminus \{i\}$ is greater than the overall similarity inside $D \setminus \{i\}$. In other words, D has higher internal similarity than $D \setminus \{i\}$. In contrast, a negative $w_D(i)$ means that D has lower internal similarity than $D \setminus \{i\}$.

By defining the total weight of D as $W(D) = \sum_{i \in D} w_D(i)$, we call a non-empty subset $D \subseteq V$ such that $W(T) > 0$ for all non-empty $T \subseteq D$ as a dominant set, if the following two conditions are satisfied

1. $w_D(i) > 0$, for all $i \in D$.
2. $w_{D \cup i}(i) < 0$, for all $i \notin D$.

In these two conditions, the first one means that D is internally coherent, whereas the second one implies that this coherency will be destroyed if any data from outside is added. These two conditions together make a dominant set a maximal subset with internal coherency. In this way, the data in a dominant set are similar to each, and those from different dominant sets are dissimilar, thereby enabling a dominant set to be regarded as a cluster.

In [17,22] the authors provided a simple method to extract dominant sets. Given the pairwise similarity matrix A, we calculate a vector $x \in R^n$ with the replicator dynamics, i.e.,

$$x_i^{(t+1)} = x_i^{(t)} \frac{(Ax^{(t)})_i}{x^{(t)\prime} A x^{(t)}} \tag{4}$$

for $i = 1, \ldots, n$. The n elements in x correspond to the weights of the n data, and the subset of data with non-zero weights form a dominant set. After extracting

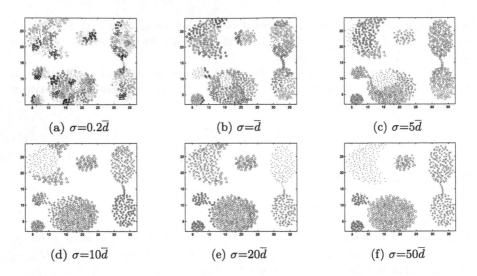

(a) $\sigma=0.2\bar{d}$ (b) $\sigma=\bar{d}$ (c) $\sigma=5\bar{d}$

(d) $\sigma=10\bar{d}$ (e) $\sigma=20\bar{d}$ (f) $\sigma=50\bar{d}$

Fig. 1. DSets clustering results on Aggregation dataset, with different σ's.

a dominant set, we extract the next one in the remaining data, until all data are included in dominant sets or some criterions are satisfied. In this way, we extract the dominant sets (clusters) sequentially and determine the number of clusters automatically.

3   Our Algorithm

Firstly we use an example to illustrate the influence of σ on DSets clustering results. The experiment is conducted on the Aggregation shape dataset [23] with 7 clusters of 2D points. With $\sigma = 0.2\bar{d}, \bar{d}, 5\bar{d}, 10\bar{d}, 20\bar{d}, 50\bar{d}$, we apply DSets clustering to the dataset and report the clustering results in Fig. 2.

It is evident from Fig. 2 that with the increase of σ, DSets clustering tends to generate larger and larger clusters, and the over-segmentation in the clustering result is switched to under-segmentation gradually. This observation can be explained as follows. From $s(x, y) = exp(-d(x, y)/\sigma)$ we see that with a small σ, all the similarity values in the similarity matrix will be driven down, as illustrated in Fig. 1(a). In this case, one data will have large similarity with only a small number of nearest neighbors. Since DSets clustering requires the data in a cluster to be similar to each other, it is difficult for DSets clustering to generate large clusters in case of small σ's. As a result, a small σ corresponds to a large amount of small clusters, i.e., over-segmentation, as illustrated in Fig. 2(a). With the increase of σ, the similarity values rise and the clusters become large, thereby relieving the over-segmentation effect, as shown in Fig. 2(b) to Fig. 2(d). If the value of σ is too large, most similarity values become quite large (Fig. 1(d)) and the obtained clusters will become over-large, and over-segmentation is switched to under-segmentation, as shown in Fig. 2(e) and 2(f). From the above analysis

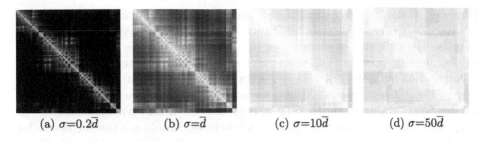

(a) $\sigma=0.2\bar{d}$ (b) $\sigma=\bar{d}$ (c) $\sigma=10\bar{d}$ (d) $\sigma=50\bar{d}$

Fig. 2. Illustration of the similarity matrices of aggregation dataset with different σ's. The similarity values in $[0, 1]$ are normalized to $[0, 255]$ for display purpose.

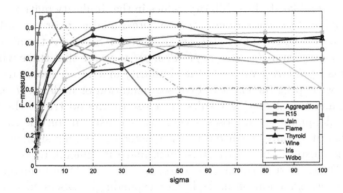

Fig. 3. DSets clustering quality (F-measure) with different σ's on eight datasets. The values of σ's are equal to the product of horizonal coordinates and $\bar{d}$.

and Fig. 2 we find that with the increase of σ, the clustering quality firstly rises and then drops, and there exists a limited σ corresponding to the best clustering quality. This observation can be validated by the following experiments. We apply DSets clustering with different σ's to eight datasets and report their clustering quality evaluated by F-measure in Fig. 3. The eight datasets include Aggregation, R15 [24], Jain [25], Flame [26], and four UCI datasets, namely Thyroid, Wine, Iris and Wdbc.

While Fig. 3 shows that for each dataset, there exist a limited σ corresponding to the best clustering quality, we also observe that these best-performing σ's vary widely. In this case, any fixed σ is not able to result in optimal clustering results for all the datasets.

As there is still no methods to determine the best-performing σ for a dataset, in this paper we present a different method to solve the dependence on σ. Specifically, we firstly use DSets clustering to generate over-segmented clustering results, and then merge clusters to improve the clustering quality. In the following we present the details of these two steps.

In the first step we generate over-segmented clustering result. As analyzed above, a small σ tends to generate over-segmentation in clustering results.

However, if the σ is too small, DSets clustering will generate a large amount of clusters of over-small sizes, as illustrated by Fig. 2(a). Even if this will not increase the difficulty of cluster merging, it will increase the computation load in the merging process. Therefore we need a suitable σ which results in over-segmentation but not too severe over-segmentation. In this paper we select $\sigma = \overline{d}$ due to the following reasons. First, we notice from Fig. 3 that the minimum of the best-performing σ's is $5\overline{d}$. This means that for these eight datasets, $\sigma = \overline{d}$ will guarantee over-segmentation, and the over-segmentation is not too severe for the dataset with the minimum best-performing σ. While it is not convincing to select $\sigma = \overline{d}$ just based on these eight datasets, we argue that it is not accidental that $\sigma = \overline{d}$ results in over-segmentation for all the eight datasets. As we analyzed above, a small σ make all the similarity values in the similarity matrix to be small (Fig. 1(a)) and lead to over-segmentation, whereas a large σ corresponds to large similarity values (Fig. 1(d)) and under-segmentation. Since $\overline{d}$ is the average of all pairwise distances, from $s(x, y) = exp(-d(x, y)/\sigma)$ we see that there will be both large and small similarity values in the similarity matrix, as shown in Fig. 1(b). In this case, we expect the effect of over-segmentation or under-segmentation is quite small, and the clustering results are close to ground truth. However, the definition of dominant set imposes a very strict constraint on the intra-cluster similarity, i.e., it requires each data in a cluster to be very similar to *all* the others in the cluster. This strict constraint make it difficult to generate large clusters by DSets clustering. In this sense, we say that DSets clustering tends to generate over-segmentation in itself. Correspondingly, DSets clustering tends to generate over-segmentation at $\sigma = \overline{d}$, and the over-segmentation can only be removed with larger σ's.

After we generate over-segmented clustering results, the next step is to merge clusters in order to improve the clustering quality. While there are some other methods proposed to merge clusters, here we show that based on the DSets clustering results, we can use a simple method to merge clusters and improve the clustering quality significantly. Our method is based on the relationship of the intra-cluster and inter-cluster similarities. Specifically, if two clusters actually belong to the same cluster and should be merged, their inter-cluster similarity is relatively large compared to the intra-cluster similarities. In contrast, if two clusters belong to two different clusters, their distance is large and the inter-cluster similarity is small compared to the intra-cluster similarities. Based on this simple observation we make decision on which clusters should be merged. The details of this method is presented below.

For a cluster D, we define its intra-cluster similarity as

$$s_{intra}(D) = \frac{1}{M(M-1)} \sum_{i \in D} \sum_{j \in D} s(i, j) \qquad (5)$$

where M is the number of data items in D. Between two clusters D_1 and D_2, their inter-cluster similarity is defined as

$$s_{inter}(D_1, D_2) = \frac{1}{M_1 M_2} \sum_{i \in D_1} \sum_{j \in D_2} s(i, j) \qquad (6)$$

where M_1 and M_2 are the numbers of data items in D_1 and D_2, respectively. With the definition of intra-cluster and inter-cluster similarities, we define the suitableness of two clusters to be merged as

$$\Psi(D_1, D_2) = \frac{s_{inter}(D_1, D_2)}{s_{intra}(D_1) + s_{intra}(D_2)} \qquad (7)$$

Evidently a large $\Psi(D_1, D_2)$ implies that D_1 and D_2 should be merged, and a small $\Psi(D_1, D_2)$ indicates that D_1 and D_2 should stay in different clusters. If we can find a threshold Ψ_{th} to differentiate between large and small Ψ, we can merge only the pairs of clusters with $\Psi > \Psi_{th}$. By studying the distribution of the pairwise Ψ's between clusters, we find that it is really possible to find such a threshold. As the pairwise Ψ's between clusters are calculated between all the pairs of clusters, and many such pairs are composed of clusters far from each other, it is reasonable to expect that most of the pairwise Ψ's correspond to the pairs of clusters not suitable to be merged, and should be labeled as *small* Ψ's. In fact, we show in Fig. 4 the histograms h of pairwise Ψ's between clusters of the eight datasets, and find that most of the Ψ's are gathered in the small part. As a result, we can define the threshold Ψ_{th} as the boundary between most of the Ψ's labeled as *small* and a small sample of Ψ's labeled as *large*. In implementation we find the largest k satisfying $h(k) < h(k-1)$ and $h(k) < h(k-2)$, and then assign the center of the k bin to the threshold Ψ_{th}. While simple, this method is found to be quite effective in experiments.

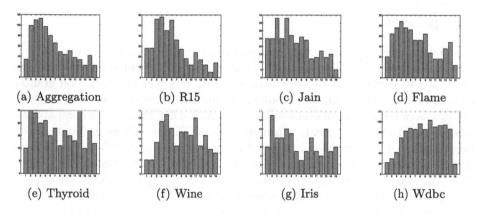

(a) Aggregation   (b) R15   (c) Jain   (d) Flame

(e) Thyroid   (f) Wine   (g) Iris   (h) Wdbc

Fig. 4. The histograms of pariwise Ψ's between clusters of eight datasets.

4   Experiments

In this section we validate the effectiveness of our algorithm in both data clustering and image segmentation experiments. For comparison, we also use the original DSets clustering algorithm and some other clustering algorithms in experiments.

Table 1. The datasets used in experiments.

	Aggregation	R15	Jain	Flame	Thyroid	Wine	Iris	Wdbc
Number of points	788	600	373	240	215	178	150	569
Data dimension	2	2	2	2	5	13	4	32
Number of clusters	7	15	2	2	2	3	3	2

Table 2. Clustering quality (F-measure) comparison of different clustering algorithms on eight datasets.

	k-means	NCuts	DBSCAN	AP	DSets-1	DSets-15	Ours
Aggregation	0.88	0.76	0.90	0.79	0.36	0.86	0.94
R15	0.85	0.98	0.73	0.66	0.86	0.77	0.89
Jain	0.80	0.70	0.85	0.58	0.18	0.57	0.74
Flame	0.84	0.79	0.96	0.74	0.25	0.76	0.69
Thyroid	0.68	0.65	0.68	0.58	0.31	0.84	0.73
Wine	0.64	0.66	0.51	0.90	0.44	0.84	0.83
Iris	0.67	0.78	0.78	0.90	0.49	0.75	0.77
Wdbc	0.84	0.84	0.69	0.79	0.14	0.62	0.84
mean	0.78	0.77	0.76	0.74	0.38	0.75	0.80

4.1   Data Clustering

The data clustering experiments are conducted on the eight datasets as used in the last section. A brief introduction of the characteristics of the eight datasets is presented in Table 1. For the four datasets with data feature dimension granter than 2, we scale the data before building the similarity matrices, in order to avoid attributes in larger ranges dominating those in smaller ranges. We compare our algorithm with DSets, k-means, normalized cuts (NCuts), DBSCAN and affinity propagation (AP). For DSets, we test the cases with $\sigma = \overline{d}$ (DSets-1) and with $\sigma = 15\overline{d}$. Since in our algorithm $\sigma = \overline{d}$, the comparison with DSets-1 is used to show the effect of the merging process. As to DSets-15, $\sigma = 15\overline{d}$ is selected as it generates the best overall clustering results. As k-means and NCuts require the number of clusters as input, we feed the ground truth number of clusters to them. With DBSCAN, the parameter $MinPts$ is manually selected to be 3 and Eps is determined based on $MinPts$ [27]. For AP, we manually select the best preference values of data. The comparison is reported in Table 2.

It is quite evident from Table 2 that our algorithm performs better than DSets-1 significantly, and this validate the effectiveness of our cluster merging method. As to the other five algorithms used in comparison, although all of them benefit from ground truth or manually selected parameters, they perform comparably to our algorithm and are outperformed by our algorithm in the mean clustering quality. This indicates that our algorithm is effective. Of course,

Fig. 5. Segmentation results. From left to right, the three columns are from DSets, NCuts and our algorithm, respectively.

we also observe that with some datasets, e.g., Jain and Flame, our algorithm is outperformed by the best ones by a large margin. In our opinion, this is attributed to the threshold determination method used in our algorithm. As can be seen from Fig. 4, for some datasets there are multiple candidates for the threshold, and our method may fail to select the optimal ones. In the next step we will try to improve the threshold determination method.

4.2   Image Segmentation

Since clustering pixels is an important image segmentation technique, we also apply our clustering algorithms to image segmentation tasks. Three examples of the segmentation results are reported in Fig. 5. For comparison, we also report the segmentation results of the original DSets algorithm and the NCuts algorithm. For each image the number of segments are selected for NCuts, and the σ are tuned for DSets, in order to obtain the best segmentation results. Although

DSets and NCuts are fed with manually selected parameters, it is evident from Fig. 5 that our algorithm performs better than or comparably to these two algorithm in segmentation effect.

5   Conclusions

In this paper we present a parameter independent clustering algorithm based on dominant sets. We firstly investigate the influence of regularization parameters on the dominant sets clustering results. As a result, we choose an appropriate regularization parameter to generate over-segmented clustering results. In the next step we use a cluster merging step to overcome over-segmentation and improve the clustering quality. The cluster merging method is based on the relationship of intra-cluster similarity with respect to inter-cluster one, and involves no parameter input. In data clustering and image segmentation experiments, our algorithm generates evidently better results than the original dominant sets algorithm. It also performs better than or comparably to other algorithms with parameter tuning.

Acknowledgments. This work is supported in part by the National Natural Science Foundation of China under Grant No. 61473045 and No. 41371425, and by the Program for Liaoning Innovative Research Team in University (LT2013023).

References

1. Ester, M., Kriegel, H.P., Sander, J., Xu, X.W.: A density-based algorithm for discovering clusters in large spatial databases with noise. In: International Conference on Knowledge Discovery and Data Mining, pp. 226–231 (1996)
2. Agrawal, R., Gehrke, J., Gunopulos, D., Raghavan, P.: Automatic subspace clustering of high dimensional data. In: International Conference on Knowledge Discovery and Data Mining, pp. 517–521 (2005)
3. Shi, J., Malik, J.: Normalized cuts and image segmentation. IEEE Trans. Pattern Anal. Mach. Intell. **22**, 167–172 (2000)
4. Brendan, J.F., Delbert, D.: Clustering by passing messages between data points. Science **315**, 972–976 (2007)
5. Rodriguez, A., Laio, A.: Clustering by fast search and find of density peaks. Science **344**, 1492–1496 (2014)
6. Zhu, X., Loy, C.C., Gong, S.: Constructing robust affinity graphs for spectral clustering. In: IEEE International Conference on Computer Vision and Pattern Recognition, pp. 1450–1457 (2014)
7. Gong, M., Liang, Y., Shi, J., Ma, W., Ma, J.: Fuzzy c-means clustering with local information and kernel metric for image segmentation. IEEE Trans. Image Process. **22**, 573–584 (2013)
8. Fowlkes, C., Belongie, S., Fan, C., Malik, J.: Spectral grouping using the nystrom method. IEEE Trans. Pattern Anal. Mach. Intell. **26**, 214–225 (2004)
9. Yan, D., Huang, L., Jordan, M.I.: Fast approximate spectral clustering. In: International Conference on Knowledge Discovery and Data Mining, pp. 907–916 (2009)

10. Niu, D., Dy, J.G., Jordan, M.I.: Dimensionality reduction for spectral clustering. In: International Conference on Artificial Intelligence and Statistics, pp. 552–560 (2011)
11. Fraley, C., Raftery, A.E.: How many clusters? Which clustering method? Answers via model-based cluster analysis. Comput. J. **41**, 578–588 (1998)
12. Monti, S., Tamayo, P., Mesirov, J., Golub, T.: Consensus clustering: a resampling-based method for class discovery and visualization of gene expression microarray data. Mach. Learn. **52**, 91–118 (2003)
13. Evanno, G., Regnaut, S., Goudet, J.: Detecting the number of clusters of individuals using the software structure: a simulation study. Mol. Ecol. **14**, 2611–2620 (2005)
14. Dueck, D., Frey, B.J.: Non-metric affinity propagation for unsupervised image categorization. In: IEEE International Conference on Computer Vision, pp. 1–8 (2007)
15. Bansal, N., Blum, A., Chawla, S.: Correlation clustering. Mach. Learn. **56**, 89–113 (2004)
16. Demaine, E.D., Emanuel, D., Fiat, A., Immorlica, N.: Correlation clustering in general weighted graphs. Theor. Comput. Sci. **361**, 172–187 (2006)
17. Pavan, M., Pelillo, M.: Dominant sets and pairwise clustering. IEEE Trans. Pattern Anal. Mach. Intell. **29**, 167–172 (2007)
18. Torsello, A., Bulo, S.R., Pelillo, M.: Grouping with asymmetric affinities: a game-theoretic perspective. In: IEEE International Conference on Computer Vision and Pattern Recognition, vol. 1, pp. 292–299 (2006)
19. Yang, X.W., Liu, H.R., Laecki, L.J.: Contour-based object detection as dominant set computation. Pattern Recogn. **45**, 1927–1936 (2012)
20. Hou, J., Pelillo, M.: A simple feature combination method based on dominant sets. Pattern Recogn. **46**, 3129–3139 (2013)
21. Hamid, R., Maddi, S., Johnson, A.Y., Bobick, A.F., Essa, I.A., Isbell, C.: A novel sequence representation for unsupervised analysis of human activities. Artif. Intell. **173**, 1221–1244 (2009)
22. Pavan, M., Pelillo, M.: A graph-theoretic approach to clustering and segmentation. In: IEEE International Conference on Computer Vision and Pattern Recognition, pp. 145–152 (2003)
23. Gionis, A., Mannila, H., Tsaparas, P.: Clustering aggregation. ACM Trans. Knowl. Discov. Data **1**, 1–30 (2007)
24. Veenman, C.J., Reinders, M., Backer, E.: A maximum variance cluster algorithm. IEEE Trans. Pattern Anal. Mach. Intell. **24**, 1273–1280 (2002)
25. Jain, A.K., Law, M.H.C.: Data clustering: a user's dilemma. In: Pal, S.K., Bandyopadhyay, S., Biswas, S. (eds.) PReMI 2005. LNCS, vol. 3776, pp. 1–10. Springer, Heidelberg (2005)
26. Fu, L., Medico, E.: Flame, a novel fuzzy clustering method for the analysis of dna microarray data. BMC Bioinform. **8**, 1–17 (2007)
27. Daszykowski, M., Walczak, B., Massart, D.L.: Looking for natural patterns in data: part 1. density-based approach. Chemometr. Intell. Lab. Syst. **56**, 83–92 (2001)

An R-CNN Based Method to Localize Speech Balloons in Comics

Yongtao Wang, Xicheng Liu[✉], and Zhi Tang

Institute of Computer Science and Technology of Peking University, Beijing,
People's Republic of China
{wyt,liuxicheng,tangzhi}@pku.edu.cn

Abstract. Comic books enjoy great popularity around the world. More and more people choose to read comic books on digital devices, especially on mobile ones. However, the screen size of most mobile devices is not big enough to display an entire comic page directly. As a consequence, without any reflow or adaption to the original books, users often find that the texts on comic pages are hard to recognize when reading comics on mobile devices. Given the positions of speech balloons, it becomes quite easy to do further processing on texts to make them easier to read on mobile devices. Because the texts on a comic page often come along with surrounding speech balloons. Therefore, it is important to devise an effective method to localize speech balloons in comics. However, only a few studies have been done in this direction. In this paper, we propose a Regions with Convolutional Neural Network (R-CNN) based method to localize speech balloons in comics. Experimental results have demonstrated that the proposed method can localize the speech balloons in comics effectively and accurately.

1 Introduction

There are people around the world enjoying reading comic books and an increasing number of them begin to read comics on digital devices, especially on mobile devices like cellular phones and tablet PCs. Though the screen size of these devices increases all the time nowadays, they are still not big enough to display comic pages on them directly. When an entire comic page is displayed on the screen, it usually becomes less clear than its printed version, especially the texts on it. This makes reading comic books on mobile devices quite inconvenient. The previous solution to this problem is to first decompose a comic page into small units called panels and then display each panel separately. This idea is demonstrated in Fig. 1. In the figure, basic comic page units like panels, texts and speech balloons are also illustrated. From Fig. 1, we can see that it is still not very convenient to browse comic books in this way.

However, recently, there is a great improvement to the size and resolution of the screens of mobile devices. Now, when an entire comic page is displayed on the screen, most elements except the texts can be displayed sharply. This enables a new way to display comic books on mobile devices. We can first locate all the text areas of a comic page and then store them in another form. After that, we remove the texts from the page. When users need to read comic books, the reading app first scales the texts according

© Springer International Publishing Switzerland 2016
Q. Tian et al. (Eds.): MMM 2016, Part I, LNCS 9516, pp. 444–453, 2016.
DOI: 10.1007/978-3-319-27671-7_37

Fig. 1. The old solution to the problem of comic page reflow. The image on the right is the original comic page and the image on the right illustrates how to display the panels separately. On the left image, we mark out different comic elements by rectangles of different colors. The red rectangle corresponds to the entire page and the blue one corresponds to a panel. The elements labeled by green and purple rectangles are a block of text and its corresponding speech balloon respectively. Image credits Inuyasha, vol. 55 p. 33.

to the screen size of the device and then redraw the scaled texts on their original positions. Figure 2 illustrates the main idea of this comic image retargeting method. The main challenge of this solution is to detect the text areas of comic pages. From our observations, most texts in comics come along with surrounding speech balloons. Given the positions of speech balloons, the detection of texts become much easier. Hence, it will make a difference if an effective method to localize speech balloons in comics is devised. From Fig. 2, it can be also seen that to avoid causing damage to reading experiences, the scaled texts should not exceed their surrounding speech balloons. Hence, speech balloons are also useful in the text scaling process, for example, serving as boundaries of the scaled texts.

To the best of our knowledge, only a few studies have been done in this direction to resolve the problem of localizing speech balloons in comics. Arai et al. in [2] propose a speech balloon detection method which is based on blob extraction. Firstly, panels of comic images are extracted and then speech balloons inside each panel are detected via blob extraction with some filtering rules. However, the filtering rules that they exploit are sensitive to noise and are mainly designed for Japanese comic books. In [3], A. Ngo ho et al. adopts a similar strategy to [2] and propose a connected component analysis based method to localize speech balloons. Both these methods are based on the assumption that a speech balloon's background is almost always white and that a speech balloon must be inside a panel. As a consequence, their methods do not have the ability to handle

Fig. 2. The main idea of the proposed comic image retargeting method. From the above images, it can be seen that only the texts are displayed vaguely and that after the text scaling process, the whole page can be displayed sharply on the screen. Image credits Rough vol. 10 p. 17.

balloons which overlap multiple panels. An active contour based method is proposed in [1] by Rigaud et al. Their method works in an unconventional way. They treat the locations of text areas as the prior knowledge. Then, for each text area, they detect its surrounding speech balloon using the active contour method proposed by them. Hence, their method cannot be used here under the circumstance that speech balloons should be detected without having to know the information about other comic elements. Figure 3 illustrates some sample speech balloons.

The speech balloon localization problem can be seen as a special case of the general object detection problem. Hence, methods for resolving the object detection problem can be used here. In this paper, we propose an R-CNN based method to localize speech balloons in comics. R-CNN is one of the state-of-art object detection methods which is proposed by Girshick et al. in [4]. The main paradigm of R-CNN is similar to the "recognition using regions" paradigm proposed in [5]. To be more specific, R-CNN first extracts many object-independent region proposals from the input image, then it computes a fixed-length feature vector for each proposal using a CNN. After that, it classifies each proposal using category-specific SVMs – to decide whether a candidate region contains objects of a specific class. To make the localization result more accurate, R-CNN uses non-maximum suppression and bounding box regression to post-process

Fig. 3. Sample speech balloons. Figure 3a is an example of the speech balloons which overlap multiple panels. Figure 3b is a normal speech balloon. Figure 3c demonstrates an example of speech balloons which are partially drawn. This kind of speech balloons is also very hard for conventional methods to localize. Figure 3d–f are used to demonstrate that the sizes and shapes of speech balloons vary from page to page and book to book. Hence, speech balloon localization is not an easy task to tackle. Figure 3a and e credit Saint Seiya vol. 5 p. 35. Figure 3b credits Edge Legend vol. 43 p. 11. Figure 3c and d credit Dr. Slump vol. 2 p. 26. Figure 3f credits Hunter vol. 2 p. 56.

candidate region proposals. Experimental results demonstrate that speech balloons in comics can be localized effectively and accurately by the proposed method.

The rest of the paper is organized as follows. Section 2 reviews the main idea of R-CNN. Experimental setup and results of the proposed method are given in Sect. 3. Finally, discussions about speech balloon localization and R-CNN are presented in Sect. 4.

2   Introduction to R-CNN

R-CNN [4] is one of the state-of-art general purpose object detection methods proposed by Ross Girshick et al. In 2012, Krizhevsky et al. [6] significantly improve the result of image classification challenge on that year's ImageNet Large Scale Visual Recognition Challenge (ILSVRC) by training a large convolutional neural network (CNN) on a huge amount of images. But this only demonstrates the effectiveness of CNNs in image classification. The question that whether CNNs can be used to significantly improve the result of object detection challenge still remains to be answered. R-CNN shows that by combining the "recognition using regions" paradigm [5], a CNN can dramatically enhance the result of the general purpose object detection challenge. In this section, we will give a brief introduction to the detection and training process of R-CNN.

2.1 Detection Process

In the detection process, R-CNN first extracts many object-independent region proposals from each input image. There are a variety of possible methods which can be used to generate region proposals, like objectness [9], selective search [8] etc. Here, R-CNN chooses selective search as the region proposal extraction method. For each region proposal, R-CNN forward propagates it through a CNN to obtain its corresponding feature vector. The length of the feature vector is fixed. As R-CNN is a general purpose object detection method, there are various kinds of objects to be detected. For each object category, R-CNN trains a SVM for it. Given a region's feature vector, each class's corresponding SVM calculates a category-specific score for the region. Then this score is used to determine whether this region contains objects of that category or not. To resolve the problem that one object may be contained or partially contained by more than one proposal, R-CNN uses non-maximum suppression to post-process the scored regions and excludes regions whose intersection-over-union (IoU) overlap rate with a higher scored region exceeds a learned threshold. The non-maximum suppression process is also category-specific. Finally, in order to make the bounding boxes of objects more accurate, R-CNN employs a linear regression model to predict a new bounding box for each candidate region proposal. The main workflow of R-CNN is illustrated in Fig. 4.

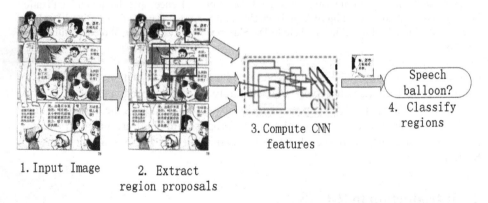

1. Input Image   2. Extract
                   region proposals

3. Compute CNN
   features

4. Classify
   regions

Speech
balloon?

Fig. 4. R-CNN detection workflow overview. R-CNN (1) takes an image as input; (2) uses a region proposal algorithm to extract many sub-regions from the input image; (3) computes a feature vector for each proposal via a CNN; (4) determines that whether a proposal contains objects of a specific class via a category-specific SVM. The sub-image of step 3 credits [4].

2.2 Training Process

The training process of R-CNN consists of three steps: supervised pre-training, domain-specific fine-tuning and object classifier training.

Supervised Pre-training and Domain-specific Fine-tuning. The CNN used in R-CNN is the same as the one proposed by Krizhevsky et al. in [6]. This is a very large neural network which contains 60 million parameters and 650,000 neurons. Hence, in order to make this neural network work properly, a huge amount of data is needed in the training process. But in most object detection problems, the amount of labeled data is often too scarce to train such a large network. To resolve this problem, R-CNN choose to train the network in this way: first pre-trains the CNN on a large auxiliary dataset (e.g. an image dataset for the task of image classification) and then fine-tunes the CNN on a relatively small domain-specific (i.e. task-specific) dataset (e.g. an image dataset for the task of speech balloon detection). Unlike conventional choice, R-CNN uses supervised learning to conduct the pre-training. In this paper, both the pre-training and fine-tuning of R-CNN are performed via the Caffe library [10].

Object Classifier Training. After the pre-training and fine-tuning, R-CNN trains one SVM classifier per category. The features of region proposals in SVM training are the outputs of layer fc_7 (i.e. the final layer of the neural network). Please refer to [6] for the details of the network's architecture. A region is regarded as a positive example if its IoU overlap with some ground truth bounding box exceeds a pre-defined threshold. When the features and labels are ready, R-CNN employs the standard hard negative mining method [11, 12] to perform the SVM training.

3   Experiment

In this section, we present the experimental result of the proposed method on the task of localizing speech balloons in comics. The implementation of R-CNN is based on the source code released on the proposer's GitHub account [7].

3.1   Dataset

Dataset for Pre-training. As mentioned previously, R-CNN uses a large CNN to extract features from proposals. But comic images with labeled speech balloons are scarce. Hence, to make the CNN work properly, we need to pre-train the CNN on an auxiliary dataset. We use the training dataset of ILSVRC 2012 for the task of image classification as the pre-training dataset.

Dataset for Fine-tuning and SVM Training. We first construct an image set consisting of comic pages from some representative comic book series like Dragon Ball, Dr. Slump, Rough etc. Then we randomly choose 900 images from the image set and label the speech balloons in these images with bounding boxes. After that, 800 images are selected randomly from the 900 images to serve as the dataset for fine-tuning and SVM training. This 800 image dataset contains 5147 speech balloons in total. The left 100 comic pages (containing 581 speech balloons) are used in the performance evaluation process.

Dataset for Performance Evaluation. We use two test datasets to evaluate the performance of the proposed method on the common measures of precision, recall and F_1 score. The first dataset consists of the 100 images mentioned previously. These 100 images originate from the same comic book series as the training dataset. To demonstrate the effectiveness of our method, we construct a larger image dataset from a much wider range of comic book series like Ranma, Twin of Brothers, Yu Yu Hakusho etc. This dataset consists of 3000 randomly selected comic pages and contains 17193 speech balloons in total.

3.2   Performance Evaluation Criteria

The output of our method for a comic image is bounding boxes of speech balloons in that image. We evaluate the correctness of a detected speech balloon according to its IoU overlap with the corresponding ground truth bounding box. IoU overlap is defined as follows:

$$\text{IoU overlap} = \frac{\text{AREA (Detected} \cap \text{Ground Truth)}}{\text{AREA (Detected} \cup \text{Ground Truth)}} \tag{1}$$

A detected speech balloon is accepted as a true positive (TP) if its IoU overlap with some ground truth bounding box is greater than 0.6. If a detected speech balloon does not match with any ground truth bounding box, then it is treated as a false positive (FP). A false negative (FN) is found if a ground truth bounding box does not have a match among the detected speech balloons. The effectiveness of R-CNN is evaluated on the common measures of precision (P), recall (R) and F_1 score which are defined as follows:

$$\text{Precision} = \frac{\text{TP}}{\text{TP} + \text{FP}} \tag{2}$$

$$\text{Recall} = \frac{\text{TP}}{\text{TP} + \text{FN}} \tag{3}$$

$$F_1 \text{ Score} = \frac{\text{Precision} \times \text{Recall} \times 2}{\text{Precision} + \text{Recall}} \tag{4}$$

3.3   Experimental Result

Explanation. As the detection of speech balloons in comics is a relatively new task, to the best of our knowledge, there are only three other speech balloon localization methods which have been reported publicly. As summarized in Sect. 1, the methods proposed in [2, 3] have to obtain the information about panels to perform the localization and are unable to localize speech balloons which overlap multiple panels (i.e. the balloons demonstrated by Fig. 3a) while the method proposed in [1] must have the prior knowledge to the positions of text areas. In this paper, we assume that speech balloons should be localized without the information about other comic elements. Hence, it is not fair or possible to compare these methods with the proposed method directly and we decide not to make a comparison between these methods and our method.

Sample Visual Result. Figure 5 demonstrates sample visual results of our method. It can be seen from Fig. 5b that even if the input is a comic page photo, our method still works properly. Figure 5a and c are examples of normal comic pages. The green bounding box in Fig. 5a corresponds to a speech balloon which overlaps three panels. And the two speech balloons labeled by green bounding boxes in Fig. 5c are examples of partially drawn speech balloons. These two kinds of balloons are very hard for conventional methods to localize.

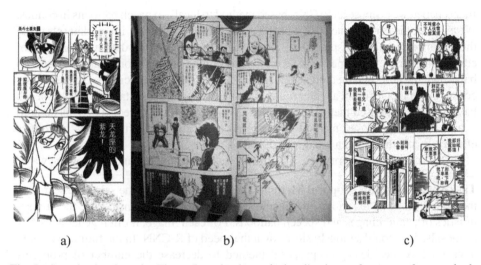

a)   b)   c)

Fig. 5. Sample visual results. Figure 5a and c shows the localization performance of our method on common comic images. The speech balloon labeled by the green bounding box in Fig. 5a is a speech balloon which overlaps three panels. Figure 5b illustrates our method's performance on less clear comic pages. Figure 5c also shows the ability of the proposed method in localizing partially drawn speech balloons (labeled by the green rectangles in Fig. 5c). Figure 5a credits Saint Seiya vol. 5 p. 35. Figure 5b credits Blue Myth Overture Chapter 2 p. 98–99. Figure 5c credits Dr. Slump vol. 2 p. 26.

Quantitative Result. The quantitative performance of the proposed method on the two test datasets are presented in Table 1. From the table, we can see that our method achieves good performance on the dataset consisting of comic pages from the same book series as the training dataset. Though the performance of our method on the dataset consisting of comic pages from different book series is not as good as it on the 100 image one, the 86.1 % precision and 84.1 % recall still can serve as a good evidence for the generalization ability of our method. We are convinced that with more training data, the performance of the proposed method will have a notable improvement.

Table 1. Quatiative results on the two datasets.

Method	Measures		
	Precision (%)	Recall (%)	F_1 score (%)
100 images	92.5	95.1	93.8
3000 images	86.1	84.1	85.1

4   Conclusion

In this paper, we propose an R-CNN based method to localize speech balloons in comics. We evaluate the performance of the proposed method on two comic image datasets using the common measures of precision, recall and F1 score. On the dataset consisting of comic pages from the same book series as the training dataset (containing 581 speech balloons), the proposed method achieves 95.1 % recall, 92.5 % precision. On the dataset consisting of images from a wider range of comic book series (containing 17193 speech balloons), the proposed method achieves 84.1 % recall, 86.1 % precision. This demonstrates that the proposed method can be used to localize speech balloons in comics effectively and that it has a good generalization ability.

In the future, the performance of our method can be improved in the following ways:

Task-specific Region Proposal Method.   In this paper, we use the selective search as the region proposal algorithm. It generates proposals without considering the characteristics of comic images and speech balloons. For each image, it often generates ~5 K proposals. This could seriously slow down the speed of R-CNN. In the future, we could devise a task-specific region proposal method to decrease the number of proposals extracted from images. This could speed up the detection and training process of R-CNN greatly.

Domain-specific Pre-training Dataset.   In this paper, we pre-train the CNN using the training dataset of ILSVRC 2012. This dataset is mainly composed of natural images. In the future, we can make a few modifications to the dataset by replacing half of the dataset with document images and setting the image-level label to document-image and non-document-image. By using this modified dataset, the performance of R-CNN for localizing speech balloons should have an improvement.

Larger Fine-tuning Dataset.   For now, we only use a dataset containing 800 comic images to serve as the fine-tuning dataset. By exploiting a larger dataset, R-CNN could gain a notable improvement on the measures of precision and recall.

Acknowledgement.   This work is supported by National Natural Science Foundation of China (Grant 61300061) and Beijing Natural Science Foundation (4132033).

References

1. Rigaud, C., Burie, J., Ogier, J., Karatzas, D., Weijer, J.: An active contour model for speech balloon detection in comics. In: International Conference on Document Analysis and Recognition, Washington, DC, pp. 1240–1244 (2013)
2. Arai, K., Tolle, H.: Method for real time text extraction of digital manga comic. Int. J. Image Process. **4**(6), 669676 (2011)
3. Ho, A.N., Burie, J., Ogier, J.: Panel and speech balloon extraction from comic books. In: International Workshop on Document Analysis Systems, Gold Cost, QLD, pp. 424–428 (2012)
4. Girshick, R., Donahue, J., Darrell, T., Malik, J.: Rich feature hierarchies for accurate object detection and semantic segmentation. In: Computer Vision and Pattern Recognition, Columbus, OH, pp. 580–587 (2014)
5. Gu, C., Lim, J.J., Arbelaez, P., Malik, J.: Recognition using regions. In: Computer Vision and Pattern Recognition, Miami, FL, pp. 1030–1037 (2009)
6. Krizhevsky, A., Sutskever, I., Hinton, G.E.: ImageNet classification with deep convolutional neural networks. In: Advances in Neural Information Processing Systems, South Lake Tahoe, Nevada, pp. 1097–1105 (2012)
7. Girshick, R.: GitHub, May 2014. https://github.com/rbgirshick/rcnn
8. Uijlings, J.R.R., van de Sande, K.E.A., Gevers, T., Smeulders, A.W.M.: Selective search for object recognition. Int. J. Comput. Vision **104**(2), 154–171 (2013)
9. Alexe, B., Deselaers, T., Ferrari, V.: Measuring the objectness of image windows. IEEE Trans. Pattern Anal. Mach. Intell. **34**(11), 2189–2202 (2012)
10. Jia, Y.: Caffe: an open source convolutional architecture for fast feature embedding, May 2013. http://caffe.berkeleyvision.org/
11. Felzenszwalb, P.F., Girshick, R.B., McAllester, D., Ramanan, D.: Object detection with discriminatively trained part based models. IEEE Trans. Pattern Anal. Mach. Intell. **32**(9), 1627–1645 (2009)
12. Sung, K.-K., Poggio, T.: Example-based learning for view-based human face detection. IEEE Trans. Pattern Anal. Mach. Intell. **20**(1), 39–51 (1998)

Facial Age Estimation with Images in the Wild

Ming Zou(✉), Jianwei Niu, Jinpeng Chen, Yu Liu, and Xiaoke Zhao

State Key Laboratory of Virtual Reality Technology and Systems,
School of Computer Science and Engineering,
Beihang University, Beijing 100191, China
{zouming,niujianwei,chenjinpeng,buaa_liuyu,allenzh}@buaa.edu.cn

Abstract. In this paper, we investigate facial age estimation with images in the wild. We aim to utilize images from the Internet to alleviate the problem of imbalance in age distribution. First, we crawl 14,283 images with their context from *Wikipedia* and infer age labels from the context for each image. After face detection, facial landmark detection and alignment, we build a set of images for facial age estimation, containing 9,456 faces with significant variations. Then, we exploit cost-sensitive learning algorithms including biased penalties SVM and Random forests for age estimation, using images in the wild as the training set. We propose to use the Gaussian function to determine varied misclassification costs. Conducted on two public aging datasets, the within-database experiments illustrate the performance improvement with the introduction of images in the wild. Furthermore, our cross-database experiments validate the generalization capability of proposed cost-sensitive age estimator.

Keywords: Age estimation · Facial images · Cost-sensitive

1 Introduction

Facial images-based age estimation has become a particularly interesting topic these years for its wide applications, such as surveillance monitoring, age-oriented advertisement and face verification for passport renewal. Many approaches have been proposed, such as RUN1 [19], RUN2 [20], AGES [11], MTWGP [21], OHRank [7] and CPNN [12]. While these methods have achieved some encouraging performance, some challenges still remain. One of the key issues is that the training data set is usually insufficient and too imbalanced in age distribution. However, collecting a large dataset with a wide age range of facial images with precise age labels is labor-intensive and time-consuming. It's not a task that could be done in a short time.

Fortunately, the Internet provides us very rich resources of human face images. Recently, Ni et al. [15] attempted to build a universal age estimator using a multi-instance regression method which was able to train on general images. Images from the Internet are usually absent from age labels while many of them are labeled with short descriptions by uploaders. These descriptions may

© Springer International Publishing Switzerland 2016
Q. Tian et al. (Eds.): MMM 2016, Part I, LNCS 9516, pp. 454–465, 2016.
DOI: 10.1007/978-3-319-27671-7_38

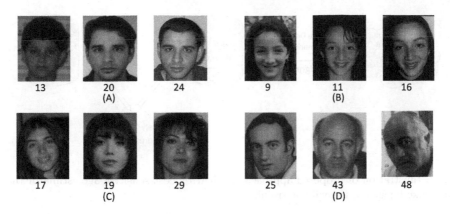

Fig. 1. Typical images with different age gaps in the FG-NET [1] database.

convey some information indicating age. Age labels can then be inferred from these information though these labels may not be 100 % correct. If we can make good use of these images with nearly-correct age labels, the imbalance problem of the training data can be alleviated. *Wikipedia*, as one of the most popular websites with collaboratively created content, is very suitable to acquire such images with context. In this paper, we crawl images from *Wikipedia* and explore the idea of improving age estimation accuracy with the crawled images. To cope with the nearly-correct labels, we turn to the characteristics of age labels for solutions.

The faces with close ages look quite similar since aging is a slow and smooth process [12]. Figure 1 gives several typical faces with different age gaps in the FG-NET database. We can observe that for each person, the difference of his face, looking between the left and the middle image is obviously not equivalent to that between the middle and the right one. Wider age gaps always lead to greater differences in look. Thus, age estimation can be modeled as a cost-sensitive classifier, where the cost varies among classes due to misclassification. State-of-the-art classifiers for age estimation like SVM treat age labels as independent. For each training image, it is assumed that only one class label that describes it, and that all others are equally bad. For the nearly-correct age labels of web crawled images, such cost-insensitive classifiers would take as noisy input which leads to performance deterioration. The cost-sensitive classifier, however, could make good use of them since the different misclassification errors provide some kind of tolerance for inaccuracy of class labels. In this paper, we utilize cost-sensitive classifiers to learn an age estimator that could take images in the wild as the training set. Illustration of our approach is shown in Fig. 2.

The contributions of this paper include: (1) we built a aging collection of people images with inferred nearly-correct age labels from *Wikipedia*; (2) we investigate cost-sensitive SVM and Random Forests for age estimation with images in the wild and propose to use the Gaussian function as the cost function to exploit the ordinal relationship among age labels; (3) both within-database and cross-database experiments are conducted to demonstrate the effectiveness and

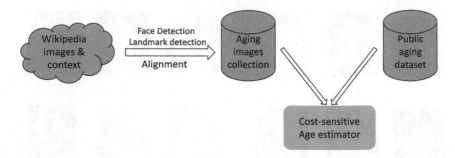

Fig. 2. Illustration of proposed approach. We learn a cost-sensitive age estimator that could take images in the wild as training set.

generalization capability of the proposed method. The rest of this paper is organized as follows. Section 2 discusses related work. In Sect. 3, our collection of *Wikipedia* images are described. We introduce cost-sensitive learning in Sect. 4. Section 5 describes our experimental results. Finally, Sect. 6 concludes.

2   Related Work

Facial Age Estimation. The existing facial age estimation systems can be roughly divided into two parts [10]: aging-related facial features extraction and age determination. For aging-related facial features extraction, many models in computer vision fields have been exploited. Representative models include anthropometric models [14], AAM [9], Aging Pattern Subspace [11] and Biologically Inspired Features [13]. As for age determination, taking age labels as individual classes or a set of sequential values, age estimation can be treated as a multi-class classification [11] or regression problem [13,15]. To fully utilize the ordinal age information of facial images, several recent work formulated age prediction as a ranking problem [7]. Geng et al. [12] proposed the label distribution to utilize the correlated classes. Most of the existing research suffers from the imbalance problem of dataset. Instead of being limited to existing few datasets, we investigate to introduce images in the wild for the training process.

Cost-Sensitive Learning. Cost-sensitive learning is concerned with the situation where certain types of error are more costly than others. Many classification systems have been proposed to solve the cost-sensitive learning problem, such as decision trees [2], support vector machines [3,6], logistic regression and Bayesian network. Cost-sensitive learning has been exploited in applications like medical diagnosis, fraud detection or risk management. Chang et al. [7] proposed an ordinal ranking algorithm with cost sensitivities for age estimation. They used the biased-penalty SVM with absolute or truncated cost for ranking and tested it on FG-NET [1] and MORPH [17]. Different from their approach, in this paper we investigate the cost-sensitive SVM and Random Forests for age estimation with images in the wild. And we propose to use the Gaussian function as the cost function to fully utilize the ordinal relationships among age labels.

3   Building an Aging Collection in the Wild

3.1   Data Collection

We built a collection of people images with inferred nearly-correct age labels from *Wikipedia*. *Wikipedia* is the 6th most popular website, the most widely used encyclopedia, and one of the best examples of truly collaboratively created content. There are more than 4.8 million articles in the English edition.[1] Besides of text, *Wikimedia Commons* is a central repository for free images, music, sound and video clips, and spoken texts and there are over 20 million images in it.[2]

In some of the biographical articles for notable people, there is an image with caption in the infobox where the birth date is usually provided, too. A typical image caption includes date and background of the photo, such as *Jobs holding an iPhone 4 at Worldwide Developers Conference 2010*. Combining the caption with the birth date information is a worth trying way to find out the actual age of the image. This motivates us to crawl biographical articles from *Wikipedia* to get images and related information. About 14,283 images with context were downloaded all together.

We first extracted birth year using simple regular expression from the birth date information like *Steven Paul Jobs, February 24, 1955, San Francisco, California, US*. The same method was then used to extract shooting date from the caption. Difference between the shooting date and the birth date was assigned to the image as age label. We excluded some obviously wrong labels such as negative ones. Images which did not contain enough information to get a label are removed from the dataset. We then did some face detection and post processing to filter out suitable images.

3.2   Face Detection and Alignment

As illustrated in Fig. 3, each image was automatically processed by a face detector [18] followed by a facial landmark detector that finds the landmarks of the face. This combination of methods excluded images without faces. Then we detected pose for remaining images. For every image, we estimated a linear transformation that transformed the detected landmarks to pre-labeled landmarks on a template model. The yaw, pitch and roll angles can be estimated from the rotation matrix. We ignored images with extreme pose. Then we aligned each image by affine transformation. Given the aligned image, we produced a cropped head or a cropped face image for future feature extraction. Finally, we got an aging image collection containing a total of 9,456 images. The age distribution is listed in Table 1.

4   Cost-Sensitive Learning for Age Estimation

As shown in Fig. 1, faces with close ages share some kind of similarity while faces with a wider age gap differ very much. Thus, the cost of misclassifying

---

[1] http://meta.wikimedia.org/wiki/List_of_Wikipedias.
[2] http://commons.wikimedia.org/wiki/Commons:MIME_type_statistics.

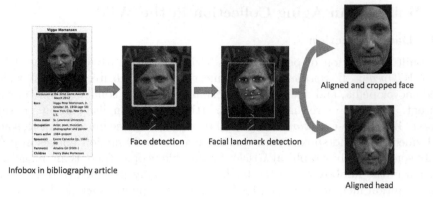

Fig. 3. The whole process of building aging collection in the wild.

Table 1. The distribution of ages in the aging collection in the wild.

Ranges	0–9	10–19	20–29	30–39	40–49	50–59	60–69	70+
No. of images	6	353	1844	2445	2357	1445	872	134

a face image in 25 to 35 should be higher than that of misclassifying it to 26. This characteristic of age labels may help us improve classification accuracy. To take this valuable information into consideration in training time, cost-sensitive learning algorithms can be utilized. The varied costs could also provide accuracy tolerance for age labels, which enables us to use images in the wild in training.

We explore the SVM and Random Forests for age estimation. SVM is fundamentally a two-class classifier while the Random Forests method is inherently a multi-class classifier. SVM has made good results in [7]. Random Forests method has the advantage of fast train and test time so it is adopted when the train and test set are large.

4.1   Biased Penalties SVM

SVM is based on the minimization of the hinge loss. To incorporate cost information into SVM, some researchers have proposed using different penalty parameters in the SVM formulation [3,6]. The proposal is known as the biased penalties method. This method introduces different penalty factors for slack variables during training, that is, assigns each instance x_i a regularization parameter C_i. It is implemented by transforming the SVM problem into

$$\min_{w,b,\xi}\quad \frac{1}{2}w^T w + C\left[\sum_i C_i\xi_i\right],\tag{1}$$

$$s.t.\quad y_i(w^T\phi(x_i)+b) \geq 1 - \xi_i,\quad \xi_i \geq 0, \forall i,$$

where $\phi(x_i)$ maps x_i into a higher-dimensional space and $C_i \geq 0$ is the regularization parameters for instance i.

As SVM is fundamentally a two-class classifier, to tackle problems involving multi classes, the one-versus-one and one-versus-the-rest approach are proposed. For age estimation, however, considering the ordinal relationship among ages, we can formulate it as a ranking problem. Let K be the number of age labels in the database, the estimation problem can be divided into $K - 1$ subproblems, and the kth subproblem is constructed from its anchor age k, by which the training set is separated into two subsets, P_k and N_k, as follows:

$$P_k = \{(x_i, +1)|y_i > k\}; \qquad N_k = \{(x_i, -1)|y_i \leq k\} \qquad (2)$$

$$s.t. \quad 1 \leq k \leq K - 1$$

This kind of formulation has been exploited in [7]. While such ranking based algorithms have shown good performance on relatively small database FG-NET and about 10 % of MORPH, the training and testing time on a larger database would greatly increase due to computational complexity. Decision trees based algorithms benefit from fast training and test time and have been used in many computer vision problems [2]. So we also explore the use of cost-sensitive decision trees for age estimation.

4.2 Random Forests

A Random Forest is an ensemble of decision trees, where each tree is trained independently on a random subset of the data. Algorithms for constructing decision trees usually work top-down from the root node, by choosing a variable that best splits the set of data points landing at each node [4]. The best split condition is quantified by split scoring functions such as information gain. The information gain is derived from information theory and uses entropy as a measure of disorder. Entropy can be computed as

$$I_{ent} = -\Sigma_j p(j|t) \log_2(p(j|t)), \qquad (3)$$

where $p(j|t)$ is the empirical frequency of class j at the node t.

To utilize the cost information in the training time, a reasonable approach is to adapt the class probabilities by including costs [4]. Instead of estimating $p(j|t)$ by N_j/N, the class probability is weighted by the relative cost, leading to an altered probability

$$p(j|t) = \frac{W_j(t)}{\Sigma_i W_i(t)}, \qquad (4)$$

where $W_j(t)$ is the summation of instance weights for class j. These cost aware posteriors can now be used in the calculation of entropy.

For age estimation in this paper, we use an improved information gain proposed in [16], which is defined as follows,

$$E_{inf} = I_{ent}(P) - \left(\frac{N_L}{N} I_{ent}(L) + \frac{N_R}{N} I_{ent}(R)\right), \qquad (5)$$

where N, N_L, and N_R are the numbers of examples landed at the parent, left and right child nodes respectively. $I_{ent}(\cdot)$ is the entropy of instances at the node. E_{inf} serves as the quality measure for spliting the set of data points landing at each node, P, into its left(L) and right(R) child nodes.

4.3 Cost Function

To apply cost-sensitive learning to age estimation, we need to determine the costs for misclassification. Considering the complexity of face aging and the good performance of Gaussian distribution used in facial age estimation by learning from label distribution [12], we use Gaussian function to calculate cost of misclassification between classes. Let cost(i, j) be the cost of misclassifying an instance belonging to class j as belonging to class i, it is defined as follows,

$$cost(i,j) = 1 - \frac{1}{\sigma\sqrt{2\pi}}e^{-(j-i)^2/2\sigma^2}, \tag{6}$$

note that for i=j, cost(i, j) =0 and the choice of standard deviation σ may describe the confidence for the priority of the class label. Concentrative cost could not fully describe the ordinal relationship between classes while dispersive cost might threat the priority of the class label.

5 Experiments

5.1 Dataset and Feature Extraction

Besides our image collection in the wild, two state-of-the-art aging databases, FG-NET [1] and MORPH [17] databases, are also included in our experiment.

The FG-NET Aging database contains 1,002 face images from 82 individuals. The age range is from 0 to 69 years. In terms of MORPH, it is a large-scale database containing more than 55,600 facial images with two to four images per person from 16 to 77 years old. The popular AAM [9] and Biologically Inspired Features(BIF) [13] are used as aging features for the FG-NET and MORPH database respectively. The first 121 model parameters of AAM are used. And the dimensionality of the BIF features are reduced to 200 using OLPP [5].

In our experiments, for biased penalties SVM, in each subproblem k, C_i in (1) is computed as $C_i = cost(k,i)$. As for Random Forests, we define W_j in (4) as $W_j = 10 * \Sigma_i cost(i,j)$.

To evaluate the performance, we adopt mean absolute error (MAE) in our experiment. The MAE is defined as the average of absolute errors between estimated age labels and the ground truth labels, namely, $MAE = \sum_{k=1}^{N} |\overline{y_k} - y_k|/N$, where y_k is the ground truth age for the kth test image, $\overline{y_k}$ is the estimated age, and N is the total number of test images.

5.2 Within-Database Experiments

Since the precise age labels are not provided in our web image collection, we use FG-NET and MORPH database for evaluation.

FG-NET and Our Collection. We adopt the popular leave-one-person-out (LOPO) test strategy, where in each fold, the images of one person are used as the test set, and the algorithm is trained on images of all other images. There are a total of 82 folds as FG-NET contains images of 82 subjects. Since FG-NET is a relatively small database, we could use algorithm that needs more computation for better results. In this experiment, we utilize biased penalties SVM in a similar ranking formulation with OHRank in [7] whereas we use Gaussian cost function for different penalties. The standard deviation in Gaussian cost function is set to 2. We use LIBSVM with weights [6] in our experiment. The RBF kernel is chosen and the parameters are determined using cross validation.

We firstly conduct an experiment on FG-NET database to demonstrate the effectiveness of cost-sensitive learning with Gaussian costs. We compare our results with established age estimation algorithms.

The results are listed in Table 2. We can see that our method achieves the lowest MAE.

To find out the main performance impairment in the FG-NET database, we record image numbers and MAEs in different age ranges, as shown in Fig. 4. We can see that the age distribution in FG-NET database is rather imbalanced. Over 67 % face images are under age 20. As images in higher age groups are rare, the MAEs in higher age groups are much higher.

Table 2. Comparisons of LOPO MAE results on the FG-NET database.

Ours	C-lsLPP [8]	CPNN [12]	OHRank [7]	MTWGP [21]	AGES [11]	RUN2 [20]	RUN1 [19]
4.34	4.38	4.76	4.48	4.83	6.77	5.33	5.78

We try to solve the imbalance problem by introducing images in the wild into training. To form a relatively balanced dataset, we randomly select images of elder people from our image collection and add them to the FG-NET database. 198 images mainly over 40 are chosen and the age distribution of the mixed dataset is listed in Table 3. A similar LOPO test is conducted on the mixed dataset while images from the web collection are included in the training set in all 82 folds. Table 3 presents the detailed results. As presented, after adding images of elder people into the training set, the MAEs in 50–59 and 60–69 reduce greatly while MAEs in 0–40 are slightly influenced. This demonstrates that cost-sensitive learning with Gaussian cost is able to handle nearly-correct age labels and thus make full use of images in the wild in the training time.

We also investigate standard SVM for this mixed dataset. The results are shown in Table 4. We can see that for both tests on FG-NET and the mixed dataset, by including cost information in the training time, the performance improves. An interesting fact is that standard SVM results on FG-NET are better than that on the mixed dataset. It might because images in the wild are taken as noisy inputs in standard SVM. Such noisy inputs lead to accuracy deterioration. However, for biased penalties SVM, the various misclassification errors

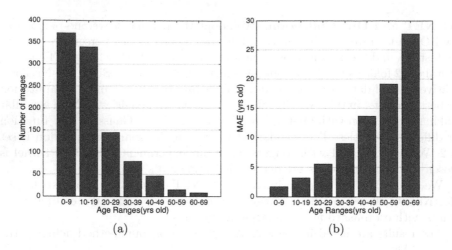

Fig. 4. Image numbers (a) and MAE (b) in different age ranges on FG-NET.

Table 3. Age distribution and LOPO MAE results on FG-NET and Mixed dataset.

Ranges	#images		LOPO MAE (years)	
	FG-NET	Mixed dataset	FG-NET	Mixed dataset
0–9	371	371	**1.69**	1.96
10–19	339	351	3.17	**3.11**
20–29	144	161	5.53	**5.02**
30–39	79	100	**9.00**	9.98
40–49	46	97	13.63	**10.66**
50–59	15	104	19.13	**13.93**
60–69	8	16	27.75	**21.25**
Global	1002	1200	4.34	**4.26**

could provide tolerance for inaccuracy of the age labels. This further demonstrates the advantage of cost-sensitive learning.

MORPH and Our Collection. The FG-NET is a small database, so only part of our collection is chosen to the training set. In this section, we use all the 9,456 images in our collection for training. As shown in Table 1, the age distribution of our collection is rather imbalanced. To alleviate the unbalanced problem, we randomly select images from MORPH in different age ranges and add them to the training set. The final age distribution of the training set is shown in Fig. 5(a). It's a rather balanced dataset. The remaining 39,581 images of MORPH are used for evaluation. Since the training set is relatively large, to save training and testing time, cost-sensitive Random Forests is adopted. As

Table 4. Comparisons of LOPO MAEs between biased penalties SVM and standard SVM. We conduct experiments on both FG-NET and the mixed dataset.

Algorithms	LOPO MAE (years)	
	FG-NET	Mixed dataset
Biased penalties SVM with Gaussian cost	4.34	**4.26**
Standard SVM	4.72	4.86

a result, we achieve an average MAE of 5.52 years. For comparison, we use standard Random Forests method as baseline. The experiments are conducted on the same training and test sets. And we get an average MAE of 7.48 years. The detailed MAEs in different age ranges are shown in Fig. 5(b). As presented, cost-sensitive method achieves lower results in almost all age ranges.

5.3 Cross-Database Experiments

To validate the generalization capability of cost-sensitive learning for age estimation, in this section we train cost-sensitive age estimator only on our aging collection in the wild and test it on MORPH. We adopt the cost-sensitive Random Forests for its fast training and test time. The standard Random Forests method is used as baseline. Our results can be compared with others in the literature that use images in the wild for training an age estimator. In [15] an age regressor was trained on their web collected database (IAD) and tested on MORPH. Detailed results are tabulated in Table 5. As shown in Table 5, we achieve a lower MAE than [15] and cost-sensitive Random Forests method significantly outperforms the standard one. The results demonstrate the generalization capability of cost-sensitive learning.

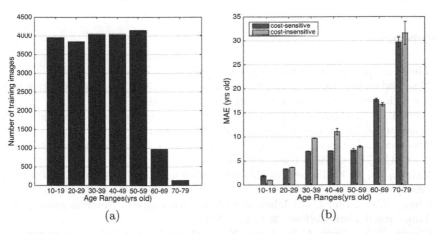

(a)   (b)

Fig. 5. Training image numbers(a) and MAE(b) in different age ranges on mixed dataset. We use 500 trees, a minimum node count of 3 and a maximum possible depth of 20 for Random Forests. The standard deviation in Gaussian cost function is 8. The results are averaged over 5 runs.

Table 5. Comparisons of MAE results on each age range in cross-database experiments. MORPH are used for evaluation. We use 500 trees, a minimum node count of 2 and a maximum possible depth of 30 for Random Forests. Standard deviation in the Gaussian cost function is set to 10. The results are averaged over 5 runs.

Age range	10–19	20–29	30–39	40–49	50–59	60–69	70–79	Avg.
No. of images	7483	15364	15511	12265	3643	324	16	–
MAE (Train on our crawled images, cost-sensitive)	13.87 ±0.05	9.40 ±0.03	6.53 ±0.01	6.62 ±0.04	5.94 ±0.01	4.73 ±0.01	12.00 ±0.19	**8.25** ±0.03
MAE (Train on our crawled images, cost-insensitive)	15.04 ±0.10	10.40 ±0.13	7.34 ±0.01	7.60 ±0.09	8.54 ±0.07	10.29 ±0.24	22.67 ±2.50	9.31 ±0.40
MAE (Train on IAD [15])	8.70	5.45	6.07	12.23	20.30	29.96	40.48	8.60

6   Conclusion

In this paper, a cost-sensitive age estimator is proposed to make good use of images in the wild. We first crawl images with context from *Wikipedia*. Age labels are then inferred from context to form an aging collection. We use cost-sensitive SVM and Random Forests to take these images as the training set. A Gaussian cost function is proposed to determine the misclassification errors. Within-database and cross-database experiments demonstrate the effectiveness and generalization capability of proposed cost-sensitive age estimator with images in the wild.

Acknowledgements. This work was supported by the National Natural Science Foundation of China (61572060, 61170296 and 61190125) and the R&D Program (2013BAH35F01).

References

1. The fg-net aging database. http://www.fgnet.rsunit.com
2. Mac Aodha, O., Brostow, G.J.: Revisiting example dependent cost-sensitive learning with decision trees. In: Proceedings of ICCV (2013)
3. Bach, F.R., Heckerman, D., Horvitz, E.: Considering cost asymmetry in learning classifiers. J. Mach. Learn. Res. **7**, 1713–1741 (2006)
4. Breiman, L., Friedman, J., Stone, C.J., Olshen, R.A.: Classification and Regression Trees. CRC Press, Boca Raton (1984)
5. Cai, D., He, X.: Orthogonal locality preserving indexing. In: Proceedings of SIGIR (2005)
6. Chang, C.-C., Lin, C.-J.: Libsvm: a library for support vector machines. ACM Trans. Intell. Syst. Technol. **2**(3), 27 (2011)
7. Chang, K.-Y., Chen, C.-S., Hung, Y.-P.: Ordinal hyperplanes ranker with cost sensitivities for age estimation. In: Proceedings of CVPR, pp. 585–592 (2011)
8. Chao, W.-L., Liu, J.-Z., Ding, J.-J.: Facial age estimation based on label-sensitive learning and age-oriented regression. Pattern Recogn. **46**(3), 628–641 (2013)

9. Cootes, T.F., Edwards, G.J., Taylor, C.J.: Active appearance models. IEEE TPAMI **23**(6), 681–685 (2001)
10. Fu, Y., Guo, G., Huang, T.S.: Age synthesis and estimation via faces: a survey. IEEE TPAMI **32**(11), 1955–1976 (2010)
11. Geng, X., Zhou, Z.-H., Smith-Miles, K.: Automatic age estimation based on facial aging patterns. IEEE TPAMI **29**(12), 2234–2240 (2007)
12. Geng, X., Yin, C., Zhou, Z.-H.: Facial age estimation by learning from label distributions. IEEE TPAMI **35**(10), 2401–2412 (2013)
13. Guo, G., Mu, G., Fu, Y., Huang, T.S.: Human age estimation using bio-inspired features. In: Proceedings of CVPR (2009)
14. Kwon, Y.H., da Vitoria Lobo, N.: Age classification from facial images. In: Proceedings of CVPR (1994)
15. Ni, B., Song, Z., Yan, S.: Web image mining towards universal age estimator. In: Proceedings of ACM MM, pp. 85–94 (2009)
16. Nowozin, S.: Improved information gain estimates for decision tree induction. In: Proceedings of ICML (2012)
17. Ricanek, K., Tesafaye, T.: Morph: a longitudinal image database of normal adult age-progression. In: Proceedings of FG (2006)
18. Saragih, J.M., Lucey, S., Cohn, J.F.: Face alignment through subspace constrained mean-shifts. In: Proceedings of ICCV (2009)
19. Yan, S., Wang, H., Huang, T.S., Yang, Q., Tang, X.: Ranking with uncertain labels. In: Proceedings of ICME, pp. 96–99 (2007)
20. Yan, S., Wang, H., Tang, X., Huang, T.S.: Learning auto-structured regressor from uncertain nonnegative labels. In: Proceedings of ICCV (2007)
21. Zhang, Y., Yeung, D.-Y.: Multi-task warped gaussian process for personalized age estimation. In: Proceedings of CVPR (2010)

Fast Visual Vocabulary Construction for Image Retrieval Using Skewed-Split k-d Trees

Ilias Gialampoukidis[✉], Stefanos Vrochidis, and Ioannis Kompatsiaris

Information Technologies Institute, CERTH, Thessaloniki, Greece
{heliasgj,stefanos,ikom}@iti.gr

Abstract. Most of the image retrieval approaches nowadays are based on the Bag-of-Words (BoW) model, which allows for representing an image efficiently and quickly. The efficiency of the BoW model is related to the efficiency of the visual vocabulary. In general, visual vocabularies are created by clustering all available visual features, formulating specific patterns. Clustering techniques are k-means oriented and they are replaced by approximate k-means methods for very large datasets. In this work, we propose a faster construction of visual vocabularies compared to the existing method in the case of SIFT descriptors, based on our observation that the values of the 128-dimensional SIFT descriptors follow the exponential distribution. The application of our method to image retrieval in specific image datasets showed that the mean Average Precision is not reduced by our approximation, despite that the visual vocabulary has been constructed significantly faster compared to the state of the art methods.

1 Introduction

Image retrieval has become very challenging over the last years, due to the large amount of images, which are produced on a daily basis. Nowadays there are many applications of image retrieval based on the query by visual example paradigm in order to support personal photo organization, shopping assistance etc. However, one of the main challenges today is the scalability and the performance in terms of time of the image indexing and retrieval methods given the fact that they have to cope with large amounts of images in small amounts of time. Searching in an image collection for similar images is strongly affected by the representation of all images. Spatial verification techniques for image representation, like RANSAC, are computationally expensive and have been outperformed by Bag-of-Words (BoW) models.

The image representation using the BoW model is based on the construction of a visual vocabulary of all visual descriptors in a dataset, in analogy to the representation of text documents, using a vocabulary of visual words [12]. Efficient construction of visual vocabularies is done by clustering the set of all available descriptors in a dataset. In the case of large datasets, the k-means clustering techniques are replaced by approximate k-means methods [10], in order to reduce the computational cost of the visual vocabulary construction, in terms

© Springer International Publishing Switzerland 2016
Q. Tian et al. (Eds.): MMM 2016, Part I, LNCS 9516, pp. 466–477, 2016.
DOI: 10.1007/978-3-319-27671-7_39

of time. However, even this approximate k-means algorithm [10] can be further elaborated in terms of its speed. For example, in [9] it is stated that for 17M descriptors and 1M clusters, a single iteration takes around 5 hours to complete (on a single CPU). Assuming that after 20 iterations, the approximate k-means algorithm is close to a solution, the visual vocabulary construction requires more than 4 days processing time.

Nowadays, the most popular way to represent an image in a set of vectors is using salient points such as SIFT descriptors [4]. However, other image representations have been proposed for image retrieval and object detection, such as VLAD [3] and GIST [7]. In this work we focus our study on SIFT descriptors [4], which is one of the most popular descriptors used nowadays for image retrieval. After observing the values of SIFT in several datasets, we found strong evidence that the values of the SIFT descriptors are exponentially distributed, sharing a similar parameter λ, which can be quickly estimated as a function of the dataset.

The main research contributions of this work are:

- Show that the values of SIFT descriptors are exponentially distributed
- Construct faster k-d trees for SIFT descriptors by introducing a novel method called "skewed-splits"
- Build visual vocabularies quickly using "skewed-splits"

Finally we evaluate our approach by performing several image retrieval tasks with well-known image datasets.

In Sect. 2 we present existing approaches in the construction of visual vocabularies. In Sect. 3 we show that, for several datasets, the values of SIFT descriptors are exponentially distributed. The non-symmetric distribution of SIFT descriptors is utilized for a modified construction of k-d trees and visual vocabularies (Sect. 4), which are applied to two collections of images, in Sect. 5, for image retrieval.

2   Related Work

The image retrieval task was tackled as a Bag-of-Words (BoW) model initially in [12], where k-means clustering was employed for the construction of a visual vocabulary, in analogy to the text retrieval techniques. The most frequent visual words (they occur in almost all images) and very sparse terms are removed, and the final results are filtered in terms of spatial consistency. The query and each image are represented as a sparse vector of term (visual word) occurrences, which are weighted using tf-idf scores. The similarity between the query and each image is calculated, using the Mahalanobis distance.

Hierarchical k-means (HKM) was the first approximate method for fast and scalable construction of a visual vocabulary [6]. Data points are clustered by $k = 2$ or $k = 10$ using k-means clustering and then k-means is applied to each one of the newly generated clusters, using the same number of clusters k. After n steps (levels), the result is k^n clusters.

Hierarchical k-means has been outperformed by approximate k-means [10] which allows for building scalable visual vocabularies. The exact k-means algorithm involves the computation of the distances between all points and cluster centers. In contrast, this computation is replaced by the computation of the distances between points and the approximately nearest cluster centers. The approximate nearest neighbor search is performed using 8 randomized k-d trees. The efficiency of k-means increases as the number of nearest centers increases in the distance computation, but the algorithm becomes slower.

Scalability issues are often tackled using distributed processing and more than 100 processing nodes [8]. The performance of offline indexing have been improved, on a semantic level, by adding semantic attributes on the set of visual descriptors [5].

Contrary to the aforementioned approaches we present a novel method for creating visual vocabularies, in which we exploit the exponential distribution of SIFT descriptors in order to provide a faster and more efficient visual vocabulary construction. After the extraction of SIFT descriptors, we fit all SIFT values to the exponential distribution. From the estimated parameter λ of the exponential distribution, we construct a k-d tree with split value the third quartile of the exponential distribution, namely "k-d tree with skewed split". Our method is comparable to the construction of visual vocabularies using approximate k-means [10], but in each k-d tree the split value is neither the median nor the mean.

3   The Exponential Distribution of SIFT Descriptors

The SIFT descriptors have been proved very reliable for the representation of an image [4]. For SIFT extraction we used the LIP-VIREO toolkit[1], where keypoints are detected using the Fast Hessian detector. After the extraction of SIFT features, we examine each one of the 128 coordinates separately as a sample of SIFT values, in order to test their fit to the exponential distribution.

For our experiments we have used the following image collections:

1. The Pascal voc 2007 dataset[2], containing 9,962 images (Pascal 10K) and its test set (Pascal 5K), containing 4952 images.
2. The Flickr logos dataset[3], containing 8240 images.
3. The Oxford buildings dataset[4] (Oxford 5K) has 5062 images.
4. The Caltech 101 dataset[5] has pictures of objects belonging to 101 categories, from which we get the category "airplanes" for the "Caltech 0.8K" dataset (800 images) and the categories airplanes, barrel, binocular, bonsai, brain, buddha, butterfly, camera, car_side, cellphone, chair, crab, faces, kangaroo, lamp, rhino, saxophone, scissors, snoopy, umbrella, water_lilly for the "Caltech 2.5K" dataset (2,516 images).

---

[1] http://pami.xmu.edu.cn/~wlzhao/lip-vireo.htm.
[2] http://host.robots.ox.ac.uk/pascal/VOC/voc2007/index.html.
[3] http://www.multimedia-computing.de/flickrlogos/.
[4] http://www.robots.ox.ac.uk/~vgg/data/oxbuildings/.
[5] http://www.vision.caltech.edu/Image_Datasets/Caltech101/.

5. The WANG dataset[6] has 1K images, belonging to 10 categories.

We test whether the values of the aforementioned SIFT descriptors are exponentially distributed. Let $x \in [0, x_{max}]$ be the SIFT values for some coordinate $j = 1, 2, \ldots, 128$, and X be the random variable that generates the sample of SIFT values. We shall show that there is strong evidence that the cumulative density function fits well (for some parameter λ) to the form:

$$Prob(X \leq x) = 1 - e^{-\lambda x} \tag{1}$$

In order to test the validity of Eq. (1) we sort the SIFT values x so as to compute $Prob(X > x)$, i.e. the fraction of them that are greater than x, for all values of x. We also define $y = Prob(X > x)$ in order to test if the logarithm $\ln y$ fits to a straight line:

$$\ln y = \alpha + \beta x \tag{2}$$

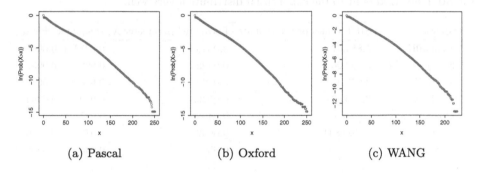

(a) Pascal   (b) Oxford   (c) WANG

Fig. 1. The linear fit of the logarithm $\ln Prob(X > x)$

Table 1. Formulation of the hypothesis tests

Hypothesis	for α	for β
Null	$H_0 : \alpha = 0$	$H_0 : \beta = 0$
Alternative	$H_1 : \alpha \neq 0$	$H_1 : \beta \neq 0$

In all datasets examined (Table 2) we formulated two hypothesis tests, as shown in Table 1. The hypothesis $H_0 : \alpha = 0$ cannot be rejected for all levels of significance because the average p-value is greater than 1 %. However, the hypothesis $H_0 : \beta = 0$, under the alternative $H_1 : \beta \neq 0$, is rejected for all levels

---

[6] http://wang.ist.psu.edu/docs/related/.

of significance because the average p-value is less than 10^{-100}. We conclude that $\alpha = 0$ and $\beta \neq 0$ and Eq. (2) is written:

$$y = e^{-\lambda x}, \quad \lambda = -\beta \tag{3}$$

Equation (1) follows from the fact that $Prob(X > x) + Prob(X \leq x) = 1$ and $y = Prob(X > x)$. The linear fit model is also evaluated by the R-squared statistic, known also as coefficient of determination [11]. We average over the 128 coordinates of the SIFT descriptors in order to have the average R-squared statistic, which is reported in Table 2 for all datasets examined. For illustrative purposes we demonstrate, in Fig. 1, the linear fit of the logarithm $\ln Prob(X > x)$ of selected coordinates for three datasets.

Table 2. The examined datasets and their corresponding fit to the exponential distribution. The average R-squared statistic is very close to 1 in all cases examined. Even for datasets which are very small, such as 800 airplane images from the Caltech dataset, the SIFT descriptors fit to the exponential distribution very well.

Dataset	SIFT values per coordinate	Estimated parameter λ	R-squared $\pm$ std
Pascal 10K	5,884,677	0.0545	0.9914 $\pm$ 0.0066
Flickr logos 8.2K	5,803,263	0.0492	0.9927 $\pm$ 0.0052
Oxford 5K	3,678,453	0.0592	0.9883 $\pm$ 0.0055
Pascal 5K	2,940,834	0.0536	0.9893 $\pm$ 0.0094
Caltech 2.5K	769,546	0.0492	0.9877 $\pm$ 0.0101
WANG 1K	505,834	0.0632	0.9789 $\pm$ 0.0096
Caltech 0.8K	174,091	0.0567	0.9461 $\pm$ 0.0402

Using the exponential distribution of SIFT descriptors we shall provide, a faster and more efficient visual vocabulary construction, in the special case of SIFT descriptors.

4 Visual Vocabulary Construction Using k-d Trees with Skewed Split

In this chapter we introduce a novel methodology for creating visual vocabularies by exploiting the exponential distribution of SIFT descriptors. In order to construct the vocabulary we apply the well established framework for extracting SIFT descriptors [4]. Initially, the keypoints are detected and SIFT descriptors are extracted, which are clustered in order to provide a set of visual words (the visual vocabulary). The clustering technique is usually approximate k-means due to the fact that exact k-means is not applicable for large datasets with billions of descriptors. A general framework is presented in Fig. 2, where k-means clustering of the SIFT descriptors is replaced by approximate k-means methods. The construction of the visual vocabulary results to the image representation using

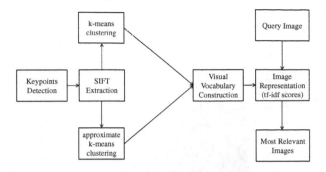

Fig. 2. Image retrieval using SIFT descriptors. After the SIFT extraction, approximate k-means methods may be used alternatively to k-means clustering for the construction of the visual vocabulary.

tf-idf scores, i.e. weighted term frequencies. The similarity between the query and each image is calculated, using the Euclidian distance.

The main novelty of our approach is on the approximate k-means part of Fig. 2. We cluster the extracted SIFT descriptors using the conjunction of 8 k-d trees with skewed split, in order to improve the construction of the visual vocabulary.

In the following, we first discuss the construction of one single k-d tree with skewed split and, secondly, we describe the approximate clustering method using the conjunction of 8 k-d trees with skewed split.

4.1   Construction of k-d Trees with Skewed Split

A k-d tree selects the coordinate with maximum variance and splits the data at the median or the mean value. A randomized k-d tree picks the coordinate to split, at random, from a set of coordinates with the highest variance and the split value s_{median} is chosen to be a value close to the median. The conjunction of 8 randomized k-d trees has been proved very efficient for approximate nearest neighbor search and approximate k-means clustering, for the construction of visual vocabularies [10].

Motivated by the lack of symmetry of the exponential distribution of SIFT values, we propose an alternative split value for the construction of a k-d tree i.e. $s_{skewed} = \ln(4)/\lambda$, which is the 3rd quartile of the exponential distribution, Eq. (1), with parameter λ. The mean value of the exponential distribution is $s_{mean} = 1/\lambda$ and the median is $\ln(2)/\lambda$.

In order to make the split more efficient, we need to take into account the mutual distances. To that end, we perform one simple k-means by $k = 2$ clusters, which results to two centers c_1, c_2. The split value, as obtained by k-means, is the border of the two clusters, i.e. $s_{kmeans} = (c_1 + c_2)/2$. Given a sample of 1K random numbers $u = \{u_1, u_2, \ldots, u_{1000}\}$ from the uniform distribution we generate a sample of 1K exponentially distributed points, using the transformation $-\ln(u)/\lambda$ [2]. After several simulations of exponentially generated datasets,

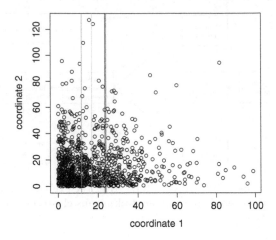

Fig. 3. The projection of 1K exponentially generated points onto the 2 dimensional plane. Choosing the coordinate 1 to split, we demonstrate the median (red line), the mean (green line), the 3rd quartile (black line) and the value s_{kmeans} (blue line) as provided by a simple k-means algorithm on the coordinate 1, by $k = 2$ clusters. We observe that it is hard to distinguish the 3rd quartile (black line) from the value s_{kmeans} (blue line), which we have chosen to be the split value (Color figure online).

we conclude that s_{kmeans} is much closer to s_{skewed} rather than to s_{mean} or s_{median}. Our statement is verified in Fig. 3, where the candidate split values $s_{mean}, s_{median}, s_{skewed}, s_{kmeans}$ are compared.

We note that we tested the correlation of all pairs of coordinates and we did not observe any high correlation. The largest observed square of the Pearson correlation coefficient is 0.75, which is far from the values 0.90–0.99.

Algorithm 1. k-d tree with skewed split

Input: Number of generations $n < 128$, $s_{skewed} = \ln(4)/\lambda$

Output: 2^n leaves

1: Sort the 128 coordinates in decreasing variance in an index set S

2: **for** $i = 1$ **to** n **do**

3: Choose the coordinate with the i-th highest variance $S[i]$ and split at s

4: **end for**

Algorithm 1 shows the construction of a k-d tree in four steps. The parameter λ is computed by Eq. (3), as $\lambda = -\beta$, where the maximum likelihood estimate for β, for any 2-dimensional set of points $(x_i, y_i), i = 1, 2, \ldots, N$, is [1,11]:

$$\beta = \frac{\sum_{i=1}^{N} x_i y_i - \frac{1}{N}(\sum_{i=1}^{N} x_i)(\sum_{i=1}^{N} y_i)}{\sum_{i=1}^{N} x_i^2 - \frac{1}{N}(\sum_{i=1}^{N} x_i)^2} \tag{4}$$

There are two main reasons that our tree is constructed faster than a k-d tree. Firstly, the variance of each coordinate is computed only once, because we sequentially split the dataset without re-calculating the variance. The variance of each coordinate is initially computed and all coordinates are sorted in decreasing variance. Starting from the coordinate with the largest variance, we split the dataset at the split value, into two leaves. For each newly generated leaf, we pick the coordinate with the second largest variance to split. The split process is repeated for all newly generated leaves, until the desired number of leaves is generated. The overall process involves the computation of all variances once for each coordinate and uses them at each step, in order to choose the coordinate to split. Secondly, the split value is fixed for all new splits and is not computed as a function of the dataset. In case a new split results to an empty new node, we set the split value to be the median, which occurs rarely. For k-d trees with 2^k leaves, it never occurred for $k < 12$. As the number of SIFT descriptors increase, it is more unlikely to get an empty new node. In all case examined, even for k-d trees with 2^{16} leaves, the split value is set to be the median for the 3 % of the newly generated leaves.

4.2   Clustering Using a Forest of 8 k-d Trees with Skewed Split

In the approximate nearest neighbors search, points close to the boarder of a leaf are likely to be assigned to incorrect clusters. Philbin et al. [10] used a conjunction of 8 randomized k-d trees to overcome this issue and expand the search area. In contrast, we propose the conjunction of 8 k-d trees with skewed split, tuning the split value s, as shown in Table 3, which results to an overlapping partition of the dataset. The overlapping regions of this partition define the search area of each point, in order to be assigned to its closest center. An illustrative example of overlapping regions is shown in Fig. 4.

Table 3. Tuning the split value s. The 3rd tree has split value the 3rd quartile and the 8th tree has split value the median.

Tree ID	Split value s	Tree ID	Split value s
1	$-\ln(0.15)/\lambda$	5	$-\ln(0.35)/\lambda$
2	$-\ln(0.20)/\lambda$	6	$-\ln(0.40)/\lambda$
3	$-\ln(0.25)/\lambda$	7	$-\ln(0.45)/\lambda$
4	$-\ln(0.30)/\lambda$	8	$-\ln(0.50)/\lambda$

The overlapping regions defined by the conjunction of 8 k-d trees with skewed split are used for clustering. For each point, we use only one search for the closest center, within each overlapping region. The number of overlapping regions determines the number of clusters because after n generations, 2^n leaves are created. Each leaf determines one region which is expanded by a collection of 8 trees with (skewed) split values close to the third quartile. Two points in each expanded region of 8 leaves are considered to be approximately close to each

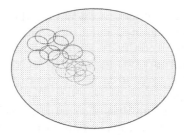

Fig. 4. Two cells of the overlapping partition. Each tree leaf is illustrated by a circle. The union of 8 blue leaves is not disjoint with the union of the 8 red leaves (Color figure online).

other and, in the case of k-means clustering, the search for the closest center is restricted to each region. Therefore, our overall clustering method coincides with one iteration of the approximate k-means algorithm [10], in terms of time, and the computational cost of our approach is $\mathcal{O}(N \log K)$, since we search over the (approximate) closest centers only once, in order to assign points to clusters.

In the following section, we test whether our significantly faster method provides also visual vocabularies of high quality.

5   Application to Image Retrieval

In order to evaluate our method we perform image retrieval experiments with the datasets Caltech 2.5K and the WANG 1K (Table 2). Since the best performing methods for fast construction of visual vocabularies are based on approximate k-means clustering and k-d trees, we create a visual vocabulary based on approximate clustering and k-d trees with skewed split.

We build one visual vocabulary using the conjunction of 8 randomized k-d trees as in [10] and another one visual vocabulary using the conjunction of 8 k-d trees with skewed-split. For the approximate k-means clustering of [10], we allow 20 iterations and the same number of nearest neighbors to search.

After the construction of each visual vocabulary, the tf-idf scores are computed [12]:

$$tfidf_{ij} = \frac{n_{id}}{n_d} \log \frac{D}{n_i}$$

where n_{id} is the number of occurrences of word i in document d, n_d is the number of words in document d, n_i is the number of occurrences of word i in the whole database and D is the total number of documents in the database.

We evaluate our method on 2 datasets of Table 2, namely the Caltech 2.5K and the WANG 1K. The 21 query images of the Caltech dataset are demonstrated in Fig. 5 and the 10 query images of the WANG dataset are demonstrated in Fig. 6. The experiments were performed on an Intel Core i7-4790K CPU at 4.00 GHz with 16GB RAM memory, using a single thread. For the statistical

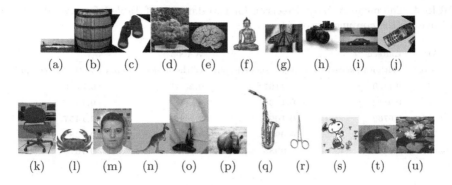

(a)   (b)   (c)   (d)   (e)   (f)   (g)   (h)   (i)   (j)

(k)   (l)   (m)   (n)   (o)   (p)   (q)   (r)   (s)   (t)   (u)

Fig. 5. The query images used for evaluation from the Caltech dataset

(a)   (b)   (c)   (d)   (e)

(f)   (g)   (h)   (i)   (j)

Fig. 6. The query images used for evaluation from the WANG dataset

analysis of SIFT descriptors and the construction of k-d trees (with and without skewed-splits) we used the R programming language[7].

For each query image, we compute the average precision, defined as the area under the precision-recall curve. Averaging for all queries, we obtain the mean Average Precision (mAP) for each visual vocabulary. In Table 4 we present the results in two selected datasets of Table 2, with different numbers of clusters, tuning the number of visual words in 2^n, $n \in \{10, 13, 14, 15, 16\}$.

The reported mAP of the WANG dataset for 1024 clusters is higher in the baseline method. In all other cases we outperform the baseline method, not only in terms of speed, but also in terms of the mean Average Precision. As the number of SIFT descriptors and clusters increase, we observe that our method performs better than the baseline approach, due to the statistical approach we have adopted. For relatively small datasets, such as the WANG dataset, and low levels of the k-d trees, we expect that the statistical laws become weak.

---

[7] https://www.r-project.org/.

Table 4. The mean Average Precision for two datasets of Table 1 for several number of leaves in the conjunction of 8 trees.

mAP	Caltech		WANG	
Clusters	Without skewed split	With skewed split	Without skewed split	With skewed split
1024	0.1100	**0.1167**	**0.2061**	0.1562
8192	0.0812	**0.1168**	0.1405	**0.1457**
16384	0.0769	**0.1009**	0.1382	**0.1457**
32768	0.0733	**0.0932**	0.1317	**0.1389**
65536	0.0731	**0.0935**	0.1314	**0.1378**

6   Conclusion

In this paper we present strong evidence that SIFT descriptors are exponentially distributed. Using the highly skewed distribution of SIFT values, we proposed an alternative split value for the construction of k-d trees. Using the BoW model for image representation, we introduced a novel method for the construction of visual vocabularies. Our tree construction of k-d trees with skewed split and the proposed clustering method are significantly faster than the corresponding baseline method. However, there are some limitations, which need to be considered, for example the fact that the number of visual words cannot be greater than 2^{128}. This is a very large number (greater than 10^{38}) and cannot easily be reached. The application of our model to the image retrieval task has shown that we obtain slightly better mAP than the baseline method, in most cases, but at a small percentage of its computational cost. Even for datasets which are very small, the SIFT descriptors fit to the exponential distribution very well and the mAP is not reduced, when compared to the baseline method.

In the future, we plan to test the statistical properties of other visual features beyond SIFT descriptors. The distribution of the visual features is crucial for fast construction of visual vocabularies.

Acknowledgements. This work was supported by the projects MULTISENSOR (FP7-610411) and KRISTINA (H2020-645012), funded by the European Commission.

References

1. Bishop, C.M.: Pattern Recognition and Machine Learning. Springer, Berlin (2006)
2. Devroye, L.: Sample-based non-uniform random variate generation. In: Proceedings of the 18th Conference on Winter Simulation, pp. 260–265. ACM, December 1986
3. Jégou, H., Douze, M., Schmid, C., Pérez, P.: Aggregating local descriptors into a compact image representation. In: 2010 IEEE Conference on Computer Vision and Pattern Recognition (CVPR), pp. 3304–3311. IEEE, June 2010
4. Lowe, D.G.: Distinctive image features from scale-invariant keypoints. Int. J. Comput. Vis. **60**(2), 91–110 (2004)

5. Luo, Q., Zhang, S., Huang, T., Gao, W., Tian, Q.: Superimage: packing semantic-relevant images for indexing and retrieval. In: Proceedings of International Conference on Multimedia Retrieval, p. 41. ACM, April 2014
6. Mikolajczyk, K., Leibe, B., Schiele, B.: Multiple object class detection with a generative model. In: 2006 IEEE Computer Society Conference on Computer Vision and Pattern Recognition, vol. 1, pp. 26–36. IEEE, June 2006
7. Mikulik, A., Chum, O., Matas, J.: Image retrieval for online browsing in large image collections. In: Brisaboa, N., Pedreira, O., Zezula, P. (eds.) SISAP 2013. LNCS, vol. 8199, pp. 3–15. Springer, Heidelberg (2013)
8. Moise, D., Shestakov, D., Gudmundsson, G., Amsaleg, L.: Indexing and searching 100 M images with map-reduce. In: Proceedings of the 3rd ACM Conference on International Conference on Multimedia Retrieval, pp. 17–24. ACM, April 2013
9. Philbin, J.: Scalable object retrieval in very large image collections. Doctoral dissertation, Oxford University (2010)
10. Philbin, J., Chum, O., Isard, M., Sivic, J., Zisserman, A.: Object retrieval with large vocabularies and fast spatial matching. In: IEEE Conference on Computer Vision and Pattern Recognition, CVPR 2007, pp. 1–8. IEEE, June 2007
11. Rawlings, J.O., Pantula, S.G., Dickey, D.A.: Applied Regression Analysis: a Research Tool. Springer Science & Business Media, New York (1998)
12. Sivic, J., Zisserman, A.: Video Google: a text retrieval approach to object matching in videos. In: Ninth IEEE International Conference on Computer Vision, Proceedings, pp. 1470–1477. IEEE, October 2003

OGB: A Distinctive and Efficient Feature
for Mobile Augmented Reality

Xin Yang[1(✉)], Xinggang Wang[1], and Kwang-Ting (Tim) Cheng[2]

[1] School of Electronics Information and Communications,
HUST, Wuhan, China
xinyang2014@hust.edu.edu, xgwang@hust.edu.cn
[2] Department of Electrical Computer Engineering, UCSB,
Santa Barbara, CA, USA
timcheng@ece.ucsb.edu

Abstract. The distinctiveness and efficiency of a feature descriptor used for object recognition and tracking are fundamental to the user experience of a mobile augmented reality (MAR) system. However, existing descriptors are either too compute-expensive to achieve real-time performance on a mobile device, or not sufficiently distinctive to identify correct matches from a large database. As a result, current MAR systems are still limited in both functionalities and capabilities, which greatly restrict their deployment in practice. In this paper, we propose a highly distinctive and efficient binary descriptor, called *Oriented Gradients Binary* (OGB). OGB captures the major edge/gradient structure that is an important characteristic of local shapes and appearance. Specifically, OGB computes the distribution of major edge/gradient directions within an image patch. To achieve high efficiency, aggressive down-sampling is applied to the patch to significantly reduce the computational complexity, while maintaining major edge/gradient directions within the patch. Comparing to the state-of-the-art binary descriptors including ORB, BRISK and FREAK, which are primarily designed for speed, OGB has similar construction efficiency, while achieves a superior performance for both object recognition and tracking tasks running on a mobile handheld device.

1 Introduction

With the proliferation of mobile handheld devices equipped with low-cost, high-quality cameras, we are witnessing a booming development of MAR systems based on visual object recognition and tracking [11–14, 24]. In such a system, an image frame, captured by a mobile camera, is first described using a set of local feature descriptors, based on which objects in the frame are recognized. Periodic recognition results are bridged by tracking the recognized contents. A distinctive and efficient local feature description could enable a high recognition rate by effectively detecting correct matches from a large database. Such a descriptor could also facilitate real-time mobile object tracking by rapidly capturing the most representative information from an image frame. Thus a high-quality local feature descriptor is critical to the user experience of a MAR system.

© Springer International Publishing Switzerland 2016
Q. Tian et al. (Eds.): MMM 2016, Part I, LNCS 9516, pp. 478–492, 2016.
DOI: 10.1007/978-3-319-27671-7_40

The success of SIFT [1] (and other similar descriptors like HOG [2]) has demonstrated that the distribution of local intensity gradients or edge directions can provide a highly distinctive and robust characterization of local object appearance and shapes. Many previous MAR systems [11–14] rely on SIFT-like descriptors. While highly distinctive and robust, SIFT-like descriptors are too costly to construct on a mobile platform such as smartphones and tablets. In addition, they are usually represented as high-dimensional floating-point vectors, thus are expensive to match and to store on a mobile handheld. As a result, systems based on a SIFT-like descriptor can only afford a small dataset for recognition and are limited in its speed for tracking.

In recent years, lightweight binary descriptors, generated by directly comparing pairs of pixel intensities, are becoming attractive options for MAR due to their small footprint and high efficiency in construction and matching. Most noticeable descriptors include Binary Robust Independent Element Feature (BRIEF) [3], Oriented Fast and Rotated BRIEF (ORB) [4], Binary Robust Invariant Scalable Keypoints (BRISK) [5] and Fast Retina Keypoint (FREAK) [6]. While efficient and compact, these binary descriptors directly utilize raw intensities and rely on ad-hoc schemes for sampling a small number of pixel pairs. The design of existing binary descriptors cursorily discards a large portion of available information resulting in their limited distinctiveness. For instance, they might fail to distinguish patches with sparse textures (e.g. Fig. 1(a)) as it is likely that most sampled pixels may reside in the blank regions. Lack of distinctiveness leads to false matches for object recognition and tracking, and in turn degrades the performance of an MAR system.

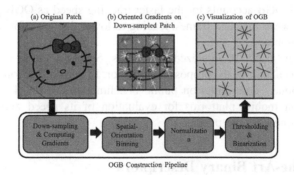

Fig. 1. Illustration OGB descriptor extraction.

In this paper we present a new binary descriptor, called *Oriented Gradient Binary* (OGB). Comparing to state-of-the-art lightweight binary descriptors ORB, BRISK and FREAK, OGB achieves higher distinctiveness while maintaining a comparable speed for descriptor construction. Thus it's more accurate in descriptor matching, particularly for large databases, yielding lower time cost for post-verification and consequently less runtime for the entire recognition and tracking task. The distinctiveness and efficiency of OGB are achieved based on the concept of local major gradient directions, which are defined as directions of strong intensity changes (i.e. having large gradient magnitudes) within a small region of an image patch. OGB encodes the distribution of local major gradient directions by dividing an image patch into grid cells and for each cell accumulating a 1D histogram of gradient directions over all pixels in the cell (see Fig. 1(b)). Then it detects the major gradient directions by thresholding the histogram and finally combines entries of thresholded-histograms to form a representation (see Fig. 1(c)). Computing the gradients for each pixel and mapping them to corresponding spatial-orientation bins are the two most time-consuming subtasks in the process. Aggressively down-sampling an image patch to reduce the amount of pixels for processing can significantly reduce the runtime. Meanwhile, sufficient information for detection of local major gradient directions can be retained by choosing a proper interpolation method to maintain the relative gradient magnitudes among image pixels (as shown by comparing Fig. 1(a) and (b)) and thus minimize the negative impact of down-sampling.

Comparing to existing lightweight binary descriptors which manually select a small set of pixels to reduce the runtime, OGB provides a more complete characterization of the local shape by intelligently aggregating neighboring pixels into one via interpolation and utilizing all aggregated pixels. Aggregation greatly reduces the computational complexity without losing the information of major gradient directions (even for sparsely textured objects like Fig. 1(a)), yielding a highly efficient yet more distinctive description than existing binary descriptors. Experimental results show that for a recognition task with 228 target objects on a mobile handheld device, OGB achieves a greater detection rate and a faster recognition speed comparing to ORB, BRISK and FREAK, while the construction time for the descriptor is about the same. For a mobile tracking task, OGB also achieves a faster tacking speed than all its competitors. Better distinctiveness of OGB accounts for the faster tracking speed as OGB produces fewer false matches between matched frames, leading to lower time cost for post-verification and pose estimation.>

The rest of the paper is organized as follows: Sect. 1 reviews the related work. Section 2 presents details of the proposed descriptor. In Sect. 3, we compare OGB with state-of-the-art binary descriptors on public benchmarks. Section 4 provides experimental results on mobile platforms for evaluation of its speed and discriminative power. Sect. 5 concludes the paper.

2   State-of-the-Art Binary Descriptors

The increasing demand of performing feature extraction and matching in real-time on mobile platforms has stimulated the development of lightweight binary descriptors. A notable one is the BRIEF descriptor [3], which directly generates bit strings by simple binary tests comparing pixel intensities in a smoothed image patch. The pixel positions are selected randomly according to a Gaussian distribution. While highly efficient, BRIEF is very sensitive to image scale and rotation changes, limiting its application to general tasks. Several efforts have been made to address these limitations and further enhance the performance of BRIEF. Rublee et al. proposed ORB [4] which incorporates image pyramids and orientation operators into BRIEF to achieve scale and rotation invariance. In addition, rather than randomly selecting pixel pairs, an ad-hoc selection scheme was proposed for selecting highly-variant and uncorrelated pixel pairs. Leutenegger et al. [5] suggested sampling pixels according to a circular sampling strategy and then selecting short-distant pairs. The resulting descriptor is called BRISK. Alahi et al. [6] further enhanced BRISK by leveraging a sampling strategy which resembles the retinal ganglion cells distribution. However, as described in Sect. 1, these descriptors directly utilize raw intensity values of pixels, sparsely sampled from an image patch based on ad-hoc rules. As a result, they are not sufficiently distinctive and robust to effectively localize matched patches on a target image or from large data-bases. Post-processing to remove false matches is usually required to ensure sufficient recognition and tracking accuracy, increasing the total runtime for MAR. Unlike these binary descriptors, OGB utilizes the distribution of major gradient directions and aggregates pixels by interpolation to reduce the computational complexity with little information loss for detection of major gradient directions. Thus it should provide a more representative and complete patch description.

Several approaches rely on machine learning to compress original floating-point feature vectors to compact binary descriptors. For instance, semantic hashing [7] trains a multi-layer neural network to learn representative and compact binary codes. Spectral hashing [8] minimizes the expected Hamming distance between similar training examples. In [9, 10], the authors explored iterative and sequential optimization strategies respectively to find projections with minimal quantization errors. These approaches aim at minimizing the footprint of the original descriptors to facilitate million-/billion-scale image retrieval, instead of improving the description efficiency. In contrast, our target is a highly efficient and distinctive feature description for real-time mobile apps.

3 ORB: Oriented Gradient Binary

Prior results [1, 2] indicated that the distribution of gradient/edge directions can well characterize a local object's appearance and shape. A common strategy to encode the distribution of gradient/edge directions is to quantize both the spatial space of an image patch and the orientation space, and, for each pixel within the patch, map its gradient magnitude to the corresponding spatial and orientation bins. We refer to descriptors constructed using such a strategy as *Oriented Gradient-based Descriptors* (OGD). The most popular OGD examples include SIFT and HOG. Most existing OGD methods incur high computational complexity, making them infeasible for real-time mobile applications such as mobile object tracking. A major reason for its high computational complexity is that existing OGD methods (e.g. HOG and SIFT) utilize every pixel (or a large portion of pixels) within an image patch for descriptor construction. As computing the gradient and spatial-orientation binning involves non-trivial operations, computing them for every pixel within a single patch is very expensive for a mobile platform.

Oriented Gradient Binary (OGB) is a lightweight version of OGD, which maintains sufficient distinctiveness of conventional OGDs and meanwhile greatly reduces their computational complexity by only encoding the local major gradient directions and the locations of their occurrences. The local major gradient directions are directions of strong intensity changes within a small sub-region (i.e. grid cell). Correspondingly, local *trivial* gradient directions are those with zero or weak intensity changes within a sub-region. As shown in Fig. 2(a), green (red) lines indicate the presence of major (trivial) gradient directions in corresponding grid cells of an image patch.

As OGB encodes only the local major gradient directions and their occurrence locations, any information loss (caused by down-sampling) on trivial gradient directions won't affect the final OGB representation. We observe that down-sampling an image patch, with a proper down-sampling rate and an interpolation method, causes little change to the relative gradient magnitudes among image pixels, and hence could still maintain sufficient information for the detection of major gradient directions (even though detection of trivial gradient directions is degraded). In Sect. 3.3, we evaluate several down-sampling rates and interpolation methods for their impacts on the final results.

In the following sub-sections, we detail the OGB extraction process and compare the distinctiveness and robustness of OGB with existing lightweight binary descriptors analytically. We then present experiments for selecting the optimal parameters and settings for OGB in Sect. 3.3.

3.1 OGB Extraction Process

Figure 1 illustrates the process of computing an OGB descriptor, which consists of 4 major steps:

(1) We down-sample an image patch and compute the intensity gradient magnitude and direction of every pixel within the down-sampled patch. In the implementation, we down-sample the entire image, instead of each image patch separately, and pre-compute the gradients of every pixel in the down-sampled image. This implementation achieves better efficiency because there are usually overlaps between detected patches; computing gradients for each patch separately may result in redundant computations. In order to achieve orientation invariance, the coordinates and gradient directions of the patch are rotated relative to its dominant

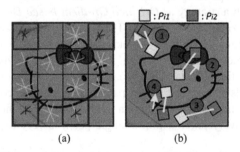

(a)   (b)

Fig. 2. (a) Green lines indicate the major gradient directions within corresponding grid cells of an image patch. Red lines in red rectangles indicate several trivial gradient directions and they are removed for the final descriptor construction. (b) Four pairs of image pixels for constructing a lightweight binary descriptor. A binary test on a pixel pair reflects the intensity change along the direction connecting the two pixels. Thus the result of each binary test encodes the presence of a gradient direction at corresponding spatial location of the patch (Color figure online).

orientation. Any dominate orientation estimation method can be used. In our implementation we use the intensity moment based method [4] due to its good performance and simplicity.

(2) We divide the image patch into $n \times n$ grid cells and quantize the entire orientation space into r bins. For each grid cell we construct a 1D orientation histogram with r bins and map the gradient magnitude of every pixel in the cell into corresponding bins of the orientation histogram. Orientation quantization may introduce undesirable boundary effects: a small change in a pixel's gradient direction may shift its vote from one orientation bin to another, consequently causing non-trivial change to the descriptor. Addressing this problem is of great importance for the OGB's robustness as OGB utilizes only a small number of pixels and thus each pixel has a relatively larger impact on the final description. Our solution uses bilinear interpolation to distribute the value of each gradient sample into its adjacent histogram bins. In other words, each entry into a bin is multiplied by a weighting factor of $1 - d$ for each dimension, where d is the distance of the sample from the central value of the bin as measured in units of the histogram bin spacing. After that, we concatenate all orientation histograms of each grid cell to form an rn^2-dimensional feature vector.

(3) We adjust the rn^2-dimensional feature vector to reduce the effects of linear/non-linear illumination change. Linear illumination changes can cause each pixel value being multiplied by a constant, which in turn will multiply the gradients by the same constant. However, as we only utilize the presence of major gradient directions (0 for absence and 1 for presence), thus changes in the absolute magnitude values won't affect the final description. Non-linear illumination changes can cause a large change in relative magnitudes for some gradients. Therefore, we reduce the influence of large gradient magnitudes by first normalizing the feature vector to a unit length, then adjusting the values in the unit-length feature vector so that no entry in the vector is greater than 0.2, and after that re-normalizing the vector to a unit length.

(4) We compute the average value of the normalized rn^2-dimensional feature vector and use it to quantize each entry of the feature vector into a binary value: a 1 (0) for a value greater (smaller) than the average value, yielding an rn^2-bit binary string.

3.2 Comparison with Lightweight Binaries

To generate a binary string, existing lightweight binary description methods (such as ORB and BRISK) perform binary test on N pixel pairs $\{(p_{i1}, p_{i2})| i_1 \neq i_2, 1 \leq i_1 \leq N, 1 \leq i_2 \leq N\}$: if $p_{i1} > p_{i2}$ the bit corresponding to this pixel pair is set to 1, otherwise is set to 0. The result of a binary test on a pixel pair reflects the intensity change along the direction connecting the two pixels: 1 indicates the presence of a gradient direction pointing from p_{i1} to p_{i2} (e.g. pixel pairs ① and ② in Fig. 2(b)), 0 indicates the presence of a gradient direction pointing from p_{i2} to p_{i1} (e.g. pixel pair ④). In other words, existing lightweight binary descriptors also encode the presence of gradient directions and their spatial distributions within the patch.

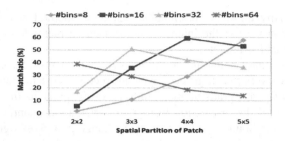

Fig. 3. This figure shows the match ratio as a function of the n × n grid cells and the number of orientations in each histogram. The results are measured by matching 5,000 test patches to a dataset of 5,000 patches. According to the results, 4 × 4 grid cells and 16 orientations give the highest match ratio.

However, existing binary description methods may sample pixel pairs in blank regions (e.g. pixel pair ③), especially for images with sparse textures. These pixel pairs will also produce "0" bits which are not distinguishable from those indicating the presence of gradient directions (e.g. pixel pair ④). In addition, existing binary description methods utilize only local intensity information of a pixel pair, thus they are not able to differentiate major gradient directions (e.g. pixel pair ②) from trivial ones (e.g. pixel pair ①). The resulting descriptor is therefore based on gradient directions that are not very reliable, making it sensitive to noises and image distortions. In comparison, our OGB excludes the non-representative information (i.e. blank regions) and unreliable directions (i.e. trivial gradient directions) and only encode the most informative and stable gradient directions and their spatial distributions. As a result, it is more robust and distinctive than existing lightweight binary descriptors.

3.3   Selection of Parameters and Settings

There are several parameters and settings that affect the performance and computation complexity of an OGB descriptor: (1) the number of orientations, r, in the histograms, (2) the number of grid-cells $n \times n$, within a patch, (3) the down-sampling rate, and (4) the interpolation method for down-sampling. In the following, we provide experimental results which evaluate the impacts of these parameters and settings.

(1)   *Evaluation Method*

We evaluate the performance of different combinations of parameters and settings using a patch matching task. More specifically, our dataset contains 5,000 patches extracted from a set of original images (i.e. the 1st images of a set of 5 image sequences whose details will be described in Sect. 4.1) and the test data consists of 5,000 patches extracted from images with some distortions from the original images (i.e. the 3rd images of the 5 image sequences). The distortions include illumination changes, viewpoint changes, image blurs and compression artifacts. We compute the OGB descriptor for all patches in the dataset and test set. Then for each test patch we performance brute-forcing matching to find its matched patch, which is the nearest neighbor among all patches in the database. We compare the matching results with the ground truth, if a matched patch is within a predefined distance range (e.g. 10 pixels in this test) to the ground truth patch we consider it as a correct match. We use the *Match Ratio*, i.e. the number of correct matches over the total number of matches, to evaluate the matching performance.

(2) *Number of Orientations and Grid Cells Selection*

The number of orientations and grid cells determine the length of the resulting OGB descriptor vector. As the length of the descriptor grows, it will be able to discriminate better in a large database, but it will also be more sensitive to shape and photometric distortions. These two parameters do not affect the computational complexity for descriptor construction, since same number of pixels is involved in the gradient computation and spatial-orientation binning for different parameter settings.

Figure 3 shows the match ratio for different numbers of orientations and grid cells. The graph shows, a larger number of orientations (#bins = 64, indicated by the purple curve) gives the best performance when the number of grid cells is small (2×2), but the results degrade as the number of grid cells increases. For an orientation count of 32 (green curve), 16 (red curve) or 8 (blue curve), the performance is poor if the number of grid cells is small (2×2), but the results peak at grid sizes of 3×3, 4×4 and 5×5 respectively for these 3 cases. These results indicate that adding more grid cells may actually hurt matching performance due to increasing its sensitivity to distortion. The results in Fig. 2 clearly indicate 16 orientations and 4×4 grid cells give the highest match ratio (59.2 %). Thus, for the rest of this paper we use this parameter setting which results in a binary feature vector of 256 bits.

(3) *Down-sampling Rate Selection*

The down-sampling rate affects both the computational complexity and the discriminating ability of the OGB description. Increasing the down-sampling rate reduces the amount of pixels for processing and consequently reduces the computational complexity. But it will also discard more detailed information of a patch, degrading its discriminating ability.

Table 1 shows the match ratio and the time cost for computing one OGB feature for different down-sampling rates. The results show that without down-sampling (i.e. the down-sampling rate = 1) the match ratio is 62.3 % and it takes 0.1213 ms to compute an OGB descriptor. Increasing the down-sampling rate to 2×2 (i.e. reducing the width and height of a patch to ½ of its original width and height respectively) slightly decreases the match ratio while reduces the description time by almost 4 times (0.034 ms per descriptor). Further increasing the down-sampling rate monotonically decreases the match ratio and the description time. In the rest of our experiments we use $8/3 \times 8/3$ as the down-sampling rate as it strikes the best balance between a high discriminating ability and a fast description speed.

(4) *Interpolation Method Selection*

Down-sampling an image may produce floating-point coordinates, resulting in unknown intensity values at certain positions. Thus, interpolation is required to estimate the values at those positions. Different interpolation methods may incur different

Table 1. Match ratio and description time when varying the down-sampling rate. We use the Area Interpolation method for down-sampling.

Down-sample rate	Match ratio (%)	Description time (ms)
1 × 1	62.3	0.121
2 × 2	61.4	0.034
8/3 × 8/3	59.2	0.022
16/5 × 16/5	56.8	0.017
4 × 4	55.1	0.013

Table 2. Match ratio and description time for different interpolation methods. The down-sampling rate is 8/3 × 8/3.

Interpolation method	Match ratio (%)	Description time (ms)
NN	50.4	0.021
Bilinear	54.6	0.022
Cubic	52.3	0.022
Area	59.2	0.022

interpolation losses in image quality and computational complexity. In this section, we evaluate 4 interpolation methods: Nearest Neighbor (NN), Bilinear, Cubic (or Bicubic) and Area-based interpolation. In the following, we briefly explain each of these methods (for more details please refer to [15]) and summarize the evaluation results in Table 2.

NN interpolation considers only one pixel — choosing the closest one as the interpolated point. It is the simplest method and requires the least processing time among all interpolation algorithms.

Bilinear interpolation considers the closest 2 × 2 neighborhood of known pixel values surrounding the unknown pixel. It computes a weighted average of these 4 pixels as the unknown pixel's final interpolated value. This results in a much smoother image than that produced by the NN interpolation.

Cubic interpolation goes one step beyond bilinear by considering the closest 4 × 4 neighborhood of known pixels — with a total of 16 pixels. Cubic interpolation produces noticeably sharper images than the previous two methods.

Area-based interpolation considers a surface area formed by three measuring points which are closest to the interpolation point. The value at the interpolation point is the average of the values in the surface area. After each point is interpolated, its value is added to the measured values. Then the interpolation process starts over again. The procedure is repeated several times until it obtains a sufficiently smooth result.

Table 2 shows the match ratio and the time cost for computing one OGB feature using different interpolation methods. Results show that the Area interpolation method gives the highest match ratio (59.2 %) while takes almost the same time for descriptor construction as other methods. Therefore, in the rest of our experiments, we use the Area interpolation method for image down-sampling.

4 Applications on Mobile Devices

In this section, we evaluate the efficiency and effectiveness of OGB for two mobile applications: mobile object recognition and real-time mobile object tracking, which will be described in Sects. 4.1 and 4.2 respectively.

4.1   Mobile Object Recognition

We apply OGB for object recognition on a mobile platform which follows a conventional object recognition pipeline: we first detect FAST keypoints and construct their OGB descriptors for a captured image frame. Then we match it to our database which returns K top-ranked database images with most matched patches as potential recognized images. Finally, we perform RANSAC for each candidate image. The image with the most and a sufficient number of matches after RANSAC validation is reported as the final recognized image.

(1) Locality Sensitive Hashing for Scalable Matching

Even though calculating the Hamming distance between two 256-bit binary descriptors is extremely fast, it becomes infeasible to performance brute-forcing matching for a large database. For a large-scale matching task, an indexing structure is usually utilized for finding approximate NNs efficiently. Locality-Sensitive-Hashing (LSH) [22] is a widely used technique for approximate NN search and we use it in our object recognition experiment.

The key of LSH is a hash function, based on which similar descriptors can be hashed into the same bucket of a hash table while dissimilar ones will be stored in different buckets. To find the NN of a query descriptor, we first retrieve its matching bucket and then check all the descriptors within this bucket using a brute-force matching.

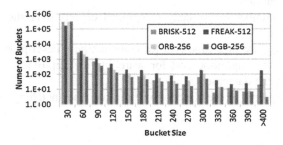

Fig. 4. LSH bucket size distribution of the 228 image dataset. OGB-256 has fewer large-sized buckets than BRISK, FREAK and ORB, thus is faster in approximate NN matching.

For binary features, the hash function is usually a subset of bits from the original bit string: buckets of a hash table contain descriptors with a common sub-bit-string. The size of the subset, i.e. the hash key size, determines a maximum Hamming distance among descriptors within the same bucket. To improve the chance that a correct NN can be retrieved by LSH, two techniques, namely multi-table and multi-probe [23], are often used. Multi-table stores the database descriptors in several hash tables, each of which uses a different hash function. In the query phase, each query descriptor is hashed into a bucket of every hash table and the matches within the bucket are checked. Multi-table improves the detection rate of NN at the cost of linearly increased memory usage and matching time. Multi-probe examines both the bucket in which a query descriptor falls and the neighboring buckets. While multi-probe would result in more matches to check, it actually requires fewer hash tables and thus lower memory usage. In addition, it enables a larger key size and therefore smaller buckets and fewer matches to check per bucket.

(2) Experiment Setup

Our database contains 119 planar objects, each of which is an image selected from the public dataset Oxford Building 5K. We added another 108 distractor images, yielding 228 database images in total. We generated our testing data, i.e. query images, by manually captured a picture for each planar object, leading to 119 query images.

We extracted around 1000 features on average from each database image and each query image. All the database features are stored in an LSH indexing structure, consisting of 5 hash tables with 1.1M + entries in total. For all the hash tables, we set the key size as 18 and the number of probes as 19. We use the OpenCV implementation for LSH. For each query, we return the top 10 database images as candidate results for which RANSAC verification was performed. The image with the most and more than 10 matches after RANSAC validation is reported as the final recognized result.

All the programs were executed in a single thread running on a 1.5 GHz processor of a Google Nexus 4 smaprtphone. The operating system is Android 4.2 Jelly Bean. Five metrics are used for evaluation:

- *Detection Rate*: the number of objects that are successfully recognized over the total number of objects. In our case, the detection rate is 1 if all the 119 planar objects are successfully identified.
- *Precision*: the number of correctly recognized objects over the total number of recognized objects.
- *Description Time*: average time cost for constructing a descriptor.
- *Matching Time:* average time cost for searching the approximate NN of a query descriptor from the database.
- *Total Time*: the sum of the description time and the matching time for a query descriptor.

(3) Results

Table 3 compares the performance of OGB with ORB, BRISK and FREAK. First, we compare the detection rate and precision of four binary descriptors in the first two columns of Table 3. We observe that OGB achieves the highest detection rate (74.0 %), which is 26.1 %, 6.7 %, and 49.6 % higher than those of BIRSK, FREAK and ORB respectively. Better distinctiveness and robustness of OGB account for its superior detection rate, as more matching keypoints can be found from a large database, increasing the probability that a correct matching image can be detected. Meanwhile, the precision of OGB is also higher than those of BRISK and ORB and very close to that of FREAK (91.7 % vs. 95.2 %). Secondly, we compare the time cost for description and matching. Even though computing an OGB descriptor is slightly slower, it is much faster for approximate NN matching than its competitors. In particular, when comparing with FREAK, which achieves the fastest description speed and a similar precision as OGB, OGB is 3.2X faster in approximate NN matching. The speedup in approximate NN matching is most likely due to the better distinctiveness of OGB. This is because non-distinctive descriptors may cluster together in feature space. As a result, when we quantize the feature space using LSH hashing functions some quantized sub-regions (i.e. buckets of a hash table) may contain a large number of descriptors, leading to a large number of checks during the approximate NN matching process. On the contrary, distinctive descriptors distribute more uniformly in the features space, and thus each bucket contains a relatively smaller number of distinctive yet similar descriptors, yielding fewer and more relevant checks. To further validate this point, we took a close look at the distribution of buckets in the hash tables, shown in Fig. 4. We observe that OGB (the purple bars) makes the buckets of the hash tables more even. Comparing to FREAK, the average bucket size of OGB is much smaller, e.g. the number of buckets whose size is smaller than 30 is 313,000+ for OGB vs. 165,000+ for FREAK. OGB also has fewer large-sized buckets than BRISK and ORB.

4.2 Real-Time Mobile Object Tracking

Object tracking on a mobile device involves matching the live frames to a previously captured frame. As consecutive frames usually share large content overlaps, it is less challenging for achieving satisfactory matching accuracy than performing a recognition task for a pair of matching images: the photometric changes (e.g. lighting differences) and geometric changes (e.g. scaling) of consecutive frames are smaller and more predictable. Therefore, for tracking we extract fewer FAST keypoints (200) than for recognition (1000). Then for each keypoint we compute its binary descriptor and find two nearest neighbors in the previous frame using a brute-force matching. This process produces a total of 400 candidate matches for a pair of consecutive frames, which are then used as input for RANSAC-based homography estimation.

We evaluated the performance of tracking using 4 public video streams "Paper"[3], "Phone"[3], "Comic"[4] and "David"[4]. The first three video streams include planar objects with unconstrained motion patterns, including viewpoint changes, translation, rotation, scaling, etc. The last video stream captures the movement of a human being with unconstrained motion patterns and lighting changes.

Table 3. Performance of object recognition on Google Nexus 4.

Desc.	Detection rate (%)	Precision (%)	Time cost (ms)		
			Desc.	Match	Total
BRISK	47.9	86.4	0.19	1.9	2.09
FREAK	67.3	95.2	0.15	6.7	6.85
ORB	24.4	80.6	0.17	2.6	2.77
OGB	74.0	91.7	0.25	1.6	1.85

We examine the total time cost for tracking, including cost for description, matching and RANSAC-based pose estimation. Figure 5 shows the time break-down for each step of the tracking process. First we compare the description time (blue bars)

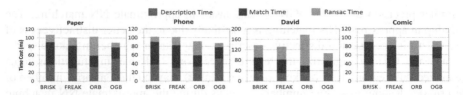

Fig. 5. Comparison of time cost for tracking. Four video streams are evaluated, including "Paper", "Phone", "David" and "Comic" (Color figure online).

Table 4. Comparison of inlier ratio between consecutive frames

Inlier ratio (%)	Paper	Phone	David	Comic
BRISK-512	51.1	57.1	45.1	50.9
FREAK-512	48.0	47.8	40.5	47.8
ORB-256	38.7	42.0	31.8	40.9
OGB-256	58.4	62.3	52.3	58.5

for all binary descriptors. The average time cost for computing 200 descriptors for a video frame is 38 ms for BRISK, 30 ms for FREAK, 33 ms for ORB and 52 ms for OGB. In terms of speed for matching (red bars), ORB and OGB are of the same speed (26 ms) and are twice faster than BRISK and FREAK (52 ms). Faster matching speed of ORB and OGB is due to the shorter descriptor length (256 bits) than those of BRISK and FREAK (512 bits). The time cost for pose estimation (green bars) varies for different video streams, but for all testing video streams OGB costs the least amount of time. Faster pose estimation speed of OGB is due to the higher inlier ratio when matching between consecutive frames. Here, the inlier ratio is defined as the number of matches that are consistent with the homography model estimated by RANSAC over the total number of matches. Table 4 reports the inlier ratio and shows that for all video streams, OGB achieves the highest inlier ratio than other binary descriptors. Please note that, in principle the inlier ratio should be smaller than 50 % as for keypoint on a current frame we find two nearest neighbors on the previous frame to guarantee a high detection rate. However, we observe that many keypoints detected by FAST detector are very close to each other, thus matching to either of these two points can be considered as a correct match. Finally, we compare the total time cost for tracking. The results show that for all video streams, OGB takes least runtime for the entire tracking process.

5   Conclusion

In this paper, we introduce a new binary descriptor, named OGB, to facilitate object recognition and tracking on a mobile platform. OGB computes the distribution of local major gradient directions within an image patch, which is a very important and representative characteristic of local shapes and appearance. Aggressively down-sampling the image patch at a rate of 8/3 × 8/3 and using the Area interpolation method can reduce the time cost by $\sim 6X$ for description, meanwhile maintaining sufficient information for detection of major gradient directions. Experimental results demonstrate that OGB achieves higher distinctiveness than ORB, BRISK and FREAK, while maintains a comparable speed for descriptor construction. Higher distinctiveness of OGB makes it more accurate in descriptor matching, leading to lower time cost for false match filtering and more efficient NN matching.

Though we apply OGB to mobile object recognition and tracking individually, it can be directly incorporated into a unified recognition-and-tracking flow for MAR. In this work we leveraged a standard LSH technique for recognizing objects on a smartphone. A more advanced indexing structure can be explored in the future to reduce the memory usage and time cost for further improving the scalability of a MAR system.

References

1. Lowe, D.G.: Distinctive image features from scale-invariant keypoints. Int. J. Compu. Vision **60**(2), 91–110 (2004)
2. Dalal, N., Triggs, B.: Histograms of oriented gradients for human detection. In: Proceedings of CVPR 2005 (2005)
3. Calonder, M., Lepetit, V., Strecha, C., Fua, P.: Brief: binary robust independent elementary features. In: Proceedings of ECCV 2010 (2010)
4. Rublee, E., Rabaud, V., Konolige, K., Bradski, G.: ORB: an efficient alternative to SIFT or SURF. In: Proceedings of ICCV 2011, Barcelona, Spain (2011)
5. Leutenegger, S., Chli, M., Siegwart, R.: BRISK: binary robust invariant scalable keypoints. In: Proceedings of CVPR 2011 (2011)
6. Alahi, A., Ortiz, R., Vandergheynst, P.: FREAK: fast retinal keypoint. In: Proceedings of CVPR 2012 (2012)
7. Salakhutdinov, R., Hinton, G.: Semantic hashing. Int. J. Approximate Reasoning, **3** (2009)
8. Weiss, Y., Fergus, R., Torralba, A.: Spectral hashing. In: Proceedings of NIPS 2009, pp: 1753–1760 (2009)
9. Gong, Y., Lazebnik, S., Gordo, A., Perronnin, F.: Iterative quantization: a procrustean approach to learning binanry codes for large-scale image retrieval. IEEE Trans. PAMI (2012)
10. Wang, J., Kumar, S., Chang, S.-F. Sequential projection learning for hashing with compact codes. In: Proceedings of ICML 2010 (2010)
11. Wagner, D., Reitmayr, G., Mulloni, A., Drummond, T., Schmalstieg, D.: Pose tracking from natural features on mobile phones. In: Proceedings of ISMAR 2008 (2008)
12. Wagner, D., Schmalstieg, D., Bischof, H.: Multiple target detection and tracking with guaranteed framerates on mobile phones. In: Proceedings of ISMAR 2009 (2009)
13. Wagner, D., Mulloni, A., Langlotz, T., Schmalstieg, D.: Real-time panoramic mapping and tracking on mobile phones. In: Proceedings of IEEE VR 2010 (2010)
14. Klein, G., Murray, D.: Parallel tracking and mapping on a camera phone. In: Proceedings of ISMAR 2009, Orlando (October 2009)
15. Parker, J., Kenyon, R., Troxel, D.: Comparison of interpolating methods for image resampling. IEEE Trans. Med. Imaging **2**(1), 31–39 (1983)

16. Rosten, E., Porter, R., Drummond, T.: Faster and better: a machine learning approach to corner detection. IEEE Trans. PAMI **32**, 105–119 (2010)
17. OpenCV. http://sourceforge.net/projects/opencvlibrary/
18. ImageSequences. http://www.robots.ox.ac.uk/~vgg/research/affine
19. Everingham, M., Van Gool, L., Williams, C.K.I., Winn, J., Zisserman, A.: The PASCAL visual object classes challenge (2009)
20. Fischler, M.A., Bolles, R.C.: Random sample consensus: a paradigm for model fitting with applications to image analysis and automated cartography. Comm. ACM **24**(6), 381–395 (1981)
21. Chum, O., Matas, J.: Matching with PROSAC – progressive sample consensus. In: Proceedings of CVPR 2005, vol. 1, pp. 220–226 (2005)
22. Gionis, A., Indyk, P., Motwani, R.: Similarity search in high dimensions via hashing. In: Proceedings of VLDB (1999)
23. Lv, Q., Josephson, W., Wang, Z., Charikar, M., Li, K.: Multi-probe LSH: efficient indexing for high-dimensional similarity search. In: Proceedings of VLDB (2007)
24. Hong, R.C., Tang, L.X., Hu, J., Li, G.D., Jiang, J.G.: Advertising object in web videos. Neurocomputing **119**, 118–124 (2013)
25. Wang, M., Li, G.D., Lu, Z., Gao, Y., Chua, T.-S.: When amazon meets google: product visualization by exploring multiple information sources. ACM Trans. Internet Technol. **12** (4), Article 2 (2013)
26. Wang, M., Li, H., Tao, D.C., Lu, K., Wu, X.D.: Multimodal graph-based reranking for web image search. IEEE Trans. Image Process. **21**(11), 4649–4661 (2012)

Learning Relative Aesthetic Quality
with a Pairwise Approach

Hao Lv and Xinmei Tian[✉]

University of Science and Technology of China, Hefei 230027, Anhui, China
haolv@mail.ustc.edu.cn, xinmei@ustc.edu.cn

Abstract. Image aesthetic quality assessment is very useful in many multimedia applications. However, most existing researchers restrict quality assessment to a binary classification problem, which is to classify the aesthetic quality of images into "high" or "low" category. The strategy they applied is to learn the mapping from the aesthetic features to the absolute binary labels of images. The binary label description is restrictive and fails to capture the general relative relationship between images. We propose a pairwise-based ranking framework that takes image pairs as input to address this challenge. The main idea is to generate and select image pairs to utilize the relative ordering information between images rather than the absolute binary label information. We test our approach on two large scale and public datasets. The experimental results show our clear advantages over traditional binary classification-based approach.

Keywords: Aesthetic quality · Binary classification · Relative ranking

1 Introduction

Image aesthetic quality assessment is a hot research topic, and has drawn much attention recent years. It is a useful technique in many real-word applications. For example, image search engine can incorporate aesthetic quality to refine its search results. Photo management system should consider aesthetic quality as an important factor when ranking photos for users. Hence, users can more easily select the photos with better aesthetic quality.

Most researchers focus their attention on aesthetic quality classification problem, which is to predict whether an image is of "high" or "low" aesthetic quality [1,2,4,5,8,11–14,17,18]. They have spent a lot of efforts on extracting effective aesthetic features, from low-level features [2,18], high-level features [4,8,12,13] to generic features [5,11,14,17]. Despite different ideas and approaches to extract aesthetic features, they share the same thought on training the binary classification model, which is to learn the mapping from aesthetic features to binary aesthetic labels of images. They utilize the absolute binary label information of images, but ignore the relative ordering information between different images.

However, the binary aesthetic labels they predicted are restrictive and unnatural. As shown in Fig. 1, it is hard to decide whether Fig. 1(b) is of "high" or

© Springer International Publishing Switzerland 2016
Q. Tian et al. (Eds.): MMM 2016, Part I, LNCS 9516, pp. 493–504, 2016.
DOI: 10.1007/978-3-319-27671-7_41

(a) "low" (2.51)   (b) "?" (5.0)   (c) "high" (8.0)

Fig. 1. The aesthetic scores (collected from popular photo sharing website DPChallenge.com [6] and scored by many different users) and labels of images. It is unnatural and restrictive to describe image quality with binary label, since it is hard to decide whether the quality of (b) is "high" or "low". However, we can express quality of (b) in a more informative and natural way: (b) is more aesthetically pleasing than (a) while less beautiful than (c).

"low" quality. However, we can describe the quality of Fig. 1(b) in a more general and natural way: Fig. 1(b) is more beautiful than Fig. 1(a) but less beautiful than Fig. 1(c). In this work, we propose to model the relative aesthetic quality, which is to focus on relative aesthetic quality ranking. It is of great practical significance, since relative comparison is a more natural way for people to describe and compare objects in real life.

To address the relative aesthetic quality ranking problem, existing methods can also estimate a probability of the learned binary classifiers prediction, which indicating the absolute aesthetic quality of an image. However, they suffer from same limitation during training. The aesthetic quality of training images is restricted to be binary, "high" or "low", which is not precise or natural. For example, it is not so reasonable to assign Fig. 1(b) with an aesthetic label of "high" or "low". Thus, this binary label description of aesthetic quality may introduce "noisy" information, while the relative supervision is more precise. For example, it is easier to define and agree on, "Is this image more beautiful than the other?" than "What the absolute aesthetic quality does this image has?". Thus, we expect the relative supervision to be more natural and precise.

How do we learn relative aesthetic quality? We propose a ranking framework based on a pairwise approach to address this problem. Traditional binary classification models learn the classifiers by utilizing the absolute binary label information. In contrast, our goal is to learn the relative ordering relationship of images with different aesthetic quality. The main idea of our approach is to capture the relative relationship of training images by generating and selecting training image pairs. We generate training image pairs which consist of images with different aesthetic quality. Furthermore, considering that not all pairs generated are useful, we select certain pairs based on proposed rules. The selecting process acts as a filter and filters out "noisy" pairs. The selected pairs contain more useful and precise relative information, which are important for improving the performance of our ranking framework. The way we generating pairs and selecting pairs is not only easy to understand but also very effective, which is verified in our experiments. We then adopt a ranking model that takes image pairs as input to learn a ranking function. It will estimate ranking scores for test-ing images. The ranking scores are used for ordering images only, which have no meaning in absolute sense.

In summary, in this paper we focus on relative aesthetic quality while most existing works are committed to traditional binary classification problem. To address relative aesthetic quality problem, we propose a pairwise-based rank-ing framework. We generate and select image pairs that contain relative order information between training images, which is essential for improving the per-formance of proposed ranking framework. The experiments on two large scale and public datasets show that our pairwise approach significantly outperforms the binary classification-based approaches.

The remainder of this paper is organized as follows. We review related works in Sect. 2 and describe the details of our proposed pairwise-based approach in Sect. 3. Then we evaluate the performance of our approach in Sect. 4. Finally, we conclude and discuss the future work in Sect. 5.

2   Related Work

In this section, we first review related works on aesthetic quality classification, and then discuss works that concern about the relative ranking problem.

Aesthetic Quality Classification. Many image aesthetic quality assess-ment approaches have been proposed in recent years. However, most existing approaches focus on aesthetic quality classification. They share the same thought on training binary classification model and they spend a lot of efforts on designing different aesthetic features. Roughly, these methods can be divided into three cat-egories: low-level feature-based approaches, high-level feature-based approaches and generic feature-based approaches.

Low-level feature-based approaches extract a set of low-level features that are commonly used in computer vision tasks [2, 18]. Tong et al. extracted blur-riness, colorfulness, saliency value and so on [18]. They achieved limited success because of these features are not specially designed for the aesthetic quality of images. Datta et al. designed a set of low-level features, which are related to user intuition and some photography literature, i.e. "rule of thirds", "simplicity" and "interestingness" [2]. After carefully designed, they extracted 56-dim features and obtained a better performance.

High-level feature-based approaches focus on designing high-level features based on photography and psychology literature [4, 8, 12, 13]. Dhar et al. proposed a set of attribute-based predictors to conduct aesthetic quality evaluation [4]. Luo et al. extracted different features for different categories of photos and then gen-erated category-specific classifiers [12]. Luo and Wang designed features mainly describing the image composition and relationship between subject region and background region [13].

Generic feature-based approaches extracted a large set of image features, which are used to describe image content [14, 17]. Marchesotti et al. extracted generic image content descriptors to conduct aesthetic quality classification and gained certain improvement [14]. Lu et al. applied three schemes to incorporate deep learning with aesthetic quality assessment, and obtained improved performance [11]. Dong et al. directly adopted the deep neural network trained on ImageNet [3] and extracted the 4096-dim output activations of the seventh layer as aesthetic features, and achieved remarkable success [5].

Relative Ranking. Many researchers focus their attention on relative ranking problems of images recent years. Kumar et al. proposed comparative facial attributes for face verification [10]. The attributes they explored are similarity-based. Wang et al. learned fine-grained image similarity with deep ranking model [19]. They proposed a deep ranking network and an efficient triplet sampling algorithm to address the fine-grained image similarity. Parikh and Grauman devised a ranking framework to learn ranking functions for image attributes, given relative similarity constraints on pairs of examples [16]. Based on relative attributes, a novel form of zero-shot learning and image describing experiments were conducted. Significant improvement was obtained by relative attributes-based approach compared with traditional binary classification-based approach. Inspired by the work related to relative ranking, we propose a pairwise-based ranking framework to address relative aesthetic quality ranking problem.

3   Relative Aesthetic Quality Ranking

Existing binary classifier-based approaches utilize absolute binary label information of training images. Unlike existing approaches, the intention of our pairwise approach is that we want to utilize the relative ordering information between training images. The architecture of proposed pairwise-based ranking framework is shown in Fig. 2. Given a set of training images, we first generate and select image pairs based on certain rules. Then we feed these selected pairs into the ranking model to learn a ranking function. During testing stage, the ranking function will estimate real-value ranking scores for images, which are used for ordering examples. In this section, we present the pairwise-based ranking model (Sect. 3.1) and explain the details of our training image pairs generation (Sect. 3.2) and selection (Sect. 3.3).

3.1   Pairwise-Based Ranking Model

We are given a set of training images $\mathcal{I} = \{I_i\}, i = 1, 2, \ldots, m$, represented in $\mathbb{R}^n$ by feature-vectors $\{x_i\}$, a set of aesthetic quality labels $\mathcal{A} = \{a_i\}, a_i \in \{0, 1\}$, and a set of image class labels $C = \{c_i\}, c_i \in \{1, 2, \ldots, C\}$. The aesthetic label "0" is for "low quality" and "1" is for "high quality". The image class labels describe the semantic content of images. Based on certain rules (described in Sects. 3.2 and 3.3), we generate a set of ordered image pairs denoted as $O = \{(I_i, I_j)\}$,

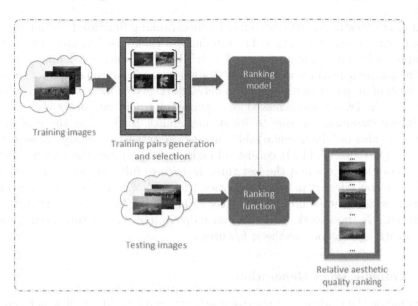

Fig. 2. The architecture of proposed pairwise-based ranking framework. We capture the relative information by training the ranking model with generated and selected image pairs.

pairs in which satisfied $(I_i, I_j) \in O \Rightarrow a_i > a_j$. Our goal is to learn a ranking function:

$$r(x_i) = w^t x_i, \tag{1}$$

by maximizing the number of the following constraints satisfied:

$$\forall (I_i, I_j) \in O : w^t x_i > w^t x_j. \tag{2}$$

This leads to an optimization problem:

$$
\begin{aligned}
\min \quad & (\frac{1}{2}\|w\|_2^2 + C\sum \xi_{ij}^2) \\
s.t. \quad & \forall (I_i, I_j) \in O : w^t x_i \geq w^t x_j + 1 - \xi_{ij}, \\
& \xi_{ij} \geq 0,
\end{aligned} \tag{3}
$$

where C is the trade-off constant.

Problem (3) can be reformulated as:

$$
\begin{aligned}
\min \quad & (\frac{1}{2}\|w\|_2^2 + C\sum \xi_{ij}^2) \\
s.t. \quad & \forall (I_i, I_j) \in O : w^t(x_i - x_j) \geq 1 - \xi_{ij}, \\
& \xi_{ij} \geq 0.
\end{aligned} \tag{4}
$$

We solve this problem using SVM^{rank} [7] with linear kernel. This learning-to-rank model explicitly enforces a desired ordering on training examples. When given a set of testing images, we apply learned ranking function to estimate real-value ranking scores for images. The ranking scores are used to order the testing examples only. The absolute values of which have no practical significance.

The aesthetic features we extracted is 4096-dim normalized output of the seventh layer of deep convolutional neural network (DCNN) designed by Krizhevsky et al. [9]. The DCNN has achieved great success in many computer vision tasks, e.g. image classification, due to its strong ability to describe the content of images. It also obtained remarkable success on aesthetic quality classification task as reported in [5,11]. It consists of eight layers in total, the first five layers are convolution layers and the last three layers are fully connected layers. The details can be referred in [9]. We extract the DCNN features as our aesthetic features, which is well-suited to the task as hand. We feed the raw RGB image to the DCNN framework, and take normalized 4096-dim output activation of the seventh layer as our aesthetic features.

3.2 Training Pairs Generation

When facing the challenge of relative aesthetic quality ranking, it is not enough to train just a binary classifier, despite that the learned classifier can also estimate a score indicating the absolute strength of images aesthetic quality. The main limitation is that the binary classifier-based approaches ignore the relative ordering information between images during training process.

To overcome the limitation shared by existing approaches, we focus on generating effective and informative image pairs to capture the relative ordering information. Considering that images with different aesthetic quality are potential to contain relative information, we generate all the possible image pairs in the training set. We generate all the possible pairs that consist of images with different aesthetic quality, which means that an image from class of "high" quality will form pairs with all images from class of "low" quality. Then the generated image pairs are denoted as:

$$O = \{(I_i, I_j)|a_i > a_j\} \tag{5}$$

Our strategy of generating pairs is easy to understand and effective, which is verified in our experiments. We feed the generated pairs to the ranking model to learn the ranking function.

3.3 Informative Training Pairs Selection

Images in the datasets are from different image classes. For example, the CUHKPQ dataset consists of seven categories (please refer Sect. 4 for detail). It has to be noticed that not all images are comparable, i.e. comparison between an image on animal and an image on architecture does not make much sense. The method described above generates all the possible pairs, which means that they contain a large set of image pairs consisting of images from different categories.

Image pairs consisting of images with similar content are much more reasonable and comparable. It is more natural to compare the aesthetic quality of images both on landscape. The ranking framework can benefit a lot from the image pairs that contain useful and comparable relative information. Therefore, it is important to keep the comparable pairs while wipe out the others. Based on this consideration, we take a selection step to reserve the comparable pairs, and the selected pairs are denoted as:

$$O_s = \{(I_i, I_j) | a_i > a_j, c_i = c_j\} \tag{6}$$

Compared with the method in Sect. 3.2, we put constraint on the image class labels during the selection step. Images from the same category are more likely to have similar content, which are more reasonable to be compared. Only image pairs consisting of images from the same category are reserved, while others are considered as "noisy" pairs and wiped out. After selection, the number of pairs are largely reduced, and the selected pairs contain less noisy information. Then we feed the selected pairs into the ranking model.

4   Experiments

In this section, we test our proposed approaches on two public datasets, CUHKPQ and a subset of AVA. Two datasets are widely used in aesthetic quality assessment field, and both contain considerable quantity of photos. We implement the-state-of-art approach as baselines [5]. We compare the performance of our approach with the baselines to verify the effectiveness of our proposed pairwise-based ranking framework.

4.1   Datasets

CUHKPQ. CUHKPQ dataset is released by Luo et al. [12]. It consists of photos downloaded from professional photography sharing websites and photos contributed by amateurs. It is divided into seven categories according to photo content. Each photo in the dataset is evaluated by ten independent viewers. Each photo is assigned with an aesthetic label "high" or "low" under the condition that eight out of ten viewers share same opinion on its assessment. It contains a total of 17690 photos from seven categories. We randomly and evenly divide it into training set and testing set, each with 8845 photos. We focus on the problem of relative aesthetic ranking. Images with similar content are more reasonable to be compared as explained in Sect. 3. Based on this consideration, during testing, we restrict relative comparison between images within the same category. Therefore, we compare the predicted relative order with original relative order within the same category to obtain test performance on seven categories.

AVA. DPChallenge.com [6] is one of the most active and popular photo sharing communities on the Internet. Users in this community are from different levels of photography enthusiasts. There are a variety of photographic challenges in the community, each defined by a title and short description. Users can upload their photos according to a specific photographic challenge. Other users can score and comment on the uploaded images. AVA is collected by Murray et al. [15] from DPChallenge.com. It consists of more than 250000 images with different tags indicating the semantic content of images. The number of aesthetic scores each image received is range from 78 to 549, with an average of 210. Average aesthetic score of the image is taken as the ground-truth value.

Murray et al. only offer the web links of images and some of them are invalid because of the update of their website. We successfully downloaded 193077 images. As aforementioned, we restrict the relative aesthetic quality comparison within the same category during testing. We extract nine categories with largest number of images in the 193077 images. We randomly and evenly divide it into training set and testing set. We adopt the same criteria to assign each image an aesthetic label "high" or "low" as reported in [15]. Images with mean score larger than or equal to $5 + \delta$ are defined as "high" quality images while those with mean score smaller than or equal to $5 - \delta$ are "low" quality images. Others are discarded. In this paper, we set $\delta = 1$. Under this setting, we have 21116 images for training and 21117 images for testing.

4.2 Experimental Settings

We implement the approach based on binary classification as baseline [5]. The aesthetic feature extraction method of baseline is the same as reported in [5]. We adopt the DCNN framework trained on ILSVRC-2012 and take the 4096-dim output activation value of seventh layer as aesthetic features. The widely used machine learning algorithm SVM is trained with the extracted features and aesthetic labels to generate the binary classifier. During testing, baseline approach estimates the probability of the binary classifiers prediction rather than a binary label for each image. We calculate the AP value within same category as the evaluation indicator. The AP value is often used in information retrieval field. We use it here to measure how well the predicted ranking is consistent with the ground truth ranking within the same category. The model parameters for all methods are determined via five-fold cross-validation on training set.

In our proposed pairwise approach, we also use extracted 4096-dim DCNN features as aesthetic features. To implement the method described in Sect. 3.2, we generate all the possible image pairs. We denote this method as "pairwise_gen". Then, we feed all these pairs to SVM^{rank}. We implement the method described in Sect. 3.3, and denote it as "pairwise_sel". We take a selecting step to wipe out some of the pairs generated in method "pairwise_gen". Then we feed the selected pairs to SVM^{rank} to learn the ranking function. Linear kernel is adopted for SVM^{rank} and the model parameters are also determined via five-fold cross-validation on training set.

Table 1. Mean AP value on CUHKPQ dataset with different approaches.

Approach	Mean AP
Baseline	0.811
Pairwise_gen	0.828
Pairwise_sel	**0.879**

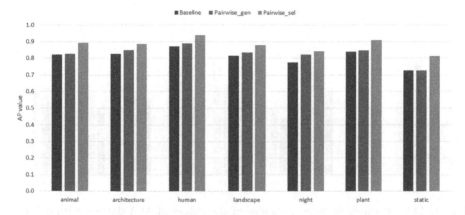

Fig. 3. The AP values on seven categories of CUHKPQ dataset with different approaches.

4.3 Experimental Results on CUHKPQ

In this experiment, we evaluate the performance of binary classification-based approach and our pairwise approaches on CUHKPQ dataset. We present the comparison of mean AP value on seven categories in Table 1 and the details of each category in Fig. 3. Among three methods, our proposed pairwise approach "pairwise_sel" achieves the best performance with mean AP value at 0.879. Method "pairwise_gen" obtains mean AP value at 0.828, while the baseline method reaches mean AP at 0.811. Although we generated image pairs in an easy way in method "pairwise_gen", we still obtain a better performance than the binary classification-based method. As shown in Fig. 3, method "pairwise_gen" performs better than baseline method on all seven categories, which indicates the robustness of our pairwise approach. Moreover, the method "pairwise_sel" significantly outperforms the method "pairwise_gen", which demonstrates that our pairwise-based ranking framework can benefit a lot from the training image pairs selection step.

The experimental results on CUHKPQ dataset show the advantage of our proposed pairwise approaches on relative aesthetic quality ranking. Although the binary classification-based method achieves an acceptable result, our proposed method "pairwise_gen" improves the performance by using a pairwise approach. Whats more, with a selecting step, proposed method "pairwise_sel" outperforms the baseline with a larger margin.

Table 2. Mean AP value on AVA dataset with different approaches.

Approach	Mean AP
Baseline	0.470
Pairwise_gen	0.531
Pairwise_sel	**0.611**

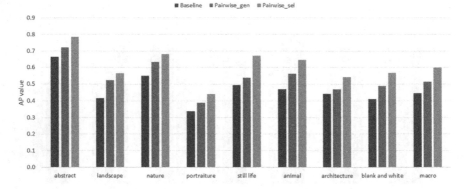

Fig. 4. The AP values on nine categories of the subset of AVA dataset with different approaches.

4.4   Experimental Results on AVA

We present the experimental results of three methods on the subset of AVA dataset in Table 2 and Fig. 3. The mean AP value over nine categories achieved by our method "pairwise_sel" is 0.611, which is the best result. The method "pairwise_gen" achieves a better result than baseline. Our proposed method "pairwise_sel" improves the performance of method "pairwise_gen" and baseline with a large margin, which indicates the effectiveness of proposed selecting step. The details on each category are shown in Fig. 4. Method "pairwise_gen" consistently outperforms baseline on all nine categories. Method "pairwise_sel" obtains better performances on nine categories over other two methods. The results on nine categories show the clear advantages of our proposed pairwise approaches.

Compared with baseline, we improves the performance by using a pairwise approach. The improvement shows the advantage of pairwise-based ranking framework at capturing the relative ranking information of images. We obtained an even larger improvement when taking a selecting step on image pairs generated in method "pairwise_gen". This verifies the contribution of proposed selecting step, which is to wipe out "noisy" pairs.

5   Conclusion and Future Work

Inspired by that it is more natural to model the relative aesthetic quality than absolute binary labels, we aim to study the aesthetic quality ranking rather than traditional aesthetic quality classification. In particular, we have proposed a pairwise-based ranking framework, which takes image pairs as input. In order to better capture the relative ordering information, we have proposed certain rules to generate and select training image pairs. We took the DCNN features as aesthetic features and SVM^{rank} as our rank model. The experimental results revealed that the proposed pairwise approach could capture the relative information of images better than traditional binary classification approach. The proposed selection step helped to wipe out "noisy" pairs and improved the performance. Despite the encouraging results achieved, this is just an attempt to study the relative aesthetic ranking problem, and there are still many open challenges. In the future, we will investigate more effective ways and more powerful models to utilize the relative order information.

Acknowledgement. This work was supported by the 973 project under Contract 2015CB351803, by the NSFC under Contracts 61390514 and 61201413, by the Youth Innovation Promotion Association CAS No. CX2100060016, by the Fundamental Research Funds for the Central Universities No. WK2100060011 and No. WK2100100021, and by the Specialized Research Fund for the Doctoral Program of Higher Education No. WJ2100060003.

References

1. Bhattacharya, S., Sukthankar, R., Shah, M.: A framework for photo-quality assessment and enhancement based on visual aesthetics. In: Proceedings of the international conference on Multimedia, pp. 271–280. ACM (2010)
2. Datta, R., Joshi, D., Li, J., Wang, J.Z.: Studying aesthetics in photographic images using a computational approach. In: Leonardis, A., Bischof, H., Pinz, A. (eds.) ECCV 2006. LNCS, vol. 3953, pp. 288–301. Springer, Heidelberg (2006)
3. Deng, J., Dong, W., Socher, R., Li, L.J., Li, K., Fei-Fei, L.: Imagenet: a large-scale hierarchical image database. In: IEEE Conference on Computer Vision and Pattern Recognition, CVPR 2009, pp. 248–255. IEEE (2009)
4. Dhar, S., Ordonez, V., Berg, T.L.: High level describable attributes for predicting aesthetics and interestingness. In: 2011 IEEE Conference on Computer Vision and Pattern Recognition (CVPR), pp. 1657–1664. IEEE (2011)
5. Dong, Z., Shen, X., Li, H., Tian, X.: Photo quality assessment with DCNN that understands image well. In: He, X., Luo, S., Tao, D., Xu, C., Yang, J., Hasan, M.A. (eds.) MMM 2015, Part II. LNCS, vol. 8936, pp. 524–535. Springer, Heidelberg (2015)
6. DPChallenge: Dpchallenge. http://www.dpchallenge.com/
7. Joachims, T.: Training linear svms in linear time. In: Proceedings of the 12th ACM SIGKDD International Conference on Knowledge Discovery and Data Mining, pp. 217–226. ACM (2006)
8. Ke, Y., Tang, X., Jing, F.: The design of high-level features for photo quality assessment. In: 2006 IEEE Computer Society Conference on Computer Vision and Pattern Recognition, vol. 1, pp. 419–426. IEEE (2006)
9. Krizhevsky, A., Sutskever, I., Hinton, G.E.: Imagenet classification with deep convolutional neural networks. In: Advances in Neural Information Processing Systems, pp. 1097–1105 (2012)

10. Kumar, N., Berg, A.C., Belhumeur, P.N., Nayar, S.K.: Attribute and simile classifiers for face verification. In: 2009 IEEE 12th International Conference on Computer Vision, pp. 365–372. IEEE (2009)
11. Lu, X., Lin, Z., Jin, H., Yang, J., Wang, J.Z.: Rapid: rating pictorial aesthetics using deep learning. In: Proceedings of the ACM International Conference on Multimedia, pp. 457–466. ACM (2014)
12. Luo, W., Wang, X., Tang, X.: Content-based photo quality assessment. In: 2011 IEEE International Conference on Computer Vision (ICCV), pp. 2206–2213. IEEE (2011)
13. Luo, Y., Tang, X.: Photo and video quality evaluation: focusing on the subject. In: Forsyth, D., Torr, P., Zisserman, A. (eds.) ECCV 2008, Part III. LNCS, vol. 5304, pp. 386–399. Springer, Heidelberg (2008)
14. Marchesotti, L., Perronnin, F., Larlus, D., Csurka, G.: Assessing the aesthetic quality of photographs using generic image descriptors. In: 2011 IEEE International Conference on Computer Vision (ICCV), pp. 1784–1791. IEEE (2011)
15. Murray, N., Marchesotti, L., Perronnin, F.: Ava: a large-scale database for aesthetic visual analysis. In: 2012 IEEE Conference on Computer Vision and Pattern Recognition (CVPR), pp. 2408–2415. IEEE (2012)
16. Parikh, D., Grauman, K.: Relative attributes. In: 2011 IEEE International Conference on Computer Vision (ICCV), pp. 503–510. IEEE (2011)
17. Su, H.H., Chen, T.W., Kao, C.C., Hsu, W.H., Chien, S.Y.: Scenic photo quality assessment with bag of aesthetics-preserving features. In: Proceedings of the 19th ACM International Conference on Multimedia, pp. 1213–1216. ACM (2011)
18. Tong, H., Li, M., Zhang, H.-J., He, J., Zhang, C.: Classification of digital photos taken by photographers or home users. In: Aizawa, K., Nakamura, Y., Satoh, S. (eds.) PCM 2004. LNCS, vol. 3331, pp. 198–205. Springer, Heidelberg (2004)
19. Wang, J., Song, Y., Leung, T., Rosenberg, C., Wang, J., Philbin, J., Chen, B., Wu, Y.: Learning fine-grained image similarity with deep ranking. In: 2014 IEEE Conference on Computer Vision and Pattern Recognition (CVPR), pp. 1386–1393. IEEE (2014)

Robust Crowd Segmentation and Counting in Indoor Scenes

Ren Yang[1](✉), Huazhong Xu[1], and Jinqiao Wang[2]

[1] School of Automation, Wuhan University of Technology, Wuhan 430070, China
yangren@whut.edu.cn, wutxhz@163.com
[2] National Laboratory of Pattern Recognition, Institute of Automation,
Chinese Academy of Sciences, Beijing, China
jqwang@nlpr.ia.ac.cn

Abstract. This paper proposes a fast counting approach to estimate the number of people in indoor scenes. Firstly, a pre-processing step is used. In order to obtain a robust gray image in complex light conditions this step includes color correlation, image smoothing and contrast stretch. Secondly, we extract foreground region by background edge modeling and contour filling. Finally, after a foreground normalization based on camera calibration, we obtain the counting results with template matching. Experimental results show that compared with the Bayesian counting approach [2], our approach is robust to illumination variation and achieves a real-time counting result in indoor scenes.

Keywords: People counting · Crowd segmentation · Template matching

1 Introduction

Crowd counting in indoor spaces and public areas is a challenging topic since the coexistence of moving crowds with stationary crowds, recurrent occlusions and complex background information. Traditional solutions rely on manual counting, which takes lots of time and money. Therefore, automatic estimation and counting the crowd number is a critical requirement for the increasing number of video cameras. The rapid development of computer vision and image processing technology makes it possible for automatic crowd counting. However, most of people counting approaches are focused on outdoor scenes such as [1–4], few works pay attention to counting in indoor scenes. Due to the coexistence of moving crowds with stationary crowds, recurrent occlusions and complex background information, existing crowd counting methods drops significantly in indoor scenes. The stationary people are missed due to moving foreground segmentation and the counting results are often disturbed by occlusions. In this paper, in order to count not only moving crowds but also stationary crowds, we proposes a method to extract human bodies and carry out template matching to estimate the crowd number.

© Springer International Publishing Switzerland 2016
Q. Tian et al. (Eds.): MMM 2016, Part I, LNCS 9516, pp. 505–514, 2016.
DOI: 10.1007/978-3-319-27671-7_42

2   Related Work

A lot of crowd counting or people counting researches have been done in the field of computer vision. The number of people was calculated by the correlation between people detection and tracking, which is acquired from multiple cameras with non-overlapping field of view [3–6]. Color histogram in RGB and HSV model can be rendered by color image acquired from cameras of three angles, and then based on feature trajectory, the number of people was counted by analyzing human face histogram [7–9]. Zhao and Nevatia [2] treated the problem of segmenting individual humans in crowd as a model-based Bayesian segmentation problem. The combination of background block-updating and head-shoulder matching is effective to avoid wrong background in people counting, Luo et al. [10] proposed a people counting approach in indoor environment. Despite the decrease of error rate in people counting, there are some drawbacks, such as the head-shoulder models are sensitive in the camera views. This paper proposes a new method for people counting based on background edge subtraction method, which is robust to light change. The crowd number is estimated from the foreground areas and normalized people template.

3   Robust Crowd Counting

In this paper, a new people counting method in indoor environment is proposed. As shown in Fig. 1, the proposed approach includes pre-processing, crowd segmentation, and crowd normalization and counting. Firstly, a background image was established when the room with normal lighting is empty. Secondly, each input frame is projected into HSV color space and pre-processed by graying and smoothing. Thirdly, taken the gray background image as a reference, the contrast of each input frame is stretched for lighting coherence. Fourthly, canny operator was used to extract edges and the foreground edge image was acquired by the difference between current edge image and background edge image. Finally, we fill the foreground edge image and the number of people was determined by the crowd area.

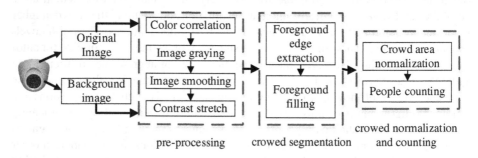

Fig. 1. The overall framework of robust crowd counting in indoor scenes

3.1 Pre-processing

Inspired by the background subtraction approaches, this paper uses background edge subtraction method to get the foreground image. Different with traditional background subtraction approaches in model update, we update the background image when there is no person in the scenes. In addition, a pre-processing step is introduced to enhance the robustness, which includes color correlation, image smoothing and contrast stretch.

Color Correlation. Each incoming frame does not always have stable values of hue, saturation and value because of the actual light conditions in the physical environment, while background image is selected appropriately. Therefore, according to the three values of hue, saturation and value in the background image, we correlate the color value in the HSV color space as follows.

$$g_h(x,y) = \begin{cases} f_h(x,y), if(\left|g_h(x,y) - f_h(x,y)\right| \leq \alpha) \\ g_h(x,y), otherwise \end{cases}$$
$$g_s(x,y) = \begin{cases} f_s(x,y), if(\left|g_s(x,y) - f_s(x,y)\right| \leq \beta) \\ g_s(x,y), otherwise \end{cases} \qquad (1)$$
$$g_v(x,y) = \begin{cases} f_v(x,y), if(\left|g_v(x,y) - f_v(x,y)\right| \leq \gamma) \\ g_v(x,y), otherwise \end{cases}$$

Here (x,y) represents the position of a pixel in the input frame. $g_h(x,y), g_s(x,y)$ and $g_v(x,y)$ are hue, saturation and value. $f_h(x,y)$, $f_s(x,y)$ and $f_v(x,y)$ represent the values of H, S, V in the background image. In order to adjust current image acquired from camera in an appropriate range, values of α, β, γ were set 30, 50 and 10 respectively.

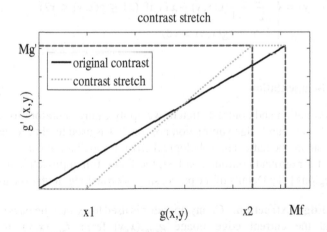

Fig. 2. Contrast stretch

Contrast Stretch. Contrast stretch can increase the contrast and adjust the pixel values for further image processing. According to grayscale of the background image, we adjust the grayscale of each incoming frame. We define $f_{gray}(x, y)$ and $g_{gray}(x, y)$ as grayscale of background image and the current frame respectively. The size of current frame is $m \times n$ and array $ff[256]$ and array $gg[256]$ were used to record the number of grey value from 0 to 255 in grayscale of the background image $f_{gray}(x, y)$ and grayscale of the current frame $g_{gray}(x, y)$ respectively. According to contrast stretch function in Fig. 2, the grayscale of the current frame $g_{gray}(x, y)$ was converted to $g'_{gray}(x, y)$.

As shown in Fig. 2, the new grayscale of the current frame of $x1$ and $x2$ was determined by Eqs. 2 and 3.

$$ff[w1] = \max(ff[0], ff[1], \ldots, ff[255])$$
$$gg[w2] = \max(gg[0], gg[1], \ldots, gg[255])$$

$$x1 = \begin{cases} \dfrac{\dfrac{\sum_{x=1}^{m} \sum_{y=1}^{n} f_{gray}(x, y)}{m \times n} - \dfrac{\sum_{x=1}^{m} \sum_{y=1}^{n} f_{gray}(x,y)}{m \times n} \times (w1 - w2)}{255}, \\ \quad if\ (w1 - w2) > 0 \\ 0, otherwise \end{cases} \tag{2}$$

$$x2 = \begin{cases} x1 + (w1 - w2), if\ (w1 - w2 > 0)\ and\ (x1 \neq 0) \\ 255, otherwise \end{cases}$$

$$g'(x, y) = \begin{cases} 0, if\ (g(x, y) < x1) \\ \dfrac{255}{x2 - x1}(g(x, y) - x1), if\ (x1 \leq g(x, y) \leq x2) \\ 255, if\ (g(x, y) > x2) \end{cases} \tag{3}$$

3.2 Crowd Segmentation

After color correlation and contrast stretch, we apply canny operators for edge extraction, then the background edge subtraction approach was used to obtain the foreground edges. After that, mathematical morphological erosion and dilation was used to improve the integrity of foreground contours. At last, the Flood Fill approach is used to fill the foreground regions [11]. Details of the processing of crowd segmentation are as follows.

Foreground Edge Extraction. Canny operator is used to extract the background edges $f_{edge}(x, y)$ and the current edge image $g_{edge}(x, y)$ from $f_{gray}(x, y)$ and $g'_{gray}(x, y)$ respectively.

$$f_{edge}(x, y) = G(x, y) * f_{gray}(x, y)$$
$$g_{edge}(x, y) = G(x, y) * g'_{gray}(x, y) \tag{4}$$

The foreground edge image is acquired by background edge subtraction, that is, the foreground edge image $h_{edge}(x, y)$ is calculated as follows.

$$h_{edge}(x, y) = \left| f_{edge}(x, y) - g_{edge}(x, y) \right| \tag{5}$$

The morphology processing [12] is widely applied to image processing, which mainly includes dilation, erosion, opening and closing operation. Since the foreground edge image got from background edge subtraction always has discrete noise and faulted edges, we apply the morphology processing to remove the noise and improve the integrity of contours. In this process, a dilation operation followed by an erosion operation is used to improve the quality of foreground edges, which provide a good foundation for the further people counting.

Foreground Filling. Based on the experimental observation, the edge image $g_{edge}(x, y)$ can't contain all background edges. There are some discrete and noisy edges in the foreground edge image, which will affect the accuracy of crowd counting. Therefore, at first we remove these useless edges. If the edge size is less than the threshold value, this edge will be removed from the foreground edge image. Then the flood fill algorithm is used to fill the foreground edges [11]. This algorithm, also called seed fill, determines the area connected to a given node in a multi-dimensional array. Through looking for all nodes in the array which are connected to the start node by a path of the target color. These nodes are given the same color and belong to the same region. As shown in Fig. 3, Fig. 3(a) is an initial image and Fig. 3(b) is the foreground image after flood fill and Fig. 3(c) is the crowd regions.

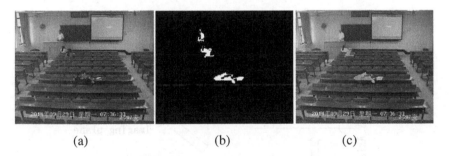

(a)   (b)   (c)

Fig. 3. An example of the foreground filling. (a) initial image; (b) foreground image; (3) crowd regions.

3.3 Crowd Normalization and Counting

As shown in Fig. 3(b), we get the size of crowd by counting the number of foreground pixels in the foreground image. However, images acquired from the digital camera have a characteristic that people far away from the camera is small on the image while people near the camera is large. Therefore, we need to normalize foreground area by camera calibration before estimating the number of people.

As shown in Fig. 4, the digital camera was installed on the roof and the monitor angle is $2\theta^{o}$. The width, length and height of the room are a, b and c respectively. The size of the image acquired from camera is $m \times n$. According to the imaging principle, the height of image contains $n1$ and $n2$ that represent the boundary of front wall and ground in the room. $n1$ and $n2$ are as follows.

$$\begin{cases} \dfrac{n2}{n1} = \dfrac{\sin(arctan\frac{b}{c})}{\sin(2\theta - arctan\frac{b}{c})} \\ n1 + n2 = n \end{cases} \tag{6}$$

As shown in Fig. 3(b), the crowd area need to be normalized with the calibrated scene. The contour area in each line of the foreground image is counted and normalized approximatively by follows.

$$s(i) = \begin{cases} s(i), if(i \leq n1) \\ \dfrac{s(i) \times n1^2}{i^2}, if(n1 < i \leq n) \end{cases} \tag{7}$$

Here $s(i)$ represents the contour area of the ith row in the foreground image. We got the normalized crowd area and then use a head-shoulder template to estimate the crowd number. As shown in Fig. 5, we use several detected results about human body and select each individual body contour area in the position of n1. The template is generated by averaging head-should models.

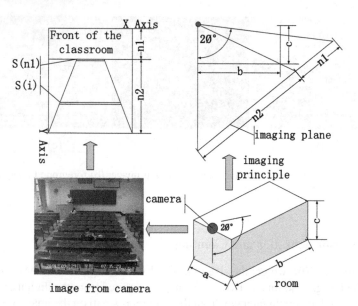

Fig. 4. Crowd normalization by calibration.

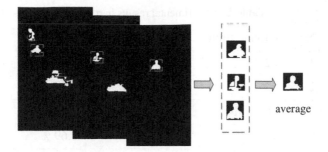

Fig. 5. The generation of the head-shoulder model

4 Experimental Results

We test our approach on the dataset collected by cameras installed on the back wall of different rooms. In room 1, experimental results are shown in Fig. 6, which present a series of results in crowd counting. The images in the first, second and third row are experiments in the case of different people. Specifically, the images in the first column are current images. The second column are grayscales which have been processed by contrast stretch. The third column are people contour area. The fourth column are crowd regions.

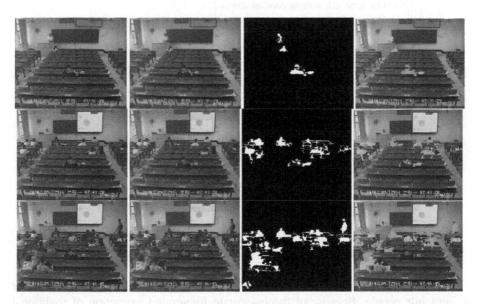

Fig. 6. Experimental results in room 1

As shown in Table 1, we give the counting results in Table 1.

Table 1. Experimental results in room1

No.	Ground truth	Counting results	Error rate
1	2	2	0 %
2	3	3	0 %
3	6	5	16.67 %
4	13	12	7.69 %
5	21	20	4.75 %

To compare with Bayesian counting approach [2], we run experiment on the dataset room2 and room3. In addition, the mobility of people in room2 is larger than room3.

The counting results of room2 are shown in Fig. 7. The result of [2] is named "Single frame". Because of the limitation of traditional background subtraction, static people can't be extracted. So the counting results of "Single frame" are below the "Ground truth" in some periods. The result of our approach without contrast stretch step is named "Our approach without contrast stretch". An excellent contrast grayscale is beneficial to edge extraction and further foreground filling, so the counting results of "Our approach without contrast stretch" are below the "Our approach" in some periods.

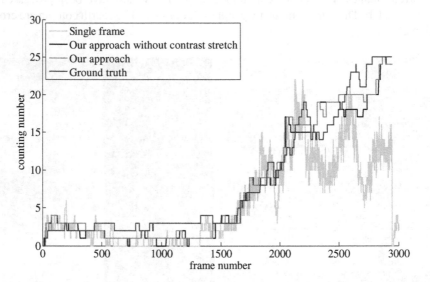

Fig. 7. Comparison results with [2] in room2.

The counting results of room3 are shown in Fig. 8. This room contains larger number of immobile people. Because of the inaccurate foreground extraction of traditional background subtraction, the counting results of "Single frame" are cluttered. The results of our approach is more accurate comparing with [2]. Sometimes the counting number of our approach is incorrect because of the serious occlusion.

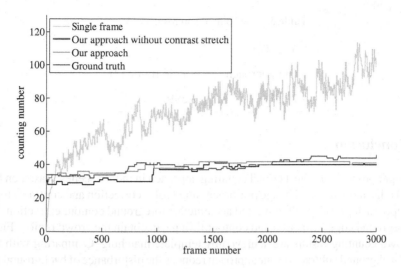

Fig. 8. Comparison results with [2] in room3.

In addition, we use *RMSE* (root mean square error) and V_{speed} (frame speed in processing) as the evaluation criterion.

$$RMSE = \sqrt{\frac{1}{n} \sum_{i=1}^{n} (num_t(i) - num_r(i))^2} \qquad (8)$$

$$V_{speed} = \frac{n}{T} \qquad (9)$$

Here n is the total number of frames in tested video. num_t is the test result of the ith frame and num_r is the ground truth of the ith frame. T is the processing time. As shown in Table 2, the comparison results shows that our approach works better in the indoor environment. As shown in Table 3, our approach have a higher speed in processing comparing with [2], which can be explained by the simple and effective algorithm in our approach.

Table 2. The comparison results of *RMSE*

No.	RMSE		
	Our approach	Without contrast stretch	Single frame [2]
room2	1.9313	2.2673	4.9058
room3	2.4740	4.3463	26.9154

Table 3. The comparison results of V_{speed}

V_{speed}(fps)	
Our approach	Single frame [2]
29.336	1.599

5 Conclusion

This paper proposes a robust crowd counting approach in indoor scenes. Based on background edge modeling and image pre-processing (color correction and contract stretch), the proposed approach is effective and accurate for foreground contour extraction. Then the foreground edge extraction and contour filling can obtain the crowd region. Finally, the crowd counting results are achieved by template matching. Comparing with traditional background subtraction our approach reduces the disturbance of background noise and improves the accuracy of crowd counting in the indoor environment.

References

1. Maddalena, L., Petrosino, A., Russo, F.: People counting by learning their appearance in a multi-view camera environment. Pattern Recogn. Lett. **36**, 125–134 (2014)
2. Zhao, T., Nevatia, R.: Bayesian human segmentation in crowded situations. In: Proceedings of the 2003 IEEE Computer Society Conference on Computer Vision and Pattern Recognition, vol. 2, p. II–459. IEEE (2003)
3. Zhao, T., Nevatia, R.: Tracking multiple humans in complex situations. IEEE Trans. Pattern Anal. Mach. Intell. **26**(9), 1208–1221 (2004)
4. Zhao, T., Nevatia, R., Wu, B.: Segmentation and tracking of multiple humans in crowded environments. IEEE Trans. Pattern Anal. Mach. Intell. **30**(7), 1198–1211 (2008)
5. Kettnaker, V., Zabih, R.: Counting people from multiple cameras. In: 1999 IEEE International Conference on Multimedia Computing and Systems, vol. 2, pp. 267–271. IEEE (1999)
6. Beyme, D.: Person counting using stereo. In: 2000 Proceedings of the Workshop on Human Motion, pp. 127–133. IEEE (2000)
7. Krumm, J., Harris S., Meyers, B., et al.: Multi-camera multi-person tracking for easyliving. In: 2000 Proceedings of the Third IEEE International Workshop on Visual Surveillance, pp. 3–10. IEEE (2000)
8. Qiang, W., Yan, F.: A fast people counting algorithm based on fusion of color and shape information. Comput. Meas. Control **9**, 068 (2010)
9. Qingming, H., Tianwen, Z., Shaojing, P.: Color image segmentation based on color learning. J. Comput. Res. Dev. **32**(9), 60–64 (1995)
10. Luo, J., Wang, J., Xu, H., Lu, H.: A real-time people counting approach in indoor environment. In: He, X., Luo, S., Tao, D., Xu, C., Yang, J., Hasan, M.A. (eds.) MMM 2015, Part I. LNCS, vol. 8935, pp. 214–223. Springer, Heidelberg (2015)
11. Burtsev, S.V., Kuzmin, Y.P.: An efficient flood-filling algorithm. Comput. Graph. **17**(5), 549–561 (1993)
12. Jang, B.K., Chin, R.T.: Analysis of thinning algorithms using mathematical morphology. IEEE Trans. Pattern Anal. Mach. Intell. **12**(6), 541–551 (1990)

Robust Sketch-Based Image Retrieval by Saliency Detection

Xiao Zhang and Xuejin Chen$^{(\boxtimes)}$

CAS Key Laboratory of Technology in Geo-spatial Information Processing
and Application System, University of Science and Technology of China, Hefei, China
leery@mail.ustc.edu.cn, xjchen99@ustc.edu.cn

Abstract. Sketch-based image retrieval (SBIR) has been extensively studied for decades because sketch is one of the most intuitive ways to describe ideas. However, the large expressional gap between hand-drawn sketches and natural images with small-scale complex structures is the fundamental challenge for SBIR systems. We present a novel framework to efficiently retrieve images with a query sketch based on saliency detection. In order to extract primary contours of the scene and depress textures, a hierarchical saliency map is computed for each image. Object contours are extracted from the saliency map instead of the original natural image. Histograms of oriented gradients (HOG) are extracted at multiple scales on a dense gradient field. Using a bag-of-visual-words representation and an inverted index structure, our system efficiently retrieves images by sketches. The experimental results conducted on a dataset of 15 k photographs demonstrate that our method performs well for a wide range of natural scenes.

Keywords: Sketch · Image retrieval · Saliency · Multi-scale

1 Introduction

With the rapid development of the Internet and social networks, the explosive growth of multimedia data asks for efficient information retrieval technologies. Sketch, as one of the most intuitive ways to describe ideas for common users, has been widely used for computer-human interaction systems. It becomes especially popular when touch screens are ubiquitous with smart phones and tablets.

Sketch-based image retrieval plays an important role in the area of content-based image retrieval and has drawn considerable attention since 1990s. Early SBIR systems append color information to strokes or regions in the query sketch to search images with similar color layouts [10,13]. However, the most fundamental challenge is to narrow the representational gap between sketches and natural images. On the one hand, hand-drawn sketches are usually very brief to describe the object of interest and lack of shape details. Moreover, there are modest distortions on sketches drawn by common users due to the lack of professional painting skills. On the other hand, natural images carry rich textures,

© Springer International Publishing Switzerland 2016
Q. Tian et al. (Eds.): MMM 2016, Part I, LNCS 9516, pp. 515–526, 2016.
DOI: 10.1007/978-3-319-27671-7_43

colors, and complex backgrounds, which are usually in small-scale and not interested in by users when search images. These distinct properties of sketches and images make SBIR significantly difficult.

Many research efforts have been spent on the investigation of an efficient matching method to simultaneously bridge the representational gap between sketches and images and search results in a large-scale database quickly. One category of methods focus on the extraction of robust local features from natural images and sketches [2,3,9]. Another category of methods compare overall shapes extracted from nature images with sketches globally [11,12,15]. However, the small-scale high-contrast patterns in texture regions of natural images bring significant difficulties to these techniques.

In this paper, we propose a novel method to extract primary contours from natural images to increase the equivalence of shape representations by sketches and natural images. A hierarchical saliency detection step is applied to extract brief object contours from images to reduce the effect of small-scale high-contrast patterns. Then a dense gradient field is interpolated from the sparse set of curves to increase the tolerance for shape varieties and sketch distortions. Multi-scale HOG are extracted from the dense gradient orientation field. Finally, with a bag-of-visual-words model and inverted index framework, our system is capable of searching similar images efficiently with query sketches and is robust to translation, scale and rotation.

2   Related Work

A series of SBIR systems have been proposed in recent years. Traditional methods focus on the extraction of robust local features from natural images and sketches [2,3,9]. A good local feature is critical for efficient image retrieval. Eitz et al. [6,7] extract an adapted edge histogram descriptor and structure tensor from evenly divided image cells. A dense gradient field is interpolated from a sparse set of edge pixels to improve the robustness on affine deformation of SBIR [8,9]. The bag-of-word model which discards the spatial information is used by these SBIR systems. However, these methods suffer from noisy edges caused by small-scale high-contrast patterns in texture regions of natural images, while textural details are not depicted in hand-drawn sketches.

Different from traditional retrieval methods, MindFinder [2] system takes image contours as image feature directly, including the position and orientation of each edge pixel. An inverted index structure using edgels is proposed to achieve a real-time SBIR system in a large-scale dataset [3]. However, the invariance of position, scale and rotation is greatly reduced due to the introduction of edgel positions into index structure.

Another category of methods compare overall shapes extracted from nature images with sketches globally. By approximating object contours as piecewise straight line segments, the overall shape is described by the relationship between line segments [11,15]. This method is suitable only for objects with a single contour. Complex images usually consist of a groups of separate edges. Saavedra and Bustos [12] propose to extract a set of primitive keyshapes such as a line, a curve or an ellipse from an image. However, the simplified keyshapes lose shape details of general objects in natural images.

In order to reduce the effect of complex backgrounds in natural images, regions of interest (ROI) are extracted to focus on retrieval of interested objects [14,17]. However, extracting ROI with low-level image cues remains a challenging problem in computer vision.

3   Algorithm

Our system consists of four main parts, as shown in Fig. 1. Our user interface (a) allows users to draw a sketch or load a sketch image on the left and presents the retrieved images on the right. Given the query sketch, our system extracts multi-scale HOG from the gradient field and generates its bag-of-words representation, as shown in Fig. 1(b). For each image in the dataset, we extract its bag-of-words representation as shown in Fig. 1(c). A saliency map is first computed to remove the small-scale complex structures in texture regions or backgrounds. Similar with the feature extraction of sketches, for each image, a gradient field is interpolated from the edges of its saliency map and its bag-of-words representation is then calculated. In order to achieve real-time retrieval, we adopt an inverted index structure as shown in Fig. 1(d) to find good image candidates first and then rank the selected image candidates by their similarities with the query sketch. Finally, the k most matched images are presented to the user.

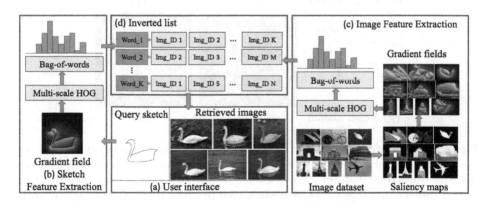

Fig. 1. System overview. Our system consists of four parts: (a) The user interface that allows users to draw query sketches and browse retrieved images; (b) The feature extraction part for the query sketch; (c) The feature extraction part for images in the dataset; (d) An inverted index structure for real-time image retrieval. The two image processing parts (blue) are computed offline while the two sketch processing parts (red) are computed online in a real-time rate (Color figure online).

3.1   Saliency Detection

Natural images usually have complex backgrounds and small-scale textures which users do not draw in their sketches. In order to more robustly compare sketches and natural images, we first extract primary object contours by detecting edges on a saliency map instead of the original image. Saliency detection has been well studied in computer vision to extract significant regions of an image. We use the hierarchical saliency detection method proposed in [16] because of its robustness on images with complex structures.

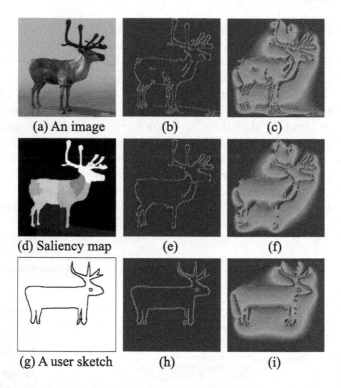

(a) An image   (b)   (c)

(d) Saliency map   (e)   (f)

(g) A user sketch   (h)   (i)

Fig. 2. Object contour detection from saliency map. The edge map (b) extracted from the original image has many noisy edges due to textures and shadows. In comparison, the edge orientation map (e) extracted from the saliency map (d) is more analogous to orientation map (h) of the hand-drawn sketch (g). As a result, the gradient field (f) computed with saliency detections is more consistent with the gradient field (i) of the hand-drawn sketch.

For each image I in the dataset, we compute its saliency map I_s. We extract edges using Canny algorithm [1] from I_s instead of the image I. In Fig. 2, we compare the edge orientation maps of an image with textures on the deer body and shadows on the ground. In the orientation map computed from the original image, there are a part of noisy edges in the shadow and texture regions, as

Fig. 2(b) shows. In contrast, the saliency map shown in Fig. 2(d) better conforms to human perception while the small scale high-contrast patterns are restrained. The orientation map (Fig. 2(e)) detected from the saliency map is more analogous with the one of the query sketch (Fig. 2(h)).

3.2 Gradient Field

Though the object contour detected from the saliency map carries the brief shape information of the primary objects in the image, the hand-drawn sketch is relatively coarse with modest distortions because of common users' lack of professional painting skills. In order to robustly match the two distinct types of shape representation, we adopt the gradient field HOG (GFHOG) descriptor [9] because of its superiority on SBIR against state-of-the-art descriptors.

From a binary Canny edge map or a query sketch I_e, we first compute its derivatives $\frac{\delta I_e}{\delta x}$ and $\frac{\delta I_e}{\delta y}$ on the horizontal and vertical directions respectively with Sobel operators. We calculate the gradient orientation $\theta \in [0, 2\pi)$ and the gradient magnitude $\rho = \sqrt{(\frac{\delta I_e}{\delta x})^2 + (\frac{\delta I_e}{\delta y})^2}$ for each pixel on I_e. By eliminating the pixels with small gradient magnitude, a sparse set of edge pixels with their orientations $\mathcal{E} = \{\theta(x,y)|_{\rho(x,y)>T_e}\}$ is obtained.

The gradient field Θ_Ω is a dense orientation field over image coordinates $\Omega \in \mathcal{R}^2$ interpolated from the sparse set of edge orientations $\mathcal{E}$. Under the assumption of smoothness in a local window, it can be solved by a Possion equation with Direchlet boundary conditions. We choose a 3×3 window to let $\delta\Theta = 0$, which leads to a linear equation at each pixel (x, y):

$$4\Theta(x,y) = \Theta(x-1,y) + \Theta(x+1,y) + \Theta(x,y-1) + \Theta(x,y+1), \quad (1)$$

with the Direchlet boundary conditions

$$\Theta(x,y) = \begin{cases} \theta(x,y), \text{ if } (x,y) \in \mathcal{E}, \\ 0, \qquad \text{ if } (x,y) \text{ is located on image boundaries.} \end{cases} \quad (2)$$

Solving Eq. 1 with Eq. 2 produces a dense gradient field Θ_Ω. Figure 2(c), (f) and (i) illustrate the three gradient fields interpolated from the edge orientation maps of the original image, the saliency map and the sketch respectively. We can see that the gradient field from the saliency map looks more similar with the gradient field of the hand-drawn sketch though there are textures and shadows in the image as well as distortions on the sketch.

3.3 Multi-scale HOG

The HOG descriptor [4] and its variants have been widely used in many object detection tasks. A HOG descriptor is computed in a local area around an interest point or a uniformly sampled point on the image. We extract HOG descriptor at multiple scales to improve the scale-invariance of our retrieval system.

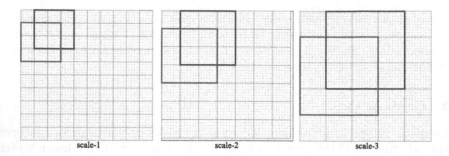

Fig. 3. Multi-scale feature extraction. A gradient field is evenly divided into grid cells in three sizes: 5×5, 7×7, and 10×10 (red grids). Then blocks consisting of 3×3 cells are sampled on each scale and a pair of adjacent blocks overlap two cells horizontally or vertically (the blue, green and purple blocks) (Color figure online).

At each scale, a gradient field Θ is uniformly divided into grid cells of $k \times k$ pixels. The orientation of each pixel in Θ is quantized into q bins in $[-\frac{1}{q}\pi, \frac{2q-1}{q}\pi)$. As a consequence, a q-D vector of HOG descriptor can be computed for each cell to represent the distribution of gradient orientations. In a larger square block consisting of $n \times n$ cells, the n^2 HOG descriptors are concatenated to form a qn^2-D vector as the shape descriptor for each block.

In our experiments, we use three scales for a tradeoff between time efficiency and scale invariance. At the three scales, the sizes of the grid cell are 5×5, 7×7, and 10×10, respectively, as shown in Fig. 3. Each block contains 3×3 cells and the blocks are sampled evenly on the gradient field by overlapping two cells horizontally or vertically. By sampling overlapping blocks, the final shape descriptor can be more robust to object translation.

3.4   Sketch-Based Image Retrieval

With the extracted local HOG descriptors at three scales on the query sketch as well as the images in the dataset, we now compare them to find the best match images with the query sketch. It is intractable to directly find one-to-one local feature matches for a large-scale image dataset. We use a widely used bag-of-visual-words model [5] for fast image retrieval.

Training Step. For each image in the dataset, hundreds of 108-D HOG descriptors can be extracted. All descriptors extracted from the entire dataset are grouped into K clusters using k-means algorithm. Each cluster is called a visual word. Then for each image, each extracted HOG feature is classified into one of the K visual words. By counting the frequency of the K visual words that occur in each image at three scales, a K-D frequency histogram $\mathbf{H}$ is obtained.

Inverted List. The image-sketch matching problem can be done by matching their K-D histograms of visual words. However, for online applications, the histogram distance calculation of tens of thousands of images with the query sketch is extremely time-consuming. In order to speed up the retrieval process to a real-time rate, we adopt an inverted index strategy of visual words for the image dataset. For each learned visual word $\mathbf{w}$, the inverted list records IDs of all images in which $\mathbf{w}$ appear as a keyword. A word in an image is defined as its keyword when it is one of the five most frequent visual words in the image.

Searching Step. Given a query sketch I_s, we extract a K-D frequency histogram $\mathbf{H}_s$ of visual words. We first choose the M most frequent visual words of the query sketch as keywords. For each keyword, we look up the inverted list and find the images containing the same keyword as image candidates. Then for each candidate image I_c, we compare its frequency histogram of visual words $\mathbf{H}_c$ with $\mathbf{H}_s$ of the query sketch by computing their histogram distance $D(\mathbf{H}_s, \mathbf{H}_c)$. By sorting the distances of all the candidate images, we finally show N images that best match the query sketch to users.

Histogram Distance. There are several widely used histogram distances: Euclidean distance, cosine distance and histogram intersection.

The cosine distance of two histograms measures the angle of them

$$D_{cosine}(\mathbf{H}_1, \mathbf{H}_2) = \frac{\sum_{i=1}^{K}(\mathbf{H}_1(i) \times \mathbf{H}_2(i))}{\sqrt{\sum_{i=1}^{K}\mathbf{H}_1^2(i) \times \mathbf{H}_2^2(i)}}. \tag{3}$$

The histogram intersection $D_{intersection}(\mathbf{H}_1, \mathbf{H}_2)$ denotes the intersection part of two histograms

$$D_{intersection}(\mathbf{H}_1, \mathbf{H}_2) = \sum_{i=1}^{K} min(\mathbf{H}_1(i), \mathbf{H}_2(i)). \tag{4}$$

Usually, hand-drawn sketches do not capture all the detailed lines as natural images, which leads to large differences between their visual word histograms. We evaluate the three distance metrics in our experiments and demonstrate the best image retrieval performance is achieved by using histogram intersection.

4   Results and Discussions

Currently, there is not any standard image database for SBIR systems. Most SBIR image databases are downloaded from the Internet. However, the content of Internet images varies widely from single objects to complex scenes. The most significant variance is caused by complex backgrounds and lighting conditions. Natural images that reflect certain variance are desired to evaluate the performance of our system. We choose the 'Flickr15K' image dataset [9] that contains 15k photographs. We manually merge categories such as 'pyramid' and 'Egypt pyramid', 'pantheon' and 'Rome pantheon' because the primary object shapes are very similar in the these categories which makes it difficult for common users to draw distinct sketches without text annotations. Finally, we use 30 categories in our experiments.

We invited ten non-expert users to draw free-hand sketches as queries to search images in the dataset. Each user is asked to draw 20 sketches of different categories to query. We finally collect 200 sketches that cover the 30 categories in the image dataset.

In saliency detection procedure, all the images in the dataset are resized to the same height (400 pixels in our system) while preserving their aspect ratios. Larger sizes will lead to much details on the detected saliency map. When computing gradient fields, we resize the edge map into a smaller size (100-pixel height) to reduce the computation complexity without significant performance decrease.

We conduct our retrieval experiments over a variety of vocabulary sizes K and three histogram distances. To evaluate the retrieval performance, we calculate the truncated Mean Average Precision (tMAP).

$$tMAP = \frac{1}{N}\sum_{i=1}^{N}\frac{rel(i)}{i}, \tag{5}$$

where N is the number of retrieved images that presented to users for each query. $N = 10$ in our experiments. $rel(i)$ denotes the relevance of ith result. If the retrieved image I_i^r is in a different category from the query sketch, $rel(i) = 0$. Otherwise, $rel(i)$ is defined as the rank of I_i^r among all the retrieved images from the same category. The value of $tMAP$ ranges from 0 to 1, and $tMAP = 0$ means that no relevant image is returned in the N retrieved images, while $tMAP = 1$ means that all the N returned images match the query sketch.

The number of the visual words in the trained vocabulary affects the retrieval performance. A large vocabulary results in that similar features are probably clustered into different visual words, while a small vocabulary results in false feature matching. Figure 4 shows the $tMAP$s of our method using different histogram distance measurements compared with GFHOG [9] without using text labels over a variety of vocabulary sizes.

Comparing the three histogram distance metrics, we can see that the histogram intersection significantly outperforms the Euclidean distance and cosine distance. The Euclidean distance of histograms performs worst when comparing natural images which have complex structures with hand-drawn sketches that exhibit shape distortions by non-expert users. Our method with the histogram intersection metric achieves the highest $tMAP \approx 0.4$ when using a vocabulary of 2000 words. In a wide range of vocabulary sizes, our method significantly improves the retrieval performance compared with GFHOG [9].

In order to evaluate the robustness of our method to translation and scale, we conduct an experiment for the sketch base. In our experiments, the size of the query sketch is 700×700. We shift the query sketch by translations $t \in [-200, 200]$ horizontally to compute the retrieval $tMAP$ for different translations. Then we scale the input sketch by a factor $\in [0.5, 1.5]$ to run another group of experiments. Figure 5 shows the $tMAP$s of our systems under different

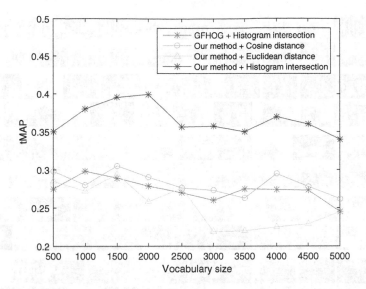

Fig. 4. The truncated MAPs of our methods using different histogram distances compared with GFHOG [9] over a range of vocabulary sizes. Our method significantly outperforms the state-of-the-art method when using histogram intersection.

translations and scalings. We can see that our system is robust to a wide range of transformations.

Figure 6 demonstrates a series of retrieval results from various query sketches drawn by non-expert users. If a retrieved image is in the same category with the query sketch, it is a true positive. Otherwise, it is a false positive. For each query sketch, we show the 10 best match images using our method with the histogram intersection metric on the upper row and the 10 best match images retrieved by GFHOG [9] without text annotations in the lower row. Both our method

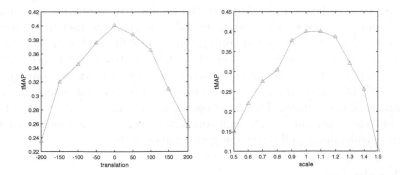

Fig. 5. The truncated MAPs of our method when we transform query sketches with a range of translations and scalings.

Fig. 6. Image retrieval results with various query sketches. For each query sketch, the 10 best match images retrieved by our method are shown in the upper row, while the 10 best match images retrieved by GFHOG [9] are shown in the lower row. The correct images in the same category with each query sketch are marked as green and false results are marked as red. The two images marked with a yellow square in (b) are not in the same category with the query sketch. However, the buildings have a very similar shape with the sketch (Color figure online).

and [9] find good matches for buildings with regular shapes, such as Fig. 6(a) and (b). For objects with curvy shapes, the gradient orientations varies from different points of view. Using our saliency detection process, primary object contours can be extracted so that the large change of orientations on small-scale structures can be depressed to improve the retrieval performance, as shown in Fig. 6(c) and (d). Figure 6(e) shows a special case when the user draws a query sketch with rich structure details on the main building. Though our method fails to sort the retrieved images well (four false matches in the ten matched images), they somehow share similar distributions of edge orientations. In comparison, GFHOG algorithm requires a Gaussian smoothing on images to reduce the number of edge pixels. This local operation will miss primary object contours. Therefore, the small-scale high-contrast textures and complex structures in images significantly pull down the performance of [9].

5   Conclusion

In this paper, we present a robust and efficient sketch-based image retrieval system using saliency detection. Primary object contours can be extracted using the hierarchical saliency detection to depress the small-scale textures and complex structures in natural images. By extracting mulit-scale HOG on the dense gradient field interpolated from sparse edges or sketches, our algorithm is robust to scale. The real-time image retrieval is achieved by using a bag-of-visual-words representation and an inverted index structure. Experimental results demonstrate that our method outperforms state-of-the-art SBIR systems for a wide variety of natural images.

While users' painting skills vary significantly, existing SBIR systems usually fail to deal with complex images by rough hand-drawn sketches. In the future, we would like to integrate the retrieval process with the sketching process so that we can use retrieved images to guide the sketching process and then refine the retrieval result by the beautified sketch. Moreover, constructing standard image datasets for SBIR systems is another significant work in the future.

Acknowledgments. We would like to thank the anonymous reviewers. This work was supported by the National Natural Science Foundation of China (NSFC) under Nos. 61472377 and 61331017, and the Fundamental Research Funds for the Central Universities under No. WK2100060011.

References

1. Canny, J.: A computational approach to edge detection. IEEE TPAMI **6**, 679–698 (1986)
2. Cao, Y., Wang, H., Wang, C., Li, Z., Zhang, L., Zhang, L.: Mindfinder: interactive sketch-based image search on millions of images. In: ACMMM (2010)
3. Cao, Y., Wang, C., Zhang, L., Zhang, L.: Edgel index for large-scale sketch-based image search. In: IEEE CVPR, pp. 761–768 (2011)
4. Dalal, N., Triggs, B.: Histograms of oriented gradients for human detection. In: CVPR, pp. 886–893 (2005)
5. Dance, C., Willamowski, J., Fan, L., Bray, C., Csurka, G.: Visual categorization with bags of keypoints. In: ECCV International Workshop on Statistical Learning in Computer Vision (2004)
6. Eitz, M., Hildebrand, K., Boubekeur, T., Alexa, M.: A descriptor for large scale image retrieval based on sketched feature lines. In: Eurographics Symposium on Sketch-Based Interfaces and Modeling, pp. 29–38 (2009)
7. Eitz, M., Hildebrand, K., Boubekeur, T., Alexa, M.: An evaluation of descriptors for large-scale image retrieval from sketched feature lines. Comput. Graph. **34**(5), 482–498 (2010)
8. Hu, R., Barnard, M., Collomosse, J.: Gradient field descriptor for sketch based retrieval and localization. ICIP **10**, 1025–1028 (2010)
9. Hu, R., Collomosse, J.: A performance evaluation of gradient field hog descriptor for sketch based image retrieval. CVIU **117**(7), 790–806 (2013)
10. Jacobs, C.E., Finkelstein, A.: Fast multiresolution image querying. In: ACM Proceedings of the 22nd Annual Conference on Computer Graphics and Interactive Techniques, pp. 277–286 (1995)
11. Parui, S., Mittal, A.: Similarity-Invariant sketch-based image retrieval in large databases. In: Fleet, D., Pajdla, T., Schiele, B., Tuytelaars, T. (eds.) ECCV 2014, Part VI. LNCS, vol. 8694, pp. 398–414. Springer, Heidelberg (2014)

12. Saavedra, J.M., Bustos, B.: Sketch-based image retrieval using keyshapes. Multimedia Tools Appl. **73**(3), 2033–2062 (2014)
13. Smith, J.R., Chang, S.F.: Visualseek: a fully automated content-based image query system. In: Proceedings of the Fourth ACM International Conference on Multimedia, pp. 87–98 (1997)
14. Tencer, L., Renakova, M., Cheriet, M.: Sketch-based retrieval of document illustrations and regions of interest. In: 2013 12th International Conference on Document Analysis and Recognition (ICDAR), pp. 728–732 (2013)
15. Wang, S., Zhang, J., Xu, T., Miao, Z.: Sketch-based image retrieval through hypothesis driven object boundary selection with HLR descriptor. IEEE Trans. Multimedia **17**, 1045–1057 (2015)
16. Yan, Q., Xu, L., Shi, J., Jia, J.: Hierarchical saliency detection. In: CVPR (2013)
17. Zhou, R., Chen, L., Zhang, L.: Sketch-based image retrieval on a large scale database. In: Proceedings of the 20th ACM international conference on Multimedia, pp. 973–976 (2012)

Image Classification Using Spatial Difference Descriptor Under Spatial Pyramid Matching Framework

Yuhui Li[1,2(✉)], Jiucheng Xu[1], Yifan Zhang[2], Chunjie Zhang[3],
Hongsheng Yin[4], and Hanqing Lu[2]

[1] School of Computer and Information Engineering, Henan Normal University, Xinxiang, China
liyuhui0224@126.com, xjch3701@sina.com
[2] National Laboratory of Pattern Recognition, Institute of Automation,
Chinese Academy of Science, Beijing, China
{yfzhang,luhq}@nlpr.ia.ac.cn
[3] School of Computer and Control Engineering, University of China Academy of Sciences,
Beijing, China
cjzhang@jdl.ac.cn
[4] China University of Mining and Technology, Xuzhou, China
xuzhouyhs@sina.com

Abstract. Spatial pyramid matching (SPM) model is an extension of the bag-of-visual words (BoW) model for local feature encoding. It firstly partitions the image into increasingly fine sub-regions, and then concatenates the histograms within each sub-region. However, the SPM model does not consider the spatial information differences between sub-regions explicitly. To make use of this information, we exploit a novel descriptor called spatial difference. In the process of promoting the performance of image classification, this descriptor is mainly used to concatenate the histograms of bag-of-visual words model under spatial pyramid matching framework. Finally, we conduct image classification experiments on several public datasets to demonstrate the effectiveness of the proposed scheme.

Keywords: Image classification · Spatial difference descriptor · Spatial pyramid matching · Bag-of-visual words · Sparse coding

1 Introduction

In recent years, the bag-of-visual words (BoW) model [1] has been very popular in various image applications, especially for image classification. Codebook generation and histogram representation are two important components for generating bag-of-visual words representation. Bag-of-visual words model has been demonstrated that combining codebook and histogram representation together can achieve good performance after being trained to predict the classes of images.

Though bag-of-visual words model has achieved good performance, there exist obvious drawbacks: both the spatial information and correlations among visual contents are neglected. Later, a spatial pyramid matching (SPM) model [2] was proposed to deal

© Springer International Publishing Switzerland 2016
Q. Tian et al. (Eds.): MMM 2016, Part I, LNCS 9516, pp. 527–539, 2016.
DOI: 10.1007/978-3-319-27671-7_44

with this problem by dividing the whole image into hierarchical sub-regions and concatenating the appropriately weighted histograms of each region. Extensive experimental results have shown that spatial pyramid matching model achieved a remarkable success on a wide range of image classification benchmarks as a fundamental model. However, spatial pyramid matching model has its own weaknesses: the finer the division, the more sensitive it is to location and orientation of visual content, as described in [3]. In view of this, many works focused on improving SPM to overcome these weaknesses, such as researchers [3–6] tried to improve the coding procedure to minimize the representation information loss. Teng et al. [3] explored a weakly spatial symmetry descriptor to boost the performance of bag-of-visual words model by combining weakly spatial symmetry (WSS) and BoW together. Despite its success in the scene image classification domain, WSS model has its own limitations. For example, after dividing the image into many sub-regions and generating histograms of bag-of-visual words model, WSS only computes spatial symmetry information inside each sub-region, rather than considering the spatial difference information between sub-regions. While spatial difference information between sub-regions is much more important than inside spatial symmetry information when dividing images into increasingly fine sub-regions.

Hence we propose a novel approach to relieve the above problems under spatial pyramid matching framework. Firstly, we compute spatial difference information in four kinds of orientations. Secondly, we combine the spatial difference descriptors with histograms of bag-of-visual words model together. Finally, experiments are conducted on several datasets, which mainly include estimating different distance measurements, evaluating the performance with different codebook sizes and comparing with other methods to prove the effectiveness of our approach.

2 Related Work

The bag-of-visual words (BoW) [1] model has been widely used for visual applications. Traditional BoW model uses the k-means clustering algorithm and considers the cluster centers as visual words. Local features are then quantized to the nearest visual word. However, this solution leads to severe information loss, which limits its discriminative power. In order to reduce the information loss in the local feature encoding process, many works [6–8] have been proposed. For instance, as a classical and typical one, Yang et al. in [6] proposed one scheme to sparse coding with spatial pyramid matching for codebook generation, and trained linear classifier to save computational cost, which was much more effective than non-linear classifier.

Further more, the traditional BoW model lacks the spatial information. Inspired by the work done by Grauman and Darrell [4], Lazebnik et al. [2] proposed the spatial pyramid matching (SPM) algorithm which was widely used by many researchers. Later, a lot of works [9–13] have been done to combine the spatial information of local features, which are motivated by the SPM algorithm. A hierarchical matching method with side information was proposed by Chen et al. [9] and it was used for image classification. A weighting scheme was used to select discriminative visual words. Randomization and discrimination was combined into a unified frame work by Yao et al. [10], which was used for fine grained image categorization. Zhang et al. [11] proposed a pose pooling kernel to recognize sub-category birds. Representing images with components and a bilinear model for object recognition was used in [12] proposed by Zhang et al., Bao and He [13] proposed an improved sparse coding model based on linear spatial pyramid matching (SPM) and scale invariant feature transform (SIFT) descriptors. Teng et al. [3] explored a weakly spatial symmetry descriptor to boost the performance of bag-of-visual words model by combining WSS and BoW together. Authors in [14, 15] explored methods of combining spectral and spatial information directly to boost the final performance. Zhu et al. [16] presented a robust semi-supervised kernel-FCM algorithm incorporating local spatial information to solve the original problem of image classification. In another way, Zhang et al. [17] proposed a novel object categorization method by using the sub-semantic space based image representation. Most of the previous works have their own superiority on some datasets. However, none of them consider the spatial difference information between sub regions. To make use of this lost information, we propose a novel descriptor named spatial difference to improve the performance for image classification.

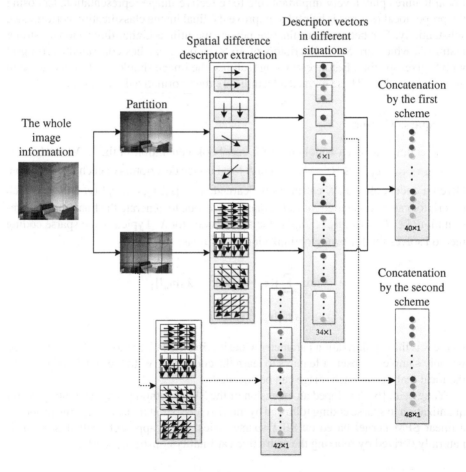

Fig. 1. Flow chart of spatial difference descriptor computation

3 Proposed Framework

Our image classification solution is derived from spatial pyramid matching model [6], which mainly includes five modules: feature extraction, sparse coding, spatial pooling, spatial difference descriptor computation, and finally linear classifier. Firstly, feature extraction accomplishes obtaining the original image representation vectors. Then sparse coding and spatial pooling are used to learn dictionary and encode the local features respectively. After this, the spatial difference descriptor is computed to complete the gist of obtaining discriminative image representation vectors by going one more step, as shown in Fig. 1. In the last module, we use linear SVM classifier to do image classification by training and testing the obtained discriminative feature the same as other common benchmarks.

3.1 Feature Extraction

Local features play a very important role for effective image representation. Choosing the proper local features is helpful to improve the final image classification performance substantially. For better discriminative power, we utilize higher-dimensional "strong features", which are SIFT descriptors of 16×16 pixel patches computed over a grid with 8-pixel spacing. Before extracting features, the image should be all processed into gray scale. At last, These extracted features are then normalized with L_2 norm.

3.2 Sparse Coding

Sparse coding has been widely used for codebook generation in the BoW model. Let $X = [x_1, x_2, ..., x_N]$ $(x_i \in \mathbf{R}^{D \times 1})$ be a set of N local image descriptors of each D dimension. Given a codebook with K entries to be learned, $V = [v_1, v_2, ..., v_K]$ $(v_i \in \mathbf{R}^{D \times 1})$, each descriptor can be converted into a K-dimensional code to generate the final image representation. Let $U = [u_1, u_2, ..., u_N]$ is the set of codes for X. Typically the sparse coding method solves the following optimization problem as:

$$\min_{U,V} \sum_{n=1}^{N} ||x_n - u_n V||^2 + \lambda ||u_n||_1$$

$$s.t. ||v_k||^2 \leq 1, \forall k$$

where λ is the regularization parameter. Considering the large amount of local features, we only sample a subset of features to learn the codebook. With the codebook in place, the local features of each image can be encoded.

Yang et al. [6] developed an extension of the SPM method [2] by generating vector quantization to sparse coding followed by multi-scale spatial max pooling, and proposed a linear SPM kernel based on SIFT sparse codes. Their approach, called ScSPM, is naturally derived by relaxing the restrictive cardinality constraint of VQ.

3.3 Spatial Pooling

In the "SPM" layer, we partition an image into $2^l \times 2^l$ spatial sub-regions, where $l = 0$, 1, 2 stands for different scales. The codes of the descriptors are pooled together to get the corresponding pooled features. These pooled features from each sub-region are concatenated and normalized to form the image feature representation. The pooling method used in this paper is max pooling:

$$z_j = \max \left(u_{ij} \right) \quad i = 1, 2, \ldots, N, \; j = K.$$

In our framework, "max pooling" combined with L_2 normalization is used. Max pooling can produce better performance than other pooling methods (i.e. *Sqrt* and *Abs*), as demonstrated by Yang et al. [6], probably due to its robustness to local spatial translation and biological plausibility.

3.4 Spatial Difference Descriptor Computation

After image representation being generated by sparse coding and spatial pooling hierarchically, spatial difference descriptors are computed according to four kinds of spatial difference information. For example, the sub-figure (a) in Fig. 2 describes left to right difference, and the sub-figure (b) describes top to down difference. As for the diagonal differences, we compare two different schemes, as shown in Fig. 1.

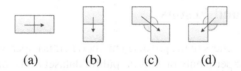

(a)   (b)   (c)   (d)

Fig. 2. Four kinds of spatial difference information

In the first scheme, diagonal spatial difference information is extracted by computing distances between sub-regions which may not be contiguous. In the second scheme, diagonal spatial difference information is extracted by computing distances between sub-regions which are all contiguous. Both of the two schemes compute spatial difference information between sub-regions rather than in each sub-region.

Because the two sub-regions to be calculated for diagonal spatial difference information may be not contiguous, their correlation is denoted by dotted line, as shown in sub-figure (c) and sub-figure (d) of Fig. 2. Especially, Fig. 1 describes the flow chart of the spatial difference descriptor computation process in the whole image level. In order to differentiate the two computing schemes, we denote the first scheme as spatial difference 1 (SD^1) and the second one as spatial difference 2 (SD^2). It is important to note that, in Figs. 1 and 2, the blocks in which the head and tail of the arrow line locate are the sub-regions we choose to compute difference descriptor.

For two computing sub-regions of the segmented image, two vectors h_1 and h_2 are built based on the size of codebook:

$$h_1 = [h_{11}, h_{12}, \ldots, h_{1K}]$$
$$h_2 = [h_{21}, h_{22}, \ldots, h_{2K}]$$

where K is the size of codebook.

If these two sub-regions are strictly the same, the distance between them would be near 0. Obviously, different distance measurements may lead to different results, and then we can use different methods to compute the distance. Finally, we will adopt *euclidean* distance measurement as the most proper one to calculate the spatial difference information between two vectors. Now we can obtain a feature vector to describe the spatial difference descriptor for the whole image:

$$D_{SD} = [d_1, d_2, \ldots, d_P]$$

where P is the number of sub-region pairs to compute according to Figs. 1 and 2.

To combine histograms of bag-of-visual words model and spatial difference information, we utilize the method as described in [3] by Teng et al. and obtain the final discriminative feature representation:

$$W_{ScSPM+SD} = [H, D_{SD}]$$
$$= [h_1, h_2, \ldots, d_m, d_1, d_2, \ldots, d_P]$$

where H is the histograms of bag-of-visual words, and m is the size of H.

4   Experiments and Results

To evaluate the effectiveness of the proposed method in this paper, we choose to conduct image classification experiments on several public datasets. The datasets are Scene 15 dataset, Caltech 101 dataset and Caltech 256 dataset.

Our approach uses the popular SIFT descriptors. The same as other common benchmarks, we randomly select the training images and use the rest images for testing. This process is repeated for ten times to get reliable results. Mean of per-class classification rates for performance measurement is used and we report the final results by mean and standard deviation of the classification rates. SPM with three pyramid and SD[1] (or SD[2]) in which spatial difference information is extracted are used to combine the hierarchical histograms and correlations between sub-regions together. Hence, each image is represented by a vector of $21 \times K$ (size of codebook) + 40 (or 48) for all datasets.

Our experiments mainly include three parts: firstly, fixing the codebook size with 1024, and for Scene 15 dataset and Caltech 101 dataset, we evaluate the performance of different distance measurements. Measuring distance includes six methods, which are *chebychev, jaccard, hamming, cosine, cityblcok* and *euclidean*. Secondly, fixing the distance measurement, we evaluate the performance by changing the codebook size. Finally, we choose to compare with other methods which are closely related with the proposed method by their reported results instead of re-implementing them and test on every datasets for fair comparison mostly.

4.1 Scene 15 Dataset

We firstly try our method on the Scene 15 dataset. There are 4485 images of 15 classes (bedroom, coast, forest, highway, industrial, insidecity, kitchen, living room, mountain, office, opencountry, store, suburb and tallbuiding) in this dataset. Each class has 200 to 400 images. We follow the same experiment setup as Yang et al. [6] did and randomly select 100 images per class for classifier training.

Then we evaluate the performance by using different distance measurements. For fair comparison, we extract WSS descriptor proposed by Teng et al. [3] and use them under the same framework with us. When, taking WSS, SD^2 and SD^1 into consideration, as shown in Fig. 3, we can see that SD^1 can achieve better performance than the other two methods, and the distance measurement of *euclidean* is the best selection to get amazing performance.

We also conduct experiments on different size of codebook to observe the effect for classification on the Scene 15 dataset. As shown in Fig. 4, when the codebook size is 2048, we can achieve the best performance on Scene 15 dataset.

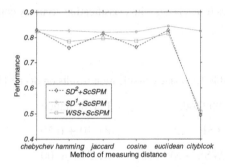

Fig. 3. Performance under different distance measurements on Scene 15

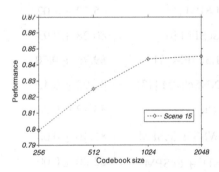

Fig. 4. Performance of codebook size on Scene 15

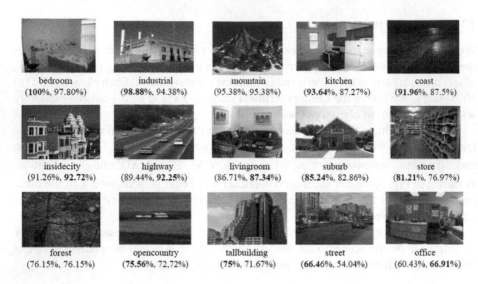

bedroom
(100%, 97.80%)

industrial
(98.88%, 94.38%)

mountain
(95.38%, 95.38%)

kitchen
(93.64%, 87.27%)

coast
(91.96%, 87.5%)

insidecity
(91.26%, 92.72%)

highway
(89.44%, 92.25%)

livingroom
(86.71%, 87.34%)

suburb
(85.24%, 82.86%)

store
(81.21%, 76.97%)

forest
(76.15%, 76.15%)

opencountry
(75.56%, 72.72%)

tallbuilding
(75%, 71.67%)

street
(66.46%, 54.04%)

office
(60.43%, 66.91%)

Fig. 5. Example images from classes with classification accuracy comparison on Scene 15

Table 1. Image classification results on Scene 15

Methods	Performance
KSPM [2]	81.40 ± 0.50
WSS + SPM [3]	81.51 ± 0.00
KCSPM [5]	76.70 ± 0.40
LSPM [6]	65.32 ± 1.02
ScSPM [6]	80.28 ± 0.93
LScSPM [7]	$\mathbf{89.75 \pm 0.50}$
NNScSPM [13]	81.92 ± 0.42
S^3R [17]	83.72 ± 0.78
WSS + ScSPM	81.46 ± 0.00
SD^2 + ScSPM	82.80 ± 0.00
SD^1 + ScSPM	84.52 ± 0.01

Finally, we give the performance of the proposed method and compare with methods proposed by [2, 3, 5–7, 13, 17] in Table 1. As shown in Table 1, LScSPM can achieve high performance on scene classification. The problem reason is that scene images contain plentiful textures in each patch, which results in the unstableness for sparse coding process. By adding Laplacian term, similar patches will be encoded into similar codes, thus the image can be accurately represented [7]. Except LScSPM, we can see that our method SD^1 descriptor under SPM framework achieves comparable results. Figure 5 shows some example images from Scene 15 dataset classes with classification accuracy in brackets. The first number in the bracket is the accuracy obtained by using SD^1 descriptor under spatial pyramid matching framework, and the second number in the bracket is the accuracy obtained by using the classic method proposed by Yang et al. in [6].

4.2 Caltech 101 Dataset

The Caltech 101 dataset contains 8144 images falling in 101 classes including animals, vehicles, flowers, etc., with significant variance in shape. The number of images per class varies from 31 to 800. Most images are medium resolution, i.e. about 300×300 pixel. We follow the common experiment setup as Yang et al. [6] did and randomly select 15 and 30 images per class for classifier training and use the rest images for testing.

On one hand, for fair comparison, as did on Scene 15 dataset, we conduct experiments to evaluate the performance of different distance measurements by taking WSS, SD^2 and SD^1 into consideration under spatial pyramid matching framework. On the other hand, we test the performance with different codebook sizes. We can see from Fig. 6, the method of *euclidean* outperforms the others. When the codebook size is 2048, we can achieve the best performance, as shown in Fig. 7.

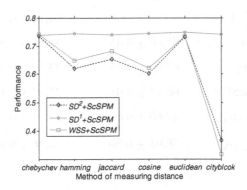

Fig. 6. Performance under different distance measurements on Caltech 101

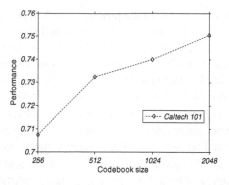

Fig. 7. Performance of codebook size on Caltech 101

At last we give the performance of the proposed method and compare with other methods described [2, 3, 5, 6] in Table 2. Figure 8 shows some typical images owning the top 18th classification accuracy in brackets. The first number in the bracket is the accuracy obtained by using SD^1 descriptors under spatial pyramid matching framework, the second one is the accuracy by using the method proposed by Yang et al. in [6]. From Table 2 and Fig. 8, we can see our method outperforms the other related methods, mainly due to the contribution of spatial difference descriptors.

Table 2. Image classification results on Caltech 101

Methods	15 training	30 training
KSPM [2]	56.40 ± 0.00	64.60 ± 0.80
WSS + SPM [3]	-	67.57 ± 0.00
KCSPM [5]	-	64.14 ± 0.18
LSPM [6]	67.00 ± 0.45	58.81 ± 1.51
ScSPM [6]	67.00 ± 0.45	73.20 ± 0.54
WSS + ScSPM	66.94 ± 0.01	73.39 ± 0.14
SD^2 + ScSPM	67.60 ± 0.00	73.15 ± 0.01
SD^1 + ScSPM	**70.01 ± 0.00**	**74.26 ± 0.01**

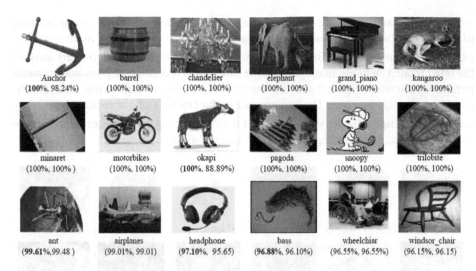

Fig. 8. Example images from classes with the top 18th classification accuracy on the Caltech 101 when using SD^1 descriptor under spatial pyramid matching framework

Table 3. Image classification results on Caltech 256

Methods	15 training	30 training	45 training	60 training
KCSPM [5]	-	27.17 ± 0.46	-	-
LSPM [6]	13.20 ± 0.62	15.45 ± 0.37	16.37 ± 0.47	16.57 ± 1.01
ScSPM [6]	27.73 ± 0.51	34.02 ± 0.35	37.46 ± 0.55	40.14 ± 0.91
LScSPM [7]	30.00 ± 0.14	35.74 ± 0.10	38.54 ± 0.36	40.43 ± 0.38
S^{3}R [17]	$\mathbf{37.85 \pm 0.48}$	$\mathbf{43.52 \pm 0.44}$	$\mathbf{46.86 \pm 0.63}$	-
KSPM [18]	-	34.10 ± 0.00	-	-
WSS + ScSPM	30.98 ± 0.00	36.90 ± 0.00	39.79 ± 0.00	41.63 ± 0.00
SD^2 + ScSPM	31.25 ± 0.00	36.84 ± 0.00	39.67 ± 0.00	41.63 ± 0.00
SD^1 + ScSPM	31.60 ± 0.00	37.04 ± 0.00	40.25 ± 0.00	$\mathbf{42.66 \pm 0.00}$

4.3 Caltech 256 Dataset

The Caltech 256 dataset has 256 classes of 29,780 images. Each class contains at least 80 images. Compared with the Caltech 101 dataset, images within the Caltech 256 dataset are more larger intra-class variant. We test our method on 15, 30, 45 and 60 training images randomly chosen in per class respectively. As shown in Table 3, S^{3}R performs better than our method. This is because it combines the visual similarity and weak semantic similarity of the training images. Furthermore it is time-costing because of learning extra classifier and space-consuming because of requiring more space for extra sub-semantic feature. Except S^3R, we can see that our approach outperforms all the other related methods, mainly due to the addition of spatial descriptors under spatial pyramid framework. Figure 9 shows some typical images owning the top 18th classification accuracy in brackets. The first number in the bracket is the accuracy obtained by using SD^1 descriptors under spatial pyramid matching framework, the second one is the accuracy by using the method proposed by Yang et al. in [6].

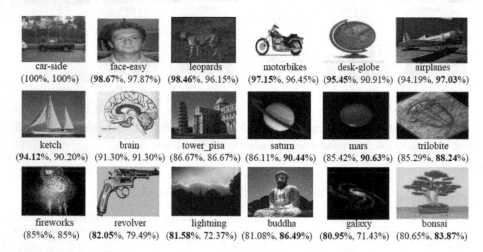

Fig. 9. Example images from classes with the top 18th classification accuracy on the Caltech 256 dataset when using SD^1 descriptors under spatial pyramid matching framework

5   Conclusion

This article focuses on boosting the performance of image classification with spatial difference information. A novel descriptor named spatial difference is proposed to describe the spatial information of differences. And this descriptor is mainly used in the combination with histograms of bag-of-visual words model under spatial pyramid matching framework, which can boost the final performance of image classification. The experimental results on the three public image datasets of the Scene 15 dataset, the Caltech 101 dataset and the Caltech 256 set demonstrate the effectiveness of the proposed method.

Acknowledgements. This work was supported by the National Natural Science Foundation of China (61202325, 61303154, 61379100, 61370169, 60873104).

References

1. Sivic, J., Zisserman, A.: Video google: a text retrieval approach to object matching in videos. In: Proceedings of ICCV, pp. 1470–1477. IEEE (2003)
2. Lazebnik, S., Schmid, C., Ponce, J.: Beyond bags of features: spatial pyramid matching for recognition natural scene categories. In: Proceedings of CVPR, pp. 2169–2178 (2006)
3. Teng, K., Wang, J., Tian, Q., Lu, H.: Improving scene classification with weakly spatial information. In: Proceedings of ICIP, pp. 3259–3263 (2013)
4. Grauman, K., Darrell, T.: Pyramid match kernels: discriminative classification with sets of image features. In: Proceedings of ICCV, pp. 725–760 (2005)
5. Smeulders, A., Gemert, J., Veenman, C., Geusebroek, J.: Visual word ambiguity. IEEE Trans. Pattern Anal. Mach. Intell. **32**(7), 1271–1283 (2010)
6. Yang, J., Yu, K., Huang, T.: Linear spatial pyramid matching using sparse coding for image classification. In: Proceedings of CVPR (2009)
7. Gao, S., Tsang, I., Chia, L.: Local features are not lonely-Laplacian sparse coding for image classification. In: Proceedings of CVPR (2010)
8. Wang, J., Yang, J., Lv, F., Huang, T., Gong, Y.: Locality-constrained linear coding for image classification. In Proceedings of CVPR (2010)
9. Chen, Q., Song, Z., Hua, Y., Huang, Z., Yan, S.: Hierarchical matching with side information for image classification. In: Proceedings of CVPR, pp. 3426–3433 (2012)
10. Yao, B., Khosla, A., Fei-Fei, L.: Combining randomization and discriminative for fine-grained image categorization. In: Proceedings of CVPR, pp. 1577–1584 (2012)
11. Zhang, N., Farrell R., Darrell, T.: Pose pooling kernels for sub-category recognition. In: Proceedings of CVPR, pp. 3665–3672 (2012)
12. Zhang, C., Liu, J., Tian, Q., Han, Y., Lu, H., Ma, S.: A boosting, sparsity-constrained bilinear model for object recognition. IEEE Multimedia **2**, 58–68 (2012)
13. Bao, C., He, L.: Linear spatial pyramid matching using non-convex and non-negative sparse coding for image classification (2015). arXiv:1504.06897v1 [cs. CV]
14. Pasolli, E., Melgoni, F., Tuia, D., Pacifici, F., Emery, W.J.: SVM active learning approach for image classification using spatial information. IEEE Trans. Geosci. Remote Sens. **52**(4), 2217–2233 (2014)
15. Jia, S., Xie, Y., Zhu, Z.: Integration of spatial and spectral information by means of sparse representation-based classification for hyper spectral imagery. In: Proceedings of the 18th Asia Pacific Symposium of Intelligent and Evolutionary Systems, Proceedings in Adaption, Learning and Optimization (2015). doi:10.1007/978-3-319-13356-0_10
16. Zhu, C., Yang, S., Zhao, Q., Cui, S., Wen, N.: Robust semi-supervised kernel-FCM algorithm incorporating local information for remote sensing image classification. J. Indian Soc. Remote Sens. **42**, 35–49 (2014)
17. Zhang, C., Chen, J., Liu, J.: Object categorization in sub-semantic space. Neurocomputing **142**, 248–255 (2014)
18. Griffin, G., Holub, A., Perona, P.: Caltech-256 object category dataset, Caltech-256 Technical report UCB/CSD-04-1366 (2007)

Exploring Relationship Between Face and Trustworthy Impression Using Mid-level Facial Features

Yan Yan[1], Jie Nie[2], Lei Huang[1], Zhen Li[1], Qinglei Cao[1], and Zhiqiang Wei[1(✉)]

[1] College of Information Science and Engineering, Ocean University of China, Qingdao China
{yanyan.azj,ithuanglei,lizhen0130,cql.levi}@gmail.com,
weizhiqiang@ouc.edu.cn
[2] Department of Computer Science and Technology, Tsinghua University, Beijing China
niejie@tsinghua.edu.cn

Abstract. When people look at a face, they always build an affective subconscious impression of the person which is very useful information in social contact. Exploring relationship between facial appearance in portraits and personality impression is an interesting and challenging issue in multimedia area. In this paper, a novel method which can build relationship between facial appearance and personality impression is proposed. Low-level visual features are extracted on the defined face regions designed from psychology at first. Then, to alleviate the semantic gap between the low-level features and high-level affective features, mid-level feature set are built through clustering method. Finally, classification model is trained using our dataset. Comprehensive experiments demonstrate the effectiveness of our method by improving 26.24 % in F1-measure and 54.28 % in recall under similar precision comparing to state-of-the-art works. Evaluation of different mid-level feature combinations further illustrates the promising of the proposed method.

Keywords: Portrait · Affective subconscious impression · Personality · Mid-level feature

1 Introduction

People could build impression including facial appearance and affective subconscious impression at first sight. The facial appearance is respect to the person's facial attributes, such as her/his eye size, face shape, etc. The affective subconscious includes personality impression primarily, e.g., "he is reliable". Many classical and famous psychology works have find that facial appearance correlated with impression of personality. How to infer relationship between face appearance and personality impression becomes an interesting and challenging problem in multimedia area, which can predict others' personality by 'reading' their faces in artificial intelligence.

Some works in recent have focused on exploring relationship between visual patterns and personality traits. Richard et al. have done some works to uncover the information driving first impression of social traits judgments, such as trustworthiness or dominance, using an attribute-based approach [1]. They create a quantitative model that can predict

© Springer International Publishing Switzerland 2016
Q. Tian et al. (Eds.): MMM 2016, Part I, LNCS 9516, pp. 540–549, 2016.
DOI: 10.1007/978-3-319-27671-7_45

first impressions of previously unseen ambient images of faces from a linear combination of facial attributes. This study shows that despite enormous variation in ambient images of faces, a substantial proportion of the variance in first impressions can be accounted for through linear changes in objectively defined features. Nie et al. presented a method to infer personality impression from portrait [2]. They used features including background, photograph color, person gender in portraits and focused on four personality impression of the whole portrait. In our work, we focus on face area. Because the appearance of face area is almost the same during ones' whole life, but clothes, photograph background and so on are changeful.

In our previous work, we have found that using facial features can predict people's trustworthiness impression at first sight [3]. From a psychological perspective, we propose novel personality-toward features combining eleven permanent facial features and five transient facial features. All sixteen features are extracted from five main facial features, consisting of eyebrow, eye, nose, mouth and face shape. For different facial regions, we used different feature extraction methods, for example, Histogram of Gradients (HOG) is used to describe eyebrow shape, and Euclidean Distance (ED) is used to describe width of eyes. This work has provided theoretical and empirical insight into finding relationship between facial appearance and trustworthiness impression. In this work, we will further explore the relationship and consider more factors of extracted features in previous work. For example, the dimension of different face regions descriptors are quite different (e.g. Scale-invariant Feature Transform (SIFT) descriptors has 127 dimensions, ED descriptors just has one dimension). We will settle the problem of vastly different dimensions of each facial trait and semantic gap between the low-level features and affective high-level features.

In this paper, we build a framework shown in Fig. 1 through proposing a new mid-level feature set from low-level features using unsupervised learning method to find relationship between facial appearance and personality impression. The mid-level feature set with mid-semantic meaning contains important discriminative information of each face region and maybe effectiveness for predicting personality impression. Relationship between face features and personality impression is inferred by the mid-level facial feature set through supervised learning method. First, we extract the low-level features from face region as face local descriptors. Then we use unsupervised feature learning algorithms translate local descriptors into a set of mid-level feature, which can represent the face in portrait. After that, Support Vector Machine (SVM) [4] is used to find relationship between face features and personality impression. In experiment section, we use K-fold cross-validation to verify the effective of our data set and then compare the effectiveness of our method with other methods. Multinomial logistic regression and Likelihood ratio test are used to compare the contribution of different mid-level feature combination. Performance of clustering low-level features into mid-level feature and correlated face colleges are demonstrated.

The remaining of the paper is organized as follows. Related works are introduced in Sect. 2. The feature extraction method is described in Sect. 3. In Sect. 4, comprehensive experiments are done and comparative results are given. Finally, some conclusions are drawn in Sect. 5.

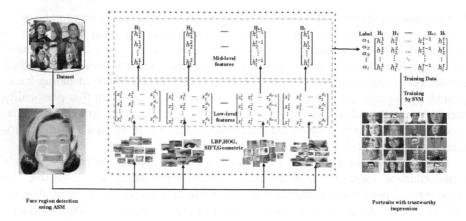

Fig. 1. Framework of the proposed method

2 Related Works

There are numerous works in current focusing on semantic learning in multimedia area using describable visual attributes. The describable visual attributes are regarded as labels that can be given to an image to describe its appearance and widely used to solve classic problems in computer vision area. Kumar et al. have introduced the usage of describable visual attributes, examples of face attributes include gender, age, jaw shape, nose size, etc., for face verification and image search [5–7]. Lin et al. tried to find the Latent Human Topic (LHT) in facial attributes comparing to Kumar et al.'s work [8]. [9, 10] have shown how mid-level semantic attributes can be used synergistically with low-level features for both identification and re-identification. These visual attribute labels in these works were learned through manual annotation, which meaning that all visual attributes are built on personal perception. Cao et al. proposed an adaptive system that allows for ties when collecting ranking data, in which each image is represented by a set of facial attributes [11]. [12, 13] focused on automatic facial attribute detection by different methods. Most of these works above, focused on extracting mid-level which related to attire, biometrics, appendix, but not natural face attributes.

Face is a distinctive region, which has settled region structure, e.g. symmetric regions, two brows located upon two eyes, one nose under middle of eyes. They give people different impression from different angles and change with different facial expressions, which can make a different impression comparing to their real facial attributes. In our work, we use a new mid-level feature set which is clustered from low-level features through unsupervised learning method to describe natural attribute of a face. Many works have applied cluster analysis method into feature classification using unsupervised learning. Sheikh et al. presented a new approach that recognizes facial expressions automatically and also to show the effectual outcome of their approach using K-Means Clustering [14].

3   Proposed Method

In this section, we extract low-level features from different face regions and cluster them into a vector of mid-level features to build a vector of facial features which is personality-toward.

3.1   Low-level Features

In this paper, we focused our work on main regions of face area. We extracted eighteen type features from main regions e.g. eye, brow, nose, mouth and relationship between regions using methods like HOG, Local Binary Patterns (LBP), SIFT and geometric descriptor [15]. Following Fig. 2 shows the workflow of low-level features extraction and description.

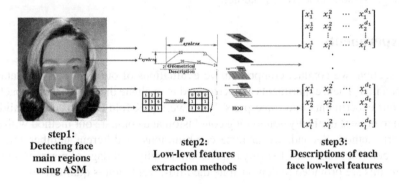

Fig. 2. Workflow of low-level features extraction and description

At first, we utilize Active Shape Model (ASM) [16] to detect main facial main regions and obtained the contours of them. Then the face area was normalized into 128×128 pixels, and converted into gray image to avert influence of illumination intensity. At last, a mixture of features descriptors consisting of HOG, LBP, SIFT and geometrical descriptions were adopted to describe personality-toward features. From each person image, we first extract eighteen kinds of face features which add up to 3412-dimensional low-level features.

3.2   Mid-level Features

As mentioned in the Sect. 3.1, each face is represented as eighteen kinds of face features which add up to 3412-dimensional low-level features. Conventionally, this extremely high dimensional vector is directly fed to an SVM for classification. However, the dimension of each attribute is quite different extracted by different methods. As result, performing classification in such feature space is susceptible to overfitting. Furthermore, each attribute may make important contribution to classi-fication. Hence, we use a vector that contains eighteen dimension face features representing a face, in which each dimension clustered from the corresponding low-level features. As one of the most popular and simple clustering algorithms, K-means algorithms [17] is used to cluster the corresponding low-level features, whose result is different from its visual attribute and we call it Mid-level feature.

For each facial portrait, suppose there are T mid-level features, we denote sample data as $X_t = [x_{t,}^1, x_{t,}^2, x_{t,}^3 \ldots, x_{t,}^{d_t}]$, where $t = 1, 2, \ldots, T$, meaning the t-th mind-level feature, d_t represents the dimension of t-th low-level feature. Then each low-level feature set X_t clustered into a single feature vector H_t valuing $C_t = \{k, k = \{1, 2, \{inK\}\}$, K is the number of features status (e.g. the number of eye status is 3 including open, close and slightly open).

For each face, we use all the mid-level features $Z = [H_1, H_2, \ldots, H_T]$ as training data and associate a training label $y_i \in [0, 1]$, where 0 means trustworthy and 1 means untrustworthy. SVM is used to classify trustworthy personality using mid-level features clustering. We select the RBF as kernel.

4   Experiments

In this section, we conduct comprehensive evaluations of our method. The dataset is described first. Then, experiments are performed to evaluate the proposed approach. We divide our experiments into three parts. First, we use K-fold (K = 5) cross-validation method to draw statistically convincing conclusion and compare our method with state-of-the-art method. Second, we analyze combinations of different number mid-level features contributions to trustworthy impression. Finally, a collage of portraits and clustering results are provided to demonstrate the mid-level features qualitatively.

4.1   Dataset

We use a challenging dataset in work of Ref. [3], which contains 2010 portraits and be labeled the ground-truth by about 250 volunteers. For each portrait, the volunteers were asked to label a tag about trustworthiness or untrustworthiness on it at their first impression. If the consistence of trustworthy by different raters achieved 70 %, the image was tagged with 'trustworthy'.

1404 images were randomly chosen as the training set and the other 606 images were used as the testing set. The details of the training set and testing set are shown in Table 1.

Table 1. Sample number of training set and testing set

Group	Training set		Testing set		Total
	Trustworthy	Untrustworthy	Trustworthy	Untrustworthy	
Portraits number	522	882	273	333	2010

4.2   Experiments and Discussions

Comparison with Other Methods. First we compare the classification accuracy of our new method with Ref. [3]. In order to draw statistically convincing conclusions, it is important to estimate the uncertainty of such estimates. In this experiment, we used K-fold (K = 5) cross-validation method, which can make it possible to reproduce the results and be unbiased to a specific data distribution.

Table 2 lists the classification results of our method and Ref. [3] using three evaluation metrics, i.e., precision, recall and F1- measure. We use the average accuracy of K-fold cross-validation method as the result. From Table 2, we can find that our method has significantly outperformed method in Ref. [3], which gains a great improvement of 54.28 % in recall with a small decrease of precision (6.14 %). Further, our method gains an improvement of 26.24 % in F1-measure.

Table 2. Performance comparison in F1-measure

Methods	Precision	Recall	F1-measure
Reference [3] method	71.10 %	37.00 %	48.70 %
Our method	64.96 %	91.28 %	74.94 %

The improvement benefits from our proposed feature set clustered from low-level features since the feature set is mid-level that can alleviate the semantic gap compared to low-level features. Moreover, feature set can alleviate the influence of the different dimensions of each low-level feature. The semantic meanings of mid-level feature set will be highlighted using qualitative experiment in Fig. 4.

To verify effectiveness of our mid-level features, we use Multinomial logistic regression model to fit the data and likelihood ratio test to demonstrate the contribution of each mid-level feature. Results are shown in following Fig. 3.

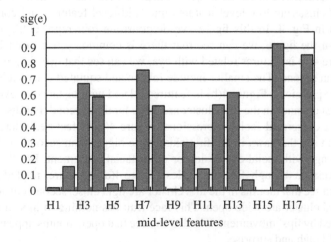

Fig. 3. Likelihood ratio test of each mid-level feature

Inferring from experience, we set a threshold as 0.05. If the likelihood of the mid-level feature is lower than threshold, this mid-level feature is correlation with personality. From Fig. 3 we can see that the likelihood of mid-level features $\{H_1, H_5, H_9, H_{15}, H_{17}\}$ are lower than 0.05, which means that these five mid-level feature are correlation with trustworthy impression.

To further represent the correlation of mid-level features and trustworthy impression, we exhibit the precision, recall and F1-measure of different number mid-level features in Table 3.

Table 3. Precision, recall and F1-measure of different number mid-level feature combination. The row of number means the combination of different number mid-level features determined by the value of likelihood ratio test. For example, combination of three mid-level features is $\{H_9, H_{15}, H_{17}\}$, which are the lowest three likelihood ratio test features. High performance is in bold.

Number	3	5	7	9	11	13	15	18
Precision	**47.37**	45.83	45.07	53.70	44.29	43.48	44.29	43.48
Recall	**72.00**	66.00	64.00	58.00	62.00	60.00	62.00	60.00
F1-measure	**57.14**	54.09	52.89	55.76	51.66	50.42	51.66	50.42

From Table 3, we can see that combination of three mid-level features perform best in all combinations, which means that we can use only three mid-level features to judge impression of a face is trustworthy or not. From the table we also can find that the combination of nine mid-level features performs better than other combinations in F1-measure and precision except for three mid-level features. It indicates that face features are not independent, such as facial expression is determined by several face AUs (Action Units).

Qualitative Demonstration of Mid-level Features. In this subsection, we demonstrate the results of clustering low-level features into mid-level features and corresponding face collages in Fig. 4. In this figure, we demonstrate four representative mid-level features. From the figure, we can see that there is common sense of each mid-level feature. Figure 4(a) is features related with eye, we can see that eyes in the first cluster are bigger than second. The smaller include inborn and squinted eyes leaded by smile or confused expressions. Figure 4(b) is features related with distance of eye and brow, we can see that distance in the first cluster are more far than second. Far distance maybe produced by exaggerated facial expressions. Figure 4(c) is features related with jaw region, we can see that jaws in the first cluster are bigger than second. This face trait could not change with facial expression or other facial movement. But they can form different impression through different angles. Figure 4(d) is features related with mouth station, we can see that mouths in the first cluster are tight close, second cluster are slight open and third cluster are wide open. This face trait has nothing to inborn feature, but just be changed by lips' movement. We also can see that open mouths appear with facial expression of laugh and surprise.

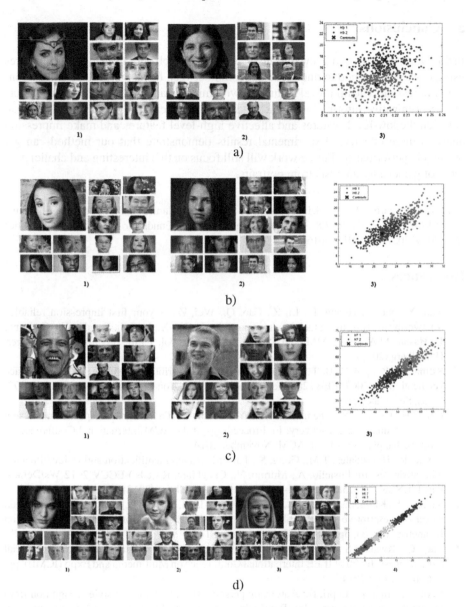

Fig. 4. Demonstrations of four representative mid-level features and corresponding face collages. (a) Represent eye region mid-level feature H_9, (b) Represent region between eye and brow mid-level feature H_8, (c) Represent jaw region mid-level feature H_7, (d) Represent mouth region mid-level feature H_{17}. (a)-3, (b)-3, (c)-3, (d)-4 show result of clustering low-level features into mid-level feature.

5   Conclusions

In this work, we proposed a new mid-level feature set which built from low-level features using clustering method to find the relationship between facial appearance and personality impression. The mid-level feature set with mid-semantic meaning contains important discriminative information of each face region that can alleviate the semantic gap between the low-level features and affective high-level features and make impression analysis more effective. Experimental results demonstrate that our method can get promising performance. Future work will still focus on this interesting and challenging topic of personality analysis from portraits.

Acknowledgments. This work is supported by the National Nature Science Foundation of China (No. 61402428, No. 61202208); the Fundamental Research Funds for the Central Universities (No. 201413021, No. 201513016).

References

1. Yan, Y., Nie, J., Huang, L., Li, Z., Cao, Q., Wei, Z.: Is your first impression reliable? trustworthy analysis using facial traits in portraits. In: He, X., Luo, S., Tao, D., Xu, C., Yang, J., Hasan, M.A. (eds.) MMM 2015, Part II. LNCS, vol. 8936, pp. 148–158. Springer, Heidelberg (2015)
2. Kumar, N., Berg, A.C., Belhumeur, P.N., Nayar, S.K.: Attribute and simile classifiers for face verification. In: 2009 IEEE 12th International Conference on Computer Vision, pp. 365–372. IEEE, September 2009
3. Lin, C.H., Chen, Y.Y., Chen, B.C., Hou, Y.L., Hsu, W.: Facial attribute space compression by latent human topic discovery. In: Proceedings of the ACM International Conference on Multimedia, pp. 1157–1160. ACM, November 2014
4. Layne, R., Hospedales, T.M., Gong, S.: Towards person identification and re-identification with attributes. In: Fusiello, A., Murino, V., Cucchiara, R. (eds.) ECCV 2012 Ws/Demos, Part I. LNCS, vol. 7583, pp. 402–412. Springer, Heidelberg (2012)
5. Klare, B.F., Klum, S., Klontz, J.C., Taborsky, E., Akgul, T., Jain, A.K.: Suspect identification based on descriptive facial attributes. In: 2014 IEEE International Joint Conference on Biometrics (IJCB), pp. 1–8. IEEE, September 2014
6. Cao, C., Kwak, I.S., Belongie, S., Kriegman, D., Ai, H.: Adaptive ranking of facial attractiveness. In: 2014 IEEE International Conference on Multimedia and Expo (ICME), pp. 1–6. IEEE, July 2014
7. Celli, F., Bruni, E., Lepri, B.: Automatic personality and interaction style recognition from facebook profile pictures. In: Proceedings of the ACM International Conference on Multimedia, pp. 1101–1104. ACM, November 2014
8. Datta, A., Feris, R., Vaquero, D.: Hierarchical ranking of facial attributes. In: 2011 IEEE International Conference on Automatic Face & Gesture Recognition and Workshops (FG 2011), pp. 36–42. IEEE, March 2011
9. Sheikh, T., Agrawal, S.: Performance comparison of an effectual approach with k-means clustering algorithm for the recognition of facial expressions. Int. J. Comput. Sci. Mob. Comput. **2**(6), 300–306 (2013)
10. Tian, Y.L., Kanade, T., Cohn, J.F.: Recognizing action units for facial expression analysis. IEEE Trans. Pattern Anal. Mach. Intell. **23**(2), 97–115 (2001)

11. Cootes, T.F., Taylor, C.J., Cooper, D.H., Graham, J.: Active shape models-their training and application. Comput. Vis. Image Underst. **61**(1), 38–59 (1995)
12. Nie, J., Cui, P., Yan, Y., Huang, L., Li, Z., Wei, Z.: How your portrait impresses people?: inferring personality impressions from portrait contents. In: Proceedings of the ACM International Conference on Multimedia, pp. 905–908. ACM, November 2014
13. Vernon, R.J., Sutherland, C.A., Young, A.W., Hartley, T.: Modeling first impressions from highly variable facial images. Proc. Natl. Acad. Sci. **111**(32), E3353–E3361 (2014)
14. Chang, C.C., Lin, C.J.: LIBSVM: a library for support vector machines. ACM Trans. Intell. Syst. Technol. (TIST) **2**(3), 27 (2011)
15. Kumar, N., Berg, A.C., Belhumeur, P.N., Nayar, S.K.: Describable visual attributes for face verification and image search. IEEE Trans. Pattern Anal. Mach. Intell. **33**(10), 1962–1977 (2011)
16. Kumar, N., Belhumeur, P.N., Nayar, S.K.: FaceTracer: a search engine for large collections of images with faces. In: Forsyth, D., Torr, P., Zisserman, A. (eds.) ECCV 2008, Part IV. LNCS, vol. 5305, pp. 340–353. Springer, Heidelberg (2008)
17. Jain, A.K.: Data clustering: 50 years beyond K-means. Pattern Recogn. Lett. **31**(8), 651–666 (2010)

Edit-Based Font Search

Ken Ishibashi[1](✉) and Kazunori Miyata[2]

[1] Prefectural University of Kumamoto,
Tsukide 3-1-100, Higashi-ku, Kumamoto-shi, Kumamoto, Japan
kishibashi@pu-kumamoto.ac.jp
[2] Japan Advanced Institute of Science and Technology,
Asahidai 1-1, Nomi-shi, Ishikawa, Japan
miyata@jaist.ac.jp

Abstract. This paper presents an interactive font search method for users who have no significant knowledge of fonts. The proposed method suggests some font candidates by deforming a displayed font image into a font resembling a user's desired font. Generally, font category names, font tags regarding human impressions such as "Cute", and a font list are used for searching a font. However, there are some issues: knowledge of font category names is required, each user gets a different impression from a font, and the font search from a font list becomes tedious as the number of font candidates increases. We expect that the proposed method can solve these problems, because it allows users to search a font easily.

Keywords: Font search · Interactive search · Deformation of letter shape

1 Introduction

Recent studies on effective image retrieval [2,12] have proposed a sketch-based retrieval method for two-dimensional (2D) or three-dimensional (3D) images. This approach is useful for a quick search because users can input their requests directly through their sketch operations. However, using the sketch-based method can be difficult for users who do not have any ideas or cannot sketch well. To overcome this problem, we propose an edit-based search method. Our proposed method has two advantages: (1) users are not required to have concrete ideas, and (2) the features of the sketch-based method are inherited. In particular, we present a font search application introducing our approach. One of the most popular smartphone graphics editors has been downloaded at least one hundred million times. Users edit their photos to enjoy decorating and sharing them with their friends using the graphics editors. In addition, creating personal presentation slides and posters is common. Fonts are an important graphic element in such graphics editing. However, there are many commercial and free fonts available on the Web. Furthermore, searching for fonts is difficult for novice users, because they do not have sufficient skills or knowledge regarding

© Springer International Publishing Switzerland 2016
Q. Tian et al. (Eds.): MMM 2016, Part I, LNCS 9516, pp. 550–561, 2016.
DOI: 10.1007/978-3-319-27671-7_46

designs. Therefore, an effective font search method can support editing tasks for many users. This paper presents our proposed font search method and reports the results of an experimental evaluation.

2   Related Work

Here, we introduce some studies that are involved in generating, learning, and searching fonts. In addition, we describe two sketch-based retrieval methods.

2.1   Generating, Learning, and Searching Fonts

Previous work has reported various manual or (semi-)automatic handling methods for fonts. Zalik [14] proposed a font editing system to easily edit a font shape. This kind of font editor enables users to create their original fonts, releasing an enormous number of different fonts on the Web. Suveeranont and Igarashi [10] implemented a semi-automatic font generation system. This system derives a skeleton and a correspondence outline. Even though the automatic font generation system creates new fonts, the font quality is lower than designers' fonts as a result of several slight distortions. However, manual font design or creation is difficult for novice users. Hence, the following studies have focused on font search to avoid the issues of font quality, and our study also focuses on a font search approach.

Recent studies have reported learning methods for fonts. O'Donovan et al. [7] developed two font-exploring interfaces. One interface uses high-level descriptive attributes, such as "warm" or "fresh". All attributes are modeled statistically through a combination of machine learning and crowdsourcing. The other uses a tree-based hierarchical menu based on perceptual font similarity. This research also provides knowledge of perceptual similarity; hence we refer to the font similarity definition. Campbell and Kautz [1] presented a manifold of fonts obtained by learning non-linear mapping. The manifold can be used to interpolate and move among existing fonts smoothly. These attribution and manifold learning methods are state-of-the art studies regarding fonts. Loviscach [6] attempted to lay out 2,000 font styles on a 2D map. Solli and Lenz [9], Cutter et al. [3], and Kataria et al. [5] proposed font image retrieval methods using font feature data. These methods search for desired fonts based on font query data. However, we expect that users might not have concrete ideas in their minds about a target font. In this case, an intuitive search method is useful. An intuitive search approach has been proposed as a sketch-based retrieval method, and our approach is similar. Next we briefly introduce the sketch-based method.

2.2   Sketch-Based Retrieval Method

Recent studies have provided sketch-based retrieval solutions as the appropriate approach for the case in which users have no concrete idea [2,12]. The image search method proposed by Chen et al. [2] enables users to create a desired image using image synthesis based on user sketch input. The 3D model search method proposed by Xu et al. [12] creates a 3D scene based on a furniture layout design that was inputted using a sketch operation. The sketch-based retrieval method is an effective and intuitive solution. Unfortunately, some users are not skilled at drawing sketches. Our solution reduces this obstacle.

3   Edit-Based Font Search Method

We illustrate our edit-based font search method. Here, we describe the framework of edit-based font search, the details relevant to font similarity, and the implementation of our proposed approach.

3.1   Framework of Edit-Based Font Search

Figure 1 shows the framework of edit-based font search.

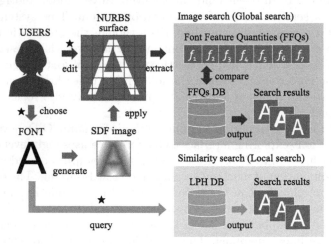

Fig. 1. Framework of edit-based font search

The method consists of two modes, global and local search. The global search mode has a function for checking a font feature quantities (FFQs) database (DB) using a user's edited font image. The global search is executed after each editing operation. This process is shown in pink in Fig. 1. The local search is a high-accuracy visual similarity search (SS). When choosing a DB font, users can search similar font candidates by querying a log-polar histogram (LPH) DB. In addition, the user's chosen font is applied as a new deformable font image using the signed distance field (SDF) of the font. The deformable image is constructed using a non-uniform rational basis spline (NURBS) surface. Users are allowed three operations, edit, choice, and query (stars in Fig. 1). They can edit to change a displayed font shape as a globally searchable system. A local search is useful if they want to check very similar font candidates. Users can obtain their desired fonts using the global and local searches.

3.2   Requirements for Font Search

This framework requires a user interface (UI) and two definitions of font features for the global and local searches. We describe the details of these requirements.

UI Implementation for Edit. In the UI of our framework, robustness and flexibility are necessary for stable user operation. To avoid any errors, we adopted a NURBS surface for a simple and robust UI. The first step is to create a NURBS surface on the display. Then, an SDF image is applied to the NURBS surface. The 7×7 control points are arranged at equal intervals in a grid on the NURBS surface. The NURBS surface size depends on the chosen font's ratio. The longer length of the font's ratio is approximately 170 pixels, and the shorter one is determined by the ratio and the actual longer length. The equations of this section are shown in a UV coordinate system.

The control points are determined by the Euclidian distances between the user's specified point and each control point. The displacements of all control points are calculated on a normalized Gaussian distribution whose center is the user's specified point. The ith control point c_i is updated as c_i' according to Eq. (1).

If $e < dist(c_i, user)$,
$$c_i' = c_i + (user' - user) \cdot \frac{1}{\sqrt{2\pi}\sigma} \cdot \exp\left(-\frac{dist(c_i, user)^2}{2\sigma^2}\right), \qquad (1)$$
otherwise, $c_i' = c_i$.

where $user$ and $user'$ are the user's specified points before and after editing, respectively, and $dist$ is the Euclidian distance function. We set $\sigma = 0.7$ empirically for comfortable user operation. In addition, we set a distance parameter e. If the parameter e has a grater value, such as 1.0, deformation will be difficult for users, because most of the control points will be moved. Thus, we set a constraint range within $[0.0, 0.1]$.

Our method has a function for overall deformation. When the control points are moved at equivalent distances, the font image might get deteriorated. Hence, our proposed method introduces an exponential function, as shown in Eq. (2).

If $0 < ID^{center} - ID_i < ID^{center}$
$$c_i' = c_i + \exp\left(-\frac{(user' - user)(ID^{center} - ID_i)}{\varphi}\right), \qquad (2)$$
Otherwise, $c_i' = c_i$.

In Fig. 2(left), a pink triangle represents a user manipulation point for overall deformation. All control points are assigned their IDs in the range of 1–7. When $ID^{center} = 4$, all control points are updated based on the number of differences between ID^{center} and each ID. The blue points indicate the update target points. When a user holds the left pink triangle down, the IDs are assigned from left to right. The effect level parameter is empirically set as $\varphi = 3.0$.

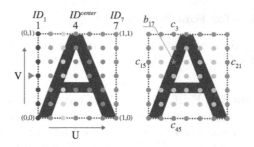

Fig. 2. (left) Example of target updating in the control points. (right) Example of a point related to a constraint point (Color figure online).

In addition, we set a force parameter ε for quick response of the UI. This is relevant to deformation smoothness and quickness of response. The high value setting provides users a quick response. However, font image deformation is not smooth. Each control point c_i' is given as follows:

$$
c_i' = \psi \cdot dist\,(a_i, c_i) \cdot \begin{bmatrix} \cos\,(\tau + \pi) \\ \sin\,(-\tau) \end{bmatrix} + \begin{bmatrix} u_i' \\ v_i' \end{bmatrix},
$$
$$
\tau = \tan^{-1}\left(\frac{v_i' - v_i}{u_i' - u_i} \right), a_i = \begin{bmatrix} u_i' \\ v_i' \end{bmatrix} = \varepsilon b_i + (1 - \varepsilon)\,c_i.
$$
(3)

where u_i and v_i are the ith coordinate values on the UV-axes. a_i and b_i are the ith constraint point and the corresponding point, respectively, calculated by dividing equally from the edge points of the same row and column. Figure 2(right) shows an example in which b_{17} is calculated as follows: $b_{17} = \begin{bmatrix} u_{15} - \frac{|u_{21} - u_{15}|}{6} \\ v_{45} + \frac{|v_3 - v_{45}|}{6} \end{bmatrix}$.
We set the force parameter as $\varepsilon = 0.3$ and the control parameter as $\psi = 2.0$ under the balance of deformation smoothness and quickness of response. The orange points in Fig. 2(right) are updated using Eq. (3).

Fonts have various features such as angle and line width. The angle or overall ratio of a font can be changed by deformation. However, the line width or swirl shape is difficult to change. Hence, we implemented thickness and swirl effect functions. The thickness effect changes the thickness of a font stroke. Green [4] proposed an improved rendering method for vector textures using an SDF image. We adopted the same method for displaying text and for this thickness effect. When a user scratches up and down in the non-display area of a font image, the thickness is changed. Scratching toward the upper side thickens the displayed text. The swirl effect creates a swirl shape at a user-specified point during a mouse down operation. These effects are implemented on a pixel shader.

Font Features for Global and Local Searches. Our proposed method requires two font features for global and local searches. First, we describe the font features for global search. Geometric font features in the previous study [7] are composed of font size and ratio with a bounding box, horizontal and vertical spacing, outline arc length, curvature histograms, angle, and stroke width. These features contain no font shape information. With the use of general graphics editors, users typically do not need to consider font size and space exactly, because they can deal with a letter flexibly. From these considerations, we defined seven FFQs, as shown in Fig. 3.

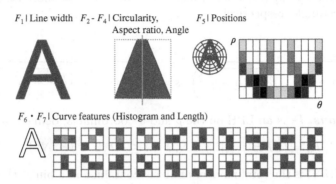

F_1 | Line width F_2 - F_4 | Circularity, Aspect ratio, Angle F_5 | Positions

$F_6 \cdot F_7$ | Curve features (Histogram and Length)

Fig. 3. Font feature quantities for global search

$\boldsymbol{F_1}|$ ***Line width.*** F_1 is the ratio of the edge pixels, given by Eq. (4), where p^{edge} and p^{all} are the number of edge and all pixels, respectively.

$$F_1 = \frac{p^{edge}}{p^{all}}. \tag{4}$$

F_2–F_4 are the geometric features of a convex hull created from a letter. Hence, these features are obtained using an image moment after creating the convex hull. The image moment is defined on a grayscale image, $I(x, y) \in [0.0, 1.0]$. The image moment of ith degree about the x-axis and jth degree about the y-axis is as follows: $M(i, j) = \sum_x \sum_y x_i y_j I(x, y)$.

$\boldsymbol{F_2}|$ ***Circularity.*** F_2 is the complexity of a convex hull, given by Eq. (5), where S is the area and L is the boundary length obtained by a contour tracing in 8-connectivity. Note that $S = M(0, 0)$.

$$F_2 = \frac{4\pi S}{L^2}. \tag{5}$$

The calculation of F_3 and F_4 uses the variables α, β, and γ, defined as follows [8]:

$$\alpha = \frac{M(2,0)}{M(0,0)} - \left(\frac{M(1,0)}{M(0,0)}\right), \quad \beta = 2\left(\frac{M(1,1)}{M(0,0)} - \frac{M(1,0)M(0,1)}{M(0,0)^2}\right),$$
$$\gamma = \frac{M(1,1)}{M(0,0)} - \left(\frac{M(0,1)}{M(0,0)}\right). \tag{6}$$

F_3| Angle. F_3 is the principal axis angle of inertia in a convex hull given by Eq. (7).

$$F_3 = \frac{1}{2} \tan^{-1} \left(\frac{\beta}{\alpha - \gamma} \right). \tag{7}$$

F_4| Aspect ratio. F_4 is the ratio of a bounding box along the principal axis of inertia given by Eq. (8), where l and s are the long and short lengths of the bounding rectangle, respectively.

$$F_4 = \frac{l}{l+s},$$
$$l = \sqrt{6 \left(\alpha + \gamma + \sqrt{\beta^2 + (\alpha - \gamma)^2} \right)}, \quad s = \sqrt{6 \left(\alpha + \gamma - \sqrt{\beta^2 + (\alpha - \gamma)^2} \right)}. \tag{8}$$

F_5| Positions. F_5 is an LPH on the ρ and θ coordinates, defined by Eq. (9).

$$F_5 = \{LPH(i,j)|i,j \in \mathbb{N}, 0 < i \le N^p, 0 < j \le N^\rho\}. \tag{9}$$

where $LPH(i,j)$ is the total number of pixels contained in a bin at the ith ρ and jth θ coordinates. N^ρ and N^θ are the maximum numbers of ρ and θ coordinates, respectively. Here, we set $N^\rho = 5$ and $N^\theta = 12$ as at the top right of Fig. 3.

$F_6 \cdot F_7$| Curve features. F_6 is a curve histogram of second-order or higher-order local autocorrelation (HLAC) features, given by Eq. (10).

$$F_6 = HLAC(i)|i \in \mathbb{N}, 0 < i \le N. \tag{10}$$

where $HLAC(i)$ is the ith HLAC pattern. All the patterns are shown at the bottom of Fig. 3, with $N = 20$.

F_7 is the total number of all curvature patterns, shown by all the pink patterns at the bottom of Fig. 3. This is defined as in Eq. (11).

$$F_7 = \sum_{i=5}^{N} HLAC(i). \tag{11}$$

The above-mentioned FFQs provide the basis for deriving a similarity function for the global search S^{global}, defined by Eq. (12).

$$S^{global} = 1.0 - \frac{\sum_{i=1}^{7} w_i F_i}{\sum_{i=1}^{7} w_i},$$
if $i \le 4$ or $i = 7$,
$$F_i = \begin{cases} \min(D, 180 - D) & \text{if } i = 3 \\ D & \text{otherwise,} \end{cases} \quad D = \frac{|F_i^{edit} - F_i^{DB}|}{F_i^{max} - F_i^{min}},$$
Otherwise,
$$F_i = 1.0 - \frac{\sum_{j=1}^{N'} \min\left(p_j^{edit}, p_j^{DB}\right)}{\max\left(\sum_{j=1}^{N'} p_j^{edit}, \sum_{j=1}^{N'} N' p_j^{DB}\right)},$$
$$N' = \begin{cases} N^\rho N^\theta & \text{if } i = 5 \\ N^{hlac} & \text{if } i = 6 \end{cases}, p = \begin{cases} LPH & \text{if } i = 5 \\ HLAC & \text{if } i = 6. \end{cases} \tag{12}$$

where w_i and F_i are the ith weight parameter and the ith FFQ, respectively. We set all weight parameters equal to 1.0. The search performance will be improved by optimizing the weight parameters through a user study. F_i^{edit} and F_i^{DB} are the ith FFQ from the user's edited font image and the ith FFQ from the DB for the global search. F^{max} and F_i^{min} are the maximum and minimum values of F_i^{DB}, respectively. p_j^{edit} and p_j^{DB} are the jth histogram patterns obtained from the edited font image and the DB, respectively. Note that f_5 and f_6 are calculated using histogram intersection [11].

In the calculation of similarity for the local search, S^{local}, we adopt a set of N^* LPHs based on different centered points [13]. The centered points are set at equal intervals. Here, we set $N^* = 10 \times 10$ points on the bounding box of the font image. This similarity DB is created by pre-processing with histogram intersection using N^* LPHs.

4   Font Search Application

We implemented a font search application to verify the effectiveness of our edit-based method. Figure 4 shows two screenshots of the implemented application. The application has two modes for global and local searches.

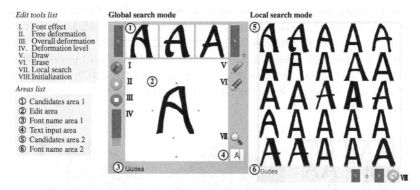

Fig. 4. Screenshots of the font search application

Global Search Mode — This mode provides four areas and seven editing tools.

Area ① displays three similar font candidates. These are updated for every user operation. Area ② is an editable area for a font image. Users can change the image by choosing a font candidate from Area ①. A user-chosen font name is displayed in Area ③. Area ④ is a text input area.

The seven editing tools have different availabilities. If Tool (V) or (VI) is available, Tools (I)–(III) are disabled. Tool (I) is related to the thickness and swirl effects. Examples are shown in Fig. 5.

In Fig. 4, the left image shows an original font image. A moving operation changes the thickness, and a staying operation creates a swirl, as shown in blue rectangles of Fig. 5. The orange rectangles indicate the top ranked font candidates using local search. Tool (II) enables free deformation of a font image in

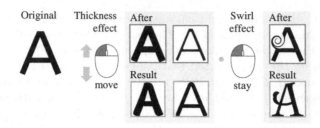

Fig. 5. Examples of the thickness and swirl effects (Color figure online)

Area ②. Tool (Ⅲ) is used to change the overall aspect ratio and angle of a font image. The four pink points and the blue triangle shown in Area ② correspond to the image aspect ratio and image rotation, respectively. Tool (Ⅳ) is a setting tool for a deformable level parameter related to e in Eq. (1). Tool (Ⅴ) is a sketch tool for drawing freely on a font image, and Tool (Ⅵ) is an erase tool. Tool (Ⅵ I) is a button for using the local search mode.

Local Search Mode — This mode shows 25 font candidates in Area ⑤. When users choose a font candidate from Area ⑤, all font candidates are updated. The current chosen font name is displayed in Area ⑥. Tool (Ⅷ) is a button for initialization. The initialization function generates 25 font candidates. In this initialization, we use a multi-dimensional scaling (MDS) based on LPHs. First, we calculate all the average values of S^{local} in the LPH DB of all upper-case letters (from "A" to "Z"). Next, we create a 3D map using the MDS based on average values. Then, we apply k-means clustering to the 3D map. Figure 6 (center) shows the clustering result.

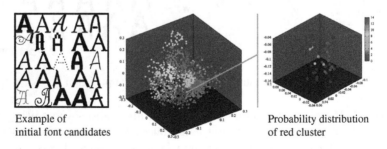

Example of initial font candidates

Probability distribution of red cluster

Fig. 6. Result of k-means clustering with a probability distribution and an example of initial font candidates (Color figure online).

Every point indicates all DB fonts, and each cluster is shown using a different color. Finally, we generated 25 probability functions based on each cluster. The probability function PF relies on an exponential function, shown in Eq. (13).

$$PF(i) = \exp\left(-\frac{d_i^2}{4d^{max}}\right). \tag{13}$$

where d_i is the Euclidian distance between the ith font and the centroid of the corresponding cluster, and d^{max} is the maximum value of the distances. An example of the probability distribution is shown in the right of Fig. 6, where red and blue indicate higher and lower probability, respectively. Each probability function generates an initial font candidate. Thus, the total number of initial font candidates is 25 (see the left of Fig. 6). We expect that checking the various types of fonts using this initialization will be useful for finding better font candidates in the early searching phase.

5   Experimental Evaluation

We conducted an experimental evaluation using our implemented font search application. This section describes the details of the experiment and the results.

5.1   Details of Experiment

Most of the existing font search methods introduced in Sect. 2.1 require font query data for their similarity searches. The method proposed by O'Donovan et al. [7] enables searching without concrete query data, because concept words are used as the query data. Their study reported the results of a font matching experiment. Hence, we conducted a font matching experiment to compare our method with the existing method. Before beginning the experiment, we first instructed every participant on the usage of the application, and then each participant used the application freely for ten minutes. However, we prohibited the use of tools (V) and (VI) to verify the search performance of the edit-based method.

The experimental procedure is as follows. When the participant clicks a button, the application displays a goal font selected randomly from 25 clusters of 3D maps as shown in Fig. 6. After selecting a cluster in a random order without duplication, a goal font is selected randomly from all font candidates belonging to the cluster. Next, the participant begins to search for a goal font by editing and using choice operations. The termination condition is either obtaining the goal font among the displayed candidates or reaching the time limit. In both cases, the target letter and the goal font are changed randomly for the next searching task. The target letter is relevant to two DBs (FFQs and LPH) and all displayed font candidates. If the target letter is set to "A," the application reads data corresponding to "A" and all font candidates are shown as "A." For the actual experiment, the target letter was set to the letters of "handgloves," one-by-one. For example, if the target letter was "h," the target letter was changed to "a" for the next search. We set two target letter types, "HANDGloves" and "handgLOVES." Eighteen undergraduate social science students (15 females and 3 males in their 20 s) participated in the experiment. The experimental equipment included laptop personal computers (Intel Core i5 2.6 GHz, 8 GB RAM, 1920 × 1080). The experiment used 1,460 font families combining Google Web Fonts and Japanese fonts of commercial software (FONT × FAN HYBRID3 produced by Font Alliance Network). The time limit of each experiment was 2 min.

Table 1. Success rate, search time, and details of user operations

Letter type	Success rate(%)	Time(s)
All	73.3	49.9
Big	75.6	51.4
Small	71.1	48.4

Case	Edit	Choice in Edit	SS	Choice in SS
Success	6.1	1.7	4.0	3.6
Failure	17.9	2.8	9.2	6.2

5.2 Experimental Results

Table 1 shows the experimental results. We recorded the success rate, search time, and number of user operations. The search time is the average time taken to find each goal font. The user operations include editing operations, candidate choices in edit mode, use of SS, and candidate choices in SS mode. In Table 1, all success rates are $>70\%$. The result of the existing method [7] had a 15 % success rate for both of the proposed interfaces. The experimental condition included a 2-min time limit and 750 fonts, including Google Web Fonts. Despite using twice as many fonts for our experiment, the experimental results indicated that the search performance of our method is superior to that of existing two interfaces. The average search time is <50 s. These results show that our method enables users to obtain their desired fonts quickly. At the bottom of Table 1, the average number of user operations is approximately 15 in successful cases. In contrast, the number of editing operations is notably large in the failure case. We suspect that users' edit operations were difficult owing to the complexity of goal fonts, or there were a lot of font candidates closely resembling the goal font.

6 Conclusion

The aim of this study was to compensate for the weakness of sketch-based retrieval methods. Hence, we proposed an edit-based search method to solve those problems. The proposed method can be used in combination with the existing sketch-based method. This paper described the details of the framework and its application. In addition, we conducted an experimental evaluation in the form of a comparison with the state-of-the-art method. The experimental results showed that our method is superior to the existing method. Using the proposed method, users obtained a specific font within 50 s with an average success rate of 70 %. Furthermore, the method requires no skill or knowledge. We expect that our method will improve the existing sketch-based method through the use of combination. In the future, we will investigate the performance with Japanese characters, because the method is available for all character set.

References

1. Campbell, N.D.F., Kautz, J.: Learning a manifold of fonts. ACM Trans. Graph. **33**(4), 91:1–91:11 (2014). http://doi.acm.org/10.1145/2601097.2601212
2. Chen, T., Cheng, M.M., Tan, P., Shamir, A., Hu, S.M.: Sketch2photo: internet image montage. ACM Trans. Graph. **28**(5), 124:1–124:10 (2009). http://doi.acm.org/10.1145/1618452.1618470
3. Cutter, M.P., Beusekom, J.v., Shafait, F., Breuel, T.M.: Unsupervised font reconstruction based on token co-occurrence. In: Proceedings of the 10th ACM Symposium on Document Engineering, DocEng 2010, pp. 143–150. ACM, New York (2010). http://doi.acm.org/10.1145/1860559.1860589
4. Green, C.: Improved alpha-tested magnification for vector textures and special effects. In: ACM SIGGRAPH 2007 Courses, SIGGRAPH 2007, pp. 9–18. ACM, New York (2007). http://doi.acm.org/10.1145/1281500.1281665
5. Kataria, S., Marchesotti, L., Perronnin, F.: Font retrieval on a large scale: an experimental study. In: 2010 17th IEEE International Conference on Image Processing (ICIP), pp. 2177–2180 (2010)
6. Loviscach, J.: The universe of fonts, charted by machine. In: ACM SIGGRAPH 2010 Talks, SIGGRAPH 2010, pp. 27:1–27:1. ACM, New York (2010). http://doi.acm.org/10.1145/1837026.1837062
7. O'Donovan, P., Lībeks, J., Agarwala, A., Hertzmann, A.: Exploratory font selection using crowdsourced attributes. ACM Trans. Graph. **33**(4), 92:1–92:9 (2014). http://doi.acm.org/10.1145/2601097.2601110
8. Shiraishi, M., Yamaguchi, Y.: An algorithm for automatic painterly rendering based on local source image approximation. In: Proceedings of the 1st International Symposium on Non-photorealistic Animation and Rendering, pp. 53–58 (2000)
9. Solli, M., Lenz, R.: FyFont: find-your-font in large font databases. In: Ersbøll, B.K., Pedersen, K.S. (eds.) SCIA 2007. LNCS, vol. 4522, pp. 432–441. Springer, Heidelberg (2007)
10. Suveeranont, R., Igarashi, T.: Feature-preserving morphable model for automatic font generation. In: ACM SIGGRAPH ASIA 2009 Sketches, p. 7:1 (2009)
11. Swain, M.J., Ballard, D.H.: Color indexing. Int. J. Comput. Vis. **7**, 11–32 (1991)
12. Xu, K., Chen, K., Fu, H., Sun, W.L., Hu, S.M.: Sketch2scene: sketch-based co-retrieval and co-placement of 3d models. ACM Trans. Graph. **32**(4), 123:1–123:15 (2013). http://doi.acm.org/10.1145/2461912.2461968
13. Xu, X., Zhang, L., Wong, T.T.: Structure-based ASCII art. ACM Trans. Graph. **29**(4), 52:1–52:10 (2010). http://doi.acm.org/10.1145/1778765.1778789
14. Zalik, B.: An interactive constraint-based graphics system with partially constrained from features. In: Presse Universitaire de Namur, pp. 129–139 (1996)

Private Video Foreground Extraction Through Chaotic Mapping Based Encryption in the Cloud

Xin Jin[1,2,4]($\boxtimes$), Kui Guo[1,2], Chenggen Song[1,2], Xiaodong Li[1,2]($\boxtimes$),
Geng Zhao[1,2], Jing Luo[1,2], Yuzhen Li[1,2,3], Yingya Chen[1,2], Yan Liu[1,2,3],
and Huaichao Wang[4]

[1] Beijing Electronic Science and Technology Institute, Beijing 100070, China
{jinxin,lxd}@besti.edu.cn
[2] GOCPCCC Key Laboratory of Information Security, Beijing 100070, China
[3] Xidian University, Xi'an 710071, China
[4] Information Technology Research Base of Civil Aviation Administration of China,
Civil Aviation University of China, Tianjin 300300, China

Abstract. Recently, storage and processing large-scale visual media data are being outsourced to Cloud Data Centres (CDCs). However, the CDCs are always third party entities. Thus the privacy of the users' visual media data may be leaked to the public or unauthorized parties. In this paper we propose a method of privacy preserving foreground extraction of video surveillance through chaotic mapping based encryption in the cloud. The client captures surveillance videos, which are then encrypted by our proposed chaotic mapping based encryption method. The encrypted surveillance videos are transmitted to the cloud server, in which the foreground extraction algorithm is running on the encrypted videos. The results are transmitted back to the client, in which the extraction results are decrypted to get the extraction results in plain videos. The extraction correctness in the encryption videos is similar as that in the plain videos. The proposed method has several advantages: (1) The server only learns the obfuscated extraction results and can not recognize anything from the results. (2) Based on our encryption method, the original extraction method in the plain videos need not be changed. (3) The chaotic mapping ensure high level security and the ability to resistant several attacks.

Keywords: Privacy preserving · Video surveillance · Foreground extraction · Chaotic mapping · Cloud computing

1 Introduction

Digital video surveillance has been equipped everywhere in our daily life and public security. Nowadays, cloud computing has changed the way of traditional video surveillance. The big data of surveillance videos are stored and automatically

© Springer International Publishing Switzerland 2016
Q. Tian et al. (Eds.): MMM 2016, Part I, LNCS 9516, pp. 562–573, 2016.
DOI: 10.1007/978-3-319-27671-7_47

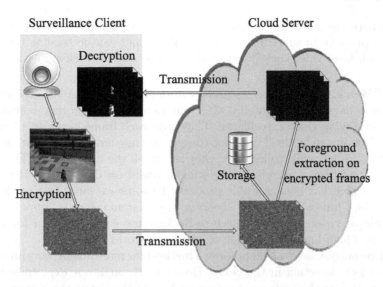

Fig. 1. The application scenario. The client captures surveillance videos, which are then encrypted by our proposed chaotic mapping based encryption method. The encrypted surveillance videos are transmitted to the cloud server, in which the foreground extraction algorithm is running on the encrypted videos. The results are transmitted back to the client, in which the extraction results are decrypted to get the extraction results in plain videos. The server learns nothing and the client knows where the foreground objects are.

analysed in the cloud server, which supports large scale video surveillance applications such as face tracking, suspect searching.

However, tremendous surveillance cameras have distributed everywhere. The privacy of the contents in the surveillance videos from the public places is being violated. One can suppose that a unauthorized person violates the cloud server and track your trajectory, your home address and the company you work for are learned by the enemy. Such a result can bring violence or crimes for you.

In this work, we focus on the fundamental problem of video surveillance application: the foreground extraction. As shown in Fig. 1, our scenario is set as that the surveillance client captures videos, which are then encrypted and transmitted to the cloud server. The cloud server run a foreground extraction algorithm in the encrypted videos and get the obfuscated results, which are sent back to the client. The client decrypts the obfuscated results and get the final foreground extraction results in plain videos. Using traditional foreground extraction methods, the contents of client videos are completely known to the server.

To protect the privacy of the public surveillance videos, we propose a method of privacy preserving foreground extraction of video surveillance through chaotic mapping based encryption in the cloud. The server learns nothing and the client knows where the foreground objects are. The foreground extraction is a fundamental problem in surveillance video analysis and can be used as the inputs several high level video surveillance tasks such as object tracking, action recognition.

Related Work. Recently, various privacy preserving computer vision algorithms in the cloud have been proposed [1–3]. Upmanyu et al. [1] propose a system for privacy preserving video surveillance. They split each frame into a set of random images. Each image by itself does not convey any meaningful information about the original frame, while collectively, they retain all the information. Their solution is derived from a secret sharing scheme based on the Chinese Remainder Theorem, which has homomorphic property to some extends. Due to the limitation of the Chinese Remainder Theorem, this system can only support several related simple vision algorithms such as background subtraction for foreground extraction, which limits their application in real world problems.

Most recently, Chu et al. [2] propose a method for real-time privacy preserving moving object detection in the cloud. However, through our experiments, it is found that this method has a main drawbacks: (1) the server can clearly see the contours of the foreground objects, which can release some privacy information of the original surveillance videos, (2) the security of their encryption method is not that well because of the less secure random scheme if the randomness functions used in their work are not based on chaotic mapping.

The particular properties of chaos [4,5], such as sensitivity to initial conditions and system parameters, pseudo-randomness, ergodicity and so on, have granted chaotic dynamics as a promising alternative for the conventional cryptographic algorithms. The inherent properties connect it directly with cryptographic characteristics of confusion and diffusion, which is presented in Shannon's works [6]. Chaotic system has been widely used in chaotic cryptography in recent years for its excellent chaotic dynamic properties, which could maintain longer periodicity in digitalization and gain good performance in cryptography. Moreover, it could be implemented in parallel by hardware, and has larger key space. Thus, it is also suitable for image encryption [7–10]. Modern image or video encryption needs to follow a rule that cipher image should be sensitive with the changes of the secret key and plain images or videos.

The Proposed Method. Thus, we employ chaotic mapping to increase the security of the video encryption scheme and we de-shuffle the extraction results in the client to avoid the leak of the contours of the foreground objects. First, the client captures surveillance videos, which are then encrypted by our proposed chaotic mapping based encryption method. The encrypted surveillance videos are transmitted to the cloud server, in which the foreground extraction algorithm is running on the encrypted videos. The results are transmitted back to the client, in which the extraction results are decrypted to get the extraction results in plain videos. The extraction correctness in the encryption videos is similar as that in the plain videos. The proposed method has several advantages: (1) The server only learns the obfuscated extraction results and can not recognize anything from the results. (2) Based on our encryption method, the original extraction method in the plain videos need not be changed. (3) The chaotic mapping ensure high level security and the ability to resistant several attacks.

2 Cryptography Primitive

In this section we briefly introduce the cryptography primitive we used in this paper. The simple but efficient logistic mapping is defined as follows:

$$x_{n+1} = \mu x_n (1 - x_n)$$
$$3.569945672... < \mu \leq 4, 0 \leq x_n \leq 1 \tag{1}$$
$$n = 0, 1, 2, ...$$

When the parameter μ and the initial value x_0 follow the Eq. 1, the outputs of this chaotic mapping $x_n \in (0, 1)$ become chaotic state and have good potential to form a random sequence.

3 Private Video Foreground Extraction

An overview of our method is shown in Fig. 2. In this section, we will describe the details of the random inverse, frame confusion, frame diffusion and the final foreground extraction.

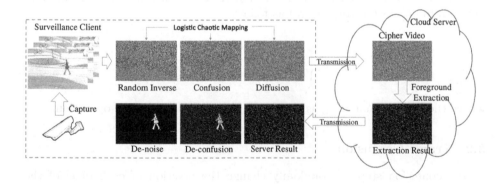

Fig. 2. The overview of our method. In the client, each frame of the captured videos is first been randomly inversed pixel by pixel using a random matrix generated by the logistic mapping. Then we use the logistic mapping to confuse the randomly inversed frame, followed by a diffuse operation to get the final cipher frame. Each are transmitted to the cloud server. The cloud server use the standard mixture Gaussian model [11] to extract foreground directly in each cipher frame. The server can only learn the cipher result and can not recognize anything in the server result. This result is transmitted back to the client. The client de-confuse the server result to get the intermediate extraction result, followed by the median filter to de-noise so as to obtain the final foreground extraction result.

3.1   Random Inverse

Firstly, we use the logistic chaotic mapping in Eq. 1 to generate a sequence of random number $x_n \in (0,1)$, whose length equals to the total number of the frame pixels $L = M \times N$, where M and N are the length and the height of the video frame. The randomness evaluation of the chaotic system shows that when selecting the eighth digit from 16 digit after the decimal point of the output of the logistic chaotic mapping, the randomness of the sequence is the best. Thus, we select the eighth digit x_n^8 of the output of the logistic chaotic mapping.

For example, when $\mu^I = 3.9, x_0^I = 0.62$, one of the subsequence of the output sequence $\{x_n^I\}$ of the logistic mapping is:

$$\{x_n^I\} = \{...,0.804374867512651,$$
$$0.613688166103959,$$
$$0.924592503462883,$$
$$0.271912703412174, \tag{2}$$
$$0.772107122027503,$$
$$0.686235085153447, ...\}_L$$

Then we select the $8th$ digit after the decimal point and get the sequence:

$$\{L_n\} = \{...,6,6,0,0,2,8,...\}_L \tag{3}$$

This sequence is converted to a $M \times N$ matrix Z. We inverse the original plain video frame F_p (8 bit gray image) as follow:

$$F_I^i = \begin{cases} F_p^i & \text{if } L_i \leq 5 \\ 255 - F_p^i & \text{if otherwise} \end{cases} \tag{4}$$

where the $M \times N$ matrix F_I is the result of the random inverse operation.

3.2   Frame Confusion

In the confusion step, we randomly change the position of each pixel of the matrix F_I. Once again, we use the logistic chaotic mapping in Eq. 1 and another pair of the parameter μ^C and the initial value x_0^C to generate a chaotic sequence $\{x_n^C\} = \{x_1^C, X_2^C, ..., X_L^C\}$ with the length of $L = M \times N$. Then we sort this sequence $\{x_n^C\}$ in ascending order:

$$\{x_n^{C'}\} = \{x_1^{C'}, X_2^{C'}, ..., X_L^{C'}\}$$
$$Ind = \{i_1, i_2, ..., i_L\}, x_{i_n}^C = x_n^{C'} \tag{5}$$

According to the index sequence Ind, the confused matrix F_C of the F_I is defined as:

$$F_C(i,j) = F_I(Ind(i \times N + j)) \tag{6}$$

where N is the height of the video frame.

3.3   Frame Diffusion

In the diffusion step, the value of each pixel in the confused matrix F_C will be changed randomly. By the logistic chaotic mapping in Eq. 1, we use the parameters $\{\mu^{D_1}, \mu^{D_2}, ..., \mu^{D_Q}\}$ and the initial value $\{x_0^{D_1}, x_0^{D_2}, ..., x_0^{D_Q}\}$ to generate Q confused matrix $\{F_{D_1}, F_{D_2}, ..., F_{D_Q}\}$ using the method in Sect. 3.2. Thus the diffused matrix F_D of the F_C is defined as:

$$F_D = \frac{1}{P+1} \cdot (F_C + \sum_{q=1}^{Q} F_{D_q}) \tag{7}$$

where the coefficient $\frac{1}{P+1}$ ensures that each pixel of the cipher frame is in the range of $[0, 255]$.

3.4   Foreground Extraction

The foreground extraction is done by the standard mixture Gaussian model [11], which can be directly conducted in the cipher frame F_D for three reasons: (1) the inverse operation, the confusion and the diffusion only shift the original mixture Gaussian distribution of each pixel to another mixture Gaussian distribution for 3 facts: (a) the inverse of a pixel only changes the mean value of the Gaussian distribution, (b) the confusion operation do not have any impact because every pixel has its own Gaussian distributions, (c) the sum of Gaussian variables is also a Gaussian) [2], (2) we use the same inverse, confusion and diffusion matrix for each video frame, thus each pixel in the cipher frame F_D still follows a Gaussian distribution, (3) the foreground extraction of the mixture Gaussian model is conducted in each pixel independently.

4   Experimental Results

We test the proposed method in various categories of surveillance videos from a large public dataset [12], which contains nearly 16000 manually annotated surveillance video frames and several subset from other public datasets. The tested surveillance videos contains categories of baseline, shadow, night videos, and intermittent object motion. We test the correctness rate of the foreground extraction, the visual results and the security analysis. In addition we compare our method with that of Chu et al. [2]. Notice that, all the foreground extraction experiments are run in cipher video frames. The experimental results reveal that:

- The correctness rate of our method are slightly higher than that of Chu et al. [2] in cipher frames.
- Our server can learn nothing about the client video while the server in Chu et al. [2] can clearly observe the contours of the foreground objects.
- Our method is secure enough in several attacks such as the brute-force attack and the statistical attack.

4.1 The Correctness Rate

The correctness rate of the foreground extraction is defined as the number of pixels correctly labelled as *foreground* or *background* against the total pixels in the video sequences. As shown in Fig. 3, we use the same standard mixture of Gaussian model as Chu et al. [2], thus the overall correctness rate in the cipher videos of our method is only slightly higher than that of [2] to some extend.

4.2 The Extraction Results

The foreground extraction algorithm of [11] is tested in 7 video sequences from the public change detection benchmark dataset [12]. In the work of Chu et al. [2], the server can completely observes the contours of the extracted foreground objects. While in our work, the server can only observes the shuffled extracted results as random white points, and can not recognize anything of the client videos. We show the foreground extraction results of our method and Chu et al. [2] in Fig. 4. **The video results of two of the test scenes are shown in the supplemental material.**

4.3 Security Analysis

A well designed image/video encryption scheme should be robust against different kinds of attacks, such as brute-force attack and statistical attack [7]. In this section, we analyse the security of the proposed encryption method in an example frame from the pedestrian video sequence of the [12].

videos/frames	Our Method	Method of [2]
backdoor/1816	0.795028	**0.807418**
busstation/1111	**0.94699**	0.939721
cubicle/2811	**0.974191**	0.973237
highway/1179	**0.936468**	0.929562
office/629	**0.901332**	0.897626
pedestrians/921	**0.988313**	0.984055
sofa/628	0.944334	**0.945863**

Fig. 3. The comparison of the correctness rate with [2] in seven video sequences. The overall correctness rate in the cipher videos of our method is only slightly higher than that of [2] to some extend.

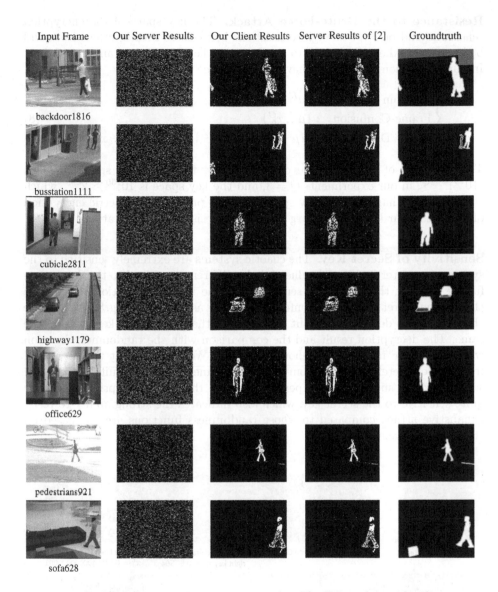

Fig. 4. Parts of the foreground extraction results. The input plain video frames are shown in the first column. The extraction results in the server using our method are shown in the second column. The extraction results after the encryption and median filter in the client are shown in the third column. The extraction results in the server using the method in [2] are shown in the fourth column. The serve can clearly observe the contours of the foreground objects and the privacy of the client video is leaked. The ground truth manually segmented and annotated in [12] are shown in the last column.

Resistance to the Brute-Force Attack. The key space of the encryption scheme should be large enough to resist the brute-force attack, otherwise it will be broken by exhaustive search to get the secret key in a limited amount of time. In our encryption method, we have the key space as follow:

$$\begin{cases} \text{Random Inverse} & : \{\mu^I, x_0^I\} \\ \text{Frame Confusion} & : \{\mu^C, x_0^C\} \\ \text{Frame Diffusion} & : \{\{\mu^{D_1}, \mu^{D_2}, ..., \mu^{D_Q}\}, \{x_0^{D_1}, x_0^{D_2}, ..., x_0^{D_Q}\}\} \end{cases} \quad (8)$$

The precision of 64-bit double data is 10^{-15}, thus the key space is about $(10^{15})^{4+2Q}$, in our experiments $Q = 3$, and the key space is $10^{150} \approx 2^{499}$, which is much lager than the max key space (2^{256}) of practical symmetric encryption of the AES. Our key space is large enough to resist brute-force attack.

Sensitivity of Secret Key. The chaotic systems are extremely sensitive to the system parameter and initial value. A light difference can lead to the decryption failure. To test the secret key sensitivity of the image encryption scheme, we change the secret key of the confusion step x_0^C as shown in Fig. 8. We use the changed key to decrypt the input frame, while the other secret keys remain the same. The decryption result and the comparison with the randomness function without chaotic mapping are shown in Fig. 5. We can see that the decrypted frame is completely different from the input frame. The test results of the other secret key are similar. The experiments show that the encryption scheme is quite sensitive to the secret key, which also indicates the strong ability to resist exhaustive attack compared to whose randomness functions are not based on chaotic mapping.

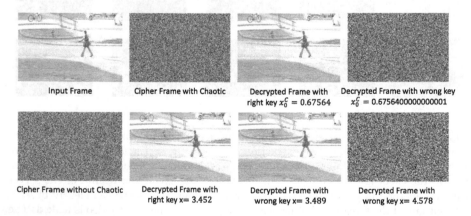

Fig. 5. Decrypted with wrong key. First line: using our method we slightly change the key and get the wrong decrypted result. Second line: using a standard random function, the sensitivity of the secret key x in the confusion step as a example is not well, and is much easier be attacked by the brute-force attack

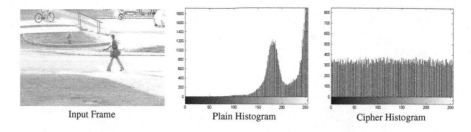

| Input Frame | Plain Histogram | Cipher Histogram |

Fig. 6. The histogram of the input frame before and after encryption.

The Histogram Analysis. The histogram is used to show the distribution of pixel values of a frame. The histogram of cipher frame should be flat enough, otherwise some information can be leaked to cause the statistical attack. This makes cipher-only attack possible through analysing the statistic property of the cipher image. Figure 6 shows the histograms of the input frame and its corresponding cipher image, respectively. Comparing the two histograms we can see that the pixel values of the original frame are concentrated on some values, but the histograms of its cipher image are very uniform, which makes statistical attacks impossible.

The Information Entropy. The information entropy [7] is used to express randomness and can measure the distribution of gray values in the image. The more uniform the distribution of pixel gray values, the greater the information entropy is. It is defined as follows:

$$\text{H}(m) = -\sum_{l=0}^{L} \text{P}(m_i) \log_2(m_i) \tag{9}$$

where m_i is the i-th gray value for an L level gray image, $L = 255$. $P(m_i)$ is the probability of m_i in the image and $\sum_{i=0}^{L} P(m_i) = 1$. The information entropy of an ideal random image is 8, which shows that the information is completely random. The information entropy of the cipher image should be close to 8 after encryption. The closer it is to 8, the smaller possibility for the scheme leaks information. The information entropy of the cipher frame using our method in Fig. 5 in our experiments is 7.9978, which is very close to 8.

The Correlation Analysis. Correlation indicates the linear relationship between two random variables. In image/video processing, it is usually employed to investigate the relationship between two adjacent pixels. Usually, the correlation of between adjacent pixels in the plain image is very high. A good encryption scheme should reduce the correlation between adjacent pixels, i.e., the less correlation of two adjacent pixels have, the safer the cipher image is. In order to test

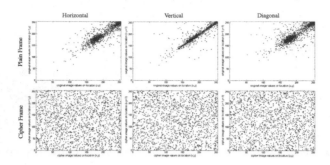

Fig. 7. The correlation of the input frame before and after encryption in 3 directions (horizontal, vertical and diagonal).

the correlation of two adjacent pixels, we test 3 directions (horizontal, vertical and diagonal) of adjacent pixels from the input frame using our method in Fig. 5 and its corresponding cipher frame (Fig. 7).

5   Conclusion and Discussion

In this paper we propose a method of privacy preserving foreground extraction of video surveillance through chaotic mapping based encryption in the cloud. We employ chaotic mapping to increase the security of the video encryption scheme and we de-shuffle the extraction results in the client to avoid the leak of the contours of the foreground objects. The experimental results show that The correctness rate of our method are slightly higher than that of Chu et al. [2] in cipher frames. Our server can learn nothing about the client video while the server in Chu et al. [2] can clearly observe the contours of the foreground objects. Our method is secure enough in several attacks such as the brute-force attack and the statistical attack.

In the future work, we will use high dimensional chaotic mapping to increase the secure level. We plan to embed most recent foreground extraction algorithms into the privacy preserving framework.

Acknowledgements. This work is partially supported by the National Natural Science Foundation of China (No. 61402021, No. 61402023, No. 61170037), the Fundamental Research Funds for the Central Universities (No. 2014XSYJ01, No. 2015XSYJ25), the Science and Technology Project of the State Archives Administrator. (No. 2015-B-10), and the Open Project Foundation of Information Technology Research Base of Civil Aviation Administration of China (No. CAAC-ITRB-201403).

References

1. Upmanyu, M., Namboodiri, A.M., Srinathan, K., Jawahar, C.V.: Efficient privacy preserving video surveillance. In: IEEE 12th International Conference on Computer Vision (ICCV), pp. 1639–1646 (2009)

2. Chu, K.Y., Kuo, Y.H., Hsu, W.H.: Real-time privacy-preserving moving object detection in the cloud. In: Proceedings of the 21st ACM International Conference on Multimedia (MM 2013), pp. 597–600. ACM, New York (2013). http://dx.doi.org/10.1145/2502081.2502157
3. Jin, X., Liu, Y., Li, X.D., Zhao, G., Chen, Y.Y., Guo, K.: Privacy preserving face identification through sparse representation. To Appear in the 10th Chinese Conference on Biometric Recognition (CCBR), Tianjin, China, 13–15 November 2015
4. Huang, C., Nien, H.: Multi chaotic systems based pixel shuffle for image encryption. Opt. Commun. **282**, 2123–2127 (2009)
5. Lian, S., Sun, J., Wang, Z.: A block cipher based on a suitable use of the chaotic standard map. Chaos Soliton Fract. **26**(1), 117–129 (2005)
6. Claude, S.: Communication theory of secrecy systems. Bell Syst. Tech. J. **28**(4), 656–715 (1949)
7. Zhen, P., Zhao, G., Min, L.Q., Jin, X.: Chaos-based image encryption scheme combining dna coding and entropy. Multimedia Tools and Applications (MTA), Published Online: 10 April 2015
8. Jin, X., Chen, Y., Ge, S., Zhang, K., Li, X., Li, Y., Liu, Y., Guo, K., Tian, Y., Zhao, G., Zhang, X., Wang, Z.: Color image encryption in CIE L*a*b* space. In: Niu, W., Li, G., Liu, J., Tan, J., Guo, L., Han, Z., Batten, L. (eds.) ATIS 2015. CCIS, vol. 557, pp. 74–85. Springer, Heidelberg (2015). doi:10.1007/978-3-662-48683-2_8
9. Jin, X., Tian, Y.L., Song C.G., Wei, G.Z., Li, X.D., et al.: An invertible and anti-chosen plaintext attack image encryption method based on DNA encoding and chaotic mapping. To Appear in the Chinese Automation Congress, Wuhan, 27–29 November 2015
10. Li, Y.Z., Li, X.D., Jin, X., Zhao, G., Ge, S.M., et al.: An image encryption algorithm based on zigzag transformation and 3-dimension chaotic logistic map. In: Niu, W., Li, G., Liu, J., Tan, J., Guo, L., Han, Z., Batten, L. (eds.) ATIS 2015. CCIS, vol. 557, pp. 3–13. Springer, Heidelberg (2015). doi:10.1007/978-3-662-48683-2_1
11. Stauffer, C., Grimson, W.E.L.: Adaptive background mixture models for real-time tracking. In: International Conference on Computer Vision and Pattern Recognition, vol. 2, p. 2246 (1999)
12. Wang, Y., Jodoin, P.M., Porikli, F., Konrad, J., Benezeth, Y., Ishwar, P.: CDnet 2014: an expanded change detection benchmark dataset. In: Proceedings IEEE Workshop on Change Detection (CDW-2014) at CVPR-2014, pp. 387–394 (2014)

Evaluating Access Mechanisms for Multimodal Representations of Lifelogs

Zhengwei Qiu$^{(\boxtimes)}$, Cathal Gurrin, and Alan F. Smeaton

Insight Centre for Data Analytics, Dublin City University, Dublin 9, Ireland
{zhengwei.qiu,cathal.gurrin,alan.smeaton}@dcu.ie

Abstract. Lifelogging, the automatic and ambient capture of daily life activities into a digital archive called a lifelog, is an increasingly popular activity with a wide range of applications areas including medical (memory support), behavioural science (analysis of quality of life), work-related (auto-recording of tasks) and more. In this paper we focus on lifelogging where there is sometimes a need to re-find something from one's past, recent or distant, from the lifelog. To be effective, a lifelog should be accessible across a variety of access devices. In the work reported here we create eight lifelogging interfaces and evaluate their effectiveness on three access devices; laptop, smartphone and e-book reader, for a searching task. Based on tests with 16 users, we identify which of the eight interfaces are most effective for each access device in a known-item search task through the lifelog, for both the lifelog owner, and for other searchers. Our results are important in suggesting ways in which personal lifelogs can be most effectively used and accessed.

Keywords: Lifelogging · HCI · Multimodal access

1 Introduction

Continuing advances in technology have led us to the point where lifelogging, the recording of the totality of life experience using digital sensors, is an activity that anyone can engage in [11]. In selecting an appropriate definition of lifelogging, we use the description by Dodge and Kitchin [5] and define it as "*a form of pervasive computing, consisting of a unified digital record of the totality of an individual's experiences, captured multi-modally using digital sensors, stored permanently as a personal multimedia archive and made accessible in a pervasive manner through appropriate interfaces*". Lifelogging can offer societal benefits in applications like memory support, diet monitoring, quality of life analysis, self-reflection, monitoring progress of degenerative conditions, and more, as well as auto-recording of work-related tasks. Yet in each application, the benefits can only be realised if the lifelog content can be easily and efficiently accessed.

Heretofore, lifelog access has typically based on a *browsing* metaphor whereby data/time, geo-location, or content filters support visual browsing of a lifelog, which has been typically structured into events. There has been little consideration of how people might *search* a lifelog across a variety of devices at any

© Springer International Publishing Switzerland 2016
Q. Tian et al. (Eds.): MMM 2016, Part I, LNCS 9516, pp. 574–585, 2016.
DOI: 10.1007/978-3-319-27671-7_48

time and in any place [11]. It is our conjecture that to be truly effective, a lifelog needs to support ubiquitous access from multiple devices, for both browsing and search tasks. Hence, when designing lifelogging tools, we believe that appropriate content modeling and representation techniques should be applied to lifelogs in order to support multi-modal access. In this work we develop a state-of-the-art holistic lifelogging solution covering capture, analysis, indexing, browsing and search. We present a first multi-modal access model for lifelog data and we consider its impact in terms of the time to find a known item across three popular access devices, smartphone, laptop and ebook reader, using eight different user interface types.

2 Background to Lifelogging and Pervasive Access

Lifelogging has been the subject of research since Mann's early work on wearable computing culminating in his CHI paper from a decade ago [17], which focused mainly on data gathering. The seminal MyLifeBits project [9] from Microsoft Research created the first database of life data from Gordon Bell's life and was one of the motivating factors in the development of the first accessible visual lifelogging device, the Microsoft SenseCam. This is a wearable camera that can automatically take up to 5,000 images from the wearer's viewpoint in a single day. It resulted in a host of lifelogging research that explored issues of data organisation for lifelogs such as how to segment a sequence of images from a day into a series of events and represent them in a storyboard manner [15] on a desktop system. Nowadays we find lifelogging applications and interfaces running on many mobile devices and even TVs [12]. At the same time as the work on general visual lifelogs, there have been domain-focussed applications of lifelogging, such as Aizawa's Foodlog [1], or any of the current generation of commercial lifelogging devices from OMG or Narrative. More recently, there have been efforts at using the smartphone as a lifelogging platform [4], which has shown to be as effective as a dedicated device such as a SenseCam [10]. In all of the above work, there is one lifelog interface running on one device. However, if lifelogging is to be a pervasive activity with anytime access as per the definition of lifelogging that we adopt, then there needs to be consideration of how to access a lifelog from across multiple devices.

Before we even begin to consider user interaction, lifelogging solutions require a certain level of basic data organisation or pre-processing, such as segmenting a day's activities into events or perhaps annotating data [11]. Studies on autobiographical memory have suggested that it is part of our nature to organise our daily experiences into events. Prior research has shown that appropriate annotations of lifelog events can significantly increase the success rate for locating known items from a lifelog [6]. Once an appropriate event segmentation model has been identified, then the challenge of how to efficiently access a lifelog needs to be considered [2,22]. Automatic or manual annotation of events in lifelogs makes them more accessible for either searching or browsing tasks [7] and this is something we need to consider when designing lifelog access tools.

With regard to multi-modal access, Hess *et al.* investigated user requirements and preferences for multiple-device media systems [13] arguing that lifelog content needs to be shared and accessed across personal and shared devices. Fleury *et al.* report an approach to transform video content from a mobile device to a television [8] and point out that more attention is needed to consider differences among devices such as size of screen and keyboard availability. Ubiquitous computing scenarios not only enable ubiquitous access, but also bring a challenge for system and user interface designers. As it becomes increasingly difficult to optimally represent content across the myriad of devices currently available due to different sizes, resolutions, interaction mechanisms and environments of use, Human-Computer Interaction (HCI) experts have realised the importance of characteristics of output devices on interface design [19]. We see that different products representing different paradigms require different solutions for accessing [21]. Since lifelog content is media-rich, it can naturally be presented in different ways and since there is no single best way to present lifelog data on every device, the best is to use different representations on different devices for different usage scenarios. That is the main point addressed in this paper.

3   An End-to-End Holistic Lifelogging Solution

In order to explore the effectiveness of different approaches to multimodal lifelog access covering both searching and browsing tasks, one needs an end-to-end lifelogging solution, from capture to access, and to evaluate this with users and real-world lifelog data. To this end, we applied state-of-the-art components to develop an holistic lifelogging process as illustrated in Fig. 1.

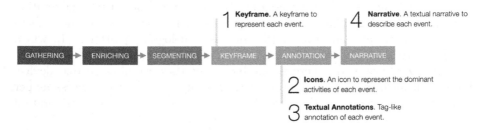

Fig. 1. Components of an holistic lifelogging process, from gathering data, through enriching, segmenting, keyframe extraction, semantic annotation and narrative generation.

We summarise the six core components of our holistic lifelogging process (as shown in Fig. 1):

- **Gathering.** The lifelog is gathered using a smartphone running our in-house developed lifelogging software [18]. Sensed data from the smartphone includes auto-captured images, location (using GPS, WIFI and base station), music listened to, photos viewed, environmental (sound levels, weather), communications (phone-call and SMS logs) and data from other sensors (accelerometer, in-range Bluetooth devices, screen, etc..).

- **Enriching.** Images gathered are annotated with metadata from the smartphone based on temporal alignment of captured data. This occurs in a semantic enrichment step in which the smartphone's sensor data is semantically enriched using empirically observed rules, machine learning techniques and an analysis of the statistical frequency and (re)occurrence of the sensed data. This produces a set of six 'contexts' forming the semantic enrichment. These are physical and social locations (e.g. addresses and meaningful descriptions such as home, work, etc.), social relationships & interactions (e.g. friends, family, stranger identification), actual and relative time (where relative time refers to concepts such as yesterday, last week, etc.), environmental factors at the time of capture such as weather, the physical and semantic activities of the wearer (e.g. sitting, running, using the phone, working, etc.), and a personal profile of the lifelogger that understands where home is and who people are.
- **Segmenting.** As with most work on visual lifelogging, an algorithm for event segmentation is used to segment each day's data into a list of discrete events based on analysing data for context changes, a typical approach in lifelogging [3]. We employ a Support Vector Machine (SVM) in the event segmentation process and consider a wider range of source features than has been done in prior work, including image, location, activity, social and environment attributes. We employed attributes extracted from raw sensor data instead of using raw sensor data itself, such as the similarity of adjacent images, semantic locations and activity changes. We evaluated the performance of this event segmentation algorithm and found that 77.4 % of 2,470 event boundaries (on the experiment dataset described later) were detected correctly vs. a manual ground truth, 22.6 % were false negatives and 0.7 % false positive instances which are not actually event boundaries but which were detected as such. This accuracy is similar though not significantly better than the state-of-the-art and gives a suitable platform to base our work on. The end result is that a lifelog becomes a series of events, each represented by a set of temporally organised images and metadata.
- **Keyframe.** To represent each event in a lifelog, a single keyframe is usually identified and used [11]. Our approach for identifying keyframes is built on two concepts, social and quality; if some photographs contain faces, choose the best quality ones as the keyframe. If no face is detected in the event, choose the best quality photograph as the keyframe.
- **Annotation.** Each event is annotated automatically to produce both an iconic representation of the dominant activity of the event as well as a textual annotation of the dominant activities, concepts and contexts of the event.
- **Narrative.** Finally, each event is represented by a narrative description of the event, similar to a diary entry. The process by which narrative is generated is a three-part process (namely fabula, sjuzet and discourse generation). Fabula is a series of sentences based on the detected contexts and segmented events, generated from the second annotation phase; sjuzet is a paragraph of

narratives generated from the fabula without the repeated sentences; and discourse is a paragraph of narrative with an illustrated picture/keyframe taken during the event.

Based on this lifelogging framework we are able to create four individual semantic elements of lifelog representation for multimodal access, as shown in Fig. 1. These represent the ways in which people communicate information in a static manner namely keyframes in storyboards, icons in storyboards, tag annotations and textual narratives, each of which we now describe.

1. **Keyframes.** The single keyframe used to represent each event is chosen as described. This allows a storyboard to be generated for each day in a lifelog and allows a user to scan a series of events based on the keyframe alone and make a decision as to whether that event may contain a known item/activity in which the user is interested.

2. **Icons.** Based on the set of identified activities within a given event, an icon is used to represent the dominant activity in that event. We focus on a single icon to represent each event on the assumption that multiple icons might be confusing. In total, we identified fifteen possible activities, based on the semantic lifestyle activities identified in [14]. The icons include activities such as working, eating, commuting, and relaxing, and could be considered to cover the majority of the range of daily activities. This representation allows a user to scan a series of events on a device that may not have the high resolution expected for keyframes or even screens which render colour.

3. **Textual Annotations.** Here we use a tag-like annotation to represent each event generated based on the context and activities of that event. This is in the form of a set of tags that describe the context of the event and on average, five such tags were generated for each event.

4. **Narrative.** A narrative description that represents the context and activities of an event is the final representation format. This can be seen as a more natural description of the event, as one might create for a diary entry. An example of a narrative description for an event would be "It was 10am and you were working at your computer with Frank for one hour". The narrative text is automatically generated based on the annotations.

4 Modeling for Multimodal Access

Based on the four semantic elements described earlier, we explored eight different interfaces to accessing a lifelog. These were interfaces that had been previously built and deployed, for example storyboards of event keyframes (with and without annotations) and animated sequences. Where the access device allowed, all keyframe-based interfaces supported drill-down analysis to show all images that are associated with the event. In addition, we designed new interfaces inspired by diaries and information visualisation, such as storyboards of icons and textual narratives (with and without keyframes). This resulted in the creation of the eight different lifelog interface types outlined below:

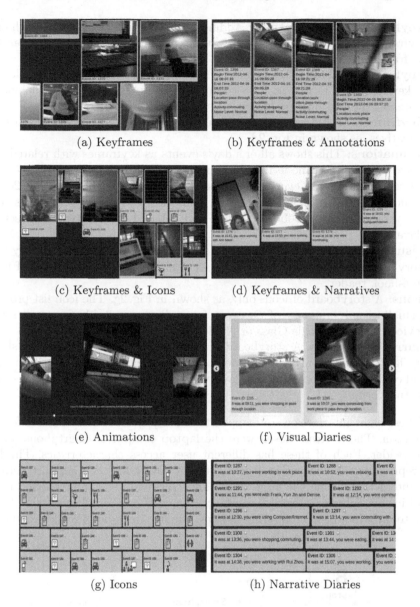

(a) Keyframes

(b) Keyframes & Annotations

(c) Keyframes & Icons

(d) Keyframes & Narratives

(e) Animations

(f) Visual Diaries

(g) Icons

(h) Narrative Diaries

Fig. 2. User interfaces employed in this researchUser interfaces employed in this research

- **Keyframes:** A traditional lifelogging interface in which all events in a day are shown on a single webpage, as in Fig. 2a with events presented chronologically. This is our baseline interface and is based on the storyboard interface used in [15].

- **Keyframes & Annotations:** Adding textual annotations underneath each keyframe, as in Fig. 2b; this represents a minimal addition to the baseline keyframe-only interface.
- **Keyframes & Icons:** Users' activities are represented by icons in addition to keyframes, as in Fig. 2c. The icon provides a quick reference visual cue to represent the 15 semantic life activities previously mentioned.
- **Keyframes & Narratives:** Containing keyframes and associated textual narratives, as shown in Fig. 2d.
- **Animations:** This shows all of a day's events as keyframes with related narrative in animation mode from beginning to end of the day. The time span between showing two events is 500 ms. The interface is shown in Fig. 2e and would be useful in a less-interactive, lean-back environment, such as on a TV or other relaxation-focused device and is based on the original SenseCam player [11].
- **Visual Diaries:** Keyframes and related narratives are shown in a diary-style storybook, as shown in Fig. 2f. The user can turn pages of events as if using an e-book reader.
- **Icons:** A storyboard of icons only, as shown in Fig. 2g. The icon list provides a small visual summary of an event for a display suitable for small screen devices, such as Google Glass or Apple watch.
- **Narrative Diaries:** A storyboard of event narrative descriptions, as shown in Fig. 2h. This narrative description replicates the concept of a written diary and could be used to summarise days, weeks or longer time periods in a simple textual form.

These eight interfaces were mapped to three physical access devices for our evaluation. The devices chosen were the laptop computer, smartphone and e-book reader. Each of these has different user access characteristics. The laptop represents the conventional lifelog access device and can support detailed user interactions using the large screen and trackpad. The smartphone (4-inch Android smartphone) integrates touch-screen access on a small screen and represents today's handheld computing devices. Finally the e-book reader represents

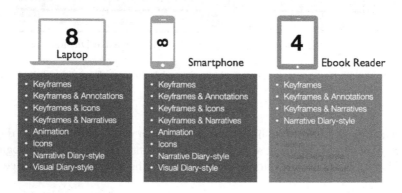

Fig. 3. Mapping Interfaces to Devices

a diary for replicating how people have traditionally logged and recalled their memories and while it can display images, it is monochrome and there is limited potential for user interaction, except for turning pages (Fig. 3).

For both the laptop and smartphone, we evaluated all eight interfaces. However, for the e-book reader, we evaluated only four of the interfaces due to limitations of the device and the lack of clarity on-screen when we have a complex interface display. Hence we focused on "Keyframes", "Keyframes & Annotations", "Keyframes & Narratives" and "Narrative Diary" on the e-book reader and we generated different PDF files to represent these interfaces.

5   User Evaluation

In order to evaluate which types of lifelog visualisations and representations are most suitable for display on different devices, we recruited one lifelogger (data gatherer and query generator) and 16 participants to evaluate the interfaces, 11 male and 5 female. Each works in, or have studied in a university, are from a broad range of disciplines, all are computer literate and would have computers/laptops and most have used e-book readers previously. All evaluators were given basic background knowledge about the lifelogger, but they were not told who the lifelogger was. In this way, any effects of some users knowing the lifelogger's lifestyle better than others were thus avoided.

5.1   Experimental Dataset

We generated a ten-day lifelog dataset of all-day lifelog use from the lifelogger (referred to as LL). This lifelog data was imported and enriched by the lifelogging process described earlier and the access interfaces were generated. From the ten days' data, the event segmentation technique identified 253 events and automatically selected appropriate keyframes. Accompanying the data, LL generated 16 information needs from the ten day lifelog, including queries, descriptions and correct answers, that could be identified by an independent third-party user. Information needs such as *"How many times was LL using the computer on the 20th ?"*, *"Find all the events involving working with AM on the 9th April"* and *"At what time was LL driving to work on the 18th ?"*.

5.2   The Interface Evaluation Process

For the evaluation, eight subjects evaluated the laptop-based interface, four evaluated the mobile interface and four evaluated the e-book interface. Interfaces were presented to subjects in a controlled order to avoid bias. For the laptop-based interface, each user attempted eight topics (one for each interface), meaning that every topic was evaluated 4 times. For the mobile and e-book interfaces, the subjects evaluated all sixteen topics across the 8 and 4 interfaces respectively.

Immediately following the evaluation, subjects completed a survey aimed at identifying their qualitative opinions about each interface. As a reminder, for

Table 1. Average time to completion (in seconds) for each interface.

Device	Keyframes	Keyframes & Annotations	Keyframes & Icons	Keyframes & Narratives	Animations	Visual Diaries	Icons	Narrative Diaries
Laptop	7.4	11.5	12.3	9.8	11.6	14.4	13.0	16.0
Phone	23.4	29.2	21.5	23.4	32.3	46.6	28.7	23.4
E-book	94.6	73	-	52.6	-	-	-	47.5

each user interface, the participant was presented with a screenshot of the user interface. A seven-point Likert scale was employed and the questionnaire took approximately 5–10 min to complete and all users completed the questionnaire without errors. This survey included four aspects: visual appeal, subjective satisfaction, potential for errors and speed of use. Visual appeal is believed to dominate impression judgments [16]. The other three aspects (effectiveness, efficiency and satisfaction) are defined by the international standard ISO/IEC 9241-11 and have been extensively employed in interface evaluation [20].

5.3   Results of the Interface Evaluation

Beginning with a quantitative analysis and the seek time to find the event of interest, the laptop computer has been shown to be the fastest approach to access information in personal lifelogs (as shown in Table 1). It took participants only 12 s on average to find the answer on the computer. On the smartphone and E-book reader it took 29 and 73 s respectively. This is not unexpected since both the smartphone and the e-book have interface limitations. Interestingly, the slowest laptop interface was faster than the best smartphone interface, and similar when comparing the smartphone and e-book reader. As can be seen from Table 1, keyframe-based interfaces were the best-performing on the laptop, with the narrative diaries performing slowest. On the smartphone, the best-performing interfaces were also keyframe based. For the e-book reader, the best-performing interfaces by a significant amount of time were narrative based, with narrative alone being the fastest. We understand that this is due to the limitations of the display technology of a monochrome e-book reader and we noted that in the PDF conversion, the images were compressed heavily, further hindering any image-based interface.

With regard to the qualitative user survey, on the laptop, the subjects preferred the Visual-Diary Style interface in terms of subjective satisfaction and

Table 2. The two suggested user interfaces for different devices.

	Optional interface	Reasons to use
Top 2 interfaces 2* for Computer	Visual-Diary Style	Highest visual appeal and subjective satisfaction, lowest error potential
	Keyframes	Fastest for laptop users
Top 2 interfaces 2* for Smartphone	Keyframes & icons	Highest accuracy, fastest UI
	Animations	Highest visual appeal and subjective satisfaction
Top 2 interfaces 2* for E-book reader	Narrative Diary Style	The fastest, most accurate, highest visual appeal and subjective satisfaction, lowest potential for errors
	Keyframes and Narratives	The second fastest, second most accurate, second highest visual appeal and subjective satisfaction, second lowest potential for errors

visual appeal. For the smartphone, the preferred interface was keyframes and icons because of its high accuracy and speed of use. Finally, due to the characteristics of the e-book reader, Narratives was selected as the preferred interface for E-book reader, because it was fastest, most accurate, highest visual appeal and subjective satisfaction, lowest potential for errors.

Based on these results, we propose the top two interfaces for each access device in Table 2. For the laptop, we suggest that a Keyframes storyboard helps locate content quickly, but that users enjoy the slower Visual-Diary style most. However, Keyframes & Icons provided a visually appealing and reasonably fast interface that provides a trade-off. For the smartphone, Keyframes & Icons would be the fastest interface, though in terms of user-appeal, the significantly slower animation interface would be best. The trade-off in this case is Keyframes & Annotations, which provided reasonable speed and user satisfaction. Finally, for the e-book reader, Narratives has shown great potential in helping users to access their lifelog data using an E-book reader, hence this is our recommendation for E-book reader users. Given the visual limitations of the e-book reader, a second option would be merging narratives with keyframes, which was reasonably fast, but also had high visual appeal and subjective satisfaction.

6 Discussion

To be effectively exploited and used across a range of usage scenarios, a personal lifelog needs to be accessible across a variety of access devices and not just from a laptop or from a mobile device. In this work we created eight lifelogging interfaces using state-of-the-art components and evaluated their effectiveness using three access devices on a holistic lifelogging framework. We identified (unsurprisingly) that the laptop is significantly faster than the other devices, with the e-book reader the slowest by far. For each device, we identified the fastest interface for users to browse for a 'known item' and identified the interfaces that the experimental subjects found most appealing.

Naturally there are some limitations of this study. Our users set were small and all were university-based and very computer-literate. We employed a small test collection and any browsing mechanism is likely to fail when a user is faced with months or years of lifelog data. In such cases, a search mechanism is required. It is our belief that many of the preliminary findings in this paper will translate into a search mechanism also. In addition, like all other work in personal lifelogging it is difficult to make our data collection openly available for use by others because of the very personal nature of the data itself, and this prevents the kind of collaborative benchmarking exercises seen in TREC. However, this is the first such study which evaluates lifelog access and will provide valuable clues for lifelog designers in future work. Finally, there are new ubiquitous-access devices, such as Google Glass and Apple watch which represent other potential interfaces to evaluate and these will be the subject of future experimentation.

Acknowledgements. This paper is based on research conducted with financial support of Science Foundation Ireland under grant number SFI/12/RC/2289.

References

1. Aizawa, K., Maruyama, Y., Li, H., Morikawa, C.: Food balance estimation by using personal dietary tendencies in a multimedia food log. IEEE Trans. Multimedia **15**(8), 2176–2185 (2013)
2. Aizawa, K., Tancharoen, D., Kawasaki, S., Yamasaki, T.: Efficient retrieval of life log based on context and content. In: Proceedings of the 1st ACM workshop on Continuous Archival and Retrieval of Personal Experiences, pp. 22–31 (2004)
3. Chen, G., Kotz, D.: A survey of context-aware mobile computing research. Technical report, TR2000-381, Dept. of Computer Science, Dartmouth College (2000)
4. De Jager, D., Wood, A.L., Merrett, G.V., Al-Hashimi, B.M., O'Hara, K., Shadbolt, N.R., Hall, W.: A low-power, distributed, pervasive healthcare system for supporting memory. In: Proceedings of the First ACM MobiHoc Workshop on Pervasive Wireless Healthcare, p. 5. ACM, May 2011
5. Dodge, M., Kitchin, R.: Outlines of a world coming into existence: pervasive computing and the ethics of forgetting. Env. Plan. B **34**(3), 431–445 (2007)
6. Doherty, A.R., Gurrin, C., Smeaton, A.F.: An investigation into event decay from large personal media archives. In: 1st ACM International Workshop on Events in Multimedia, EIMM 2009, pp. 49–56, October 2009
7. Doherty, A.R., Pauly-Takacs, K., Caprani, N., Gurrin, C., Moulin, C.J.A., O'Connor, N.E., Smeaton, A.F.: Experiences of aiding autobiographical memory using the sensecam. Hum.-Comput. Interact. **27**(1–2), 151–174 (2012)

8. Fleury, A., Pedersen, J.S., Bo Larsen, L.: Evaluating user preferences for video transfer methods from a mobile device to a TV screen. Pervasive Mob. Comput. **9**(2), 228–241 (2013)
9. Gemmell, J., Bell, G., Lueder, R.: MyLifeBits: a personal database for everything. Commun. ACM **49**(1), 88–95 (2006)
10. Gurrin, C., Qiu, Z., Hughes, M., Caprani, N., Doherty, A.R., Hodges, S.E., Smeaton, A.F.: The smartphone as a platform for wearable cameras in preventative medicine. Am. J. Prev. Med. **44**(3), 308–313 (2013)
11. Gurrin, C., Smeaton, A.F., Doherty, A.R.: LifeLogging: personal big data. Found. Trends Inf. Retrieval **8**(1), 1–125 (2014)
12. Gurrin, C., Zhang, Z., Lee, H., Caprani, N., Carthy, D., O'Connor, N.E.: Gesture-based personal archive browsing in a lean-back environment. In: The 16th International Conference on Multimedia Modelling, MMM 2010, January 2010
13. Hess, J., Ley, B., Ogonowski, C., Wan, L., Wulf, V.: Jumping between devices and services: towards an integrated concept for social TV. In: Proceedings of the 9th International Interactive Conference on Interactive Television, pp. 11–20. ACM (2011)
14. Kahneman, D., Krueger, A.B., Schkade, D.A., Schwarz, N., Stone, A.A.: A survey method for characterizing daily life experience: the day reconstruction method. Science **306**, 1776–1780 (2004)
15. Lee, H., Smeaton, A.F., O'Connor, N.E., Jones, G.J., Blighe, M., Byrne, D., Doherty, A., Gurrin, C.: Constructing a sensecam visual diary as a media process multimedia systems. Multimedia Syst. J. **14**(6), 341–349 (2008)
16. Lindgaard, G., Dudek, C., Sen, D., Sumegi, L., Noonan, P.: An exploration of relations between visual appeal, trustworthiness and perceived usability of homepages. ACM Trans. Comput.-Hum. Interact. (TOCHI) **18**(1), 1 (2011)
17. Mann, S., Fung, J., Aimone, C., Sehgal, A., Chen, D.: Designing eyetap digital eyeglasses for continuous lifelong capture and sharing of personal experiences. In:Proceedings of CHI, ALT.CHI (2005)
18. Qiu, Z., Gurrin, C., Doherty, A.R., Smeaton, A.F.: A real-time life experience logging tool. In: Schoeffmann, K., Merialdo, B., Hauptmann, A.G., Ngo, C.-W., Andreopoulos, Y., Breiteneder, C. (eds.) MMM 2012. LNCS, vol. 7131, pp. 636–638. Springer, Heidelberg (2012)
19. Robertson, S., Wharton, C., Ashworth, C., Franzke, M.: Dual device user interface design: PDAs and interactive television. In: Proceedings of the SIGCHI Conference on Human Factors in Computing Systems, pp. 79–86. ACM (1996)
20. Shneiderman, B.: Designing the User Interface: Strategies for Effective Human-Computer Interaction, vol. 2. Addison-Wesley, Reading (1992)
21. Tambe, M.: TV human interface: different paradigm from that of PC and Mobile. IEEE Code of Ethics (2012)
22. Tancharoen, D., Yamasaki, T., Aizawa, K.: Practical life log video indexing based on content and context. In: Electronic Imaging 2006, pp. 60730E–60730E. International Society for Optics and Photonics (2006)

Analysis and Comparison of Inter-Channel Level Difference and Interaural Level Difference

Tingzhao Wu[1,2,3], Ruimin Hu[1,2]($\boxtimes$), Li Gao[1,2], Xiaochen Wang[1,3], and Shanfa Ke[1,3]

[1] National Engineering Research Center for Multimedia Software, School of Computer, Wuhan University, Wuhan, China
785860285@qq.com
[2] Collaborative Innovation Center of Geospatial Technology, Wuhan, China
hurm1964@gmail.com, gllynnie@126.com
[3] Research Institute of Wuhan University in Shenzhen, Shenzhen, China
{clowang,kimmyfa}@163.com

Abstract. The directional perception of human ear for the sound at horizontal plane mainly depends on binaural cues, Interaural Level Difference (ILD), Interaural Time Difference (ITD) and Interaural Correlation (IC). And ILD plays a leading role for human to locate the position of sound with frequency above 1.5 KHz. In spatial audio applications, Inter-Channel Level Difference (ICLD) between loudspeaker signals are used to represent the location information of phantom sources generated by two loudspeakers. For headphone application, ILD and ICLD are approximate, so the perceptual characteristics of ILD can be used as a replacement for that of ICLD. But due to the attenuation influence of the transfer procedure from loudspeakers to humans ears, ICLD between loudspeakers signals are no longer the same with ILD between signals arrive at two ears. And these differences are always ignored in current spatial audio applications such as the perceptual coding of spatial parameters. So in this paper we focus on the analysis and comparison of ICLD and ILD from their formation and their values with different loudspeaker configurations. Experimental results showed that the difference of ILD and ICLD could be up to 55 dB, and the research of this paper may be an important part or reference for further research about spatial audio applications such as coding, reconstruction, etc.

Keywords: Binaural cues · ILD · Spatial parameters · ICLD · Phantom source

1 Introduction

As known to all, when we hear a sound, we can easily distinguish the direction of the sound source. In 1838, Johannes Muller has put forward that the directional perception on sound of human ears mainly depends on the correlation and differences of binaural signals [1]. J. Blauert further proposed the signal relations

© Springer International Publishing Switzerland 2016
Q. Tian et al. (Eds.): MMM 2016, Part I, LNCS 9516, pp. 586–595, 2016.
DOI: 10.1007/978-3-319-27671-7_49

between two ears represented as Interaural Level Difference (ILD), Interaural Time Difference (ITD), and Interaural Correlation (IC), called binaural cues [2]. Because of the differences and correlation between binaural signals, people can determine the location of the sound source. And when the sound frequency is higher than 1.5 KHz, ILD plays a main role in people's perception of the sound source [3].

Corresponding to binaural cues between two ears, spatial cues between two speaker channels includes Inter-channel Level Difference (ICLD), Inter-channel Time Difference (ICTD) and Inter-channel Correlation (ICC). The signals arrive our ears after influenced by the acoustical transfer function (ATF) during the sound transmission from sound source (such as speaker) to ears. So Binaural cues and spatial cues are different. When binaural cues change (such as ILD), we can perceive sound come from different positions, but it is hard to precisely change binaural cues directly by speakers. However we can change spatial cues (such as ICLD) to indirectly change binaural cues (such as ILD). Owing to different ATF under different playback environment, the relation between ICLD and ILD would be different when the speaker configuration changes. For example, the relation between ICLD and ILD of headphone playback environment is obviously different from that of loudspeaker playback environment.

The sound played by headphone go directly into our ears when we use headphone to listen, so the sound signal we heard and the sound signal played by headphones are almost the same. Thus, we may believe that the ILD of the signals received by our ears is the same as ICLD of the signals played by headphone.

When we use loudspeakers for listening, the signals of loudspeakers are transferred to ears during the ATF from the loudspeaker to the eardrums. To simulate the acoustical transfer pathway, the Head Related Transfer Function (HRTF) is introduced [4]. That is to say, the loudspeaker signals reach our ears after the function of HRTF, so it is to see that the ILD of the signals our ears received are different from the ICLD of the loudspeaker signals. Especially in surround sound, and 3D audio loudspeaker system, the difference is more significant.

However, the difference of binaural cues and spatial cues (such as ICLD and ILD, which are the most important cues for directional perception) did not get enough attention, many existing encoding methods ignored the difference between the ILD and ICLD, and use the perception characteristics of ILD to formulate ICLD quantitative codebook. For example, the MPEG-Surround [5,6] still used the same ICLD quantitative method to extract parameters and code in the multichannel signal space. It is obviously not accurate for the ICLD quantization.

ICLD is the level difference of loudspeaker signals, while ILD is the level difference of signals received by ears, so the two parameters are clearly not the same. There are few literatures make a detailed comparison and analysis between the loudspeaker signals and the corresponding binaural signals. Breebaart [7] compared the binaural signals respectively generated by phantom source and real source, with a part of the experiment results to indicate the difference between ILD and ICLD, but the he did not specifically discuss the difference and the effect factors. In addition, Gao [8] pointed out the difference between ICLD and ILD in the quantification of ICLD, but she did not make a qualitative and quantitative analysis of the difference between the two parameters.

Therefore, in order to meet the lack of detailed difference between ILD and ICLD, this paper made the more comprehensive and more specific analysis and comparison among the two parameters, as a result, it can modify the quantization of ICLD and act as a reference of the new precise spatial audio coding method.

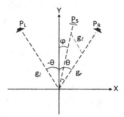

Fig. 1. Tangent amplitude panning law. Two loudspeaker signals $P_L(t)$ and $P_R(t)$ are derived from the phantom source signal $P_S(t)$. Two azimuth angles of loudspeakers are respectively given by $\theta_1 = -\theta$, $\theta_2 = \theta$. Phantom source azimuth is marked for $\theta_S = \varphi$.

2   Compare ICLD with ILD Theoretically

2.1   Calculation of ICLD

When we use two loudspeakers to rebuild a phantom source or decompose a sound source into two loudspeakers, we can use tangent amplitude panning law [7] to calculate the gains of loudspeakers with phantom source azimuth and loudspeaker deviation angle, as shown in Fig. 1. g_l and g_r are respectively the gain of P_L and P_R, they are calculated by [9]:

$$g_l = \frac{\tan\theta - \tan\varphi}{\sqrt{2\tan^2\theta + 2\tan^2\varphi}}, \quad g_r = \frac{\tan\theta + \tan\varphi}{\sqrt{2\tan^2\theta + 2\tan^2\varphi}} \tag{1}$$

Then the ICLD of loudspeaker signals is given by [9]:

$$ICLD = 20\log_{10}\left(\frac{g_l}{g_r}\right) = 20\log_{10}\left(\frac{\tan\theta - \tan\phi}{\tan\theta + \tan\phi}\right) \tag{2}$$

With regard to different loudspeaker configurations, calculation method of ICLD is the same. So we can learn from the Eq. 2, ICLD only related to two factors: the loudspeaker deviation angle and phantom source azimuth (relative deviation angle between loudspeakers).

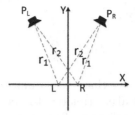

Fig. 2. Sound pressure and particle velocity schematic diagram. Point L and R respectively represent left ear and right ear.

2.2   Generation and Estimation of ILD

It is can be considered that voice is transmitted from sound source to human ears after two filtering processes, the first is air attenuation from source to the head position, the second is the diffraction of sound by head shape and auricle. Therefore, for ILD calculation process, the influence of these two parts need to be considered at the same time. Here we use the sound wave equation to simulate the first filtering process in order to calculate the sound energy at the head position, then we use a transfer function to simulate the second filtering process.

Assuming that the human's head is a rigid sphere, then the sound pressure at point L and point R produced by loudspeaker P_L and P_R are marked for $pressure_L$ and $pressure_R$, they can be expressed as [10]:

$$pressure_L = \frac{g_l}{r_1}e^{-iKr_1} + \frac{g_r}{r_2}e^{-iKr_2}, \quad pressure_R = \frac{g_l}{r_2}e^{-iKr_2} + \frac{g_r}{r_1}e^{-iKr_1} \quad (3)$$

Where g_l and g_r are the gains of P_L and P_R, r_1 and r_2 are the distance from loudspeakers to ears, and K is the number of wave forms.

The particle velocity at point L and point R can be shown as [10]:

$$v_L = \frac{g_l}{\rho_0 c_0}\frac{e^{-iKr_1}}{r_1} + \frac{g_r}{\rho_0 c_0}\frac{e^{-iKr_2}}{r_2}, \quad v_R = \frac{g_l}{\rho_0 c_0}\frac{e^{-iKr_2}}{r_2} + \frac{g_r}{\rho_0 c_0}\frac{e^{-iKr_1}}{r_1} \quad (4)$$

Where ρ_0 is air density, and c_0 is voice speed in the air. Then we can calculate the sound energy density by sound pressure and particle velocity, given as [11,12]:

$$w_L = \frac{pressure_L^2}{4\rho_0 c_0^2} + \frac{\rho_0 v_L^2}{4}, \quad w_R = \frac{pressure_R^2}{4\rho_0 c_0^2} + \frac{\rho_0 v_R^2}{4} \quad (5)$$

So the sound energy are [12]:

$$I_L = c_0 w_L, \quad I_R = c_0 w_R \quad (6)$$

We mark the filter function of human head and auricle as *head_function* $\{I, A\}$, where A is a parameter related human head. So the signals received by ears can be written as:

$$E_L = head\_function\{I_L, A\}, \quad E_R = head\_function\{I_R, A\} \quad (7)$$

At last, the ILD is given by:

$$ILD = \frac{E_L}{E_R} \tag{8}$$

The derivation process shows that the parameters which influence the ILD include: two gains of loudspeakers (related to the loudspeaker deviation and phantom source azimuth), the distance from loudspeakers to ears, air density, sound velocity in the air, the frequency of the sound signal and etc.

Therefore, ILD and ICLD have essential difference, and they cannot be treated as the same. In other words, designing ICLD quantitative codebook according to ILD directly is not accurate.

3 Experiments

We do simulation experiments based on existing HRTF database in order to further analyze the difference between the ICLD and ILD qualitatively and quantitatively.

With regard to a given loudspeaker configuration, we produce loudspeaker signals by control the ICLD of loudspeaker signals. Then we take the loudspeaker as source and simulate binaural signals by utilizing HRTF database, ILD can be derived through the analog signals. In the experiments, we compare the influence introduced by different loudspeaker configurations, different frequencies, different phantom source azimuths and etc. More details are described next.

3.1 Experimental Data Generation

We use three single frequency audio sequences as test sequences, the frequencies are 7 K, 13 K and 20 K. The HRTF database used in this paper is CIPIC database, we can elect the HRTF data we need according to subject, elevation and azimuth [13]. And we define three configurations, the first configuration is refer to Standard 5.1 Channels, in the second and the third configuration, the deviation angles are adjusted, as shown in Fig. 3.

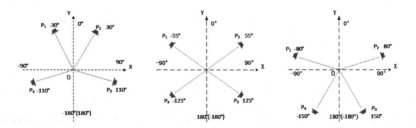

Fig. 3. Three configurations with different deviation angle. The left one is config.1, the middle one is config.2, and the right one is config.3.

We conduct experiments on four intervals of each configuration respectively, and in total 110 phantom source azimuth angles are selected (on the basis of the first configuration), as shown in Table 1.

Table 1. Phantom source azimuths.

Index	Interval	Azimuth:°
1	$-30° - 30°$	$-28 - 26 - 24 \ldots 24\ 26\ 28$
2	$30° - 110°$	$31\ 34\ 37 \ldots 103\ 106\ 109$
3	$-110° - 110°$	$-115 - 120 - 125 \ldots 125\ 120\ 115$
4	$-110° - -30°$	$-109 - 106 - 103 \ldots -37 - 34 - 31$

We write deviation angle of each interval as $\theta\,(C_I, I_I)$, where C_I is the index of configurations ($C_I = 1, 2, 3$), I_I is the index of intervals ($I_I = 1, 2, 3, 4$). For each phantom source azimuth angle φ_i ($i = 1, 2\ldots110$), then we can get the corresponding ICLD by Eqs. 1 and 2:

$$g_l(C_I, I_I, \varphi_i) = \frac{\tan\theta\,(C_I, I_I) - \tan\varphi_i}{\sqrt{2\tan^2\theta\,(C_I, I_I) + 2\tan^2\varphi_i}} \tag{9}$$

$$g_r(C_I, I_I, \varphi_i) = \frac{\tan\theta\,(C_I, I_I) + \tan\varphi_i}{\sqrt{2\tan^2\theta\,(C_I, I_I) + 2\tan^2\varphi_i}} \tag{10}$$

$$ICLD(C_I, I_I, \varphi_i) = 20\log_{10}\left(\frac{\tan\theta\,(C_I, I_I) - \tan\varphi_i}{\tan\theta\,(C_I, I_I) + \tan\varphi_i}\right) \tag{11}$$

Suppose the phantom source signal which we want to reconstruct is $P_S\,(n, k)$, then the two loudspeaker signals which form the phantom source are as follows:

$$P_L\,(n, k) = g_l P_S\,(n, k), \quad P_R\,(n, k) = g_r P_S\,(n, k) \tag{12}$$

Where n is the index of frame and k is index of sampling point.

Then we need to use the loudspeaker signals and the HRTF of corresponding position to make a convolution in order to simulate signals that reach to the human ears. There are four channels signals reach to the human ears, respectively are the left ear signal $P_{LL}\,(n, k)$ and the right ear signal $P_{LR}\,(n, k)$ generated by left loudspeaker signal $P_L\,(n, k)$, and the left ear signal $P_{RL}\,(n, k)$ and the right ear signal $P_{RR}\,(n, k)$ generated by right loudspeaker signal $P_R\,(n, k)$, they can be calculated by:

$$P_{Ll}\,(n, k) = left\_channel\,\{P_L\,(n, k) * HRTF\,(sub, \theta_i, e)\} \tag{13}$$

$$P_{Lr}\,(n, k) = right\_channel\,\{P_L\,(n, k) * HRTF\,(sub, \theta_i, e)\} \tag{14}$$

$$P_{Rl}\,(n, k) = left\_channel\,\{P_R\,(n, k) * HRTF\,(sub, \theta_j, e)\} \tag{15}$$

$$P_{Rr}\,(n, k) = right\_channel\,\{P_R\,(n, k) * HRTF\,(sub, \theta_j, e)\} \tag{16}$$

Where $left\_channel\{*\}$ and $right\_channel\{*\}$ respectively are methods to obtain the left channel data and right channel data, θ_i and θ_j are the loudspeaker azimuth angles in the original coordinate system, $HRTF(*)$ represent for the selected HRTF data.

Then calculate the energy sum of left channels and right channels, the energy E_L and E_R are derived by:

$$E_L = \sum_{n=1}^{N} \sum_{k=1}^{K} \{fft\,[P_{Ll}\,(n,k)] + fft\,[P_{Rl}\,(n,k)]\}^2 \tag{17}$$

Where $fft[*]$ is the function of Fast Fourier Transform.

Finally, the parameter ILD is given by:

$$ILD = 10\log_{10}\left(\frac{E_L}{E_R}\right) \tag{18}$$

3.2   Experimental Comparison

With different azimuth angles φ_i and different deviation angles $\theta\,(C_I, I_I)$ of each configuration, we can calculate the ICLD and ILD of the configuration. And then applying the cubic spline interpolation to these data, so we can get ICLD and ILD of 300° range. The result is shown in follow figures.

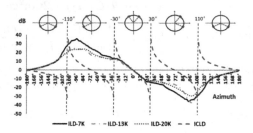

Fig. 4. Comparison of ICLD and ILD based on different frequencies sequences (config.1).

Corresponding to different phantom source position, ICLD and ILD features are different. When phantom source is located in the front interval (such as the −30° to 30° interval of config.1), ILD and ICLD are very approximate, the maximum difference in config.1 is 15.0141 dB, but if the deviation angle is larger, ILD and ICLD are almost overlap completely. When phantom source is located in the lateral interval (such as −30° to −110° interval of config.1), there is a greater difference between ICLD and ILD, the maximum difference is up to 55.8059 dB. When phantom source is located in the rear interval (such as −110° to 110° interval of config.1), ILD and ICLD difference is biggest, the maximum difference is up to 42.6577 dB. But separately analyze each interval, the closer

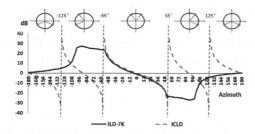

Fig. 5. Comparison of ICLD and ILD based on config.2.

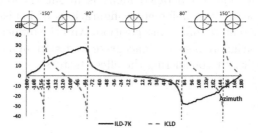

Fig. 6. Comparison of ICLD and ILD based on config.3.

to loudspeaker, the greater difference between ILD and ICLD is, for example, in the $-150°$ to $-80°$ interval of config.3, the maximum difference value appears near $-80°$.

The tendency of ICLD and ILD are also changed with phantom source azimuth. In the front interval, ILD trend is consistent with ICLD trend. In lateral interval, when the phantom source azimuth angle less than $90°$, ILD trend and ICLD trend are consistent, while when the phantom source azimuth angle is greater than $90°$, ILD trend is contrary to ICLD trend. And in the rear interval, ILD and ICLD have opposite change tendency.

Different frequencies of sequences also influence ILD, as shown in Fig. 4. In front and rear intervals, coincidence degree of ILD curves with different frequencies is higher, it shows that the effect introduced by frequency is small. In lateral interval, the difference of ILD curves is more obvious, it shows that frequency effect is more apparent, and the maximum difference is up to 12 dB. But there is no obvious change of ILD curve trends for different frequencies.

Besides, the ICLD itself is different between different configurations. Take the config.1 as the standard, to calculate the difference between config.1 and config.2 or config.3. The result is shown in the Table 2.

The data in Table 2 contains the average difference and the maximum difference of ICLD which is calculated based on different configurations. It shows that the average difference introduced by change of configuration is larger than 13 dB, even the maximum difference can be up to 45 dB. While the biggest quantization step length of ICLD is only 5 dB in MPEG Surround technology (when

Table 2. Difference of ICLD between configurations.

Configuration	Average difference:dB	Max difference:dB
config.2-config.1	13.1561	45.0516
config.3-config.1	15.3408	38.6646

the ICLD is less than 45 dB) [5], so the rebuilding will lead to the change of phantom source position.

In conclusion, we make a detailed analysis about the difference of ICLD and ILD from the aspects of different configurations of loudspeakers, different frequencies and different virtual sound position. And the influence introduced by the factors cannot be ignored, so it is inaccurate to design quantization codebook according to ILD without considering the effect factors.

4   Conclusion

There are two filtering process through sound traveling from source to ears. The first one is attenuation in transmission route to head position. Then the second filtering is caused by head shape and auricle. From the aspects of sound pressure and particle velocity, we theoretically deduce the ILD and ICLD in the first process. Then we conclude that ICLD is a physical quantity related to the loudspeaker derivation angle and phantom source position, while ILD is not only related to the loudspeakers derivation angle and phantom source position, but also related to the distance between loudspeakers and listening position. As a result, there is an essential distinction between ILD and ICLD. The second filtering process is similar, thus the distinction mainly reflects in the first filtering process.

In addition, we conduct experiments to verify the distinction between ILD and ICLD. The experiments show that different configuration of loudspeakers, frequency and phantom source position all contribute to the change of ILD and ICLD.

When the phantom source lies in the dead ahead position, the values of ILD and ICLD are similar, the maximum of the difference is 15.0141 dB in config.1, but in config.3, where the angle between two front loudspeakers is bigger, the values of ILD and ICLD are nearly the same. When the phantom source lies in the side space, there is a great difference between values of ILD and ICLD, which maximum is 55.8059 dB. When the phantom source lies in the back space, the difference of values is the largest, which reaches 42.6577 dB.

The average variation of ICLD reaches 13 dB as the change of loudspeakers configuration, is larger than the biggest quantization step length (5 dB) of ICLD in MPEG Surround technology. So the rebuilding will lead to the change of phantom source position.

As a consequence, improving the quantitative method is necessary in order to get better performance of rebuilding. Our research and analysis provides reference to 3D audio coding method of more accurate and higher quality. The further work is to put forward more accurate ICLD quantitative method.

Acknowledgments. The research was supported by National Nature Science Foundation of China (No. 61231015); National High Technology Research and Development Program of China (863 Program) No. 2015AA016306; National Nature Science Foundation of China (No. 61201169, 61201340); Science and Technology Plan Projects of Shenzhen (ZDSYS2014050916575763).

References

1. Boring, E.G.: Sensation and Perception in the History of Experimental Psychology. D. Appleton Century Company, New York (1942)
2. Blauert, J.: Spatial Hearing: The Psychophysics of Human Sound Localization. MIT Press, Cambridge (1997)
3. Baumgarte, F., Faller, C.: Binaural cue coding-Part I: psychoacoustic fundamentals and design principles. IEEE Trans. Speech Audio Process. 11(LCAV–ARTICLE–2005–032), 509–519 (2003)
4. Pulkki, V., Karjalainen, M.: Localization of amplitude-panned virtual sources I: stereophonic panning. J. Audio Eng. Soc. 49(9), 739–752 (2001)
5. ISO/IEC JTC1/SC29/WG11 (MPEG) Document N7947, Text of ISO/IEC 23003–1:2006/FCD, MPEG Surround, Bangkok (2006)
6. Jiang, W., Wang, J., Zhao, Y., et al.: Multi-channel audio compression method based on ITU-T G. 719 Codec. In: 2013 Ninth International Conference on Intelligent Information Hiding and Multimedia Signal Processing, pp. 293–297. IEEE (2013)
7. Breebaart, J.: Comparison of interaural intensity differences evoked by real and phantom sources. J. Audio Eng. Soc. 61(11), 850–859 (2013)
8. Gao, L., Hu, R., Yang, Y., Wang, X., Tu, W., Wu, T.: Azimuthal perceptual resolution model based adaptive 3D spatial parameter coding. In: He, X., Luo, S., Tao, D., Xu, C., Yang, J., Hasan, M.A. (eds.) MMM 2015, Part I. LNCS, vol. 8935, pp. 534–545. Springer, Heidelberg (2015)
9. Gao, L., Hu, R., Yang, Y.: A spatial priority based scalable audio coding. In: 2014 IEEE International Conference on Acoustics, Speech and Signal Processing (ICASSP), pp. 3670–3674. IEEE (2014)
10. Pierce, A.D.: Acoustics: An Introduction to Its Physical Principles and Applications. McGrawHill, New York (1981)
11. Fahy, F.J.: Measurement of acoustic intensity using the cross spectral density of two microphone signals. J. Acoust. Soc. Am. 62(4), 1057–1059 (1977)
12. Everest, F.A., Pohlmann, K.C.: The Master Handbook of Acoustics. McGraw-Hill, New York (2001)
13. Algazi, V.R., Duda, R.O., Thompson, D.M., Avendano, C.: The CIPIC HRTF database. In: Proceedings of the 2001 IEEE Workshop on Applications of Signal Processing to Audio and Electroacoustics, pp. 99–102. Mohonk Mountain House, New Paltz, 21–24 October 2001

Automatic Scribble Simulation for Interactive Image Segmentation Evaluation

Bingjie Jiang[1,2], Tongwei Ren[1,2(✉)], and Jia Bei[1,2]

[1] State Key Laboratory for Novel Software Technology, Nanjing University,
Nanjing, China
[2] Software Institute, Nanjing University, Nanjing, China
bjie.jiang@gmail.com, rentw@nju.edu.cn, beijia@software.nju.edu.cn

Abstract. To provide comprehensive evaluation of interactive image segmentation algorithms, we propose an automatic scribble simulation approach. We first analyze the variety of scribbles labelled by different users and its influence on segmentation result. Then, we describe the consistency and inconsistency of scribbles with normal distribution on superpixel level and superpixel group level, and analyze the effect of connection in scribble for interactive segmentation evaluation. Based on the above analysis, we simulate scribbles on foreground and background respectively by randomly selecting superpixel groups and superpixels with the previously determined coverage values. The experimental results show that the scribbles simulated by the proposed approach can obtain similar evaluation results to manually labelled scribbles and avoid serious deviation in precision and recall evaluation.

Keywords: Scribble simulation · Interactive image segmentation evaluation · Scribble variety · Superpixel group

1 Introduction

As the foundation of numerous multimedia applications, interactive image segmentation has been widely utilized in object recognition [1], image retrieval [2] and annotation [3], scene understanding [4], visual tracking [5], social media mining [3], surveillance analysis [6] and so on. It can effectively extract the desired objects from images with the assistance of manual labels, which is used to approximately outline the regions of objects [7]. There have been several types of manual labels applied in the existing interactive image segmentation algorithms, including triple map, boundary box and scribble, in which scribble is commonly used for its simplicity and flexibility in labelling [8].

For different users may provide various scribbles in labelling, an effective interactive image segmentation algorithm should be robust to scribbles, i.e., its performance should not be obviously influenced by the difference of scribbles labelled by different users. It requires to evaluate interactive image segmentation algorithms on the scribbles provided by numbers of users with sufficient variety.

© Springer International Publishing Switzerland 2016
Q. Tian et al. (Eds.): MMM 2016, Part I, LNCS 9516, pp. 596–608, 2016.
DOI: 10.1007/978-3-319-27671-7_50

However, in the evaluations of the existing algorithms [7,8], only the scribbles provided by one specific user are used to reduce human labor. Obviously, such evaluations cannot provide a comprehensive comparison for they contain high randomness in scribble labelling and segmentation. Moreover, if the scribbles are intentionally selected, the evaluations may have a bias to some algorithms, which influences the fairness of the evaluations.

To overcome the above problem, we propose an automatic scribble simulation approach for interactive image segmentation evaluation. Based on the ground truths of segmentation results, a number of scribbles with sufficient variety are automatically generated to simulate the labelling results provided by different users, which can provide a comprehensive evaluation of interactive image segmentation algorithms (Fig. 1).

The rest of the paper is organized as follows: In Sect. 2, we briefly review the existing interactive image segmentation algorithms and evaluation strategies. In Sect. 3, we introduce the data set used in our experiments. In Sect. 4, we analyze the variety of scribbles labelled by different users and its influence on segmentation results. In Sect. 5, we present the details of our scribble simulation approach. In Sect. 6, the proposed scribble simulation approach is validated by comparing manual labelling and other two automatic simulation approaches. Finally, the paper is concluded in Sect. 7.

2   Related Work

Interactive Image Segmentation. Amounts of interactive image segmentation algorithms have been proposed in decades. As one of the most representative algorithms, Graph cuts [7] converts each image to a graph and formulates image segmentation as a min-cut energy minimization problem. Grabcut [9] allows to label a rectangle around foreground and improves segmentation performance by iteration strategy. Random walker [10] assigns each unlabelled pixel a maximal probability that a random walker could reach it starting from the existing labels. Geodesic distance [11] uses star-convexity prior and replaces Euclidean rays with geodesic path to exploit the structure of shortest paths. Recently, some researchers extend interactive segmentation from monocular image to other media types, such as binocular image [8], RGB-D image [12] and video [13].

Interactive Image Segmentation Evaluation. In evaluation of interactive image segmentation algorithms, a key problem is how to effectively generate sufficient user labels. Manual labelling with several participants has high labor cost and time consumption even with a facilitate tool [14], and the labelled scribbles cannot be utilized to handle new test images [15]. To overcome the limitations of manual labelling, automatic interactive image segmentation evaluation is studied as in image and video compression [16], resizing [17] and summarization [18]. Some existing approaches attempt to generate the scribbles similar to manual labelling results in appearance [19–21], for example, extracting the sketches of foreground objects and labelling the background around the objects. However, the appearance of scribbles generated by these approaches are constricted by pre-defined rules, which reduces the variety of the generated scribbles. To increase representation flexibility, another approaches use pixel sets instead of connected curves in representing scribbles. Nevertheless, the existing approaches usually sample the pixels by manual defined strategies without considering the characteristics of manual labelling [22].

3   Data Set

We construct our data set with 96 images from Berkeley Segmentation Dataset [15]. Each image contains at least one obvious object, which could be unambiguously explained to users. And these images are also representative of some major challenges of image segmentation, including fuzzy boundary, complex texture and complex lighting conditions. The ground truths of segmentation results are precisely hand-labelled for each image to avoid biases.

To analyze the rules for scribble simulation, we invite five users to manually label the images with The K-Space Segmentation Tool Set [14]. All the users are the students with basic computer operating skills but limited knowledge about interactive image segmentation. Each user is given a clear guidance and enough time to familiarize themselves with the labelling software, and all the labelling operations are carried out by mouse.

4   Analysis of Scribble Variety

4.1   Scribble Difference

An instinctive observation is that different users cannot keep high consistency in labelling images with scribbles. In order to validate the observation, we analyze the variety of scribbles labelled by different users. To facilitate the following description, we indicate the five users with A, B, C, D and E. To the kth image, the scribbles labelled by user n are represented as s_n^k, here $n \in \{A, B, C, D, E\}$. And the scribbles labelled by user n on all the images are represented as S_n.

We first analyze the difference of scribbles by pair-wise intersection rate. To the kth image, the intersection rate of the scribbles labelled by user m and n is calculated as $\phi_{m,n}^k = (s_m^k \cap s_n^k)/(s_m^k \cup s_n^k)$. And the average intersection rate of all the scribbles labelled by user m and n is calculated as $\bar{\phi}_{m,n} = \frac{1}{K} \sum_{k=1}^{K} \phi_{m,n}^k$, here $K = 96$ is the number of images in our data set. Table 1 shows the average interaction rates between the scribbles labelled by all the users on foreground and background, respectively. It is observed that the values of average intersection rates between the scribbles labelled by different users are quite low.

4.2   Influence on Segmentation Result

We further evaluate the influence of different scribbles on segmentation results. We utilize the scribbles labelled by different users as the inputs of interactive

Table 1. Pixel level average intersection rates of different scribbles on foreground and background.

Foreground						Background				
	S_A	S_B	S_C	S_D	S_E		S_A	S_B	S_C	S_D	S_E
S_A	–	2.4 %	3.5 %	2.0 %	1.7 %	S_A	–	0.9 %	0.9 %	0.1 %	0.6 %
S_B	2.4 %	–	2.7 %	2.2 %	2.3 %	S_B	0.9 %	–	1.0 %	0.2 %	0.7 %
S_C	3.5 %	2.7 %	–	2.1 %	2.0 %	S_C	0.9 %	1.0 %	–	0.2 %	0.7 %
S_D	2.0 %	2.2 %	2.1 %	–	1.8 %	S_D	0.1 %	0.2 %	0.2 %	–	0.7 %
S_E	1.7 %	2.3 %	2.0 %	1.8 %	–	S_E	0.6 %	0.7 %	0.7 %	0.7 %	–

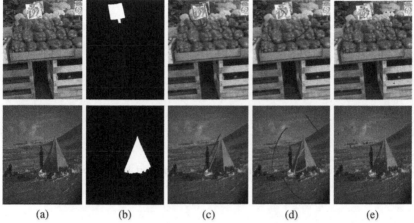

    (a)   (b)   (c)   (d)   (e)

Fig. 1. Examples of manually labelled scribbles and automatic simulation results. (a) Original image. (b) Ground truth of segmentation result. (c)–(d) Scribbles labelled by different users. (e) Automatic simulation result.

image segmentation algorithms. In our experiments, four interactive image segmentation algorithms, including graph cuts (GC) [7], geodesic star convexity (GSC) [11], random walker (RW) [10], and geodesic shortest path (GSP) [13], are used, whose implementations are all provided in [11]. Totally, $96 \times 5 \times 4$ segmentation results are generated by the four algorithms initialized with all the scribbles. To the five segmentation results generated by one algorithm on each original image, we calculate the percentages of pixels which occur as foreground in one, two, three, four or five segmentation results, respectively.

Figure 2 shows the boxplots of pixel co-occurrence percentages in different numbers of segmentation results for each segmentation algorithm. We can observe that the pixel co-occurrence percentages decline greatly when the numbers of segmentation results increase for all the algorithms. It means the segmentation results generated by the scribbles from different users are quite inconsistent. Moreover, to some segmentation algorithms which can generated relatively

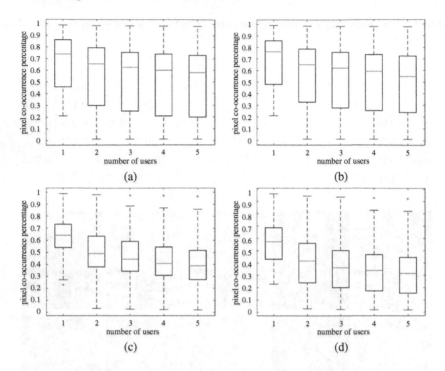

Fig. 2. Percentages of pixel co-occurrence as foreground in different numbers of segmentation results. (a) GC. (b) GSC. (c) RW. (d) GSP.

higher consistent results, such as GC and GSC, we can also find that their segmentation results contain high variability (larger lengthes of the boxes in Fig. 2(a) and (b)). Therefore, multiple scribbles with sufficient variety is required for comprehensively evaluating the performance of interactive segmentation algorithms.

5   Automatic Scribble Simulation

5.1   Scribble Consistency on Superpixel and Superpixel Group Levels

For the scribbles labelled by different users have high inconsistence on pixel level, we analyze the scribbles on superpixel levels for expecting high consistency. we use the simple linear iterative clustering algorithm [23], which is implemented by VLFeat open source library [24], to cluster image pixels into compact and nearly uniform superpixels. To each superpixel, we consider it to be labelled by a scribble if one or more pixels within it are labelled by the scribble. Similar to pixel-level scribble consistency analysis, we analyze the scribble consistency on superpixel level by calculating the pair-wise intersection rates of the scribbles labelled by different users.

Table 2. Superpixel level average intersection rates of different scribbles on foreground and background.

Foreground					Background				
$\widehat{S}_A$	$\widehat{S}_B$	$\widehat{S}_C$	$\widehat{S}_D$	$\widehat{S}_E$	$\widehat{S}_A$	$\widehat{S}_B$	$\widehat{S}_C$	$\widehat{S}_D$	$\widehat{S}_E$
$\widehat{S}_A$ –	49.3%	49.0%	38.9%	37.2%	$\widehat{S}_A$ –	19.1%	17.8%	4.4%	11.9%
$\widehat{S}_B$ 49.3%	–	52.7%	43.5%	39.8%	$\widehat{S}_B$ 19.1%	–	21.2%	7.7%	15.2%
$\widehat{S}_C$ 49.0%	52.7%	–	41.6%	40.2%	$\widehat{S}_C$ 17.8%	21.2%	–	6.3%	16.1%
$\widehat{S}_D$ 38.9%	43.5%	41.6%	–	34.3%	$\widehat{S}_D$ 4.4%	7.7%	6.3%	–	16.1%
$\widehat{S}_E$ 37.2%	39.8%	40.2%	34.3%	–	$\widehat{S}_E$ 11.9%	15.2%	16.1%	16.1%	–

Table 2 shows the intersection rates of the scribbles labelled by different users on superpixel level. Here, $\widehat{S}_n$ denotes the scribbles labelled by user n on all the images on superpixel level. Compared to pixel level analysis in Table 1, the scribbles have higher consistency on superpixel level than on pixel level, but the values of intersection rates are not high enough especially on background.

To further explore the consistency of different scribbles, we divide the superpixels into groups by quantifying them on RGB color space. We uniformly decompose R, G, B channels into eight parts and the whole RGB color space is decomposed into $8 \times 8 \times 8$ subspaces. All the superpixels whose average color belong to the same subspace are considered as a superpixel group. Similarly, if one or more pixels in a superpixel group are labelled by a scribble, we consider the superpixel group to be labelled by the scribble.

Table 3 shows the intersection rates of the scribbles labelled by different users on superpixel group level. Here, $\widetilde{S}_n$ denotes the scribbles labelled by user n on all the images on superpixel group level, and intersection rates of different scribbles are calculated in a similar way to intersection rates on pixel level and superpixel level. We can find that the consistency of scribbles on superpixel group level keeps increasing, and the values of intersection rates are rather high on both foreground and background.

Based on the observation in Tables 2 and 3, we conclude that the scribbles labelled by different users are consistent on superpixel group level but keep

Table 3. Superpixel group level average intersection rates of different scribbles on foreground and background.

Foreground					Background				
$\widetilde{S}_A$	$\widetilde{S}_B$	$\widetilde{S}_C$	$\widetilde{S}_D$	$\widetilde{S}_E$	$\widetilde{S}_A$	$\widetilde{S}_B$	$\widetilde{S}_C$	$\widetilde{S}_D$	$\widetilde{S}_E$
$\widetilde{S}_A$ –	77.7%	78.3%	73.2%	77.6%	$\widetilde{S}_A$ –	66.9%	67.1%	58.2%	68.1%
$\widetilde{S}_B$ 77.7%	–	81.5%	78.4%	79.8%	$\widetilde{S}_B$ 66.9%	–	68.6%	61.7%	70.2%
$\widetilde{S}_C$ 78.3%	81.5%	–	78.4%	78.8%	$\widetilde{S}_C$ 67.1%	68.6%	–	60.6%	70.5%
$\widetilde{S}_D$ 73.2%	78.4%	78.4%	–	75.3%	$\widetilde{S}_D$ 58.2%	61.7%	60.6%	–	66.1%
$\widetilde{S}_E$ 77.6%	79.8%	78.8%	75.3%	–	$\widetilde{S}_E$ 68.1%	70.2%	70.5%	66.1%	–

some varieties on superpixel level. The reason is that the key characteristics of foreground and background in each image are limited, and each key characteristic is represented by multiple superpixels especially on background. When a user labels an image, he/she usually tries to cover all the key characteristics of foreground and background to avoid further providing more interaction. Hence, most superpixel groups are labelled on both foreground and background. Nevertheless, the selection of superpixels to represent each key characteristic highly depends on personal habits. When a key characteristic is represented by multiple superpixels, the scribbles will appear obvious inconsistency.

5.2   Distribution of Superpixel Group Coverage

To effectively simulate the scribbles generated by different users, one important problem is how much image content should be covered by scribbles on superpixel group level, i.e., which percentage of superpixel groups should be labelled in scribble simulation. To answer this question, we calculate the percentages of superpixel groups labelled as foreground and background by different users. Table 4 shows the mean, variance and coefficient of variation (CV) of content coverage rate by superpixel groups on foreground and background, respectively. We can find that the content coverage rates of the scribbles labelled by different users only have small differences from similar mean values, and the content coverage rates of the scribbles labelled by the same user are stable from low values of variance and CV. Hence, in simulating the scribbles labelled by one user, the content coverage rate can be randomly selected in a small range but it should be keep consistent in scribble simulation on all images.

Table 4. Content coverage rate by superpixel groups on foreground and background.

Foreground						Background				
	$\tilde{S}_A$	$\tilde{S}_B$	$\tilde{S}_C$	$\tilde{S}_D$	$\tilde{S}_E$		$\tilde{S}_A$	$\tilde{S}_B$	$\tilde{S}_C$	$\tilde{S}_D$	$\tilde{S}_E$
Mean	74.9 %	81.9 %	83.6 %	83.2 %	76.5 %	Mean	64.9 %	64.5 %	65.4 %	61.2 %	74.1 %
Variance	0.03	0.03	0.02	0.02	0.03	Variance	0.03	0.02	0.02	0.03	0.02
CV	0.22	0.17	0.17	0.18	0.22	CV	0.27	0.21	0.22	0.31	0.17

To describe the distribution of superpixel group coverage, we test its normality with normal Q-Q plot. Figure 3(a) and (b) show the normal Q-Q plots of superpixel group coverage on foreground and background, respectively. It shows that data points in the plots are both close to the diagonals, which indicates that the distribution of superpixel group coverage on foreground and background are both normal distribution. We analyze the parameters of these two normal distributions in Fig. 3(c) and (d), and use them to describe the distribution of superpixel group coverage on foreground and background.

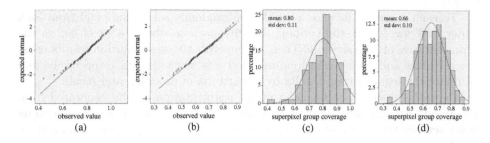

Fig. 3. Distribution of superpixel group coverage on foreground and background.

5.3 Distribution of Superpixel Coverage

Another problem in scribble simulation is how to describe the distribution of superpixel coverage since it has obvious inconsistency among the scribbles labelled by different users.

We analyze the distribution of superpixel coverage in a similar way to superpixel group coverage. Figure 4(a) and (b) show the normal Q-Q plots of superpixel coverage on foreground and background, respectively. It shows that the distribution of superpixel coverage on foreground and background are also both normal distribution. And Fig. 4(c) and (d) show the parameters of these two normal distributions, which are used to describe the distribution of superpixel coverage on foreground and background.

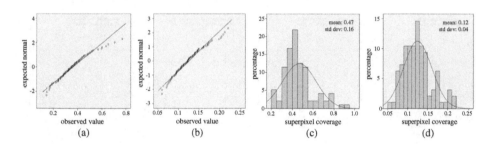

Fig. 4. Distribution of superpixel coverage on foreground and background.

5.4 Effect of Connection in Scribble

The third problem in scribble simulation is how to generate the smooth curves to represent the scribbles which should look natural and cover the prescribed percentage of superpixel groups. It is a difficult and complex problem though it has been researched for decades [19–21]. To simplify the problem, we analyze the effective elements in scribble for interactive image segmentation.

To each manually labelled scribble, we randomly select one pixel from each superpixel covered by the scribble. In this way, we obtain a pixel set as the representative of each scribble. Then, we generate the segmentation results using the scribbles and their corresponding pixel sets as the inputs respectively, and compare the segmentation results by the criteria of precision and recall.

Figure 5 shows the comparison of the segmentation results generated from the inputs of manually labelled scribbles (blue bins) and their corresponding pixel sets without connections (orange bins). It shows that the segmentation results generated by these two types of inputs are very similar in performance. It means that the effect of connection in scribbles is weak to interactive image segmentation. Hence, we can use arbitrary connections between pixels in scribble simulation when keeping the stability of superpixel group coverage, or even completely ignore the connections. In our experiments, we directly use pixel sets without connections to simulate scribbles to simplify processing procedure and reduce computational cost.

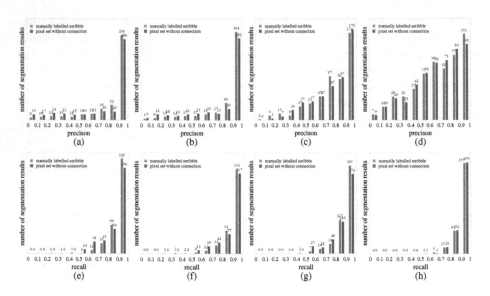

Fig. 5. Comparison of the segmentation results generated by manually labelled scribbles and their corresponding pixel sets. (a)–(d) Precision comparison using SC, GSC, RW, and GSP algorithms, respectively. (e)–(h) Recall comparison using SC, GSC, RW, and GSP algorithms, respectively (Color figure online).

5.5   Scribble Simulation

Based on the above analysis, we simulate scribbles on foreground and background separately according to the segmentation ground truth of each image. We first generate the superpixels and superpixel groups according to the corresponding segmentation ground truth. Then, we determine the values of superpixel group

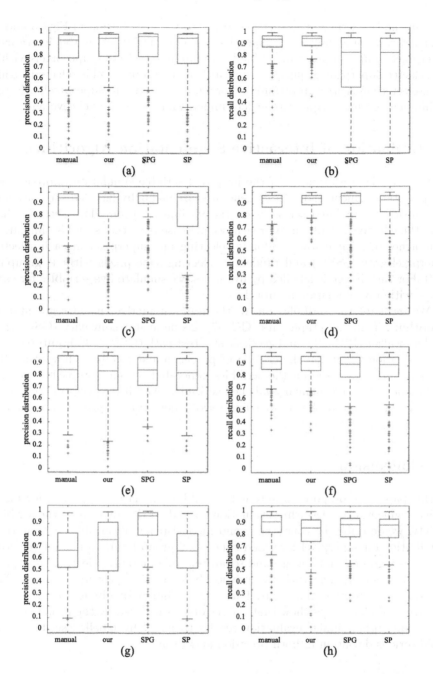

Fig. 6. Comparison of the evaluation results. (a)(c)(e)(g) Precision evaluation the segmentation results generated by manually labelled scribbles and the scribbles simulated by our approach, SPG, SP, respectively. (b)(d)(f)(h) Recall evaluation the segmentation results generated by manually labelled scribbles and the scribbles simulated by our approach, SPG, SP, respectively.

coverage and superpixel coverage according to the distribution in Figs. 3 and 4. Thereafter, we randomly select superpixels groups until reaching the determined superpixel group coverage value. Finally, we randomly selected superpixels within the selected superpixel groups, in which at least one superpixel is selected within each selected superpixel group and the total number of all the superpixels selected within in all groups is equal to the determined value of superpixel coverage.

6   Evaluation of Interactive Segmentation Algorithms

To validate the performance of the proposed scribble simulation approach, we compare the segmentation results generated by manually labelled scribbles and automatically simulated scribbles on the 96 images in our data set. To illustrate the effectiveness of our approach, we also use two other scribble simulation approaches in comparison, randomly selecting one superpixel in each selected superpixel group (SPG) and randomly selecting superpixels without grouping (SP). For each image is labelled by five users, we simulate five scribbles for each image with any simulation approach.

We evaluate the scribbles on all the four interactive image segmentation algorithms, including graph cuts (GC) [7], geodesic star convexity (GSC) [11], random walker (RW) [10], and geodesic shortest path (GSP) [13]. Figure 6 shows the evaluation results on the criteria of precision and recall. It shows that the evaluation results based on the scribbles generated by our approach is similar to the ones based on manually labelled scribbles. And it outperforms the other two simulation approaches in avoiding serious deviation in precision and recall evaluation (Fig. 6(b)(c)(g)).

7   Conclusion

In this paper, we propose an automatic scribble simulation approach for interactive segmentation algorithm evaluation. Based on the analysis of the scribble variety, we describe the consistency and inconsistency of scribbles with normal distribution on superpixel level and superpixel group level, and simulate scribbles on foreground and background separately by randomly selecting superpixel groups and superpixels with the previously determined coverage values. The experimental results evaluated by four existing interactive image segmentation algorithms on 96 images show that the scribbles simulated by the proposed approach can obtain similar evaluation results to manually labelled scribbles and avoid serious deviation in both precision and recall.

Acknowledgments. This work is supported by the National Science Foundation of China (61321491, 61202320), Research Project of Excellent State Key Laboratory (61223003), National Undergraduate Innovation Project (G1410284075) and Collaborative Innovation Center of Novel Software Technology and Industrialization.

References

1. Bao, B.-K., Liu, G., Hong, R., Yan, S., Changsheng, X.: General subspace learning with corrupted training data via graph embedding. IEEE TIP **22**(11), 4380–4393 (2013)
2. Xu, X., Geng, W., Ju, R., Yang, Y., Ren, T., Wu, G.: OBSIR: object-based stereo image retrieval. In: IEEE ICME, pp. 1–6 (2014)
3. Sang, J., Changsheng, X., Liu, J.: User-aware image tag refinement via ternary semantic analysis. IEEE TMM **14**(3–2), 883–895 (2012)
4. Li, T., Chang, H., Wang, M., Ni, B., Hong, R., Yan, S.: Crowded scene analysis: a survey. IEEE TCSVT **25**(3), 367–386 (2015)
5. Ren, T., Qiu, Z., Liu, Y., Tong, Y., Bei, J.: Soft-assigned bag of features for object tracking. MMSJ **21**(2), 189–205 (2015)
6. Bao, B.-K., Liu, G., Changsheng, X., Yan, S.: Inductive robust principal component analysis. IEEE TIP **21**(8), 3794–3800 (2012)
7. Boykov, Y.Y., Jolly, M.-P.: Interactive graph cuts for optimal boundary & region segmentation of objects in N-D images. In: IEEE ICCV, pp. 105–112 (2001)
8. Ju, R., Ren, T., Wu, G.: StereoSnakes: contour based consistent object extraction for stereo images. In: IEEE ICCV (2015)
9. Rother, C., Kolmogorov, V., Blake, A.: Grabcut: interactive foreground extraction using iterated graph cuts. ACM TOG **23**(3), 309–314 (2004)
10. Grady, L.: Random walks for image segmentation. IEEE TPAMI **28**(11), 1768–1783 (2006)
11. Gulshan, V., Rother, C., Criminisi, A., Blake, A., Zisserman, A.: Geodesic star convexity for interactive image segmentation. In: IEEE CVPR, pp. 3129–3136 (2010)
12. Ge, L., Ju, R., Ren, T., Wu, G.: Interactive RGB-D image segmentation using hierarchical graph cut and geodesic distance. In: Ho, Y.-S., Sang, J., Ro, Y.M., Kim, J., Wu, F. (eds.) PCM 2015. LNCS, vol. 9314, pp. 114–124. Springer, Heidelberg (2015)
13. Bai, X., Sapiro, G.: A geodesic framework for fast interactive image and video segmentation and matting. In: IEEE ICCV, pp. 1–8 (2007)
14. McGuinness, K., O'Connor, N.E.: A comparative evaluation of interactive segmentation algorithms. PR **43**(2), 434–444 (2010)
15. Martin, D., Fowlkes, C., Tal, D., Malik, J.: A database of human segmented natural images and its application to evaluating segmentation algorithms and measuring ecological statistics. In: IEEE ICCV, pp. 416–423 (2001)
16. Wang, Z., Bovik, A.C., Sheikh, H.R., Simoncelli, E.P.: Image quality assessment: from error visibility to structural similarity. IEEE TIP **13**(4), 600–612 (2004)
17. Ren, T., Wu, G.: Automatic image retargeting evaluation based on user perception. In: IEEE ICIP, pp. 1569–1572 (2010)
18. Ren, T., Liu, Y., Wu, G.: Full-reference quality assessment for video summary. In: IEEE ICDM Workshops, pp. 874–883 (2008)
19. Fu, Y., Cheng, J., Li, Z., Lu, H.: Saliency cuts: an automatic approach to object segmentation. In: ICPR, pp. 1–4 (2008)
20. McGuinness, K., O'Connor, N.: Toward automated evaluation of interactive segmentation. CVIU **115**(6), 868–884 (2011)
21. Kohli, P., Nickisch, H., Rother, C., Rhemann, C.: User-centric learning and evaluation of interactive segmentation systems. IJCV **100**(3), 261–274 (2012)
22. Moschidis, E., Graham, J.: A systematic performance evaluation of interactive image segmentation methods based on simulated user interaction. In: IEEE ISBI, pp. 928–931 (2010)

23. Achanta, R., Shaji, A., Smith, K., Lucchi, A., Fua, P., Susstrunk, S.: Slic superpixels compared to state-of-the-art superpixel methods. IEEE TPAMI **34**(11), 2274–2282 (2012)
24. Vedaldi, A., Fulkerson, B.: VLFeat: an open and portable library of computer vision algorithms. In: ACM MM, pp. 1469–1472 (2010)

Multi-modal Image Re-ranking
with Autoencoders and Click Semantics

Chaohui Tang[1,2], Qingxin Zhu[1], Chaoqun Hong[2(✉)], and Jun Yu[3]

[1] School of Information and Software Engineering,
University of Electronic Science and Technology of China, Chengdu 611731, China
[2] School of Computer and Information Engineering,
Xiamen University of Technology, Xiamen 361024, China
`cqhong@xmut.edu.cn`
[3] School of Computer Science, Hangzhou Dianzi University, Hangzhou 310018, China

Abstract. Image re-ranking is effective in improving text-based image retrieval experience. However, to construct an efficacious algorithm to achieve such a target is limited by two important issues: one is that visual features extracted for image re-ranking from images are too superficial to represent the whole information contained within images; the other is that the corresponding text information often mismatches semantics of images. In this paper, we utilize autoencoders to extract deeper features of images and exploit click data to bridge the semantic gap between query words and image semantics. A graph-based algorithm(MIR-AC) is proposed to adaptively integrate features from autoencoders and click information by constructing two manifolds with updating weights. In particular, MIR-AC completes image re-ranking by conducting an iterative optimization process in which image ranking scores and weights of manifolds are updated alternatively. Experiments are conducted on a real world dataset and results demonstrate that MIR-AC outperforms given state-of-arts in image re-ranking.

Keywords: Image re-ranking · Autoencoders · Click data · Patch alignment framework

1 Introduction

Currently, due to the rapidly increasing number of images in the internet, image retrieval has become a hot topic [19, 20]. However, most current searching engines actually retrieve text labeled images, and ambiguities exists since the labeling textual information often mismatches contents of images.

Researchers have devote themselves into exploiting methods of retrieving images according to their contents, which is called Content-based Image Retrieval (CBIR). A well-designed feature should be discriminative with respect to images. Until now, quite a lot of features have been proposed for image description. Generally speaking, feature descriptors can be divided into two types. The first one is named as global features. They represent one aspect of image

© Springer International Publishing Switzerland 2016
Q. Tian et al. (Eds.): MMM 2016, Part I, LNCS 9516, pp. 609–620, 2016.
DOI: 10.1007/978-3-319-27671-7_51

characteristics with a single vector, such as color features [19], shape features [2], textural features [7], edge features [3] and so on. The second one is named as local features. They represent image characteristics with a set of vectors by detecting feature points (regions) and describing these features. SIFT (Scale Invariant Feature Transform) [12] and SURF (Speeded-Up Robust Features) [1] are the widely used.

Since a single type of features could not completely describe one image, researchers have proposed learning methods by combining different types of features [10,17,18,23]. These methods are called multiview learning methods, which have attracted plenty of attention. Graph-based approaches could describe different features in a unified form [9,17], so they are one of the most popular solution. Representative approaches include Laplacian Regularization [27], Normalized Laplacian Regularization [26], Local Learning Regularization [21], Markov random walk explanation [15] and multi-task low-rank affinity pursuit [6].

In the past few years, click data have become another type of important information in measuring the relationship between queries and retrieved images [4]. Therefore, image re-ranking with click data has been proven to be effective since it comes from feedback of users and is more reliable than textual information [8]. Hua et al. argued that the massive amount of click data from commercial search engines provides a data set that is unique in the bridging of the semantic and intent gap [11]. Yu et al. proposed a multi-view hypergraph-based learning method that adaptively integrates click data with varied visual features [24].

In this paper, a novel approach is proposed for image re-ranking using autoencoders and click semantics. Contributions of this paper are as follows:

- We effectively utilize autoencoders to obtain deep representation of images for image re-ranking, rather than simple visual features. Furthermore, click data of each single query are exploit by splitting the corresponding query results into the relevant and irrelevant parts, and a pairwise strategy is adopted to explore more complicated click information.
- Graph-based learning framework, PAF [25], are used to construct two manifolds from autoencoder-produced features and click information, named L_{au} and L_{click} respectively. We propose an image re-ranking algorithm, which can effectively integrate L_{au} and L_{click} and obtain the refined images scores by updating weighs of the two manifolds in a iterative procedure.
- Experiments are fully conducted to comprehensively analyze the performance of our method on real world image datasets with user query generated click information, which is extracted from a commercial search engine. The results demonstrate the effectiveness of our method.

The remainder of this paper is organized as follows. The proposed image re-ranking method is presented in Sect. 2. Then, experimental results on a public dataset and comparisons with state-of-the-art methods are presented in Sect. 3. Finally, we conclude the paper in Sect. 4.

2 Multimodal Learning with Autoencoders and Click Semantics

2.1 Marginalized Denoising Autoencoders

Deep learning architectures [14] have been useful for exploring hidden representations in natural images and have proven success in a variety of vision tasks. Based on it, autoencoders [22] are a type of unsupervised feature-learning scheme in which the internal layer acts as a generic extractor of inner image representations. Theoretically speaking, the training objective of an autoencoder is to minimize the reconstruction error:

$$\sum_i^n \| x_i - \hat{x}_i \|^2, \tag{1}$$

In denoising autoencoders, inputs $x_1, ..., x_n$ are corrupted by random feature removal. $\hat{x}_i$ is denoted as the corrupted version of x_i and $W : \mathbb{R}^d \rightarrow \mathbb{R}^d$ is denoted as the mapping of reconstructing the corrupted inputs. In this way, we can defined the squared reconstruction loss as:

$$\frac{1}{2n} \sum_{i=1}^n \| x_i - W\hat{x}_i \|^2. \tag{2}$$

The solution to (2) depends on which features of each input are randomly corrupted. To lower the variance, MDA [5] perform multiple passes over the training set, each time with different corruption. In this way, the overall squared loss is defined as:

$$\frac{1}{2mn} \sum_{j=1}^m \sum_{i=1}^n \| x_i - W\hat{x}_{i,j} \|^2, \tag{3}$$

where $\hat{x}_{i,j}$ represents the jth corrupted version of the original input x_i and m is the number of layers.

For the matrix form, $X = [x_1, ..., x_n] \in \mathbb{R}^{d \times n}$ is denoted as the data matrix, $\overline{X} = [X, .., X]$ is denoted as the m-times repeated version and $\hat{X}$ is defined as the corrupted version of $\overline{X}$. Then, the loss in (3) is reduced to:

$$\frac{1}{2mn} tr[(\overline{X} - W\hat{X})^T (\overline{X} - W\hat{X})]. \tag{4}$$

The minimization to (4) can be expressed as the well-known closed-form solution for ordinary least squares:

$$W = PQ^{-1} \text{ with } Q = \hat{X}\hat{X}^T \text{ and } P = \overline{X}\hat{X}^T. \tag{5}$$

Table 1. Notations adopted in the paper

N	The number of images in dataset
NQ	The number of results to a query
$f = [f_1, ..., f_N]^T \in R^N$	The re-ranking scores of all images to a query
RS and IRS	The positive and negative results to a query
NR and NI	The number of image in RS and IRS
$f_{init} \in R^N$	The initial ranking scores to a query
$f(u)$ or f_u	A score function to image u
D_v	The vertex degree matrix
D_e	The hyperedge degree matrix
H	The adjacency matrix of hypergraph
Ω	$\Omega_{(i,i)}$ is the weight of the ith hyperedge
L_{au}	The manifold of feature space from autoencoder
L_{click}	The manifold of feature space from click data
γ	The weights of two manifolds
λ	The trade-off parameter

2.2 Manifold Learning with Click Semantics

To clearly describe the details of our proposed method, some important notations are given as shown in Table 1.

To re-rank image ranking scores f_{init}, which is provided by a verified image retrieval algorithm for a query, graph-based framework is adopted to exploit the geometrical structures among all the images. Then a better ranking scores f can be obtained by adopting a strategy that images, which have stronger connections, have closer ranking scores. The objective function for image re-ranking can be formulated as [26]:

$$\underset{f}{\text{argmin}}\{Regularizer(f) + \lambda Loss(f)\} \qquad (6)$$

According to [36], $Regularizer(f)$ can be written as

$$Regularizer(f) = fLf^T \qquad (7)$$

where fLf^T is a kind of smoothness measurement of f to the hypergraph laplacian matrix L,which is defined as

$$I - \frac{1}{D_v^{1/2}}H\Omega D_e H^T \frac{1}{D_v^{1/2}} \qquad (8)$$

$Loss(f)$ is used to measure the difference between the initial ranking scores f_{init} of query results and re-ranked scores f, specially in our case, $Loss(f) = ||f - f_{init}||^2$. Then (6) can be rewritten as

$$\begin{aligned}&\underset{f}{\text{argmin}}\{Regularizer(f) + \lambda Loss(f)\}\\&= \underset{f}{\text{argmin}}\{fLf^T + \lambda||f - f_{init}||^2\}\end{aligned} \qquad (9)$$

Patch alignment framework(PAF) was proposed by Zhang et al. [25]. It unifies spectral analysis based manifold learning approaches, including LLE/NPE/ONPP, ISOMAP, LE/LPP, LTSA/LLTSA, HLLE, PCA and LDA. It is proposed as a powerful analysis and development tool for manifold learning. It consists of two stages:local patch construction stage and global alignment stage.

In local patch construction stage, PAF maps high dimension feature space HD, in our case the image features extracted by autoencoders, to a lower one LD while keeping the local geometrical structure L_{ae},the laplacian matrix. To obtain L_{ae}, local patch of each image in high dimension feature space is calculated. First, for any instance $x_i \in HD$, the corresponding local patch $patch_{x_i}$ is constructed by x_i itself and its k nearest neighbors in HD in this paper, and $patch_{x_i}$ forms a hyperedge, which can be used to construct a local laplacian matrix L_i according to (8). To map $patch_{x_i}$ to $patch_{y_i}$ of $y_i \in LD$, the objective function of local patch optimization is formulated as [25]:

$$\underset{patch_{y_i}}{\operatorname{argmin}} \, trace(patch_{y_i} * L_i * patch_{y_i}{}^T) \tag{10}$$

In the global alignment stage, PAF uses projection matrix $Proj_i$, also known as selection matrix, to map the global low dimension feature space LD to the local geometrical structure $patch_{y_i}$, and $patch_{y_i} = LD * Proj_i$. Finally, PAF cumulates over all the local patch and the objective function turns out to be [25]:

$$\underset{LD}{\operatorname{argmin}} \, trace(LD * L_{ae} * LD^T) \tag{11}$$

where L_{ae} is defined as:

$$L_{ae} = \sum_i^N Proj_i * L_i * Proj_i{}^T \tag{12}$$

N is the number of images in the data set.

But for the click semantics, in this paper, the local patch construction is much more complicated. To re-rank the query results of a certain query effectively, the process of importance but hard to achieve is to determine which one is relevant to the query semantic and which is not. Fortunately this could be achieved by investigating the query logs provided by a commercial search engine. For any query in this search engine, the query results, QR, would be split into two discriminative parts: relevant set with high click counts and irrelevant set with low click counts, noted by RS and IRS respectively, and $QR = RS \bigcup IRS$. Then a pairwise constraints analysis method [13] is adopted to get more discriminative relationship among $items \in QR$. To apply the pairwise constraints analysis method effectively in image re-ranking, the objective function can be defined as:

$$\operatorname{argmin} \sum_{r_i \in RS} (\sum_{r_j \in RS} ||f_{r_i} - f_{r_j}||^2 - \sum_{ur \in IRS} ||f_{r_i} - f_{ur}||^2) \tag{13}$$

Then for any relevant query result $r \in RS$, the optimization function is

$$\text{argmin} \sum_{r_j \in RS} ||f_r - f_{r_j}||^2 - \sum_{ur \in IRS} ||f_r - f_{ur}||^2 \tag{14}$$

To improve the calculation efficiency of (14), its vectorization form should be provided. Let $f_{loc^i} = \{f^i, f_{RS}^1, f_{RS}^2, ..., f_{RS}^{Num_{RS}}, f_{IRS}^1, f_{IRS}^2, ..., f_{IRS}^{Num_{IRS}}\}$ be ranking scores of $r_i \in RS$ and all the other query results, wherein each f_{RS}^{ri} is the ranking score of $ri \in RS$ and f_{IRS}^{ur} the ranking score of $ur \in IRS$ respectively; Let $w^i = [1_1, 1_2, ..., 1_{NR}, -1_1, -1_2, ... - 1_{NI}]$ be the corresponding coefficient vector, where NR and NI represent the number of items in RS and IRS respectively. Then (14) can be rewritten as

$$\text{argmin} \sum_{r_j \in RS} ||f_r - f_{r_j}||^2 - \sum_{ur \in IRS} ||f_r - f_{ur}||^2$$
$$= \text{argmin}\, f_{loc^i} * (Q_{loc^i}) * diag(w^i) * (Q_{loc^i})^T * f_{loc^i}{}^T \tag{15}$$
$$= \text{argmin}\, f_{loc^i} L_i f_{loc^i}{}^T$$

where $Q_{loc^i} = \begin{bmatrix} I_{1 \times NQ} \\ -1 * eye(NQ) \end{bmatrix}$, and $L_i = (Q_{loc^i}) * diag(w^i) * (Q_{loc^i})^T$ represents the pairwise discriminative relationship between $ri \in RS$ and all other query results, and $NQ = NR + NI$.

To align f_{local^i} to a consist global coordinate f, selection matrix $S^i \in R^{N \times (NQ+1)}$ is constructed as following, where N is the number of images in the dataset adopted.

$$S_{m,n}^i = \begin{cases} 1, indic(m, n) = 1 \\ 0, else. \end{cases} \tag{16}$$

where $indic(m, n) = 1$ means $m \in \{RS \bigcup IRS\}$ and m is the nth bin of $local^i$. Then f_{local^i} can be rewritten as $f_{local^i} = fS^i$.

Cumulating all the local patches of $r \in RS$ results in

$$\text{argmin}_f \sum_{r_i \in RS} (\sum_{r_j \in RS} ||f_{r_i} - f_{r_j}||^2 - \sum_{ur \in IRS} ||f_{r_i} - f_{ur}||^2)$$
$$= \text{argmin}_f \sum_{r_i \in RS} f_{local^i} L_i f_{local^i}{}^T$$
$$= \text{argmin}_f \sum_{r_i \in RS} fS^i L_i (fS^i)^T \tag{17}$$
$$= \text{argmin}_f f(\sum_{r_i \in RS} S^i L_i S^{iT}) f^T$$
$$= \text{argmin}_f f L_{click} f^T$$

where L_{click} is defined as

$$L_{click} = \sum_{r_i \in RS} S^i L_i S^{i^T} \tag{18}$$

In our manifold learning approach, we combine image geometrical structure L_{ae} and click data semantic structure L_{click} inherent in click data by assigning weights $\gamma = [\gamma_1, \gamma_2]$ to them respectively. Therefore, L in (9) can take advantages of L_{ae} and L_{click} at the meanwhile, i.e.,

$$L_{weighted} = \gamma_1 L_{ae} + \gamma_2 L_{click} \\ s.t. \, \gamma_1 + \gamma_2 = 1, \gamma_1 > 0, \gamma_2 > 0 \tag{19}$$

To fully explore the complementary property of L_{ae} and L_{click}, a trick in [16] is adopted: $\gamma_i := \gamma_i{}^r$ with $r > 1$.

Then we can rewrite the objective function (9) as

$$\underset{f \in R^N, \gamma \in R^2}{\arg\min} \{f L_{weighted} f^T + \lambda \|f - f_{init}\|^2\} \\ s.t. \, \gamma_1 + \gamma_2 = 1, \gamma_1 > 0, \gamma_2 > 0, r > 1 \tag{20}$$

We can solve (20) in an alternating style between f and γ.

First, we fix γ, and (20) degenerates to (9), which is a convex optimization problem with respect to f and has a globally optimal solution to a fixed γ as

$$f = 2\lambda f_{init}(L_{weighted} + L_{weighted}{}^T + 2\lambda I)^{-1} \tag{21}$$

where $L_{weighted}$ is positive semi definite, so $(L_{weighted} + L_{weighted}{}^T + 2\lambda I)$ is positive definite and invertible. We then fix f and solve (20) with respect to λ. Since it is a condition constrained optimization problem, we transform it by adopting a Lagrange multiplier β to a unconstrained optimization problem as

$$\underset{\beta, \gamma}{\arg\min} \, \gamma_1^r f L_{ae} f^T + \gamma_2^r f L_{click} f^T + \beta(1 - \gamma_1 - \gamma_2) \tag{22}$$

By applying lagrange optimal necessary conditions, a minima of the objective function will be achieved by setting the derivative of it with respect to β and γ to 0 respectively, then we obtain a relatively better γ as

$$\gamma_i = \frac{\left(\frac{1}{f L_i f^T}\right)^{1/(r-1)}}{\sum_{i=1}^{2} \left(\frac{1}{f L_i f^T}\right)^{1/(r-1)}} \tag{23}$$

where $L_1 = L_{ae}$, $L_2 = L_{click}$ and $i \in \{1, 2\}$. Since L_i is positive semi definite, then $f L_i f^T >= 0$ and $\gamma_i >= 0$.

With the increasing of iteration number, the value of the objective function reduces and f will converge with a certain iteration number. The algorithm framework is described as Algorithm 1.

Algorithm 1. Algorithm of Multi-modal Image Re-ranking with Autoencoders and Click Semantics

---

Input:
    A image data set with click information, X;
    The initial ranking score for optimization, f_{init};
Output:
    The re-ranked image scores for a given query,f;
1: Calculate L_{ae} and L_{click} according to (12) and (18);
2: Initialize $\gamma = [0.5, 0.5]$ for L_{ae} and L_{click};
3: **while** f has not converged **do**
4:    Update f as shown in (21);
5:    Update γ described in (23);
6: **end while**
7: **return** f;

---

3   Experimental Evaluation

3.1   Experiment Introduction

We conduct experiments on a large scale real-world image dataset provided by a commercial search engine, which contains 147634 images and each of them consists of click information created by millions of people while surfing the internet. For each single query, there are roughly 196 images retrieved, and some of them are relevant to the query and the others are not. To describe the quality of the query results, each query result are marked with 'Excellent', 'Good' or 'Bad' to indicate that it is closely, slightly or barely relevant to the query.

We adopt this dataset mainly for two reasons. First, it contains a large number of images which are close to our daily lives, so it is of great significance to explore it; Second, it provides the initial ranking information for each query, so the practicability of our re-ranking algorithm can be easily verified.

The proposed "Multi-modal Image Re-ranking with Autoencoders and Click Semantics" utilizes Autoencoders to obtain deep information of images and explores click data to narrow semantic gap between the query and query results. We name it "MIR-AC".

The data set is split into evaluation part *EvalSet* and validation part *ValiSet* in our experiments. The partition is conducted randomly among all the images in it, and *EvalSet* and *ValiSet* are 9 to 1 roughly in number. To evaluate the re-ranking performance of our algorithm, the Normalized Discounted Cumulated Gain(NDCG) is adopted, which is widely used in image retrieval researches. For a certain query, NDCG at a depth d, specially $d = 50$ in our work, is formulated as

$$NDCG(d) = (\sum_{i=1}^{d} \frac{2^{rel(i)} - 1}{log(i + 1)})/Optimal(d) \tag{24}$$

where $rel(i) \in \{0, 1, 2\}$ represents the relevant degree to the query, and $Optimal(d)$ is the Discounted Cumulated Gain(DCG) at the depth of d to the best ranking of the query results, which means that, in the ranking list, 'Bad' ones follow 'Good' ones and 'Excellent' ones are followed by 'Good' ones. The bigger the values of NDCG, the better the image ranking quality.

3.2   Experimental Results

We conduct experiments to compare the performance of the proposed MIR-AC with baseline, MHL-Click [24] and some other referenced state-of-arts in [24], and results are presented as shown in Table 2. NDCG@50 is adopted without loss of generality, and we can observe that MIR-AC outperforms all other methods. It shows that bringing in the autoencoder and click data can effectively exploit the semantics of images to improve user's query experience.

Table 2. Comparison of NDCG@50 of our augmented approach with other state-of-arts and baseline

Metric	Baseline	Aver-Rwalk	GP	MHL	MHL-Click	MIR-AC
NDCG@50	0.910	0.921	0.932	0.928	0.936	**0.939**

Since there are three common parameters both in MIR-AC and MHL-Click, the performance comparison is further conducted between them. We adopt a strategy that all parameters are tested one at a time by varying its value, while the other two are fixed to given values. Experimental results of varying k is demonstrated in Fig. 1.

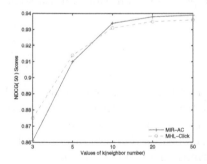

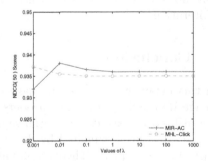

Fig. 1. NDCG(50) of varying k, with $\lambda = 0.01$, $r = 3$

Fig. 2. NDCG(50) of varying λ, with $k = 20$, $r = 3$

As we can see from Fig. 1, for both MIR-AC and MHL-Click, the larger the value of k, the bigger the value of $NDCG(50)$. The values of $NDCG(50)$ of MIR-AC are slightly smaller than those of MHL-Click while k is set to be 3 and 5, because MHL-Click extracts 6 different kinds of visual features from

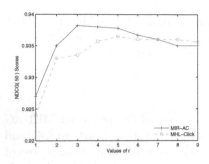
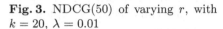
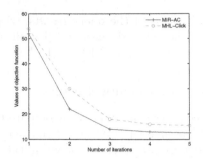

Fig. 3. NDCG(50) of varying r, with $k = 20$, $\lambda = 0.01$

Fig. 4. Curve of convergence, with $k = 20$, $\lambda = 0.01$, $r = 3$

images while MIR-AC extracts only one kind of image feature representation, and MHL-Click can obtain more effective relationship among images than MIR-AC when the number of neighbors is small, such as 3 and 5. But MIR-AC has better performance with k at 10, 20, 50 for the reason that MIR-AC has extracted more effective feature representation than MHL-Click though number in disadvantage, and the quality of feature representations has more significant impact on $NDCG(50)$ than the quantity when $k \in \{10, 20, 50\}$.

Figure 2 shows that our proposed method MIR-AC outperforms MHL-Click when λ is larger than 0.01, and the performance of MIR-AC turns out to be stable when λ is larger than 0.1. Figure 3 demonstrates that MIR-AC and MHL-Click have the best performance with $r = 3$ and $r = 5$ respectively. We also trace the change of values of the objective function with the iteration of MIR-AC and MHL-Click, and the curve of convergence is shown in Fig. 4, in which we observe that both MIR-AC and MHL-Click can converge to a certain value after 5 steps of iteration.

4   Conclusion

In this paper, we propose an image re-ranking algorithm named 'Multi-modal Image Re-ranking with Autoencoders and Click Semantics. First, Autoencoders are utilized to extract a essential representation of raw images, and click data of users has been exploited by departing query results into two discriminative parts for each query. Then graph-based learning algorithm framework, Patch Alignment Framework, is adopted to construct two manifolds representing structures of images and query results to a query. Furthermore, these structures are integrated together effectively in our proposed algorithm for image re-ranking. Finally, Experiments have been carried out on a real world image dataset, and the experimental results demonstrate the effectiveness of our proposed algorithm.

Acknowledgement. This work is supported by the Natural Science Foundation of China (No. 61202145, No. 61300192, No. 61472110), the Program for New Century

Excellent Talents in University (No. NECT-12-0323), Zhejiang Provincial Natural Science Foundation of China for Distinguished Young Scholars (No. LR15F020002), the Natural Science Foundation of Fujian Province of China under Grants (No. 2014J01256), and the education and research Foundation of Fujian Province of China under Grants (No. JB14082, JB12252S).

References

1. Bay, H., Ess, A., Tuytelaars, T., Gool, L.V.: Speeded-up robust features (surf). Computer Vis. Image Underst. **110**, 346–359 (2008)
2. Belongie, S., Malik, J., Puzicha, J.: Shape matching and object recognition using shape contexts. IEEE Trans. Pattern Anal. Mach. Intell. **24**(4), 509–522 (2002)
3. Cai, X., Han, G., Xiao, S.: An image registration method based on similarity of edge information. In: Proceedings of the IEEE International Symposium on Industrial Electronics, pp. 1111–1115. IEEE Press, May 2012
4. Carterette, B., Jones, R.: Evaluating search engines by modeling the relationship between relevance and clicks. In: Advances in Neural Information Processing Systems, pp. 217–224. MIT Press (2009)
5. Chen, M., Weinberger, K.Q., Sha, F., Bengio, Y.: Marginalized denoising autoencoders for nonlinear representations. In: IEEE International Conference on Machine Learning, pp. 1476–1484. IEEE (2014)
6. Cheng, B., Liu, G., Wang, J., Huang, Z., Yan, S.: Multi-task low-rank affinity pursuit for image segmentation. In: IEEE International Conference on Computer Vision, pp. 2439–2446. IEEE (2011)
7. Dalal, N., Triggs, B.: Histograms of oriented gradients for human detection. In: Proceedings of the IEEE International Conference on Computer Vision and Pattern Recognition, pp. 886–893. IEEE Press (2005)
8. G. Smith, H.A.: Evaluating implicit judgments from image search interactions. In: WebSci 2009: Society On-line (2009)
9. Gao, Y., Wang, M., Tao, D., Ji, R., Dai, Q.: 3-d object retrieval and recognition with hypergraph analysis. IEEE Trans. Image Process. **21**, 4290–4303 (2012)
10. Hong, C., Yu, J., Li, J., Chen, X.: Multi-view hypergraph learning by patch alignment framework. Neurocomputing **118**(22), 79–86 (2013)
11. Hua, X.S., Yang, L., Wang, J., Wang, J., Ye, M., Wang, K., Rui, Y., Li, J.: Clickage: towards bridging semantic and intent gaps via mining click logs of search engines. In: ACM International Conference on Multimedia, pp. 243–252. ACM (2013)
12. Lowe, D.G.: Distinctive image features from scale-invariant key points. Int. J. Comput. Vis. **60**, 91–110 (2004)
13. Baghshah, M.S., Shouraki, S.B.: Metric learning for semi-supervised clustering using pairwise constraints and the geometrical structure of data. Intell. Data Anal. **13**, 887–899 (2009)
14. Srivastava, N., Hinton, G., Krizhevsky, A., Sutskever, I., Salakhutdinov, R.: Dropout: a simple way to prevent neural networks from overfitting. J. Mach. Learn. Res. **15**, 1929–1958 (2014)
15. Szummer, M., Jaakkola, T.: Partially labeled classification with markov random walks. In: Advances in Neural Information Processing Systems, vol. 14, pp. 945–952. MIT Press (2001)
16. Wang, M., Hua, X., Yuan, X., Song, Y., Dai, L.: Optimizing multi-graph learning: towards a unified video annotation scheme. In: International Conference on MultiMedia Modelling, pp. 862–871. ACM MM (2007)

17. Wang, M., Hua, X.S., Hong, R., Tang, J., Qi, G.J., Song, Y.: Unified video annotation via multigraph learning. IEEE Trans. Circuits Syst. Video Technol. **19**, 733–746 (2009)

18. Wang, M., Li, G., Lu, Z., Gao, Y., Chua, T.S.: When amazon meets google: product visualization by exploring multiple information sources. ACM Trans. Internet Technol. **12**(4), 12 (2013)

19. Wang, M., Li, H., Tao, D., Lu, K., Wu, X.: Multimodal graph-based reranking for web image search. IEEE Trans. Image Process. **21**(11), 4649–4661 (2012)

20. Wang, M., Yang, K., Hua, X.S., Zhang, H.J.: Towards a relevant and diverse search of social images. IEEE Trans. Multimedia **12**, 829–842 (2010)

21. Wu, M., Scholkopf, B.: Transductive classification via local learning regularization. In: International Conference on Artificial Intelligence and Statistics, pp. 628–635. Microtome (2007)

22. Yoshua, B.: Learning deep architectures for AI. Found. Trends Mach. Learn. **2**(1), 1–127 (2009)

23. Yu, J., Liu, D., Tao, D., Seah, H.S.: On combining multiple features for cartoon character retrieval and clip synthesis. IEEE Trans. Syst. Man Cybern. Part B **42**(5), 1413–1427 (2012)

24. Yu, J., Rui, Y., Chen, B.: Exploiting click constraints and multi-view features for image re-ranking. IEEE Trans. Multimedia **16**(1), 159–168 (2014)

25. Zhang, T., Tao, D., Li, X., Yang, J.: Patch alignment for dimensionality reduction. IEEE Trans. Knowl. Data Eng. **21**, 1299–1313 (2009)

26. Zhou, D., Bousquet, O., Lal, T.N., Weston, J., Scholkopf, B.: Learning with local and global consistency. In: Advances in Neural Information Processing Systems, vol. 16, pp. 321–328. MIT Press (2004)

27. Zhu, X., Ghahramani, Z., Lafferty, J.: Semi-supervised learning using Gaussian fields and harmonic functions. In: International Conference on Machine Learning, vol. 2, pp. 912–919. ACM, August 2003

Sketch-Based Image Retrieval with a Novel BoVW Representation

Cheng Jin[1], Chenjie Li[1], Zheming Wang[1], Yuejie Zhang[1(✉)], and Tao Zhang[2]

[1] School of Computer Science, Shanghai Key Laboratory of Intelligent Information Processing,
Fudan University, Shanghai 200433, China
{jc,14210240046,13210240035,yjzhang}@fudan.edu.cn
[2] School of Information Management and Engineering,
Shanghai University of Finance and Economics, Shanghai 200433, China
taozhang@mail.shufe.edu.cn

Abstract. A novel Bag-of-Visual-Word (BoVW) based approach is developed in this paper to facilitate more effective Sketch-based Image Retrieval (SBIR). We focus on constructing the visual vocabulary based on the BoVW representation with both the spatial distribution and inter-relationship of descriptors. To optimize the sketch-image matching, the weighting quantization is created by integrating both the neighbor and spatial feature information to quantify features as visual words. We emphasize on an inverted indexing by converting an image to a trigram representation with visual words and their spatial information. Our experiments have obtained very positive results.

Keywords: Sketch-based image retrieval · Bag-of-visual-word representation · Visual vocabulary generation · Weighting quantization for matching · Inverted indexing structure construction

1 Introduction

With the massive explosion of image data on the Web, methods to seamlessly handle the complex structures of image data to achieve more effective Sketch-based Image Retrieval (SBIR) have become an important research focus, especially due to the prevalence of devices with touchable screens [1, 2]. While existing descriptors which have been used in Content-based Image Retrieval (CBIR) like Color Histogram, Edge Histogram Descriptor (EHD), Histograms of Oriented Gradients (HOG), Scale-invariant Feature Transform (SIFT), and others can also be used in SBIR, such descriptors from each cell of an image are concatenated to form a global image feature, which is effective for SBIR but involves a computationally expensive scaling optimization with the dataset size [3, 4]. Therefore, the Bag-of-Visual-Word (BoVW) based SBIR approach is becoming more popular than traditional content-based ones to alleviate the above problems, which have been proved to be successful in SBIR in more recent works.

The general BoVW approach for SBIR attempts to describe an image as a set of visual words and then create a codebook of visual words [5, 6]. Cao *et al.* [7] proposed the contour-based EdgelIndex method, in which the matching process seemed similar

Q. Tian et al. (Eds.): MMM 2016, Part I, LNCS 9516, pp. 621–631, 2016.
DOI: 10.1007/978-3-319-27671-7_52

to the BoVW model. Eitz *et al.* [8] developed new descriptors based on the bag-of-features approach for large-scale SBIR systems, which explored the standard HOG within a BoVW framework. Hu *et al.* [9] incorporated GF-HOG into a BoVW retrieval framework, which was harnessed both for robust SBIR and localizing sketched objects within an image. Unfortunately, all these existing approaches have not yet provided good solutions for the following key issues: (1) intermediate descriptor generation to bridge the representational gap; (2) pairwise matching optimization to identify better correspondences between sketches and images; and (3) proper index structure to speed up the retrieval process. To address the first issue, it's important to develop a robust algorithm that can achieve more accurate visual feature expression. To address the second, it's critical to develop a new matching algorithm with high accuracy but low cost. To address the third, it's interesting to leverage large-scale images to establish a more efficient indexing mechanism.

Based on these observations, a novel BoVW-based approach is developed in this paper to facilitate more effective SBIR. Our scheme significantly differs from other earlier work in: (a) The visual vocabulary is constructed based on visual words from discriminative features local to points within regions of images, in which the BoVW representation can capture both the spatial distribution and inter-relationship of descriptors. (b) The weighting quantization is created to achieve more precise characterization of the correlations between sketches and images, in which both the neighbor and spatial feature information are integrated to quantify features as visual words. (c) The inverted indexing strategy is designed to face scalability problems when scaling up to large-scale images, in which a specific indexing structure is established by converting an image to a trigram representation with visual words and their spatial information. (d) A new real-time SBIR framework is built by fusing the above image representation, matching and indexing patterns, which not only enables users to present on the query panel whatever they imagine in their mind, but also returns the most similar images to the picture in users' mind. Our experiments on a large number of public image data have obtained very positive results.

2   Visual Vocabulary Generation

In a typical BoVW framework, interest points are usually first detected and represented by feature descriptors, and then the unsupervised grouping is performed over all the descriptors to generate a set of k clusters. Each cluster is equivalent to a *"visual word"*, and the general visual vocabulary is composed of all these visual words. However, such a vocabulary may ignore the important spatial information involved in visual features, which is a quite crucial factor for the SBIR performance. Thus it's necessary to discover which spatial information may possibly co-occur between sketches and images and then build the visual vocabulary with such information.

For the initial query sketch, it can be represented as multiple strokes, and each stroke consists of many edge pixels. For each original natural image, it is first converted into a shape image by the edge detection. Since the selection of interest point and feature representation has notable impact on the visual word generation and the sketch-image similarity measurement, a tradeoff on complexity and descriptiveness is to sample randomly from all the possible dense multi-scale features in the sketch/image. Unlike most existing work, our interest point selection is local to detected Canny edges (for images) and pixels comprising strokes (for sketches). A set of HOG descriptors is considered as the optimal visual expression and computed to represent interest points.

Subsequently, the extracted features from the same region/ position of images are clustered to generate the visual vocabulary, and each region/position will generate a visual vocabulary with spatial properties of local feature descriptors of images. Given an image set with N images, $IMG = \{img_1, ..., img_i, ..., img_N\}$, all these images are first resized to the same size $m \times m$, in which each image img_i is represented by an feature collection, $VF_i = \{f_i(1, 1), ..., f_i(j, k), ..., f_i(m, m)\}$, where $f_i(j, k)$ represents the feature expression for the interest point $IP_i(j, k)$ in the region/position of (j, k). Thus we can get the following feature matrix VF:

$$VF = \begin{pmatrix} VF_1 \\ VF_2 \\ \cdots \\ VF_i \\ \cdots \\ VF_N \end{pmatrix} = \begin{pmatrix} f_1(1,1) & f_1(1,2) & \cdots & f_1(j,k) & \cdots & f_1(m,m) \\ f_2(1,1) & f_2(1,2) & \cdots & f_2(j,k) & \cdots & f_2(m,m) \\ \cdots & \cdots & \cdots & \cdots & \cdots & \cdots \\ f_i(1,1) & f_i(1,2) & \cdots & f_i(j,k) & \cdots & f_i(m,m) \\ \cdots & \cdots & \cdots & \cdots & \cdots & \cdots \\ f_N(1,1) & f_N(1,2) & \cdots & f_N(j,k) & \cdots & f_N(m,m) \end{pmatrix} \quad (1)$$

where each row in VF represents an feature set for each img_i. To integrate the spatial information into the visual vocabulary, the features from the same region/position of different images are grouped into the same feature vector, $VF(j, k) = \{f_1(j, k), ..., f_i(j, k), ..., f_N(j, k)\}$ (i.e., each column in VF) and thus we can get $m \times m$ vectors for $m \times m$ regions/positions. Each vector will be further processed through k-means clustering independently, and each cluster can be regarded as a visual word with the spatial attribute. This aims at partitioning each feature vector $VF(j, k)$ into $r (\leq N)$ sets, $VFS(j, k) = \{s_1(j, k), ..., s_l(j, k), ..., s_r(j, k)\}$, so as to minimize the within-cluster sum of squares, as shown in Formula (2).

$$\arg\min_{IS(j,k)} \sum_{l=1}^{r} \sum_{xq \in sl(j,k)} \|xq - \mu l\|^2 \quad (2)$$

where μ_l denotes the mean of interest points in $s_l(j, k)$. All the visual words related to each (j, k) form the visual vocabulary $VV(j, k)$ related to this region/position, and therefore $m \times m$ independent visual vocabularies with the spatial attribute can be generated for the whole image set, as shown in Formula (3).

$$VV = \begin{pmatrix} VV(1,1) \\ VV(1,2) \\ \cdots \\ VV(j,k) \\ \cdots \\ VV(m,m) \end{pmatrix} = \begin{pmatrix} vw_{11} & vw_{12} & \cdots & vw_{1l} & \cdots & vw_{1r} \\ vw_{21} & vw_{22} & \cdots & vw_{2l} & \cdots & vw_{2r} \\ \cdots & \cdots & \cdots & \cdots & \cdots & \cdots \\ vw_{i1} & vw_{i2} & \cdots & vw_{il} & \cdots & vw_{ir} \\ \cdots & \cdots & \cdots & \cdots & \cdots & \cdots \\ vw_{m\times m,1} & vw_{m\times m,2} & \cdots & vw_{m\times m,l} & \cdots & vw_{m\times m,r} \end{pmatrix} \quad (3)$$

With such a generation mechanism, multiple vocabularies with the valuable spatial attributes are constructed to better interpret the visual appearances in images, i.e., visual characterization enrichment, and the side-effect of the spatial information loss for sketch-image association can be greatly mitigated.

3   Sketch-Image Matching via Quantization

In our BoVW-based framework, we consider the sketch-image matching as a weighting quantization problem for searching the optimal sketch-image linkages. Our weighting scheme is designed to quantize features into visual words by fusing the surrounding neighborhood information, in which the noise interference can be minimized as possible. Considering an interest point $IP_i(j, k)$ in the image img_i, the corresponding visual vocabulary $VV(j, k)$ consists of N visual words, $VV(j, k) = \{vw_1, \ldots, vw_l, \ldots, vw_N\}$, the distance between $IP_i(j, k)$ and each vw_l can be computed as:

$$Dist'\left(IP_i(j,k),vwl\right) = \sum_{(j',k')\in\varphi(IPi(j,k))} \omega(IP_i(j,k),IP_i(j',k')) * Dist(j',k'),vwl \qquad (4)$$

where $\phi(IP_i(j, k))$ represents a square window of size $r \times r$ centered at $IP_i(j, k)$; $Dist(IP_i(j', k'), vw_l)$ denotes the Euclidean distance; and $\omega(IP_i(j, k), IP_i(j', k'))$ is a weighting function for the distance measurement between $IP_i(j, k)$ and each vw_l, which is defined as:

$$\omega\left(IP_i(j,k),IP_i(j',k')\right)$$
$$= \frac{1}{\rho(j,k)} \exp\left(-\frac{(IP_i(j,k) - IP_i(j',k'))}{\sigma_S^2}\right) \exp\left(-\frac{(\theta\left(IP_i(j,k)\right) - \theta\left(IP_i(j',k')\right))}{\sigma_{GO}^2}\right) \qquad (5)$$

where $\rho(j, k)$ is a normalization coefficient to guarantee $\sum_{IPi(j,k)\in\phi(IPi(j,k))}\omega(IP_i(j, k), IP_i(j', k')) = 1$; σ_S and σ_{GO} are two preset coefficients for adjusting the difference of spatial attribute and orientation between the central point $IP_i(j, k)$ and the surrounding point $IP_i(j', k')$; and $\theta(IP_i(j, k))$ and $\theta(IP_i(j', k'))$ represent the orientations for $IP_i(j, k)$ and $IP_i(j', k')$. It's worth noting that with the spatial position and gradient orientation changing, the weight for the related interest point can be adjusted dynamically by fusing such significant information. Thus our mathematical model for the weighting quantization can be formalized as Formula (6), in which the local distance between each interest point and its final associated visual word can reach the minimum value.

$$\arg\min_{vwl\in VV(j,k)} Dist'\left(IP_i(j,k),vwl\right)$$
$$= \arg\min_{vwl\in VV(j,k)} \sum_{(j',k')\in\varphi(IPi(j,k))} \omega\left(IP_i(j,k),IP_i(j',k')\right) * Dist\left(IP_i(j',k'),vwl\right) \qquad (6)$$

For the sketch-image matching, it can also be regarded as a restricted similarity or correlation measurement. For each sketch, we want to know which image has the most possible association with it. Therefore, our BoVW-based matching cost evaluation concerns more on the integral consistency for each sketch-image pair. It's better for each sketch-image pair to have the same visual words in the corresponding regions/positions as many as possible.

Firstly, it's necessary for each interest-point pair in the same region/position to appear more locally similar, which can be evaluated based on the common visual words for the specific region/position, as shown in Formula (7).

$$Sim_{IP}\left(IP_{skt}\left(j,k\right),IP_{img}\left(j,k\right)\right) = \left\{ \begin{array}{ll} 1, & vw_{skt}\left(j,k\right) = vw_{img}\left(j,k\right) \\ 0, & vw_{skt}\left(j,k\right) \neq vw_{img}\left(j,k\right) \end{array} \right. \tag{7}$$

where $IP_{skt}(j, k)$ and $IP_{img}(j, k)$ are two interest points for the region/position (j, k) in the sketch skt and image img; and $vw_{skt}(j, k)$ and $vw_{img}(j, k)$ are the visual words after quantization. If $vw_{skt}(j, k)$ and $vw_{img}(j, k)$ are same, $IP_{skt}(j, k)$ and $IP_{img}(j, k)$ are similar.

Secondly, the global similarity is utilized to measure the correlation for each sketch-image pair. A remarkable phenomenon can be observed that when the image resolution is low, many edge interest points that reflect the visual intension are ignored. Such points represent the outer boundaries of images, which play more significant roles than those inner ones. Thus a probability weight is appended to the global similarity, which indicates the likeliness to be an outer boundary interest-point pair, as shown in Formula (8). The higher weight for a point pair implies its higher importance to the global similarity for the sketch-image pair.

$$Sim_{S-I}\left(skt, img\right) = \sum_{\substack{IP_{skt}(j,k) \in skt \\ IP_{img}(j,k) \in img}} P\left(j, k\right) Sim_{IP}\left(IP_{skt}\left(j, k\right), IP_{img}\left(j, k\right)\right) \tag{8}$$

where $P(j, k)$ is the normalized weight to indicate the importance of interest-point pair. The higher value the $Sim_{S-I}(skt, img)$ has, the more correlative the sketch and image are.

Thirdly, a refinement mechanism is proposed to further optimize the sketch-image pairs with less similar interest-point pairs to be more correlative. When two images have the equal similarity to a sketch, the one with less interest points is better matched. Thus the global similarity measurement can be refined as:

$$Sim_{S-I}\left(skt, img\right) = \frac{\sum_{\substack{IP_{skt}(j,k) \in skt \\ IP_{img}(j,k) \in img}} P\left(j, k\right) Sim_{IP}\left(IP_{skt}\left(j, k\right), IP_{img}\left(j, k\right)\right)}{\left|skt_{IP}\right| + \left|img_{IP}\right|} \tag{9}$$

where $\left|skt_{IP}\right|$ and $\left|img_{IP}\right|$ denote the numbers of reasonable edge interest points in sketch and image respectively.

It is important to note that if an image has the larger proportion of similar interest points to the sketch, it will have the higher similarity to the sketch. As the sum of the "normalized" local similarity values for interest-point pairs in a sketch-image pair, the refined global similarity can be well representative of the correlation between sketch and image.

4   Indexing Structure Construction

The general BoVW-based model does not concern the region/position information of edge interest points, but simply utilizing the frequency of visual word. The common way to establish the indexing structure is that each visual word is taken as an index to indicate the images containing it, which may guarantee the acceptable efficiency at the sacrifice of retrieval performance. However, in our BoVW-based framework, the relative spatial layout information of visual words for each sketch-image pair is well integrated into the visual representation and matching of sketches and images. Thus an effective inverted indexing structure is still necessary to prune the obviously irrelevant images and optimize the realtime retrieval response speed.

In this work, we build a trigram-based inverted indexing structure for images with the fusion of spatial information. Given an edge interest point $IP_i(x, y)$ in the image img_i, it can be viewed as a particular index trigram $(x, y, vw_l(x, y))$, in which the former two items denote the spatial position of $IP_i(x, y)$ in img_i, and the last represents the visual word associated with $IP_i(x, y)$. Such index trigrams are considered as the entries to construct the inverted index list, and the images with the interest points corresponding to the trigrams are merged into these entries. Suppose the size of query sketch is unified as 128×128, the size of visual vocabulary for each region/position is set as 200, thus the trigram-based inverted index list will contain $128 \times 128 \times 200 = 3{,}276{,}800$ entries. If the scale of image database is not very large, a certain number of entries can be combined to form a compound entry, e.g., merging 100 trigrams into an entry. The trigram-based indexing structure and its instantiation are shown in Fig. 1.

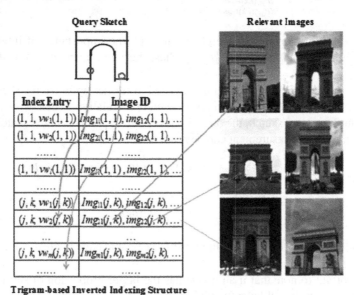

Query Sketch   **Relevant Images**

Index Entry	Image ID
$(1, 1, vw_1(1, 1))$	$Img_{11}(1, 1), img_{12}(1, 1), \ldots$
$(1, 1, vw_2(1, 1))$	$Img_{21}(1, 1), img_{22}(1, 1), \ldots$
$\ldots\ldots$	$\ldots\ldots$
$(1, 1, vw_l(1, 1))$	$Img_{l1}(1, 1), img_{l2}(1, 1), \ldots$
$\ldots\ldots$	
$(j, k, vw_1(j, k))$	$Img_{11}(j, k), img_{12}(j, k), \ldots$
$(j, k, vw_2(j, k))$	$Img_{21}(j, k), img_{22}(j, k), \ldots$
$\ldots$	$\ldots$
$(j, k, vw_m(j, k))$	$Img_{m1}(j, k), img_{m2}(j, k), \ldots$
$\ldots\ldots$	$\ldots\ldots$

Trigram-based Inverted Indexing Structure

Fig. 1. An instantiation of trigram-based indexing structure.

Based on the above indexing structure construction strategy, for each image in the whole image database, all its valid visual words are assigned to the associated trigram entries in the inverted index list, and then all the images can be well organized to facilitate more effective realtime search. Especially, with the help of such trigram-based indexing, the subsequent ranking mechanism can be efficiently implemented to get both fast and precise sketch-image correlation matching. We built the SBIR system with a common server to support a realtime search on 200,000 images. The response time for each query sketch is 0.5 ~ 1 s.

5   Experiment and Analysis

Dataset and Evaluation Metrics.   Our dataset consists of 200,000 images from *Flickr*. We have recruited 10 non-expert sketchers to draw 600 sketches by their imagination or by roughly tracing the objects in color images. The official criteria of Precision (P), Rank (R) and Median Average Precision (MAP) are introduced to evaluate our algorithm.

Experiment on Visual Feature Descriptors.   The visual feature descriptors define the visual attributes of sketches and images, which is very beneficial to the performance improvement for local matching and global retrieval. To exhibit the significant effect of feature representation, different popular feature descriptors are introduced to make a comparison. The related experimental results are shown in Fig. 2.

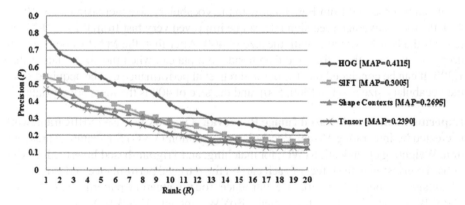

Fig. 2.   The experimental results on different visual feature descriptors.

It can be seen from Fig. 2 that the best MAP value of 0.4115 can be obtained by utilizing the HOG descriptor in our SBIR system, which is much higher than that based on the SIFT descriptor. The MAP values on the descriptors of shape contexts and Tensor appear even worse than that of SIFT, in which Tensor exhibits the worst effect on SBIR. Due to the lack of texture and color information in a sketch, SIFT and shape contexts cannot yield very good results. We can observe that when integrating Tensor into our BoVW-based framework, it performs poorly, which may be due to the relative low-dimensionality of feature representation. However, the HOG descriptor counts the occurrences of gradient orientation in localized portions of a sketch/image, thus it is quite suitable for representing the visual features of sketch and image, especially the sketch that is only represented by the spatial information of strokes.

Experiment on Visual Vocabulary. As the visual vocabulary size increases, the feature representation based on vocabulary becomes more discriminative, but less generalizable and constrained to noises. Using a larger vocabulary will also gain the higher computational cost for interest point clustering, feature representation, and matching. Thus we set the size in [100, 2,000] to investigate the influence of different size settings. The related statistical results are given in Fig. 3.

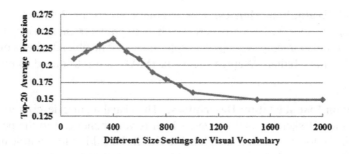

Fig. 3. The statistical results on different sizes for visual vocabulary.

It can be observed from Fig. 3 that with the vocabulary size increasing from 100 to 400, the top-20 average precision rate can be improved very fast from 0.21 to 0.24 and reaches the best MAP value with the size of 400. After that, the MAP value begins to drop with the growth of size, and then trends to a plateau when the size is larger than 1,200. It can be concluded that the optimal retrieval performance can be acquired when the vocabulary size is set in [300, 500], and the size of 400 is the most appropriate.

Experiment on Sketch-Based Image Retrieval. Our BoVW-based SBIR framework is created by integrating Visual Vocabulary Generation (VVG) for feature representation, Weighting Quantization (WQ) for matching, and Trigram-based Inverted Indexing (TII). To investigate the effect of each part on the retrieval performance and explore the advantages of our framework, we introduce four evaluation patterns: (1) *Baseline (TBoVW)* (that is, the traditional BoVW model *(TBoVW)*); (2) *Baseline (TBoVW) + VVG*; (3) *Baseline (TBoVW) + WQ*; and (4) *Baseline (TBoVW) + VVG + WQ + TII* (our framework). The experimental results are shown in Fig. 4.

It can be viewed from Fig. 4 that the best run is *Baseline (TBoVW) + VVG + WQ + TII* under our framework, and we can obtain the best MAP value of 0.4115. In comparison with the baseline model based on the traditional BoVW framework, the whole retrieval performance could be promoted to a great degree by adding *VVG*, *WQ* and *TII* under our framework. We can even achieve 166 % of the *Baseline (TBoVW)* effectiveness, 128 % of the *Baseline (TBoVW) + VVG* effectiveness, and 127 % of the *Baseline (TBoVW) + WQ*

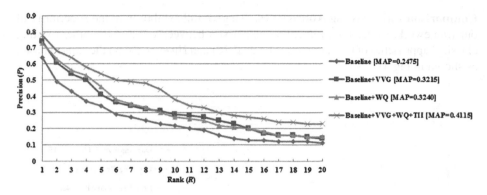

Fig. 4. The experimental results on our BoVW-based SBIR framework.

effectiveness. From the results for the former three runs, we can further verify that our *VVG* and *WQ* manners are the optimal visual vocabulary generation and feature quantization mechanisms for facilitating the whole retrieval performance. When *VVG* or *WQ* is separately integrated into the traditional BoVW framework, the retrieval performance increases and becomes pretty good. Although compared to the results of the last run with our framework, the results from two runs of *Baseline* (*TBoVW*) + *VVG/WQ* appear less performant, we can still observe the promising performance exhibition, and their MAP values can still reach a relatively high value. Through the comparison among four runs, it can be found that our framework is obviously superior to the traditional one and more suitable for solving such a sketch-image association optimization problem.

With all the experimental results above, we could gain more explicit insights about our BoVW-based SBIR framework. It can be noticed that the in-depth analysis and mining for visual feature representation and sketch-image similarity measure is available and presents more impactful ability for discovering the most important correlative feature information from sketch-image pairs. Furthermore, the high retrieval efficiency can be achieved through constructing the trigram-based inverted indexing structure, which can not only reduce the search space, but also increase the realtime response speed significantly. An instantiation of some query sketches and their relevant images is shown in Fig. 5.

Fig. 5. An instantiation of some query sketches and their relevant images.

Comparison with Existing Approaches. To give full exhibition to the superiority of our framework, we have performed a comparison between our and the other existing classical approaches of Cao *et al.* [7], Etiz *et al.* [8] and Hu *et al.* [9] on the same dataset, as shown in Fig. 6.

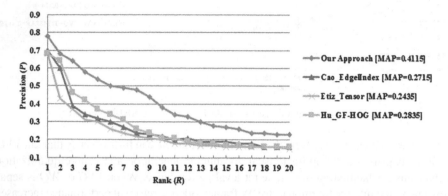

Fig. 6. The comparison results between our and the other approaches.

It can be found from Fig. 6 that for the SBIR on our dataset by Cao/Etiz/Hu *et al.*'s approaches, we can obtain the MAP values of 0.2715, 0.2435 and 0.2835 respectively. Compared the results of Cao/Etiz/Hu *et al.*'s approaches and our baseline model with *VVG* or *WQ*, we can find both the MAP values of ours are obviously higher than those of Cao/Etiz/Hu *et al.*'s, and the best MAP value of 0.3240 appears in the results of our model. This implies that both our visual vocabulary generation and weighting quantization mechanisms are feasible for facilitating more effective sketch-image correlation judgement. By fusing *VVG*, *WQ* and *TII*, our BoVW-based SBIR framework can present the significant performance superiority, and the best MAP value of 0.4115 differs greatly from those of Cao/Etiz/Hu *et al.*'s. This does not only verify the prominent roles of our *VVG*, *WQ* and *TII* once again, but also confirms that our approach is exactly a better way for determining sketch-image associations and supporting SBIR.

Analysis and Discussion. Through the analysis for the error or failure in the retrieval results, it can be found that our BoVW-based SBIR quality is highly related to the following aspects. (1) Our approach relates very closely to the BoVW representation for sketches and images. It's easier for sketch and image preprocessing to produce some errors or missing detections for valid edge interest points or visual words, which will seriously affect the whole retrieval performance. (2) The other edge interest points or visual words that co-occur in sketch-image pairs maybe helpful for facilitating the sketch's relevant image finding and judgement. Such co-occurrence information can be utilized to further improve the retrieval relevance measurement. (3) Instead of only concerning the visual appearances for images, it's helpful to exploit the appropriate multimodal information involved in both image and the corresponding caption/annotation text for optimizing the correlation measurement between sketches and images.

(4) Some particular sketches or images present an extreme vision of vague contours and cluttered lines. With very limited useful information and too many noises, it's hard for such sketches or images to successfully implement precise SBIR. This may be the most stubborn problem in our approach.

6   Conclusions and Future Work

A new BoVW-based framework is introduced in this paper to support more precise SBIR for large-scale images. The visual vocabulary is established for characterizing the visual information in sketches and images. The weighting quantization is utilized to make sketch-image matching more accurately. The trigram-based inverted indexing structure is constructed to acquire more ideal efficiency. Our future work will focus on making our system available online, so that more users can benefit from our research.

Acknowledgements. This work is supported by National Natural Science Fund of China (No. 61572140), National Science & Technology Pillar Program of China (No. 2012BAH59F04), National Natural Science Fund of China (No. 61170095; 71171126), Shanghai Philosophy Social Sciences Planning Project (No. 2014BYY009), and Zhuoxue Program of Fudan University. Yuejie Zhang is the corresponding author.

References

1. Datta, R., Joshi, D., Li, J., Wang, J.Z.: Image retrieval: ideas, influences, and trends of the new age. ACM Comput. Surv. **40**(2), 5:1–5:60 (2008)
2. Cao, Y., Wang, H., Wang C.H., Li, Z.W., Zhang, L.Q., Zhang, L.: MindFinder: interactive sketch-based image search on millions of images. In: Proceedings of MM 2010, pp. 1605–1608 (2010)
3. Furuya, T., Ohbuchi, R.: Visual saliency weighting and cross-domain manifold ranking for sketch-based image retrieval. In: Gurrin, C., Hopfgartner, F., Hurst, W., Johansen, H., Lee, H., O'Connor, N. (eds.) MMM 2014, Part I. LNCS, vol. 8325, pp. 37–49. Springer, Heidelberg (2014)
4. Sun, X.H., Wang, C.H., Xu, C., Zhang, L.: Indexing billions of images for sketch-based retrieval. In: Proceedings of MM 2013, pp. 233–242 (2013)
5. Lazebnik, S., Schmid, C., Ponce, J.: Beyond bags of features: spatial pyramid matching for recognizing natural scene categories. In: Proceedings of CVPR 2006, vol. 2, pp. 2169–2178 (2006)
6. Springmann, M., Al Kabary, I., Schuldt, H.: Image retrieval at memory's edge: known image search based on user-drawn sketches. In Proceedings of CIKM 2010, pp. 1465–1468 (2010)
7. Cao, Y., Wang, C.H., Zhang, L.Q., Zhang, L.: Edgel index for large-scale sketch-based image search. In: Proceedings of CVPR 2011, pp. 761–768 (2011)
8. Eitz, M., Hildebrand, K., Boubekeur, T., Alexa, M.: Sketch-based image retrieval: benchmark and bag-of-features descriptors. IEEE Trans. Visual Comput. Graphics **17**(11), 1624–1636 (2011)
9. Hu, R., Collomossea, J.: A performance evaluation of gradient field HOG descriptor for sketch based image retrieval. Comput. Vis. Image Underst. **117**(7), 790–806 (2013)

Symmetry-Aware Human Shape Correspondence Using Skeleton

Zongyi Xu and Qianni Zhang[✉]

Queen Mary University of London, Mile End, London E1 4NS, UK
{zongyi.xu,qianni.zhang}@qmul.ac.uk

Abstract. In this paper, we propose a symmetry-aware human shape correspondence extraction method. We address the symmetric flip problem which exists in establishing correspondences for intrinsically symmetric models and improve the accuracy of the final corresponding pairs. To achieve this goal, we extended the state-of-the-art approach by using skeleton information to further remove symmetric flipped shape correspondences. Traditional approaches that only rely on surface geometry information can hardly discriminate surface points which are symmetric. With the appearance of inexpensive RGB-D camera, such as Kinect, skeleton information can be easily obtained along with mesh. Therefore, after the initial correspondences are achieved, we extend the candidate sets for each point on the template, followed by making use of skeleton to remove the symmetric flipped false candidates. In the remaining candidates, final correspondences are achieved by choosing those with minimum geodesic distortion from base vertex set, which is formed by sampling on the mesh. Experiments demonstrate that the proposed method can effectively remove all the symmetric flipped candidates. Moreover, the final correspondence pair is more accurate than those of the state of the arts.

Keywords: Kinect · Shape correspondence · Symmetric flip problem

1 Introduction

Shape correspondence is a fundamental problem in many research topics such as 3D mesh retrieval, shape registration and mesh deformation. 3D shape correspondence is a mapping from one point set on the source mesh to another on the target mesh. There exist three kinds of mapping: one-to-one, one-to-many and many-to-one. In this paper, we aim to address the problem of establishing the accurate one-to-one correspondence between intrinsic-symmetrically isometric human models.

The target of shape correspondence is to find the point pairs that are similar or semantically equivalent. Isometric shapes appear in various contexts such as different poses of an articulated human model or two shapes presenting different but semantically similar objects [16]. It is highly demanded to find isometric

© Springer International Publishing Switzerland 2016
Q. Tian et al. (Eds.): MMM 2016, Part I, LNCS 9516, pp. 632–641, 2016.
DOI: 10.1007/978-3-319-27671-7_53

shape correspondence since most real world deformations are isometric. Moreover, shape correspondences between isometric shapes have practical values. For instance, the deformation based on isometric template will be much more efficient benefiting from their similar shapes. If two shapes are totally isometric, the geodesic distance between two points on one shape is the same as the geodesic distance between their correspondences on the other shape [16].

Embedding-based methods are popular techniques for 3D shape correspondences problem. In these methods, original mesh is embedded into a new domain where isometric deviation can be measured and optimized. Euclidean embedding can be achieved by using various techniques such as classic MDS(Multi-Dimensional Scaling) [14], least-square MDS [6], heat kernel embedding [10] and spectral embedding [7]. Besides embedding methods, other approaches [15,16] minimize the isometric distortion directly in the 3D Euclidean space. However, most existing algorithms tend to be confused by the intrinsically symmetric features and suffer from symmetric flip problems. They can hardly discriminate symmetric points on the surface even if the mesh to be matched is not perfectly symmetric. Therefore, it is common that the correspondence of the point on the right hand of the source mesh is established on the left hand of the target mesh, as shown in Fig. 1.

We proposed a method to find correspondences for human isometric shape model which is able to solve the symmetric flipping problem. Our idea is to combine skeleton information to distinguish intrinsic symmetry. Given two meshes with their skeletons, we first utilize local features to find one-to-many correspondences between two meshes. The candidate set for each feature point presents symmetric property on the mesh. A skeleton segment associated with surface points is capable of discriminating symmetry. The final correspondence is located and refined by minimize the isometric distortion with respect to based vertex set.

In summary, our contributions are: (a) we integrate skeleton information to robustly address symmetric flip problem which still exist in state-of-the-art techniques; (b) we take advantage of the base vertex set to refine the final one-to-one correspondence and achieve better accuracy.

2 Related Work

Shape correspondence is a long- and well-studied problem. In areas such as shape matching, 3D shape retrieval, and mesh registration, 3D reconstruction, many recent efforts are made on finding shape corresponding points on two meshes.

SCAPE model [1] use markers to manually locate correspondences between two meshes. Besides manual assignment, plenty of works develop local surface descriptors to automatically establish correspondences. Some works extend local descriptors in 2D images for triangulated meshes, such as MeshHOG [18] or 3D shape context [9]. The embedding-based method is a more reliable approach when it comes to isometric deformation. Multidimensional Scaling(MDS) [14] approximate geodesic distance with Euclidean distance in embedding space.

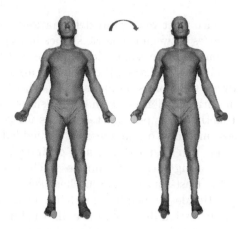

Fig. 1. Flipping correspondences. Each correspondence pair is labelled with the same colour (Color figure online).

Dey et al. [5] uses the Global Point Signature(GPS) [13] for spectral embedding of meshes and thereby find the mesh extremities. Sahillioglu et al. [15] also transfers vertices into spectral domain and optimize the result using expectation-maximization algorithm. However, these above methods sometimes provide false correspondence due to the presence of model symmetries. Ovsjanikov et al. [11] firstly identify the intrinsic symmetry of object in a quotient space, and then factor it out. Zhang et al. [19] differentiate the intrinsic symmetric points by calculating a signed angle field from the gradient fields of the harmonic field which is derived from four points on the hands and feet. For scan data produced by RGB-D cameras, e.g. Kinect, many imperfections make it harder to find the correspondences. Holes and non-smooth mesh result in difficulties for calculating geodesic distance. Noises and missing data have negative influence on the matrix structure which is the basis of embedding-based methods. Jiang et al. [8] and Zheng [20] detect the intrinsic symmetry of point clouds using skeleton but the skeleton they use is produced according to the surface or point clouds.

Sahillioglu et al. [16] propose a coarse-to-fine scheme to track symmetric flips. Although this method is more accurate than the previous ones based on embedding approach, it still has the symmetric flipping false pairs even in the final level. We will compare our method with them in both accuracy and addressing problem of symmetric flips. The strength of our approach is that it takes into account the skeleton information.

3   Skeleton-Based Symmetry-Aware Approach

The intrinsic symmetry leads to the symmetric flipped correspondence between two meshes. Neither embedding-based methods like MDS, GMDS nor local descriptors can differentiate them effectively. Previous works which are solely replying on surface-related information, i.e. geodesic distance, face normal, are unable to solve symmetric problems completely. However, with the help of a set of skeleton information where different skeleton segments have different labels and surface point and skeleton segment are associated, it is possible to address the symmetric flipping problem with skeleton. Moreover, along with the appearance of Kinect camera, we are able to obtain skeleton of mesh with ease. To perform the skeleton attachment process, we use the algorithm based on the work by [2], in which the input is the joints positions tracked from Kinect. The output is the skeleton attached the human model. In the following, we firstly discuss how to obtain candidate set for source point, followed by our method to address symmetric flip problem as well as to refine the final correspondence which is more accurate in terms of both visual effect and semantics.

3.1   Correspondence Candidate Set

As mentioned before, in order to make sure that the candidate set includes the correct correspondence as much as possible, we first compute the one-to-many correspondences using Heat Kernel Signature(HKS) [17], we select the top N similar points to construct the candidate set which is shown in Fig. 2(a). To compute the heat of point i at time t_i, we firstly perform the Laplace-Beltrami operator L on the mesh. Let Λ be the diagonal matrix of the eigenvalues of L, and Φ be the matrix with the corresponding eigenvectors, the heat kernel of the mesh is computed as Eq. 1:

$$K_t = \Phi exp(-t\Lambda)\Phi^T \tag{1}$$

Each entry in $k_t(i,j)$ represents the heat diffusion between point i and j. The diagonal elements of this matrix is composed of HKS. Thus, HKS feature is a vector whose entry $k_{t_j}(p_i, p_i)$ is the heat at point i at time of t_j:

$$\{k_{t1}(p_i, p_i), k_{t2}(p_i, p_i), \ldots, k_{tn}(p_i, p_i)\} \tag{2}$$

When the dissimilarity of HKS between the template point and target point in Eq. 3 is less than a threshold t, the target point is selected as candidate for the template point.

$$\Delta s = \|HKS(p_t) - HKS(p_s)\|, \tag{3}$$

where $HKS(p)$ is the heat kernel signature at point p, p_t and p_s are the points on the template and target respectively. Here, we apply the scale-invariant HKS(si-HKS) [3] to get feature for meshes.

After the initial correspondence is achieved by si-HKS, an expanded set of candidate points are obtained as shown in Fig. 2(a). As it can be observed, the expanded candidates for the point on the right foot of the source model distribute on both feet of the target model, presenting symmetric property.

3.2   Skeleton-Based Symmetry-Aware Shape Correspondence

To locate the single correspondence for template point, the next step is to remove those symmetric flipping points. Skeleton is an important clue for filtering flipped

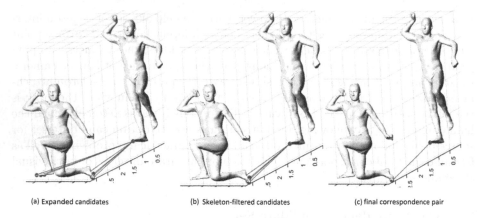

(a) Expanded candidates   (b) Skeleton-filtered candidates   (c) final correspondence pair

Fig. 2. Overflow of proposed method: (a) the expanded candidate set for one point on the template; (b) the after-filtered candidate set using skeleton filtering; (c) the final one candidate point on the source based on base vertex set

correspondences. As shown in Fig 3, skeleton divides mesh into 17 parts and each mesh part attached a segment has a unique label and the right extremity and its left counterpart have different labels. Therefore, our method is able to discriminate the right points and their counterparts on the left, addressing symmetric flip problems. When the template point and candidate points are on the same skeleton segment, they are kept; otherwise, the candidates are removed. The filtered candidate set for template point is shown in Fig. 2(b).

3.3 One-to-One Correspondence

After the symmetric flip problem is solved, the remaining candidates need to be further filtered to find the one-to-one correspondence pair. Therefore, the next step in our method uses the sum of relative distances from candidates to the base vertex set to filter invalid candidates.

The base vertex set [15] is selected based on Gaussian curvatures. This process is illustrated in Fig. 4. Initially, at each vertex of the original mesh, we compute the Gaussian curvatures using a simple way proposed in [12] with Eq. 4.

$$gc(p) = 3(2\pi - \sum \alpha_i)/\sum A(f_i), \tag{4}$$

where $A(f_i)$ is the area of the face f_i that adjacent to the vertex and the angle α_i is the angle of f_i at the vertex. Then we sort the vertices into a list in descending order with respect to their curvature values like in Fig. 4(a) and choose the top vertex as the first base vertex, e.g. marked point (x_1, y_1, z_1) in Fig. 4(b). Then, as shown in Fig. 4(c), we compute the geodesic distance from this vertex and mark all its neighboring points lying within a radius r. In our experiment, we adopt the Dijkstra's shortest path algorithm to compute the geodesic distance

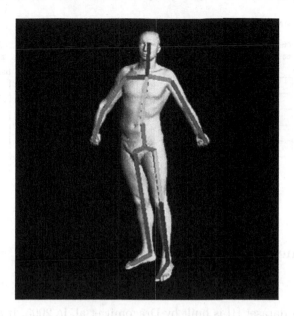

Fig. 3. Mesh division by skeleton; each colour represents a skeleton segment (Color figure online)

between two vertices as Eq. 5. The weight of each edge of Dijkstra's path is the Euclidean distance between neighboring vertices by Eq. 6.

$$g(i,j) = \sum_{i \in P} \omega_i \tag{5}$$

$$\omega_i = \min_{v_k \in N_i} ||v_i - v_k||, \tag{6}$$

where N_i is the neighbors of point i. The next base vertex is the first unmarked vertex in the list like (x_3, y_3, z_3) in Fig. 4(d). This process is repeated until all points are marked and based vertex set is built. The final base vertex set is illustrated in Fig. 4(f). Given base vertex set ϕ, we compute the relative surface distance from each candidate to ϕ with Eq. 7. The candidate C with the minimum relative distance to ϕ is regarded as the final correspondence as shown in Fig. 2(c).

$$D_{iso}(c_i, \phi) = \sum_{(v_j \in \phi, c_i \in \Theta)} g(c_i, v_j) \tag{7}$$

$$C = arg \min_{c_i \in \Theta} (D_{iso}(c_i, \phi) - D_{iso}(p, \phi)) \tag{8}$$

Here, $g(.,.)$ is the geodesic distance between two vertices. p is the point on the template. After the distances from the candidates to *base vertex set* are acquired, we select the candidate with minimum distance to base vertex set as the final correspondence as illustrated in Eq. 8.

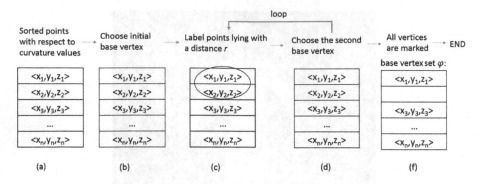

Fig. 4. The process of base vertex set

4 Experiments

4.1 Dataset

SCAPE human dataset [1] is built by Dragomir et al. in 2005. It is composed of pose dataset and shape dataset. In pose dataset, it contains scans of 70 different poses of a particular person. The shape model consists of 45 different people in a similar but un-identical pose. In the pose dataset, one mesh is chosen as the template mesh and others are denoted as instance meshes. Each mesh has 25000 triangle faces and 12500 vertices. Although the original work made use of both shape and pose data, only the pose data is distributed together with its skeleton information. Meshes in SCAPE model are hole-filled using the algorithm by Davis et al. [4]. SCAPE model also constructs a skeleton for the template mesh based on the fact that vertices on the same skeleton joint are spatially contiguous and exhibit similar motion across different scans. Thus, after scanning the pose instance for a particular person, authors decompose the mesh into several approximately rigid parts and get the location of the parts in different pose instances as well as the articulated object skeleton linking the parts. Based on the pose dataset, a tree-structured articulated skeleton is automatically constructed with 16 parts. Since SCAPE model contain both symmetric and various deformed shapes, we evaluate the performance of our proposed method with respect to symmetric flipped correspondence as well as the accuracy of final correspondences with SCAPE dataset.

4.2 Performance Evaluation

We compare our method with latest shape correspondence algorithm [16]. Firstly, we intuitively compare our result with those in coarse-to-fine combinatorial matching (C2FCM) algorithm [16] in Fig. 5. Our method outperforms C2FCM [16] with respect to semantical equivalence. For the same point on the third toe of template, C2FCM method finds its correspondence on the foot bottom. However, our method locates its correspondence almost on the third toe of foot. Secondly, the average

Fig. 5. Compare our method with C2FCM. Left: C2FCM Right: Our method

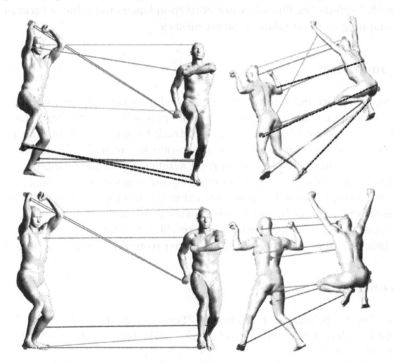

Fig. 6. Comparison between two methods: Top: results obtained using method in [16]; Bottom: our results. Matched point pairs are connected by lines. Symmetry flips are connected by black dash lines.

geodesic error is compared with C2FCM algorithm in Table 1. We can see that for different proportions of correspondences, the geodesic errors of our method are less than those of C2FCM. The average of geodesic error of all correspondences, shown in the column of 100 % correspondences, our method outperforms C2FCM,

Table 1. Comparison of our method with C2FCM method

GeoErr %Corr Method	10	20	30	40	50	60	70	80	90	100
C2FCM	0.122	0.103	0.093	0.093	0.092	0.097	0.098	0.097	0.097	0.098
Ours	0.097	0.074	0.070	0.074	0.081	0.084	0.082	0.081	0.080	0.083

which means our method is able to find the correspondences more accurately. More results are shown in Fig. 6, we can see that in coarse-to-fine algorithm correspondences which are shown in the top line present symmetry (both on the left and right foot) for template point. Our method is able to find the unique correspondences which is more accurate than [16] in terms of semantics. Moreover, it successfully removes that symmetric flipped invalid correspondences and achieved accurate one to one mapping from template to target meshes.

5 Conclusion

In summary, we present a robust method to address the symmetric flip problem in shape correspondence research area. This approach can effectively remove the flipped correspondences by introducing skeleton information and through minimising distortion error. It can locate the one-to-one semantically similar correspondence more accurately. Experimental results indicate that the proposed approach outperformed traditional approaches that rely only on surface information.

In the future, we hope to investigate other mesh data which is obtained from cheap scanners such as Kinect to find correspondence between template mesh and scanned mesh. The focus will be on tracking the intrinsic challenges posed by the incomplete and noisy data that is used to build such scanned mesh.

References

1. Anguelov, D., Srinivasan, P., Koller, D., Thrun, S., Rodgers, J., Davis, J.: Scape: shape completion and animation of people. ACM Trans. Graph. (TOG) **24**, 408–416 (2005). ACM
2. Baran, I., Popović, J.: Automatic rigging and animation of 3d characters. ACM Trans. Graph. (TOG) **26**, 72 (2007). ACM
3. Bronstein, M.M., Kokkinos, I.: Scale-invariant heat kernel signatures for non-rigid shape recognition. In: 2010 IEEE Conference on Computer Vision and Pattern Recognition (CVPR), pp. 1704–1711. IEEE (2010)
4. Davis, J., Marschner, S.R., Garr, M., Levoy, M.: Filling holes in complex surfaces using volumetric diffusion. In: Proceedings of the First International Symposium on 3D Data Processing Visualization and Transmission, pp. 428–441. IEEE (2002)
5. Dey, T.K., Fu, B., Wang, H., Wang, L.: Automatic posing of a meshed human model using point clouds. Comput. Graph. **46**, 14–24 (2015)

6. Elad, A., Kimmel, R.: On bending invariant signatures for surfaces. IEEE Trans. Pattern Anal. Mach. Intell. **25**(10), 1285–1295 (2003)
7. Jain, V., Zhang, H.: Robust 3d shape correspondence in the spectral domain. In: IEEE International Conference on Shape Modeling and Applications, SMI 2006, pp. 19–19. IEEE (2006)
8. Jiang, W., Xu, K., Cheng, Z.Q., Zhang, H.: Skeleton-based intrinsic symmetry detection on point clouds. Graph. Models **75**(4), 177–188 (2013)
9. Körtgen, M., Park, G.J., Novotni, M., Klein, R.: 3d shape matching with 3d shape contexts. In: The 7th Central European Seminar on Computer Graphics, vol. 3, pp. 5–17 (2003)
10. Ovsjanikov, M., Mérigot, Q., Mémoli, F., Guibas, L.: One point isometric matching with the heat kernel. Comput. Graph. Forum **29**, 1555–1564 (2010). Wiley Online Library
11. Ovsjanikov, M., Mérigot, Q., Pătrăucean, V., Guibas, L.: Shape matching via quotient spaces. Comput. Graph. Forum **32**, 1–11 (2013). Wiley Online Library
12. Rugis, J., Klette, R.: Surface curvature extraction for 3d image analysis or surface rendering
13. Rustamov, R.M.: Laplace-beltrami eigenfunctions for deformation invariant shape representation. In: Proceedings of the Fifth Eurographics Symposium on Geometry Processing, pp. 225–233. Eurographics Association (2007)
14. Sahillioğlu, Y., Yemez, Y.: 3d shape correspondence by isometry-driven greedy optimization. In: 2010 IEEE Conference on Computer Vision and Pattern Recognition (CVPR), pp. 453–458. IEEE (2010)
15. Sahillioglu, Y., Yemez, Y.: Minimum-distortion isometric shape correspondence using em algorithm. IEEE Trans. Pattern Anal. Mach. Intell. **34**(11), 2203–2215 (2012)
16. Sahillioğlu, Y., Yemez, Y.: Coarse-to-fine isometric shape correspondence by tracking symmetric flips. Comput. Graph. Forum **32**, 177–189 (2013). Wiley Online Library
17. Sun, J., Ovsjanikov, M., Guibas, L.: A concise and provably informative multi-scale signature based on heat diffusion. Comput. Graph. Forum **28**, 1383–1392 (2009). Wiley Online Library
18. Zaharescu, A., Boyer, E., Varanasi, K., Horaud, R.: Surface feature detection and description with applications to mesh matching. In: IEEE Conference on Computer Vision and Pattern Recognition, CVPR 2009, pp. 373–380. IEEE (2009)
19. Zhang, Z., Yin, K., Foong, K.W.: Symmetry robust descriptor for non-rigid surface matching. Comput. Graph. Forum **32**, 355–362 (2013). Wiley Online Library
20. Zheng, Q., Hao, Z., Huang, H., Xu, K., Zhang, H., Cohen-Or, D., Chen, B.: Skeleton-intrinsic symmetrization of shapes. Comput. Graph. Forum **34**, 275–286 (2015). Wiley Online Library

XTemplate 4.0: Providing Adaptive Layouts and Nested Templates for Hypermedia Documents

Glauco F. Amorim[✉], Joel A.F. dos Santos, and Débora C. Muchaluat-Saade

MidiaCom Lab, Computer Science Department,
Fluminense Federal University - UFF, Niterói, Brazil
{gamorim,joel,debora}@midiacom.uff.br,
http://www.midiacom.uff.br

Abstract. A hypermedia composite template defines generic structures of nodes and links that can be reused in different hypermedia compositions. XTemplate is an XML-based language for the definition of hypermedia composite templates. XTemplate can currently be used to create templates for NCL documents, but other hosting languages can also be used.

In current versions of hypermedia document template languages, including XTemplate, there is no facility for defining template layouts. This work extends XTemplate, incorporating the concept of adaptive layouts. *Adaptive layouts* enable the definition of generic presentation characteristics for multimedia documents that are instantiated at processing time and adapted to the number of media objects declared in a given document that uses a template.

Another important facility that this work incorporates in XTemplate is hypermedia composite template nesting. *Template nesting* enables the inclusion of template components inside other hypermedia composite templates, thus making the use of multiple nested templates transparent to the document author that uses templates.

1 Introduction

Currently, XTemplate 3.0 [1] and TAL [2] are the main languages used to build hypermedia composite templates (or just template for simplicity) for multimedia documents specified with Nested Context Language (NCL) [3]. Both XTemplate and TAL are used to reduce the complexity of creating multimedia applications, mainly when the document contains a large number of media objects and synchronization relationships between these objects.

Although the use of templates reduce the authoring effort in defining the structure and synchronization of NCL documents, there are some points that need to be improved.

Hypermedia composite templates are used to define the generic structure of nodes and links for one single hypermedia composition. Therefore, whenever one wants to use multiple composite templates inside a document, it has to be done explicitly by both importing all desired templates and associating each (desired) composition to one imported template. The drawback of such a scenario is that

© Springer International Publishing Switzerland 2016
Q. Tian et al. (Eds.): MMM 2016, Part I, LNCS 9516, pp. 642–653, 2016.
DOI: 10.1007/978-3-319-27671-7_54

the author using a group of templates must have the knowledge about each template and how they can be put together. TAL allows a main template to declare inner templates, but the communication interface between them is not well-defined. XTemplate 3.0 allows template extension, but does not provide template nesting in the main template definition. Then, the first contribution of this work is to enable templates to be nested in another template, maintaining compositionality. Thus, for the scenario previously described, an author would just need to use one template that nests all other templates.

The second contribution of this work is related to multimedia documents layout. Typically, NCL applications involve the presentation of various types of media objects in devices ranging from smartphones to digital TVs. For each media object to be presented at a given device screen, its presentation characteristics include its spatial coordinates (x, y) on the player device screen, along with its size. One may notice that, as well as for relationships, the amount of presentation characteristics to be declared tends to grow when the number of media objects in a given application grows. The process of defining presentation characteristics in NCL in such a scenario is both cumbersome and prone to errors. Template authoring languages for NCL, such as XTemplate 3.0 and TAL, do not provide a generic way to define visual presentation characteristics. So, the second contribution of this work is to provide an approach for the generic definition of document presentation characteristics through the so-called *adaptive layouts*.

This work presents an extension to the XTemplate 3.0 language, called XTemplate 4.0. In this new version of XTemplate, we provide both facilities for defining *adaptive layouts* and *nested templates*. Through testing it was observed that this version reduces the authoring effort in creating and using templates created with the XTemplate language. This paper extends the work presented in [4], where an XML language for defining adaptive layouts was proposed.

The remaining of the paper is structured as follows. Section 2 describes both adaptive layout and template nesting features. Section 3 discusses related work together with comparison with XTemplate previous version. Section 4 presents XTemplate 4.0, and the features it adds to the language. Section 5 describes evaluation tests done with XTemplate 4.0 and discusses its results. Section 6 finishes the paper with conclusions and future work.

2   Adaptive Layouts and Nested Templates

2.1   Adaptive Layouts

NCL documents are (typically) composed by several media objects, where a (great) part of those objects are presented on a device screen. How media objects are presented in the screen is specified in an NCL document by *region-descriptor* pairs in the following way, a media object in NCL is represented by a *media* element. A media element refers to a *descriptor* element for defining how it will be presented on a device screen, such as its transparency, sound level (if applicable), navigation specification, and the screen area it will occupy. The latter is defined by referring to a *region* element. Although such separation is intended to improve definition reuse - several *media* nodes may refer to the same *descriptor* and several *descriptors* may refer to the same *region* - when specifications are different and no reuse is possible, the number of *region* and *descriptor* elements is the same as the number of *media* elements.

What we call **adaptive layout** is an approach that enables document authors to create generic presentation characteristics that adapt itself to the number of media objects in a given document, thus diminishing authoring effort regarding how media objects will be displayed.

An adaptive layout relies on (possibly several) **layout models**, where a layout model specifies general directives for media object presentation. For example, a layout model may specify to place objects inside a grid.

Layout model instances are declared by **layout components**. A layout component specifies the device screen area where media objects referring to it should be placed, besides specific information about its layout model. For example, a layout component using a grid layout model specifies its number of lines and columns. A layout component can also have as children one or more **layout items**. A layout item declares a subset of presentation characteristics to be associated to media objects, for example a specific size. Media objects, therefore, may refer to specific layout items inside a layout component or to the layout component as a whole.

A **layout template** defines one or more layout components. Hypermedia composite templates define *components* that represent groups of media objects. While using a composite template, an author associates specific media objects to generic template components by labeling media objects (*xlabel* attribute) with a given template component identification. Likewise, specific media objects can be associated to layout components by labeling them (*layout* attribute) with a given layout component identification. Alternatively, a specific media object can be associated to a template component that, by its turn, is associated to a layout component.

2.2   Template Nesting

Composite templates define generic structures to be inherited by a composition that uses it. It means that synchronization specification provided by a composite template will take into account the child nodes of a given composition. Suppose that a composition c_1 declares three elements e_1, e_2 and e_3 inside it. Composition c_1 uses composite template T_1, which defines components G_1 and G_2 along with a synchronization specification relating both components.

Suppose, for example, element e_3 is another composition nested in c_1. If a document author also wants to embed semantics into e_3, this has to be done by making e_3 use another composite template (T_2). This approach is depicted in Fig. 1a.

An NCL author creating a document with multiple compositions, therefore, is supposed to associate all compositions that are intended to use templates to their corresponding templates. The drawback of such a scenario is that an author

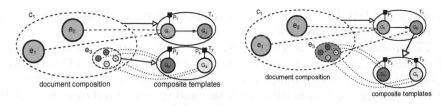

a) Nested composition using composite templates   b) Template component using nested template

Fig. 1. Composite templates without and with nesting template

using a group of templates must have knowledge about each template separately and how they can be put together. Once the user of composite templates is not supposed to have previous knowledge about the template authoring language (in our case XTemplate), using more than one template in the same document can be difficult for him.

What we call **template nesting** is an approach where template components can be associated to other templates. Therefore when a document composition is associated to a given template component, it will have the embedded semantics specified in the composite template that component uses. This idea is depicted in Fig. 1b. Notice that a hypermedia composite template cannot nest itself.

In order to ensure communication between two templates (let's call them the external and the internal templates), it is necessary to establish a communication interface. This interface is defined by *port* elements. The internal template must declare in its vocabulary one or more *port* elements to work as its communication interface. In the external template, the same *port* elements are declared inside the component using the internal template.

3 Related Work

Some works that address the creation of documents based on templates are discussed in LimSee3 [5] and Stamp [6].

The LimSee3 [5] model uses templates for multimedia document authoring. It defines an authoring language, independent of the existing languages, focusing on the logical structure of a document and not in the target language element semantics. With this approach, LimSee 3 provides great flexibility and reuse. The template created with that language describes a generic hierarchy among its components. A template has to be edited in order to become a complete document or another template.

The STAMP model [6] proposes a solution for the adaptation of multimedia presentations. That model is applied when the presentation content comes from a database, focusing on web systems. STAMP uses templates for the automatic generation of presentations. The presentation model and structure can be adapted according to the number of elements retrieved from the database.

However, STAMP does not work with template extension and cannot embed spatio-temporal semantics into a composition.

XTemplate 3.0 [1] is the current version of XTemplate. It is an expressive template authoring language, and it also allows for the definition of presentation (layout) characteristics for template components. However, XTemplate 3.0 does not have features such as nesting templates and generic specifications of media object presentation characteristics, as the concept of adaptive layouts. This work extends XTemplate 3.0 including both new facilities.

TAL [2] is described by its authors as a modular declarative language that supports the specification of templates: incomplete hypermedia compositions. As XTemplate, TAL can define a set of documents that share the same specification for their compositional structure. However, TAL does not allow the specification of layout information for its components, nor provides any facility to create generic definition of document presentation characteristics. TAL has nesting templates as one of its features, but it does not define template interfaces for nesting templates, so template compositionality is not satisfied.

In [7], the authors propose a system for creating and presenting grid-based documents to suit various viewing conditions and content selection. The system can display static content or dynamic content from several different sources. A set of templates is proposed for layouts that were inspired by models of traditional newspapers. Each template organizes a collection of content in the region determined by the model. In another work [8], the authors present a solution for publishing interactive digital documents, which is based on document authoring instead of programming. The solution is generic and describes how the definitions of templates and variable content elements can be used to decrease redundancy and increase the flexibility for those applications. Some ideas proposed in that related work to define adaptive layouts were used in the model proposed in our work.

There are tools that facilitate the definition of layout features in web documents, such as the CSS Regions Module [9]. The CSS Regions module allows content from one or more elements to flow through one or more boxes called CSS Regions. Besides, it allows dynamic magazine layouts that are flexible in placement of boxes for content flows.

Although those layout managers are proposed for different types of applications, the idea of organizing the layout suggested by them matches the dynamic management of layouts and, thus, were used as a basis for the adaptive layout modeling proposed in this paper.

4   XTemplate 4.0

XTemplate 4.0 keeps the facilities added to the template authoring language by XTemplate 3.0 and incorporates two new facilities for defining *adaptive layouts* for a given template and also *nesting templates*, so that multiple templates can be used together (in a main template).

Different interactive multimedia applications created by the NCL community, available at the NCL Club[1], were analyzed in order to determine different useful layout models to be provided. After analyzing them, we identified presentations that resembled the characteristics described by *FlowLayout* and *GridLayout* managers in Java language. Other layouts found did not have a well-defined structure or could be represented by compositions of *FlowLayout* and *GridLayout*. XTemplate 4.0 currently provides two layout models, which are:

- **FlowLayout:** layout items are placed from left to right, row by row. When there is no more space in a row considering the size of a given layout item, another row is created below it and the same principle is used again.
- **GridLayout:** layout items are placed from left to right, row by row, in a grid format. The grid is always built considering the whole area specified by the layout component.

XTemplate 4.0 adds a new element called *layoutBase* where the layout components for a given template are declared. A layout component declaration is done through element *layout*. The *layout* element has a child element called *format* that defines the area the layout component covers by its attributes *width*, *height*, *top* and *bottom*. Layout item arrangement is specified with attribute *align*, while items spacing is specified by attributes *hspace* and *vspace* of element *format*.

The layout model a given layout component instantiates is declared through attribute *type*. The *layout* element has specific attributes for each kind of layout model it instantiates. A *layout* element with type *flowLayout* does not need to define a number of rows and columns, on the other hand, it has to define the alignment for layout items. A *layout* element with type *gridLayout* defines a number of rows and columns, but does not define layout item alignment or size. Layout items are declared by the *item* element. Whenever a layout component instantiates a *flowLayout* model, it is possible to include different sized items in the same layout component. So each *item* element can define its own *width* and *height*.

Besides size and positioning, a layout component is capable of describing the navigation behavior among its items and between components. Each *layout* element may define a *focus* child element with a *focusIndex* attribute that determines a unique navigation index for that element. Whenever the *focus* element is defined for a given *layout* element, it establishes the possibility of navigation among its items. Therefore, media objects referring to a given *layout* element will be associated to navigational definition, depending on their relative position to each other.

Navigation among layout components is declared through attributes *moveUp*, *moveDown*, *moveLeft* and *moveRight* of element *focus*. It is not mandatory to have all those attributes, but at least one is necessary for establishing navigation among layout components. Each attribute indicates the *focusIndex* of the layout component to receive focus when the corresponding remote control

---

[1] http://clube.ncl.org.br.

(or keyboard) key is pressed while that layout component is in focus. Whenever navigation among layout components is established, layout items in the border of a given layout component will inherit a given navigation attribute from the layout component, thus allowing navigation to items in other layout components. For example, items in the bottom border of a layout component will inherit the *moveDown* attribute. The ones in the left border will inherit the *moveLeft* attribute and so on. It is important to notice, however, that when navigation among layout components is not established, the default behavior takes place, i.e. navigation from one border moves to the opposite border of the same layout component (cyclic navigation).

In addition, it is possible define media exhibition parameters in the layout component. Suppose, for example, that a media object should be reproduced with 90 % transparency. Then, it is necessary to include a child element, called **layoutParam**, in the related layout item. This parameter is represented by a tuple $< name, value, item >$.

While creating a layout template, the author associates layout components to template components through its new *layout* attribute, whose value is the layout component *id*.

Template nesting is achieved by extending the *component* element with a new *xtemplate* type. Thus, a *component* element with *xtemplate* type contains the same structure of the internal template it references. The internal template a *component* element references is indicated along with its type as *xtemplate/alias*, where *alias* is the unique identification of a given template in the template base. To enable using components of a nested template, the *component* element declares *port* child elements with the same *xlabels* as ports defined in the vocabulary of the internal nested template. An example of template nesting is presented in Sect. 4.2.

It is possible to define more than one level of nesting among templates. The template processor will process nesting in a recursive way, from the most inner to the root template element.

4.1 Template Processing

Template processing is done in two main steps. At the first processing step, elements referencing a given template component will be associated to synchronization relationships related to that component and declared in the template *body*. If a given template component represents a nested template, composite elements referring to that component are associated to the given template and processed before continuing the main template processing. At the end of the step, media objects inherit from template components their reference to layout components. At the second processing step, each layout component is translated to an NCL *region* representing the whole area declared by it. Each layout item is translated into a *region* with the same size and positioning attributes and an NCL *descriptor* with the same navigational attributes. Media objects are associated to *region-descriptor* pairs representing layout items in the order they are declared in the NCL document, i.e. the first media object declared in the NCL document will be associated to the first *region-descriptor* pair and so on. When no more *region* can be created inside the *region* representing the whole layout component, media objects are associated to existing *region-descriptor* pairs starting from the beginning. If a specific *item* element of the layout component is indicated, media objects are associated to a specific *region-descriptor* pair.

4.2   Template Example

To illustrate the new facilities provided by XTemplate 4.0, we modified the template "quiz.xml" presented in [1]. The "quiz.xml" template helps creating an interactive quiz that will be presented during a video presentation. "quiz.xml" uses the "screen.xml" template for presenting a question and its possible answers on the screen. Between a screen component and its successor, there are "change_screen" links. Those links are responsible for checking if one of the color keys of the remote control is pressed (red, green, yellow and blue). Once one of those keys is pressed, the link stops the presentation of the current screen and starts its successor. That link also passes a value representing the key pressed to a program (Lua counter node), which is a script written in the Lua language. That node will be responsible for testing if the answer is correct or not. If so, it updates one of its variable values, counting the correct answers.

Using XTemplate 3.0, where no layout components are available, "screen.xml" had to declare four answer types, because each answer will be presented in a different screen region (one below the other). Using XTemplate 4.0, just one answer component has to be defined in "screen.xml" using the *flowLayout* model, regardless of the number of answers for a question. This template was modified in the following way: (i) each *screen* component now represents the nested template "screen.xml"; (ii) "change_screen" links where modified to check if a given answer was selected and pass its position to the Lua counter node; and (iii) template "screen.xml" declares a *layout* element with type *flowLayout* for presenting possible answers one bellow the other together with their navigation attributes, regardless of the number of answers provided.

File "layout.xml" defines two layout components to be used for a question and its related answers. The question is presented on top of the screen (centralized) and answers are presented one bellow the other (centralized). Listing 1.1 presents a fragment of the definition of answer layout components.

```
1  ...
2  <layout id="ansFl" type="flowlayout">
3      <format align="center" height="640" hspace="10" left="10" top="300" vspace="0"
           width="200" zIndex="3"/>
4      <item id="c1" height="150" width="200" />
5      <focus focusIndex="1"/>
6  </layout>
7  ...
```

Listing 1.1. "layout.xml" layout components

File "screen.xml" defines how the question and answers for each question are presented. This template is nested inside the *quiz* template, thus it defines port *portAnswer* to work as its communication interface. A fragment of this template vocabulary is presented in Listing 1.2. As aforementioned, the template will be processed in a recursive way, from the most inner to the root element. Then, when the "screen.xml" is processed, each *media* element with *answer* label will receive a *port* element. This element will be related to a *portAnswer* label to identify the communication interface.

```
1   <vocabulary>
2      <port xlabel="portAnswer"/>
3      <component xlabel="question" layout="lay#qstF1"/>
4      <component xlabel="answer" layout="lay#ansF1"/>
5   </vocabulary>
6   <body >
7      ...
8      <variable name="i" select="1"/>
9      <for-each select="child::media[@xlabel='answer']">
10        <port id="port" select="current()" xlabel="portMenu"/>
11        <variable name="i" select="$i + 1"/>
12     </for-each>
13     ...
14  </body>
```

Listing 1.2. "screen.xml" vocabulary fragment

File "quiz.xml" represents the main template. It nests template *screen* (alias "scn") for its *screen* component. To enable referencing nested template components, the *screen* component declares a *port* child element also named *portAnswer*. A Fragment of "quiz.xml" is presented in Listing 1.3.

```
1   <component xlabel="screen" xtype="xtemplate/scn">
2      <port xlabel="portAnswer"/>
3   </component>
```

Listing 1.3. "quiz.xml" fragments

An example of NCL document using the *quiz* template is presented in Listing 1.4. Notice that, different from the example in [1], the NCL author only has to use the *quiz* template. The *context* element refers the nested template declaring a *screen* label.

```
1   <body xtemplate="quiz">
2      <context id="screen_01" xlabel="screen">
3        <media id="question_01" xlabel="question"/>
4        <media id="answer_01_A" xlabel="answer"/>
5        <media id="answer_01_B" xlabel="answer"/>
6        ...
7      </context>
8   </body>
```

Listing 1.4. NCL document (fragment) using *quiz* template

5 XTemplate 4.0 Evaluation

We performed tests to evaluate the *adaptive layout* feature. The test involved five activities, where in each activity authors should write code for one or more layout components and NCL code using layout components to define the application's presentation characteristics. The authors had to specify the layout template document in each activity and use this template within the NCL document.

Activity 1: authors should place media objects in a grid with three columns and two rows in the center of the screen; *Activity 2:* authors should place media objects in a flow with one item in the bottom of the screen; *Activity 3:* authors should place media objects in a flow with two different item sizes at the top of the screen and a grid with four columns and one row in the bottom; *Activity 4:* authors should place media objects along two grids, one in the left and one in the right side of the screen, both with one column and four rows, and one flow in the bottom; *Activity 5:* authors should create any placement they wanted.

A total of 20 students participated in the current study: seven students of a high-school technical course in Telecommunications (Group 1) and thirteen students of a computer science graduation course in (Group 2). All students in Group 1 were enrolled in a digital TV applications course, therefore, authors had a good knowledge about NCL. No student in Group 2 had any knowledge about NCL, but had good knowledge about HTML language, which is also XML-based.

After completing all five activities, authors filled out a questionnaire with ten questions, nine related to the following cognitive dimensions [10]: *visibility, role-expressiveness, closeness of mapping, verbosity, premature commitment, hidden dependencies, error-proneness, consistency* and *viscosity*, and one to define a final score. Each question should be answered with a score of one to ten, which was used to evaluate the facility. The overall result of the tests can be seen in Fig. 2a.

The first interesting conclusion about test results is that, although the two groups had different knowledge about NCL, results are quite similar in both groups. This result shows that the author experience with the adaptive layout authoring language is not influenced by the author expertise about the hosting multimedia authoring language.

One main concern was not to increase *verbosity* when using adaptive layouts. As it can be seen by test results, this goal was achieved. Another important result is in *role-expressiveness* and *visibility* dimensions. Average results indicate that the *adaptive layout* feature is self-explaining and therefore easy to use. Combined with the results for *error-proneness* and *viscosity* dimensions, we can conclude that the new feature does not insert any difficulty for those who want to use it, leading to new authoring mistakes, even with non-expert authors.

Although the evaluation results for both *hidden dependencies* and *premature commitment* dimensions were not very high, it was better than expected. *Hidden dependencies* is explained because of the navigation feature provided by layout components. Both the *id* and *focus* attributes of a layout component are used to define NCL descriptor focus index. This information needs to be provided to authors using that feature. *Premature commitment* is explained because authors did not know that the relative position among layout components in the layout template does not interfere in the final positioning of media objects in the device screen.

In general, the *adaptive layout* feature has been well evaluated. However, some adjustments can be made to improve its use. Suggestions given by test subjects include providing a graphical tool to create layout components and to provide other more sophisticated layout models. Both suggestions are work in progress.

A graphical editor that supports templates is very useful for helping document authors using templates [11]. When template nesting is available, a graphical editor can make the use of several nested templates transparent to document authors.

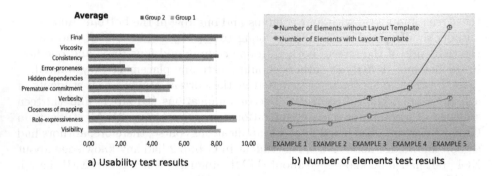

a) Usability test results   b) Number of elements test results

Fig. 2. Test results

Using only textual edition, the document author will not be able to understand clearly the difference between using separate templates and using one main template with nested templates. Because of this, we did not apply usability tests to evaluate the nested template facility with subject groups 1 and 2. However, we intend to run those tests when a graphical editor that supports nested template is available.

In addition, another evaluation about the number of language elements used for creating templates with and without adaptive layouts was made. The five scenarios used in the previous tests were considered and the results can be seen in Fig. 2b.

The number of template language elements increases considerably when the complexity of the example document increases. This result is expected because they involve the creation of more region-descriptor pairs. With the use of layout models, this effort is mitigated because the author only needs to set the proper model and, for each model, define a few required elements.

In order to evaluate nested template authoring, several XTemplate examples using more than one template were redesigned to use the template-nesting feature. Given the small changes necessary to use that feature for those examples and our experience about authoring templates, we state that nesting templates do not increase the complexity when authoring templates.

6   Conclusions

This paper presented a new version of the XTemplate language, called XTemplate 4.0. In this new version, two new features are provided, *adaptive layouts* and *template nesting*. Both features represent the main contribution of this paper.

Adaptive layouts enable the definition of generic presentation characteristics for multimedia documents that are instantiated at processing time and adapted to the number of media objects declared in a given document. As a test case, we provided two layout models, *GridLyout* and *FlowLayout*, that are implemented for the NCL language. It is important to notice that this feature is not provided by any other NCL template language. XTemplate 4.0 usability test results indicate that the *adaptive layout* feature is intuitive and does not increase verbosity.

Template nesting enables the association of template components to other templates. Our solution for template nesting specifies interface points (ports) for nesting templates, which does satisfy composionality for authoring hypermedia composite templates.

A future work is to develop a graphical editor that supports nested templates to help the NCL document author when using templates. With textual edition, the difference between using several different templates and one main template with nested templates is subtle and might not be clearly understood by template users. That explains why we have not run usability tests about using nested templates yet.

Another future work is to include new layout models and perform evaluations for each new model provided and develop a graphical editor to help specifying adaptive layouts for multimedia documents.

References

1. dos Santos, J.A.F., Muchaluat-Saade, D.C.: XTemplate 3.0: spatio-temporal semantics and structure reuse for hypermedia compositions. MTAP **61**(3), 645–673 (2012)
2. Neto, C.S., Pinto, H.F., Soares, L.F.G.: TAL processor of hypermedia applications. In: DocEng, pp. 69–78. ACM (2012)
3. Recommendation ITU-T H.761, Nested Context Language (NCL) and Ginga-NCL for IPTV Services (2011)
4. Amorim, G.F., dos Santos, J.A.F., Muchaluat-Saade, D.C.: Adaptive layouts for authoring ncl programs. In: 19th WebMedia. ACM (2013) (in Portuguese)
5. Deltour, R., Roisin, C.: The limsee3 multimedia authoring model. In: DocEng, pp. 173–175. ACM (2006)
6. Bilasco, I.M., Gensel, J., Villanova-Oliver, M.: STAMP: a model for generating adaptable multimedia presentations. MTAP **25**(3), 361–375 (2005)
7. Schrier, E., Dontcheva, M., Jacobs, C., Wade, G., Salesin, D.: Adaptive layout for dynamically aggregated documents. In: 13th IUI, pp. 99–108. ACM (2008)
8. Signer, B., Norrie, M.C., Weibel, N., Ispas, A.: Advanced authoring of paper-digital systems. MTAP **70**(2), 1309–1332 (2014)
9. W3C, CSS Regions Module Level 1. http://www.w3.org/TR/css-regions-1/
10. Blackwell, A., Green, T.: HCI Models, Theories, and Frameworks: Toward an Interdisciplinary Science. Morgan Kaufmann, San Francisco (2003)
11. Mattos, D., Silva, J., Muchaluat-Saade, D.: Next: graphical editor for authoring NCL documents supporting composite templates. In: EuroITV (2013)

Level Ratio Based Inter and Intra Channel Prediction with Application to Stereo Audio Frame Loss Concealment

Yuhong Yang[1]([✉]), Yanye Wang[1], Ruimin Hu[1,2], Hongjiang Yu[1], Li Gao[1], and Song Wang[1]

[1] National Engineering Research Center for Multimedia Software, Computer School, Wuhan University, Wuhan, China
ahka_yang@yeah.net, hrm1964@163.com,
{944708668,1739593607,5420655,395584404}@qq.com
[2] Collaborative Innovation Center of Geospatial Technology, Wuhan, China

Abstract. The problem of side signal frame loss concealment for Mid/Side (M/S) stereo audio is addressed in this paper. The proposed level ratio based inter and intra channel (LRIIC) prediction method is designed to overcome the main challenge due to the diversified stereophonic nature of audio signals. To identify the stereophonic nature, we employ the time varying level ratio between mid (also called monophonic) signal and side signal. In the first phase, one Winer filter is designed from the monophonic signal and side signal of previous frame. And the other one is designed from the side signal of previous two frames. The two filters allow reconstruction of the current loss frame with inter and intra channel prediction respectively. In the second phase, available current frame of monophonic signal is used as the input of inter-channel Winer filter. While long-term prediction filter is employed to find the periodic components of previous frames as the input of intra-channel Winer filter. Finally, a level ratio based linear combination of inter and intra channel prediction output is employed to get the current lost side signal. Objective and subjective evaluation results for the proposed LRIIC approach, in comparison with existing techniques, all incorporated within a 3GPP AMR-WB+ decoder, provide evidence for gains across a variety of stereophonic signals.

Keywords: Index terms: frame loss concealment · M/S stereo · Stereophonic signals · Level ratio · AMR-WB+

1 Introduction

Highly efficient stereo audio coding methods are required for new compelling and commercially interesting applications such as mobile audio on demand, audio messages and music on cellphones. These services use audio media in mobile communication systems. But the compressed audio bit-stream which had been corrupted by transmission errors or permanent packet losses could result in very annoying artifacts to users. Audio coding for mobile applications has to cope with hard requirements due to the nature of

© Springer International Publishing Switzerland 2016
Q. Tian et al. (Eds.): MMM 2016, Part I, LNCS 9516, pp. 654–661, 2016.
DOI: 10.1007/978-3-319-27671-7_55

mobile wireless transmission. In order to minimize the above audible effect of missing frames, error recovery approaches must be employed in audio codecs to improve the overall playback quality.

Error recovery techniques are generally classified into sender-based and receiver-based reproduction [1]. Sender-based approaches include retransmission, forward error correction (FEC) [2–4] and interleaving. These approaches generally require higher complexity in the encoder, and produce bit streams with much larger sizes, which are not suitable for transmission over bandwidth-limited channels. Receiver-based frame loss concealment approaches include insertion, interpolation and regeneration at the decoder. Receiver-based approaches don't increase the compressed audio size, have been of interest for mobile applications [5]. The simple techniques of insertion just replace the lost packet by silence, noise or the previous frame [6]. The insertion approaches are easy to implement, result in poor performance. The interpolation methods find the waveform substitution by some pattern matching scheme [7]. These methods are with better performance. The regeneration approaches rebuild a loss frame by codec parameters derived from source modelling of audio [8–11]. The advanced regeneration approaches are more complex, but with the best results.

Above mentioned receiver-based approaches are also applied in frame loss concealment of multi-channel audio. Rishi [12] proposed the first work on inter-channel error concealment for multi-channel streaming audio. An insertion based waveform substitution for a missing interval in a channel is derived from the sample values of other channels during the same interval. AMR-WB+ [13] codec also adopts the inter-channel prediction approach in frame loss concealment of side signal. While a novel time-domain filter is obtained from monophonic and side signals of the latest good frame. Then the filter is employed in side signal prediction of current lost frame. And current mono signals either concealed or correctly received are used as the filter inputs. AMR-WB + codec only use inter-channel prediction of side signals without referring to the intra-channel correlation. In [14, 15], Koji Yoshida has shown that intra channel prediction is also effective for side signal frame loss concealment.

Due to the fact that both inter and intra channel prediction approaches contribute to frame loss concealment of side signal, we adopt a level ratio based inter and intra channel (LRIIC) prediction method. Different from above mentioned methods, we employ the level ratio between mid (also called monophonic) signal and side signal of latest good frame to handle the unification weight problem. Experiments on AMR-WB+ test datasets demonstrate the effectiveness of the proposed method.

2   Level Ratio Based Inter and Intra Channel Prediction

The stereo audio in the decoder is composed by an monophonic signal $M(n)$ which is the average of the left channel and right channel and a side signal $S(n)$ which is the difference of the two channels. The $M(n)$ can be concealed as the single channel loss concealment method when the current frame is lost. Here we present the proposed LRIIC prediction method for side signal frame loss concealment.

2.1   Inter-channel Prediction

Presented in the Fig. 1, a wiener filter is designed with a transfer function as formulation (1) for short term prediction. The filter coefficients are calculated as formulation (2) where R_{mm} is a Toeplitz matrix of autocorrelations of the monophonic signal $M(n)$ and R_{ms} is a vector of cross-correlations of the current windowed monophonic signal and the side signal. Save the filter coefficients in a buffer. If the next frame is lost, using the recovered monophonic signal of the next frame as excitation of the filter and the filter coefficients to reconstruct the lost side signal.

$$H_1(z) = \sum_{i=0}^{P} h_1(i) \cdot z^{-i}, \quad P = 8 \tag{1}$$

$$R_{mm} \cdot h_1 = R_{ms} \tag{2}$$

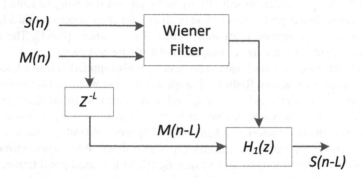

Fig. 1.   Inter-channel prediction

2.2   Intra-channel Prediction

In this part, we design a cascade model by long term prediction and short term prediction. The short term prediction is also based on a Wiener filter. And the excitation of filter is interpolated based on pattern matching, searches the received waveform for a match on the shape of the part of the waveform just before the loss. The waveform substitution method is analogous to long term prediction. As shown in Fig. 2, L is the frame length, find the most correlated periodic segment $S'(n)$ in the range of $(n + L - r)$.

Then intra-channel prediction Wiener filter is presented in formulation (3). The excitation and the response of the filter is $S'(n)$ and $S(n)$ respectively. Compute and save the coefficients h2 of the filter as formulation (4) where $R_{s's'}$ is a Toeplitz matrix of autocorrelations of the derived signal $S'(n)$ and the $R_{s's}$ is a vector of cross-correlation of the derived signal $S'(n)$ and current side signal $S(n)$. The periodic substitution of the next frame $S'(n - L)$ is saved in buffer for future use. When the next frame is lost, use $S'(n - L)$ and the filter coefficients h2 to reconstruct the side signal of the next frame.

$$H_2(z) = \sum_{i=0}^{P} h_2(i) \cdot z^{-i}, \quad P = 8 \tag{3}$$

$$R_{s's'} \cdot h_2 = R_{s's} \tag{4}$$

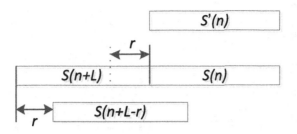

Fig. 2. Periodic waveform substitution

2.3 Level Ratio Based Inter and Intra Channel Prediction

Presented in the Fig. 3, the final reconstruction of the side signal $S(n)$ is the combination of the reconstructed side signals in Sects. 2.1 and 2.2 as described in formulation (5). $S_1(n)$ is the side signal reconstructed by the inter-channel and $S_2(n)$ is the side signal reconstructed by the intra-channel, w_1 and w_2 are the weights of the two reconstructed side signals. To make the $S_1(n)$ and $S_2(n)$ in the same cope, we normalize the $S_1(n)$ with a gain which meets the formulation (6), w_1 and w_2 meets the formulation (7)

$$S(n) = w_1 \cdot S_1(n) + w_2 \cdot S_2(n), \quad n = 0, 1, \dots L - 1 \tag{5}$$

$$\sum_{n=0}^{L-1} S_1^2(n) \cdot gain^2 = \sum_{n=0}^{L-1} S'^2(n) \tag{6}$$

$$w_1 = (1 - w_2) * gain \tag{7}$$

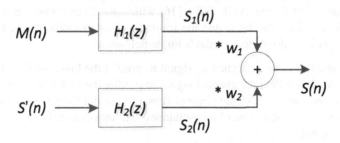

Fig. 3. The union of inter-channel and intra-channel

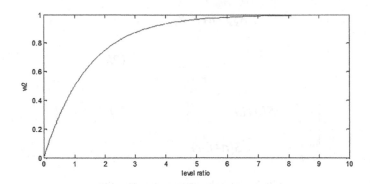

Fig. 4. Curve of w_2 with level ratio

The channel level ratio of monophonic signal and side signal reflects the difference of the left and right channel. So we assumed that weights are changed as the level ratio's changing. When stereo image of the audio source is in the middle, the level ratio is infinite, the intra-channel take the main part in the prediction of side signal and w_2 is equal to zero. As the stereo image shift from the middle, the level ratio decreases, the influence of the inter-channel prediction increase. Here we assume that w_2 is a increased with level ratio monotonically. According to the empirical tests we found that it takes the best result in when w_2 meets the formulation (8) whose curve is presented in Fig. 4.

$$w_2 = 1 - e^{-0.6931 * ratio} \tag{8}$$

3   Experiments

In our experiment, AMR-WB+ is operated at 32 kbps to generate the bit-streams. 19 test audio files are selected from AMR-WB+ [16] which are divided into 3 groups: speech, music and mixed. Here we integrated the following four methods in AMR-WB+, and compare the performance of four methods implementation:

1. Inter-channel: use the inter channel signal to predict the loss side signal.
2. Intra-channel: use the intra channel signal to predict the loss side signal.
3. Selective: use the more related channel signal to predict the loss side signal.
4. LRIIC: use the combination of inter-channel and intra-channel signal to predict the loss side signal.

For fair comparison, we set the random 10 % Frame Lost Rate (FLR).

3.1   Objective Evaluation

We evaluate Segmental Signal to Noise Ratio (SNR) as an objective measure. Segmental SNR (SSNR) is the average of SNR in dB at each of the lost frame. For SSNR the signal energy is of the originally decoded side signals and noise energy is of the difference between originally decoded side signals and the side signals generated by an error concealment module. The frames are lost randomly at the rate of 10 % using the four different error concealment methods. The files are classified into speech, music and mixed audio. Calculated the average SSNR of each file type and average of all. The results of SSNR for all files in the different schemes are shown in Table 1.

Table 1. SSNR at 10 % FLR

File type	Inter-channel	Intra-channel	Selective	LRIIC
Music	−4.276	−2.767	−3.168	−2.661
Speech	−5.241	−3.135	−4.082	−2.868
Mixed	−4.139	−2.957	−3.263	−2.818
average	−4.552	−2.953	−3.504	−2.782

As shown in Table 1, the intra-channel prediction method is better than the inter-channel method. This indicates that the cascade model for intra-channel is more robust than the inter-channel which uses the short term prediction. Since the inter-channel adopts the shape-gain constrained time-domain filter approach. Small deviation of prediction filer coefficients might result in quite low SSNR in the pre-diction outputs. Selective method is better than inter-channel prediction but worse than intra-channel prediction which shows that using inter-channel and intra-channel separately according to the cross-correlation has little increase for the performance. This is mainly due to the incorrect selection by using correlation criterion. The performance might be improved with revised criterion. LRIIC pre-diction is the best one which shows that the intra-channel prediction and inter-channel prediction have a jointly effect on the prediction of side signal.

3.2 Subjective Evaluation

For subjective test, we use score of Comparison Mean Opinion Score (CMOS) for quality evaluation. LRIIC method is compared with Intra-channel method with the highest SNR among others.

We choose 10 graduated students majored in audio processing and trained to focus on the frame loss parts before the formal subjective tests. And only 10 test files from the above mentioned 19 files are used to reduce workload in subjective test. The test files contain Ref/A/B in which Ref is the original audio, A is decoded by Intra-channel method and B is decoded by LRIIC method. The files were presented to the listeners in a random order. The scores have 7 levels showed in Table 2.

Table 2. CMOS scores

Comparison of the Stimuli	Score
B is much better than A	+3
B is better than A	+2
B is slightly better than A	+1
B is the same as A	0
B is slightly worse than A	−1
B is worse than A	−2
B is much worse than A	−3

The results of the subjective evaluation presented in the Fig. 5 show that LRIIC prediction is better than intra-channel prediction in most cases. The average improvement is 0.5875. The subjective evaluation and objective evaluation both verify the effectiveness of the unification model.

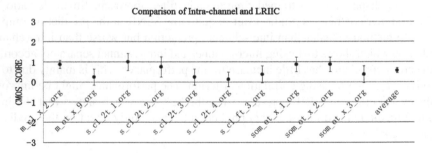

Fig. 5. Comparison of Intra-channel and LRIIC prediction

4   Conclusion and Future Work

This paper proposed a frame error concealment method for stereo audio with combination of inter-channel and intra-channel based on channel level ratio. Contrary to the current methods which use inter-channel or intra-channel, the proposed method address the jointly effect on the concealing of the lost frame. The results of objective and subjective evaluation show that the proposed algorithm has the best performance. Further direction include enhancing the pro-posed technique to include factors which have effect on the weights of the inter-channel and intra-channel. Future work also include generalization of the frame loss concealment algorithm for multi-channel audio or 3D audio.

Acknowledgement. This work is supported by National Natural Science Foundation of China (No. 61102127, 61231015, 61201340, 61201169, 61272278, 61471271) and National High Technology Research and Development Program of China (863 Program) No. 2015AA016306.

References

1. Perkins, C., Hodson, O., Hardman, V.: A survey of packet loss recovery techniques for streaming audio. IEEE Network **12**(5), 40–48 (1998)
2. Westerlund, M., Nohlgren, A., Svedberg, J., et al.: Forward error correction in speech coding: U.S. Patent 6,757,654, 29 June 2004
3. Merazka, F.: Forward error correction concealment method for celp-based coders in packet networks. In: Proceedings of the World Congress on Engineering, vol. 1 (2010)
4. Midya, A., Ranjan, R., Sengupta, S.: Scene content driven FEC allocation for video streaming. Sig. Process. Image Commun. **29**(1), 37–48 (2014)
5. Carvalho, L.S.G.D., Mota, E.D.S.: Survey on application-layer mechanisms for speech quality adaptation in VoIP. ACM Computing Surv. (CSUR) **45**(3), 36 (2013)
6. Hardman, V., Sasse M.A., Handley, M., et al.: Reliable audio for use over the Internet. In: Proceedings of INET, vol. 95, pp. 171–178 (1995)
7. Wasem, O.J., Goodman, D.J., Dvorak, C.A., et al.: The effect of waveform substitution on the quality of PCM packet communications. IEEE Trans. Acoust. Speech Signal Process. **36**(3), 342–348 (1988)
8. Ryu, S.U., Choy, E., Rose, K.: Encoder assisted frame loss concealment for MPEG-AAC decoder. In: IEEE International Conference on Acoustics, Speech and Signal Processing, ICASSP 2006 Proceedings, vol. 5, p. V. IEEE (2006)
9. GPP TS 26 190 V7.0.0, Adaptive Multi-Rate -Wideband (AMR-WB) speech codec: Transcoding functions (Rel. 7), June 2007
10. Nanjundaswamy, T., Rose, K.: Bidirectional cascaded long term prediction for frame loss concealment in polyphonic audio signals. In: 2012 IEEE International Conference on Acoustics, Speech and Signal Processing (ICASSP), pp. 417–420. IEEE (2012)
11. Yang, Y., Dong, S., Hu, R., et al.: An inter-frame correlation based error concealment of immittance spectral coefficients for mobile speech and audio codecs. In: 2014 IEEE International Conference on High Performance Computing and Communications, 2014 IEEE 6th International Symposium on Cyberspace Safety and Security, 2014 IEEE 11th International Conference on Embedded Software and Systems (HPCC, CSS, ICESS), pp. 436–441. IEEE (2014)
12. Sinha, R., Papadopoulos, C., Kyriakakis, C.: Loss concealment for multi-channel streaming audio. In: Proceedings of the 13th International Workshop on Network and Operating Systems Support for Digital Audio and Video, pp. 100–109. ACM (2003)
13. GPP TS 126 290 V10.0.0,Extended Adaptive Multi-Rate -Wideband (AMR-WB+) codec: Transcoding functions (Rel. 7), April 2011
14. Yoshida, K.: Stereo decoder that conceals a lost frame in one channel using data from another channel: U.S. Patent 8,209,168, 26 June 2012
15. Yoshida, K.: Stereo sound decoding apparatus, stereo sound encoding apparatus and lost-frame compensating method: U.S. Patent 8,359,196, 22 January 2013
16. GPP TS 26.274. Audio codec processing functions; Extended Adaptive Multi-Rate -Wideband (AMR-WB+) speech codec; Conformance testing (2005)

Depth Map Coding by Modeling the Locality and Local Correlation of View Synthesis Distortion in 3-D Video

Qiong Xue, Xuguang Lan$^{(\boxtimes)}$, and Meng Yang

Institute of Artificial Intelligence and Robotics,
Xi'an Jiaotong University, Xi'an, China
joan95@163.com, xglan@mail.xjtu.edu.cn,
myang59@gmail.com

Abstract. We propose depth map coding method by modeling the locality and local correlation of view synthesis distortion (VSD) in three dimensional (3D) video. Taking into account local characteristics of both depth map quantization error and color video, we start by dividing depth map into two kinds of regions: error sensitive area (ESA), where depth map quality is sensitive to quantization and the synthesized view quality is sensitive to depth map distortion, and error resilient area (ERA) otherwise. The locality of VSD is established by modeling the relationship between synthesized view distortion and depth map distortion for different regions separately. Then the local correlation of VSD is exploited in terms of synthesized view distortion propagated from other non-corresponding regions of depth map due to correlation of regions during temporal/spatial prediction in depth map coding. The final VSD model is obtained by considering both the locality and local correlation and implemented in the rate-distortion (RD) optimized mode selection for depth map coding. Experimental results show that our solution can achieve 12.67 % and 11.22 % bit rate saving as well as 1.07 dB and 1.94 dB improvement of synthesized view quality on average at high and low bit rate, respectively, compared with H.264/AVC.

Keywords: Depth map coding · View synthesis · Rate-distortion optimization · H.264/AVC

1 Introduction

In the past decades, three dimensional (3D) video has become popular due to its capability in providing realistic 3D viewing experience comparing to general 2D video. Among different 3D video systems, multi-view-plus-depth (MVD) 3D video [1] has been a more and more promising solution due to its low data storage and view flexibility. In this solution, both color videos and depth maps are captured from different views. At the terminal, view synthesis techniques, such as depth-image-based-rendering (DIBR) are used to construct virtual videos of other views that are not captured, such that any desired views can be presented to the users.

Like color video, depth map in MVD 3D video system also needs to be compressed such that the limited bandwidth can be utilized efficiently. Considering that depth map

© Springer International Publishing Switzerland 2016
Q. Tian et al. (Eds.): MMM 2016, Part I, LNCS 9516, pp. 662–674, 2016.
DOI: 10.1007/978-3-319-27671-7_56

can be treated as gray scale picture, depth map coding can be directly solved with general 2D video coding solutions, such as H.264, HEVC, etc. However, depth map often has specific characteristics such as regionally flat with clear and sharp boundaries separating different areas comparing to general color image. Therefore it is necessary to design coding method for depth map itself instead of using the standard video codec. Many efforts have been concentrated on exploiting the specific characteristics of depth map [15–18]. Moreover, it has been acknowledged that, depth map in 3D video system is used to generate virtual views in view synthesis process rather than displayed in client side. So the coding performance of depth map should be measured by the quality of virtual synthesized view instead of depth map quality itself. To achieve this goal, many research works have focused on estimating the synthesized view distortion with the depth map distortion [3–8, 14]. Considering the local characteristics of relationship between view synthesis distortion and depth map distortion, Chung et al. [10] and Zhang et al. [9] decomposed depth map into multiple regions and then well estimated their synthesized view distortion separately. With the estimated synthesized view distortion above, the depth map coding can be achieved by replacing the conventional distortion metric with the estimated synthesized view distortion in practical application.

When estimating the synthesized view distortion with depth map distortion, the relationship between depth errors and rendering errors has been the key point. It has been acknowledged that depth map error will cause position shift when generating the virtual view with the color video in view synthesis, which is proportional to depth map error [6]. The larger the depth map error is, the larger the rendering position error is and then the worse the quality of the synthesized view may become. In addition, given depth error of the same level, the quality of the synthesized view may also be affected by the local characteristics of the color video. For example, depth error will cause more severe degradation of the synthesized view quality for the regions with rich texture in color video than that for flat regions. So the effect of depth error on synthesized view quality varies for different local regions, which is determined by the local characteristics of both the depth map quantization error and color video information. Works [15–18] were all devoted to preserve the quality of depth map due to its specific characteristics without considering its effect on synthesized view quality. Kim et al. [6, 7, 14] and Oh et al. [5] exploited the corresponding color video information for view synthesis distortion model based on depth distortion, without considering the regional difference of depth map quantization distortion. Cheung et al. [4] and Fang et al. [3] jointly considered the depth map distortion and color view characteristic to model the view synthesis distortion, while the locality of view synthesis distortion due to depth quantization error have not been exploited. Zhang et al. [9] simply divided depth map into two regions by performing canny operator on the corresponding color video, which can't be proved to be the real measurement of the properties of relationship between depth error and view synthesis distortion. Moreover, the rendering error caused by depth error is supposed to be uniform for all pixels in view synthesis. However, in practical application, depth error due to lossy compression is generally not uniform. In this paper, we aim to exploit the locality of view synthesis distortion due to depth map distortion by jointly considering the depth map quantization error and color view information for each local region.

In addition, when exploring the locality of view synthesis distortion due to depth quantization distortion, it is generally assumed that different regions of depth map are independent from each other, and their synthesized view distortion are estimated separately. However, when encoding depth map with standard video coding solutions (H.264 or HEVC), the corresponding synthesized distortion of different regions may interact with each other. To be specific, due to the spatial/temporal prediction techniques in depth map coding, the distortion of a block may also affect the quality of the neighboring blocks of the same frame and the blocks of the following frames, which will in turn lead to synthesized view distortion. However, the existing relevant works which attempt to exploit the view synthesis distortion with depth distortion have not taken into account the view synthesis distortion propagated from other regions due to correlation of regions in depth map coding. In this paper, we aim to further consider the local correlation of different regions when modeling the regional view synthesis distortion due to depth map quantization distortion.

In this paper, we model the view synthesis distortion (VSD) in terms of locality and local correlation of different regions. We start by dividing depth map into two kinds of regions: error sensitive area (ESA), where depth map is sensitive to quantization and the synthesized view quality is sensitive to depth map distortion, and error resilient area (ERA) otherwise. The locality of VSD is established by statistically modeling the relationship between synthesized view distortion and depth map distortion for the local regions separately. Then the local correlation of VSD is exploited in terms of synthesized view distortion propagated from non-corresponding regions of depth map due to temporal/spatial prediction in depth map coding. The proposed VSD model is used in the rate-distortion optimization (RDO) process for depth map coding. The solution is achieved by adjusting Lagrangian multiplier, and is verified in standard video coding platform H.264/AVC.

The rest of this paper is organized as follows. In Sect. 2, we model the VSD by exploiting both the locality and local correlation of it. In Sect. 3, the VSD model is incorporated in the RDO process for depth map coding. Experimental results are show in Sect. 4. Finally, we conclude the paper in Sect. 5.

2 Proposed VSD Model

2.1 Previous Work on View Synthesis Distortion

In MVD based 3D video system, the virtual synthesized view is obtained by warping the reference color video to the virtual view base on DIBR technique, where depth map is used to calculate the warping position for the synthesized view [2]. Therefore, depth map error will cause rendering position shift during view synthesis. When the cameras are rectified and linearly arranged, there exists only horizontal position error during rendering. Since the virtual synthesized view is the warped version of the reference color video, the synthesis distortion SSD_{vir} in terms of sum of squared Distance (SSD) can be approximated as [11]

$$SSD_{vir} = \sum_{x,y} \left(C(x,y) - C(x - \Delta P(x,y), y) \right)^2 \qquad (1)$$

where $C(x, y)$ denotes the reference color pixel at position (x, y), and $(x - \Delta P(x, y), y)$ is shifted pixel position during rendering. Clearly, the synthesis distortion depends on the local characteristics of the reference color video. For example, the synthesized view quality is more sensitive to rendering position shift for the textural regions than that for the flat regions. It has been proved that rendering position shift $\Delta P(x, y)$ is proportional to depth error $\Delta D_{depth}(x, y)$ in [6], i.e.

$$\Delta P(x, y) = \frac{f \cdot l}{255} \cdot \left(\frac{1}{z_{near}} - \frac{1}{z_{far}} \right) \cdot \Delta D_{depth}(x, y) \tag{2}$$

Where f is the focal length of the camera with the unit of pixels, l is baseline distance between two cameras of the reference view and virtual view, z_{near} and z_{far} are the nearest and farthest depth values, respectively.

We also note that the depth quantization error we are concerned with is unevenly distributed in different regions of depth map. Consequently, it can be concluded that the effect of depth map error on the synthesized view quality varies for different local regions, which is determined by both local characteristics of depth map quantization error and the effect of color video on the relationship between depth map distortion and view synthesis distortion.

2.2   Locality of VSD

A. Depth Map Classification for Modeling the Locality of VSD. In this section, we explore the locality of VSD taking into account local characteristics of both depth quantization distortion and the effect of color video on the relationship between depth map distortion and synthesized view distortion. We pre-encode the depth map with a set of quantization parameters (QPs) and synthesize virtual views for statistical modeling. We start by dividing a depth map frame and the corresponding virtual synthesized view frame into blocks, and then compute depth map distortion and synthesized view distortion for each block. To be specific, we exploit the local characteristics of depth map quantization distortion by statistically modeling the relationship between QP and depth map distortion for each block. Likewise, the local characteristics of color video's effect on the relationship between depth map distortion and synthesized view distortion are exploited by statistically modeling their relationship for each block. The distortion is measured by mean squared error (MSE). Figure 1 shows the empirical results of the statistical modeling for one block in a frame of sequence Kendo.

We observe experimentally that depth distortion $D_{depth,B}$ varies linearly with QP in each block as shown in Fig. 1(a), and there relationship can be linearly formulated as,

$$D_{depth,B} = \alpha \cdot QP + \gamma \tag{3}$$

where the slope α indicates the intensity of the relationship between QP and depth map distortion within a block, and it can represent the local characteristics of depth map quantization error. Similarly, as shown in Fig. 1(b), the relationship between depth map

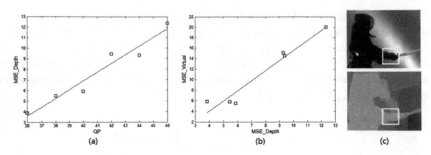

Fig. 1. Empirical modeling results for one block of sequence kendo. (a) Relationship between QP and depth map distortion. (b) Relationship between depth distortion and synthesized view distortion. (c) Location of the tested block in color/depth frame.

distortion $D_{depth,B}$ and synthesized view distortion $D_{vir,B}$ can also be approximated to be linear for each block,

$$D_{vir,B} = \beta \cdot D_{depth,B} + \eta \qquad (4)$$

where the slope β indicates the intensity of relationship between depth map distortion and synthesized view distortion within a block, which differs with the local color video characteristics. γ and η are fitting coefficients.

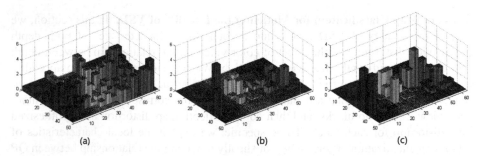

Fig. 2. Value maps of (a) α, (b) β, (c) ψ, for a frame of blocks in sequence kendo.

Then we draw the value map of α and β for a frame of blocks to observe intuitively the local characteristics of depth quantization distortion and that of color video during view synthesis. We can then draw the map of ψ by multiplying α and β, which exploits the local characteristics of synthesized view distortion in terms of depth distortion considering both depth quantization and view synthesis processes. Figure 2 shows the empirical value maps of α, β and ψ for a frame of sequence Kendo. Figure 3 shows the corresponding distribution histogram of ψ for the frame, where we can see that a large part of ψ approaches zero which means that a majority of blocks in a frame of depth map contribute to negligible errors to synthesized view after quantization. Therefore, it is sensible to classify the depth map into regions of the same property to efficiently exploit the locality of VSD due to depth map quantization distortion according to ψ.

Proper threshold T of ψ can be determined referring to the distribution histogram so as for depth map classification. A block is labeled as error sensitive block if the corresponding ψ is larger than T, or error resilient block otherwise. All the error sensitive blocks constitute the error sensitive area (ESA), where the depth map quality is sensitive to quantization and the synthesized view quality is sensitive to depth map distortion. The other blocks make up the error resilient area (ERA).

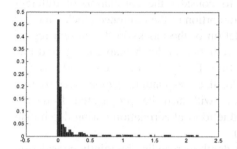

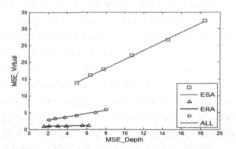

Fig. 3. Distribution histogram of ψ for a frame of sequence Kendo

Fig. 4. Relationship between depth map distortion and view synthesis distortion in ESA, ERA and the entire frame.

B. Locality Model of VSD. In last section, we have divided depth map into two kinds of regions ESA and ERA. Since the relationship between depth distortion and synthesized view distortion has been statistically modeled for each block in Eq. (4), the synthesized view distortion can be estimated with the model coefficients β and η given different depth map distortion for each block. Furthermore, we can obtain view synthesis distortion and depth map distortion for each region by aggregating the distortion of blocks in the same region. Experimentally, we observe that the relationships between depth map distortion and synthesized view distortion for the ESA, ERA and the entire image are well linear fitted, respectively. Figure 4 shows the empirical results for different regions in a frame of sequence Kendo. Therefore the locality of VSD can be modeled by linearly formulating the relationship between synthesized view distortion $D_{vir,\phi}$ and depth map distortion $D_{depth,\phi}$ for the corresponding regions,

$$D_{vir,\phi} = K_\phi \cdot D_{depth,\phi} + B_\phi \tag{5}$$

where $\phi \in \{ESA, ERA, ALL\}$, ALL detotes the entire frame. K_ϕ and B_ϕ are the fitting coefficients for different areas.

As we can see in Fig. 4, the depth map distortion of ESA is much larger than that of ERA and the slope K_{ESA} is much larger than K_{ERA}, which verifies the locality of depth map quantization error.

2.3   VSD Model Based on Locality and Local Correlation

When exploring the local characteristics of relationship between synthesized view distortion and depth distortion, it is generally assumed that different regions of depth map are independent from each other, and the synthesized view distortion is estimated only in terms of the depth map distortion of the corresponding regions. However, note that we estimate the view synthesis distortion to better facilitate depth map coding, and the depth map distortion we are concerned about for VSD modeling is from depth map quantization. Consequently, it is important to consider the correlation of different regions in depth map coding. To be specific, distortion of the reference blocks in other regions will indirectly lead to quality degradation of the blocks in the current region during both temporal prediction and spatial prediction in depth map coding, and this will in turn result in synthesized view distortion. Except for the view synthesis distortion due to depth map distortion of the local corresponding regions (referred as locality of VSD), view synthesis distortion will also be propagated from the non-corresponding regions of the depth map due to local correlation during depth map coding (referred as local correlation of VSD).

   Therefore, we propose the VSD model for depth map coding by jointly considering the locality and local correlation of view synthesis distortion due to depth map quantization distortion as illustrated in Fig. 5. Specifically, the VSD model is obtained as weighted sum of depth map distortion and the locality model of VSD for each region. Since we have verified experimentally in Sect. 2.2 that depth quantization distortion is unevenly distributed as shown in Fig. 2(a), different weighting coefficients are employed for different regions of depth map. The final proposed VSD metric $D_{J,\phi}$ can be represented in terms of depth map distortion $D_{depth,\phi}$ for each region,

$$D_{J,\phi} = p_\phi \cdot D_{depth,\phi} + q_\phi \cdot D_{vir,\phi} = p_\phi \cdot D_{depth,\phi} + q_\phi \cdot \left(K_\phi \cdot D_{depth,\phi} + B_\phi \right) \qquad (6)$$

Where $\phi \in \{ESA, ERA, ALL\}$, p_ϕ and q_ϕ are coefficients which satisfy $p_\phi + q_\phi = 1$.

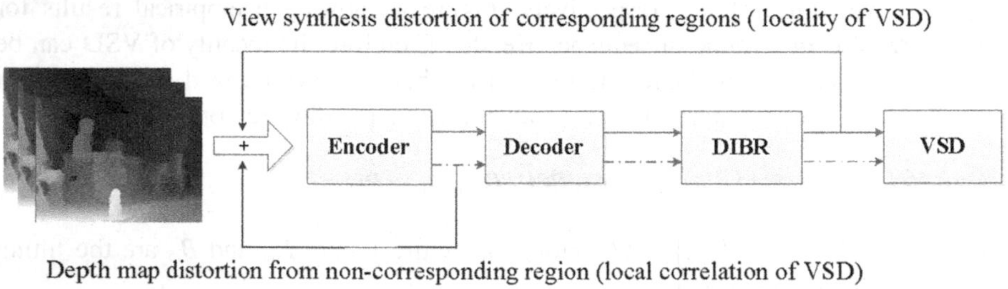

View synthesis distortion of corresponding regions (locality of VSD)

Depth map distortion from non-corresponding region (local correlation of VSD)

Fig. 5. VSD in terms of locality and local correlation of view synthesis distortion due to depth map quantization distortion.

3   Rate-Distortion Optimization Using the VSD Model

In the traditional video codec, such as those based on H.264/AVC, Lagrangian optimization has been extensively used to select the optimized mode for video coding,

$$J_{depth} = D_{depth} + \lambda_{depth} \cdot R_{depth} \tag{7}$$

The optimal Lagrangian multiplier λ_{depth} is determined by taking derivative of (7) and setting it to zero [12, 13], from which we have,

$$\frac{dR_{depth}}{dD_{depth}} = -\frac{1}{\lambda_{depth}} \tag{8}$$

In the case of the proposed approach for depth map coding, the depth distortion metric is replaced by the VSD in (6), and the corresponding bit rate is,

$$R_{J,ALL} = p_{ALL} \cdot R_{depth} + q_{ALL} \cdot R_{vir} = p_{ALL} \cdot R_{depth} + q_{ALL} \cdot \left(R_{depth} + R_{color}\right) \tag{9}$$

Note that we concentrate on the improvement of depth map coding in this paper, we assume that depth and color video are independently encoded and the rate of color video is considered as a constant. The modified optimal Lagrangian multiplier λ_J for the whole frame can be obtained referring to (8)

$$\frac{dR_{J,ALL}}{dD_{J,ALL}} = \frac{d\left(p_{ALL} \cdot R_{depth} + q_{ALL} \cdot \left(R_{depth} + R_{color}\right)\right)}{d\left(p_{ALL} \cdot D_{depth,ALL} + q_{ALL} \cdot D_{vir,ALL}\right)}$$
$$= \frac{1}{p_{All} + q_{ALL} \cdot K_{ALL}} \cdot \frac{dR_{depth}}{dD_{depth}} = -\frac{1}{\lambda_J} \tag{10}$$

Applying (8) to (10), the modified multiplier can be obtained,

$$\lambda_J = \left(p_{ALL} + q_{ALL} \cdot K_{ALL}\right) \cdot \lambda_{depth} \tag{11}$$

When we consider the rate distortion optimization for the ESA and ERA respectively, the Lagrangian cost function $J_{J,\phi}$ can be written as,

$$J_{J,\phi} = D_{J,\phi} + \lambda_{J,\phi} \cdot R_{J,\phi} \tag{12}$$

Where $\phi \in \{ESA, ERA\}$, $D_{J,\phi}$ is the VSD for area ϕ which can be derived from (6), and bit rate $R_{J,\phi}$ can be obtained as the same way in (9). Therefore, Eq. (12) can be further written as

$$J_{J,\phi} = \left(p_\phi + q_\phi \cdot K_\phi\right) \cdot D_{depth} + \lambda_J \cdot R_{depth} + \lambda_J \cdot q_\phi \cdot R_{color} + q_\phi \cdot B_\phi \tag{13}$$

Where the terms of $\lambda_J \cdot q_\phi \cdot R_{color}$ and $q_\phi \cdot B_\phi$ can be regarded as constants. Therefore, by substituting (11) the Lagrangian cost function in (13) can be further rewritten as,

$$J_{J,\phi} = D_{depth} + \frac{p_{All} + q_{ALL} \cdot K_{ALL}}{p_\phi + q_\phi \cdot A_\phi} \cdot \lambda_{depth} \cdot R_{depth} \tag{14}$$

Hence the Lagrangian multiplier is modified by a product factor for the ESA and ERA, respectively, which can be represented as

$$\lambda_\phi = \frac{p_{ALL} + q_{ALL} \cdot K_\phi}{p_\phi + q_\phi \cdot K_\phi} \tag{15}$$

Therefore, our solution is finally achieved by adjusting Lagrangian multiplier relating to weighting factors p_{ESA}, p_{ERA} and p_{ALL} for different regions of depth map.

4 Experimental Results

The proposed VSD metric and the associated RD optimized mode selection are applied based on H.264/AVC (JM18.0). Various test sequences including Kendo, Balloons, BookArrival and Ballet are used to verify the solution. QP values are set as 36, 38, 40, 42, 44 and 46, and the intra period is set as 4. Since we concentrate on the synthesized view quality based depth map coding in this paper, the color video is not coded. View Synthesis Reference Software, VSRS 3.0 [20] is then used to synthesize the virtual view with depth reconstruction and original color video. The synthesized view distortion is measured with PSNR. We verify the Rate-Distortion (R-D) performance of our solution, in which rate refers to the depth map and the distortion refers to the synthesized view.

In the experiments, we implement our solution (denoted as 'Proposed'), as well as the other two schemes, the original H.264/AVC and the most relevant work of the regional RDO scheme proposed by Zhang et al. [9] (denoted as 'RDO_Zhang') as comparison. Figure 6 shows the RD performance comparison results. The ground truth is obtained by synthesizing virtual view with the original depth map and color video of the reference view. From the results, we can easily observe that our proposal can achieve constant and sufficient improvement of RD performance compared with the original H.264/AVC and Zhang's RDO scheme. Though Zhang's RDO method also attempts to exploit the regional characteristics of synthesized view distortion due to depth map distortion, they classify depth map simply by performing canny operator on the corresponding color image. It may correspond to the local characteristics in some cases, but not in all cases. More importantly, Zhang's proposal exploits only the synthesized view distortion of the corresponding local regions without considering the correlation of different regions, while our VSD metric exploits both synthesized view distortion of the local corresponding regions and that propagated from non-corresponding regions due to correlation of regions. Consequently, our solution can efficiently improve the comprehensive synthesized view quality.

Table 1 presents the BDBR and BRPSNR [20] comparison of the proposed RDO scheme and that of Zhang's at high and low bit rate, where high rate refers to $QP \in \{36, 38, 40, 42\}$, and low bit rate refers to $QP \in \{40, 42, 44, 46\}$.

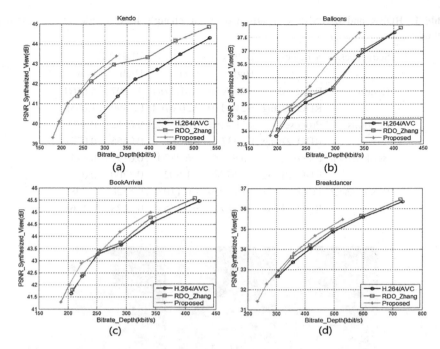

Fig. 6. RD curves of the original H.264/AVC, Zhang's proposed RDO scheme, and our proposed scheme for different sequences

Negative BDBR means percentage of bit rate reduction and positive BDPSNR indicates improvement of image quality compared to H.264/AVC. The results show that our method can achieve 12.67 % and 11.22 % bit rate saving as well as 1.07 dB and 1.94 dB improvement of synthesized view quality on average at high and low bit rate, respectively, compared with H.264/AVC. Besides, the proposed approach can achieve constant improvement of depth map coding.

Figure 7 presents the subjective quality of the synthesized view for a local region of a frame of sequence Balloons. Compared with the result using the 'H.264/AVC' method in (b), our method in (c) can largely preserve the original quality of the virtual synthesized view as in (a), especially the local properties. That is because our approach efficiently exploits the locality of VSD based on a reasonable depth map classification method, which can preserve the regions which are really sensitive to distortion both during quantization and view synthesis. Therefore, the local properties of the synthesized view can be well preserved.

In terms of computational complexity, it's important to note that our method is finally derived as a modifying factor for the Lagrangian multiplier for different regions of depth map, which is easy to apply. And the weighting factors and the coefficients relating to local color and depth characteristics included in the Lagrangian multiplier can be calculated off-line with several QPs for each sequence. Compared with the relevant work, our method is no more complex than Zhang's RDO proposal. And it reduces much computation complexity compared to Chung's method in [10].

Table 1. BDBR and BDPSNR comparison between Zhang's RDO metric ad the proposed method

Sequences	Schemes	BDBR		BDPSNR	
		High	Low	High	Low
Kendo	Proposed	−22.7060	−31.2784	3.1297	5.8169
	RDO_Zhang	−24.3110	−20.6704	0.7490	1.6638
Balloons	Proposed	−11.4161	−0.62522	0.6252	0.8700
	RDO_Zhang	0.0337	0.06548	0.0655	0.0731
BookArrival	Proposed	−5.9698	−6.7830	0.01925	0.6749
	RDO_Zhang	−3.6778	−1.5027	0.1700	0.0901
Breakdancer	Proposed	−10.5930	−6.1107	0.5126	0.3983
	RDO_Zhang	−2.7872	−3.6055	0.1178	0.1565

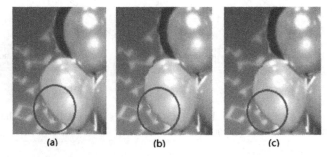

(a)   (b)   (c)

Fig. 7. Subjective quality evaluation of the synthesized view of sequence Balloons. (a) with original depth map, (b) when depth map is coded using H.264/AVC (c) when depth map is coded using the proposed method.

5   Conclusion

In this paper, we propose depth map coding method by modeling the locality and local correlation of view synthesis distortion (VSD) in three dimensional (3D) video. Taking into account local characteristics of both depth map quantization error and color video, we start by dividing depth map into two kinds of regions: error sensitive area (ESA), where depth map quality is sensitive to quantization and the synthesized view quality is sensitive to depth map distortion, and error resilient area (ERA) otherwise. The locality of VSD is established by modeling the relationship between synthesized view distortion and depth map distortion for the different corresponding regions separately. Then the local correlation of VSD is exploited in terms of synthesized view distortion propagated from other non-corresponding regions of depth map during temporal/spatial prediction of depth map coding. The final VSD model is obtained by considering both the locality and local correlation. The proposed VSD metric is implemented in the rate-distortion (RD) optimized mode selection for depth map coding. Experimental results show that our solution can achieve significant improvement of depth map coding.

Acknowledgement. This work was supported in part by the Program 973 No. 2012CB316400, and NSFC No.61175010, 61573268, and International Cooperation Research funding.

References

1. Mülle, K., Merkle, P., Wiegand, T.: 3-D video representation using depth maps. Proc. IEEE **99**(4), 643–656 (2011)
2. Fehn, C.: Depth-image-based rendering (DIBR), compression, and transmission for a new approach on 3D-TV. In: Proceedings of SPIE 5291, Stereoscopic Displays and Virtual Reality Systems, vol. XI, p. 93 (2004)
3. Fang, L., Cheung, N.M., Tian, D., Vetro, A.: An analytical model for synthesis distortoin estimation in 3D video. IEEE Trans. Image Processing **23**(1), 185–199 (2014)
4. Cheung, N.-M., Tian, D., Vetro, A., Sun, H.: On modeling the rendering error in 3D video. In: Proceedings of IEEE International Conference on Image Processing (ICIP) (2012)
5. Oh, B.T., Lee, J., Park, D.-S.: Depth map coding based on synthesized view distortion function. IEEE J. Sel. Top. Sign. Process. **5**(7), 1344–1352 (2011)
6. Kim, W.-S., Ortega, A., Lai, P., Tian, D., Gomila, C.: Depth map distortion analysis for view rendering and depth coding. In: Proceedings of IEEE International Conference on Image Processing ICIP 2009, Cairo, Egypt (2009)
7. Kim, W.S., Ortega, A., Lai, P., Tian, D., Gomila, C.: Depth map coding with distortion estimation of rendered view. In: Proceedings of SPIE, vol. 7543, pp. 75430B-1–75430B-10, January 2010
8. Oh, B.T., Lee, J., Park, D.S., Tech, G., Muller, K., Wiegand, T.: 3D-CE8.h results on view synthesis optimization. ITU-T SG 16 WP 3 and ISO/IEC JTC 1/SC 29/WG 11 Document JCT2-A0033 (2012)
9. Zhang, Y., Kwong, S., Xu, L.: Regional bit allocation and rate distortion optimization for multiview depth coding with view synthesis distortion model. IEEE Trans. Image Process. **22**(9), 3497–3512 (2013)
10. Chung, T.Y., Sim, J.Y., Kim, C.S.: Bit allocation algorithm with novel view synthesis distortion model for multiview video plus depth coding (2014). doi:10.1109/TIP.2014.2327801
11. ISO/IEC JTC1 SC29 WG11 MPEG: Description of 3-D video coding technology proposal by Qualcomm incorporated. Doc. m22583 (2011)
12. Wiegand, T., Girod, B.: Lagrange multiplier selection in hybrid video coder control (2001). doi:10.1109/ICIP.2001.958171
13. Wiegand, T., Schwarz, H., Joch, A., Kossentini, F., Sullivan, G.J.: Rate-constrained coder control and comparison of video coding standards. IEEE Trans. Circ. Syst. Video Technol. **13**(7), 688–703 (2003)
14. Kim, W.S., Ortega, A., Lai, P., Tian, D.: Depth map coding optimization using rendered view distortion for 3-D video coding. IEEE Trans. Image Process. **24**(11), 3534–3545 (2015)
15. Morvan, Y., Farin, D., de With, P.H.N.: Depth-image compression based on an R-D optimized quadtree decomposition for the transmission of multiview images. In: Proceedings of of IEEE International Conference on Image Processing ICIP (2007)
16. Oh, K.J., Vetro, A., Ho, Y.S.: Depth coding using a boundary reconstruction filter for 3-D video systems. IEEE Trans. Circ. Syst. Video Technol. **21**(3), 350–359 (2011)

17. Ekmekcioglu, E., Mrak, M., Worrall, S., Kondoz, A.: Utilization of edge adaptive upsampling in compression of depth map videos for enhanced free-viewpoint rendering. In: Proceedings of IEEE International Conference on Image Processing, pp. 733–736, November 2009
18. Zhao, Y., Zhu, C., Chen, Z., Yu, L.: Depth no-synthesis-error model for view synthesis in 3-D video. IEEE Trans. Image Process. **20**(8), 2221–2228 (2011)
19. Bjontegaard, G.: Calculation of average PSNR differences between RDCurves. ITU-T VCEG (ITU-T SG16 Q.6), Austin TX, US, Document VCEG-M33 (2001)
20. Tanimoto, M., Fujii, T., Suzuki, K.: View synthesis algorithm inview synthesis reference software 3.0 (VSRS3.0), MPEG (ISO/IECJTC1/SC29/WG11), Lausanne, Switzerland, Technical Report M16090 (2009)

Discriminative Feature Learning
with an Optimal Pattern Model
for Image Classification

Lijuan Liu, Yu Bao, Haojie Li$^{(\boxtimes)}$, Xin Fan, and Zhongxuan Luo

School of Software, Key Laboratory for Ubiquitous Network and Service
Software of Liaoning Province, Dalian University of Technology, Dalian, China
lijuanliu@mail.dlut.edu.cn, baoyu_dlut@foxmail.com,
{hjli, zxluo}@dlut.edu.cn, xin.fan@ieee.org

Abstract. The co-occurrence features learned through pattern mining methods have more discriminative power to separate images from other categories than individual low-level features. However, the "pattern explosion" problem involved in mining process prevents its application in many visual tasks. In this paper, we propose a novel scheme to learn discriminative features based on a mined optimal pattern model. The proposed method deals with the "pattern explosion" problem from two aspects, (1) it uses selected weak semantic patches instead of grid patches to substantially reduce the database to mine; (2) the adopted optimal pattern model can produce compact and representative patterns which make the resulted image code more effective and discriminative for classification. In our work, we apply the minimal description length (MDL) to mine the optimal pattern model. We evaluate the proposed method on two publicly available datasets (15-Scenes and Oxford-Flowers17) and the experimental results demonstrate its effectiveness.

Keywords: Weak semantic patches · Pattern mining · MDL principle

1   Introduction

Image classification, as one of the most fundamental issues in computer vision, has been widely studied [1, 25, 26]. A general method to deal with this problem is to represent an image with bag-of-words model [1] and then feed it into a classifier. In the bag-of-words model each image is represented as the distribution of the first order of the low-level features, such as SIFT [2] and HOG [3]. However, low-level features like SIFT extracted from small rigid cells contain no semantic information and lead to a non-discriminative representation. A lot of recent works have been proposed to alleviate this problem. There has been a trend to make use of the well-established frequent pattern mining methods to discover the frequently co-occurred groups which can capture more discriminative information for the image classification. By taking leverage of pattern mining methodology, the relevance dependences between the visual words can be discovered which provide more discriminative property than single visual words [12, 27]. However, the major challenge which is called as *"pattern explosion"*

© Springer International Publishing Switzerland 2016
Q. Tian et al. (Eds.): MMM 2016, Part I, LNCS 9516, pp. 675–685, 2016.
DOI: 10.1007/978-3-319-27671-7_57

makes the pattern mining methods less practical in image classification. To deal with this problem, a lot of auxiliary procedures have been proposed in recent works.

To solve the "*pattern explosion*" problem, reducing the size of the searching space is a popular solution. One hand, a lot of recent works focus on constructing a smaller database of transactions. In [6, 7] transactions are restricted to pairs or triplets for sake of some objects or scenes. [8] divides images into small sub-regions densely and each region is represented as a transaction while [9] reduces the dimension of low-level image representation by a set of random projection matrix. It is observed that image representation with good performance usually capture the meaningful image content and variants, a meaningful database of transactions constructed on low-level features with rich semantic contents is preferred. However, these works have neglected semantic information due to the strong constraints or manually settings in patches extraction procedure. At the opposite extreme, high-level semantic components [4, 5] or stronger compositional features [10, 11] are used to construct the database of transactions, but it is still an open question to capture explicit semantic components.

Actually pattern mining methods can work well in text fields on account of the grammar and sentences in documents but cannot be able to obtain competitive results in standard public image datasets. One of the most important reasons is that the input database of transactions contains less semantic information by the strict pretreatment processes such as mentioned before. In our work, by taking the analogy to the structure in documents (*documents* $\rightarrow$ *paragraphs* $\rightarrow$ *sentences* $\rightarrow$ *words*), we re-read images in a new structure, as *images* $\rightarrow$ *weak semantic patches* $\rightarrow$ *local descriptive regions* $\rightarrow$ *pixels*. The *weak semantic patches* [22], denoted as mid-level concept here, can be detected with a fast method under some rules [24]. Such detected patches are representative and discriminative for containing certain semantic information. We propose to construct the basic input database from these weak semantic patches. To benefit from the semantic property, we can reduce the size of input space in pattern mining phrase semantically and obtain more understandable mid-level patterns.

On the other hand, as the size of mined patterns trough exhaustive searching methods is an exponential relationship to the size of input space, the mined patterns cannot be used as the mid-level model directly. Therefore, smaller subsets should be selected firstly. However this is another exhaustive searching problem with high computation complexity that we cannot obtain the optimal solution in practice in most cases. To solve this problem, some recent works focus on avoiding exhaustive search to decrease the computational complexity. [12, 13] use sequential forward selection method to reduce the search time, while [14, 15] propose to mine both conjunction and disjunction patterns. In these works the pattern mining methods are aiming at finding all individual frequent patterns that satisfy the mining rules, which may lead to redundant pattern models.

We propose to apply the minimal description length (MDL) principle [19] to our problem to find the optimal pattern model, which is a small set of informative patterns while sufficiently describes the structure of the database. Following the MDL principle, we can mine the frequent pattern model that yields the best lossless compression of the database, as MDL ensures that the model will not be overly elaborate or simplistic w.r. t. the complexity of the data [20]. In practice, we propose to use KRIMP algorithm [20] that aims at selecting specific patterns according to MDL principle to mine the mid-level pattern model. The pattern model mined through this way is non redundant and can be used directly.

In this paper, we propose a novel scheme to learn discriminative features based on a mined optimal pattern model and our contribution can be detailed in three aspects. (1) We first propose to construct the input database of transactions based on weak semantic patches. To benefit from the semantic property, we can reduce the size of input space in pattern mining phrase semantically. (2) Secondly, we adopt the minimal description length principle to find the optimal model. This mined pattern model can be used as our mid-level feature model with no further procedures. (3) Lastly, we quantize each image into a mid-level feature vector according this mined model. To take the label information into account further, we propose to apply a new term weighting scheme named term relevance ratio (TRR) on these vectors. The proposed image representation vector reaches the state-of-art performance on image classification task.

2   The Proposed Framework

In this section, we first introduce some notations, then depict our proposed method in detail. As mentioned before, our work has three main steps: (i) construct the transaction database from the input image dataset, (ii) mine the mid-level pattern model with KRIMP algorithm based on the MDL Principle that "the best set of patterns is the set that compresses the database best", (iii) translate images into histograms based on the bag-of-patterns model with a new pooling scheme referred as Term Relevance Ratio.

Notations. In our case, each image I_{c_i} is first described by a set of selected weak semantic patches $P_{c_i} = \left\{ p_{c_{i1}}, p_{c_{i2}}, \ldots, p_{c_{il}} \right\}$ with a class label c, $c \in \{1 \ldots C\}$, C is the number of classes in the dataset, where I_{c_i} means the ith image in the cth class, $p_{c_{ij}}$ means the jth selected patch from image I_{c_i}, l is a non-fixed constant and it depends on images. We represent each patch into a histogram according to the generated visual vocabulary $V = \{v_1, v_2, \ldots, v_D\}$ in a standard bag-of-words scheme, D is the length of visual vocabulary. While translating histograms into transactions, let $E = \{e_1, e_2, \ldots, e_D\}$ be the set of items in which each *item* e_k is an index label of a visual word. A *transaction* t is represented as a subset of items, $t \in PowerSet(E)$ in which $PowerSet(E)$ means the power set of E. The database of transactions with class label c is donated as $db_c = \{t_1, t_2, \ldots, t_n\}$, and the *pattern model* mined from this database is M_c, which is a set of frequent *patterns*, each *pattern* is denoted as x, $x \in PowerSet(PowerSet(E))$.

2.1   From Images to the Database of Transactions

As mentioned in the introduction, we re-read the images in a new structure as *images* $\rightarrow$ *semantic patches* $\rightarrow$ *local descriptive regions* $\rightarrow$ *pixels*. Here we define the second level of images as semantic patches that are both representative and discriminative defined in [22]. The question is that what kind of patches is preferred? *The first selection of course is high-level semantic component (e.g. object, scene etc.).* However, to our knowledge it is still an open question to capture explicit semantic components and for another reason that the high-level semantic components are not representative enough to cope with the deformations or other image noise. Therefore in our work, we prefer to select *flexible meaningful patches under certain rules from each image.*

We propose to use selective search [24] which employs multiple hierarchical segmentations based on a variety of color spaces to detect the weak semantic patches. As showed in [24], this method can generate discriminative object-orient patches with a very high recall and a reasonable number of patches. In our work, as the Fig. 1 shows, we use this selective search on general images and it seems that this method can also work very well with that the retuned patch set overlaps the original image with a high ratio and in which each patch seems discriminative and representative.

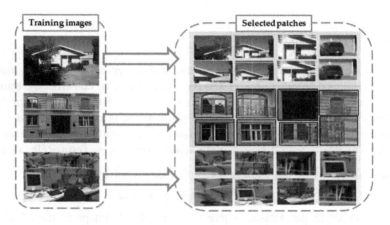

Fig. 1. The selected patches from image dataset (15-Scenes) based on selective search [24]

After we selected the set of patches $P_c = \left\{ P_{c_1}, P_{c_2}, \ldots, P_{c_{N_c}} \right\}$ from the set of images with class label c, in which $P_{c_i} = \left\{ p_{c_{i1}}, p_{c_{i2}}, \ldots, p_{c_{il}} \right\}$, N_c is the number of training images, we quantize each patch into a representative vector in a standard bag-of-words scheme referred as local BOW (LBOW). We extract SIFT descriptors from the selected patches, then create visual vocabulary $V = \{v_1, v_2, \ldots, v_D\}$ using K-Means clustering and lastly assign each SIFT descriptor to the nearest visual word (Euclid space). By keeping the non-zero bins of each LBOW vector we can translate the patches into a set of transactions donated as a database $db_c = \{t_1, t_2, \ldots, t_n\}$, in which $t_i \in PowerSet(E)$.

2.2   Pattern Mining Based on MDL Principle

For a given database of transactions, we can use any standard data mining methods to find the patterns. Here we propose the MDL principle based pattern mining algorithm, KRIMP [20]. MDL principle is based on the insight that "any regularity in the data can be used to compress the data". Equating learning about data with finding regularity from data, MDL methods can be interpreted as searching for a model that compress the data best and the mined model can be regarded as the best description of the input database. Moreover, the pattern mining method based on MDL principle can avoid searching the whole input database exhaustively and the mined pattern model is non-redundant.

According to the MDL principle, for a given potential set of models M_*, the best model $M_c \in M_*$ is the one that satisfies

$$M_c = \underset{M \in M_*}{\mathrm{argmin}}\, L(M) + L(db_c|M) \tag{1}$$

where $L(M)$ is the length of the description the model M in bits and $L(db_c|M)$ is the length of the description of the database when encoded with model M in bits.

To solve this function, KRIMP algorithm defines two functions. One is a cover function of model M and transaction t, $cover{:}M \times t \mapsto PowerSet(PowerSet(E))$, which identifies which patterns of the pattern model M are used to encode the transaction t in the input database db_c. The other is a usage function, $usage(x) = |\{t \in db_c|x \in cover(M,t)\}|$, which counts the number of transactions t from database db_c where x is used to cover.

Here Shannon entropy is used to compute the length. Firstly $L(M)$ is computed as follows,

$$L(M) = \sum_{x \in M,\, usage_{db}(x) \neq 0} L(x|M_{st}) + L(x|M) \tag{2}$$

where M_{st} is the simplest model that only contains the singleton patterns. For a given model M, the length of a pattern x [20] is

$$L(x|M) = -\log P(x|db_c) \tag{3}$$

where $P(x|db_c)$ is a probability distribution of $x \in M$ for db_c, given by

$$P(x|db_c) = \frac{usage_{db_c}(x)}{\sum\limits_{y \in M} usage_{db_c}(y)} \tag{4}$$

Then the length of the database db_c encoded with a given pattern model is the sum of all transactions encoded with the current model

$$L(db_c|M) = \sum_{t \in db_c} L(t|M) \tag{5}$$

$$L(t|M) = \sum_{x \in cover(M,t)} L(x|M) \tag{6}$$

By solving the above optimal function we obtain a pattern model M_c that describes the original database best. In practice, we select K patterns with the shortest description length from each class and then union all of them to serve as a mid-level feature model $\mathbf{M_f} = \{x_1, x_2, \ldots, x_L\}$, L is the dimension of the resultant representation of images.

2.3 Term Relevance Ratio Weighting

Based on the mined mid-level feature model, we can then quantize each image into a histogram vector with bag-of-patterns model. However, this is not a wise choice here as we only mine the most frequent patterns from each class to avoid over-fitting. We propose a new term weighting scheme to take label information into account further. TF-IDF is a widely used scheme in image retrieval and image classification. TF (Term Frequency) measures the importance of terms in the current image and IDF (Inverse Document Frequency) measures the importance of terms overall the database which is a general importance that neglect label information.

Here we propose to use a new term weighting scheme which is popular in text classification in recent years [21]. In the new term weighting scheme the IDF part is replaced by the ratio of the probability estimations on the relevant class and other classes, which is referred as Term Relevance Ratio (TRR). The ratio is formulated into a log-odds form,

$$trr_{ij} = \log\left(\frac{P(x_i|c_j)}{P(x_i|\bar{c}_j)} + \alpha\right) \tag{7}$$

where c_j is the relevant image class and $\bar{c}_j$ are the rest image classes in the database, α is a constant value to make sure that logarithmic value is positive. The probabilities are estimated as follows,

$$P(x_i|c_j) = \frac{\sum_{k=1}^{|\mathbf{I}_{c_j}|} tf_{ik}}{\sum_{l=1}^{|\mathbf{M_f}|} \sum_{k=1}^{|\mathbf{I}_{c_j}|} tf_{lk}} \tag{8}$$

$$P(x_i|\bar{c}_j) = \frac{\sum_{k=1}^{|\mathbf{I}_{\bar{c}_j}|} tf_{ik}}{\sum_{l=1}^{|\mathbf{M_f}|} \sum_{k=1}^{|\mathbf{I}_{\bar{c}_j}|} tf_{lk}} \tag{9}$$

where $\mathbf{I}_{c_j}$ is the set of images in relevant class and $\mathbf{I}_{\bar{c}_j}$ is the set of images from the rest classes, tf_{ik} means the term frequency of the ith pattern in the kth image. Now the integrated term weighting scheme is

$$tw_{ik} = (\log(tf_{ik} + 1)) \cdot trr_{ij}, \quad \forall I_k \in \mathbf{I}_{c_j} \tag{10}$$

After we obtained the quantized histogram for each image, we use this new term weighting scheme to reweight the histogram and then obtain a new vector which is more specified to classification task.

In our work, we also propose a new rectangle based spatial matching scheme to capture the spatial information. By the observation that most of the meaningful patches locate at the center of images, the proposed match scheme focuses on the image center and is more adaptive to our selected flexible located semantic patches than the traditional SMP approach [18]. In our experiment, the image is partitioned into the center located rectangle and the rest surroundings of the image, the resultant representation is $h = [(h_{st}),(h_{cen1},h_{sur1}),(h_{cen2},h_{sur2})]$, h_{st} is the quantized vector based on all selected patches within the image, h_{cen} is calculated on the patches matched in the center space while h_{sur} is calculated on the patches matched in the surrounding space, the ratio of center space and surrounding space is $cen1/sru1 = 1/3$, $cen2/sru2 = 1/1$, "match" means the overlapped area $area_o$ between patches and partitioned space satisfies $area_o/area_p > 0.5$, $area_p$ is the size of patches.

3   Experiments

In this section, we first validate our proposed approach on two public datasets. We also conduct a group of experiments to evaluate the effect of parameters.

3.1   Datasets and Experimental Setups

We evaluate the performance of proposed approach on two public datasets: 15-Scenes [18] and Oxford-Flowers17 [17].

The 15-Scenes dataset is used to evaluate scene classification and contains totally 4485 images falling into 15 categories. Following the experimental procedures in the literature, 100 randomly selected images per category are used for training and the rest are used for testing. The Oxford-Flowers17 dataset contains 17 flower categories and each category contains 80 images. According to [8], we randomly take 60 images per category for training and use the rest for testing.

In our implementations we densely extracted SIFT descriptors from 16×16 pixels patches from each image on a grid with the stepsize 8 pixels. Then we use standard K-Means clustering to train the visual vocabulary and fix the size as 1024 according to [9]. We use the default parameters [24] in the weak semantic patch detection phrase. In the pattern mining stage, we keep the top shortest 1200 patterns and 1000 patterns for each category in 15-Scenes dataset and Oxford-Flowers17 dataset respectively. The classifiers are trained using SVM and we choose the kernel as histogram intersection kernel. The parameters used in classification are selected by 5-cross validation and we use mean classification accuracy to evaluate our results on the two datasets.

3.2   Results and Discuss

Comparing Several State-of-the-Art Methods. In Table 1 We compare our results on 15-Scenes dataset against several recent results on the same dataset. [16, 23] are deformable part based methods and each part is represented as a HOG [3] vector. [8] is the most similar method to ours that the database of transactions is constructed from fixed rigid patches. Table 2 shows the comparisons between the state-of-the-art approaches and our proposed image representation on Oxford-Flower-17 dataset. In [9] the database of transactions are created from the subspace generated with random projection matrix on the low-level features.

The experimental results demonstrate the effectiveness of our proposed approach. By archiving more than 2 % improvement over both the deformable part based methods and the approach using pattern mining on fixed rigid patches on 15-Scenes dataset, we have verified the two assumptions introduced in introduction, (1) the co-occurrence features capturing the relevance dependence between low-level features are more discriminative and representative than single low-level feature, furthermore, the mined optimal pattern model which applied MDL principle in practice can work well without any extra procedures; (2) the database of transactions constructed from weak semantic patches selected under certain rules is more understandable than that constructed by complex mathematical methods or just simply dividing the images. Most of all, our proposed scheme is proved to be an effective approach in image understanding for the task of image classification.

Table 1. Comparison with state-of-the-art results on the 15-Scenes dataset

Methods	Results
FLH + GRID [8]	86.2
D_Parts [23]	86.0
DSFL [16]	84.2
The proposed	**88.41**

Table 2. Comparison with state-of-the-art results on the Oxford-Flower-17 dataset

Methods	Results
FLH + GRID [8]	92.9
HoPs [9]	93.8
FLH + SPM [8]	92.6
The proposed	**94.1**

Detailed Analysis. We also conduct a group of experiments on 15-Scenes dataset to evaluate the effect of parameters.

The size of the pattern model. In our work, we first mine one pattern model from the set of one class images and then union them to a mid-level feature model. According to the MDL principle, the larger the model is, the more regularities the method mines. Then intuitively, the large pattern model is discriminative and the small pattern model will be representative. From Table 3 we can see that, the mean accuracy decreases when the size of the pattern model is small, which may be due to the fact that the representative mid-level feature model is not discriminative enough. As the size becomes larger, the mean accuracy also goes down, the cause of this phenomenon may be the data overfitting.

Weak semantic patches or fixed rigid patches? To investigate whether the weak semantic patches improve the result of classification or not, we conduct a group of comparative experiments between propose weak semantic patches and fixed rigid patches. The result showed in Table 4 confirmed our assumption and the result based on weak semantic patches outperforms that based on fixed rigid patches.

Table 3. The results on 15-Scenes with different number of mined patterns per class

K	800	1000	1200	1400
Results	86.8	88.1	**88.4**	88.04

Table 4. The results on 15-Scenes with different kind of patches

Kinds of patches	Fixed rigid patches	Weak semantic patches
Results	87.3	**88.4**

Rectangle based spatial matching scheme. In our work, we also propose a rectangle based spatial matching scheme to capture the spatial information rather than traditional SMP scheme. From the results showed in Table 5, we find that the SPM scheme can barely improve the ability of our image representation, however, our proposed rectangle based spatial matching scheme has archived a better performance.

Table 5. The results on 15-Scenes with different kind of spatial matching schemes

Matching scheme	No spatial info	SPM scheme	Rectangle SM
Results	85.9	86.3	**88.4**

4 Conclusions

In this paper we proposed a novel mid-level semantic feature learning method base on pattern mining method. We re-read the image into a new structure that the weak semantic patches dominate the second level structure and address the "pattern explosion" problem through taking the weak semantic patches into use. Moreover, we prefer to mine the optimal pattern set that compress the database with best lossless used as our mid-level pattern model according to the MDL principle. We also propose to use a new pooling scheme to take the class label into account to make the feature representations of images more discriminative. The experimental results on two public datasets have demonstrated the effectiveness of the proposed approach.

Acknowledgements. This work was partially supported by National Natural Science Funds of China (61173104, 61472059, 61428202).

References

1. Sivic, J., Zisserman, A.: Video google: a text retrieval approach to object matching in videos. In: 2003 IEEE Conference on Computer Vision and Pattern Recognition (CVPR), pp. 1470–1477. IEEE (2003)
2. Lowe, D.G.: Distinctive image features from scale-invariant keypoints. Int. J. Comput. Vision **60**(2), 91–110 (2004)

3. Dalal, N., Triggs, B.: Histograms of oriented gradients for human detection. In: 2005 IEEE Conference on Computer Vision and Pattern Recognition (CVPR), pp. 886–893. IEEE (2005)

4. Savarese, S., Winn, J., Criminisi, A.: Discriminative object class models of appearance and shape by correlatons. In: 2006 IEEE Conference on Computer Vision and Pattern Recognition (CVPR), pp. 2033–2040. IEEE (2006)

5. Yao, B., Fei-Fei, L.: Grouplet: A structured image representation for recognizing human and object interactions. In: 2003 IEEE Conference on Computer Vision and Pattern Recognition (CVPR), pp. 9–16. IEEE (2010)

6. Liu, D., Hua, G., Viola, P., Chen, T.: Integrated feature selection and higher-order spatial feature extraction for object categorization. In: 2008 IEEE Conference on Computer Vision and Pattern Recognition (CVPR), pp. 1–8. IEEE (2008)

7. Yang, Y., Newsam, S.: Spatial pyramid co-occurrence for image classification. In: 2011 IEEE International Conference on Computer Vision (ICCV), pp. 1465–1472. IEEE (2011)

8. Fernando, B., Fromont, E., Tuytelaars, T.: Mining mid-level features for image classification. Int. J. Comput. Vision 108, 186–203 (2014)

9. Voravuthikunchai, W., Crémilleux, B., Jurie, F.: Histograms of pattern sets for image classification and object recognition. In: 2014 IEEE Conference on Computer Vision and Pattern Recognition (CVPR), pp. 224–231. IEEE (2014)

10. Jiang, Y., Meng, J., Yuan, J.: Randomized visual phrases for object search. In: 2012 IEEE Conference on Computer Vision and Pattern Recognition (CVPR), pp. 3100–3107. IEEE (2012)

11. Zhang, S., Huang, Q., Hua, G., Jiang, S., Gao, W., Tian, Q.: Building contextual visual vocabulary for large-scale image applications. In: 2010 ACM International Conference on Multimedia, pp. 501–510. ACM (2010)

12. Mita, T., Kaneko, T., Stenger, B., Hori, O.: Discriminative feature co-occurrence selection for object detection. In: 2008 IEEE Transactions on Pattern Analysis and Machine Intelligence, pp. 1257–1269. (2008)

13. Torralba, A., Murphy, K.P., Freeman, W.T.: Sharing visual features for multiclass and multi-view object detection. In: 2007 IEEE Transactions on Pattern Analysis and Machine Intelligence, pp. 854–869. (2007)

14. Yuan, J., Yang, M., Wu, Y.: Mining discriminative co-occurrence patterns for visual recognition. In: 2011 IEEE Conference on Computer Vision and Pattern Recognition (CVPR), pp. 2777–2784. IEEE (2011)

15. Weng, C., Yuan, J.: Efficient mining of optimal AND/OR patterns for visual recognition. IEEE Trans. Multimedia 17(5), 626–635 (2015)

16. Zuo, Z., Wang, G., Shuai, B., Zhao, L., Yang, Q., Jiang, X.: Learning discriminative and shareable features for scene classification. In: Fleet, D., Pajdla, T., Schiele, B., Tuytelaars, T. (eds.) ECCV 2014, Part I. LNCS, vol. 8689, pp. 552–568. Springer, Heidelberg (2014)

17. Nilsback, M.E., Zisserman, A.: Automated flower classification over a large number of classes. In: 2008 Indian Conference on Computer Vision, Graphics and Image Processing (ICVGIP), pp. 722–729. IEEE (2008)

18. Lazebnik, S., Schmid, C., Ponce, J.: Beyond bags of features: spatial pyramid matching for recognizing natural scene categories. In: 2006 IEEE Conference on Computer Vision and Pattern Recognition (CVPR), pp. 2169–2178. IEEE (2006)

19. Grünwald, P.: A tutorial introduction to the minimum description length principle. In: Advances in Minimum Description Length: Theory and Applications, pp. 23–81 (2005)

20. Vreeken, J., Van Leeuwen, M., Siebes, A.: Krimp: mining itemsets that compress. Data Min. Knowl. Disc. 23(1), 169–214 (2011)

21. Ko, Y.: A study of term weighting schemes using class information for text classification. In: Proceedings of the 35th International ACM SIGIR Conference on Research and Development in Information Retrieval, pp. 1029–1030. ACM (2012)
22. Singh, S., Gupta, A., Efros, A.A.: Unsupervised discovery of mid-level discriminative patches. In: Fitzgibbon, A., Lazebnik, S., Perona, P., Sato, Y., Schmid, C. (eds.) ECCV 2012, Part II. LNCS, vol. 7573, pp. 73–86. Springer, Heidelberg (2012)
23. Sun, J., Ponce, J.: Learning discriminative part detectors for image classification and cosegmentation. In: 2013 IEEE International Conference on Computer Vision (ICCV), pp. 3400–3407. IEEE (2013)
24. Van de Sande, K.E., Uijlings, J.R., Gevers, T., Smeulders, A.W.: Segmentation as selective search for object recognition. In: 2011 IEEE International Conference on Computer Vision (ICCV), pp. 1879–1886. IEEE (2011)
25. Wang, M., Liu, X., Wu, X.: Visual Classification by l_1 -Hypergraph Modeling. In: 2015 IEEE Transactions on Knowledge and Data Engineering, pp. 2564–2574. IEEE (2015)
26. Lu, D., Weng, Q.: A survey of image classification methods and techniques for improving classification performance. Int. J. Remote Sens. **28**, 823–870 (2007)
27. Wang, J., Wang, M., Li, P., Liu, L., Zhao, Z., Hu, X., Wu, X.: Online feature selection with group structure analysis. IEEE Trans. Knowl. Data Eng.

Sign Language Recognition Based on Trajectory Modeling with HMMs

Junfu Pu, Wengang Zhou[✉], Jihai Zhang, and Houqiang Li

University of Science and Technology of China, Hefei, Anhui,
People's Republic of China
{pjh,jihzhang}@mail.ustc.edu.cn, {zhwg,lihq}@ustc.edu.cn

Abstract. Sign language recognition targets on interpreting and understanding the sign language for convenience of communication between the deaf and the normal people, which has broad social impact. The problem is challenging due to the large variations for different signers and the subtle difference between sign words. In this paper, we propose a new method for isolated sign language recognition based on trajectory modeling with hidden Markov models (HMMs). In our approach, we first normalize and re-sample the raw trajectory data and partition the trajectory into multiple segments. To represent each trajectory segment, we proposed a new curve feature descriptor based on shape context. After that, hidden Markov model is used to model each isolated sign word for recognition. To evaluate the performance of our proposed algorithm, we have built a large isolated Chinese sign language vocabulary with Kinect 2.0. The dataset contains 100 unique isolated sign words, each of which is performed by 50 signers for 5 times. Experimental results demonstrate that the proposed method achieves a better performance compared with normal coordinate feature with HMM.

1 Introduction

Sign language is one of the most important ways for communication between the deaf and the normal people. With broad social impact, this problem has attracted considerable attention from many researchers around the world. The target is to build a system to automatically translate sign language into text or interpret it into spoken language [17,18]. The research methods of sign language recognition (SLR) can also be applied in general human-computer interaction systems.

Sign language conveys semantic meaning through hand shapes, trajectories, and facial expressions. Most existing recognition methods for sign language are based on gestures and trajectories of sign words. For gesture recognition, Murakami and Taguchi [9] proposed a gesture recognition method for Japanese sign language using recurrent neural network, but they used the data gloves, which were expensive and inconvenient in real application for the signers. They developed a posture recognition system which could recognize a finger alphabet of 42 symbols and achieved a recognition rate of 98 % for registered people.

© Springer International Publishing Switzerland 2016
Q. Tian et al. (Eds.): MMM 2016, Part I, LNCS 9516, pp. 686–697, 2016.
DOI: 10.1007/978-3-319-27671-7_58

Also, an alphabet gesture recognition system was designed for American sign language (ASL) with ANN by Oz and Leu [10]. Hidden Markov models (HMMs) [13] show an extremely good performance in temporal pattern recognition, especially in speech recognition [2,12]. Schlenzig et al. [14] used a single universal HMM and a finite state estimator for the determination of gestures. Their proposed method also achieved a high recognition rate. Huang et al. [6] built a deep neural network based on postures captured by Real-Sense.

The above works are commonly based on gesture recognition. However, these methods are not able to handle the recognition for signs with only trajectory information. To address this problem, more and more researchers focus on sign language recognition by trajectory matching [16]. Yushun et al. [8] presented a new method of curve matching from the views of manifold for sign language recognition. They divided the curve of the sign word into a set of several linear segments, and defined the distance between two segments. Thus, the matching of two curves was transformed into the matching between two sets of linear segments. Their method achieved a recognition rate of 78.3 % in a dataset with 370 daily sign words. In addition, the combination of both trajectory and hand shape features for sign language recognition was proposed by Grobel and Assan [3], and data gloves were also necessary in there experiments. Although sign language recognition with data gloves [4] achieved a high recognition rate, it's inconvenient to be applied in SLR system for the expensive device. Kinect developed by Microsoft [15] is capable of capturing the depth, color, and joint locations easily and accurately. Hence, more and more researchers use Kinect for sign language recognition [5,20–22].

The method we proposed in this paper is based on trajectories of sign words. The data captured by Kinect consists of a set of 3D points, which are the axis locations of joints in each temporal stamp. We use the trajectories of both hands for recognition. However, recognition only based on trajectories suffers difficulties in some specific cases. Take the Chinese sign words "You" and "Good" for example, the trajectories of both words are shown in Fig. 1. These two trajectories are quite similar, which makes it tough to realize the similar signs' accurate recognition.

The rest of this paper is organized as follows. Section 2 gives an overview of our system. The method of feature extraction is discussed in Sect. 3. Section 4 introduces the details of sign words modeling with HMMs. The experiments are carried out in Sect. 5. Finally, in Sect. 6 we make a summary and brief discussion for future work.

2   System Overview

As shown in Fig. 2, the trajectory matching system for sign language recognition consists of 5 modules, i.e., data preprocessing, discrete contour evolution (DCE) [7] for segmentation, feature extraction for HMMs, codebook training and quantization, and trajectory matching with HMMs. The aim of preprocessing is to normalize the data and decrease the negative effect of noise. We re-sample the

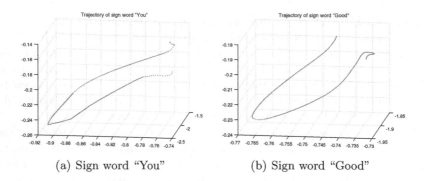

(a) Sign word "You"   (b) Sign word "Good"

Fig. 1. The illustration of two trajectories for Chinese sign words "You" and "Good". These two trajectory curves are quite similar with subtle difference.

raw data path into a new curve with fixed length. For the re-sampled curves, the DCE algorithm [7] is used for segmentation. Then we extract the shape context [1] of each point and randomly choose some of them for training a codebook. The curve features introduced in Sect. 3 are extracted for each trajectory. With these features, we build a HMM model for each isolated sign word. Then the label with maximum posterior probability is regarded as the recognition result.

3   Curve Feature Extraction

In this section, we will introduce how to build robust features with an effective representation of the 3D trajectory curve. These features will be used in trajectory curve recognition and retrieval. First of all, we extract the shape context of all points in the curve. For all word samples, we choose some of them randomly for training codebooks, without taking account of their labels. Then for each word sample, we quantize the shape context features of all points in the trajectory curve with the pre-trained codebooks. An illustration of curve feature extraction based on shape context is shown in Fig. 3.

3.1   Shape Context

In literature, shape context [1] has been wildly used in shape recognition. Shape context is a feature that describes the distribution of other points in the neighborhood of a reference point. For a point p_i on the curve, shape context is defined as a histogram h_i of the relative coordinates of the remaining $n - 1$ points, where n is the number of points in the curve. Equation 1 gives the definition of h_i, where # means the cardinality of a set. In order to make the descriptor more discriminative to nearby points, we use $log - polar^2$ coordinate system. A brief illustration is shown in Fig. 4.

$$h_i(k) = \#\{q \neq p_i : (q - p_i) \in bin(k)\}. \tag{1}$$

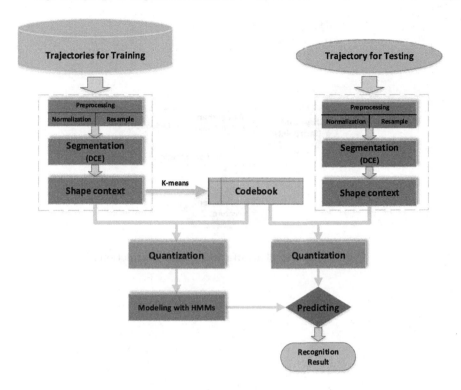

Fig. 2. An illustration of proposed method. The steps in the left column consist the stage of training. For a testing trajectory of sign word, the HMMs that we've trained are used for recognition.

Considering that our sign word trajectory is 3D, we project it along three orthogonal planes, i.e., x-y, y-z, and x-z, and obtain three 2D curves. Then, for each curve, we extract a shape context feature and concatenate them into a single feature vector to represent the 3D sign word trajectory. We use 2D shape context feature since they are efficient for extraction and well capture the data structure.

3.2 Codebook Training

After getting the shape context features of all sample points in a sign word trajectory, we train the codebook for quantization. Here, we use K-means algorithm for codebook training. The main steps are as follow: First, we randomly sample a set of training features. Then we extract shape context features of these chosen samples and use K-means clustering algorithm to generate a set of cluster centers. The cluster centroids constitute our codebook. Suppose the codebook size is K, then the codebook can be described as $B = [b_1, b_2, \ldots, b_K]^T \in R^{K \times d}$, where b_i is the cluster centroid vector and d denotes the dimension of shape context feature.

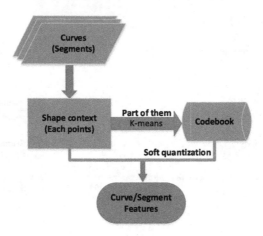

Fig. 3. The flow chart of curve feature extraction

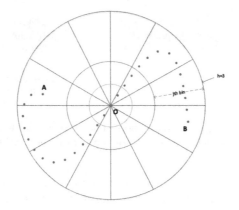

Fig. 4. An illustration of shape context. For point O, we count the points dropped into each bin. As is marked in this figure, there are 3 points in the j-th bin, so $h_j = 3$.

3.3   Quantization

A 3D trajectory is formed with a set of sequential 3D points. For a N-point curve $C = (p_1, p_2, \cdots, p_N)$, we use the codebook discussed in Sect. 3.2 to quantize the shape context feature of each point p_i. In our experiment, soft quantization [11] is used to get curve feature. The illustration of soft quantization is shown in Fig. 5.

As is shown in Fig. 5, $B = [b_1, b_2, \ldots, b_K]^T$ is the codebook, which b_i is the cluster centroid feature. SC_i denotes the shape context of point p_i in curve C. We calculate the Euclidean distance between SC_i and $b_q(q = 1, 2, \cdots, K)$.

$$d_{i,q} = ||SC_i - b_q||. \tag{2}$$

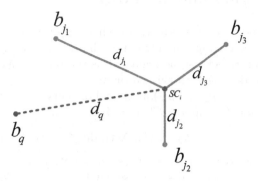

Fig. 5. An illustration for soft-quantization. b_{j_1}, b_{j_2} and b_{j_3} are 3 nearest centroid features far from SC_i in feature space. So, we use the distance d_{j_1}, d_{j_2} and d_{j_3} to calculate the weights for quantization.

Denote $d_i = [d_{i,1}, d_{i,2}, \ldots, d_{i,k}]^T$, then we select Q smallest distances $d_{j_1}, d_{j_2}, \ldots,$ d_{j_Q}, and the weights of quantization is given by Eq. 3 [11].

$$\omega_i = \frac{exp^{-\frac{d_i^2}{\sigma^2}}}{\sum\limits_{j=1}^{Q} exp^{-\frac{d_j^2}{\sigma^2}}}, \quad i = 1, 2, \cdots, Q, \tag{3}$$

where σ is a constant. Then the curve feature $f = [f_1, f_2, \ldots, f_K]$ could be generated as Algorithm 1.

Algorithm 1. Extraction of Curve Feature

Input: The 3D trajectory points $C = (p_1, p_2, \cdots, p_N)$;
    Codebook $B = [b_1, b_2, \cdots, b_K]^T \in R^{K \times d}$;
Output: The curve feature with a good enough description of curve C;
 1: n=1;
 2: $f = [f_1, f_2, \cdots, f_K]^{K \times 1}$, set $f_i = 0$ for all $i = 1, 2, \cdots, K$;
 3: **for** $n = 1 : Q$ **do**
 4:    Extract the shape context SC_n of point p_n;
 5:    $d = [d_1, d_2, \cdots, d_K]^{K \times 1}, d = B \times SC_n$;
 6:    Make d sorted by increasing and get Q minimum values $d_{i_1}, d_{i_2}, \cdots, d_{i_Q}$ and its
       corresponding index $idx_{i_1}, idx_{i_2}, \cdots, idx_{i_Q}$;
 7:    **for** $j = 1 \rightarrow Q$ **do**
 8:       Get ω_j refers to Eq. 3;
 9:       $f_{idx_j} = f_{idx_j} + \omega_j$;
10:    **end for**
11: **end for**
12: Output the curve feature f;

---

4   Character Modeling by HMM

Hidden Markov models are well known for their application in temporal patten recognition such as speech and handwritten characters. In sign language recognition, the trajectories can also be regarded as temporal sequences, and HMMs are appropriate for dealing with this problem.

Let's denote $C = (p_1, p_2, \ldots, p_N)$ as the whole curve with N points, which is divided into M segments. Suppose $C^{(i)} (i = 1, 2, \cdots, M)$ denotes the segment set of C, then $C = \bigcup\limits_{i=1}^{M} C^{(i)} (i = 1, 2, \ldots, M)$. Actually, $C^{(i)}$ can be regarded as a set of sub-motions and the context between these sub-motions is going to be modeled with HMMs. In our experiments, we use DCE (Discrete Contour Evolution) algorithm to partition the curve path into M segments. Data preprocessing is necessary before segmentation.

4.1   Preprocessing

The procedure of data preprocessing includes two aspects: normalization and resampling. The trajectories should be normalized since they are quite different in scales when performed by different signers. Specifically, we use the location of the signer's head and width of shoulder to realize normalization. Resampling also plays an important role in preprocessing. It will make the trajectories more smooth and remove the noise. Generally, the velocities of different parts can be much different when a signer is playing a sign language word, so the sample points are not uniformly distributed, resampling will solve this problem much more better. We use the $1 algorithm proposed by Wobbrock et al. [19] for resampling.

4.2   DCE Algorithm

In sign language recognition, $C = (p_1, p_2, \ldots, p_N)$ is composed of digital line segments $s_1, s_2, \ldots, s_{N-1}$, where s_i is the line segment joining p_i to p_{i+1} for $i = 1, 2, \ldots, N - 1$. We normalize s_i by the length of curve C. Let's denote $l(s)$ as the length of s, and $\beta(s_1, s_2)$ as the angle between s_1 and s_2. The cost function $Z(s_1, s_2)$ for a pair of segments s_1 and s_2 is defined in Eq. 4 [7].

$$Z(s_1, s_2) = \frac{\beta(s_1, s_2) l(s_1) l(s_2)}{l(s_1) + l(s_2)}. \tag{4}$$

The main procedure of DCE is given by Algorithm 2 [7].

Algorithm 2. DCE Algorithm

Input: The 3D trajectory points $C - (p_1, p_2, ..., p_N)$;
   The number of segments M;
Output: The segmentation $C^{(i)}(i = 1, 2, ..., M)$ of curve C;
 1: k=N;
 2: **while** k>M+1 **do**
 3:    Find a pair s_i, s_{i+1} such that $Z(s_i, s_{i+1})$ is minimal;
 4:    Replace s_i, s_{i+1} by the segment s' joining the endpoints of arc $s_i \bigcup s_{i+1}$;
 5:    k=k-1;
 6: **end while**
 7: Use the index of remaining points to get segmentation $C^{(i)}, i = 1, 2, ..., M$;

4.3   HMM Modeling

After getting the partition of curve C, we can extract curve feature from each sub-motion $C^{(i)}(i = 1, 2, ..., M)$ for HMMs training. For each word, we can use some of these curve features as training samples to build a hidden Markov model. Each character HMM is structured as left-to-right model.

The HMM is modeled by the parameter vector $\lambda = (A, B, \pi)$, where A is the transition probability matrix, B is the emission probability matrix, and π is the initial state probability distribution. For each word, we can use our training samples to get the parameter vector λ by Baum-Welch algorithm. Suppose there are N HMMs to be trained, that is to say, the whole data set contains N different isolated sign language words. Given an unknown testing sequence $O = [o_1, o_2, ..., o_n]$, we classify it to class C_p with the following decision rule:

$$C_p = \max_{C_i} \log p(O|\lambda_{C_i}), \quad i = 1, 2, ..., N, \tag{5}$$

where $\log p(O|\lambda)$ is the logarithm of the probability of sequence O, given the model parameter λ. The sum of $\log p(O|\lambda)$ for both hands is used for recognition.

5   Experiments

5.1   Datasets and Experimental Setup

Our dataset is built by ourselves with Kinect and will be released to the public. It contains 100 isolated Chinese sign language words in daily life. Each word is played by 50 signers for 5 times. As a result, the dataset consists of $100 \times 50 \times 5$ samples. We divide the whole dataset into 2 subsets for training, validation, and testing. The details about these 2 subsets are shown as follow:

Subset A. Subset A contains 100 words, each of which is performed by 14 signers for 5 times. So there are 7,000 samples in total. In the experiments introduced in Sect. 5.2, we choose 60 samples of each word for training, and the rest 10 for validation.

Subset B. The vocabularies for both dataset are the same. Each word in Subset B is performed by another 36 signers for 5 times. In our experiments, we use Subset B as a large testing set to evaluate the effectiveness and stability of our method.

The Kinect developed by Microsoft is able to detect the joints and we can obtain the locations of joints in real-time. In sign language recognition, the locations of both hands will make sense for recognition. Hence, the trajectories of left hand and right hand will be modeled by HMM independently.

5.2 Optimal Parameters Setting

For the step of preprocessing, we re-sample the raw trajectory into a new path with 300 points. Suppose there are V observations and Q hidden states, and the trajectory is divided into M segments. It's tough to determine the optimal parameters at the same time, making the models fit the sign words well. So, we vary one parameter with the other parameters fixed. Hence, we can get approximate optimal parameters by experiments in this way. For each word, we choose 60 samples of each word in Subset A for training and the rest in Subset A for validation.

As shown in Fig. 6, when we fixed the segment number M as 15, the optimal V and Q are 10 and 8, respectively, and it can reach the accuracy rate up to 60.1 %. Hence, it reasonable to use the optimal parameters in the final recognition. In the same way, we can fix the V and Q randomly and find the optimal segment number. The results with fixed $V = 20$ and $Q = 6$ are shown in Fig. 7. We can find that when we choose the segment number as 30, we obtain the best performance. Hence, in the following test, we set the parameters as $V = 10$, $Q = 8$ and $M = 30$.

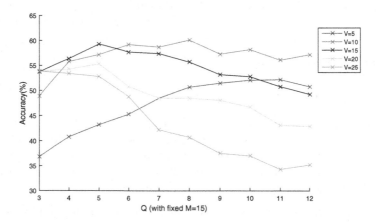

Fig. 6. Recognition rates with various V and Q. We fixed the number of segments as 15.

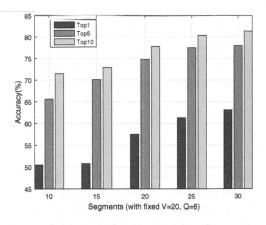

Fig. 7. Recognition rates with virous M. With fixed $V = 20$ and $Q = 6$, we find that $M = 30$ is the optimal parameter.

5.3 Results and Analysis

After getting the optimal parameters of V, Q and M, we use the dataset built by ourselves to evaluate the performance of our method. HMM with normal coordinate and curve matching from the view of manifold (CM_VoM) proposed in [8] are our baseline methods. In our experiments with Subset A, we choose 60 samples of each word for training while the rest 10 for testing. The recognition rates on Subset A for different methods are shown in Table 1. Our method can get the accuracies of 67.3%, 86.6%, 89.8% in top 1, top 5, and top 10, respectively. While HMM with normal coordinate features gets the accuracies of 32.2%, 54.8%, 65.3% in top 1, top 5 and top 10, respectively. CM_VoM [8] obtains the accuracies of 57.6%, 82.4%, 89.4%, respectively. In order to show the effectiveness and stability of our method, we also conduct the experiments on large Subset B. We use the pre-trained models with Subset A to recognize the words in Subset B. That is to say, the training and testing samples are from different signers. Table 2 gives the accuracies of different methods on Subset B. Our method can get the recognition rates at 54.4%, 77.3%, and 82.7% in top 1, top 5, top 10, respectively, which performs better than the other 2 methods.

Table 1. The recognition rates for different methods on Subset A

Subset A	Top1	Top5	Top10
Normal HMM	0.322	0.548	0.653
CM_VoM	0.576	0.824	0.894
Our method	0.673	0.866	0.898

Table 2. The recognition rates for different methods on Subset B with unseen signers

Subset B	Top1	Top5	Top10
Normal HMM	0.125	0.271	0.386
CM_VoM	0.451	0.719	0.819
Our method	0.544	0.773	0.827

6   Conclusion

In this paper, we propose a new approach for Chinese sign language recognition based on trajectory modeling. The method is inspired from the shape recognition with shape context. We partition the projected curve of sign word trajectory into multiple segments and represent each segment into histogram feature by shape context quantization. With these features, the HMMs are applied for modeling the sign words. The experiments show that our method outperforms the comparison methods by a large margin on a large dataset containing 100 sign words with over 25,000 samples. For the future work, we will integrate both trajectory of sign word and hand shapes for more accurate SLR recognition.

Acknowledgement. This work was supported in part to Dr. Zhou by the Fundamental Research Funds for the Central Universities under contract No. WK2100060014 and WK2100060011 and the National Science Foundation of China under contract No. 61472378, and in part to Prof. Li by the National Science Foundation of China under contract No. 61272316.

References

1. Belongie, S., Malik, J., Puzicha, J.: Shape matching and object recognition using shape contexts. IEEE Trans. Pattern Anal. Mach. Intell. **24**(4), 509–522 (2002)
2. Gales, M., Young, S.: The application of hidden markov models in speech recognition. Found. Trends Sig. Process. **1**(3), 195–304 (2008)
3. Grobel, K., Assan, M.: Isolated sign language recognition using hidden markov models. In: Proceedings of the IEEE International Conference on Systems, Man and Cybernetics, pp. 162–167. IEEE (1997)
4. Hienz, H., Kraiss, K.-F., Bauer, B.: Continuous sign language recognition using hidden markov models. In: International Conference on Multimodal Interfaces, vol. 4, pp. 10–15 (1999)
5. Huang, J., Zhou, W., Li, H., Li, W.: Sign language recognition using 3D convolutional neural networks. In: IEEE International Conference on Multimedia and Expo (ICME), pp. 1–6. IEEE, 2015
6. Huang, J., Zhou, W., Li, H., Li, W.: Sign language recognition using real-sense. In: IEEE China Summit and International Conference on Signal and Information Processing (ChinaSIP), pp. 166–170. IEEE (2015)
7. Latecki, L.J., Lakämper, R.: Convexity rule for shape decomposition based on discrete contour evolution. Comput. Vis. Image Underst. **73**(3), 441–454 (1999)

8. Lin, Y., Chai, X., Zhou, Y., Chen, X.: Curve matching from the view of manifold for sign language recognition. In: Shan, S., Jawahar, C.V., Jawahar, C.V. (eds.) ACCV 2014 Workshops. LNCS, vol. 9010, pp. 233–246. Springer, Heidelberg (2014)
9. Murakami, K., Taguchi, H.: Gesture recognition using recurrent neural networks. In: Proceedings of the SIGCHI Conference on Human Factors in Computing Systems, pp. 237–242. ACM (1991)
10. Oz, C., Leu, M.C.: Recognition of finger spelling of American sign language with artificial neural network using position/orientation sensors and data glove. In: Wang, J., Liao, X.-F., Yi, Z. (eds.) ISNN 2005. LNCS, vol. 3497, pp. 157–164. Springer, Heidelberg (2005)
11. Philbin, J., Chum, O., Isard, M., Sivic, J., Zisserman, A.: Lost in quantization: improving particular object retrieval in large scale image databases. In: Proceedings of the IEEE Conference on Computer Vision and Pattern Recognition (2008)
12. Rabiner, L.R.: A tutorial on hidden markov models and selected applications in speech recognition. Proc. IEEE **77**(2), 257–286 (1989)
13. Rabiner, L.R., Juang, B.-H.: An introduction to hidden markov models. IEEE ASSP Mag. **3**(1), 4–16 (1986)
14. Schlenzig, J., Hunter, E., Jain, R.: Vision based hand gesture interpretation using recursive estimation. In: Conference Record of the Twenty-Eighth Asilomar Conference on Signals, Systems and Computers, vol. 2, pp. 1267–1271. IEEE (1994)
15. Shotton, J., Sharp, T., Kipman, A., Fitzgibbon, A., Finocchio, M., Blake, A., Cook, M., Moore, R.: Real-time human pose recognition in parts from single depth images. Commun. ACM **56**(1), 116–124 (2013)
16. Wang, H., Chai, X., Zhou, Y., Chen, X.: Fast sign language recognition benefited from low rank approximation. In: IEEE International Conference and Workshops on Automatic Face and Gesture Recognition, pp. 1–6. IEEE (2015)
17. Wang, M., Hua, X.-S., Hong, R., Tang, J., Qi, G.-J., Song, Y.: Unified video annotation via multigraph learning. IEEE Trans. Circuits Syst. Video Technol. **19**(5), 733–746 (2009)
18. Wang, M., Ni, B., Hua, X.-S., Chua, T.-S.: Assistive tagging: a survey of multimedia tagging with human-computer joint exploration. ACM Comput. Surv. (CSUR) **44**(4), 25 (2012)
19. Wobbrock, J.O., Wilson, A.D., Li, Y.: Gestures without libraries, toolkits or training: a $1 recognizer for user interface prototypes. In: Proceedings of the 20th Annual ACM Symposium on User Interface Software and Technology, pp. 159–168. ACM (2007)
20. Zafrulla, Z., Brashear, H., Starner, T., Hamilton, H., Presti, P.: American sign language recognition with the kinect. In: Proceedings of the 13th International Conference on Multimodal Interfaces, pp. 279–286. ACM (2011)
21. Zhang, J., Zhou, W., Li, H.: A threshold-based hmm-dtw approach for continuous sign language recognition. In: Proceedings of International Conference on Internet Multimedia Computing and Service, p. 237. ACM (2014)
22. Zhang, J., Zhou, W., Li, H.: A new system for chinese sign language recognition. In: IEEE China Summit and International Conference on Signal and Information Processing (ChinaSIP), pp. 534–538. IEEE (2015)

MusicMixer: Automatic DJ System Considering Beat and Latent Topic Similarity

Tatsunori Hirai[1]([⊠]), Hironori Doi[2], and Shigeo Morishima[3,4]

[1] Waseda University, Tokyo, Japan
tatsunori_hirai@asagi.waseda.jp
[2] Dwango, Tokyo, Japan
[3] Waseda Research Institute for Science and Engineering, Tokyo, Japan
[4] JST CREST, Tokyo, Japan

Abstract. This paper presents *MusicMixer*, an automatic DJ system that mixes songs in a seamless manner. MusicMixer mixes songs based on audio similarity calculated via beat analysis and latent topic analysis of the chromatic signal in the audio. The topic represents latent semantics about how chromatic sounds are generated. Given a list of songs, a DJ selects a song with beat and sounds similar to a specific point of the currently playing song to seamlessly transition between songs. By calculating the similarity of all existing pairs of songs, the proposed system can retrieve the best mixing point from innumerable possibilities. Although it is comparatively easy to calculate beat similarity from audio signals, it has been difficult to consider the semantics of songs as a human DJ considers. To consider such semantics, we propose a method to represent audio signals to construct topic models that acquire latent semantics of audio. The results of a subjective experiment demonstrate the effectiveness of the proposed latent semantic analysis method.

Keywords: DJ system · Song mixing · Latent topic analysis · Beat similarity · Machine learning

1 Introduction

Many people enjoy listening to music. The digitalization of music content has made it possible for many people to carry their favorite songs on a digital music player. Opportunities to play such songs are frequent, e.g., at a house party or while driving a car. At some parties, an exclusive DJ performs for the attendants. DJs never stop playing the music until the party ends. They control the atmosphere of the event by seamlessly mixing songs[1]. However, it is not always realistic to personally hire a DJ. Thus, we present *MusicMixer*, an automatic DJ system that can mix songs for a user.

One of the most important things in a DJ's performance is to mix songs as naturally as possible. Given a list of songs, a DJ selects a song with beats and

---

[1] The word "mix" here refers to the gradual transiton of one song to another.

© Springer International Publishing Switzerland 2016
Q. Tian et al. (Eds.): MMM 2016, Part I, LNCS 9516, pp. 698–709, 2016.
DOI: 10.1007/978-3-319-27671-7_59

sounds that are similar to a specific point in the currently playing song such that the song transition is seamless. Consequently, the songs will be mixed as a consecutive song. The beats are particularly important and should be carefully considered. Maintaining stable beats during song transition is the key to realizing a seamless mix. The time to select the next song is limited and the songs are numerous; therefore, many DJs intuitively select a song to connect. However, this might not be the best song. The innumerable possibilities of mixing songs make performing difficult for the DJ.

Computers are good at searching for the best pairs of beats from innumerable possibilities. It is possible to solve this problem using a signal processing technique to extract beats and rely on a computer to retrieve a similar beat for effective mixing. However, computers handle audio signals numerically without considering the underlying song semantics; thus, the resulting mix will be mechanical if the system only considers beat similarity. The latent semantics of songs must be considered in addition to the beats. The DJ attempts to switch to a new song when the two songs sound similar. To consider the latent semantics, we propose a method to analyze the latent topics of a song from the polyphonic audio signal. These topics represent latent semantics about how chromatic sounds are generated. In addition to beat similarity, the proposed system considers the similarity of latent topics of songs. In particular, by employing a machine learning method called latent Dirichlet allocation (LDA) [1], the proposed system infers latent topics that generate chromatic audio signals. This process corresponds to consideration of how sound is generated from latent topics in a given song. By inferring similarity among song topics, higher level song information can be considered.

MusicMixer takes advantage of computational machine power to retrieve a good mix of songs. To make mixing more seamless as the DJ mix, the system focuses on the similarity of latent topics in addition to beat similarity and realizes natural song mixing (Fig. 1).

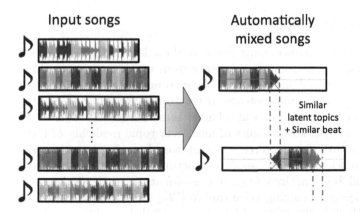

Fig. 1. Conceptual image of mixing songs with similar latent topics and beats using MusicMixer

2   Related Work

2.1   Music Mixing and Playlist Generation

Ishizaki *et al.* proposed a DJ system that adjusts the tempo of songs [2]. They defined a measurement function for user discomfort relative to tempo adjustment based on a subjective experiment. However, that system only considers tempo and beat. In addition, their system does not retrieve a mixing point but forcibly changes the tempo of songs.

Several studies have focused on generating a music playlist [3–6]. AutoDJ generates a playlist based on one or more seed songs using Gaussian process regression [3]. The AutoDJ project team has also proposed a method to infer the similarity between music objects and have applied this to playlist generation [6]. However, these approaches focused on playlist generation, and the importance of mixing (connecting) songs was not considered.

Goto *et al.* proposed *Musicream* [7], which provides a novel music listening experience, including sticking, sorting, and recalling musical pieces. It also provides a playlist generation function; however, mixing is not considered.

There is another approach to mixing songs, referred to as mashup. Mashup creates a single song from multiple songs. AutoMashUpper [8] generates a mashup according to a mashability measure. Tokui proposed an interactive music mashup system called Massh [9]. Mashups are a DJ track composition style; however, not all DJs can perform mashups live without using a pre-recorded mashup song. The mainstay of a DJ performance is mixing.

However, there has been little research on DJ mixing comparing to the research on playlist generation. We believe that a mixing method combined with playlist generation methods could be a powerful tool. In this paper, we propose a DJ system for mixing songs that considers beats and the higher level information of a song. During a DJ performance, the higher level music information that should be considered is the semantics of songs.

2.2   Topic Modeling

In natural language processing research, there is a method called topic modeling which estimates the topic of a sentence from words that appear in the sentence. Those words depend on the topic of the sentence; thus, the topic can be estimated by observing the actual sentence. If the topics are the same for two sentences, the sentences will be similar at a higher semantic level. Sasaki *et al.* proposed a system to analyze latent topics of music via topic modeling of lyrics [10]. They proposed an interface to retrieve a song based on the latent topics of lyrics.

Topic modeling can be applied to actual features. In this case, feature vectors should be quantized (e.g., a bag-of-features) [11]. Nakano *et al.* applied topic modeling to singing voice timbre [12]. They defined a similarity measure based on the KL-divergence of latent topics and showed that singers with similar singing voices have similar latent topics. However, it is difficult to understand the meaning of each topic explicitly using feature vectors rather than words.

Hu *et al.* used the note names of a song as words to estimate the musical key of a song using topic modeling [13]. This shows that topic analysis using note names is effective for inferring the latent semantics of a song. Hu *et al.* also proposed an extended method to estimate musical keys from an audio signal using a chroma vector (i.e., audio features based on a histogram of a 12 chromatic scale) rather than note names [14]. This approach shows that topic modeling using a chroma vector is useful for inferring the latent topic of a song.

3   System Overview

MusicMixer requires preprocesses to analyze song beats and latent topics. The beat analysis is performed using an audio signal processing approach. Figure 2 shows the system flow.

First, a low-pass filter (LPF) is applied to the input song collection to extract low-frequency signals. In the low-frequency signal, beat information such as bass and snare drums and bass sounds is prominent. Thus, beat information can be acquired by detecting the peaks of the envelope in the low-frequency signal.

The latent topic analysis is realized using LDA [1]. First, the system constructs a topic model using a music database that includes various music genres. The latent topics for a new input song can be estimated using the constructed

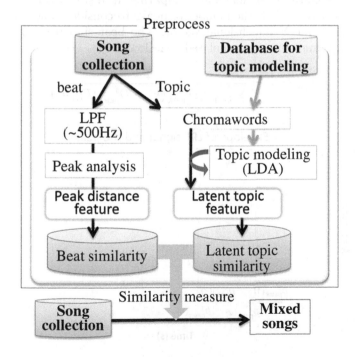

Fig. 2. System flow

model. Our goal is to find a good mixing point rather than analyze the topic of a whole song; thus, we analyze the topics of segmented song portions.

Finally, the system retrieves the most similar song fragments based on the combination of beat similarity and latent topic similarity. Once a similar pair of song fragments is retrieved, the system mixes songs at the fragment by cross-fading (i.e., fading in and out). Thus, the songs are mixed naturally. To mix more songs for endless playback, the similarity-based retrieval is applied to the mixed song.

4   Beat Similarity

In particular, the sound of the bass drum plays a significant role (e.g., the rhythm pattern called four-on-the-floor is composed of bass drum sounds). In addition to the bass drum, the snare drum and other bass sounds are important to express detailed rhythm. Note that we assume that all the other sounds do not affect to the beat. To ignore other audio signals, we apply an LPF, which passes signals with frequency below 500 Hz. The LPF passes the attack sounds of a general snare drum. By analyzing the peaks of the envelope of a low-frequency signal, dominant sound events in the low-frequency spectrum, such as the attack of a bass drum, can be detected. The distances between peaks correspond to the length of the beat (Fig. 3).

The beat similarity is calculated by comparing the distances between N peaks of the envelope. Here, N is the number of peaks to consider. The peak distance feature D_{peak} is an N dimensional vector. The beat similarity S_{beat} between fragment i and fragment j is calculated as follows:

$$S_{beat}(i,j) = \frac{1}{\sum_{k=1}^{N-1} ||D_{peak}^i(k) - D_{peak}^j(k)|| + 1}. \tag{1}$$

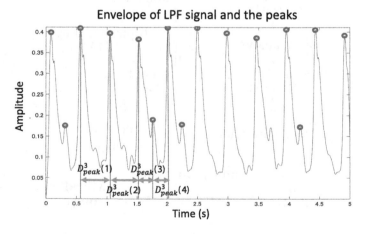

Fig. 3. Extracting peak distance features from the envelope of low frequency audio signal

Here, larger N values realize better matching relative to beats. However, the number of candidate songs to be mixed will be excessively reduced if the N value is too large. At present, this parameter is user-defined.

5   Latent Topic Similarity

This section describes the method to analyze a latent topic of a song using topic modeling. In particular, we propose a topic modeling method that considers the latent topic of a song by expressing the audio signal symbolically.

5.1   Topic Modeling

The topic model is constructed by extracting the features of songs and applying LDA [1] to the features. We extract the chroma vector from the audio signal and represent the feature symbolically, which we refer to as "ChromaWords."

Extraction of ChromaWords. Latent topics typically include semantics. However, topic analysis using audio feature values makes it difficult to understand the meaning of topics explicitly. It is difficult to determine meaning from a high dimensional feature value; therefore, previous topic modeling methods could not describe the meaning of each topic clearly. To avoid losing topic meaning, we express the audio signal symbolically. There are symbols in music that are represented in a musical score, e.g., note names. Since a letter is assigned to each note, we can use the letters to construct a word for topic modeling.

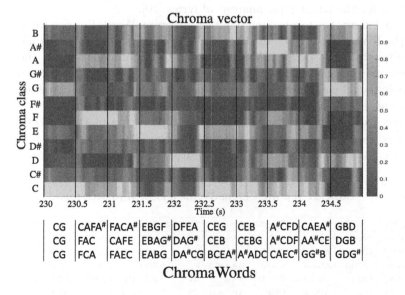

Fig. 4. Extraction of ChromaWords from chroma vector

Here, we employ an audio feature referred to as a chroma vector, which is a histogram of 12 notes. Each bin of the chroma vector represents a musical note. By sorting the chroma vector by dominant notes, a word can be generated (e.g., [CADE], [BAD#]) which we refer to as ChromaWords. Typically, a chroma vector includes noise caused primarily by inharmonic sounds. To avoid the effects of the noise, we use the top 70 % power of notes. Here, we set the maximum length of the word to four letters. Thus, we can represent polyphonic audio signals symbolically with natural language processing. Figure 4 shows an example of ChromaWords (bottom) acquired from the chroma vector of an actual song (top). Note that "#" is not counted as a single letter. Because of space limitations, we only display three ChromaWords per 0.5 s. These three ChromaWords are sampled at equal interval (0.5 s). The leftmost letter is the most dominant component, and less dominant components are to the right.

ChromaWords are acquired per audio frame. Here, the audio sampling rate is 16000 Hz monaural, and the frame length is 200 ms, shifting every 10 ms. One hundred words can be acquired from a 1-s audio signal.

Latent Dirichlet Allocation. By acquiring ChromaWords from a song, topic modeling can be applied similar to methods in natural language processing. MusicMixer employs LDA [1] for training of latent topic analysis. The number of topics is set to 100 in order to express semantics more complex than those of basic western tonality. The vocabulary of ChromaWords is 13345 ($= {}_{12}P_4 + {}_{12}P_3 + {}_{12}P_2 + {}_{12}P_1 + 1$), including perfect silence.

Training is required prior to latent topic analysis. We use 100 songs from the RWC music database [15], which comprises of songs of various genres. The parameters and algorithm for LDA is the same as the topic modeling method employed for the latent topic analysis of lyrics [10].

Table 1 shows the top-five representative ChromaWords for each topic learned from the RWC music database (10 topics out of 100, sorted by probability). The leftmost letter in a ChromaWord indicates the dominant note in the sound. Because many initial letters in ChromaWords for the same topic are the same, the topic model constructed by LDA reflects the semantics of chromatic notes, which was difficult for previous methods to explicitly express.

Using the constructed model, the latent topics for a new input song can be estimated by calculating a predictive distribution. The latent topics for the new input song are represented as a mixing ratio of all 100 topics. Figure 5 shows

Table 1. Top-five ChromaWords allocated to each topic.

Topic22	Topic90	Topic7	Topic98	Topic78	Topic52	Topic9	Topic79	Topic43	Topic80
CBC#A	Silent	AG	CFG	AA#BC	ED	AFB	Silent	AEA#	DAA#C
CC#BA	GDCA	AA#G	CFGD	ABA#C	E	AA#BF	GA	AEA#B	DACA#
CC#AB	GCDA	AGA#	CGFA	BAA#C	F	Silent	CF	Silent	ADA#C
CBAA#	GDAC	ADGA#	CGFD	AA#CB	DE	AA#GB	CFAG	AEA#G	DACC#
CABA#	DA	ADA#G	CGF	ABCA#	EF	AA#FB	FC	AEA#C	AA#BC

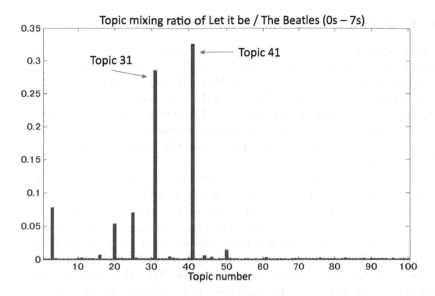

Fig. 5. Results of latent topic analysis applied to a 7-s fragment of the song "Let it Be"

an example of topic analysis for a fragment of the song "Let it Be" by The Beatles. In this result, topic 41 includes ChromaWords "DB," "GB," and "AC" as dominant words, and the dominant letters in the ChromaWords of topic 31 are "F," "C," and "A." In fact, the chord progression of this part of the song is "C, G, Am, F," which shows that topics mostly reflect the notes consisted in these chords. This indicates the relevance between chords and ChromaWords. Note that chords or harmony effect the ChromaWords, but the topics themselves do not directly represent chords or harmony.

Here, our goal is to find a good mixing point rather than analyze the topics of an entire song. Therefore, MusicMixer analyzes latent topics every 5 s to acquire the temporal transition of the topic ratio.

5.2   Calculation of Latent Topic Similarity

The mixing ratio of latent topics for each 5-s song fragment is acquired by the above-mentioned method. The mixing ratio is extracted as 100-dimensional feature vectors, and we use the mixing ratio as the latent topic feature f.

The latent topic similarity S_{topic} between fragment i and fragment j is calculated in the same form as the beat similarity:

$$S_{topic}(i,j) = \frac{1}{\sum_{k=1}^{K} ||f_i(k) - f_j(k)|| + 1},\tag{2}$$

where K is the number of topics (100).

5.3   Evaluation

We performed a subjective evaluation experiment to evaluate the effectiveness of latent topic analysis using ChromaWords. We compared the proposed method to a latent topic analysis method using MFCC feature values and chroma vector feature values. The topic modeling method for the compared methods is based on the method proposed by Nakano *et al.* [12], which uses k-means clustering to describe feature values in a bag-of-features expression. Note that we do not use similarity calculated from raw feature values; thus, we can focus on the effects of ChromaWords.

Fifty pop, rock, and dance songs were used in the experiment, and 2192 segments were generated by cutting the songs into 5-s fragments. We calculated the latent topic similarity of all pairs between the 2192 fragments.

The subjects were asked to listen to two pairs of song and indicate which pair was more similar. A pair was generated based on the latent topic similarity of each method. We selected three pairs of songs per method. The three pairs were selected from the top-30 latent topic similarity. Note that song repetition was avoided in this experiment. To avoid the effects of beat, we did not mix songs but played each song separately.

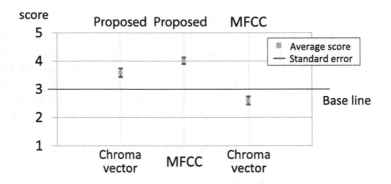

Fig. 6. Result of subjective evaluation experiment

Eight students (ages 22 to 24) with no DJ experience participated in the experiment. The subjects listened to a song pair generated by method A first, and then a pair generated by method B. They then rated the pairs from 1 (Pair A is more similar than B) to 5 (Pair B is more similar than A). A score of 3 indicates that "both pairs are equal in terms of similarity." The pairs were presented randomly.

Figure 6 shows the results of the experiment. The score is the average of all eight subjects and all nine compared pairs. Comparing the proposed topic modeling method with the topic modeling method using raw chroma vector feature, the score was 3.58, which indicates that the proposed method expresses similarity better. Comparing the proposed method with the topic modeling with

MFCC, the score was 4.01, which also indicates that the proposed method performs better. The rightmost plot shows a comparison of MFCC and chroma vector methods, which are not related to the proposed method. As can be seen, the topic model using the chroma vector outperforms the MFCC method. These results indicate that the proposed method using ChromaWords outperformed the other methods in terms of music fragment similarity.

6   Mixing Songs

MusicMixer mixes songs based on the similarity measurements described above. The combined similarity S between fragments i and j can be calculated as follows:

$$S(i,j) = w \times S_{beat}(i,j) + (1 - w) \times S_{topic}(i,j), \tag{3}$$

where w denotes the weight parameter used to change the balance of the beat and latent topic similarity ($w = 0.5$ in the current implementation).

The length of each song fragment depends on the number of peaks in the beat similarity calculation. Although the length of each fragment differs, beat similarity ensures that fragments with similar length are selected as similar beat fragments. In addition, the fragment lengths are not 5 s (the length for topic analysis). Therefore, we assume that the fragment in a 5-s fragment is similar relative to the latent topic feature. Thus, we use the same latent topic feature even though the length of the fragment is not 5 s.

There is a function to specify the scope of when mixing can occur. For example, we do not want to switch to a new song during the beginning of the previous song or start a song at the end. In the current implementation, a song will not change until the latter half, and a song will start no later than the first half.

7   Discussion

7.1   Limitations

MusicMixer considers both beat and latent semantics. However, latent semantics are limited to the chromatic audio signal. Therefore, other types of high level information such as variation of instrument or dynamics within a song cannot be considered in the current implementation. In future, we will explore the possibility of semantic topic analysis using the symbolic representation acquired from the audio signals.

MusicMixer does not consider lyrics or their semantics. Therefore, a summer song may be selected after a winter song, which is undesirable. It is possible that a user compensates for such flaws by introducing user interface.

7.2   Applications of MusicMixer

MusicMixer's applications are not limited to an automatic DJ tool. There is a style of DJ performance referred to as "back-to-back" which is collaborative play among multiple DJs. In a back-to-back session, a partner DJ selects the next song while one DJ's song is playing. Thus, the partner DJ's play may be unpredictable. Although the back-to-back style cannot be performed alone, a DJ system such as MusicMixer could act as a partner DJ for a back-to-back performance. This is similar to playing a video game against the computer, which can improve the player's technique. Furthermore, collaboration with a computer might produce new or unexpected groove.

It is also possible for inexperienced people to practice DJ performance using MusicMixer. For example, mixing songs is the difficult part of a DJ's performance, but song selection might be easier for inexperienced people. In this case, the connection of songs could be performed by the system, and the user can focus on song selection. Conversely, the user can focus on song mixing without worrying about song selection by allowing system to select songs. DJ performance requires significant skill that can be only acquired from practical experience.

7.3   Conclusion

We have presented MusicMixer, an automatic DJ system. We have proposed a method to mix songs naturally by considering both beat and latent topic similarity. Our main contribution is the application of topic modeling using Chroma-Words, which are an audio signal-based symbolic representation. Previous topic modeling methods have analyzed the latent topics of audio or images using features represented as a bag-of-features, so the meaning of topic was not clear. We have achieved topic modeling with understandable topic meanings using Chroma-Words. Furthermore, the results of a subjective evaluation indicate that our topic modeling method outperforms other methods in terms of music similarity. Topic modeling is primarily used to analyze latent semantics in observed data. The proposed method makes it possible to employ the latent semantics of chromatic sounds. However, the semantics of chromatic sounds do not cover all the semantics of a song. Thus, in future, we plan to consider other semantics such as timbre.

This study has focused on song mixing without changing the original songs. In a future implementation, the proposed system will perform modulation of songs to allow free connection of any type of song pairs. For example, by using a song morphing method [16], it may be possible to embed such a function. In addition, we plan to consider the structure of songs. In this manner, a new song will not be played until the verse of the current song ends. We will explore the possibility of human and computer collaborative DJ performance.

Acknowledgments. This work was supported by OngaCREST, CREST, JST and JSPS Grant-in-Aid for JSPS Fellows. This work was inspired by Tonkatsu DJ Agetaro.

References

1. Blei, D., Ng, A., Jordan, M.: Latent dirichlet allocation. J. Mach. Learn. Res. **3**, 993–1022 (2003)
2. Ishizaki, H., Hoashi, K., Takishima, Y.: Full-automatic DJ mixing system with optimal tempo adjustment based on measurement function of user discomfort. In: Proceedings of ISMIR, pp. 135–140 (2009)
3. Platt, J. Burges, C., Swenson, S., Weare, C., Zheng, A.: Learning a gaussian process prior for automatically generating music playlists. In: Proceedings of NIPS, pp. 1425–1432 (2001)
4. Aucouturier, J.J., Pachet, F.: Scaling up music playlist generation. In: Proceedings of ICME, pp. 105–108 (2002)
5. Pampalk, E., Pohle, T., Widmer, G.: Dynamic playlist generation based on skipping behavior. In: Proceedings of ISMIR, pp. 634–637 (2005)
6. Ragno, R., Burges, C., Herley, C.: Inferring similarity between music objects with application to playlist generation. In: Proceedings of MIR, pp. 73–80 (2005)
7. Goto, M., Goto, T.: Musicream: integrated music-listening interface for active, flexible, and unexpected encounters with musical pieces. Inf. Media Technol. **5**(1), 139–152 (2010)
8. Davies, M., Hamel, P., Yoshii, K., Goto, M.: AutoMashUpper: automatic creation of multi-song music mashups. Trans. Audio Speech Lang. Process. **22**(12), 1726–1737 (2014)
9. Tokui, N.: Massh!: a web-based collective music mashup system. In: Proceedings of DIMEA, pp. 526–527 (2009)
10. Sasaki, S., Yoshii, K., Nakano, T., Goto, M., Morishima, S.: LyricsRadar: a lyrics retrieval system based on latent topics of lyrics. In: Proceedings of ISMIR, pp. 585–590 (2014)
11. Sivic, J., Zisserman, A.: Video Google: a text retrieval approach to object matching in videos. In: Proceedings of ICCV, pp. 1470–1477 (2003)
12. Nakano, T., Yoshii, K., Goto, M.: Vocal timbre analysis using latent Dirichlet allocation and cross-gender vocal timbre similarity. In: Proceedings of ICASSP, pp. 5202–5206 (2014)
13. Hu, D., Saul, L.: A probabilistic topic model for unsupervised learning of musical key-profiles. In: Proceedings of ISMIR, pp. 441–446 (2009)
14. Hu, D., Saul, L.: A probabilistic topic model for music analysis. In: Proceedings of NIPS (2009)
15. Goto, M.: Development of the RWC music database. In: Proceedings of ICA, pp. 553–556 (2004)
16. Hirai, T., Sasaki, S., Morishima, S.: MusicMean: fusion-based music generation. In: Proceedings of SMC, pp. 323–327 (2015)

Adaptive Synopsis of Non-Human Primates' Surveillance Video Based on Behavior Classification

Dongqi Cai[1(✉)], Fei Su[1,2], and Zhicheng Zhao[1,2]

[1] School of Information and Communication Engineering,
Beijing University of Posts and Telecommunications, Beijing 100876, China
{caidongqi,sufei,zhaozc}@bupt.edu.cn
[2] Beijing Key Laboratory of Network System and Network Culture,
Beijing University of Posts and Telecommunications, Beijing 100876, China

Abstract. Non-human primates (NHPs) play a critical role in biomedical research. Automated monitoring and analysis of NHP's behaviors through the surveillance video can greatly support the NHP-related studies. However, little research work has been undertaken yet. There are two challenges in analyzing the NHP's surveillance video: the NHP's behaviors are lack of regularity and intention, and serious occlusions are brought by the fences of the cages. In this paper, four typical NHPs' behaviors are defined based on the requirement in pharmaceutical analysis. We design a novel feature set combining contextual attributes and local motion information to overcome the effects of occlusions. A hierarchical linear discriminant analysis (LDA) classifier is proposed to categorize the NHPs' behaviors. Based on the behavior classification, an adaptive synopsis algorithm is further proposed to condense the NHPs' surveillance video, which offers a mechanism to retrieve any NHP's behavior information corresponding to specified events or time periods in the surveillance video. Experimental results show the effectiveness of the proposed method in categorizing and condensing NHPs' surveillance video.

Keywords: Non-human primate · Behavior classification · Video synopsis · Occlusions

1 Introduction

Since NHPs are similar to humans in the biological characteristics and way of action, they play an important role in biology, medical science and psychology research. The current research of NHP mainly focuses on building noninvasive recording system of NHP's biometric signals [1] (such as electrodermal activity (EDA), electrocardiography (ECG), 3-axis acceleration and temperature), analyzing the recorded biometric signals using traditional signal processing tools according to some specific applications (such as autonomic activity and social behavior studying [1], epileptic seizure detection [2] and dynamic studies on the

© Springer International Publishing Switzerland 2016
Q. Tian et al. (Eds.): MMM 2016, Part I, LNCS 9516, pp. 710–721, 2016.
DOI: 10.1007/978-3-319-27671-7_60

HRRT [3]). Besides the direct biometric signals recording, a video surveillance system of NHPs is usually established to provide reference information for the target application. The automatic analysis and synopsis of the NHP's surveillance video can help interpreting the NHP's detailed behaviors and some unusual bio-signal data, saving storage space and thus reducing the workload.

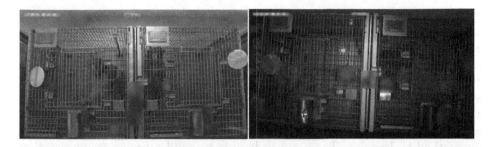

Fig. 1. Example frames of NHPs' surveillance video in RGB (left) and IR (right) modes.

However, little research work of analyzing NHP's behaviors through surveillance video has been undertaken yet. In the pharmaceutical research, NHPs are usually in cages to facilitate monitoring the effects of medicine. In this paper, the NHP's surveillance video is captured in a 24/7 fashion, including RGB and Infra-Red (IR) modes, as shown in Fig. 1. There are some challenges in the video analysis in this kind of scenario. Firstly, the fences of the cages seriously block the surveillance cameras and make the video objects difficult to be distinguished. Second, being easily affected by environmental factors and suddenly happening, NHPs' behavior patterns are totally different from human beings', so the traditional video analysis methods are not fit here. Finally, the occlusions and the changing in lighting conditions also make it difficult, especially in the phase to categorize the behaviors at a fine-grained level. Through discussions with biomedical scientists, in this paper, we define the NHPs' behaviors under normal circumstances into four typical categories: still, mild, normal and excitation. Except for these cases, any other behavior is categorized into an extra type as "others". This automatic classification of NHPs' behaviors is very useful for pharmacological analysis. To overcome the effect of occlusions, a feature set combining the contextual attributes and local motion information is proposed. Finally, a hierarchical behavior classifier including linear discriminant analysis (LDA) and temporal grouping is designed. Based on the classification results, an adaptive video synopsis algorithm is proposed to condense the NHP's surveillance video which offers a mechanism to retrieve any NHP's behavior information corresponding to specified events or time periods in NHPs' surveillance video. Experimental results show the effectiveness of behaviors classification, and the condensed videos keep the significant information and show good rationality as well. The remaining parts of this paper are organized as follows. Section 2 gives the related work on video synopsis, behavior classification and occlusion handling. Section 3 introduces the proposed adaptive synopsis algorithm in detail, including preprocessing, feature extraction, behavior classification and synopsis. The experimental results on the NHP's video surveillance data are shown in Sect. 4, which is followed by the conclusions and future work in Sect. 5.

2   Related Work

Synopsis methods condense the video by selecting a series of keyframes [5,6] or subshots [7,8] that best represent the original video [4]. According to specific type of video content, available methods use selection criteria based on diversity (selected frames should not be redundant), anomalies (unusual events ought to be included), and temporal spacing (coverage ought to be spread across the video). They typically rely on low-level cues such as motion or color [7], or else track pre-trained objects of interest [6]. In this paper, according to the real scenario of using fixed camera to record the video, the sampling-based synopsis method is desirable. The problem here is to build a proper frame-selection criterion reflecting the significances of frames to provide useful information for pharmacological analysis. It is desirable to adjust the subsampling rates in accordance with the NHPs behavior information. However, behavior classification of NHP's surveillance video under heavy occlusions is a new challenging problem. To our knowledge, there has been an enormous amount of research on analyzing video data of human beings, but relatively little on visual analysis of other organisms. Visual observation and analysis of insects [9], flying animals [10], ground animals [11] and marine animals [12] behaviors has been an increasingly popular and important area of study recently. These works tried to model the animals' motion style through analyzing the movement path. In this paper, the local motion information is also incorporated into the feature set for behavior recognition. Sandikci et al. present a two-stage method for behavior recognition of laboratory mice [11]. The first stage is used to discriminate still actions such as sleeping from the others. The second stage classifies the remaining four actions, namely, drinking, eating, exploring and grooming. Inspired by this work, a hierarchical classifier which takes into account the importance of behaviors and their relations is proposed to classify the NHPs' different behaviors. Referring to occlusions, there are mainly two ways to deal with it. One is to remove the occlusions directly by object detection and region filling [13–15]. The other is to design robust features for visual representation [16–18]. Because of the irregular distribution of cages' mental bars and variation of camera angles and lighting conditions, cage removal in NHPs' surveillance videos is too costly. Referring to the concepts in [18], influence of the occlusions can be subtly avoided by developing mid-level visual attributes. So, here we focuses on extracting robust feature sets by combining mid-level contextual attributes with local motion information.

3   Adaptive Synopsis

The scheme of the proposed algorithm is illustrated in Fig. 2. Preprocessing is first applied to locate the boundaries of the cages and determine whether

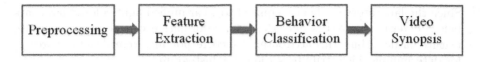

Fig. 2. The overall diagram of the proposed algorithm.

the video is within normal field of view (FOV). Then feature set combining contextual attributes and local motion information is extracted and fed to the behavior classifier. Finally, video synopsis is performed based on the obtained behavior labels.

3.1 Preprocessing

As shown in Fig. 1, two NHPs separately in two cages are being monitored simultaneously. There are no interactions between those two NHPs. In order to distinguish them and analyze each NHP separately, the boundary between two cages should be located. It should be mentioned that the cages may be moved during daily cleaning or routine operations and no extra markers can be attached to the cages in order to avoid making NHPs discomfort. General statistics approaches based on color information or texture information fail to perform well here due to the variability of lighting and the positions and angles of the cages.

Since the cages are approximately the same in structure, symmetry detection algorithms can be applied to give the cages' middle boundary. Similar to [17], SIFT detection and matching are used to select the keypoints pairs. Differences between adjacent frames are calculated to filter out the keypoints falling on the moving NHPs. Symmetry axis is obtained through Hough space voting on the matched SIFT pairs. The detected symmetric axis can be denoted as a two-point-vector $l_{sym} = (x_1, y_1, x_2, y_2)$ and the left and right areas divided by l_{sym} correspond to the left and right NHPs respectively.

3.2 Feature Extraction

Attributes are visual concepts that can be detected by machines and understood by humans [18]. To adequately represent the NHPs' behaviors under occlusions, both mid-level contextual attributes and lower-level local motion information should be included.

3.2.1 Contextual Attributes

Taking into account factors that may influence the NHP's behaviors, a meaningful attribute vocabulary is constructed as follows.

- **_Attribute 1:_** Biomedical scientists' long-term observation on NHPs' behavior patterns concludes that NHPs tend to get excited when people pass by or show up to feed them. So the first attribute is defined as $a_1 = \{$Are there people shown up?$\}$.

Since the laboratory staffs are required to wear unified light-colored lab coats before entering the NHPs' lab, the color information can be an appropriate feature for describing this attribute. The corresponding lower-level features f_1 for determining this attribute is the color histogram.

- **Attribute 2:** NHPs are more likely to be still when the lights are off according to statistical analysis. The video turns to IR mode after turning off the lights. So the second attribute is defined as $a_2 = \{$Is it in RGB mode?$\}$.

  In IR mode, the output of R, G and B color channels are the same theoretically. There is a slight difference among channels due to the noise brought by the hardware constraint. The corresponding lower-level features f_2 for a_2 is computed by the intensity variance of R, G and B components.

- **Attribute 3:** During daily cleaning or routine operations, the cages may be moved and the NHP's behaviors will be influenced. $a_3 = \{$Are the cages in the proper area?$\}$ is defined to indicate this contextual cue. For this attribute, the horizontal location of the boundary edge l_sym obtained in the preprocessing phase is calculated as $f_3 = 1/2(x_1 + x_2)$.

These contextual attributes can be viewed as binary features determined by the trained SVM separately using the corresponding lower-level features.

3.2.2   NHP's Motion Information

The NHPs tend to move fast without regular exercise in the behavior of "excitation", for example, turning somersaults and circling around. While being in the state of "mild", only part of the NHPs' limbs move, such as scratching and grooming. To distinguish these behaviors, frame difference and optical flow based motion features are extracted.

Frame difference reflects not only the NHPs' motion scales but also the locations. Let the distortion between two frames i and j be denoted by $d(f_i, f_j)$. The motion area is the subset of pixels with intensity higher than a threshold:

$$\mathcal{M} = \left\{ (x, y) \,|abs\left(d\left(f_i, f_j\right)\right) \geq T_1 \right\} \tag{1}$$

where T_1 is a predefined threshold to suppress the noise introduced by the change of lighting and the slight movement of cages. The percentage of the motion area on the frame and the corresponding regional center are as follows:

$$d_1 = \frac{|\mathcal{M}|}{|d(f_i, f_j)|} \tag{2}$$

$$d_2 = \frac{1}{|\mathcal{M}|} \sum_{(x_i, y_i) \in \mathcal{M}} (x_i, y_i) \tag{3}$$

where $|\cdot|$ denotes element counting operator.

As we previously noted, the behavior "excitation" involves fast movement. And the type "normal" indicates a whole body movement at normal speed, for example, walking and exploring. To distinguish these behaviors, the optical flow based feature is preferred. Let the optical flow vector at pixel (x, y) with velocity $(\dot{x}, \dot{y})$ be denoted by $\dot{o} = (x, y, \dot{x}, \dot{y})$. The subset of optical flow which effectively represents the NHP's movement is defined by:

$$\mathscr{O} = \left\{ \dot{o} \mid \sqrt{(\dot{x})^2 + (\dot{y})^2} \geq T_2 \right\} \tag{4}$$

T_2 is a threshold of velocity amplitude to filter out the slight movement of cages. The location and amplitude of the mean optical flow $\overline{O}$ are extracted:

$$d_2 = \frac{1}{\mathscr{M}} \sum_{(x_i, y_i, \dot{x}_i, \dot{y}_i) \in \mathscr{O}} (x_i, y_i, \dot{x}_i, \dot{y}_i) \tag{5}$$

Then the contextual attributes $[a_1, a_2, a_3]$ and motion information $[d_1, d_2, d_3]$ are combined to form the final feature set.

3.3 Behavior Classification

Let $\mathscr{B} = \{b_i \mid i = 0(others), 1(excitation), 2(normal), 3(mild), 4(still)\}$ represents the behavior set. According to biomedical statistics, different behaviors contribute different significances in pharmacological analysis. Comparing to the other three behaviors, "excitation" plays an important role. Therefore, a hierarchical behavior classifier is designed to guarantee not to miss the important behaviors, as shown in Fig. 3. The sub-classifier at each layer solves a binary classification problem progressively using LDA and temporal grouping. LDA is used to classify the feature set into two temporary subsets and temporal grouping merges the temporary labels by temporal continuity. The mini plots in Fig. 3 show the labeling results at each layer after temporal grouping.

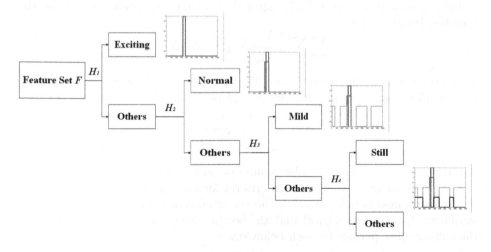

Fig. 3. Diagram of the hierarchical behaviors classifier.

For the ith layer of the classifier, the binary classifier based on LDA is formulated as:

$$H_i(F_i) = w_i^T F_i \tag{6}$$

where w_i is the learned coefficients which project the feature set to a linearly separable space. Let L_i denote the subset of frame index with temporary labels b_i:

$$\mathscr{L}_i = \{j | H_i(f_{ij} = 1)\} \tag{7}$$

f_{ij} is a feature vector with frame index j in F_i. Given $m, n \in \mathscr{L}_i$, the temporal continuity based grouping is denoted by:

$$m - n \leq T_{\Delta i} \rightarrow m : n \in \mathscr{L}_i \tag{8}$$

$T_{\Delta i}$ is a predefined threshold indicating the minimum durations for behavior b_i. The colon operator means from m to n. Through the order of the subclassifiers and temporal grouping, the prior information about the NHP's motion pattern as mentioned above is subtly integrated into the hierarchical behavior classifier.

3.4   Adaptive Synopsis

After preprocessing, feature extraction and behavior classification, behavior labels obtained for the left and right NHPs are denoted by Y_L and Y_R respectively. The proposed adaptive synopsis algorithm aims to subsample the original video according to the significance of frames in providing useful information for pharmacological research, which is reflected by the behavior types.

Assuming the length of a video clip is N, the number of frames with the left and right behaviors being b_i and b_j is denoted by $N_{(b_i, b_j)}$ and its corresponding subsampling period is $t_{(b_i, b_j)}$. Therefore, given the compression ratio R, the resulted length of the video clip is required to be approximately the same as the specified length.

$$\sum_j \sum_i \left\lceil \frac{N_{(b_i, b_j)}}{t_{(b_i, b_j)}} \right\rceil \approx R \times N \tag{9}$$

where $\lceil \cdot \rceil$ is a rounding up operation. That is, the ratio of the resulted length to the specified length of the video clip should be as close to 1 as possible.

$$S = \frac{R \times N}{\sum_j \sum_i \left\lceil \frac{N_{(b_i, b_j)}}{t_{(b_i, b_j)}} \right\rceil} \tag{10}$$

S can be used to adjust the subsampling periods iteratively and the difference $\varepsilon = (S - 1)$ is used as a termination criteria for the algorithm.

As mentioned before, although the occurrence of being excited is low, it is significant for pharmacological analysis. So the occurrences are used to initialize the subsampling periods for each behavior.

$$t_{(b_i, b_j)} = \left\lceil \frac{N_{(b_i, b_j)}}{\alpha \times N} \right\rceil \tag{11}$$

where α is the coefficient for normalization:

$$\alpha = \frac{1}{\sum_j \sum_i \frac{N}{N_{(b_i, b_j)}}} \tag{12}$$

With the subsampling rates integrating the significances of the behaviors, the resulted synopsis is the essence of the original video.

4   Experimental Results

4.1   Datasets

The NHP surveillance system monitors two separate NHPs in cages simultaneously as shown in Fig. 1. The camera resolution is 1920×1080 and the frame rate is 30 fps. The video data is stored as a separate file at regular intervals for storage security and management convenience. Therefore, the videos can be processed in an off-line manner.

The dataset used in the experiments is constructed by 40 NHPs' surveillance video clips with an average length of 8 min including RGB and IR modes. They are randomly selected and manually labelled with contextual attributes and predefined behaviors by well-trained professionals. Half of these videos are used as training set and the others are used for testing.

The behavior labels are treated as ground truth in evaluating the performance of the hierarchical classifier, denoted as $Y_G T$. The subsampling rate used in calculating the optical flow is 4. The predefined thresholds T_1 and T_2 are set to 10 and 2 respectively. And the minimum duration thresholds $T_{\Delta i}(i = 1, \ldots, 4)$ are set to $(2, 2, 2, 5)$ seconds based on the prior knowledge about NHPs' behavior patterns. The SVM classifiers for determining the contextual attributes and the projection vectors $w_i(i = 1, \ldots, 4)$ of the hierarchical classifier can be learned using the training video clips.

4.2   Behavior Classification

To evaluate the performance of the behavior classifier, a standard accuracy is defined as the percentage of the video frames with correct labels, that is

$$Score = \frac{\sum \delta(Y_{GT}, Y)}{N} \tag{13}$$

where $\delta(\cdot, \cdot)$ is an indicator function which takes 1 when the two arguments are equal and takes 0 otherwise.

Figure 4 shows the behaviors classification scores of the left and right NHPs in the 20 testing video clips. As can be seen, the proposed hierarchical behaviors classifier can effectively classify the NHPs' behaviors with an average score of 0.8391. The videos with scores approached full marks 1 are usually the ones in IR modes when the possibility of NHPs' being still is high. On the other hand, very

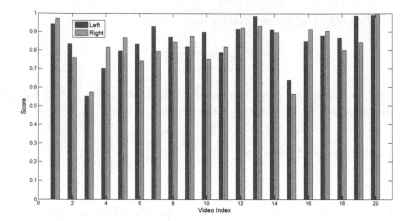

Fig. 4. Behavior classification scores of the testing videos.

few videos with scores below 0.6 are those when the NHPs are frequently alternating between "excitation" and "normal". In reality, the boundaries between these two behaviors are even difficult to be clearly distinguished with human beings. It would be more reasonable to use a soft weighted score which considers the levels of difficulties in classifying different behaviors. However, this standard score can still objectively reflect the performance of the classifier.

To compare the performance of behavior classification in RGB and IR modes, Fig. 5 gives an example of both situations. The behaviors are encoded by integers from 0 to 4, specifically $b_1 = 4$, $b_2 = 3$, $b_3 = 2$, $b_4 = 1$ and $b_0 = 0$. The blue solid line indicates the output labels and the red dashed line is the ground truth. As can be seen, the categorized results are basically consistent with the ground truth. In Fig. 5(a), the inconsistency occurs before and after the change of states. This subtle error is inevitable due to the gap between the predefined minimum durations of each behavior and the subjectively labeling boundaries, and it is

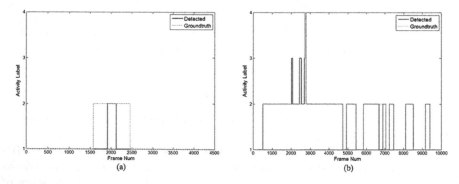

Fig. 5. Comparisons of the detected behavior labels and the ground truth in (a) IR and (b) RGB modes (Color figure onlline).

acceptable for the following synopsis algorithm. In Fig. 5(b), the classification results take two small segments of "being mild" as "being normal". This kind of "emphasizing the more active behaviors" accords with the idea when designing the hierarchical classifier.

4.3   Video Synopsis

Since there are no previous algorithms worked on condensing animal surveillance video under experimental environment, comparative experiments cannot be conducted. To better evaluate the performance of the proposed synopsis algorithm, 3D visualizations of the results are provided.

The traditional background-subtraction based tracker is applied to obtain the NHPs' motion trajectories in one video clip. And manual correction is also applied to avoid the interference of occlusions. The left and right NHPs' positions at each frame and the detected behavior labels are shown in Fig. 6(a) and (b). The corresponding ones after synopsis resulted with a compression ratio of 0.01 are shown in Fig. 6(c) and (d).

Intuitively, the dispersion degree of the positions reflects the intensity of motion. As can be seen, in Fig. 6, "being excitation" appears to be a dispersed subset of points in a small consecutive spatial-temporal cube as shown by red stars. And "being still" appears to be an approximately straight linesegment in the spatial-temporal domain as shown by magenta triangles.

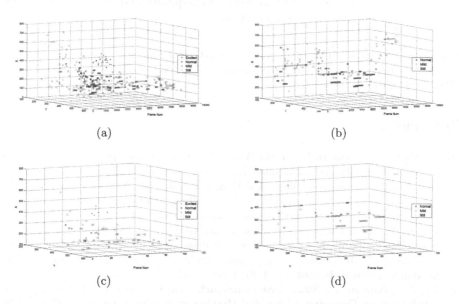

Fig. 6. The left and right NHPs' positions and their corresponding behaviors categorized by the proposed hierarchical classifier. (c) and (d) correspond to the synopsis of (a) and (b) when the compression ratio is 0.01 (Color figure onlline).

As can be seen from Fig. 6(c) and (d), the proposed synopsis algorithm selects the frames in "dispersed spatial-temporal cube" with greater probability than those in "straight line segments", which demonstrates the effectiveness and reasonability of the proposed synopsis algorithm.

After behavior classification and video synopsis, a mechanism to retrieve any NHP's behavior information corresponding to specified events or time periods in NHP experiments is implemented as well, using appropriate video frame labels obtained and the corresponding indexes.

5   Conclusions and Future Work

In this paper, four typical NHP's behaviors are defined based on biomedical research. A feature set combining the contextual attributes and local motion information is designed to overcome the interferences of the occlusions. A hierarchical behaviors classifier including LDA and temporal grouping is applied to categorize the NHPs' behaviors. Based on the classification results, an adaptive synopsis algorithm reflecting the significances of the NHPs' behaviors is proposed to condense the surveillance video, which offers a mechanism to retrieve any NHP's behavior information corresponding to specified events or time periods in NHP surveillance video. Experimental results show the accuracy and effectiveness of the proposed algorithms.

For future work, we will explore more complicated NHPs' behaviors by utilizing image processing tools to enhance the video quality and designing occlusion-robust feature for behavior representation.

Acknowledgements. This work is supported by Chinese National Natural Science Foundation (61372169, 61471049), and National Key Technology R&D Program (2012BAH63F01, 2012BA-H41F03).

References

1. Fletcher, R., Amemori, K., et al.: Wearable wireless sensor platform for studying autonomic activity and social behavior in non-human primates. In: IEEE Conference on Engineering in Medicine and Biology Society, pp. 4046–4049. IEEE (2012)
2. Sauter-Starace, F., Torres-Martinez, N., et al.: Epileptic seizure recordings of a non-human primate using carbon nanotube microelectrodes on implantable silicon shanks. In: IEEE Conference on Neural Engineering, pp. 589–592. IEEE (2011)
3. Sossi, V., Camborde, M.L., et al.: Dynamic imaging on the high resolution research tomograph (HRRT): non-human primate studies. In: IEEE Nuclear Science Symposium Conference Record, pp. 1981–1985 (2005)
4. Lu, Z., Grauman, K.: Story-driven summarization for egocentric video. In: IEEE Conference on Computer Vision and Pattern Recognition, pp. 2714–2721, 23–28 June 2013
5. Goldman, D.B., Curless, S., et al.: Schematic storyboarding for video visualization and editing. In: ACM Transactions on Graphics, vol. 25, no. 3, pp. 862–871. ACM (2006)

6. Liu, D., Hua, G., Chen, T.: A hierarchical visual model for video object summarization. IEEE Trans. Pattern Anal. Mach. Intell. **32**(12), 2178 2190 (2010)
7. Ngo, C.W., Ma, Y.F., Zhang, H.J.: Automatic video summarization by graph modeling. In: Ninth IEEE Conference on Computer Vision, pp. 104–109. IEEE (2003)
8. Laganire, R., Lambert, P., Ionescu, B.E.: Video summarization from spatio-temporal features. In: Proceedings of the 2nd ACM TRECVid Video Summarization Workshop, pp. 144–148. ACM (2008)
9. Miranda, B., Salas, J., Vera, P.: Bumblebees detection and tracking. In: Visual Observation and Analysis of Animal and Insect Behavior (2012)
10. Martinez, F., Manzanera, A., Romero, E.: Analysing the hovering flight of the hummingbird using statistics of the optical flow field. In: Visual Observation and Analysis of Animal and Insect Behavior (2012)
11. Sandikci, S., Duygulu, P., et al.: HMM based behavior recognition of laboratory animals. In: Visual Observation and Analysis of Animal and Insect Behavior (2012)
12. Kawasue, K., Nagatomo, S., Oya, Y.: Three-dimensional behavior measurements of small aquatic lives using a single camera. In: Visual Observation and Analysis of Animal and Insect Behavior (2012)
13. Liu, Y., Belkina, T., Hays, J.H., Lublinerman, R.: Image defencing. In: IEEE Conference on Computer Vision and Pattern Recognition, pp. 1–8, 23–28 June 2008
14. Park, M., Brocklehurst, K., Collins, R.T., Liu, Y.: Image defencing revisited. In: Kimmel, R., Klette, R., Sugimoto, A. (eds.) ACCV 2010, Part IV. LNCS, vol. 6495, pp. 422–434. Springer, Heidelberg (2011)
15. Mu, Y., Liu, W., Yan, S.: Video defencing. IEEE Trans. Circuits Syst. Video Technol. 1 12 (2013)
16. Bettadapura, V., Schindler, G., Plotz, T., et al.: Augmenting bag-of-words: data-driven discovery of temporal and structural information for activity recognition. In: IEEE Conference on Computer Vision and Pattern Recognition (2013)
17. Loy, G., Eklundh, J.-O.: Detecting symmetry and symmetric constellations of features. In: Leonardis, A., Bischof, H., Pinz, A. (eds.) ECCV 2006. LNCS, vol. 3952, pp. 508–521. Springer, Heidelberg (2006)
18. Duan, K., Parikh, D., Crandall, D., et al.: Discovering localized attributes for fine-grained recognition. In: IEEE Conference on Computer Vision and Pattern Recognition, pp. 3474–3481 (2012)

A Packet Scheduling Method for Multimedia QoS Provisioning

Jinbang Chen[1]([✉]), Zhen Huang[2], Martin Heusse[3],
and Guillaume Urvoy-Keller[4]

[1] East China Normal University, Shanghai, China
jbchen@cs.ecnu.edu.cn
[2] National University of Defense Technology, Changsha, China
[3] Laboratoire LIG CNRS UMR 5217, Grenoble, France
[4] University of Nice Sophia Antipolis, CNRS, I3S, UMR 7271,
Sophia Antipolis, France

Abstract. Size-based scheduling is an appealing solution to manage bottleneck links as the interactive (short) flows of users are offered almost constant service times, whatever the level of congestion of the link is. Size-based schedulers like LAS, Run2C or LARS offer different additional features, for instance the ability to protect low/medium rate long lasting multimedia transfers like VoIP, for the case of LARS. However, all these solutions have a significant memory footprint as they require to keep one state per flow, or alternatively to modify the TCP implementation of every end host. This constitutes a significant hindrance to the deployment of size-based scheduling in the wild. In this paper, we propose Early Flow Discard (EFD), a new size-based scheduler, which keeps the salient properties of state-of-the-art size based schedulers like LARS, with a bounded memory requirement. We demonstrate its efficiency by comparing it with several size-based and size-oblivious schedulers. We further demonstrate that size-based scheduling offers an interesting solution to the so-called bufferbloat phenomenon. This is achieved with a completely different approach than the one advocated in PIE or CoDel for instance, as EFD does not strive to keep the queue occupancy low but controls per flow response times, which increases with the flow size.

Keywords: Packet scheduling · Multimedia · QoS · VoIP

1 Introduction

To improve the Quality of Experience of end users, ISPs have, at their disposal, a toolbox of traffic management mechanisms that they can deploy in their switches and routers. These mechanisms include scheduling and buffer management policies like *e.g.*, FIFO and FQ (*Fair Queuing*) which are popular scheduling policies;

J. Chen—The main part of this work was conducted when the first author was at Eurecom, France.

© Springer International Publishing Switzerland 2016
Q. Tian et al. (Eds.): MMM 2016, Part I, LNCS 9516, pp. 722–737, 2016.
DOI: 10.1007/978-3-319-27671-7_61

while droptail, RED (*Random Early Discard* [6]) , ARED (*Adaptive RED* [5]), and more recently, CoDel (*Controlled Delay* [13]) and PIE (*Proportional Integral controller Enhanced* [15]) are examples of buffer management policies. As highlighted by Sivaraman *et al.* [19], the exact combination of policies to use depends on the set of applications, their QoS requirements as well as on the underlying access network, *e.g.*, wired or wireless, cellular networks. There is no one size fits all solution. Still, a consensus exists concerning the (conflicting) objectives to reach: a highly utilized network with minimally utilized buffers. Indeed, while buffers aim at acting as shock absorbers, they lead to delay increase if they are continuously utilized [13]. Failure to meet these requirements can lead to the so-called bufferbloat phenomenon [3], as observed in wired and mobile networks [9].

RED, ARED, CoDel or PIE are solutions to keep buffers occupancy low [10]. The two key challenges faced by these active queue management (AQM) mechanisms is to assess the queue utilization and decide which flow to penalize in case the queue utilization is deemed too high. Penalizing a short flow might be detrimental to the end users as those flows are often generated by interactive applications (Web browsing, email, DNS queries, tweets, chats, . . .). Penalizing a high rate flow might be a good strategy if this flow requires a low RTT during its transfer, while it can be under optimal in the case of a download to a mobile device, see [19]. An approach to solve this dilemma (which flow to penalize) is to use a size-based scheduler, *e.g.*, SRPT (*Shortest Remaining Processing Time* [11]), LAS (*Least Attained Service* [18]), LARS (*Least Attained Recently Serviced* [8]) or Run2C (*Running Number differentiation mechanism* [2]). A size-based scheduler is flow aware as it takes the flow size or flow rate into account in its scheduling (and also buffer management) decision. When applied to the case of Internet traffic that consists of a mix of short and large flows, the idea behind size-based scheduling is to give priority to short flows and long flows in their early stage (which we call young flows hereafter) so as to favor interactive applications and, more generally, TCP flows in their infancy. Indeed, TCP is more sensitive to losses and delay in its slow start phase than during the later congestion avoidance phase. While size-based scheduling does not try to limit the queue size, it alleviates the bufferbloat phenomenon by granting short/young flows access to the head of the queue, which in general results in a response time almost independent of the physical queue size and actual load condition.

There is however a price to pay to use size-based scheduling, which is the bookkeeping cost of tracking each and every active flow at the switch/router. Indeed, a flow table is required that needs to be updated for every arriving packet[1]. In addition, the number of flows in progress can grow to a large value under high load. Recent advances in SDN (Software Defined Networking) demonstrate that bookkeeping of flows in commercial routers is possible. However, SDN aims at exposing the control plane and not the data plane of a switch [19].

---

[1] The authors of [2] proposed an elegant solution to work around this issue by encoding the flow size into the TCP sequence number. However, such a solution requires modifications of (and trust in) every end hosts.

In this work, we explore a radical approach to the scalability issue of size-based schedulers by limiting the memory of the scheduler to the packets in the buffer. Our scheduler, that we call EFD (*Early Flow Discard*) features the following properties: (i) Low response time to small and young flows (*i.e.*, long flows in their start-up phase), irrespectively of queue size and load; (ii) Low bookkeeping cost, *i.e.*, the number of flows tracked at any given time instant remains consistently low, and upper bounded by the physical queue size; (iii) Differentiation of flows based on volumes but also on rate.

The last property is known deficiency of formerly proposed size-based scheduler. It occurs if a low rate flow, *e.g.*, a VoIP transfer, runs for a long time and appears fat to the scheduler that simply accumulates transferred volume of flows like LAS or Run2C.

Our contributions are as follows:

- We present in detail the internal dynamics of EFD, *i.e.*, the way connections are handled by the scheduler and broken into so-called segments that are treated independently from each other.
- We compare the EFD strategy to reduce memory footprint with other strategies that rely on an on the fly classification of short and long flows and show its superiority.
- Through extensive simulations for the case of full duplex access links (Ethernet), we demonstrate that EFD behaves similarly to state-of-the art size-based schedulers like LARS, with the ability of effectively protecting low/medium rate multimedia transfers.

The remaining of this paper is organized as follows. Section 2 reviews the previous work on size-based scheduling and the methods proposed to delineate small flows from large ones. The detailed description of EFD discipline are given in Sect. 3. We then present the performance evaluation of EFD in full duplex in Sect. 4. Section 5 provides a summary of results.

2   Related Works

Size-based scheduling has received a lot of attention from the research community with applications to Web servers, Internet traffic or 3G networks – as they can significantly improve the quality of experience of users by favoring short flows. The representatives consist of SRPT, LAS, Run2C and LARS. Although appealing, SRPT is impractical for routers and switches as it requires prior knowledge of flow sizes - which is not achievable for most networking elements (router, access point, *etc.*). In addition, most of the size-based schedulers proposed so far, including the aforementioned disciplines, feature a common drawback – they either need to track the ongoing sizes of all flows, or require changes at the end-hosts. We term "statelessness" the property of a scheduler to not keep any state concerning the ongoing flows it is servicing. Run2C achieves this property albeit at the cost of a modification of TCP - which not only makes the scheme TCP dependent, but also reduces the randomness of initial sequence numbers.

Recall that a small amount of long flows contribute to the majority of the Internet traffic load. As such, if we are able to properly identify long flows so as to maintain flow states for these long flows only which are limited instead of all flows, a significant overhead-saving for flow state keeping will be naturally obtained, retaining the desirable property of providing low response time to short flows as long as short flows are preferentially served over long ones. A simple way to achieve this is to apply a probabilistic method to detect long flows[2] [12,16]. SIFT [16] uses such a probabilistic scheme along with a PS+PS scheduler. In SIFT, a flow is "short" as long as none of its packets passes the probabilistic test. All its packets go to high priority queue until this flow is identified as "long". There are however *false positives* induced with such strategy, which is detrimental as short flows identified as long ones by mistake will be sent to low priority queue.

Summary: Size-based scheduling policies can significantly improve the quality of experience of users although at the price of keeping one state per flow, which constitutes a major drawback in today's Internet. On the other hand, the mechanisms proposed to detect long flows suffer from false positives, which is an issue due to the huge number of short flows in typical traffic aggregates. Our proposal, EFD manages to keep the desirable properties of size-based schedulers with the ability of protecting low/medium rate multimedia transfers, while bounding the number of flows that need to be tracked, as we will see in the next sections.

3   Early Flow Discard Scheduling

EFD belongs to the family of Multi-Level Processor Sharing scheduling policy [11]. EFD features two queues. The low priority queue is served only if the high priority queue is empty. Both queues are drained in a FIFO manner at the packet level. In practice, a single physical queue for packet storage is divided into two virtual queues. The first part of the physical queue is dedicated to the virtual high priority queue while the second part is the low priority queue. A pointer is used to indicate the position of the last packet of the virtual high priority queue. This idea is similar to the one proposed in the Cross-Protect mechanism [12]. Algorithm 1 presents the EFD algorithm in pseudo-code.

A last point to mention is that each of the scheduling policies that we consider in this paper are paired with a buffer management scheme. For FIFO or SCFQ (an implementation of Processor Sharing for packet networks [7]), this is drop tail. In contrast, for all size-based scheduling policies, when the queue is full, the newly arriving packet is assigned a priority according the scheduling policy and this is the packet with the smallest priority that is discarded.

---

[2] Note that this mechanism is proposed in X-protect [12] to do admission control and not for scheduling.

Algorithm 1. Early Flow Discard algorithm

```
1: function packet_arrival(p)
2: # A new packet p of flow F arrives
3: if no packets of F are present in the queue then
4:    create a flow entry R(F) for F;
5:    # p is a high priority packet
6:    if the queue is full then
7:       if only high priority packets in the queue then
8:          p is dropped;
9:          return;
10:      else
11:         the last packet of low priority queue is dropped;
12:         p is inserted at the end of high priority queue;
13:      end if
14:   else
15:      p is inserted at the end of high priority queue;
16:   end if
17: else
18:    # at least one packet of F reside in the queue, so that a flow entry for F exists in the
          table
19:    if number of bytes already served of flow F < threshold th  then
20:       # p is a high priority packet
21:       if the queue is full then
22:          if only high priority packets in the queue then
23:             p is dropped;
24:             return;
25:          else
26:             the last packet of low priority queue is dropped;
27:             p is inserted at the end of high priority queue;
28:             update the flow entry R(F) in the table;
29:          end if
30:       else
31:          p is inserted at the end of high priority queue;
32:          update the flow entry R(F) in the table;
33:       end if
34:    else
35:       # p is a low priority packet
36:       if the queue is full then
37:          p is dropped;
38:          return;
39:       else
40:          p is put at the end of low priority queue;
41:          update the flow entry R(F) in the table;
42:       end if
43:    end if
44: end if
45:
46: function packet_departure(p)
47: # A packet p of flow F leaves due to the end of service or dropping
48: if no more packets of flow F are in the queue after p leaves then
49:    the flow entry R(F) is deleted from the table;
50: else
51:    update the flow entry R(F) in the table;
52: end if
```

4   EFD for a Full-Duplex Bottleneck Link

In this section, we investigate the performance of EFD in wired networks. Such a scenario enables to consider relatively large buffer sizes, which might be a plus for EFD as its memory is proportional to depth of the buffer. Specifically, we fix the buffer size at the bottleneck to be 300 packets, which means 450 KB. It is still fairly small, as buffer size can be a few MB or more[3].

4.1   Simulation Methodology

We present the network set up - network topology and workload - used to evaluate the performance of EFD and to compare it to other scheduling policies. All simulations are done using QualNet 4.5 [1].

Network Topology. We consider the case of a single bottleneck network, using the classical dumbbell topology depicted in Fig. 1. The buffer size is fixed and equal to 300 packets. A group of senders (nodes 1 to 5) are connected to a router (node 6) by 100 Mb/s bandwidth links and a group of receivers (nodes 8 to 12) are connected to another router (node 7) with a 100 Mb/s bandwidth link. The traffic is conveyed over the link connecting the two routers with a capacity of 10 Mb/s - which is therefore the bottleneck link. All links have a propagation delay of 1 ms.

Workload Generation. Suppose that the bottleneck link has a capacity of C bits/s. The traffic demand, expressed as a bit rate, is the product of the flow arrival rate λ and the average flow size $E[\sigma]$. The load offered to the link is then defined as:

$$\rho = \frac{\lambda E[\sigma]}{C} \qquad (1)$$

In all cases, the global load is controlled by tuning the arrival rate of requests. Thus, the congestion level increases as a function of the load. The flow arrivals

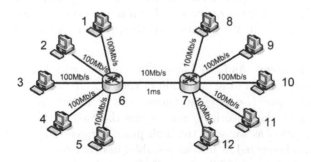

Fig. 1. Wired network topology

---

[3] see http://people.ucsc.edu/~warner/buffer.html.

follow a Poisson process, and the content requested is distributed according to a bounded Zipf distribution. Note that a bounded Zipf distribution is a discrete analog of the continuous bounded Pareto distribution, and Pareto is a heavy-tailed distribution usually adopted for modeling flows in the Internet.

Transfers are performed over TCP or UDP depending on the simulation. For each simulation set-up, we consider an underload and an overload regime, which correspond respectively to workloads of 8 and 15 Mb/s (80 % and 150 % of the bottleneck capacity). For TCP simulations, we use the GENERIC-FTP model of Qualnet, which corresponds to an unidirectional transfer of data. For UDP transfers, we use a CBR application model where one controls the inter-packet arrival time. The latter enables to control the exact rate at which packets are sent to the bottleneck. In both TCP and UDP cases, IP packets have a fixed size of 1500 bytes.

4.2 EFD Internal Dynamics

In this section, we present a detailed analysis of the way EFD manages flows. This is of utmost interest as it is a key distinguishing property of EFD as compared to other size-based scheduling policy. We focus on the following aspects:

- The evolution of the flow table size;
- How traffic is split between the low and high priority queue.
- How connections are fragmented by the scheduler due to the insertion/removal process within the flow table;

Flow Table. Due to the discarding mechanism of flow entries within the flow table in EFD, a flow entry exists in the flow table only if at least one packet of the flow is present in the queue. As stated in Sect. 3, an important consequence is that the flow table size is bounded by the physical queue size in packets[4].

For the TCP workload, we plot in Fig. 2 the instantaneous queue size and the instantaneous flow table size in underload and overload. Remind that the buffer size is 300 packets in our experiments. Figure 2 reveals that both flow table and packet queue grow up as the traffic intensity increases, but the table size is consistently below the queue size. Even in the overload case, the flow table size remains fairly small. We further investigate this issue in Sect. 4.3.

Virtual Queue Sizes. As EFD features two queues with high and low priority respectively, Fig. 3 depicts the evolution of the two virtual queue sizes, together with the overall queue (*i.e.* physical queue) size in underload and overload. One clearly sees that the low priority queue carries the bulk of the traffic. It is in line with our expectation as we want the high priority queue to be lightly loaded so that packets can be served as fast as possible, in order to grant short flows with low mean response times. While the bufferbloat phenomenon is often presented

---

[4] In most if not all active equipments – routers, access points – queues are counted in packets and not in bytes.

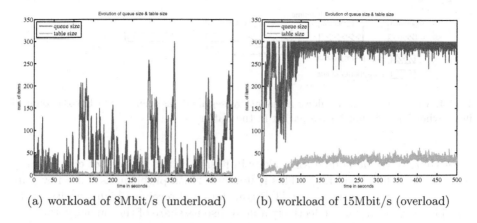

(a) workload of 8Mbit/s (underload)   (b) workload of 15Mbit/s (overload)

Fig. 2. Queue size & Table size

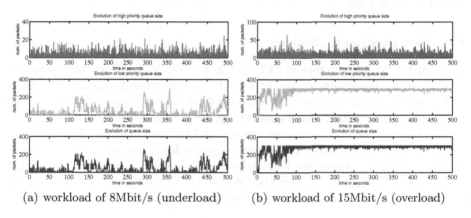

(a) workload of 8Mbit/s (underload)   (b) workload of 15Mbit/s (overload)

Fig. 3. Evolution of high and low priority queue size in both underload and overload.

as the persistence of large queues that degrade the performance of every flow, we see that using size-based scheduling, we can simultaneously sustain a high buffer occupancy and grantlow response times to young/short flows.

Flow Fragmentation. With EFD, a connection can be fragmented into many flows - each one is treated as fresh by the scheduler. In addition, the packets of one of these flows might be partly serviced in high priority or low priority queue: the first *th* packets are serviced by the high priority queue and the rest by the low priority queue. We call this phenomenon "flow fragmentation". It is in clear contrast with FIFO, LAS and RuN2C.

In practice, several phenomena can lead to break a connection into many fragments. For instance, during connection establishment, the TCP slow start

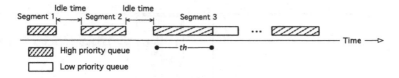

Fig. 4. EFD considers the flows segment by segment, as there is no bookkeeping of flows when they do not have a packet in the buffer

algorithm limits the number of packets in flight so that it does not continuously occupy the buffer (see Fig. 4). This is however not a problem, as those flows are smaller than th and thus young TCP transfers will receive a high priority. If the flow lasts longer and it is effectively able to use its share of the capacity, then the connection will eventually occupy the buffer without interruption and therefore stay in the flow table.

4.3   Performation Evaluation

We compare the performance of EFD to other scheduling policies. Our objective is to illustrate the ability of EFD to fulfill the following 3 objectives, namely (i) low bookkeeping cost, (ii) low response time to small flows, (iii) protecting long lasting delay sensitive flows, such as VoIP like multimedia transfers.

To illustrate the first 2 items, we consider a TCP workload with homogeneous transfers, *i.e.*, transfers that take place on paths having similar characteristics. For the last item - protecting long lived delay sensitive flows - we add a UDP workload to the TCP workload in the form of a CBR traffic, in order to highlight the behavior of each scheduler in presence of long lasting delay sensitive flows.

Overhead of Flow State Keeping. The approaches to maintain the flow table in the size-based scheduling policies proposed so far can be categorized as: (i) full flow table approach as in LAS [17], (ii) no flow table approach – an external support is provided to the scheduler, either at the end-hosts by modificaitons of the transport layer [2] or by some intermediate boxes, as in the case of a DiffServ scheme [14], that marks the packets, (iii) probabilistic approaches: a test is performed at each packet arrival for flows that have not already be incorporated in the flow table [4,12,16], (iv) EFD deterministic approach – the EFD approach is fully deterministic as flow entries are removed from the flow table once they have no more packet in the queue.

In this section, we compare all the approaches presented except the "No flow table approach" for our TCP workload scenario. We consider one representative of each family: LAS, X-Protect and EFD. We term X-Protect a Multi-Level Processor Scheduling policy that maintains two queues, similarly to Run2C, but uses the probabilistic mechanism proposed by Kortebi [12] to track long flows.

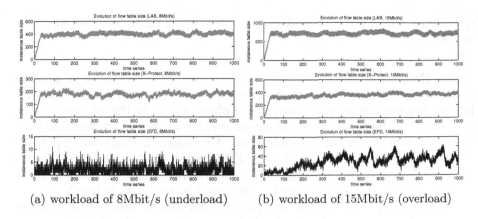

(a) workload of 8Mbit/s (underload)   (b) workload of 15Mbit/s (overload)

Fig. 5. Evolution of flow table size over time

As for the actual scheduling of packets, X-Protect mimics Run2C based on the information it possesses. If the packet does not belong to a flow in the flow table nor passes the test, it is put in the high priority queue. If it belongs to a flow in the flow table, it is put either in the high priority queue or in the low priority queue, depending on the amount of bytes sent by the flow. We use a threshold of 30KB, similar to the one used for EFD.

The evolution of flow table size over time for load of 8 Mbit/s (underload) and 15 Mbit/s (overload) are shown in Fig. 5. For LAS and X-Protect, the flow table is visited every 5 s and the flows that have been inactive for 30 s are removed. We also report the mean value and the 95 % level confidence interval of the flow table size over 1000 s simulation for both load conditions in Table 1.

Table 1. Statistics - flow table size

		LAS	X-Protect	EFD
8 Mbit/s (underload)	mean	392.49	176.84	1.99
	95 %-CI	[392.37, 392.61]	[176.78, 176.89]	[1.9962, 2.0032]
15 Mbit/s (overload)	mean	704	352.95	27.90
	95 %-CI	[703.8, 704.2]	[352.86, 353.05]	[27.88, 27.92]

We observe how X-Protect roughly halves the number of tracked flows, compared to LAS. By contrast, EFD reduces it by one order of magnitude. The reason why X-Protect offers deceptive performance is the race condition that exists between the flow size distribution and the probabilistic detection mechanism. Indeed, even though a low probability, say 1 %, is used to test if a flow is long, there exists so many short flows that the number of false positives becomes quite large, which prevents the flow table from being significantly smaller than in LAS.

Mean Response Time. Response time is a key metric for a lot of applications, especially interactive ones. An objective of EFD and size-based scheduling policies in general is to favor interactive applications, hence the emphasis put on response time. Response times are computed only for flows that complete their transfer before the end of the simulation. We consider five scheduling policies: FIFO, LAS, Run2C, EFD and LARS. FIFO is the current de facto standard and it is thus important to compare the performance of EFD to this policy. We do not consider the X-protect policy discussed in Sect. 4.3, as Run2C can be considered as a perfect version of X-protect since Run2C distinguishes packets of flows below and above the threshold th (we use the same threshold th for both EFD and Run2C).

We first turn our attention to the aggregate volumes of traffic per policy for the underload and overload cases. We observe no significant difference between the different scheduling policies in terms both of number of complete and incomplete connections. The various scheduling policies lead to a similar level of medium[5] utilization.

Distributions of the response times for the (complete) short and long transfers in underload and overload conditions are presented in Fig. 6, in which we plot the mean value together with the 95 % confidence interval of the response time over the flow size. Remember that the distribution of flow sizes generated exhibits high variability - meaning that small number of longest flows carry the majority of traffic load. Thus, it is problematic when calculating the confidence interval of flow response time, especially for long flows as the number of long flows collected from the workload is limited. To handle this issue, we accumulate the samples by starting from a certain flow size and spanning adjacent flow sizes in an ascending order until the number of samples reaches a threshold value given (threshold value equals to 200 for example), during which the mean value of all flow sizes traversed is taken as the flow size to pair with the confidence interval.

Under all load conditions, LAS, EFD and Run2C manage to significantly improve the response time of the short flows as compared to FIFO. EFD and Run2C offer similar performance. They both have a transition of behavior at about th value ($th = 20$ MSS). Still, the transition of EFD is smoother than the one of Run2C. This was expected as Run2C applies a strict rule: below or above th for a given transfer, whereas EFD can further cut a long transfer into fragments which individually go first to the high priority queue. Overall, EFD provides similar or slightly better performance than Run2C with a minimal price in terms of flow bookkeeping. LAS offers the best response time of size-based scheduling policies in our experiment for small and intermediate size flows. For large flows its performance is equivalent to what other policies obtain for the

---

[5] The medium is the IP path as those policies operate at the IP level.

underload case and significantly better for overload. However, one has to keep in mind that in overload conditions, LAS deliberately killed a large set of long flows, hence its apparent better performance. Also note how LARS behaves similarly to LAS in underload and degrades to fair queueing –which brings it close to FIFO in this case– when the networks is overloaded.

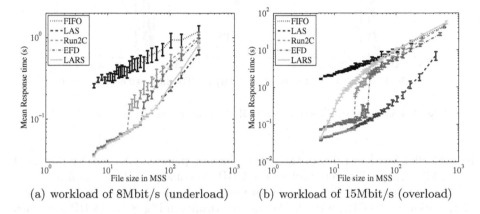

(a) workload of 8Mbit/s (underload)   (b) workload of 15Mbit/s (overload)

Fig. 6. Confidence interval of response time over flow size

The Case of Multimedia Traffic. In the TCP scenario considered above, FTP servers were homogeneous in the sense that they had the same access link capacity and the same latency to each client. The transfer rate was controlled by TCP. In such conditions, it is difficult to illustrate how EFD takes into accounts the actual transmission rate of data sources. In this section, we have added a single CBR flow to the TCP workload used previously.

We consider two rates 64 Kb/s and 500 Kb/s for the CBR flow, representing typical audio (*e.g.*, VoIP) and video stream (*e.g.*, YouTube video - even though YouTube uses HTTP streaming) respectively. The background load also varies - 4, 8 and 12 Mbps- which correspond to underload/moderate/overload regimes as the bottleneck capacity is 10 Mbps. To avoid the warm-up period of the background workload, the CBR flow is started at time $t = 10$ s and keeps on sending packets continuously until the end of the simulation. The simulation lasts for 1000 seconds. Since small buffers are prone to packet loss, we assign to the bottleneck a buffer of 50 packets, instead of 300 packets previously. The loss rates experienced by the CBR flow are given in Fig. 7, in which a well-known fair scheduling scheme called SCFQ [7] is added for the comparison, in addition to the disciplines mentioned before.

As we can see from the figure, for the case of a CBR flow with rate of 64 Kbps, LAS discards a large fraction of packets even at low load. This was expected as LAS only considers the accumulated volume of traffic of the flow and even at 64 kbps, the CBR flow has sent more than 8 MB of data in 1000 s (without taking the Ethernet/IP layers overhead into account). In contrast, FIFO, SCFQ and

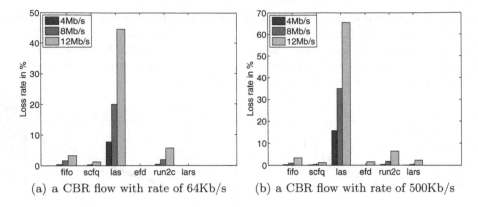

(a) a CBR flow with rate of 64Kb/s   (b) a CBR flow with rate of 500Kb/s

Fig. 7. Loss rate experienced by a CBR flow in different background loads

Run2C offer low loss rates in the order of a few percents at most. As for EFD and LARS, they effectively protect the CBR flow under all load conditions.

To further analyze this behavior, we next examine the inter-departure time distribution of the CBR flow [8]. We present results in Fig. 8 for a CBR flow with rate of 64 Kbps for two background loads: 4 Mbps and 12 Mbps. We observe from Fig. 8 that, LAS serves packets in batch (many packets have short delay between them) while EFD and LARS forward packets in a much more regular way as most packets have almost the same delay between them in both background load regimes. In addition, the jitter apparently ramps up under LAS and Run2C as the background traffic grows from 4Mbps (underload) to 12 Mbps (overload).

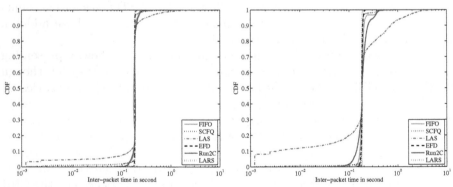

(a) CDF of inter-departure times for a CBR flow with 4 Mb/s background traffic   (b) CDF of inter-departure times for a CBR flow with 12 Mb/s background traffic

Fig. 8. Jitter of a CBR flow with rate of 64 Kb/s

As the rate of the CBR flow increases from 64 Kbps to 500 Kbps, no packet loss is observed for EFD in underload/moderate load conditions, similarly to SCFQ, whereas the other scheduling disciplines (FIFO, LAS, Run2C and LARS) are hit at various degrees. In overload, EFD and LARS blow up similarly to LAS (which still represents an upper bound on the loss rate as the CBR flow is continuously granted the lowest priority). EFD behaves slightly better than LARS as the load in the high priority queue is by definition lower under EFD than under Run2C.

When looking at the above results from a high level perspective, one can think at first sight that FIFO and SCFQ do a decent job as they provide low loss rates to the CBR flow in most scenarios (under or overload). However, those apparently appealing results are a side effect of a well-known and non desirable behavior of FIFO. Indeed, under FIFO, the non responsive CBR flow adversely impacts the TCP workload, leading to high loss rates. This is especially true for the CBR flow working at 500 kbps. SCFQ tends to behave similarly if not paired with an appropriate buffer management policy [7]. In contrast, LARS and EFD offer a nice trade-off as they manage to simultaneously grant low loss rates to the CBR flow with a low penalty to the TCP background workload. Run2C avoids the infinite memory of LAS but still features quite high loss rates since the CBR flow remains continuously stuck in the low priority queue.

Overall, EFD manages to keep the desirable properties of size-based scheduling policies and in addition manages, with a low bookkeeping cost, to protect multimedia flows as it implicitly accounts for the rate of this flow and not only its accumulated volume.

5   Conclusions

In this paper, we have investigated a radical approach to the bookkeeping cost by constraining it to the buffer size of router/switch. The proposed policy, *Early Flow Discard* (EFD), removes a flow entry from the table once its last packet currently residing in the queue leaves, and gives time and space priority to small/young. Through extensive simulations with Ethernet, we have demonstrated that EFD meets our expectations: it gives small response time to small flows, and is able to protect low/medium rate multimedia traffic, especially if it operates at low rate due to its limited memory. The latter means that as long as a flow requires a rate that is such that it does not create a significant backlog in the queue, it will be granted a high priority.

Acknowledgments. This work was supported by the French National Research Agency (ANR) project ELAN under contract ANR-08-VERS-008, the Science and Technology Commission of Shanghai Municipality under research grant no. 14DZ2260800, China Postdoctoral Science Foundation under grant no. 2014M561438, the National Basic Research Program of China (973) under Grant No. 2014CB340303,

National Natural Science Foundation of China (NS-F) under Grant No. 61402490 and Excellent Ph.D Dissertation Foundation of Hunan.

References

1. Qualnet 4.5. Scalable Networks (2010). http://www.scalable-networks.com
2. Avrachenkov, K., Ayesta, U., Brown, P., Nyberg, E.: Differentiation between short and long TCP flows: predictability of the response time. In: Proceedings of IEEE INFOCOM (2004)
3. Cerf, V., Jacobson, V., Weaver, N., Gettys, J.: BufferBloat: what's wrong with the Internet? ACM Queue **9**(12), 10 (2012)
4. Divakaran, D.M., Carofiglio, G., Altman, E., Primet, P.V.-B.: A flow scheduler architecture. In: IFIP Networking 2010, pp. 122–134 (2010)
5. Floyd, S., Gummadi, R., Shenker, S.: Adaptive red: an algorithm for increasing the robustness of red's active queue management. Technical report (2001). http://www.icir.org/floyd/
6. Floyd, S., Jacobson, V.: Random early detection gateways for congestion avoidance. IEEE/ACM Trans. Netw. **1**(4), 397–413 (1993)
7. Golestani, S.J.: A self-clocked fair queueing scheme for broadband applications. In: 13th Proceedings IEEE INFOCOM 1994, Networking for Global Communications, vol. 2, pp. 636–646, June 1994
8. Heusse, M., Urvoy-Keller, G., Duda, A., Brown, T.X.: Least attained recent service for packet scheduling over wireless LANs. In: WoWMoM 2010 (2010)
9. Jiang, H., Wang, Y., Lee, K., Rhee, I.: Tackling bufferbloat in 3G/4G networks. In: Byers, J.W., Kurose, J., Mahajan, R., Snoeren, A.C. (eds.) Internet Measurement Conference, pp. 329–342. ACM (2012)
10. Khademi, N., Ros, D., Welzl, M.: The new aqm kids on the block: an experimental evaluation of codel and pie. In: 2014 IEEE Conference on Computer Communications Workshops (INFOCOM WKSHPS) (2014)
11. Kleinrock, L.: Computer Applications, Volume 2: Queueing Systems, 1st edn. Wiley-Interscience, New York (1976)
12. Kortebi, A., Oueslati, S., Roberts, J.: Cross-protect: implicit service differentiation and admission control. In: IEEE HPSR 2004 (2004)
13. Nichols, K.M., Jacobson, V.: Controlling queue delay. ACM Queue **10**(5), 20 (2012)
14. Noureddine, W., Tobagi, F.: Improving the performance of interactive TCP applications using service differentiation. In: Computer Networks Journal, p. 354. IEEE (2002)
15. Pan, R., Natarajan, P., Piglione, C., Prabhu, M.S., Subramanian, V., Baker, F., VerSteeg, B.: Pie: a lightweight control scheme to address the bufferbloat problem. In: 2013 IEEE 14th International Conference on High Performance Switching and Routing (HPSR) (2013)
16. Psounis, K., Ghosh, A., Prabhakar, B., Wang, G.: Sift: a simple algorithm for tracking elephant flows, and taking advantage of power laws. In: The 43rd Annual Allerton Conference on Control, Communication and Computing (2005)
17. Rai, I.A., Biersack, E.W., Urvoy-Keller, G.: Size-based scheduling to Improve the performance of short TCP flows. IEEE Netw. **19**, 12–17 (2004)

18. Rai, I.A., Urvoy-Keller, G., Biersack, E.W.: Size-based scheduling with differentiated services to improve response time of highly varying flow sizes. In: Proceedings of the 15th ITC Specialist Seminar, Internet Traffic Engineering and Traffic Management (2002)
19. Sivaraman, A., Winstein, K., Subramanian, S., Balakrishnan, H.: No silver bullet: extending SDN to the data plane. In: HotNets-XII, College Park, MD, November 2013

Robust Object Tracking Using Valid Fragments Selection

Jin Zheng[1,2(✉)], Bo Li[1,2], Peng Tian[1], and Gang Luo[3]

[1] Beijing Key Laboratory of Digital Media, School of Computer Science and Engineering,
Beihang University, Beijing 100191, China
{JinZheng,boli,TianPengDigMedia}@buaa.edu.cn
[2] State Key Laboratory of Virtual Reality Technology and Systems, Beihang University,
Beijing 100191, China
[3] Schepens Eye Research Institute, Mass Eye and Ear, Harvard Medical School,
Boston, MA 02114, USA
Gang_Luo@MEEI.HARVARD.EDU

Abstract. Local features are widely used in visual tracking to improve robustness in cases of partial occlusion, deformation and rotation. This paper proposes a local fragment-based object tracking algorithm. Unlike many existing fragment-based algorithms that allocate the weights to each fragment, this method firstly defines discrimination and uniqueness for local fragment, and builds an automatic pre-selection of useful fragments for tracking. Then, a Harris-SIFT filter is used to choose the current valid fragments, excluding occluded or highly deformed fragments. Based on those valid fragments, fragment-based color histogram provides a structured and effective description for the object. Finally, the object is tracked using a valid fragment template combining the displacement constraint and similarity of each valid fragment. The object template is updated by fusing feature similarity and valid fragments, which is scale-adaptive and robust to partial occlusion. The experimental results show that the proposed algorithm is accurate and robust in challenging scenarios.

Keywords: Fragments-based tracking · Structured fragments · Valid fragment selection · Harris-SIFT filter · Template update

1 Introduction

Object tracking is one of the important areas in computer vision, and it has a wide range of applications in intelligent surveillance, activity analysis, content understanding, video compression, human-computer interaction etc. [1]. At present, object tracking still remains challenging for situations in which the tracked object undergoes large and unpredictable changes in its visual appearance due to occlusions, deformation, scales change, varying illumination, as well as the dynamic and cluttered environments.

Object tracking has been largely formulated in a match-and-search framework. In this framework, the process of tracking usually includes two steps: appearance modeling and motion searching. It is critical for robust tracking to handle appearance changes of objects [2]. In general, appearance features include global features and local features. For global features, the lacks of spatial constraints cause them sensitive to the changes in object itself

© Springer International Publishing Switzerland 2016
Q. Tian et al. (Eds.): MMM 2016, Part I, LNCS 9516, pp. 738–751, 2016.
DOI: 10.1007/978-3-319-27671-7_62

and the surrounding background. When there are appearance changes in tracked objects, global features also undergo great variations and oftentimes are hard to determine where the appearance changes happen. On the contrary, local features may provide the spatial information, and are more robust to object deformation and partial occlusion. For example, ALIEN (Appearance Learning In Evidential Nuisance [3]) proposes a novel object representation based on the weakly aligned multi-instance local features, and resides on local features to detect and track. FoT (Flock of Trackers [4]) estimates the object motion using local trackers covering the object. Matrioska [5] also resides on local features and uses a voting procedure, and the detection module uses multiple key point-based methods (ORB, FREAK, BRISK, SURF, etc.) inside a fallback model to correctly localize the object. Similarly, Yi et al. [6] describes the object using feature points, and utilizes motion saliency and description discrimination to choose local feature for tracking. Goferman et al. [7] proposes a Context-Aware Saliency Detection (CASD) method, which is based on four principles observed in the psychological literature. The most popular one is multiple-patch or fragment-based tracking. In this type of methods, the tracked object is divided into several fragments and each fragment is tracked independently based on features matching. The whole object is tracked using linear weighting scheme, vote map, or maximum similarity of the fragment location.

The advantages of constructing multi-fragments to represent the object include: (1) Multi-fragments provide the spatial information of the object template based on the relative spatial arrangements of different object parts. This offers an important advantage over the conventional region-based trackers in which only the region of interest is modeled by a single histogram with the loss of spatial information. (2) The use of fragments makes tracking more robust to partial occlusion and deformation etc. For example, even if some of the fragments are lost due to occlusion, the object could be located using the un-occluded fragments. (3) The structured template composed of multi-fragments is similar to the voting mechanism, and it can effectively prevent wrong decisions due to a large change of an individual fragment. (4) The dimensions of fragments are much lower than the dimensions of global features, therefore, it can minimize dimensionality computation effectively.

Some recent fragment-based tracking algorithms [8–16] are summarized. Almost all these algorithms derive features from color histogram, except [10] combines edge histogram and [13] combines gradient histogram. Using either overlapping [8–11, 14, 16] or non-overlapping [12, 13, 15] fragment partitions, the position of each fragment with respect to the center of the object is fixed and known in most of these algorithms [8, 10–16]. As a result, they may not deal with deformation and rotation well. Besides, each fragment is equally treated and just assigned a different weight based on its contribution. For estimating the contribution of each fragment, some algorithms combine the background [10, 12, 14], but still ignore the differences among these object fragments. Actually, similar object fragments of an object may cause confusion and therefore result in wrong tracking. Even more, some algorithms locate the object just using one maximal similarity fragment [8, 10], but an individual fragment probably has a large drifting due to the unreliable similarity scores, especially in cluttered background. Even though a linear weighting location scheme of multi-fragments are used [9, 11–16], they may still fail as the number of unreliable fragments increases in complex environments, such as serious occlusion or deformation, cluttered background etc. In these situations, the color histogram similarity scores of incorrectly tracked fragments are unreliable, and the use of many such fragments would negatively affect tracking. Therefore, the selection of reliable patches, especially for the object enduring large deformation, is particularly important. It is worth mentioning that Kown [17] analyzes the landscape of local mode of the patch to estimate its robustness, and evolves the topology between local patches as the appearance of the object changes geometrically. Moreover, Kown emphasizes the robustness of patch, as well as the discrimination ability and the distribution of each patch [18].

As for motion searching, exhaustively searching [8, 11, 15], mean-shift [9, 10, 12, 14] and particle filter [13, 16] are commonly used. Exhaustively searching is time-consuming, and mean-shift probably falls into local optimization, and the original mean-shift formulation cannot handle orientation and scale. Particle filter [19] is a robust approach due to the ability to maintain multiple hypotheses of the object state. However, the complexity of object tracking based on particle filter can linearly increases along with the dimension of features and the number of particles. The feature selection and dimensionality reduction, as well as scale estimation are important for particle filter.

This paper proposes a fragment-based tracking algorithm using multiple validation measures to prevent erroneous tracking. The first validation uses a discrimination and a uniqueness measure of the local fragments, which describe the discriminative difference between the fragment and background, and the unique difference between a given fragment and the other fragments for the tracked object, respectively. This algorithm builds a pre-selection mechanism to determine discriminative and unique fragments (DUFrags), which can help reduce the fragment searching costs and overcome the problem of confusing fragments. In a further validation procedure, valid DUFrags (V_DUFrags) are selected using Harris-SIFT filter. Thus, V_DUFrags construct a valid structured object. The robustness of Harris-SIFT to scale changes, rotation and illumination changes can help improve the robustness of tracking, and the local characteristic of Harris-SIFT make the overall object tracking robust to partial occlusion. Importantly, Harris-SIFT allows us to make use of highly localized features in tracking. Thus, our method utilizes object features at 3 levels: (a) overall level – a group of structured fragments to describe the tracked object; (b) a medium level – each V_DUFrag is tracked based on features in local histogram; (c) a low level – Harris corners within each V_DUFrag. Furthermore, V_DUFrags are searched with certain spatial constraints. With these validation procedures, the flexible fragment grid does not necessarily result in object drifting.

2   The Proposed Algorithm

Assuming the object appearance is explicit and can be expressed in the detection phase, the algorithm selects DUFrags firstly. In the tracking process, it uses Harris-SIFT filter to exclude the current invalid fragments, which may be due to occlusion or too large appearance change. The remaining V_DUFrags are used for locating the object. The framework of the algorithm is illustrated in Fig. 1.

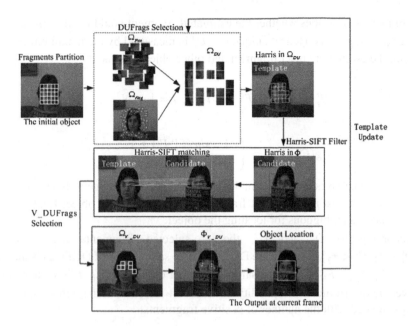

Fig. 1. The framework of the proposed algorithm

2.1 DUFrags Selection Based on Discrimination and Uniqueness

Given an tracked object, firstly we divide the bounding box of the object template into several fragments. Our algorithm adopts adjacent non-overlapping partition. The fragment must be large enough to provide a reliable histogram. On the other hand, the number of fragments must be large enough to provide sufficient spatial information. Thus, an object is empirically divided into squares of approximately 20-pixel wide. HSV color histogram with 85 bins [20] is adopted, and Bhattacharyya distance is used to calculate the color histogram similarity.

Assume that the object template T is divided into N fragments, which is represented as positive samples set $\Omega_{Pos} = \{F_1^t, F_2^t, \cdots, F_N^t\}$, and there are negative samples set $\Omega_{Neg} = \{\tilde{F}_1^t, \tilde{F}_2^t, \cdots \tilde{F}_M^t\}$ existing in the neighboring background. N is the number of positive samples, M is the number of negative samples, and t is the time. Discrimination is defined as D_i^t as formula (1).

$$D_i^t = \max(sim(F_i^t, \tilde{F}_j^t)) \quad i = 1 \ldots N, \ j = 1 \ldots M \tag{1}$$

Here, F_i^t denotes the ith object fragment at the tth frame, and $\tilde{F}_j^t$ denotes the jth background fragment at the tth frame. $sim(\cdot)$ is the similarity measure function. Searching the most similar negative fragment with the object fragment F_i^t, and the maximal similarity is denoted by D_i^t. If D_i^t is big enough, it means that the object fragment F_i^t is not discriminative from the background.

Uniqueness describes whether the estimated fragment could be distinguished from other object fragments. Hence, uniqueness U_i^t is measured by the maximum difference between the estimated object fragment and other object fragments.

$$U_i^t = \sum_{j=1}^{N} g(F_i^t, F_j^t, \tau) \quad i \neq j, \ i = 1 \ldots N \tag{2}$$

$$g(F_i^t, F_j^t, \tau) = \begin{cases} 1 & sim(F_i^t, F_j^t) \geq \tau \\ 0 & else \end{cases} \tag{3}$$

When U_i^t increases, it means that there are many object fragments having the similar features with the estimated object fragment, so the estimated object fragment is not unique, and it is ambiguous for locating the object.

According to D_i^t and U_i^t, the pre-selected fragments are decided by $\{F_i : D_i^t < \tau_1, U_i^t < \tau_2\}$, and they are DUFrags. The samples set of DUFrags is marked as $\Omega_{DU} = \{F_1, F_2, \cdots F_{N_{DU}}\}$, and the samples number of Ω_{DU} is N_{DU}. τ_1 is set according to the average discrimination capability in the worst selection at the initialization frame, and τ_2 is related with the number of positive fragments.

$$\tau_1 = \alpha_1 \cdot \frac{1}{N} \sum_{i=1}^{N} D_i^0, \quad \tau_2 = \alpha_2 \cdot N, \ \alpha_1, \alpha_2 \in [0, 1] \tag{4}$$

For convenience, time t is omitted in the following.

Fragment selection is actually a process of features selection and dimension reduction. The obtained DUFrags allow the proposed algorithm to better handle distractive environments and interference from similar objects. It also could focus its computational resources on more informative regions.

2.2   V_DUFrags Selection Based on Harris-SIFT Filter and Spatial Constraint

In the tracking process, particle filter is used to get the candidate fragments at each frame, and the searching area is restrained in a small range. If the highest color-histogram similarity is used to locate the object (Fig. 2(a)), it is obvious that some of them are basically correct (such as fragments 1–5), the others are mismatched obviously (such as fragments 6–10). Such deformed fragments or noisy fragments sometimes drift off the object completely, and this could lead to completely wrong localization of the object. How to exclude the mismatching in flexible fragment grid is important. Actually, applying histogram calculation to validate fragments and then using those valid fragments with histogram similarity weight to track is analogous to circular reasoning, if validity is not clearly defined. This is the essential limitation for the existing weighted fragment-based tracking methods only using color histogram.

Thus, Harris-SIFT filter and spatial constraint are proposed for excluding the mismatching and achieving accurate tracking. For each DUFrags, the top-rank N_P locations which have higher color histogram similarities are selected. We believe that color histogram similarity is appropriate for confirming the candidates, but

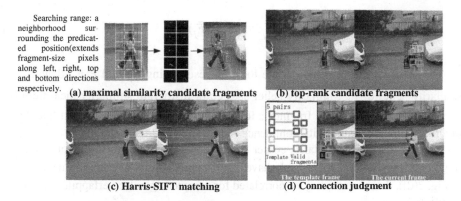

<table>
<tr><td>Searching range: a neighborhood surrounding the predicated position(extends fragment-size pixels along left, right, top and bottom directions respectively.</td></tr>
</table>

(a) maximal similarity candidate fragments   (b) top-rank candidate fragments

(c) Harris-SIFT matching   (d) Connection judgment

Fig. 2. Float fragments with structure constraint (Color figure online)

cannot locate the unique correct location accurately. For N_{DU} fragments in Ω_{DU}, there are $N_{DU} \times N_P$ candidate fragments marked as $\Phi = \{\bar{F}_1^1, \bar{F}_1^2, \cdots, \bar{F}_1^{N_P}, \bar{F}_2^1, \bar{F}_2^2, \cdots, \bar{F}_2^{N_P}, \cdots, \bar{F}_{N_{DU}}^1, \bar{F}_{N_{DU}}^2, \cdots, \bar{F}_{N_{DU}}^{N_P}\}$. Then we use Harris-Sift filter to identify the valid DUFrags. A Harris-SIFT feature is defined as a Harris corner and its SIFT feature vector, and it uses the efficiency of Harris corners, as well as the robustness of SIFT features. In detail, Harris corners in Ω_{DU} and Φ are firstly extracted respectively, and their SIFT feature vectors are computed. Then, the Euclidean distances of SIFT feature vectors between Ω_{DU} and Φ are calculated for matching. Random Sample Consensus (RANSAC) is used [21] to exclude the mismatched points. To realize the spatial constraint, it is required that the valid candidate fragments are 8-neighbor connected with template fragment in Ω_{DU}.

As can be seen in Fig. 2(b), for each DUFrag, 10 candidate fragments having top-rank color histogram similarities are found using particle filter. Then, Harris-SIFT are computed for these candidate fragments, and five matched pairs are obtained (Fig. 2(c)). The five fragments in Ω_{DU} are marked by red, red, blue, yellow and orange squares. According to Harris-SIFT, the corresponding matched fragments in Φ are red; red and blue; blue and yellow; yellow and blue; orange and violet ones (Fig. 2(d)). Take the orange DUFrags on the person's knee as an example. Its matched points at the current frame are covered by one orange fragments and one violet fragment. Because the two candidate fragments are all 8-neighbor connected with the orange fragment in the template, they are valid fragments.

Thus, a candidate fragment in Φ is considered valid when there exists SIFT matching-pair and connectivity relationship for this fragment and its corresponding object template fragment in Ω_{DU}. All the valid fragments in Φ are marked as Φ_{V_DU}, which keeps the stable and structured parts for the object.

2.3 Object Location Based on V_DUFrag Fusion

After Harris-SIFT filtering, all the fragments in Φ_{V_DU} are fused to determine the object location. Suppose the matched set is $\Phi_{V\_DU} = \{\bar{F}_1', \bar{F}_2', \cdots \bar{F}_L'\}$ at the current frame, the object tracking likelihood function using multi-fragment is define as

$$\hat{P}(T) = \sum_{j=1}^{L} P(T|\bar{F}'_j)P(\bar{F}'_j) \tag{5}$$

The mode of $P(T|\bar{F}'_j)$ denotes the location result using $\bar{F}'_j$, and $P(\bar{F}'_j)$ denotes the similarity belief of $\bar{F}'_j$. Similarity belief uses color histogram similarity measurement. As a voting process, all the fragments in Φ_{V_DU} participate to form a joint likelihood function using color similarity confidences and fragment displacements.

Because one matched Harris corner can be covered by several different candidate fragments, thus several $\bar{F}'_i$ probably have a substantial spatial overlap in images (such as Fig. 2(d)), their beliefs may be correlated for conquering non-overlapping fragments partition.

2.4   Feature Fusion Update

Object drift is largely due to imperfect template updating. It is a common updating strategy depending on the feature similarity between the best-matched candidate and the template. However, oftentimes the similarity measure is not reliable, for instance, histogram similarity is prone to error due to occlusion, similar background and noises etc. Moreover, the threshold of similarity for making updating decision is not easy to determine.

This paper sets a rule that the template should not be updated in cases of occlusion or large deformation, otherwise it can be updated in cases of zoom in/out or rotation. Because Harris-SIFT is sensitive to occlusion and can provide some information about rotation, scale change, we propose an object template update strategy based on Harris-SIFT and color histogram measures, which is described as follow.

Update condition: If there is SIFT matched feature point pair in the fragment, or its color histogram similarity is higher than the average color histogram similarity of the fragment computed based on the previous frames, the fragment is regarded as in updating state. If all the fragments of the object are in updating state, the object update condition is satisfied.

Scale estimation: The object size estimated by particle filter is not accurate sometimes, especially when the object has irregular scale change. Therefore, our algorithm estimates scale change based on readily available SIFT matching-pairs. Define the bounding boxes of the matched SIFT points in the object template and the current frame as C_1 and C_2, and the bounding box of object template as R_1. Then calculate the distance between C_1 and R_1, which is marked as $Gap_1 = [top, bottom, left, right]$. Then,

$$Gap_2 = \sqrt{\frac{Area(C_2)}{Area(C_1)}} \times Gap_1 \tag{6}$$

Add Gap_2 to C_2, and the bounding box of the object in the current frame is gotten, which is denoted as R_2.

For each tracking frame, once the update condition is satisfied, scale estimation is processed, and the object template is updated. Afterwards, DUFrags are chosen again.

3   Experimental Results

We first qualitatively tested the selection of DUFrags and V_DUFrags to demonstrate the validity of this procedure for object tracking. And then, tracking results with qualitative and quantitative evaluation are presented. Finally, computational cost and limitation are also analyzed in this section.

3.1   DUFrags and V_DUFrags Selection

Detecting discriminative and unique fragments, as well as occluded or deformed fragments of the object are the important steps for the proposed tracking algorithm. Figure 3 shows some results of DUFrags and V_DUFrags selection.

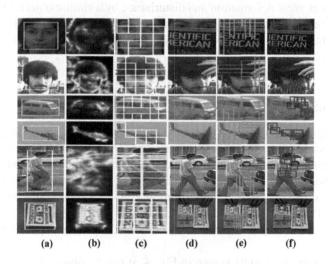

| (a) | (b) | (c) | (d) | (e) | (f) |

Fig. 3. DUFrags & V_DUFrags selection (a) object and background, (b) CASD, (c) DUFrags, (d) occluded/deformed candidate, (e) RFT, (f) V_DUFrags (Coloe figure online)

Figure 3(a) show the object and its neighboring background, and Fig. 3(c) show DUFrags found by our pre-selection method (labeled by yellow rectangles). As most existing fragment-based methods [8–16] do not include pre-selection mechanism, we illustrate Context-Aware Saliency Detection (CASD) results [7] for reference. Similar to our method, CASD doesn't need prior knowledge or training database, and the detected saliency regions (shown as bright areas in Fig. 3(b)) is based on local low-level consideration, global considerations, visual organizational rules, and high-level factors. The experimental results show our DUFrags and CASD are well coincide with each other not only for high discrimination regions, such as eye, nose and mouth, but also for low uniqueness regions, such as hair and cheek in face images. Especially, when the background is simple, such as plane and dollar images, CASD and DUFrags both behave the similar results. However, when the background is complex, such as walker images, CASD is not exclusive to the object, for instance, the leg is darker than some neighboring background. On the contrary, DUFrags regard all the object fragments as discriminative and unique fragments. In case of lack of local feature, such as the vehicle roof and window in the minibus image, CASD may ignore the regions lacking small scale features, while DUFrags appear to be tracking relevant.

For V_DUFrags selection, the fragment selection method in Robust Fragments-based Tracking (RFT) [8] is used for comparison because it is similar to our algorithm. In RFT, the object is divided into 36 overlapping fragments and those with top 25 % ranked similarity are selected to represent the current candidate. The color histogram is used in similarity measure. As can be seen in Fig. 3(e), some occluded fragments or background fragments are selected incorrectly by RFT, especially when the occlusions are large or have the similar color as the object, while some useful fragments are ignored. On the contrary, the proposed Harris-SIFT filter could filter out occluded or highly deformed fragments. A typical example is shown by the dollar images. The tracked object undergoes great deformation and disturbance by a similar object. Our method is able to capture the corrected fragments of the tracked dollar bill. These examples demonstrate that overall V_DUFrags are robust and reliable fragments to describe the tracked objects.

3.2   Tracking Results

For evaluating the tracking performances, eight challenging video sequences used in previous publications [8, 19] were tested. These videos include partial occlusion, non-rigid deformation, scale change, background disturbances and motion blur. For each sequence, the location of the tracked object is manually labeled at the first frame. We evaluate our tracking algorithm against five state-of-the-art algorithms, including RFT [8], DFT [22], LOT [23], RCT [24] and LSST [25]. We used the source codes provided by the authors and ran them using optimal parameters for each algorithm.

From the exemplar results shown in Fig. 4, it can be observed that the proposed algorithm was among the best methods, as it could track objects with the highest accuracy for most of the sequences. Especially in case of very large scale change, such as Highway2. According to the quantitative evaluation in Table 1 (the average center location error) and Table 2 (the average overlap rate), LSST tracker gave better results than ours for DavidOutdoor, DavidIndoor and Face sequences. In these sequences, the appearances of the objects vary too much due to non-rigid activity. Due to limitations of valid fragment updating, our method could not adapt to too severe deformation.

Table 1. Average center location error (in pixels)

Sequence	RFT	DFT	LOT	RCT	LSST	Ours
Occlusion1	6.2	22.4	22.0	21.5	7.1	1.5
DavidOutdoor	65.3	58.6	64.6	103.3	5.1	5.5
Caviar1	5.2	9.9	2.5	100.4	2.5	2.5
Highway	49.4	8.8	44.7	11.0	13.4	3.2
DavidIndoor	52.1	17.2	84.7	12.5	4.8	7.9
Deer	91.0	255.9	89.9	201.4	8.9	5.2
Face	49.6	135.7	34.1	158.8	4.3	8.5
Dollar	11.6	3.7	68.9	15.4	3.0	1.2
Avg.	**41.3**	**64.0**	**51.4**	**78.0**	**6.1**	**4.4**

Table 2. Average overlap rate

Sequence	RFT	DFT	LOT	RCT	LSST	Ours
Occlusion1	0.89	0.73	0.51	0.70	0.88	0.96
DavidOutdoor	0.41	0.57	0.50	0.24	0.77	0.78
Caviar1	0.70	0.66	0.85	0.23	0.83	0.85
Highway	0.16	0.34	0.30	0.35	0.32	0.57
DavidIndoor	0.25	0.44	0.15	0.51	0.75	0.52
Deer	0.11	0.07	0.12	0.07	0.59	0.64
Face	0.37	0.36	0.48	0.16	0.89	0.79
Dollar	0.69	0.88	0.20	0.66	0.89	0.95
Avg.	**0.45**	**0.51**	**0.39**	**0.37**	**0.74**	**0.76**

Fig. 4. Sample tracking results for challenging sequences

The tracking results on those challenging sequences show that the proposed algorithm was among the top 2 and mostly the top 1 in terms of average center location error and the average overlap rate.

3.3   Computational Cost

The proposed tracking algorithm was implemented in Matlab 2013 code, and was tested on a PC with a AMD Athlon (tm) II X4 635 Processor (2.9 GHz) and 4 GB RAM. For the aforementioned video sequences, the processing time of each sequence is listed in Table 3. The processing time of the algorithm mainly spent on color similarity computation in particle filter framework, SIFT features computation and matching. Apparently, the processing time was different depending on the size and texture of the object. In our experiments, the number of positive samples were determined by the size of the object, and the number of negative $M = 200$. In addition, the color histogram is computed using d-dimensional feature vectors, $d = 85$, and the size of fragment is $h \times w$, $w = h = 20$, $\alpha_1 = 0.3$, $\alpha_2 = 0.2$. The particle number for each DUFrag was $M_p = 100$, and $N_P = 10$.

Table 3. Processing time

Sequence	Image size (pixels)	Object size (pixels)	Length	Average processing time (s/f)
Occlusion1	352×288	116×146	898	0.6674
DavidOudoor	640×480	39×131	252	0.3340
Caviar1	384×288	31×80	382	0.0798
Highway	320×240	43×63	45	0.1433
DavidIndoor	320×240	60×93	462	0.2891
Deer	704×400	101×71	71	0.0884
Face	640×480	94×110	492	0.4949
Dollar	320×240	58×90	327	0.3892

It should be noted that, since each patch is tracked individually, parallel hardwires (such as multiple/multi-core processors or graphics programmable units) can be explored to further increase running speed.

3.4   Limitations and Future Work

Even if the tracked object has a large area of homogeneous color, as long as the object is distinct from the background, the proposed algorithm can use the peripheral discriminative fragments for tracking, such as the person in Caviar1 sequence. In rare cases where our algorithm may fail, the reasons could be that, all fragments are too homogeneous and do not contain unique fragments, or the object is blurry and does not contain discriminative fragments from the background, or Harris-SIFT tracking failed. Another limitation is that small objects (less than 20×20 pixels) are not suitable for fragment-based tracking or fragment selection. How to design more efficient local features, especially for homogeneous and blurry object, as well as small object, is our future work.

More experimental works should be implemented on benchmark sequences (e.g., OTB2013, VOT 2013, VOT2014), as well as comparing with more state-of-the-art tracking algorithms, to examine the algorithm's adaptive capability to severe, non-rigid deformation. This is also our future work.

4   Conclusion

The conventional features selection usually directly use the local features selected from previous frames to locate object while ignoring their validity, and this can easily lead to feature degradation and impair tracking robustness and accuracy. Unlike many existing algorithms, this paper proposes a robust fragments-based object tracking algorithm that uses discriminative and unique fragments pre-selection mechanism to determine DUFrags, and then uses Harris-SIFT filter and spatial constraint to further identify the current valid fragments (VDUFrags). This process helps to exclude the occlusion or transformed fragments. Finally, the object is localized using a structured fragment template combining the displacement and similarity of each valid fragment.

We don't consider the fragments tracking directly determined by SIFT matching. The reason is mainly that our method is possible to use the multiple overlapped candidate fragments, which is coincident with multiple hypotheses of particle filter. In addition, in our approach SIFT deal with Harris corners at a low level, a fragment describes an object at a medium level, and the combination of multiple structured fragments represents the object at a high level. Such a multi-level framework can be more robust than methods that only rely on single level. The experimental results show that the proposed algorithm is accurate and robust in challenging scenarios that include severe occlusions and large scale changes.

Acknowledgements. This work is supported by the National Science Foundation of China (No. 61370124), China 863 Program (Project No. 2014AA015104), the National Science Foundation of China for Distinguished Young Scholars (No. 61125206), China Scholarship Foundation (No. 201303070205), and NIH/NIA grant AG041974.

References

1. Li, X., Hu, W., Shen, C., et al.: A survey of appearance models in visual object tracking. ACM Trans. Intell. Syst. Technol. (TIST) 4(4), 58 (2013)
2. Salti, S., Cavallaro, A., Di Stefano, L.: Adaptive appearance modeling for video tracking: survey and evaluation. IEEE Trans. Image Process. 21(10), 4334–4348 (2012)

3. Pernici, F., Del Bimbo, A.: Object tracking by oversampling local features. IEEE Trans. Pattern Anal. Mach. Intell. **99**(PrePrints), 1 (2013)
4. Vojir, T., Matas, J.: Robustifying the flock of trackers. In: Computer Vision Winter Workshop, pp. 91–97. IEEE (2011)
5. Maresca, M.E., Petrosino, A.: MATRIOSKA: a multi-level approach to fast tracking by learning. In: Petrosino, A. (ed.) ICIAP 2013, Part II. LNCS, vol. 8157, pp. 419–428. Springer, Heidelberg (2013)
6. Yi, K.M., Jeong, H., et al.: Initialization-insensitive visual tracking through voting with salient local features. In: IEEE International Conference on Computer Vision (ICCV), pp. 2912–2919. IEEE (2013)
7. Goferman, S., Zelnik-Manor, L., Tal, A.: Context-aware saliency detection. IEEE Trans. Pattern Anal. Mach. Intell. **34**(10), 1915–1926 (2012)
8. Adam, A., Rivlin, E., Shimshoni, I.: Robust fragments-based tracking using the integral histogram. In: IEEE Computer Society Conference on Computer Vision and Pattern Recognition, vol. 1, pp. 798–805. IEEE (2006)
9. Srikrishnan, V., Nagaraj, T., Chaudhuri, S.: Fragment based tracking for scale and orientation adaptation. Comput. Vision Graph. Image Process., 328–335 (2008)
10. Jeyakar, J., Babu, R.V., Ramakrishnan, K.R.: Robust object tracking with background-weighted local kernels. Comput. Vision Image Underst. **112**(3), 296–309 (2008)
11. Naik, N., Patil, S., Joshi, M.: A fragment based scale adaptive tracker with partial occlusion handling. In: TENCON 2009–2009 IEEE Region 10 Conference, pp. 1–6. IEEE (2009)
12. Wang, F., Yu, S., Yang, J.: Robust and efficient fragments-based tracking using mean shift. AEU-Int. J. Electron. Commun. **64**(7), 614–623 (2010)
13. Nigam, C., Babu, R.V., Raja, S.K., et al.: Fragmented particles-based robust object tracking with feature fusion. Int. J. Image Graph. **10**(1), 93–112 (2010)
14. Li, G., Wu, H.: Robust object tracking using kernel-based weighted fragments. In: 2011 International Conference on Multimedia Technology (ICMT), pp. 3643–3646. IEEE (2011)
15. Dihl, L., Jung, C.R., Bins, J.: Robust adaptive patch-based object tracking using weighted vector median filters. In: 24th SIBGRAPI Conference on Graphics, Patterns and Images (Sibgrapi), pp. 149–156. IEEE (2011)
16. Erdem, E., Dubuisson, S., Bloch, I.: Fragments based tracking with adaptive cue integration. Comput. Vision Image Underst. **116**(7), 827–841 (2012)
17. Kwon, J., Lee, K.M.: Tracking of a non-rigid object via patch-based dynamic appearance modeling and adaptive basin hopping monte carlo sampling. In: Proceedings of IEEE Conference on Computer Vision and Pattern Recognition (2009)
18. Kwon, J., Lee, K.M.: Highly nonrigid object tracking via patch-based dynamic appearance modeling. IEEE Trans. Pattern Anal. Mach. Intell. **35**(10), 2427–2441 (2013)
19. Isard, M., Blake, A.: Condensation—conditional density propagation for visual tracking. Int. J. Comput. Vision **29**(1), 5–28 (1999)
20. Yang, J., Wang, J., Liu, R.: Color histogram image retrieval based on spatial and neighboring information. Comput. Eng. Appl. **43**(27), 158–160 (2007)
21. Chen, H.Y., Lin, Y.Y., Chen, B.Y.: Robust feature matching with alternate hough and inverted hough transforms. In: IEEE Conference on Computer Vision and Pattern Recognition (CVPR), pp. 2762–2769. IEEE (2013)
22. Sevilla-Lara, L., Learned-Miller, E.: Distribution fields for tracking. In: IEEE Conference on Computer Vision and Pattern Recognition (CVPR), pp. 1910–1917. IEEE (2012)
23. Oron, S., Bar-Hillel, A., Levi, D., et al.: Locally orderless tracking. In: IEEE Conference on Computer Vision and Pattern Recognition (CVPR), pp. 1940–1947. IEEE (2012)

24. Zhang, K., Zhang, L., Yang, M.-H.: Real-Time compressive tracking. In: Fitzgibbon, A., Lazebnik, S., Perona, P., Sato, Y., Schmid, C. (eds.) ECCV 2012, Part III. LNCS, vol. 7574, pp. 864–877. Springer, Heidelberg (2012)
25. Wang, D., Lu, H., Yang, M.H.: Least soft-threshold squares tracking. In: IEEE Conference on Computer Vision and Pattern Recognition (CVPR), pp. 2371–2378. IEEE (2013)

Special Session Poster Papers

Exploring Discriminative Views for 3D Object Retrieval

Dong Wang[1,2], Bin Wang[1(✉)], Sicheng Zhao[2], Hongxun Yao[2], and Hong Liu[1]

[1] State Key Laboratory of Robotics and System, Harbin Institute of Technology,
Harbin, China
{dwang89,wbhit,hong.liu}@hit.edu.cn
[2] School of Computer Science and Technology, Harbin Institute of Technology,
Harbin, China
{zsc,h.yao}@hit.edu.cn

Abstract. View-based 3D object retrieval techniques have become prevalent in various fields, and lots of ingenious studies have promoted the development of retrieval performance from different aspects. In this paper, we focus on the 2D projective views that represent the 3D objects and propose a boosting approach by evaluating the discriminative ability of each object's views. Different from previous works on selecting representative views of query object, we investigate the discriminative information of each view in dataset. By employing the proposed reverse distance metric, we utilize the discriminative information for many to many view set matching. The proposed algorithm is then employed with various features to boost the multi-model graph learning method. We compare our approach with several state of the art methods on ETH-80 dataset and National Taiwan University 3D model dataset. The results demonstrate the effectiveness of our method and its excellent boosting performance.

Keywords: Discriminative view · Reverse sum-min distance · 3D object retrieval

1 Introduction

The rapid advance of computer techniques lead to the swift growth of 3D data nowadays. With rich information contained, 3D models have been widely used in various domains [4,18,24]. Meanwhile, efficient retrieval and recognition technologies are urgently needed.

There are two main types of retrieval paradigms: model-based methods [8,19] and view-based methods [1,21]. Early object retrieval techniques mainly belong to model-based methods, which require explicitly virtual 3D models. However, it is not convenient to build 3D models in many practical applications. In addition, the model-based methods are always computational expensive. To overcome these difficulties, researchers consider to use a set of 2D pictures to represent a 3D object. This alternative approach, namely view-based method, gives rise to a

© Springer International Publishing Switzerland 2016
Q. Tian et al. (Eds.): MMM 2016, Part I, LNCS 9516, pp. 755–766, 2016.
DOI: 10.1007/978-3-319-27671-7_63

profitable and intensive research in computer vision community. In this scheme, one single view or multiple views are captured to represent a 3D object. Therefore, the retrieval of 3D object becomes the problem of multiple views matching.

It is commonly agreed in the 3D retrieval field that view-based paradigm has superior performance to the model-based framework [6,11,26]. Nevertheless, the plenty of representative views bring problems.

As the fundamental step in view-based retrieval, view capturing and selection is important for both computational efficiency and retrieval precision [6]. A typical view selection method is Adaptive Views Clustering (AVC) [1], which employed Bayesian information criteria to cluster representative views from the view pool. Giorgi et al. [13] used single semantically grounded 2D views for 3D retrieval by geometrical characteristics. These methods typically refined the intra information and selected the most representative view subset as the query object representation. In the further work, Gao et al. [12] proposed a discriminative view selection scheme based on users' relevance feedback. The system incrementally selected the most informative view in each round of relevance feedback. Hamid [16] formulated the view selection procedure as a classification and feature selection problem, which trained a classifier and measured the classification performance of each view to acquire the best views of each 3D shape.

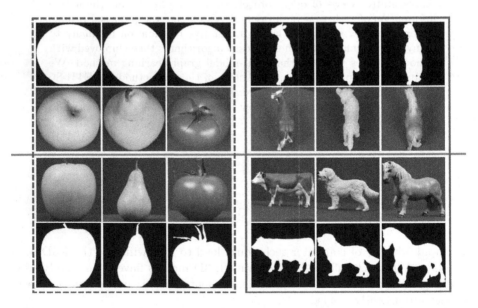

Fig. 1. View examples of six 3D objects with different discriminative information in ETH dataset. There is a fruit group (in left dashed box) and an animal group (in right solid box). From left to right column are apple, peer, tomato, cow, dog, and horse. We select two views of each object and its corresponding shape map. It can be seen from the upper two lines that views in each group present similar appearance, in contrast, the shape maps in the bottom two lines have object information that is able to distinguish one from another within each group.

An excellent retrieval algorithm implies that the higher number of related objects belong to the same category meanwhile less irrelevant objects from other categories. We will try to boost the retrieval performance by investigating the role of each view, which mainly comes from the fact that the views representing a 3D object possess different information of that instance. Figure 1 reveals that some views of the objects from different categories present similar shape property, which will confuse the retrieval system to distinguish objects from each other. In this paper, we propose a method to evaluate the discriminative importance of the object's views and a new distance metric that can utilize the information for retrieval. Contrary to most of the methods which select representative views in an unsupervised manner, we compute the weight of views for each object in the dataset based on their distinction performance of inter category.

In summary, the main contributions of this paper are three-fold.

1. We evaluate the discriminative performance of each view in a simple but effective manner for the objects in dataset rather than selecting the representative views which are difficult to measure. This is an off-line training process and does not need any prior knowledge of the camera settings.
2. We propose a new distance metric to integrate the weight information of views for object matching. The proposed method employs the discriminative information and can be more robust for similar views from different categories.
3. The proposed method can be easily integrated into learning based framework and can improve the retrieval results substantially. The added bottom information can be regarded as the basic components of some learning based approaches to boost their performance.

The remainder of this paper is organized as follows. Section 2 briefly reviews the related work of view-based 3D object retrieval. The proposed view weights and distance metric are presented in Sect. 3. Experimental results and analysis are provided in Sect. 4. Finally, the conclusion and discussion are given in Sect. 5.

2   Related Works

View-based 3D retrieval is a special task of content-based image retrieval (CBIR), where each 3D object is represented by a set of 2D pictures captured from various views. There are two key issues in this technology: view representation and model matching. In the past few decades, great efforts have been dedicated to addressing the two issues. LightField Descriptor (LFD) [3] is a widely used shape descriptor, in which ten views are captured from vertices of dodecahedron over a hemisphere to represent one object, and described by Zernike Moments and Fourier descriptors. Another famous representation method is Compact Multi-View Descriptor (CMVD) [5], which uses Polar-Fourier Transform, Zernike Moments and Krawtchouk Moments to construct a set of 2D rotation-invariant shape descriptor vectors. Gau et al. proposed spatial structure circular descriptor (SSCD) [8], which is one of the synthesized view methods and invariant to rotation and scaling. Camera constraint-free view-based method

(CCFV) [9] builds query models by clustering views which are free from camera array settings. Visual features are important for view representation, and the bag-of-visual-feature algorithm [20] is a classical method on this issue. In this method, the SIFT features are extracted to generate a bag-of-words feature for each 3D object. In addition, the fairly popular features [26] include Zernike Moments, Krawtchouk Moments, and Fourier descriptors. We extract the three features from each view for view representation in this work.

For model matching, some typical distance metrics, including Average distance, Minimal distance, Hausdorff distance and Sum distance [21], are usually used to measure the similarity of two models. However, these conventional approaches ignore the higher order information among multiple views. Intrinsically, the model matching is a many-to-many problem and some advanced methods have been proposed in recent years, such as bipartite graph matching [7], probabilistic matching [1] and learning-based methods [11]. [7] models the two compared view sets as a bipartite graph, and the corresponding weights are computed with random walk. The weighted bipartite graph matching is then conducted to measure the distance between the comparing objects. In [1], a Bayesian probabilistic method is constructed to compute the matching scores. The hypergraph formulation is introduced for 3D retrieval in [11]. By clustering the views, the edges in the hypergraph are generated and the semi-supervised learning process is conducted to estimate the query relevance score. Hong et al. [14] model image features with support vector machine and employ hypergraph to capture the connectivity among views. They improved retrieval accuracy substantially. Multigraph learning method has shown its superior performance by fusing multiple graphs and has been applied in various fields [22,25,26]. Zhao et al. [26] employed multigraph fusing features and acquired excellent retrieval results.

The existing matching methods achieve excellent retrieval performance by estimating the relationship among multiple views or learning the underneath structure among all compared 3D objects. However, they are inadequate to take advantage of the description ability of the views. In this work, we adopt a simple but effective distance approach to measure the views' discriminative performance among inter category. Then, a reverse sum-min distance metric is proposed to integrate the views' discriminative information. Utilizing the graph-based learning framework, we achieved superior performance to the previous methods.

3   Proposed Method

In this section, we introduce how to measure the view weight of each object and how to make use of this information to boost the retrieval performance. For clarity, the algorithms are divided into two stages: training and querying. We illustrate the notations and their definitions used in this paper in Table 1.

We extract three different kinds of visual features (2D Zernike Moments [15], 2D Krawtchouk Moments [23] and 2D Fourier descriptor [2]) from each view. We adopt 49 dimensional Zernike moment features as in [11], 78 dimensional Fourier and 78 dimensional Krawtchouk descriptors according to the optimal experiment in [5].

Table 1. Notations and Definitions

$\mathcal{D}$	The 3D object dataset
$\mathcal{C}_i = \{\mathcal{O}_{i,j}, \cdots, \mathcal{O}_{i,n_c}\}$	The i-th category which contains n_c objects in dataset
Q	The query object
q_i	The i-th view of query object Q
$\mathcal{V}_i$	The view set of i-th category in dataset
$\mathcal{V}_{i,j} = \{v_{i,j}^1, \cdots, v_{i,j}^{n_v}\}$	The view set of object $\mathcal{O}_{i,j}$, which contains n_v views
$w_{i,j}^k$	The weight of k-th view that belong to $\mathcal{O}_{i,j}$
$d(v_i, v_j)$	The Mahalanobis distance between view v_i and v_j
$d(v, \mathcal{V})$	The minimum distance between view v and the view set $\mathcal{V}$

3.1 Training

The retrieval system will return the ranking objects from the dataset according to the relevance score between the query object and the objects in dataset. In view of the 2D views that are used to represent both the query object and the objects in dataset, the views do matter with their representative and discriminative ability. As initially mentioned, the views that represent a 3D object do not have equal contribution in object representation as well as discrimination. What's more interesting is that some confusing views (as demonstrated in Fig. 1) even will degrade retrieval accuracy. In [12], Gao et al. have investigated the role of views and selected informative view through relevance feedback. They named their work which employed informative set of view instead of all views as "less is more". However, it depends on the user's feedback. Without any intervention, we propose a simple but effective training procedure to measure the importance of each view in this subsection.

To alleviate the mismatching problem, we consider that the views which can distinguish the indicative object from objects of other categories should have higher weights when comparing two objects. And those similar views of inter categories need to be eliminated due to their confusing characteristics. Instead of setting a threshold to pick out those useful object views, we give the views discriminative weights by measuring their shortest distance with views from inter categories. It should be noted that this stage can be trained off-line.

The shortest distance between the k-th view of object j in category i and all the views contained in other categories is defined as

$$\min_{\mathcal{V}_m \in \mathcal{D}/\mathcal{V}_i} d(v_{i,j}^k, \mathcal{V}_m) \tag{1}$$

where

$$d(v_{i,j}^k, \mathcal{V}_m) = \min_n d(v_{i,j}^k, \mathcal{V}_{m,n}) = \min_n \{min_l d(v_{i,j}^k, v_{m,n}^l)\} \tag{2}$$

The objective in Eq. (1) indicates the shortest distance of nearest view from inter category objects to $v_{i,j}^k$. A larger distance means that view $v_{i,j}^k$ is

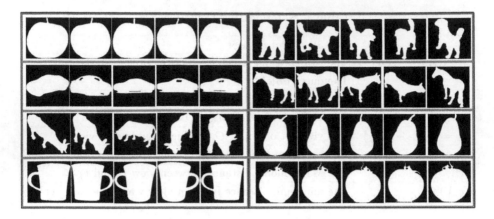

Fig. 2. The top 5 discriminative views of objects from each category in ETH dataset.

more discriminative compared to other irrelevant objects in dataset. After acquiring the nearest distance of each view of the object $\mathcal{O}_{i,j}$, the weight $w_{i,j}^k$ is caculated as

$$w_{i,j}^k = \frac{exp(\frac{d(v_{i,j}^k, V_{D/c_i})}{\sigma})}{\sum_{k=1}^{n_v} exp(\frac{d(v_{i,j}^k, V_{D/c_i})}{\sigma})} \tag{3}$$

where σ is the parameter to reduce the effect of small values of the exponential term and expand the weights' difference.

Figure 2 list the top 5 discriminative views according to the discriminative weights of some objects from ETH dataset. We can see from the figure that the views of high weights contain semantic information of the specific object, which is somewhat consistent with the views in the bottom row of Fig. 1.

3.2 Querying

(A) Distance Based Matching

For a given query 3D object, conventional retrieval approaches usually match view pairs between the query object and the objects in dataset. Among the many-to-many distance measurements, Hausdorff distance has shown its effectiveness in the existing retrieval works [10,26]. The distance between query object Q and $O_{i,j}$ can be formulated as:

$$D_{Hausdorff}(Q, \mathcal{O}_{i,j}) = \max\{\max_{q_t \in Q}\{\min_{v_{i,j}^k \in \mathcal{V}_{i,j}} d(q_t, v_{i,j}^k)\}, \max_{\min_{v_{i,j}^k \in \mathcal{V}_{i,j}}}\{\min_{q_t \in Q} d(q_t, v_{i,j}^k)\}\} \tag{4}$$

Hausdorff distance adopts a max-min scheme to measure the distance between two view sets, which may not be able to integrate the weight $w_{i,j}^k$ of each view because of its extreme distance formula mode. Instead, the Sum-min distance makes use of the information of each view from one object, which is defined as:

$$D_{sum\_min}(Q, \mathcal{O}_{i,j}) = \frac{1}{|Q|} \sum_{q_t \in Q} \min_{v_{i,j}^k \in \mathcal{V}_{i,j}} d(q_t, v_{i,j}^k) \tag{5}$$

Although it is a simple distance metric, we would like to present more details for introducing our method. The traditional Sum-min distance measurement [11] is composed of two steps: (1) compute the Mahalanobis distance between each view of the query object to all the views from a candidate object in dataset, and select the nearest distance for each view of the query object; (2) after acquiring the nearest distance of each query view, average them to get the mean distance between the two view set. An intuitive idea is to replace the average operation with weighted coefficients to give more reasonable distance measurement. Unfortunately, it is hard, if not impossible, to decide which view should be considered more important without much prior information about the query 3D object. Considering that we have weighted each view of objects in dataset, conducting a reverse comparison will integrate the weight information. This is to say that we regard the objects in dataset as the query object, while the original query object is considered as a retrieval candidate. Our reverse weighted sum-min distance metric (RWSM) can be formalized as

$$D_{RWSM}(\mathcal{O}_{i,j}, Q) = \sum_{k=1}^{n_v} w_{i,j}^k \min_{q_t \in Q} d(v_{i,j}^k, q_t) \tag{6}$$

(B) Graph Based Retrieval

However, the simple distance measures are not able to make sufficient use of the structure relevance information among the multiple views. Many learning based methods have been proposed to investigate the higher order information, such as bipartite [7], Hypergraph [11], multi-modal graph [26] learning methods. Based on the weighted view set, we try to boost the retrieval performance by integrating the weighted information.

The proposed reverse weighting algorithm can be seamlessly used in various features, and in this paper, we extrack Zernike, Fourier and Krawtchouk features from each view. To take full advantage of the construction and relevance information, we integrate our algorithm into the multi-modal graph learning method [26].

The cost function for single graph with similarity matrixes from each feature $c \in \{Zernike, Fourier, Krawtchouk\}$ is computed by

$$J_c(f) = \sum_{i,j} W_{c,ij} \left| \frac{f_i}{\sqrt{D_{c,ii}}} - \frac{f_j}{\sqrt{D_{c,jj}}} \right|^2 + \mu_c \sum_i |f_i - Y_i|^2 \tag{7}$$

where $W_{c,ij}$ is the similarity between the i-th and j-th sample, $D_{c,ii} = \sum_j W_{c,ij}$ and f_i can be regarded as a relevance score. Y_i is set to 1 if the i-th sample is labeled positive, and 0 otherwise. The final optimization problem is defined as

$$[f, \alpha] = \arg\min_{f,\alpha} \{\alpha_Z J_Z(f) + \alpha_F J_F(f) + \alpha_K J_K(f)\},$$

$$\text{s.t. } \alpha_Z, \alpha_F, \alpha_K > 0, \ \alpha_Z + \alpha_F + \alpha_K = 1. \tag{8}$$

The relevance score f and weight α_c are computed in alternating minimization manner and the regularization parameters μ_c is selected by grid search. The details of the implementation procedure can be found in [26].

4   Experiments

In this section, we compare the proposed approach with several baseline methods on two 3D datasets. The selected ETH 3D object collection (ETH) [17] contains 80 objects that evenly belong to 8 categories. Each object has 41 views. The National Taiwan University 3D Model dataset (NTU) [3] include 330 objects evenly belonging to 33 categories. Each object is represented by 216 views. We adopt 7 widely used performance evaluation metrics [11]:

- Precision-recall curve (PR). PR is a comprehensive performance measurement.

$$Precision = \frac{\#(\{relevant\ objects\} \cap \{retrieved\ objects\})}{\#\{retrieved\ objects\}}$$

$$Recall = \frac{\#(\{relevant\ objects\} \cap \{retrieved\ objects\})}{\#\{relevant\ objects\}}$$

- Nearest neighbor rate (NN). It calculates the ratio that the first returned object belongs to the same category of the query object.
- First tier (FT) and Second tier (ST). They are defined as the recall of the top τ and 2τ relevant results respectively that account for the percent of the corresponding number of samples in dataset.
- F1 score (F1). It is computed as 2·Precision·Recall/(Precision+Recall), which is used to measure the integration accuracy based on precision-recall value.
- Discounted cumulative gain (DCG). DCG is a statistic that assigns higher weights at the top ranking results to cater for the users' selection bias toward the frontal results.
- Average normalized modified retrieval rank (ANMRR). It is a rank based metric by considering the ranking information of relevant objects within the retrieved objects.

Higher values of NN, FT, ST, F1, DCG, indicate better performance, while lower ANMRR represents better result.

We firstly compare our reverse weighted distance measurement with two conventional metrics. We extract Zernike features of each view, and measure the similarity of two objects by matching their corresponding feature set using different distance measurements. Then the retrieval results are sorted by computing the distance of query object with objects from dataset in descending order.

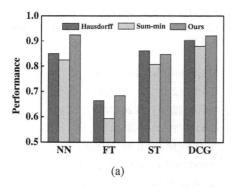
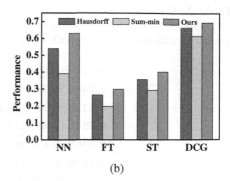

(a)   (b)

Fig. 3. Performance comparison of different distance measurements on (a) ETH and (b) NTU datasets. We compare their performance using 4 measurements.

The results in Fig. 3 reveal that the Hausdorff matching method surpasses the Sum-min overall. By reversing pairs of Sum-min algorithm with the weights of views, the proposed method can achieve great improvement over the traditional Sum-min method. Meanwhile, the new approach performs better than Hausdorff in general.

Table 2. Performance comparison of different methods for several measures on the ETH dataset, where the values in () are the gains achieved by the proposed method compared with each of other methods in terms of %.

	Ours	Multigraph	Hypergraph	Hausdorff
NN	**0.9250**	0.8750(5.4)	0.9125(1.35)	0.8500(8.1)
FT	**0.7569**	0.6917(8.6)	0.7079(6.5)	0.6639(12.29)
ST	0.9111	0.8986(1.4)	**0.9194**(-0.9)	0.8611(5.5)
F1	0.6175	0.6083(1.5)	**0.6292**(-1.9)	0.5925(4)
DCG	**0.9410**	0.9064(3.7)	0.9267(1.5)	0.9022(4.1)
ANMRR	**0.1926**	0.2525(3.1)	0.2341(2.2)	0.2777(4.4)

Graph based learning algorithms have presented excellent performance in 3D object retrieval work due to its powerful information utilizing ability. We integrate the proposed method into multigraph learning framework. In the following experiments, we compare the boosting multigraph with traditional multigraph and hypergraph learning methods, Hausdorff is also considered as a baseline method.

We extract Zernike, Fourier and Krawtchouk features from each view, and compute the object similarity of each feature by the reverse weighting distance measurement. Based on multigraph learning method, we regard the query object itself as the positive sample, and iteratively update the parameters for combining

Table 3. Performance comparison of different methods for several measures on the NTU dataset, where the values in () are the gains achieved by the proposed method compared with each of other methods in terms of %.

	Ours	Multigraph	Hypergraph	Hausdorff
NN	**0.7485**	0.6970(6.9)	0.5939(2.1)	0.5394(27.9)
FT	**0.4128**	0.3697(10.4)	0.3051(26.1)	0.2643(36)
ST	**0.5246**	0.4684(10.7)	0.4175(20.4)	0.3576(31.8)
F1	**0.3869**	0.3531(8.7)	0.3226(16.6)	0.2875(25.7)
DCG	**0.7695**	0.7386(4.0)	0.6951(9.7)	0.6662(13.4)
ANMRR	**0.4856**	0.5272(8.6)	0.5870(20.9)	0.6245(28.6)

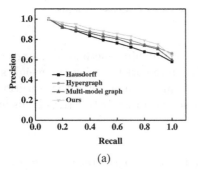
(a)

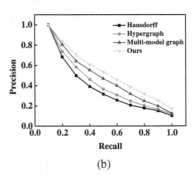
(b)

Fig. 4. Precision-recall curve comparison of different methods on (a) ETH and (b) NTU datasets.

three single feature graphs and relevance score between query object and samples in dataset. Hypergraph based 3D retrieval method [11] is another excellent work which can explore the higher order relationship among objects and does not need to calculate the distance between objects.

Tables 2 and 3 compare the proposed boosting method with three baselines on ETH and NTU dataset respectively, and the precision-recall curves are shown in Fig. 4. From the tables and the figures, we can see that the proposed approach outperforms the other methods in general. Compared with original multi-model method, we boost the retrieval performance in all rounds. In comparison with Hypergraph, the proposed method is better in general, except the second tier and F1 score which are slightly lower on ETH dataset. On the more challenging NTU dataset, our method perform the best on all evaluation metrics.

5   Conclusion

This paper studies the role of views for 3D object retrieval. By learning the discriminative information of each view in dataset, the views are given corresponding confidence based on their discriminative ability for matching. With the reverse

Sum-min distance metric, we can take full advantage of the discriminative information of views when computing the many-to-many distance of two objects. In this way, the effectiveness of detrimental views of objects in dataset can be weaken to a large extend when comparing with the query object, and the retrieval performance based on distance comparison can be improved remarkably. Experiments on the ETH and NTU dataset show that the proposed method obtains better matching performance compared with traditional distance metrics. In addition, combing three different features with our methods within multiple graph learning approach, the boosted algorithm outperforms the state of the art approaches.

Acknowledgments. This work is supported by State Key Development Program of Basic Research of China (973 Program) (No. 2013CB733105), the National Natural Science Foundation of China (No. 61472103) and Key Program (No. 61133003).

References

1. Ansary, T.F., Daoudi, M., Vandeborre, J.P.: A bayesian 3-D search engine using adaptive views clustering. IEEE Trans. Multimedia 9(1), 78–88 (2007)
2. Brigham, E.O., Brigham, E.: The fast Fourier transform and its applications, vol. 1. Prentice Hall Englewood Cliffs, NJ (1988)
3. Chen, D.Y., Tian, X.P., Shen, Y.T., Ouhyoung, M.: On visual similarity based 3D model retrieval. In: Computer Graphics Forum, vol. 22, pp. 223–232. Wiley Online Library (2003)
4. Cheng, J., Bian, W., Tao, D.: Locally regularized sliced inverse regression based 3D hand gesture recognition on a dance robot. Inf. Sci. **221**, 274–283 (2013)
5. Daras, P., Axenopoulos, A.: A 3D shape retrieval framework supporting multimodal queries. Int. J. Comput. Vis. **89**(2–3), 229–247 (2010)
6. Gao, Y., Dai, Q.: View-based 3-d object retrieval: challenges and approaches. IEEE Trans. Multimedia (2014)
7. Gao, Y., Dai, Q., Wang, M., Zhang, N.: 3D model retrieval using weighted bipartite graph matching. Sig. Process. Image Commun. **26**(1), 39–47 (2011)
8. Gao, Y., Dai, Q., Zhang, N.Y.: 3D model comparison using spatial structure circular descriptor. Pattern Recogn. **43**(3), 1142–1151 (2010)
9. Gao, Y., Tang, J., Hong, R., Yan, S., Dai, Q., Zhang, N., Chua, T.S.: Camera constraint-free view-based 3-D object retrieval. IEEE Trans. Image Process. **21**(4), 2269–2281 (2012)
10. Gao, Y., Wang, M., Ji, R., Wu, X., Dai, Q.: 3-D object retrieval with hausdorff distance learning. IEEE Trans. Ind. Electron. **61**(4), 2088–2098 (2014)
11. Gao, Y., Wang, M., Tao, D., Ji, R., Dai, Q.: 3-D object retrieval and recognition with hypergraph analysis. IEEE Trans. Image Process. **21**(9), 4290–4303 (2012)
12. Gao, Y., Wang, M., Zha, Z.J., Tian, Q., Dai, Q., Zhang, N.: Less is more: efficient 3-D object retrieval with query view selection. IEEE Trans. Multimedia **13**(5), 1007–1018 (2011)
13. Giorgi, D., Mortara, M., Spagnuolo, M.: 3D shape retrieval based on best view selection. In: Proceedings of the ACM Workshop on 3D Object Retrieval, pp. 9–14 (2010)
14. Hong, C., Yu, J., You, J., Chen, X., Tao, D.: Multi-view ensemble manifold regularization for 3D object recognition. Inf. Sci. (2015)

15. Khotanzad, A., Hong, Y.H.: Invariant image recognition by zernike moments. IEEE Trans. Pattern Anal. Mach. Intell. **12**(5), 489–497 (1990)
16. Laga, H.: Semantics-driven approach for automatic selection of best views of 3D shapes. In: Proceedings of the 3rd Eurographics Conference on 3D Object Retrieval, pp. 15–22 (2010)
17. Leibe, B., Schiele, B.: Analyzing appearance and contour based methods for object categorization. In: Computer Vision and Pattern Recognition, vol. 2, p. II-409 (2003)
18. Liu, Q., Yang, Y., Ji, R., Gao, Y., Yu, L.: Cross-view down/up-sampling method for multiview depth video coding. Sig. Process. Lett. **19**(5), 295–298 (2012)
19. Liu, Y., Wang, X.L., Wang, H.Y., Zha, H., Qin, H.: Learning robust similarity measures for 3D partial shape retrieval. Int. J. Comput. Vis. **89**(2–3), 408–431 (2010)
20. Ohbuchi, R., Osada, K., Furuya, T., Banno, T.: Salient local visual features for shape-based 3D model retrieval. In: IEEE International Conference on Shape Modeling and Applications, pp. 93–102 (2008)
21. Shih, J.L., Lee, C.H., Wang, J.T.: A new 3D model retrieval approach based on the elevation descriptor. Pattern Recogn. **40**(1), 283–295 (2007)
22. Wang, M., Hua, X.S., Hong, R., Tang, J., Qi, G.J., Song, Y.: Unified video annotation via multigraph learning. IEEE Trans. Circ. Syst. Video Technol. **19**(5), 733–746 (2009)
23. Yap, P.T., Paramesran, R., Ong, S.H.: Image analysis by Krawtchouk moments. IEEE Trans. Image Process. **12**(11), 1367–1377 (2003)
24. Zhao, S., Chen, L., Yao, H., Zhang, Y., Sun, X.: Strategy for dynamic 3D depth data matching towards robust action retrieval. Neurocomputing **151**, 533–543 (2015)
25. Zhao, S., Yao, H., Yang, Y., Zhang, Y.: Affective image retrieval via multi-graph learning. In: ACM International Conference on Multimedia, pp. 1025–1028 (2014)
26. Zhao, S., Yao, H., Zhang, Y., Wang, Y., Liu, S.: View-based 3D object retrieval via multi-modal graph learning. Sig. Process. **112**, 110–118 (2015)

What Catches Your Eyes as You Move Around? On the Discovery of Interesting Regions in the Street

Heng-Yu Chi[1,2]([✉]), Wen-Huang Cheng[2], Chuang-Wen You[3],
and Ming-Syan Chen[1,2]

[1] Department of Electrical Engineering, National Taiwan University, Taipei, Taiwan
{hengyuchi,mschen}@citi.sinica.edu.tw
[2] Research Center for IT Innovation, Academia Sinica, Taipei, Taiwan
whcheng@citi.sinica.edu.tw
[3] Intel-NTU Connected Context Computing Center, Taipei, Taiwan
cwyou@citi.sinica.edu.tw

Abstract. Interesting regions are defined as parts of street view scenes that can attract people's interests when they are moving on the road, and play an important role in various daily-life scenarios. In this paper, therefore, we propose a framework for locating interesting regions in the street and explore the potential use for advanced multimedia applications. Based on the psychological findings and cognitive processes, we proposed and quantified three properties for modeling interesting regions, including attractive, unique, and familiar. Also, a spatial-temporal fusion scheme is developed to combine the multiple properties for discovering the presence of interesting regions. We conduct a set of user studies to demonstrate the effectiveness of our approach. The results support that most users agreed with the interesting regions found by the proposed approach. Finally, a novel application based on interesting regions is also presented for offering an improved navigation experience to vehicle drivers.

Keywords: Interesting regions · Spatio-temporal fusion · Attractive · Unique · Familiar · Vehicle navigation applications

1 Introduction

Given a street view scene, what attracts your interest at your first glance? Figure 1 is a capture of the street view in New York's Times Square. In an informal survey, most people would see the McDonald's business signs at first sight. Other parts of high interestingness include the yellow school bus on the road and the outdoor electronic advertising boards. Generally, all these entities that can attract high interest on the street are named as *Interesting Regions*.

Interesting regions play an important role in our daily-life scenarios. For example, when you try to meet up with your friends in a cluttered and crowded

© Springer International Publishing Switzerland 2016
Q. Tian et al. (Eds.): MMM 2016, Part I, LNCS 9516, pp. 767–779, 2016.
DOI: 10.1007/978-3-319-27671-7_64

Fig. 1. People tend to show higher interest in certain *interesting regions* when moving on the street. For example, the McDonald's business signs, the yellow school bus, and the outdoor electronic advertising boards (Color figure online).

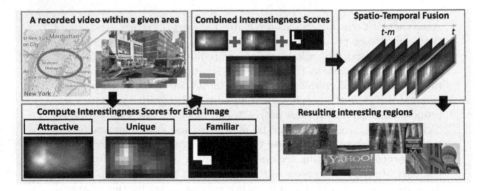

Fig. 2. The proposed framework for locating interesting regions in street view images.

place, you might probably find an interesting region nearby (e.g., the business sign with a distinct look) as an easy-to-find location to tell them where you are exactly [1,2]. Interesting regions might also help enhance driver comprehension of the navigation guidance while driving [3–5]. By showing interesting regions as a visual prompt (cf. Fig. 5), it can make drivers easier to determine when to make a turn on the road when entering into a visually cluttered area like the city center district (cf. Sect. 3.4). Moreover, the discovered properties of interesting regions in the street are important cues that can help people design the street objects, such as buildings, traffic signs, and business logos [2,6], with a better visual appearance for special purposes.

To be precise, interesting regions on the street can be defined by several properties. First, according to psychological findings [7,8], interesting regions in a scene are largely determined by human observers based on low-level visual properties, i.e., the visual appearance is *attractive* [9]. For example, the red electronic advertising board around the center of Fig. 1 is relatively attractive on the street because of its noticeable colors. In addition, inspired by cognitive processes [10,11], interesting regions should not repeatedly or frequently appear in an area, i.e., the interesting regions are *unique*. For example, the yellow cabs on the road in Fig. 1 are not unique since there are lots of yellow cabs with similar appearances literally everywhere in New York City. Moreover, people are interested in the regions with *familiar* icons (e.g., logos), texts, and architectures (e.g., landmarks) on the street [2,12]. For example, most people are interested in the McDonald's store in Fig. 1 since the McDonald's store is a famous fast-food restaurant worldwide. Finally, for people moving on the street (e.g., driving, biking, or walking), interesting regions should appear for a long-enough time such that people can recognize them [13]. For example, electronic advertising boards with fast-changing visual contents cannot be interesting regions because most people do not have enough time to recognize them.

For finding interesting regions, previous works focus on only one or two properties mentioned above, e.g., attractive [8,9,14], unique [10,11], and familiar [2,12,15]. However, it is arguable that identifying interesting objects on the street is a complex interplay between the multiple properties. For example, people would be more interested in the McDonald's than the red electronic advertising board in Fig. 1, since they both are attractive but the McDonald's is also a familiar brand.

Therefore, in this work, we proposed a framework for locating interesting regions from the sight of people moving on the road, cf. Fig. 2. Specifically, we specify and quantify the defining properties of interesting objects, including attractive, unique, and familiar, whereby a spatio-temporal fusion scheme is developed to encode the presence of interesting regions based on the combined interestingness scores. Beyond discovering interesting regions within a single image, the proposed spatio-temporal fusion finds the interestingness scores of regions by taking into account both the spatial and temporal dimensions and can ensure that people have long-enough time to recognize those interesting regions. In the experiments, we conducted a set of user studies to explore the usability of interesting regions and demonstrate the effectiveness of our approach. The results shed light on new directions for the development of advanced multimedia applications, e.g., an application prototype based on interesting regions for providing vehicle drivers clearer instructions on the road is also showcased, cf. Fig. 5 and Sect. 3.4.

The remainder of the paper is organized as follows. In Sect. 2, we define the properties of interesting regions and describe the approach for locating interesting regions. We conduct a set of user studies to demonstrate the effectiveness of our approach in Sect. 3. Finally, we conclude our work in Sect. 4.

2   Interesting Regions

In this section, we define three properties, i.e., *attractive, unique,* and *familiar,* for interesting regions on the street and propose a set of extraction algorithms for detecting the adopted three main properties of interesting regions, respectively. The found properties are individually quantized as a corresponding score and then combined as the interestingness score. Finally, we present the spatial-temporal fusion to retrieve those regions which have higher interestingness scores as the resulting interesting regions.

2.1 Attractive

An interesting region should be visually attractive to users. That is, it may consist of salient colors, shapes, or textures with respect to its surroundings on the street. For example, a red building is visually salient when standing in a street that is full of green trees. Generally, the visual saliency of a position in an image refers to its relative attractiveness with respect to the whole image. The saliency map is commonly used for finding the visual saliency within a single image [9, 16]. Therefore, in this work, we adopt the Graph-Based Visual Saliency (GBVS) scheme [9,17] to generate the local saliency map for each image. The values of an output saliency map are normalized to a range between 0 (low attractive) and 1 (high attractive). A region which consists of image pixels with higher attractive values relatively is regarded as a more attractive region.

2.2 Unique

An interesting region should be unique, i.e., it does not commonly appear within an area. For example, a street lamp is a common object on the street and would receive less interest. For computing the uniqueness, we adopted the TF-IDF technique [11,18] to suppress the non-unique image regions, i.e., those are generally more common than others in an area. In the learning stage, given a set of training images within an area, a color based structure descriptor, i.e., the OpponentSIFT [19], is adopted to extract visual features. Then the extracted visual features were used to generate a codebook, i.e., word $w_1, \ldots, w_i, \ldots, w_l$, by using the Bag-of-Words (BoW) model [20]. Afterwards, each input image I is segmented into K grids, i.e., $I_1, I_2, \ldots, I_K$, at first. Then each grid I_j can be represented as a histogram V_{I_j} of visual words:

$$V_{I_j} = (W_1, \ldots, W_i, \ldots, W_l), \text{ where } W_i = \frac{E(w_i, I_j)}{N_{I_j}} log(\frac{K}{K_i}), \tag{1}$$

where $E(w_i, I_j)$ is whether the word w_i is in grid I_j, i.e., $E(w_i, I_j) = 1$, or not, i.e., $E(w_i, I_j) = 0$, N_{I_j} is the total number of words in grid I_j, and K_i is the number of grids containing word w_i. Finally, the unique score of each grid can be obtained by the summation over all W_i, i.e., $S_{I_j} = \sum_i W_i$. All the scores are normalized to a range between 0 (low unique) and 1 (high unique). A region which consists of image pixels with higher unique values relatively is regarded as a more unique region.

2.3 Familiar

An interesting region had better to be familiar by users. Generally, when users come to a *new* place, they often can quickly identify something *old* (i.e., familiar) to them because of the cognitive processes, such as landmarks or business signs. Previous work has shown that the salient portions often correspond to semantic objects in an image [12]. On the street, it is especially true for business signs since they are purposely designed to be distinct in look for attracting people's attention. Without loss of generality, we initially adopt the method, MOSRO [2], to extract the business signs and landmarks for an image. Finally, the output pixel-level masks of business signs and landmarks will be normalized to a range between 0 (low likelihood of presence of the specified objects) and 1 (high likelihood of presence of the specified objects) and regarded as the familiar score.

2.4 Spatio-Temporal Fusion

In this section, we present the spatial-temporal fusion to retrieve the interesting regions based on the combined interestingness scores in consecutive time instances. Given an image subset of an area with geo-tagged information, we first partition each image into K grids, i.e., $I_1, \ldots, I_j, \ldots, I_K$, for computing the interestingness score for each grid in the image. The interestingness score S is defined as a weighted value for the attractive score S^{At}, unique score S^{Un}, and familiar score S^{Fa} as follows.

$$S_{I_j} = \alpha S_{I_j}^{At} + \beta S_{I_j}^{Un} + \gamma S_{I_j}^{Fa}, \tag{2}$$

where α, β, and γ are control parameters (with equal weights as the default setting).

In addition, according to the findings in [13], it takes a period of time for people to recognize interesting objects after seeing them. The prior work [21] also reported that the duration of human's short term memory is between 15 and 30 s. Therefore, we can thus apply a time decay function, a Gaussian filter $F(t) = \frac{1}{2\pi\sigma}e^{-\frac{t^2}{2\sigma^2}}$, as a weighted moving average for each grid I_j, i.e., $S_{I_j}(t)$, to obtain the final interestingness values, $S_{I_j}{}^F(t)$:

$$S_{I_j}{}^F(t) = \frac{\sum_{u=-m}^{0} F(t+u) \times S_{I_j}(t+u)}{\sum_{u=-m}^{0} F(t+u)} \tag{3}$$

where $S_{I_j}{}^F(t)$ is the interestingness values of grid I_j in t^{th} second after spatial-temporal fusion according to the original interestingness values S_{I_j} from $(t-m)^{th}$ second to t^{th} second, respectively, m is set to be 15 as suggested by [21], and σ is set to be 0.2 empirically. With the fusion, a grid I_{j_1} might be regarded as an interesting region if the scores of the grid $S_{I_{j_1}}$ retain relatively high for a long-enough time from $t - m$ to t. In contrast, if the score of another grid $S_{I_{j_2}}$ is only sparsely high at certain time instances, it might not be regarded as an interesting region.

Table 1. The collected video clips used in the experiments (T: Taipei city in Asia, NY: New York city in America)

	v_1	v_2	v_3	v_4	v_5	v_6	v_7	v_8	v_9	v_{10}	v_{11}	v_{12}
Period (s)	32	27	24	37	30	36	33	25	30	38	47	40
Location	T	T	T	T	T	T	T	NY	NY	NY	NY	NY
Video quality	720p	720p	720p	720p	720p	360p	360p	720p	720p	720p	720p	720p

3   User Study

To demonstrate the effectiveness of our approach, two sets of experiments involving 30 participants are conducted. The goal of the first objective evaluation (Sect. 3.3) is to assess if the system-detected interesting regions match the participant-reported ones. The second user study (Sect. 3.4) is to further explore the potential use of our interesting regions for advanced applications. That is, interesting regions are used as a novel navigation prompt to help drivers better identify right turning points during driving.

3.1   Data Set

To emulate scenes seen by people moving on the road, we collected 12 YouTube video clips that were recorded by in-car dash cameras. These clips are the street view scenes in urban cities with constant recording. To encompass visual diversity, 7 clips are from Taipei city in Asia and 5 clips are from New York city in America. The detailed information of the collected video clips are shown in Table 1.

3.2   Participants

In the experiments, 30 participants (22 males and 8 females) were recruited as unpaid adult volunteers. These participants age from 22 to 38 years old (with an average of 28.77 and the standard deviation of 4.14 years old). All of the participants were licensed drivers and had used GPS navigation systems during driving in recent months. Among them, 13 participants have more than five years of driving experiences.

3.3   Evaluation of Detected Interesting Regions

Benchmark. Participants were asked to label interesting regions as the ground truth, called *P regions* hereafter, which attract their focus in twelve video clips from the data set. Through a developed user interface, each of the twelve video clips is playbacked in turn and the participants can click the mouse to indicate the occurrence of an interesting region. After each participant p_i finished labeling interesting regions in the jth video clip, the corresponding frames containing those P regions were immediately extracted for participants to localize the P regions by drawing rectangular boxes to bound the P regions in a frame.

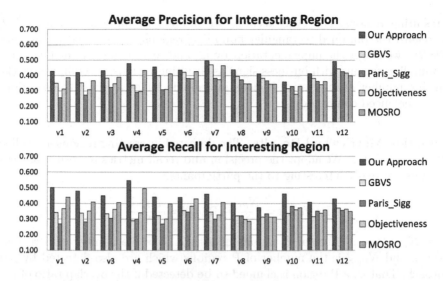

Fig. 3. The average precision (AP) and average recall (AR) for interesting region.

(a) Sample image from v_5 (b) Sample image from v_{12}

Fig. 4. Sample interesting regions detected by the proposed approach and our comparisons (Black: P region; blue, orange, red, yellow, and green: S region detected by our approach, GBVS, Paris_Sigg, Objectiveness, and MOSRO, respectively) (Color figure online)

Methods. The screenshot images are regularly sampled from twelve video clips as the input images. We use the proposed approach in Sect. 2 to obtain interesting regions. The control parameters, i.e., α, β, and γ, of Eq. (2) are set with equal weights and the m of Eq (3) is set to be 15 by [21]. To assure that obtained interesting regions can appear for large enough in size, the number of grids, i.e., K, is set to be 100 empirically. Since the input data are videos, we can thus sample the screenshot images of the videos per seconds to do the spatial temporal fusion. There are four previous methods, i.e., GBVS, Paris_Sigg, Objectiveness, and MOSRO, in comparison with our approach. For GBVS, we adopt the implementation by [17] and regard those regions which consists of image pixels with higher value, i.e., more than 0.6, in saliency map as resulting regions. For Paris_Sigg, we adopt the implementation by [11] and regard the obtained visual elements as resulting regions. For *Objectiveness*, we adopt the implementation by [12] and regard the obtained rectangular bounding regions as resulting regions. For *MOSRO*, we adopt the implementation by [2] and regard the obtained composite bounding regions with business signs as resulting regions. Overall, we set the resulting number of interesting regions in a single image at most three regions to keep balanced comparison.

Evaluation Metrics and Results. The detected interesting regions are called S regions hereafter. We adopt the precision and recall metrics to verify that the S regions are really interesting to the participants.

$$\text{Precision} = \frac{N_{P \cap S}}{N_S}, \text{Recall} = \frac{N_{P \cap S}}{N_P} \tag{4}$$

where N_P is the number of P regions in total, N_S is the number of S regions in total, and $N_{P \cap S}$ is the number of P regions which are also detected by our approach. That is, a P region is claimed to be detected if the overlap ratio of the P region and an S region, i.e., $\frac{Area(P \cap S)}{Area(P \cup S)}$, is higher than a predefined threshold, say 0.5. The experimental results are shown in Fig. 3.

Overall, the experimental results show that the average APs and the average ARs of the proposed approach are 0.438 and 0.450, respectively. Considering the fact that interesting objects are diverse and subjective, the average APs of our approach can outperform our comparisons (0.386 for *GBVS*, 0.339 for *Paris_Sigg*, 0.338 for *Objectiveness*, and 0.388 for *MOSRO*) with a relative improvement from a minimum of 12.7 % to a maximum of 29.4 %. In addition, the average ARs of our approach also outperform our comparisons (0.332 for *GBVS*, 0.318 for *Paris_Sigg*, 0.342 for *Objectiveness*, and 0.388 for *MOSRO*) with a significant relative improvement from a minimum of 15.6 % to a maximum of 41.3 %.

Sample P regions and S regions are shown in Fig. 4. For clearly showing the sample results, we show only P regions, i.e., black rectangle, and the **Top-1** S region, i.e., the blue, orange, red, yellow, and green rectangles, respectively, for each approach, i.e., our approach, *GBVS*, *Paris_Sigg*, *Objectiveness*, and *MOSRO*, respectively. By Fig. 4(a), the P region can be detected by our approach and object recognition methods, i.e., *Objectiveness* and *MOSRO*, since the McDonald's business sign is relatively clear in comparison with other objects in the image. However, as shown in Fig. 4(b), the P region may not be detected by most methods since there are a variety of visual salient objects in the image. Our approach can detect the P region in Fig. 4(b) since the proposed spatial-temporal fusion (cf. Sect. 2.4) can consider the detected interesting regions in previous images and apply a weighted moving average to calculate the combined interestingness score for each region and their related region in neighbor images spatially. The results also support our assumption that when people moving on the street, interesting regions should appear for a long-enough time such that people can recognize them [13].

In addition, for different cities, the performance in Asia (v_1 to v_7), i.e., 0.449 in average AP and 0.474 in average AR, is better than that in America (v_8 to v_{12}),

Table 2. The average AP and AR with different setting of the control parameters (α, β, and γ) of Eq. (2) (α for S^{At}, β for S^{Un}, and γ for S^{Fa})

	$\frac{1}{3}(1,1,1)$	$\frac{1}{2}(0,1,1)$	$\frac{1}{2}(1,0,1)$	$\frac{1}{2}(1,1,0)$	(0,0,1)	(0,1,0)	(1,0,0)
Average AP	0.438	0.409	0.419	0.402	0.392	0.326	0.391
Average AR	0.450	0.411	0.431	0.392	0.391	0.302	0.355

i.e., 0.422 in average AP and 0.414 in average AR. One reason might be that the familiar score can contribute more in Asia than in America, since Asian streets usually exhibit a larger number of business signs, billboards, and electronic signs (cf. Fig. 4(b)). In general, the experimental results are very encouraging and show that most users are interested in the interesting regions found by our proposed approach.

Moreover, we also evaluate whether each adopted property has positive contribution or not by varying the control parameters (α, β, and γ) of Eq. (2). The experimental results are shown in Table 2. It is worthy noting that the attractive property (visual saliency), i.e., 0.391 in average AP and 0.355 in average AR, has long been recognized as the most effective modeling for visual attention, but our results show that the familiar property, i.e., 0.392 in average AP and 0.391 in average AR, could be an even more important factor to attract people's interest. In other words, the results imply that people might show more interest in semantic objects than those of visual attractiveness on the street. In addition, all of the adopted properties do have positive contributions, i.e., the lack of any property will decrease the accuracy of the resulting interesting regions with degradations of 6.62 %, 4.34 %, and 8.22 % in average AP and 8.67 %, 4.22 %, and 12.89 % in average AR, respectively. Finally, as shown in Fig. 3 and Table 2, the experimental results also still show that the approach using single property with spatial temporal fusion will improve the performance in comparison with the approach using single property without spatial-temporal fusion. For example, the attractive score with spatial temporal fusion (0.391 in average AP and 0.355 in average AR) can outperform the *GBVS* (0.386 in in average AP and 0.332 in AR) with a 1.30 % and 6.93 % relative improvement in average AP and average AR, respectively. The results can verify the feasibility of the proposed spatial-temporal fusion.

3.4 iNavi for Driving

According to a 2013 survey involving car drivers in the United States, more than 63 % of them who have used GPS say they have been led astray because of receiving "complex, confusing and incorrect routes" [4]. Therefore, in this section, as a demonstration of the potential use of the proposed interesting regions for advanced multimedia applications, we develop an application, iNavi, with the purpose of providing vehicle drivers clearer instructions based on interesting regions (cf. Fig. 5) when driving on the road. The user study is conducted as

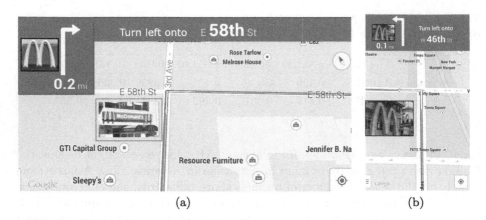

Fig. 5. An illustration of the iNavi's navigation instruction (interesting-region guidance) for vehicle drivers.

follows. First, participants were required to answer the pre-study questionnaire survey to collect their personal experiences of using traditional navigation systems. Then an emulated driving environment, which displays videos (from the data set in Sect. 3.1) and suggests turns before the participants are going to pass an intersection recorded in the videos, was built to emulate conditions that drivers receive turning cues to follow a suggested route during driving. All participants were asked to make turns (by pressing an assigned button on a keyboard) during the emulated driving when receiving the iNavi instructions, i.e., the street name and the associated interesting region are prompted, cf. Fig. 5. Finally, a post-study interview is applied to gather user feedbacks from the participants after using iNavi.

The results of the questionnaire for the experiences of using general navigation systems are shown in Fig. 6. According to the responses to Questions 1-1 and 1-2, most of the drivers have experiences of making a wrong turn when they were driving because they have difficulties on recognizing suggested turning intersections in time after receiving the visual and audible cues raised by traditional navigation systems. The remaining question, Questions 1-3, shows that most drivers have had wasted 10 to 20 min after being astray. Among them, 36.67 % drivers have wasted a significant amount of time, i.e., over 30 min, to return to the original route after making a wrong turn.

Figure 6 also lists questions of the post-study questionnaire to collect opinions from the participants after using our system, iNavi. The results show that 73.33 % of the participants can effectively identify the corresponding regions in real-world scenes after iNavi prompted interesting regions (Question 2-1). Overall, iNavi can help 96.66 % of the participants to timely make correct turning at intersection through recognizing interesting regions in comparison to current navigation systems (Question 2-2). Finally, 90 % of the participants are willing to use iNavi in the future (Question 2-3). Generally, the results demonstrate

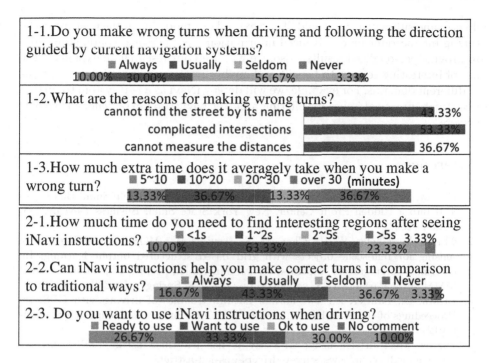

Fig. 6. Personal experiences of using traditional navigation systems (1-1 to 1-3) and post-study interviewing results (2-1 to 2-3).

the effectiveness of our novel navigation system over the traditional navigation systems for practical applications.

Moreover, there are several advanced suggestions from participants after using iNavi. First, most participants recommend that it is better to add additional hints in a voice message so that the voice message will be in accordance with the prompted visual message. For example, as shown in Fig. 5, the voice message are recommended to be "turn left onto E58th street after McDonald's store", instead of "turn left onto E58th street in 0.2 miles" only. We will adopt the techniques of automatic annotation to name all the detected interesting regions based on their semantic meanings and revise the voice message in the future. In addition, several participants point out that it is better to select personalized interesting regions which is familiar in their daily life. In the future, we will collect the records whether drivers make right turns after seeing iNavi visual messages or not, and learn the personalized familiar score for each user.

4   Conclusions and Future Work

In this paper, we proposed a novel framework for locating interesting regions from the sight of people moving on the road. A set of user studies were conducted and verified the effectiveness of our approach. Since interesting regions are essentially diverse and subjective, we would like to extend our modeling in the near future by taking into account more advanced information (e.g., personal context) in order to provide personalized results. Also, we will keep our research on exploiting the use of interesting regions to beneift and enable creative multimedia applications in different domains. For example, we will deploy iNavi as a real system to further investigate practical issues for offering a better navigation experience for vehicle drivers.

References

1. Conroy, D.: What's Your Signage?: How On-Premise Signs Help Small Businesses Tap Into a Hidden Profit Center. New York State Small Business Development Center (2004)
2. Chi, H.-Y., Cheng, W.-H., Chen, M.-S., Tsui, A.W.: MOSRO: enabling mobile sensing for real-scene objects with grid based structured output learning. In: Gurrin, C., Hopfgartner, F., Hurst, W., Johansen, H., Lee, H., O'Connor, N. (eds.) MMM 2014, Part I. LNCS, vol. 8325, pp. 207–218. Springer, Heidelberg (2014)
3. Brown, B., Laurier, E.: The normal natural troubles of driving with GPS. In: Proceedings of the SIGCHI Conference on Human Factors in Computing Systems (2012)
4. Survey: Most drivers using GPS say it has led them astray (2013). http://michelinmedia.com/news/survey-drivers-gps-led-astray/
5. Forlizzi, J., Barley, W.C., Seder, T.: Where should i turn: moving from individual to collaborative navigation strategies to inform the interaction design of future navigation systems. In: Proceedings of the SIGCHI Conference on Human Factors in Computing Systems (2010)
6. Tsai, T.-H., Cheng, W.-H., You, C.-W., Hu, M.-C., Tsui, A.W., Chi, H.-Y.: Learning and recognition of on-premise signs from weakly labeled street view images. IEEE Transactions on Image Processing (2014)
7. Elazary, L., Itti, L.: Interesting objects are visually salient. J. Vis. (2008)
8. Liu, T., et al.: Learning to detect a salient object. IEEE Trans. Pattern Anal. Mach. Intell. (2011)
9. Harel, J., Koch, C., Perona, P.: Graph-based visual saliency. In: NIPS (2007)
10. Parikh, D., et al.: Interactively building a discriminative vocabulary of nameable attributes. In: CVPR (2011)
11. Doersch, C., et al.: What makes paris look like Paris. In: SIGGRAPH (2012)
12. Alexe, B., Deselaers, T., Ferrari, V.: Measuring the objectness of image windows. IEEE Trans. Pattern Anal. Mach. Intell. (2012)
13. Short Term Memory. http://www.simplypsychology.org/short-term-memory.html (2009)
14. Huang, Y.-M., et al.: Advances in Multimedia Information Processing. In: PCM (2008)
15. Ross, T., May, A.: Design advice for the inclusion of landmarks in vehicle navigation systems. Loughborough University (2002)
16. Itti, L., Koch, C., Niebur, E.: A model of saliency-based visual attention for rapid scene analysis. IEEE Trans. Pattern Anal. Mach. Intell. (1998)
17. Hou, X., Harel, J., Koch, C.: Image signature: highlighting sparse salient regions. IEEE Trans. Pattern Anal. Mach. Intell. (2012)

18. Sivic, J., Zisserman, A.: Efficient visual search of videos cast as text retrieval. IEEE Trans. Pattern Anal. Mach. Intell. (2009)
19. van de Sande, K.E.A., Gevers, T., Snoek, C.G.M.: Evaluating color descriptors for object and scene recognition. IEEE Trans. Pattern Anal. Mach. Intell. (2010)
20. Nister, D., Stewenius, H.: Scalable recognition with a vocabulary tree. In: CVPR (2006)
21. Atkinson, R.C., et al.: The Control Processes of Short-term Memory. Stanford University, Stanford (1971)

Bag Detection and Retrieval in Street Shots

Chong Cao, Yuning Du, and Haizhou Ai$^{(\boxtimes)}$

Tsinghua National Laboratory for Information Science and Technology,
Department of Computer Science and Technology,
Tsinghua University, Beijing 100084, China
mcc.ygz@gmail.com, ahz@mail.tsinghua.edu.cn

Abstract. In recent years, e-commerce has become an important way people shop. Among this, clothes and bags are extraordinarily important for customers. However, traditional online shopping modes only allow users to search with key words. Sometimes users may find it very hard to precisely describe what they want in words. Moreover, even if a user gives a detailed description, it may not agree with the description provided by the seller. Therefore, search-by-image without the help of semantic descriptions becomes a research focus in computer vision and multi-media processing. In this paper, we address the problem of object detection and retrieval and focus particularly on bags in street shots. First, we locate the bag region in an image by Pairwise Context based Convolutional Neural Network (PC-CNN). After that, we learn high-level descriptions of bag images based on attributes and build a retrieval system allowing for image search. We test our approach on the publicly available Fashionista Benchmark (FB) and a Pedestrian with Bags dataset (PB) collected by ourselves to demonstrate the effectiveness of the proposed method.

1   Introduction

With the rapid growth of the Internet and mobile applications, e-commerce is playing an increasingly important role in daily life. In some online stores of certain brands, users can find what they want with keywords or attribute categories. However, un-professional users may be unfamiliar with the keywords a website uses. For example, some brands name items with the designer or celebrities such as Hermes Kelly bag. It is hard for a user to find the bag on the website if he/she does not know the category "Kelly bag". In fact, it may be hard for users to name out some commonly used attributes of an item such as its color. There are hundreds of colors in a designer's color book[1] while most ordinary people can only name out a few of them, not to mention more sophisticated attributes such as shape, pattern and material. Such problems are more obvious on all-brands websites such as eBay, where the number of items is extremely large and descriptions from sellers are always inconsistent. In this case, simple semantic descriptions can not meet people's needs.

---

[1] http://html-color-codes.info/color-names/.

© Springer International Publishing Switzerland 2016
Q. Tian et al. (Eds.): MMM 2016, Part I, LNCS 9516, pp. 780–792, 2016.
DOI: 10.1007/978-3-319-27671-7_65

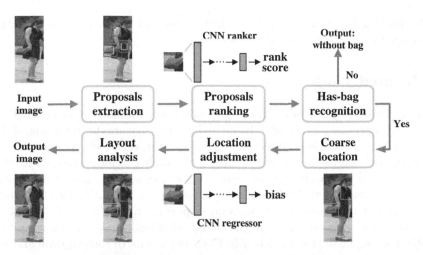

Fig. 1. Illustration of the framework of our bag detection approach. The cyan boxes are the original proposals, the fuchsia ones are the context proposals, the green box is the ground truth and the yellow box is the final detection result of our approach (Color figure online).

In this paper, we develop a framework for object detection and retrieval and apply it to bags. This is because bags are common in daily life and studies on bags have wide application prospects. For example, bags can provide a lot of information for person re-identification and abnormal event detection in surveillance and bag retrieval can bring huge economic benefits in online shopping and fashion recommendations. Recent works on cloth retrieval [19–21] segment out the cloth region with the help of face detection or pose estimation and then calculate the visual similarity between the query image and images in the gallery. However, the segmentation step is always very time-consuming and does not fit for online retrieval. Besides, compared with clothes, which can be aligned via pose estimation, bags are more flexible in size, pose and position. Therefore, we extract high-level semantic descriptions of bag images using Convolutional Neural Network (CNN) supervised by attribute information. To quickly locate bag regions in images, we start by generating bag proposals in the form of bounding boxes by selective search [17]. After that, we train a CNN to find bag region candidates from these proposals and estimate their bias to the ground truth bounding box. Finally, we confirm bag positions by filtering and merging the proposals. The framework of our bag detection method is shown in Fig. 1.

The contribution of this paper can be summarized as below:

1. We propose a PC-CNN structure for object detection that builds the relation between the original proposal and context proposal with a deep model.
2. We combine classification and regression together to improve detection.
3. We extract object features based on semantic attributes, which keep image appearance information and contain semantic meanings.

4. We build a bag retrieval system which allows for automatic bag detection, retrieval and person re-identification *etc.*

2   Related Work

Object Detection: Traditional object detection methods scan through the image with a pre-trained classifier based on low-level features to judge each sliding window whether there is an object in it [18]. However, such methods limit the size of the window thus the height-width ratio of the sliding window is always fixed for a certain object. To handle the cases where the object is deformable, Felzenszwalb *et al.* propose more flexible classifiers for detection [7]. Recently, with the growth of computer performance, CNN plays an important role in research [3,5]. CNN has a more complicated structure compared with traditional classifiers and can handle more sophisticated models. But CNN needs a lot of computation and usually are very slow, which motivates some pre-processing method to accelerate the network, for example, generating some proposals instead of scanning through the whole image [8]. The PC-CNN we propose for bag detection uses selective search [17] to decrease the number of bounding boxes and is very similar to this work. Rather than considering the region inside the object bounding box, Ding and Xiao improve the performance of pedestrian detection by context [4]. But the relation between the object bounding box and context is unknown. In this paper, we build the relation with a deep model and use context information to improve detection accuracy.

Image Retrieval: Most existing object retrieval methods focus on two aspects: feature extraction and similarity measurement. A simple but effective way to solve the problem is to extract low-level features from all images and choose the K nearest neighbors (KNN) of the probe from the gallery set [11]. However, this only works for images where the background is clean and the object occupies most of the image. In addition, such methods cannot handle environmental changes, pose changes, occlusions and deformations which are very common in daily photos. Zhao *et al.* [22] calculate saliency scores for densely sampled image patches to describe how important a patch is in retrieval and measure the similarity of two images with patch weights based on the saliency. They apply the approach to person re-identification. However, calculating the saliency scores consumes much time. A few recent works on cloth parsing and retrieval infer the tag for each super pixel with the help of face detection or pose estimation and use the segmentation result to help retrieval [16,19–21]. However, these methods are hard to apply to bags since the position and size of bags are very flexible and hard to align by human pose. Besides, the segmentation and parsing process takes a lot of time. Some works [1,2,10,15] train semantic attribute classifiers that can handle above problems. But these binary attributes only cover limited information about the image, and each binary attribute value needs a separate classifier. Instead, we train a multi-value classifier with CNN, which decides the attribute class in a single classifier and avoids the messy process of score fusion.

3   Bag Detection with PC-CNN

Given an input street-shot image, we first use selective search [17] to generate original proposals and corresponding context proposals. Then we put the proposal pairs into the Pairwise Context based Convolutional Neural Network (PC-CNN) and rank the proposals according to confidence scores. At the same time, we train a PC-CNN regressor with bag proposals and their bias to the ground truth. After that, we select a small number of top-ranking proposals as the coarse bag locations and estimate fine bag locations with the estimated bias we learned from the regressor. Last, we filter out the isolate ones and merge the overlapping original proposals to obtain final bag locations. Figure 1 shows the framework of our approach for bag detection.

3.1   Generating Bag Proposals with Selective Search

Selective search [17] exploits the structure of an image and generates proposals from super pixels. It uses a variety of grouping criteria to cover as many object appearances as possible. The overlap ratio a_o between the bounding box of a proposal B_p and the ground truth bounding box B_{gt} is defined as:

$$a_o(B_p, B_{gt}) = \frac{area(B_p \cap B_{gt})}{area(B_p \cup B_{gt})} \tag{1}$$

A proposal is considered as a correct proposal if its overlap ratio with a ground truth bounding box is over 50 %. For bag images in the pedestrian dataset we build, selective search will yield 95.4 % recall rate by generating about 552 proposals on average for a 150×400 pedestrian image.

Denote the bounding box of an original proposal s_i as $\{t_i, b_i, l_i, r_i\}$, where t_i, b_i, l_i, r_i are its top, bottom, left and right border positions, as shown in Fig. 1, we extract the context proposal by extending its border. The bounding box of the context proposal of s_i will be $\{t_i - o_i^1, b_i + o_i^1, l_i - o_i^2, r_i + o_i^2\}$, where o_i^1 and o_i^2 are the vertically and horizontally extended margins. If o_i^1 and o_i^2 are too small, the context proposals are similar to the original proposals and contain little context information. If o_i^1 and o_i^2 are too large, the context proposals may contain much noise. In our experiments, we set o_i^1 and o_i^2 as one third of the height and width of the original proposal s_i, i.e., $o_i^1 = h_i/3$, $o_i^2 = w_i/3$. To facilitate the following PC-CNN design, similar to the method in [8], we warp all original and context proposals to a fixed size of 48×48 pixels.

3.2   PC-CNN

Girshick et al. utilize the structure of CNN from ImageNet classification [14] to learn feature representations of a proposal for object detection in R-CNN [8]. However, the ImageNet structure is very complex and time-consuming. In this paper, we use a simpler structure of CNN similar to CIFAR-10 classification [13] to build the proposed PC-CNN and reduce time consumption.

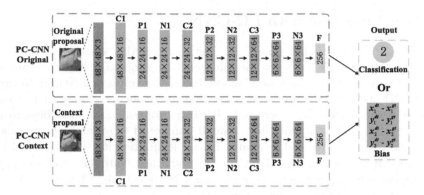

Fig. 2. The proposed PC-CNN structure. The network layers are visualized with their corresponding output dimensions. The output layer can be a 2 value classification probability (classification PC-CNN) or a 4-dimension vector (regression PC-CNN) indicating the relative bias of the proposal to the bounding box.

An overview of the structure of the PC-CNN is shown in Fig. 2. This structure incorporates two sub CNNs we learn from the original and context proposals respectively. PC-CNN utilizes a fully connected classification layer to combine the above features. The output layer of the classification PC-CNN is a softmax layer with 2 output values indicating the probabilities of an original proposal belonging to bag or not. The structures of two sub CNNs are identical and each sub CNN contains three C-P-N layers, where C denotes a convolutional layer, P denotes a pooling layer, N denotes a normalization layer and F denotes a fully connected layer. We denote the sub CNN built from the original proposal as PC-CNN_ori and the one built from the context proposal as PC-CNN_con. For C, P, N layers, the output dimension is defined as width × height × depth, where the first two dimensions have a spatial meaning while the depth defines the number of filters or channels. For F layer, the output dimension is the number of activation units. For further details, please refer to [13,14]. The regression PC-CNN has a similar structure with the classification PC-CNN by changing the output layer to a 4D vector indicating bias of the top left and bottom right corners of a bounding box to the ground truth. We add the bias to the positions of the bounding boxes classified as containing bags to get more accurate positions.

A traditional method to train a PC-CNN is stochastic gradient descent (SGD) with randomly initialized parameters. However, this is not a good strategy because it ignores the discriminative abilities of two sub CNNs, PC-CNN_ori and PC-CNN_con. Instead, we use a two-stage strategy to train PC-CNN. At first, we train PC-CNN_ori and PC-CNN_con individually by using SGD with the parameters randomly initialized. This step can obtain better appearance features of the original and context proposal. During the training, a softmax classification layer with 2 output values follows the fully connected layer in PC-CNN_ori and PC-CNN_con in the classification PC-CNN. Secondly, we continue SGD with the initialization parameters from the above pre-trained two sub CNNs.

Fig. 3. We show some coarse-to-fine post-processing results. Each group of figures consists of a coarse detection result (left) and a fine detection result (right).

The structure of the proposed PC-CNN is much simpler than the ImageNet classification CNN [14] used in R-CNN [8] due to the smaller input size, the fewer filters in the convolutional layers and the fewer activation units in fully connected layers. Experiments show that although PC-CNN is simple, its performance for bag detection is competitive compared with the original R-CNN and it consumes less time for bag detection.

3.3   Post Processing

After calculating the rank of the original proposals from PC-CNN, Du *et al.* [6] consider the top-1 proposal as the bag location. However, from the training dataset, we observe that each bag image contains about 4 bag proposals on average and most of the top-4 proposals are likely to contain bags. Therefore we use the top-4 proposals as coarse locations of a bag and utilize these proposals to seek for a more accurate bag location. We adopt two strategies: (1) Given the bounding boxes B_i and B_j of two proposals s_i and s_j, when $a_o(B_i, B_j) \geq 0.2$, we assume s_i and s_j are connected. Then, the isolate proposals are considered as non-bag proposals that are filtered out. (2) The connected proposals are merged and the average bounding box is considered as a fine candidate bag location. The confidence of a fine candidate bag location is the average confidence of its component proposals. Finally, whether a candidate bounding box contains bags is decided by comparing the confidence value to a threshold. We show some coarse-to-fine post-processing results in Fig. 3.

4   Attribute Learning and Retrieval

Attribute is a very common way to describe objects both in research and daily-life. Most shopping websites provide keyword search to help customers find what they want. Usually, bag attributes can be summarized in four classes: color, shape, material and pattern. We assume such semantic categorizations are important in comparing the similarity between bags. Different from existing works [1, 15] that regard each value in each categorization (*e.g.* red in color) as an attribute and train binary or relative classifiers for each attribute, we use the category itself as an attribute and train a multi-class classifier for each attribute.

The structure of the attribute classification CNN is similar to the bag detection CNN using the original proposal. The only difference is that the output of the classification CNN is the probability of the bag region with a certain class.

Table 1. A brief review of the two datasets we use in the experiment.

Name	Image number	Source	Image size	Annotation
FB	685	[21]	400 × 600	Super-pixel level annotation
PB	9794	Surveillance video	150 × 400	Pedestrian and bag positions

For example, for the attribute "color", we classify all the bags into eight classes. Each class represent a color name (*e.g.* red). Due to the resolution limitation of the dataset we use, we only annotate color and shape of the bags.

Kumar *et al.* [15] directly use the output of each attribute classifier to describe human faces and measure the similarity between different faces. However, bags do not have as many mature semantic categorizations as human faces. Instead we use the F layer as feature representations. The CNN feature we extract can keep original bag appearance and contain potential semantic information since we learn the feature from bag attributes. Each bag region is described with a 256-dimension vector and the similarity between two bag regions are measured by their Euclidean distance. In object image retrieval, we sort all the images in the gallery set according to their distance to the query image and show the user the first a few images as retrieval results. Such a retrieval system can be applied to online shopping, person re-identification, *etc.*

5   Experiment

In this section, we introduce the PB Dataset we build and demonstrate the effectiveness of our bag detection and retrieval approach on PB and another publicly available FB dataset. We compare our detection result to state-of-the-art detection and parsing results [7–9, 20, 21] and show results of attribute learning and bag retrieval.

5.1   Dataset

Some basic information of the two datasets are shown in Table 1. The FB Dataset is a publicly available benchmark for fashion parsing that contains 685 images of street shot images. Existing work on human parsing [20, 21] and apparel detection [9] already shows promising results of bag detection on this dataset. The experimental setup we use for this dataset is similar to [9] (456 images for training and 229 images for testing). The PB Dataset is a relative large dataset we collect from a few surveillance video shots. We manually cut out pedestrians from video frames, and annotate the position of bags in each pedestrian image. Among these images, there are 4734 images with bags in them, and the other 5060 images without. The appearance of bags in this dataset changes greatly and including backpack, shoulder bag, handbag and waist bag in different colors. 3230 bag images and 3385 non-bag images are selected randomly for training and the other images are for testing.

Table 2. Bag detection accuracy and time consumption of different approaches on PS dataset.

	DPM [7]	R-CNN [8]	CNN_ori [6]	PC_cls	PC_cls+reg	PC_full
Acc	0.725	0.751	0.581	0.703	0.758	**0.791**
Time (s)	6.429	1.492	0.496	0.517	0.579	0.581

5.2   Bag Detection

Since our goal is to detect bags in street-shot images, some non-bag proposals can be filtered out according to the prior information of a bag. We utilize two strategies to decrease the number of proposals for reducing the burden of the classification and saving the time consumption on detection. (1) We assume that the size of a bag is within a certain range. A proposal that is too large or too small can be considered as a non-bag proposal. From our training dataset, 99.8 % bag proposals satisfy the following criteria: $h_i/h \in [0.1, 0.43]$, $w_i/w \in [0.15, 0.9]$, where h_i and w_i are the height and width of a proposal s_i, and h and w are the height and width of the pedestrian image. (2) Selective search only removes absolute overlapping proposals and considers proposals with large overlapping as different ones. Nevertheless, these proposals contain almost the same information. So we only randomly reserve one of the proposals with large overlapping. Given the bounding boxes B_i and B_j of two proposals s_i and s_j, when $a_o(B_i, B_j) \geq 0.85$, we consider s_i and s_j overlap largely. After we adopt the above two strategies, we observe that the residual proposals can still yield 94.7 % recall rate. Although the strategies lead to a slight reduction of the recall rate, the average number of proposals reduces a lot and is only about 80 for each pedestrian image, which saves a lot of time for further classification.

We commit experiments on different methods and different training and testing sets. First, we use the training set of PS to train a CNN bag detector same as [6] with only original proposals (CNN_ori) as baseline result. To test the effect of context proposals, regression, and post processing, we also train a PC-CNN detector (PC_cls), a PC-CNN regressor (PC_cls+reg) and add afterwards layout analysis (PC_full). Finally, we compare the accuracy of our method to some state-of-art public available methods [7,8] trained with original proposals. Detection accuracy and time consumption are shown in Table 2. All experiments are carried out on a Dell $T6500$ workstation with an Intel $2.0GHz$ CPU, $16GB$ memory and a Quadro $K4000$ GPU. We try different number of models n from 1 to 4 in DPM and report the best accuracy with $n = 3$. From the results we can see that our approach increases performance even though it has simpler structure than R-CNN. PC_full achieves a speedup rate of about 2.57x over R-CNN with an accuracy increase of 4 %.

To compare with other parsing and detection based methods and evaluate the influence of training set used, we also carry out experiments on FB dataset. As mentioned above, FB has super-pixel level bag annotations. We consider

Table 3. Bag detection accuracy of different approaches on the FB dataset.

	CRF [21]	PD [20]	Apparel [9]	DPM [7]			PC-CNN		
	FB	FB	FB	FB	PB+FB	PB	FB	PB+FB	PB
Accuracy	0.041	0.049	0.225	0.370	0.361	0.319	**0.426**	0.417	0.379

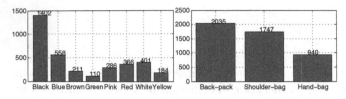

Table 4. Attribute classification accuracy on PS dataset.

	Color	Shape
Class #	8	3
Acc (%)	81.2	67.3

Fig. 4. Final color and shape values used for training and the number of images collected with each attribute value

all bag-related tags (bag, purse and wallet). However, due to the inaccuracy of super pixel boundaries and human annotation noises, some annotations are not accurate. Thus simply using a bounding box to cover these super pixels as bag regions is not reasonable. We use connected component analysis, region merging and region size limits to convert the super-pixel level annotations to bounding boxes. After a few tries, we choose a set of parameters that best suits our needs. The first three rows of Fig. 6(b) show some converted results. The red translucent area is ground truth super-pixel annotations provided by humans or fashion parsing results [20,21], and the green box is the converted bounding box used in our experiment. Noting that the converted ground truth results still have some inaccurate or missing annotations, we manually correct them (*e.g.* the last two examples).

The detection accuracy of different methods using different training sets are shown in Table 3. Although we do not know exactly how much time consumed using parsing-based approaches [20,21], according to their online demo it is much

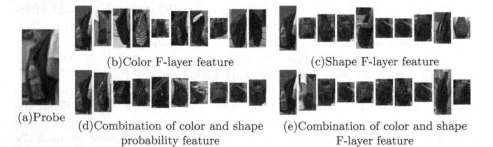

(a)Probe

(b)Color F-layer feature

(c)Shape F-layer feature

(d)Combination of color and shape probability feature

(e)Combination of color and shape F-layer feature

Fig. 5. Top-10 bag retrieval results using different features. (a) to (e) show the query image, retrieval results using the color F-layer feature, the Shape F-layer feature, the combination of color and shape probability and F-layer feature respectively (Color figure online).

(a)Top to bottom are results on the PB dataset using our approach and DPM.

(b)Top to bottom are results on the FB dataset using ground truth annotation,
CRF[21], PD[20], our approach and DPM. For the first three rows, the red
translucent area is super-pixel annotations and the green box shows converted result.

Fig. 6. Bag detection results of different approaches on PB and FB datasets. Ground truth is shown in green, correct detection (overlap more than 50 %) in yellow and wrong detection (overlap less than 50 %) in red (Color figure online).

slower than our method. In addition, enlarging training sets will improve performance in general but for a small dataset taken under certain scenario (such as FB) using a training set that only contains images from the same scenario obtains a higher accuracy. In Fig. 6 we show some results of bag detection using different methods.

5.3   Attribute Learning and Bag Retrieval

Attribute is a very common way to describe objects both in research and daily-life. Most shopping websites provide keyword search to help customers find what they want. Usually, bag attributes can be summarized in four classes: color, shape, material and pattern. As is mentioned previously, due to the limitations of the dataset and human labor, we only annotate color and shape of the bags. There are plenty of vocabulary to describe the color of an object. For example, the RGB color space differentiate colors into $256 \times 256 \times 256$ values. And the designer's color book mentioned in Sect. 1 provides semantic descriptions of hundreds of color names. However, most ordinary people cannot distinguish so many colors, especially in natural scenes where the appearance of an object is largely influenced by illumination and the environment. We simply provide ten color options and four shape options for annotators to choose from. After collecting the attributes for all the images in the PB dataset, we give up three color values (orange, purple and grey) and one shape values (waist bag) since images collected in these classes are far from enough for training. And the final color and shape values used for training and the number of images collected under each attribute value are shown in Fig. 4.

The train-test split we use in attribute learning is the same as the one we use in detection. We train a multi-class CNN classifier for each attribute and use the class with the highest probability as the learned attribute value. The classification accuracy is shown in Table 4.

With weak supervision of attribute label, the multi-class CNN classifier can distinguish bag images of different attribute value and we assume this is very important in comparing the similarity between bags. However, the attribute information is far from enough to completely describe a bag. Therefore, we use the last F layer as feature representations. We apply the bag feature extracted from the trained CNN and use Euclidean distance to measure the similarity between two bags. Retrieval results are shown in Fig. 5. We can see that using the F-layer feature keeps more appearance information of the image and the top-10 retrieval result are all similar to the probe image in color and shape.

6   Conclusions

We build a bag retrieval system that automatically detects bags in a street shot and retrieves for bags with similar styles in a photo gallery. We combine the original and context proposals together to train a PC-CNN to classify bounding box proposals generated by selective search into bag or non-bag classes and estimate the bias of the proposals to the ground truth. To compare bag regions with different size in street shots, we train an attribute classifier using CNN and use the F layer to describe bag regions. The CNN feature is applied to the bag retrieval system and experiments validate its effectiveness.

Acknowledgements. This work is supported in part by the 973 Program of China under Grant No. 2011CB302203.

References

1. Berg, T.L., Berg, A.C., Shih, J.: Automatic attribute discovery and characterization from Noisy Web Data. In: Daniilidis, K., Maragos, P., Paragios, N. (eds.) ECCV 2010, Part I. LNCS, vol. 6311, pp. 663–676. Springer, Heidelberg (2010)
2. Chen, H., Gallagher, A., Girod, B.: Describing clothing by semantic attributes. In: Fitzgibbon, A., Lazebnik, S., Perona, P., Sato, Y., Schmid, C. (eds.) ECCV 2012, Part III. LNCS, vol. 7574, pp. 609–623. Springer, Heidelberg (2012)
3. Deng, J., Dong, W., Socher, R., Li, L.-J., Li, K., Fei-Fei, L.: Imagenet: a large-scale hierarchical image database. In: IEEE Conference on Computer Vision and Pattern Recognition, CVPR 2009, pp. 248–255. IEEE (2009)
4. Ding, Y., Xiao, J.: Contextual boost for pedestrian detection. In: 2012 IEEE Conference on Computer Vision and Pattern Recognition (CVPR), pp. 2895–2902. IEEE (2012)
5. Donahue, J., Jia, Y., Vinyals, O., Hoffman, J., Zhang, N., Tzeng, E., Darrell, T.: Decaf: a deep convolutional activation feature for generic visual recognition. arXiv preprint arXiv:1310.1531 (2013)
6. Du, Y., Ai, H., Lao, S.: A two-stage approach for bag detection in pedestrian images. In: Cremers, D., Reid, I., Saito, H., Yang, M.-H. (eds.) ACCV 2014. LNCS, vol. 9006, pp. 507–521. Springer, Heidelberg (2015)
7. Felzenszwalb, P., McAllester, D., Ramanan, D.: A discriminatively trained, multiscale, deformable part model. In: IEEE Conference on Computer Vision and Pattern Recognition, CVPR 2008, pp. 1–8. IEEE (2008)
8. Girshick, R., Donahue, J., Darrell, T., Malik, J.: Rich feature hierarchies for accurate object detection and semantic segmentation. In: 2014 IEEE Conference on Computer Vision and Pattern Recognition (CVPR), pp. 580–587. IEEE (2014)
9. Hara, K., Jagadeesh, V., Piramuthu, R.: Fashion apparel detection: the role of deep convolutional neural network and pose-dependent priors. arXiv preprint arXiv:1411.5319 (2014)
10. Huang, J., Liu, S., Xing, J., Mei, T., Yan, S.: Circle & search: attribute-aware shoe retrieval. ACM Trans. Multimedia Comput. Commun. Appl. (TOMM) **11**(1), 3 (2014)
11. Kim, H.-C., Kim, D., Bang, S.Y.: Face retrieval using 1st-and 2nd-order pca mixture model. In: 2002 International Conference on Proceedings of the Image Processing, vol. 2, pp. II-605. IEEE (2002)
12. Krizhevskey, A.: Cuda-convnet (2014). https://code.google.com/p/cuda-convnet/
13. Krizhevsky, A., Hinton, G.: Learning multiple layers of features from tiny images. Computer Science Department, University of Toronto. Technical Report, vol. 1, no. 4, p. 7 (2009)

14. Krizhevsky, A., Sutskever, I., Hinton, G.E.: Imagenet classification with deep convolutional neural networks. In: Advances in Neural Information Processing Systems, pp. 1097–1105 (2012)
15. Kumar, N., Berg, A.C., Belhumeur, P.N., Nayar, S.K.: Attribute and simile classifiers for face verification. In: 2009 IEEE 12th International Conference on Computer Vision, pp. 365–372. IEEE (2009)
16. Liu, S., Liang, X., Liu, L., Lu, K., Lin, L., Yan, S.: Fashion parsing with video context. In: Proceedings of the ACM International Conference on Multimedia, pp. 467–476. ACM (2014)
17. Uijlings, J.R., van de Sande, K.E., Gevers, T., Smeulders, A.W.: Selective search for object recognition. Int. J. Comput. Vis. **104**(2), 154–171 (2013)
18. Viola, P., Jones, M.: Robust real-time object detection. Int. J. Comput. Vis. **4**, 34–47 (2001)
19. Wang, X., Zhang, T.: Clothes search in consumer photos via color matching and attribute learning. In: Proceedings of the 19th ACM International Conference on Multimedia, pp. 1353–1356. ACM (2011)
20. Yamaguchi, K., Kiapour, M.H., Berg, T.L.: Paper doll parsing: retrieving similar styles to parse clothing items. In: 2013 IEEE International Conference on Computer Vision (ICCV), pp. 3519–3526. IEEE (2013)
21. Yamaguchi, K., Kiapour, M.H., Ortiz, L.E., Berg, T.L.: Parsing clothing in fashion photographs. In: 2012 IEEE Conference on Computer Vision and Pattern Recognition (CVPR), pp. 3570–3577. IEEE (2012)
22. Zhao, R., Ouyang, W., Wang, X.: Unsupervised salience learning for person re-identification. In: 2013 IEEE Conference on Computer Vision and Pattern Recognition (CVPR), pp. 3586–3593. IEEE (2013)

TV Commercial Detection Using Success Based Locally Weighted Kernel Combination

Raghvendra Kannao[(✉)] and Prithwijit Guha

Department of Electronics and Electrical Engineering,
Indian Institute of Technology Guwahati, Guwahati 781039, Assam, India
{raghvendra,pguha}@iitg.ernet.in

Abstract. Classification problems using multiple kernel learning (MKL) algorithms achieve superior performance on account of using a weighted combination of base kernels on feature sub-sets. Each of the base kernels are characterized by the similarity measures defined over the feature sub-sets. Existing works in MKL have mostly used fixed weights which are shown to be related to the overall discriminative capability of corresponding base kernels. We argue that this class discrimination ability of a kernel is a local phenomenon and thus, advocate the necessity of using instance dependent functions for weighing the kernels. We propose a new framework for learning such weighing functions linked to ability of kernels to discriminate in the local regions of the feature space. During training, we first identify the regions of success in the feature sub-spaces, where the base kernels have high likelihood of success. These regions are identified by evaluating the performance of support vector machines (SVM) trained using corresponding (single) base kernels. The weighing functions are then estimated by using support vector regression (SVR). The target for SVRs is set to 1.0 for the successfully classified patterns and to 0.0, otherwise. The second contribution of this work is the construction and public domain release of a commercial detection dataset of 150 hours, acquired from 5 different TV news channels. Empirical results on 8 standard datasets and our own TV commercial detection dataset have shown the superiority of the proposed scheme of multiple kernel learning.

1 Introduction

Automatic identification of commercials in television news videos finds applications in the domain of broadcast analysis and monitoring. Existing works on TV commercial detection use either frequentist or knowledge based approaches. The *frequentist approaches* [22] assume the TV commercials to have higher reappearance frequency compared to the non-commercials. The repeating video segments are efficiently identified by finger-printing and hashing techniques. These methods have shown excellent performance on archival video data but, can't be deployed for on-the-run commercial detection [3, 22].

Knowledge based methods on the other hand, learn discriminative [11] or generative models [23] using features extracted from video segments for segregating

© Springer International Publishing Switzerland 2016
Q. Tian et al. (Eds.): MMM 2016, Part I, LNCS 9516, pp. 793–805, 2016.
DOI: 10.1007/978-3-319-27671-7_66

the TV commercials. These features represent either audio-visual properties or meta information of the video segments [11,13,24]. Knowledge based methods are comparatively popular due to their superior performance and applicability in commercial detection for both real-time and off-line applications. However, these methods require judicious selection and efficient fusion of multi-modal features to maintain the superiority. The feature fusion in case of TV commercial detection is non-trivial and requires a special attention due to following reasons. First, most of the existing approaches use assumptions specific to particular channel or broadcasting standards and hence, discriminative abilities of features are local in the feature space. Second, different features have different notions of similarity and hence, requires the use of separate similarity measure for every feature.

Fusing the information available from heterogeneous sources to solve classification problems is well studied in the literature. These fusion schemes can be categorized as either "Early Fusion", "Late Fusion" or "Intermediate Fusion" based on the level at which multiple information sources are combined [1,4,16]. In early fusion, information available from different sources is fused at the feature level by concatenating all the available features. A single classifier is then trained on the concatenated feature vector. Early fusion has simple and comparatively inexpensive training and testing procedure. However, usage of a single similarity measure or classifier architecture for all the features and presence of non-discriminative features restricts the performance early fusion schemes [1]. This mandates a pre-selection or pre-processing stage to select optimal set of features and a suitable similarity measure or classifier architecture [1,16].

Late fusion involves the combination of information sources at the decision level. On each of the available features, a separate classifier is trained. Decisions of these classifiers are rationally aggregated using a decision combiner stage. Each of the trained classifier has discriminative abilities only in local regions of the feature space and thus, the base classifiers may have diversity in errors. Decision combiners harness this diversity in errors to reach at a more accurate decision [11,16]. Fusion at the decision level enables the use of suitable classifier architecture and the similarity measure for features. Moreover, the irrelevant features are suppressed at the decision combiner stage. Thus, late fusion overcomes the drawbacks of early fusion to an extent but, at an additional cost of complex and rigorous training procedure [11,16]. The upper bound on the possible improvement in performance due to feature fusion is determined by the ability of the fusion scheme to generate a set of diverse base classifiers and the ability of the decision combiner to suppress the false decisions [1].

The third family of information fusion schemes (intermediate fusion), combines different features at the kernel level using the multiple kernel learning framework (MKL, henceforth) [4]. In MKL, weighted additive or multiplicative combination of different kernels is used to train a single SVM classifier. Each base kernel may be defined either on a sub-set of features or individual features with a suitable similarity measure. The weight of each kernel determines the influence of feature-similarity measure pair on the final decision. The kernel weights are selected such that only discriminative kernels have non-zero weights, while non-discriminative kernels are suppressed. Intermediate fusion overcomes the drawbacks of both early and late fusion schemes. Several recent works have used and empirically established the superiority of intermediate fusion schemes over traditional feature fusion techniques [2,15,19,21]. Various methods are proposed in the literature for estimating the kernel weights (see review on multiple kernel learning methods by Gonen and Alpaydin [4]). For example, Moguerza et al. [14] have proposed to use linear combination of different kernels with conditional class probabilities estimated using nearest neighbor approach for weighing the kernels. Tanabe et al. [20] have used the F-score of the classifier trained on individual kernels as weights. Kernel weights are also estimated by formulating it as constrained optimization problems [7,17]. The objective of these optimization problems is either to maximize the similarity between the combined kernel and the ideal kernel [17] or to minimize the structural risk of the MKL SVM [7,19]. These optimization problems are formulated as a quadratically constrained quadratic programming (QCQP) problem and are solved to get kernel weights as well as SVM parameters. One of the popular and computationally efficient technique for solving the QCQP problem is proposed by Sonnenburg et al. [18,19]. They have proposed to use semi-infinite linear programming for optimizing the kernel weights and quadratic programming for computing the SVM parameters.

Most of the works on MKL have proposed to use fixed weights. However, kernels behave as local experts and have discriminative abilities in local regions of the feature space. Gonen and Alpaydin [5] (L-MKL, henceforth) have proposed to use instance dependent functions instead of fixed weights. A gating model defined by a combination of perceptrons decides the weights of the kernels and hence, their corresponding regions of influence. The weights are estimated using a two step optimization process. In the first step, the Lagrange multipliers of the canonical SVM are estimated by keeping the weights of perceptrons in the gating model fixed. In second step, the parameters of the gating model (perceptron weights) are re-estimated. This two step process is continued till convergence. The number of perceptrons used in the gating model are equal to the number of kernels. The gating model divides the feature space into non-overlapping regions separated by perceptrons and in each region a single kernel is active. This particular approach has shown the significant improvement in performance over existing methods using fixed weights. However, the discriminative regions of the kernels may not be continuous in the feature space. Thus, the assumption of this approach to have regions of influence of the kernels to be non-overlapping, is not feasible for real life datasets.

First major contribution of this work is a novel framework for learning instance dependent kernel weighing functions. These kernel weighing functions are linked to the discriminative abilities of the kernels. We define the local regions in the feature space, where a particular kernel has discriminative abilities as "Regions of Success" (RoS, henceforth) of the kernel. The proposed kernel weighing functions have high weights in the RoS of corresponding kernels. We call these weighing functions as "Success Prediction Functions" (SPF, henceforth).

This weighing scheme ensures that, only locally successful kernels are allowed to contribute in the final kernel combination while, suppressing the failed ones. For learning success prediction functions, we use a two stage non-iterative procedure. In the first stage, we identify the regions of success of each of the base kernels. These regions are identified by training SVMs with individual base kernels and evaluating their performance on the cross-validation set. The correctly classified instances are identified as instances from the regions of success. The success prediction functions are estimated using regression models trained on the cross-validation set. The target for regression models is set to 1.0 for the correctly classified instances and to 0.0, otherwise. In our experimentation, we have used the support vector regression (SVR) to learn the SPFs. For a given test pattern, SPFs provide higher weights for kernels having high likelihood of success in the neighborhood of test pattern, while reducing the importance of the probably erroneous kernels. Thus, increasing the classification performance. The empirical results have shown the superiority of our proposal.

Our second contribution is the creation of a TV News Commercial Dataset of approximately 150 h of TV news videos from 5 different news channels [9]. To the best of our knowledge this is the first publicly available dataset for TV news commercial detection which will enable benchmarking of different machine learning algorithms. We have used existing audio-visual features viz. shot length [11], scene motion distribution [6], scene text distribution [11], frame difference distribution, zero crossing rate [24], short time energy [24], fundamental frequency, spectral centroid, spectral flux, spectral roll-off frequency [11] and MFCC bag of audio words [12] for characterizing commercial shots.

The rest of the paper is organized in the following manner. In Sect. 2, we briefly describe the different audio-visual features used for characterizing commercials. The proposal of the success based locally weighted kernel combinations is explained in Sect. 3. The results of experimentation in terms of comparative F-score and generalization performances are presented in Sect. 4. Finally, we conclude in Sect. 5 and outline the future extensions.

2   Audio-Visual Features

We choose a video shot as basic unit for commercial detection as transition to and from commercials will result in a shot change. The television video broadcast is first segmented into shots based on simple color distribution consistency [11]. We extract following eleven audio-visual features from every video shots to segregate commercials.

Video Shot Length. (SL) [6] is a discriminating feature as the commercial shots are of short durations compared to that of news reports. **Overlay Text Distribution** is an important clue for identifying the commercials [11]. We observe that the major ticker text bands situated in the upper and lower portions of the scene are generally present during news and other programs. Nevertheless, during commercials only the lower most band remains. Although, product specific small text patches appear throughout the frame during commercials. Following existing work [11], we have divided the frame into a 5×3 grid and have constructed a 30 dimensional feature vector (TD) storing mean and variance of the fractions of text area in each grid block of every frame over the entire shot. The **Motion Distribution** (MD) is a significant feature as previous works have indicated that commercial shots have high motion content as they try to convey maximum information in minimum possible time. This motivates us to compute dense optical flow (Horn-Schunk formulation) between consecutive frames and construct a distribution of flow magnitudes over the entire shot with 40 uniformly divided bins in range of $[0, 40]$ [6]. Optical flow misses sudden changes in pixel intensities of a region if the boundaries of the region are unchanged. Thus, **Frame Difference Distribution** (FD) is also computed along with flow magnitude distributions. We obtain the frame difference by averaging absolute frame difference in each of 3 color channels and the distribution is constructed with 32 bins in the range of $[0, 255]$ [6].

To attract viewer's attention TV commercials usually have loud background music and non-pure speech. Whereas, pure speech in moderate volume dominates regular programs [6]. This motivated us to use low level audio features like **Short Time Energy** (STE, henceforth), **Zero Crossing Rate** (ZCR) [24], **Spectral Centroid** (SC), **Spectral Flux** (SF), **Spectral Roll-Off Frequency** (SR) and **Fundamental Frequency** (FF) [11] to discriminate between speech, music and non-pure speech. For all the above mentioned audio features, we have used the non-overlapping frames of 20 msec duration and sampling frequency of 8 kHz. The Mean and standard deviation of each audio feature over the shot results in a $2D$ vector for each feature.

The **MFCC Bag of Audio Words** (BoAW) have been successfully used in existing speech/audio processing applications [12]. The MFCC coefficients along with Delta and Delta-Delta Cepstrum from 150 h of audio tracks are clustered (using K-means) into 4000 (empirically determined) groups to form our audio word dictionary. Every shot is then represented as a Bag of Audio Words by forming the normalized histograms of the MFCC co-efficients extracted from overlapping windows in the shots.

Existing approaches have experimented with classifiers (mainly SVM, AdaBoost etc.) learned on different combinations of the above mentioned features to detect commercials. We observe that at different locations of the feature space, a particular combination of features is generally successful in identifying the commercial shots (Fig. 1). This motivated us to propose a spatially varying composition of kernels trained on a particular feature. These locally varying weights effectively work as feature selectors in multiple kernel learning framework. Our proposed methodology of "success based locally weighted kernel combination" is described next.

3   Success Based Locally Weighted Kernel Combination

Consider the problem of designing a binary classifier over the training set $S = \{(x_i, y_i); i = 1, \ldots n\}$ containing n independent and identically distributed instances, where $y_i \in \{-1, +1\}$ is the class label of the instance $\boldsymbol{x}_i$. Let, each training instance consists of m independent features, such that $\boldsymbol{x}_i = [^1\boldsymbol{x}_i, \ldots^j \boldsymbol{x}_i, \ldots^m \boldsymbol{x}_i]^T$. Each feature may have multiple dimensions. Let us consider that, q base kernels can be defined on a single feature based q different similarity measures (e.g. $RBF, \chi^2, Linear$). Thus, for m features, we have a total of $P = m \times q$ base kernels. Let, $\boldsymbol{k}_r(\cdot, \cdot)$ $(r = 1, \ldots P)$ be the rth kernel in the combination.

One of the simplest formulation for multiple kernel learning is proposed by Tanabe et al. [20]. They have used the F-scores (on cross-validation sets) of the single kernel SVMs trained with the corresponding base kernels as weights. The F-score weighted kernel combination is given by $\boldsymbol{k}_F(\boldsymbol{x}_i, \boldsymbol{x}_l) = \frac{1}{\gamma_f} \sum_{r=1}^{P} f_r \boldsymbol{k}_r(\boldsymbol{x}_i, \boldsymbol{x}_l)$. Here f_r is the F-score of the classifier $\boldsymbol{C}_r$ and $\gamma_f = \sum_{r=1}^{P} f_r$ is a normalizing factor. In most practical cases, it is observed that kernels don't have uniform performance throughout the feature space. Thus, the use of fixed kernel weights degrades the performance [5]. We note that, each kernel has their own regions of success in the feature space (Fig. 1) where they exhibit their discriminative capabilities. In MKL framework, domination of non-discriminative kernels over discriminative ones, often leads to misclassification [20] for certain test patterns. This motivates us to learn a set of adaptive weighing functions $\eta_r(\boldsymbol{x}_i, \boldsymbol{x}_l) = g_r(\boldsymbol{x}_i) \cdot g_r(\boldsymbol{x}_l)$ such that, only discriminative kernels are allowed to contribute in the kernel combination. Note that the proposed weighing function $\eta_r(\cdot, \cdot)$ is instance dependent as opposed to fixed weight formulations [2, 15, 19, 21] (e.g. f_r in Tanabe et al. [20]).

The first step in the estimation of $g_r(\cdot)$ is to identify the regions of success of kernel $\boldsymbol{k}_r(\cdot, \cdot)$ in the feature space. These regions are identified by training a SVM $\boldsymbol{C}_r$ using the rth base kernel. SVM $\boldsymbol{C}_r$ is then evaluated on the cross-validation set to identify the regions of success in the feature space. The function $g_r(\cdot)$ is estimated by Support Vector Regression (SVR) using the cross-validation set of classifier $\boldsymbol{C}_r$. The target for $g_r(\cdot)$ is set such that, $g_r(\boldsymbol{x}_i) = \delta[\boldsymbol{C}_r(\boldsymbol{x}_i) - y_i]$ where, $\delta[\cdot]$ is the Kronecker delta function. Such a success based weighing scheme will assign high weights (near 1.0) to discriminative kernels, while suppressing the

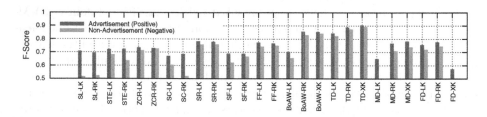

Fig. 1. Figure shows the performance of individual base kernels on TV news commercial dataset. Base kernels are defined on individual features (Sect. 2) with linear (L-K), RBF (R-K) and the χ^2 (X-K) similarity. Note the varying capabilities of the base kernels and their biases towards positives and negatives. We observe that base kernels defined on text distribution (TD) and MFCC Bag of Words (BoAW) with RBF or χ^2 similarity has the best performance.

irrelevant ones (weights near 0.0). The resulting success based locally weighted kernel combination $k_S(x_i, x_l)$ is given by

$$k_S(x_i, x_l) = \frac{1}{\gamma_{il}} \sum_{r=1}^{P} \eta_r(x_i, x_l) k_r(x_i, x_l) = \frac{1}{\gamma_{il}} \sum_{r=1}^{P} g_r(x_i) k_r(x_i, x_l) g_r(x_l) \quad (1)$$

where, $\gamma_{il} = \sum_{r=1}^{P} \eta_r(x_i, x_l)$ is the normalizing factor. This linear combination is weighted by the success level predictions $(g_r(x_i), g_r(x_l))$ of both the inputs x_i and x_l of the kernel function thereby, enhancing the contributions from successful kernels at a particular instance. The performance of the proposed approach was found to be superior than baseline methods on our TV news commercial dataset and on 5 out of 8 standard benchmark datasets in terms of balanced F-score. The results of our experimentation are presented in the next section.

4   Experimentation

Real life datasets often have intraclass variability and interclass imbalances in terms of number of positive and negative instances. Classifiers trained on imbalanced datasets will either lead to biased classification (biased towards majority class) or over fitting on minority class. To avoid ill effects of interclass and/or intraclass imbalances of the training data, we have used cluster based over sampling (CBO, henceforth) scheme proposed in [8]. The testing set is used as it is. For each dataset we have used Linear(L-K) and RBF(R-K) similarity functions and RBF kernel for Regression. The χ^2 similarity is also used in commercial detection for distribution like features (e.g. scene text, optical flow, frame difference distributions and MFCC BoAW). For each feature similarity function pairs, kernels are created and success prediction functions are trained.

The results are reported by dividing the available datasets into testing (40 %) and training sets (60 %) with stratification. We have also reported the results on five other methods apart from the proposed method (S-MKL). These baseline methods are as follows – Concatenation (CONCAT) of all features (early fusion) with single SVM; F-score Weighted ensemble (F-EC) of classifiers trained on each Feature Kernel combination; optimization based MKL (SG-MKL) [18,19], data dependent Localized MKL (L-MKL) [5] and F-score weighted multiple kernel learning (F-MKL) [20]. The number of kernels for SG-MKL and L-MKL are same as that of S-MKL. We have implemented feature extraction codes in C++. We have used LibSVM for support vector classification (C-SVC), L-MKL and regression (ϵ-SVR) and Shogun Library [18] for SG-MKL. All the datasets are scaled to the range $[-1, 1]$. The hyper parameters for C-SVC and ϵ-SVR (C, ϵ and γ for RBF kernel) are obtained by a grid search with the objective of maximizing the balanced accuracy and minimizing the MSE respectively. Use of balanced accuracy instead of accuracy of a single class ensures the unbiasedness of the classifier.

Table 1. Table shows shot wise (S) and broadcast time wise (T) performance of different methods on TV news commercial dataset. Our Proposed method **S-MKL** outperforms all baseline methods. Note the improvement in time wise performance. (Higher numbers are better except in training time)

Methods ↓	Commercial (Positive)						Non-commercial (Negative)						Train Time	Test Time
	Precision		Recall		F-score		Precision		Recall		F-score			
	S	T	S	T	S	T	S	T	S	T	S	T	(h)	(msec)
CONCAT	0.94	0.82	0.90	0.82	0.92	0.84	0.93	0.89	0.89	0.88	0.91	0.89	**18.4**	19
F-EC	0.91	0.85	0.95	0.85	0.93	0.84	0.92	0.88	0.90	0.87	0.91	0.87	38.6	**14**
SG-MKL	0.96	0.81	0.83	0.82	0.89	0.83	0.88	0.86	0.94	0.86	0.91	0.86	67.8	45
L-MKL	0.97	0.83	0.95	0.83	0.96	0.84	0.5	0.62	0.81	0.72	0.62	0.67	75.1	**14**
F-MKL	0.94	0.91	0.92	0.92	0.93	0.91	0.97	0.91	0.95	0.91	0.96	0.91	43.1	28
S-MKL	**0.99**	**0.98**	**0.99**	**0.99**	**0.99**	**0.99**	1	1	**0.98**	**0.99**	**0.99**	**0.99**	48.6	27

4.1 TV News Commercial Dataset

The works of Liu et al. [11] and Wu and Satoh [22] are current state of art methods for TV commercial detection. However, it is difficult to benchmark the performance of our proposed method for commercial detection with current state of art as both the works have used channel and/or region specific heuristics and benchmarked the results on their own datasets which are not available in public domain. We have created a TV News commercial dataset [9] of approximately 150 h of TV news broadcast with 30 h of news broadcast from each of the 5 television news channels – *CNN-IBN, TIMES NOW, NDTV 24X7, BBC WORLD* and *CNN*. The dataset consists of a total of 1, 29, 676 shots with 33, 117, 39, 252, 17, 052, 17, 720, and 22, 535 shots from each of the channels respectively. The dataset is dominated by commercial shots with roughly 63 % positives. From each of the shots 11 audio visual features described in Sect. 2 are extracted. The TV News Commercials Dataset can be download from UCI machine learning repository [9].

In our experimentation, we have used 11 linear kernels (one kernel for each feature), 11 RBF kernels (one kernel per feature) and 4 χ^2 kernels (one each with text distribution, motion distribution, frame difference distribution and audio bag of words). Hence, for commercial detection we use SVM with a combination of 26 kernels. The performance of classifiers trained with individual kernel functions is presented in Fig. 1. Out of these, text distribution and MFCC bag of audio words with χ^2 and RBF similarity turned out to be the best performing classifiers. The classification results of different methods on TV News Commercial dataset are tabulated in Table 1. Our proposed method outperforms all other baseline methods in terms of shot-wise performance as well as broadcast time-wise performance.

During training, L-MKL turns out to be the most expensive approach followed by SG-MKL and S-MKL. L-MKL assumes the linear separability between the regions of use of each kernel and locates these regions by gradient descent [5]. This is not practical and hence, convergence takes comparatively more time. The SVM trained on concatenation of all features with RBF kernel took least

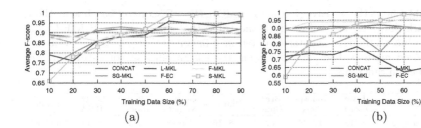

Fig. 2. Figure shows the variations in F-scores for (a) positive and (b) negative categories with respect to changing training set size.

training time. L-MKL and F-EC are found to be the fastest during testing due to reduced number of kernel calculations. In L-MKL, kernel computations are reduced as for every support vector only a single kernel is active. In F-EC, kernel computations are reduced as individual classifiers are trained on features with small dimension. Our proposed method stands third in terms of training and testing time. Comparatively long time taken by our proposed method may be attributed to number of classifiers and regressors involved. However, longer training and testing time is justified by the gain in performance.

The results of experiments are presented by varying the training data size in Fig. 2. Intraclass variability preserved by CBO based data balancing is reflected in the consistent performance of classifiers even after varying the training data size. All the methods except our proposed method (S-MKL) and localized MKL (L-MKL) exhibit the consistent performance over varying training data sizes. S-MKL becomes consistent after sufficient data is available for training. However, L-MKL shows consistency in F-score for positive class only and hence, results in highly biased classification. Poor performance of our proposed method on smaller datasets is attributed to the imperfect learning of success prediction function due to unavailability of sufficient data since, SVRs have larger MSE for smaller training data sizes.

4.2   Benchmark Datasets

To demonstrate the performance of S-MKL, we have benchmarked our results on 8 datasets from UCI machine learning repository [10]. S-MKL have outperformed other baseline methods on 5 datasets in terms of balanced F-score (Table 2). We have used Linear and RBF similarity with each of the feature. In almost all the cases our method produced balanced output in terms of F-scores of positive and negative classes while, other methods showed strong bias towards either of the classes.

4.3   Discussions

In the proposed method, we have used a weighted linear combination of the kernels for training SVM instead of a predefined single kernel. The weights for the kernels are adaptively estimated using kernel weighing functions. Out of the

Table 2. Table shows the positive (F+) and negative (F−) F-scores of baseline methods on benchmark datasets. It is clear from the table that our method (**S-MKL**) out performs baseline methods on 5 datasets in terms of balanced F-score. The performance of our proposed method degrades if training data is insufficient for estimating the success prediction functions

Datasets		Liver Disorder	Ionosphere	Breast Cancer	Diabetes	German Numeric	Mushrooms	COD-RNA	Adult
Features/Classifiers		6/12	17/34	10/20	8/16	24/48	21/42	8/16	14/28
Positive Instances (%)		42.09	64.1	34.99	65.1	30	64.1	33.33	24.84
Total Instances		345	351	683	768	1000	8124	244109	270000
CONCAT	F+	0.52	0.68	0.69	0.75	0.67	0.49	0.76	0.28
	F−	0.71	0.77	0.87	0.49	0.43	0.56	0.64	0.82
F-EC	F+	0.34	0.59	0.71	0.78	0.65	0.3	0.79	0.2
	F−	0.81	0.62	0.76	0.34	0.63	0.79	0.71	0.79
SG-MKL	F+	0.62	0.72	0.74	0.81	0.71	0.52	0.62	0.58
	F−	0.76	0.79	0.69	0.58	0.69	0.69	0.54	0.49
L-MKL	F+	0.63	0.94	0.69	0.72	0.79	0.52	0.4	0.6
	F−	0.75	0.87	0.79	0.69	0.78	0.72	0.51	0.3
F-MKL	F+	0.58	0.82	0.74	0.71	0.71	0.73	0.79	0.58
	F−	0.56	0.86	0.86	0.79	0.69	0.75	0.82	0.62
S-MKL	F+	0.54	0.65	0.89	0.79	0.71	0.87	0.9	0.79
	F−	0.51	0.69	0.94	0.82	0.76	0.83	0.89	0.84

existing methods, S-MKL most closely relates to F-MKL [20] and L-MKL [5]. In F-MKL, kernels have fixed weights throughout the feature space whereas, L-MKL uses locally varying weights. In [5] it was reported for some datasets that L-MKL performs better than F-MKL but, in our experimentation we have observed that in most cases F-MKL have outperformed L-MKL. This reduction in performance of L-MKL is due to the violation of assumption of L-MKL on linear separability of regions of use of kernels [5]. On the other hand, our proposed S-MKL does not make any assumptions on linear separability of regions of use of kernels and hence, outperforms L-MKL.

Domination of non-discriminative kernels (redundant information) hampers the performance of multiple classifier systems (F-EC) and multiple kernel classifiers. It may be noted that, F-MKL (intermediate fusion) and F-EC (late fusion) uses the same weighing function (F-score). F-MKL has comparatively unbiased and better performance. Hence, it may be concluded that multiple kernel learning framework (F-MKL) takes care of redundant information to an extent. Moreover, in our proposed scheme, only successful kernels have sufficient weights to contribute to the final decision. Our weighing scheme ensures that fewer number of correct kernels won't be dominated by larger number of failed ones. We have observed that in most cases, even with a single successful kernel, S-MKL can predict correct labels.

S-MKL has comparatively balance behavior for positive and negative classes despite interclass imbalances in datasets (For example note the performance of different methods on Adult dataset having only 25 % positives). This balanced performance is mainly the result of learning process of success prediction function which treats successful prediction of both the positive and negative categories equally.

5   Conclusion

We have proposed a "success based local weighing" scheme for the selection of kernels in the context of commercial detection in news broadcast videos. The video shots are characterized by eleven different (existing) audio-visual features like shot length, motion and scene text distribution, ZCR, STE, spectral features, fundamental frequency and MFCC Bag of Audio Words. We have trained SVM based classifiers with linear and RBF kernel for all the features and χ^2 kernels for distribution like features, resulting in a total of 26 kernels. Our first proposition involves the use of weighted linear combination of kernels instead of a single kernel in SVM where, the weighing functions are estimated (using support vector regression with RBF kernel) from the zones of success of the classifiers trained with individual kernels. Success prediction functions are designed to have values closer to 1.0 where, the corresponding kernel functions had success and to 0.0, otherwise. Our second contribution is the creation of a TV News commercial dataset of 150 h from 5 different channels and is the first publicly available dataset [9]. MKL with local success based weighing scheme has outperformed other baseline methods on TV News Commercials dataset as well as on 5 standard datasets.

In the present work, we have proposed a single stage weight prediction algorithm from multiple kernel combination. However, we have not experimented with the possibilities of kernel combinations in the support vector regression stage and have only used the RBF kernel. We believe that the simultaneous estimation of weighing functions for kernel combinations in both classifier and regressors will require a reformulation of the problem involving stages of iterative optimization. Also, in this work, we have only contributed in the classifier stage while using existing features. This work can be extended further to include text/audio content and style as features whose combination with the proposed classifier will definitely lead to better performances.

Acknowledgement. This work is part of the ongoing project on "Multi-modal Broadcast Analytics – Structured Evidence Visualization for Events of Security Concern" funded by the Department of Electronics & Information Technology (DeitY), Govt. of India.

References

1. Atrey, P.K., Hossain, M.A., El, S.A., Kankanhalli, M.S.: Multimodal fusion for multimedia analysis: a survey. Multimedia Syst. **16**(6), 345–379 (2010)
2. Ben-Hur, A., Noble, W.S.: Kernel methods for predicting protein-protein interactions. Bioinformatics **21**(1), i38–i46 (2005)

3. Duygulu, P., yu Chen, M., Hauptmann, A.: Comparison and combination of two novel commercial detection methods. In: International Conference on Multimedia and Expo, vol. 2, pp. 1267–1270 (2004)
4. Gonen, M., Alpaydin, E.: Multiple kernel learning algorithms. J. Mach. Learn. Res. **12**, 2211–2268 (2011)
5. Gonen, M., Alpaydin, E.: Localized algorithms for multiple kernel learning. Pattern Recogn. **46**, 795–807 (2013)
6. Hua, X.S., Lu, L., Zhang, H.J.: Robust learning-based TV commercial detection. In: International Conference on Multimedia and Expo (2005)
7. Jawanpuria, P., Varma, M., Nath, J.S.: On p-norm path following in multiple kernel learning for non-linear feature selection. In: International Conference on Machine Learning, June 2014
8. Jo, T., Japkowicz, N.: Class imbalances versus small disjuncts. SIGKDD Explor. Newslett. **6**(1), 40–49 (2004)
9. Kannao, R., Soni, R.S., Guha, P.: TV news channel commercial detection dataset (2015). http://archive.ics.uci.edu/ml/datasets/TV+News+Channel+Commercial+Detection+Dataset
10. Lichman, M.: UCI machine learning repository (2013). http://archive.ics.uci.edu/ml
11. Liu, N., Zhao, Y., Zhu, Z., Lu, H.: Exploiting visual-audio-textual characteristics for automatic tv commercial block detection and segmentation. IEEE Trans. Multimedia **13**(5), 961–973 (2011)
12. Mühling, M., Ewerth, R., Zhou, J., Freisleben, B.: Multimodal Video Concept Detection via Bag of Auditory Words and Multiple Kernel Learning. In: Schoeffmann, K., Merialdo, B., Hauptmann, A.G., Ngo, C.-W., Andreopoulos, Y., Breiteneder, C. (eds.) MMM 2012. LNCS, vol. 7131, pp. 40–50. Springer, Heidelberg (2012)
13. Meng, L., Cai, Y., Wang, M., Li, Y.: Tv commercial detection based on shot change and text extraction. In: International Congress on Image and Signal Processing, pp. 1–5 (2009)
14. Moguerza, J.M., Muñoz, A., de Diego, I.M.: Improving support vector classification via the combination of multiple sources of information. In: Fred, A., Caelli, T.M., Duin, R.P.W., Campilho, A.C., de Ridder, D. (eds.) SSPR 2004 and SPR 2004. LNCS, pp. 592–600. Springer, Heidelberg (2004)
15. Natarajan, P., Wu, S., Vitaladevuni, S., Zhuang, X., Tsakalidis, S., Park, U., Prasad, R., Natarajan, P.: Multimodal feature fusion for robust event detection in web videos. In: Computer Vision and Pattern Recognition, pp. 1298–1305. IEEE (2012)
16. Rokach, L.: Ensemble-based classifiers. Artif. Intell. Rev. **33**(1–2), 1–39 (2010)
17. Shawe-Taylor, N., Kandola, A.: On kernel target alignment. Adv. Neural Inf. Process. Syst. **14**, 367 (2002)
18. Sonnenburg, S., Ratsch, G., Henschel, S., Widmer, C., Behr, J., Zien, A., de Bona, F., Binder, A., Gehl, C.: The shogun machine learning toolbox. J. Mach. Learn. Res. **11**, 1799–1802 (2010)
19. Sonnenburg, S., Ratsch, G., Schafer, C., Scholkopf, B.: Large scale multiple kernel learning. J. Mach. Learn. Res. **7**, 1531–1565 (2006)
20. Tanabe, H., Ho, T.B., Nguyen, C.H., Kawasaki, S.: Simple but effective methods for combining kernels in computational biology. In: RIVF, pp. 71–78. IEEE (2008)
21. Vahdat, A., Cannons, K., Mori, G., Oh, S., Kim, I.: Compositional models for video event detection: a multiple kernel learning latent variable approach. In: International Conference on Computer Vision, pp. 1185–1192. IEEE (2013)

22. Wu, X., Satoh, S.: Ultrahigh-speed tv commercial detection, extraction, and matching. IEEE Trans. Circ. Syst. Video Technol. **23**(6), 1054–1069 (2013)
23. Wang, X., Guo, Z.: A novel real-time commercial detection scheme. In: International Conference on Innovative Computing Information and Control, pp. 536–536 (2008)
24. Zhang, L., Zhu, Z., Zhao, Y.: Robust commercial detection system. In: International Conference on Multimedia and Expo, pp. 587–590 (2007)

Frame-Wise Continuity-Based Video Summarization and Stretching

Tatsunori Hirai[1]([✉]) and Shigeo Morishima[2,3]

[1] Waseda University, Tokyo, Japan
tatsunori_hirai@asagi.waseda.jp
[2] Waseda Research Institute for Science and Engineering, Tokyo, Japan
[3] JST CREST, Tokyo, Japan

Abstract. This paper describes a method for freely changing the length of a video clip, leaving its content almost unchanged, by removing video frames considering both audio and video transitions. In a video clip that contains many video frames, there are less important frames that only extend the length of the clip. Taking the continuity of audio and video frames into account, the method enables changing the length of a video clip by removing or inserting frames that do not significantly affect the content. Our method can be used to change the length of a clip without changing the playback speed. Subjective experimental results demonstrate the effectiveness of our method in preserving the clip content.

Keywords: Flexible video · Video analysis · Video summarization · Video stretching

1 Introduction

Video content exists everywhere, on televisions (TVs), digital versatile disks (DVDs), the Internet, and personal devices. Such video content is complete, and people watch the content as it was created. People watch video content as it is, and the time consumed watching it is driven by the video clip length. On the other hand, spare time to watch such video clips is limited, and the time available and clip length rarely match. Here we propose a method that enables changing the length of a video clip rather than adjusting the watching time to the length by adding flexibility to the clip length while preserving the content.

Because video sharing services are widespread, the number of video clips uploaded on the Web is greater than the number that a person could watch in a lifetime. From the beginning of the first TV broadcast, multiple stations have been broadcasting incessantly. Thus, the amount of video clip is increasing monotonically, and the difficulty in finding a clip that a user wants to watch is increasing correspondingly. In such an environment, developing technologies for efficient video retrieval and browsing is gaining attention. Such methods are being studied from various viewpoints, for example, video retrieval and recommendation methods for deciding what to watch and video summarization and fast-forwarding methods for improving how to watch. Improvement in

© Springer International Publishing Switzerland 2016
Q. Tian et al. (Eds.): MMM 2016, Part I, LNCS 9516, pp. 806–817, 2016.
DOI: 10.1007/978-3-319-27671-7_67

information technology, particularly computer memory, is making it possible to freely handle the video clips as a research subject, and such research is gaining attention.

In this paper, we propose a novel video summarization method for watching a video clip efficiently, which can also be applied to other applications such as video editing. The objective of video summarization is to shorten a video clip while preserving its essence. It is to maximize the amount of information per unit time and the summarization amount and an understandable degree are trade-off. Our method achieves this objective by thinning out video frames. Focusing on audio and video frame transitions, our method removes frames that do not affect the content in a frame-wise manner. Thus, the length of a clip can be reduced, while the content and continuity of audio and video frames are preserved.

Furthermore, our method can stretch the length of a video clip by inserting video frames that do not significantly affect the content. In this manner, the length of a video clip can be flexible. Video summarization methods are an important technology for an efficient video browsing experience. Our method also supports video editing by enabling changes to the video clip length. In the video editing process, an editor must use trial and error to determine shot change points to fit the resulting video clip to a fixed time. If the length of the video fits the fixed time perfectly, editors can refrain from editing the video. Our method helps such editors by enabling changes to the length of a video clip without requiring changes to its content. We aim to help people browse or edit video clips by increasing the flexibility of a video clip.

2   Related Work

Video summarization is one of the most important and well-established video processing research topics. There are many video summarization approaches in various domains [1]. In particular, summarization methods for sports video are actively pursued. For example, Kawamura et al. proposed a method for summarizing racquet sports video by extracting rally scenes that are important for understanding the game [2]. Tjondronegoro et al. proposed a method for detecting and summarizing sports video highlights from whistle sounds, audience cheers, and textual information [3].

DeMenthon et al. proposed a method for summarizing videos by expressing a video as a curve in multi-dimensional feature space and then simplifying the curve [4]. Their method enables frame-wise thinning out of video frames but does not consider audio information. Smiths et al. designed an indicator for thinning out video segments by integrating scene changes, camera motions, object recognition, and keywords in the audio [5]. This method eliminates video segments that are less important for understanding the content. However, the elimination of video segments results in video sequence discontinuity, and the original content cannot be preserved. Therefore, the summarized result is equivalent to that of edited videos that do not preserve the original information.

Most video summarization either extracts important video segments or eliminates unimportant segments. However, such methods eliminate portions of video using their own criteria, preventing a user from knowing what types of scenes are eliminated. Therefore, a fast-forwarding video browsing approach that does not eliminate information but enables efficient video watching has been proposed. Kurihara *et al.* proposed a fast-forwarding method that changes the playback speed in speech and non-speech segments to enable understanding of the speech content while fast forwarding a video [6]. This method does not take visual information into account and is thus not well suited to videos, such as sports videos, in which motion is important. Moreover, this method changes the playback speed at a fixed rate regardless of the content. There are methods for changing the playback speed depending on factors such as motions in a clip [7] or user preferences [8]. However, those methods do not take audio information into account and are thus not suitable for videos, such as conversation videos, in which audio plays an important role. On the other hand, there is a method for changing the speed of audio media to enhance its intelligibility. Imai *et al.* proposed a method, speech rate conversion, for slowing the speech speed and shrinking the silent sections in audio media [9]. In this manner, the entire audio length does not change but intelligibility improves. In summary, fast forwarding and speed conversion techniques for video and audio have been proposed separately, however, no method considering both modalities simultaneously has been proposed.

In this paper, we present a method for removing and inserting video frames considering both video and audio, while preserving the content. Although the video frame removal and insertion can be regarded as a video playback speed conversion, we focus on the cost of frame thinning rather than on changing the playback speed on the basis of specific criteria. What we focus on to thin out video frames is the continuity of video frames with respect to video and audio. Because video frames in a clip have time continuity, it is important to preserve the continuity. Therefore, our method removes a video frame that does not affect transition, thereby preserving continuity in the resulting frames. At this point, before removing a frame that does not affect transition, we also consider the continuity after removing the frame. In other words, we consider both the transition of a video frame and continuity between anteroposterior video frames. Consequently, our method can preserve more continuity than by focusing only on the transition of corresponding video frames. Another difference between our method and fast forwarding is that our result can be used in video editing applications because our resulting video clips retain both their content and their continuity, thereby ensuring the naturalness of the results while preserving the content. To distinguish our method from fast forwarding, we call our video frame thinning out method a "frame-wise video summarization" method.

The advantage of video summarization and fast-forwarding methods is primarily related to the efficiency of browsing video content, however, our goal is not restricted to efficient browsing and includes video editing support as well. In the video editing workflow, the video clip material is cut and pasted by an editor to match a fixed time. The lengths of material clips and the fixed time rarely match without editing. However, editing is time consuming because there are many possibilities for cutting and pasting, and the editor must consider the length and the content at the same time, which requires substantial trial and error. There are methods for resizing the length of background music to fit with the corresponding video. Sato *et al.* proposed a method for reconstructing bars of music and resizing them to match a user's preference [10], and Wenner *et al.* proposed a method for resizing music by calculating natural jumping points within a musical piece [15]. These methods help a video editor by providing background music with flexibility. Using flexible music reduces a restriction in video editing because the editor does not need to consider the background music length. This flexibility is also helpful for the editor if we can change the video length as well. Berthouzoz *et al.* proposed a system for seamlessly cutting and pasting interview video clips that an editor wants to use [11]. The quality of edited video is sufficiently natural that people who watch the interview video cannot detect the cut point easily. However, this method can be applied to a video that contains minimal transitions between camera and objects, such as interview video. This system focuses on supporting video editing of interview video. We aim to support video editing more extensively by achieving flexible video, frame-wise video summarization, and stretching.

3   Frame-Wise Video Summarization

Digital video media display multiple images called frames from moment to moment and express motion and dynamics. The general video displays 30 frames per second (FPS), with an interval between frames of approximately 0.033 s. The number of FPS is called the frame rate, which is 29.97 FPS for general video content of NTSC, such as that on TV broadcasting in the United States. Actually, video media do not capture images uninterruptedly, however, the human cognitive system interpolates the visual information between the frames such that a human perceives the content as consecutive visual information. The frame rate corresponds to the temporal resolution, and it need not always be 30 FPS or more. For example, most movie films and animations are 24 FPS, and some video content uploaded on the Web is 15 FPS to minimize size. Although a higher frame rate achieves smoother motion and a lower frame rate results in jumpy motion, the human brain interpolates the motion as long as a minimum frame rate is maintained.

Removing one frame per pair in a 30 FPS video clip doubles the playback speed. Consequently, the amount of visual information received from the video clip will be equivalent to that of a video recorded at 15 FPS. Here we focus on transitions in a video clip. For example, if an object in a 30 FPS video remains stationary during capture, thinning out of video frames has no effect. The amount of visual information that can be received from the clip will not change, however, the clip will be shorter. In the extreme, keeping only one frame and eliminating all the other frames will not change the visual information as long as the objects are stationary, but only the time required for its playback. Thus, transition in video frames corresponds to the amount of visual information, and lower transition frames have less affect after thinning out is performed.

The main idea of this research is to focus on transition of video frames and to remove low transition frames in a frame-wise manner in order to reduce the length of a video clip, while preserving the amount of information (i.e., remaining the content). The same can be said with regard to the audio portion. If the same sound occurs in consecutive frames, removing some frames has little effect on the content. Using this feature, we propose a method that thins out video frames considering audio-visual transition.

3.1   Frame-Wise Thinning Based on Visual Transition

Human eyesight is sensitive to sudden change, hence, temporal continuity is important. The significance of video frame thinning depends on transitions. If the frame to be removed contains drastic change, the resulting video clip will cause discomfort compared with one resulting from removing a stable frame. For example, if the object to be captured by a camera is moving fast and can be captured for only a few frames, one frame plays an important role. Removing such a frame destroys object motion continuity. In contrast, if an object is moving very slowly, the moving distance during one frame is sufficiently short such that the loss of continuity from removing it is lesser than that from removing a frame containing a fast-moving object.

Consequently, we focus on transitions that can handle object motion comprehensively. In particular, we use the sum of squared differences (SSD) corresponding to the difference between adjacent video frames. The SSD value $s(t)$ at frame t can be calculated as

$$s(t) = \sum_{y=1}^{height} \sum_{x=1}^{width} (f_t(x,y) - f_{t-1}(x,y))^2, \tag{1}$$

where $f_t(x,y)$ is the pixel value at the coordinates (x,y) in frame t. Figure 1 shows the SSD transition in a video of a person walking past a stable camera twice. In this graph, the SSD value increases sharply at an important event in the video (viz., a person walking by).

In this type of video, frames with stable objects will be preferentially removed when the lower value frames are thinned out. Thus, thinning out frames with low SSD values shortens a video clip, while preserving its content.

To some extent, this method is effective. However, only thinning out the frames with low SSD values sometimes creates new discontinuities. Therefore, new SSD values to be inserted after thinning must also be considered. To take new SSD values into account, we set the thinning cost C_{video} as the summation of SSD values of the frame and the new SSD value of anteroposterior video frames. Here the cost C_{video} is normalized to enable combining it with an audio thinning cost later. Our method thins out video frames according to the cost, preserving the continuity of the entire video clip as much as possible. To achieve this aim, SSD values of removed frames must be recalculated after the first round of thinning. At this point, the new SSD value of thinned out frames is not normalized, creating a difference between normalized costs. To avoid this

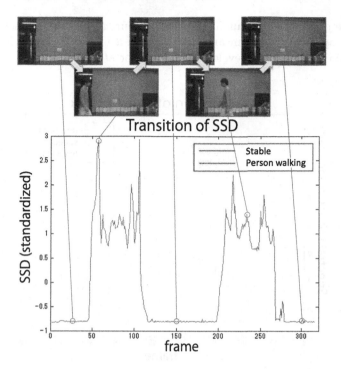

Fig. 1. Transition of SSD value in a video clip.

difference, we retain the value for the normalizing process and apply the same parameter to rescale the new SSD value. We prohibit thinning out the first or last frame of a video clip to ensure that the thinning cost can always be calculated.

Avidan *et al.* proposed an image resizing (retargeting) method called seam carving [12]. This method searches for a path that is not important and thins out pixels in a row- or column-wise manner. Rubinstein *et al.* expanded this method and proposed an improved seam carving that considers insertion cost in addition to thinning cost [13]. The improved seam carving can be applied to video retargeting. Our method can be considered as a form of seam carving applied to entire video frames for video summarization instead of video retargeting.

3.2   Frame-Wise Thinning Out Based on Audio Transition

The sampling rate of audio is generally 44,100 or 48,000 samples per second, which is very different from that of video frames. We designed an audio frame for analysis to adjust the time step between audio and video. When the video frame rate is r, we set the audio frame length to $2/r$ seconds and the step length for the audio frame to $1/r$ seconds. Hence, the audio and video time steps can be synchronized.

As an audio feature expressing audio continuity, we use spectral flux, which represents local temporal transition of the audio spectrum. It takes a high feature value at the point when an audio transition occurs (e.g., sound onset or offset). We extract spectral flux from an audio part of a video clip using MIRtoolbox 1.5, an audio analysis tool developed by Lartillot *et al.* [14].

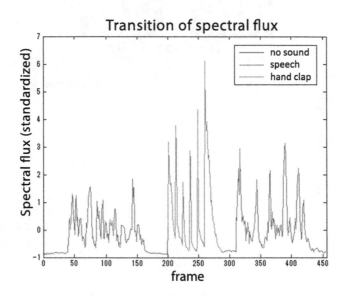

Fig. 2. Transition of spectral flux value.

Figure 2 shows the transition of spectral flux for an audio sample that includes speech and hand claps. The audio sample is recorded in an indoor environment where particular sounds are not observed other than the speech and claps. In the graph in Fig. 2, the spectral flux value reflects the auditory events. Therefore, we can detect a section in which an audio event occurs by focusing on the spectral flux value.

Thinning out audio frames with a low spectral flux value shortens the audio without losing the audio event content. Here we use audio frame thinning cost C_{audio} and thin out frames exactly as in Sect. 3.1, considering the new insertion cost in addition to the spectral flux value of a frame. An audio can be shortened by calculating the spectral flux value per thinning round.

3.3 Frame-Wise Thinning Based on Audio-Visual Transition

We combine video and audio frame thinning. Video and audio frame thinning consider visual and audio continuity, respectively. To consider audio-visual continuity, we design the audio-visual thinning cost $C(t)$ for removing frame t as

$$C(t) = \alpha C_{video}(t) + (1 - \alpha)C_{audio}(t). \tag{2}$$

Here the parameter α is a weight for audio-visual balance. When $\alpha = 0.5$, audio and visual continuity is considered equally. Both C_{video} and C_{audio} are normalized in advance to have mean zero and variance one in order to achieve uniformity. Frame-wise video summarization can be achieved by removing a video frame with minimum cost using Eq. 2. Figure 3 shows the transition of audio-visual thinning cost calculated for the video with a person walking by used in Fig. 1 in Sect. 3.1. In this video, the visual event is a human walking across the video twice, and the audio event is the sound of footsteps. Before and after the walking, there are sections with only the footsteps can be heard. In such a section, the cost remains high, indicating that the cost reflects not only visual but also audio events. The parameter α is an important parameter to add weight to audio and video, which is highly dependent to the content. For example, auditory cost is not as important in action scenes as it is in conversation scenes. At the current implementation, this parameter is set by a user. Automatic optimization of α is our future work.

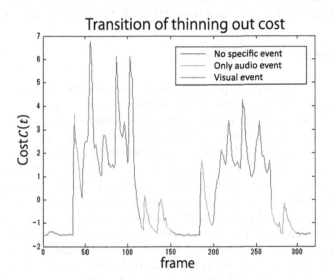

Fig. 3. Transition of audio-visual thinning cost.

4   Video Stretching via Frame Insertion

We also propose a method for stretching a video clip to provide it with flexibility. Video stretching can be achieved by using the same cost (Eq. 2) as in the video summarization. Inserting video frames with low cost at the same point can stretch a video clip while preserving its content. There are two differences between video summarization and stretching. One is that we do not consider the new inserted cost after frame insertion because the inserted cost incurred from the insertion of the same frame is zero. Another difference is that we do not re-calculate the cost for each frame insertion round because the video frame with minimum cost remains the minimum despite frame insertion. Instead of re-calculating cost, we insert video frames in the ascending order of cost. Hence, we begin video frame insertion with the least-cost frame and then insert the frame with the second-least cost.

However, inserting video frames in an ascending order leads to disregarding content. If we double the length of a video clip by inserting video frames in an ascending order, all the frames will be doubled and the playback speed will be halved. It is possible to preserve the clip content with slow playback, however, the motion in the clip will be slowed and changes its meaning. Therefore, we set a maximum percentage of video frames to insert and do not use all the frames for insertion. For example, if the maximum percentage is set to 50 %, only half of the video frames will be used for insertion. This value corresponds to the percentage of frames that do not play important roles for content. The value depends on the content, and it must be set differently for each video clip. However, for the current implementation, we leave the setting of this value to the user.

5   Subjective Experiment

We performed a subjective experiment to evaluate the results of our video summarization and stretching method with respect to naturalness.

We compared our results with video clips produced using playback speed conversion. Seven subjects were asked to watch video clips generated using our proposed method and the playback speed conversion method and to determine which were more natural. The subjects watched a total of 20 video clips and scored them from "1: the video has been shortened" to "5: the video has been stretched" by comparing the clips generated using each method. If the score was close to "3: the video is natural," summarization or stretching succeeded in changing the length of the video clip without introducing artifacts. We used two types of video clips and five clip lengths, with shrinking rates of 50 %, 75 %, 100 %, 125 %, and 150 %. The video clips included "No.1: video of fireworks with narration included" and "No.2: speech video with gesture included." The subjects watched five clips with different lengths produced using our method, followed by five clips produced using the comparison method. The clips were displayed to the subjects in random clip length order, and subjects were asked to assign each score independently, regardless of the previous clip displayed. The video clips included the original clip, with a shrinking rate of 100 %, however, the subjects were not told which clip it was. Although the 100 % shrinking rate clips are exactly the same in the proposed and comparison methods, we displayed both clips in order to check the variance of the evaluation scores. We set the audio-visual balance parameter α in Eq. 2 to 0.5 and the maximum percentage for frame insertion to 20 %.

Table 1 shows the evaluation score for each video clip. The score 3.00 indicates naturalness, and scores closer to that value reflect better results. The score in

Table 1 is the average of seven subjects, and the row average indicates average scores of the two video clips. From this result, we can conclude that our proposed method scores more closely to 3.00 and is thereby shown to be more natural than the speed conversion method for shrinking rates of 75 %, 125 %, and 150 %. However, the result for the 50 % shrinking rate is the same as that for the speed conversion method. This is because thinning out 50 % of frames is excessive and causes the removal of meaningful frames. From the questionnaire provided to the subjects, we learned that discontinuity, particularly in speech video, was the major cause of unnaturalness. This experiment handled only two types of video clips and four shrinking rates, and we are planning to further evaluate the relation between shrinking rate and naturalness to understand the effective range of our method. In addition, the audio-visual balance parameter α should be optimized in the future.

Table 1. The result of subjective evaluation experiment

Video	Method	Shrinking rate				
		50 %	75 %	100 %	125 %	150 %
No.1 fireworks	Comparison	1.00	**1.86**	3.00	4.14	4.86
	Proposed	1.00	1.71	3.00	**4.00**	4.86
No.2 speech with gestures	Comparison	1.14	1.86	3.00	4.00	4.43
	Proposed	1.14	**2.29**	2.86	**3.71**	**4.00**
Average	Comparison	1.07	1.86	3.00	4.07	4.64
	Proposed	1.07	**2.00**	2.93	**3.86**	**4.43**

6   Applications of the Proposed Method

In this section, we introduce some further applications of our proposed method that we are exploring beyond summarization and stretching.

In the video summarization method, video frames are thinned out in an ascending order. Here we consider the thinning out of video frames in a descending order and the elimination of moving objects in a video clip. It is possible to eliminate moving objects by thinning out video frames at a high cost if all the other objects are stable. However, if the camera is moving or the scene is complex and many objects are moving, it is difficult to eliminate frames using this approach. In such a condition, pixel-wise elimination rather than frame-wise elimination is more effective, and we are investigating that possibility.

Another possible application is audio-visual synchronization for a music video clip. In Sects. 3 and 4, we considered audio and visual continuity simultaneously. In contrast, if we consider them separately, it may be possible to synchronize music and video. Sato *et al.* proposed a method to synchronize music to a video clip by recomposing a musical piece [10]. Similarly, our method has the potential of synchronizing music and video by resizing video. When a user specifies the points of video and audio to be synchronized, our method enables shrinking and stretching the video frames between those points. Because synchronizing audio and video is difficult, it is effective for a user to support audio-visual synchronization at the frame level.

Our video summarization method is effective for content-based video retrieval as well. Since the method shortens the video while preserving the content, content-based video retrieval can be more efficient and fast without considering verbose video frames. In addition, data storage can be used more efficiently if the content can be reversible. We can discuss the reversibility of our summarization method by referring the theory of information compression. Our method is related to the theory of information compression, and we will further explore the relation in the future research.

7   Conclusion

In this paper, we present a novel method for summarizing and stretching video clips by focusing on transition and continuity. The proposal of frame-wise removal and insertion and the cost to preserve continuity are our main contributions. The subjective experimental results demonstrated effectiveness in a reasonable manner.

Our method thins out video frames regardless of the meaning of silence or stability. Therefore, if a silence is meaningful, our method may have a negative effect. The method summarizes video considering the content, however, if the silence or stable state is the content, our method eliminates the content. It is a limitation of our method that it is difficult in the current implementation to consider the semantics of the content. Most but not all the semantics included in audio-visual transition can be considered using our method. A further limitation is that our method is not well suited to apply to a video that includes music. Because every audio frame in music has meaning, our method must not remove any frames. In such a case, resizing music using the method proposed by Wenner et al. [15] may be effective, or perhaps, simply thinning out the frames equally would produce a better result.

There are limits to the number of frames that can be removed, while still preserving the content. We will explore the limit further and enable adaptive and automatic setting of parameters such as *alpha* in Eq. 2 by quantitatively defining an amount of information for a video clip. The idea of our method is similar to that of video compression. We are planning to apply those theories used in video compression technique to achieve effective frame thinning. In addition, we will evaluate the effectiveness of our method further.

The time we can assign to watching the video content is limited. We can watch the video content efficiently by scheduling what and when to watch. However, we would like to enjoy freely without letting such an annoying schedule restrain our daily lives. Instead of adjusting our schedules to the video content, the video content should be adjusted to our schedules. Our method has the possibility of enabling a user to watch a video regardless of how much time he or she has available for video watching, and we will further explore that possibility.

Acknowledgments. This work was supported by OngaCREST, CREST, JST and partially supported by JSPS Grant-in-Aid for JSPS Fellows.

References

1. Money, A.G., Agius, H.: Video summarisation: a conceptual framework and survey of the state of the art. J. Vis. Commun. Image Represent. **19**, 121–143 (2008)
2. Kawamura, S., Fukusato, T., Hirai, T., Morishima, S.: Efficient video viewing system for racquet sports with automatic summarization focusing on rally scenes. In: Proceedings of SIGGRAPH 2014, Article 62 (2014)
3. Tjondronegoro, D., Chen, Y.-P.P., Pham, B.: Integrating highlights for more complete sports video summarization. IEEE Multimedia **11**(4), 22–37 (2004)
4. DeMenthon, D., Kobla, V., Doermann, D.: Video summarization by curve simplification. In: Proceedings of ACM Multimedia 1998, pp. 211–218 (1998)
5. Smith, M., Kanade, T.: Video skimming and characterization through the combination of image and language understanding. In: Proceedings of CBAIVL1998, pp. 61–70 (1998)
6. Kurihara, K.: CinemaGazer: a system for watching videos at very high speed. In: Proceedings of AVI, pp. 108–115 (2012)
7. Peker, K., Divakaran, A., Sun, H.: Constant pace skimming and temporal subsampling of video using motion activity. In: Proceedings of ICIP2001, vol. 3, pp. 414–417 (2001)
8. Cheng, K.-Y., Luo, S.-J., Chen, B.-Y., Chu, H.-H.: SmartPlayer: user-centric video fast-forwarding. In: Proceedings of CHI 2009, pp. 789–798 (2009)
9. Imai, A., Seiyama, N., Mishima, T., Takagi, T., Miyasaka, E.: Application of speech rate conversion technology to video editing: allows up to 5 times normal speed playback while maintaining speech intelligibility. In: Proceedings of Audio Engineering Society Conference (2001)
10. Sato, H., Hirai, T., Nakano, T., Goto, M., Morishima, S.: A music video authoring system synchronizing climax of video clips and music via rearrangement of musical bars. In: Proceedings of SIGGRAPH 2015, Article 42 (2015)
11. Berthouzoz, F., Li, W., Agrawala, M.: Tools for placing cuts and transitions in interview video. ACM Trans. Graph. **31**(4), 67:1–67:8 (2012)
12. Avidan, S., Shamir, A.: Seam carving for content-aware image resizing. ACM Trans. Graph. **26**(3), 10:1–10:9 (2007)
13. Rubinstein, M., Shamir, A., Avidan, S.: Improved seam carving for video retargeting. ACM Trans. Graph. **27**(3), 16:1–16:9 (2008)
14. Lartillot, O., Toiviainen, P., Eerola, T.: A Matlab toolbox for music information retrieval. In: Preisach, C., Burkhardt, H., Schmidt-Thieme, L., Decker, R. (eds.) Data Analysis, Machine Learning and Applications, p. 261. Springer, Heidelbeg (2008)
15. Wenner, S., Bazin, J.C., Sorkine-Hornung, A., Kim, C., Gross, M.: Scalable music: automatic music retargeting and synthesis. Comput. Graph. Forum **32**(2), 345–354 (2013)

Respiration Motion State Estimation on 4D CT Rib Cage Images

Chao Xie$^{(\boxtimes)}$, Wengang Zhou, Weiping Ding, Houqiang Li, and Weiping Li

EEIS Department, University of Science and Technology of China, Hefei, China
{chaoxie,zhwg,wpdings,lihq,wpli}@ustc.edu.cn

Abstract. Respiration motion state is an important indicator for disease diagnose in clinical practice. In this paper, we approach this problem with 4D CT rib cage images and target on identifying the end-inhalation and end-exhalation phrases. Observing that the motion of rib bones well reflect the respiration motion state, we transform this problem into a rib bone segmentation problem. Firstly, we propose a novel steerable filter enhanced level set method for rib bone segmentation. We formulate the level set segmentation problem as a variational optimization problem. To address the blurry edge issue, we enhance the image with the classic steerable filter. After that, by comparing the positions of rib bones in sequential frames, we present an criterion to determine the end-expiration and end-inspiration phrases. We validate our approach with real 4D CT rib cage images and demonstrate the effectiveness of our approach.

Keywords: 4D CT imaging · Respiration motion stage · Image segmentation · Level set · Active contour · Steerable filter

1 Introduction

In clinic study, respiration motion phase is important indicator to diagnose disease and investigate the impact of medicines on human body. To approach it, 4D CT imaging is popularly applied to take images over the 3D rib cage of patients in regular time phase, resulting in 4D CT rib cage images. In such a scenario, the problem becomes how to determine the end-inhalation and end-exhalation phases [1]. Such expiration and inspiration information will help to study the state of tumors [9] or organs [11] in the rib cage and assist in the pharmaceutical research.

In 4D CT rib cage images, the most notable feature for the respiration motion state is the rib bones, which exhibits distinctive contrast with the surrounding region and moves periodically in specific directions. Motivated by such observation, we approach the respiration motion state estimation problem by segmentation of rib bones. Once the rib bones are segmented out, the end-inhalation and end-exhalation phases can be readily determined with some heuristic criteria by checking the motion extent of rib bones. Therefore, the problem is transformed to the rib bone segmentation from the 4D CT images.

© Springer International Publishing Switzerland 2016
Q. Tian et al. (Eds.): MMM 2016, Part I, LNCS 9516, pp. 818–828, 2016.
DOI: 10.1007/978-3-319-27671-7_68

There are many image segmentation algorithms proposed for medical images in literature. For instance, ISODATA algorithm has been successfully applied to cell image segmentation [8,10]. Watershed is also a valuable tool to solve such problems because it can segment touching objects, but over-segmentation would be likely to happen simultaneously [7]. Level set is another important method to address biomedical image segmentation [3,14] by evolving active contours. The advantage of level set is that it allows flexible interaction by initializing the active contour and implicit embedding of bio-image characteristics to lead the active contour to converge on the boundary of object of interest [6]. However, it is still infeasible to directly apply the existing level set technique to our rib segmentation problem due to the prevalent hot noise in the image domain and the intensity variations.

To address the problems mentioned above, we propose an active contour technique based on steerable filter to segment the bleb cells. With the proposed method, satisfactory segmentation results can be obtained. After the segmentation of rib bones, we compare the relative positions of rib bones in sequential frames and assign motion marks. Finally, the end-inhalation and end-exhalation phases are determined based on the motion marks.

The rest of this paper is organized as follows. In Sect. 2, we discuss our approach in details. Section 3 provides the experimental results. Finally, we conclude the paper in Sect. 4.

2   Our Approach

In our approach, we estimate the end-inhalation and end-exhalation for the 4D CT rib cage images based on results of rib bone segmentation. In this section, we first discuss the rib bone segmentation with an active contour model in Sect. 2.1. Before the description of our segmentation method, we first briefly recall the background of active contour models and discuss the limitations of direction application to our data. Then we propose to embed the steerable filter into the active contour model for our task. After that, we present a scheme to estimate the respiration motion state via the segmented rib bones in Sect. 2.2.

2.1   Bone Segmentation

Two sample slice images of the 3D CT rib cage are sown in Fig. 5. It is observed that in the 4D CT rib cage images, there are two features that can be considered to detect the respiration cycle. The first one is the organ area, and the second is the motion of bones. However, the feature of organ area is not as credible. Because there exists a motion along the z-direction perpendicular to the image plane and it may compensate for the area expansion or shrinkage. Although each scan is performed in eight sequential positions in 3D, it only covers 20 mm which is too small to provide helpful 3D information. This is also validated with our experiments.

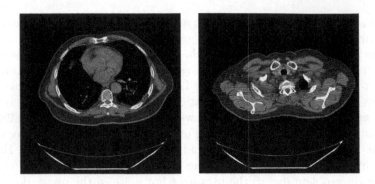

Fig. 1. Two rib cage slice images obtained through the CT scan.

The motion state of the bone lends itself as an effective feature to determine the respiration cycle. Before analyzing the bone motion, it is necessary to first segment the bones. However, the segmentation of bone is a difficult task for a number of reasons. First, hot noise is prevalent in the image domain. Second, intensity variations are present.

To address the problems mentioned above, we propose an active contour technique based on steerable filter to segment the white bone. Before the description of our own method, we first briefly recall the background of active contour models and discuss the limitations of direction application to out data. Then some modification is proposed to make it more adaptive for our task.

Level Set Segmentation. Level set is a versatile tool for analyzing and computing the motion of an active contour, and has been applied successfully in image segmentation [12,13]. Generally, active contour models can be classified into two categories: edge-based models and region-based models, both of which possess their own pros and cons [2,3,5,6].

As mentioned above, our cell image suffers from hot noise and intensity inhomogeneity. Simple thresholding methods, such as Otsu and ISODATA, cannot yield satisfactory results. Besides, the intensity distribution is complex and it is difficult to apply region-based active contour methods [3,6] to obtain satisfactory segmentation result. In this paper, we adopt an edge-based approach based on Laplacian of Gaussian (LoG) image with dominate orientation of local image. Our method can effectively tackle the problems mentioned above.

In [4], R. Kimmle have proposed to use regularized Laplacian zero crossings as optical edge integrators. In this paper, we deal with it in the variational view. Moreover, we extend it by introducing dominant orientation obtained through steerable filter [4]. Our method is formulated as follows.

Let C denote a closed contour and Ω be image domain. In level set methods $C \subset \Omega$ is represented by the zero level set of a Lipschitz function $\phi : \Omega \to \Re$, such that ϕ is negative for point inside C, positive outside C and zero at C.

Define ϕ as a signed distance function, satisfying $|\nabla\phi| = 1$. In an ideal case, assuming the intensity of object in an image is greater than that of the background, the object boundary will lie on local maxima of intensity gradient. Therefore, the negative integration of image intensity gradient along a closed curve in the image domain should achieve minimum when the curve locates in the image with local maximum gradient value. Such a curve is desired to be a zero level set curve, which can be guided by the following energy functional:

$$E(\phi) = \oint_{\Gamma(\phi=0)} \nabla(G_{\sigma_1} * I) \cdot n d\Gamma, \tag{1}$$

where G_{σ_1} is the Gaussian kernel with standard deviation σ_1 and n denotes the outer normal direction of the zero level set. In the formulation of $E(\phi)$, I is convolved with a Gaussian kernel for two considerations: suppress the noise present in the image and extend the capture range. To regularize the zero level set contour of ϕ, we also need the area of the zero level set surface of ϕ and the weighted volume inside the zeros level set surface, which are given respectively by

$$L(\phi) = \iiint_{\Omega} \delta(\phi)|\nabla\phi|dxdy, \tag{2}$$

and

$$V(\phi) = \iiint_{\Omega} g(\nabla I)H(-\phi)dxdy, \tag{3}$$

where δ is the Dirac function, H is the Heaviside function and g is a positive non-increasing function defined as

$$g(\nabla I) = \frac{1}{1 + |\nabla(G_{\delta_2} * I)|^p}, \tag{4}$$

in which G_{δ_2} is the Gaussian kernel with standard deviation δ_2.

Finally, the entire energy functional is defined as follows,

$$F(\phi) = \lambda E(\phi) + \alpha L(\phi) + \gamma V(\phi). \tag{5}$$

By minimizing $F(\phi)$ with respect to ϕ, we deduce the associated Euler-Lagrange equation for ϕ. Through calculus of variations, the Gateaux derivative of the functional $F(\phi)$ can be written as

$$\frac{\partial F}{\partial \phi} = -\delta(\phi)[\lambda\Delta(G_\delta * I) + \alpha\nabla[\frac{\nabla\phi}{|\nabla\phi|}] + vg(\nabla I)]. \tag{6}$$

With steepest descent algorithm, we obtain the evolution equation of ϕ:

$$\frac{\partial \phi}{\partial t} = \delta(\phi)[\lambda\Delta(G_\delta * I) + \alpha\nabla[\frac{\nabla\phi}{|\nabla\phi|}] + vg(\nabla I)], \tag{7}$$

where λ, α and γ are constant for weighting the different evolution force. It should be pointed out that, in [4], an equation similar to Eq. (7) is given, except that the Dirac function $\delta(\phi)$ is replaced with $|\nabla\phi|$, with the implication that the evolution is extended to all level sets of ϕ. However, that extension may suffer from instability on evolution.

On the right of Eq. (7), the first term indicates Laplacian of Gaussian (LoG) image. LoG acts like a doublet, which aligns the curve to be closer to the zero crossings along the edge. When the evolution front is around the zero crossings of LoG image, it will be pulled toward the zero crossings whether it locates either side of the zero crossings, until it stops on the zero crossings. The second term indicates the curvature evolution the level set, with which the surface of the zero level set stays smooth all along the evolution. The third term is a balloon force, which acts to speed up the evolution in those regions where image gradient is low or slow down the evolution when where there is high gradient.

In terms of the challenges mentioned above, direction application of Eq. (7) to our data set suffers from some disadvantages. Considering that our image is polluted with heavy hot noise, it is difficult to select a proper standard variation σ of the Gaussian kernel for the LoG. On the one hand, small σ may make the LoG term sensitive to noise. On the other hand, large σ may over-smooth the image and make the zero-crossing the image intensity filed deviate. The reason is that the Gaussion filter is an isotropic and all directions are treated equally. Thus, the response of the LoG for the noise-polluted image is not great enough to distinguish the image edge. An illustration on this dilemma is shown in Fig. 2(b) and (c). To address this problem, we take account of dominant orientation at local image and replace the LoG term with a direction-adaptive one based on steerable filters, which is explained in the next section.

Steerable Filter Enhancement. Steerable filters make use of quadrature pairs of filters to allow adaptive control over phase and orientation [4]. One important application of Steerable filters is to detect the local orientation of images. For a second derivative of a Gaussian rotated along to an angle θ, it can be written as

$$G_2^\theta = k_a(\theta)G_{2a} + k_b(\theta)G_{2b} + k_c(\theta)G_{2c}, \qquad (8)$$

where

$$G_{2a} = \frac{0.9213(2x^2 - 1)}{\sqrt{2}\sigma} exp(\frac{x^2 + y^2}{2\sigma^2}),$$

$$G_{2b} = \frac{0.9213xy}{\sqrt{2}\sigma} exp(\frac{x^2 + y^2}{2\sigma^2}),$$

$$G_{2c} = \frac{0.9213(2y^2 - 1)}{\sqrt{2}\sigma} exp(\frac{x^2 + y^2}{2\sigma^2}),$$

$$k_a(\theta) = cos(\theta)^2,$$

$$k_b(\theta) = -2cos(\theta)sin(\theta),$$

$$k_c(\theta) = sin(\theta)^2.$$

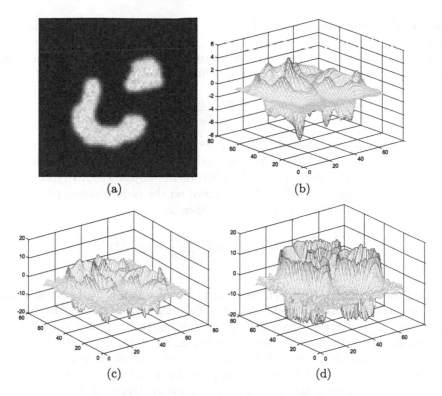

Fig. 2. Comparison between LoG image and G_2^θ convolved image under dominant orientation. (a) A noisy image; (b) the mesh of LoG with standard variation 3; (c) the mesh of LoG image with standard variation 1.5; (d) the mesh of G_2^θ image with standard variation 1.5.

Convolving both sides of Eq. (8) with image I, we can obtain that

$$G_2^\theta * I = k_a(\theta)G_{2a} * I + k_b(\theta)G_{2b} * I + k_c(\theta)G_{2c} * I. \tag{9}$$

If the local image dominant orientation is known, we can obtain the greatest response of the second derivative of Gaussian steerable filters. Now, the problem is left to how to find the proper angle. In [4], Freeman describes an approach to address it as follows. Define oriented energy $E(\theta)$ as the orientation strength along a particular direction θ by the squared output of a quadrature pair of band pass filters steered to the angle θ. With the nth derivative of a Gaussian and its Hilbert transform as the band pass filter, it can be obtained that

$$E_n(\theta) = [G_n^\theta]^2 + [H_n^\theta]^2. \tag{10}$$

Rewrite G_n^θ and H_n^θ as the sum of basis filter outputs, $E_n(\theta)$ simplifies to a Fourier series in angle, where only even frequencies are present due to the squaring operation

$$E_n(\theta) = C_1 + C_2 cos(2\theta) + C_3 sin(2\theta) + O(\theta), \tag{11}$$

where $O(\theta)$ is higher order terms. For the details about C_1, C_2 and C_3, we refer readers to [4].

With the lowest frequency term to approximate the strength S of the dominant orientation, the direction θ_d can be obtained as

$$\theta_d = \frac{atan(C_2, C_3)}{2}. \tag{12}$$

Thus, the image convolved with the steerable filter G_2^θ along the dominant orientation θ_d can be obtained. An illustration of the impact of G_2^θ is shown in Fig. 1(d). It can be observed that the edge of the object is greatly enhanced, which will facilitate the level set to converge on the object boundary. Finally, the level set evolution Eq. (7) can be rewritten as

$$\frac{\partial \phi}{\partial t} = \delta(\phi)[\lambda G_2^{\theta_d} * I + \alpha \nabla (\frac{\nabla \phi}{|\nabla \phi|} + vg(\nabla I))]. \tag{13}$$

Segmentation Procedure. Through the CT scan, different parts in the human body are reflected as different intensity. To accurately segment the bone, a threshold T_{low} is defined. Pixels with intensities less than T_{low} are considered as background. The remains are candidates belonging to the bone and are left to the active contour methods to distinguish. T_{low} is determined as follows.

A typical histogram of a rib cage image is illustrated in Fig. 3(a). To remove the noise of the histogram, a de-noising algorithm based on wavelet can be adopted and the de-noised result is shown in Fig. 3(b). Since we are interested in the bone with greater intensity, our attention is only focused on the two right bumps in the histogram. Denote the most right peak as T_p and the most right trough as T_t, as illustrated with green circle and red circle respectively. Then T_{low} is defined as $T_{low} = \frac{T_p + T_t}{2}$. With T_{low}, the front of the active contours can be prevented from leaking through the weak boundary of the bone.

Finally, the main steps for segmentation are summarized as follows (Fig. 4).

- Step 1. Calculate the second derivative of a Gaussian image rotated along θ_d in Eq. (12) to the dominate orientation.
- Step 2. Calculate T_{low}, and initialize the contour for level set function ϕ.
- Step 3. Evolve the level set function according to Eq. (13).
- Step 4. Re-initialize ϕ locally to be the signed distance function for the evolution curve.
- Step 5. Check convergence. If the active contour no longer mover or a maximum iteration number is reached, stop evolution. Otherwise, go to step 3.

2.2   Motion State Estimation

The respiration is a periodic motion consisting of a series of phases. In fact, if we can determine the end-inhalation phase and end-exhalation phase, a whole respiration course can be obtained.

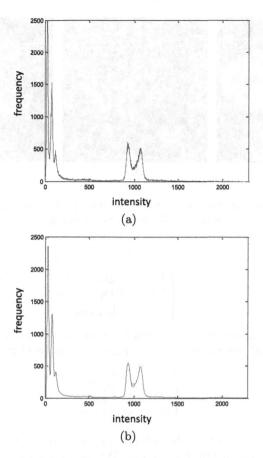

Fig. 3. Intensity histogram of rig cage image slice (a) before and (b) after wavelet de-noising (Color figure online).

Generally, in each CT image, the white bone may include vertebra, ribs, sternum and clavicle. The ribs and vertebra exist in all the CT images, while the other two types of white bone appear only in some of the scans. The vertebra nearly stays still in all the scans. During the course of respiration, the ribs move regularly. To be detailed, ribs move up towards the vertebra on exhalation and outwards the vertebra on inhalation. Consequently, the ribs are good alternative to be used to check the respiration motion.

Considering that the motion between each two sequential images is very small, the decision of motion state is made between each image and the one before its previous one. Not all the ribs are adopted. Instead, we only select those with a hole inside.

For each rib, the corresponding one in the one before last time point is found first. Then the position of their centroids are compared. If the difference D is greater than a pre-defined threshold T, a motion mark s_b for the state of the bone is assigned.

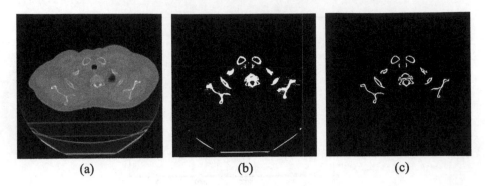

Fig. 4. Illustration of the bone segmentation results. (a) Original image slice; (b) Segmentation result with the Otsu's algorithm; (c) Segmentation results with our proposed approach.

$$s_b(i) = \begin{cases} 1, & \text{if } D(i) > T \\ -1, & \text{if } D(i) < -T \\ 0, & \text{otherwise} \end{cases} \tag{14}$$

For each slice image at a time point, there are several valid ribs and the overall motion mark for the state of the image s_I can be obtained.

$$s_I(j) = \begin{cases} 1, & \text{if } \sum_i s_b(i) > 0 \\ -1, & \text{if } \sum_i s_b(i) < 0 \\ 0, & \text{otherwise} \end{cases} \tag{15}$$

For each time point, eight slice images are scanned at each sequential time stamp along the human trunk direction. We aggregate the states of all those images to estimate the state of each time stamp as follows.

$$s_t(k) = \begin{cases} 1, & \text{if } \sum_j s_I(j) > 0 \\ -1, & \text{if } \sum_j s_I(j) < 0 \\ 0, & \text{otherwise} \end{cases} \tag{16}$$

Once the states of all time stamps are calculated, we regard the time stamps with peak values as the end-inhalation phase and end-exhalation phases. Figure 5 illustrates the result of a scan containing 46 3D images. From the curve, it is easy to locate the end-inhalation phase and end-exhalation phases, which correspond to the local peak and trough, respectively.

3   Experiments

We validate our method with 32 scans of 4D-CT files involving 6,024 image slices. Of them, there are three scans each containing 46 time points, 13 scans each containing 18 time points, and 16 scans each containing 23 time points. For

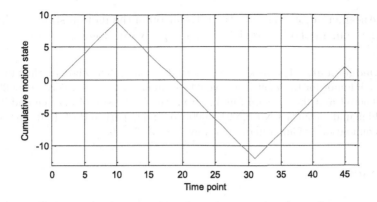

Fig. 5. Cumulative motion state for 46 3D images in series. The horizontal axis denotes the time stamp while the vertical axis denotes s_t.

each time stamp, each CT scan captures data in eight sequential positions along the human trunk direction and covers 20 mm physically on the patient chest.

To evaluate our approach, for each scan, end-inhalation phase and end-exhalation phase are estimated and compared with ground truth. To quantitatively evaluate the performance, we compute the absolute difference between end-inhalation/end-exhalation phases with the ground truth values. The results are reported in Table 1. From Table 1, it is observed that the maximum error for both end-inhalation and end-exhalation is two time points (frames). The end-inhalation is accurately identified for 62.50 % scans, with 25.00 % error rate for 1-frame absolute error and 12.5 % for 2-frame error. On the other hand, the end-exhalation estimation is accurate for 68.78 % frames, with 21.88 % for 1-frame absolute error and 9.38 % for 2-frame error.

Table 1. The error rate of estimation on end-inhalation/end-exhalation phases with the 32 scans of 4D-CT rib cage image data.

Absolute error (frames)	0 (true)	1	2	≥ 3
Error rate (scan number) on end-inhalation	62.50 %(20)	25.00 %(8)	12.50 %(4)	0
Error rate (scan number) on end-exhalation	68.78 %(22)	21.88 %(7)	9.38 %(3)	0

4   Conclusions

In this paper, we propose a novel approach to determine the respiratory phases. The motion bone is selected to locate the end-inhalation and end-exhalation phase. Bones are segmented through a level set method based on steerable filter.

Accurate segmentation facilitate the estimation of the motion state of rib bones. We validate our method on real 4D-CT images with promising result.

Acknowledgement. This work was supported in part to Dr. Zhou by the Fundamental Research Funds for the Central Universities under contract No. WK2100060014 and WK2100060011 and the start-up funding from the University of Science and Technology of China under contract No. KY2100000036, and in part to Prof. Li by the National Science Foundation of China under contract No. 61272316.

References

1. Abdelnour, A.F., Nehmeh, S.A., Pan, T., Humm, J.L., Vernon, P., Schöder, H., Rosenzweig, K.E., Mageras, G.S., Yorke, E., Larson, S.M., et al.: Phase and amplitude binning for 4D-CT imaging. Phys. Med. Biol. **52**(12), 3515 (2007)
2. Caselles, V., Kimmel, R., Sapiro, G.: Geodesic active contours. Int. J. Comput. Vis. **22**(1), 61–79 (1997)
3. Chan, T.F., Vese, L.A.: Active contours without edges. IEEE Trans. Image Process. **10**(2), 266–277 (2001)
4. Freeman, W.T., Adelson, E.H.: The design and use of steerable filters. IEEE Trans. Pattern Anal. Mach. Intell. **13**(9), 891–906 (1991)
5. Kimmel, R., Bruckstein, A.M.: Regularized laplacian zero crossings as optimal edge integrators. Int. J. Comput. Vis. **53**(3), 225–243 (2003)
6. Li, C., Kao, C.-Y., Gore, J.C., Ding, Z.: Implicit active contours driven by local binary fitting energy. In: 2007 IEEE Conference on Computer Vision and Pattern Recognition, CVPR 2007, pp. 1–7. IEEE (2007)
7. Lim, F.Y., Chiam, K.-H., Mahadevan, L.: The size, shape, and dynamics of cellular blebs. EPL (Europhys. Lett.) **100**(2), 28004 (2012)
8. Norman, L.L., Brugés, J., Sengupta, K., Sens, P., Aranda-Espinoza, H.: Cell blebbing and membrane area homeostasis in spreading and retracting cells. Biophys. J. **99**(6), 1726–1733 (2010)
9. Shirato, H., Seppenwoolde, Y., Kitamura, K., Onimura, R., Shimizu, S.: Intrafractional tumor motion: lung and liver. Semin. Radiat. Oncol. **14**, 10–18 (2004)
10. Spangler, E.J., Harvey, C.W., Revalee, J.D., Sunil Kumar, P.B., Laradji, M.: Computer simulation of cytoskeleton-induced blebbing in lipid membranes. Phys. Rev. E **84**(5), 051906 (2011)
11. Werner, R., Ehrhardt, J., Schmidt, R., Handels, H.: Patient-specific finite element modeling of respiratory lung motion using 4D CT image data. Med. Phys. **36**(5), 1500–1511 (2009)
12. Yang, X., Gao, X., Tao, D., Li, X., Li, J.: An efficient MRF embedded level set method for image segmentation. IEEE Trans. Image Process. **24**(1), 9–21 (2015)
13. Zhang, K., Zhang, L., Song, H., Zhou, W.: Active contours with selective local or global segmentation: a new formulation and level set method. Image Vis. Comput. **28**(4), 668–676 (2010)
14. Zhou, W., Li, H., Zhou, X.: 3D dendrite reconstruction and spine identification. In: Metaxas, D., Axel, L., Fichtinger, G., Székely, G. (eds.) MICCAI 2008, Part II. LNCS, vol. 5242, pp. 18–26. Springer, Heidelberg (2008)

Location-Aware Image Classification

Xinggang Wang[1]([⊠]), Xin Yang[1], Wenyu Liu[1], Chen Duan[2],
and Longin Jan Latecki[3]

[1] School of EIC, Huazhong University of Science and Technology, Wuhan, China
{xgwang,xinyang2014,liuwy}@hust.edu.cn
[2] Wuhan Second Ship Design and Research Institute, Wuhan, China
duanchen@hust.edu.cn
[3] Department of CIS, Temple University, Philadelphia, PA, USA
latecki@temple.edu

Abstract. Currently, the most popular image classification methods are based on global image representations. They face an obvious contradiction between the uncertainty of object position and the global image representation. In this paper, we propose a novel location-aware image classification framework to address this problem. In our framework, an image is classified based on local image representation, and the classifier is learned using an iterative multi-instance learning with a latent SVM, *i.e.*, we infer object location using latent SVM to improve image classification. Our method is very efficient and outperforms the popular spatial pyramid matching (SPM) method and the Region Based Latent SVM (RBLSVM) method [1] on the challenging PASCAL VOC dataset.

Keywords: Image classification · Latent SVM · Spatial pyramid matching

1 Introduction

We consider the problem of classifying images based on the presence of certain objects in the image. For training images, we only know the image label but do not know the locations of objects. For testing images, both image label and object locations are unknown, and therefore, we need to infer them. This problem is very challenging, not only because of the large appearance variation of objects within the same class, but also due to the variation of object locations in images. In this paper, we aim to improve image classification performance by inferring object locations. The latent SVM is used to iteratively infer the location of objects and produce a more discriminatively classifier, as illustrated in Fig. 1.

To explain our idea we begin with an review of some global image representation methods. The *bag-of-features* (BoF) method [2] representing an image as an unstructured collection of local features has been widely used. The BoF methods discard all spatial layout of local features within the image. Obviously the spatial layout information of local features within and around the objects is very important for recognition. To overcome this problem, the most efficient

© Springer International Publishing Switzerland 2016
Q. Tian et al. (Eds.): MMM 2016, Part I, LNCS 9516, pp. 829–841, 2016.
DOI: 10.1007/978-3-319-27671-7_69

Fig. 1. Classical image classifiers use global image representation. Instead of training a classifier based on global image representation, we train a more discriminative classifier based on local image representation. As illustrated by the cars, the key challenge is the fact that locations of the target object are unknown and vary significantly. We address this problem by iterative multi-instance learning in the framework of the latent SVM. We utilize a local feature context (LFC) to represent local image regions. The locations of the cars automatically learned in our framework are shown in the second row as log-polar coordinates of the LFC descriptor.

and effective extension of BoF is the *Spatial Pyramid Matching* (SPM) method proposed by Lazebnik *et al* in [3]. Motivated by [4], the SPM method partitions an image into increasingly finer spatial sub-regions, and computes histograms of local features for each sub-region. Typically, $2^l \times 2^l$ sub-regions for $l = 0, 1, 2$ are used. The main idea of SPM is to represent an image in different levels of resolution, and then let SVM decide which levels are useful to recognize the images. Many state-of-the-art image classification systems [5–8] have used SPM, which clearly demonstrates its usefulness.

However, SPM, like BoF and many other image classification approaches, are still based on a global image representation, because it not only represents the features of the target objects but also of other parts of images. In contrast, we propose a method that classifies images based on local image representation. In other words, our approach detects object location and classifies a given image based on the detected location.

To deal with the object locations variation issue, the proposed approach requires two components: one is a rich object-level local image representation method and the other is a good multiple instance learning [9] algorithm. Rich object-level local image representation should capture spatial layout information within and around the object. In this paper we modified the image representation method proposed in [10] as a local object descriptor. Since we only know the image label in the training phase, in each image we extract many candidate object-level features. Only one or few of these are located on objects of interest. This learning problem is a typical example of multiple instance learning, and we choose the excellent latent SVM algorithm proposed by Felzenszwalb *et al* [11] to solve it. We provide more details in the following paragraphs.

An image representation called *feature context* (FC) is introduced in [10]. FC captures the spatial layout information of local image features by dividing the image into different regions using a log-polar coordinate system like Shape Context [12]. Given a reference point located at the center of the log-polar coordinate system, a FC histogram can be computed. In [10] all FC histograms at different image locations are concatenated to a long vector for image classification, while in this paper, we treat each FC histogram as an individual instance, so we call the representation in this paper *local feature context* (LFC). There are at least three advantages of using LFC: First, LFC is a location-aware image representation, so given a reference point, LFC is focused on describing the region around the reference point. This contrasts the global representation of SPM, which is not location-aware. Second, LFC inherits from FC the rich spatial layout information of local descriptors. Third, LFC does not require segmenting the image into regions containing objects. This kind of segmentation is even harder than image classification. Instead, we compute LFC at many locations in a given image. Then the latent SVM framework allows us to learn the true locations of the objects.

The *Latent SVM* (LSVM) [11] algorithm is a kind of multiple instance SVM [13]. It has been successfully used for learning a deformable part model when given object bounding boxes. In our framework, LSVM is used in a different way, since our latent variables describe the position of the object in the image as opposed to a relative locations of the parts with respect to a center of the bounding box. We use the latent position variable to iteratively identify the location of the target object, and consequently, learn its local representation. Our approach also differs in the feature descriptor. In [11] each part of the model is a HOG filter and locations of HOG filters are latent variables. In the training process, the LSVM algorithm is used to infer these latent variables. This approach has achieved the *state-of-the-art* performance on the PASCAL VOC detection task. In addition, multiple instance learning is also applied for effective image annotation [14].

In our case, we have image labels but do not know bounding boxes of objects. Therefore, we need to discover the object locations if we want to represent them with correct local features. HOG filter is not suitable to be our local image representation for two reasons: First, HOG filter is too precise in that the value of HOG filter will change a lot if the location changes a little. Second, HOG filter is very sensitive to the size of the region. Since the values of LFC features are significantly more stable under changes of location and scale, we selected LFC as our local feature. We choose LSVM as our multiple instance learning algorithm, because it can iteratively infer the object locations and produce more discriminative local image classifiers. Once LSVM converges, we not only obtain the final classifier, but at the same time, we obtain a rough object location in each training image. In the testing phase, the LSVM classifier detects the most likely object location and classifies the image based on this location. We also observe that the LSVM learning process is efficient, since for each training image only one instance is selected (The term "instance" means a location-aware feature in our framework). The large number of negative instances can be handled using the bootstrapping method or the "data mining" method proposed in [11].

We test our method on the challenging PASCAL VOC 2007 dataset [15], and both image classification and object localization performance are reported.

The experimental results show that our method can outperform SPM and RBLSVM methods on the image classification task and can get a very promising object localization accuracy.

2   Related Work

Image classification is a hot topic in computer vision and has been widely applied in many applications [16–19]. In this section we describe related works on topics such as common object discovery, image classification using multiple instance learning algorithms, and combining object detection with image classification.

Weakly supervised learning (WSL) for object detection has recently received growing attentions. Various methods have been proposed to address this difficult problem, such as [14, 20–24]. The most related one to our method is [25], where the objectness measure from [26] is used to extract lots of boxes in images that form an object candidate set. The boxes are described using texture and shape descriptors. Then sMIL [27] is used to autonomously learn an object detector. This method can obtain a reasonable object detection performance on the challenging PASCAL VOC 2007 dataset [15].

Latent SVM, which belongs to multiple instance learning methods, has been proven very powerful in [11] in many applications. Some recent approaches [1, 28] use it for recognition when only image labels are given. [28] uses a deformable part model in [11] learned by latent SVM to capture visual structure of images, and can obtain *state-of-the-art* weakly supervised object localization performance. However, the image classification performance of [28] has not been studied yet. In [1], images are divided into many non-overlapping regions. Latent SVM is used to search a combination of regions that contain an object. This method can train a classifier for image classification that is less affected by the background cluster. Since in our method, each object is associated with a local image descriptor, we do not need to solve the combinatorial optimization problem, so time complexity of our method is much lower. In addition, the method in [1] only compares to the BoF method, while we also compare to the more powerful SPM method. Moreover, the result of our method is significantly better than the method in [1].

Another very related topic is combining object detection with image classification [29]. The systems in [6, 30] show that the information from detected objects can significantly boost image classification performance. Both methods [6, 30] and our method are based on an obvious observation that knowing object locations can lead to more discriminative image classifiers. In contrast to the methods in [6, 30], our method do not need object bounding boxes for training. Hence, our method stays in the classical framework of image classification, where only the class labels are known for the training images.

3   Approach

Our approach is composed of two main parts. The first part is the location-aware image representation, local feature context, described in Sect. 3.1. The second part is the latent SVM algorithm formulated in Sect. 3.2.

3.1   Local Feature Context

Local feature context image representation is based on the descriptor proposed in [10]. To calculate local feature context in a given image I, we extract a set of local image descriptors, such as densely computed SIFT [31] features, and we denote locations of features as a set $L = \{l_1, \ldots, l_L\}$. Each feature located at $l \in L$ is encoded as vector $C(l) = (w_1^l, \ldots, w_K^l)$, where K is the size of codebook generated using the k-means algorithm. We use the coding method introduced in [5]. Given a reference point z in image I, following the way of calculating the shape context descriptor [12], the area around z is divided into regions $Region_r^z$ in log polar coordinate system for $r = 1, \ldots, R$. The **feature context** around point z is defined as a matrix

$$fc(z, r, i) = \mathcal{M}\{w_i^l \mid (l - z) \in \Delta_s(Region_r^z)\}, \tag{1}$$

where $i = 1, \ldots, K$ indexes the codewords and $\mathcal{M}$ is a pooling function, which extracts the most relevant codewords present in the region $Region_r^p$. The function $\mathcal{M}$ can be *max*, *sum*, *mean* or some other functions. We use the *max* pooling function as $\mathcal{M}$ in our experimental results, since the max pooling method is more robust to local transformation than mean statistics in a histogram [32]. $\Delta_s(Region_r^z)$ denotes a neighborhood of region $Region_r^z$ of radius s. This fact is useful to compensate for spatial uncertainty of local descriptors, because the local descriptors near the boundaries of regions may belong to multiple regions. The LFC of image I at position z is a tensor of dimension $R \times K$ given by

$$lfc(I, z) = (fc(z, r, i))_{r,i} \in \mathbb{R}^{R \times K}. \tag{2}$$

As customary, $lfc(I, z)$ is vectorized. We also compute a simple bag of words histogram of codewords with max pooling in the whole image I and append it to each vector. Hence each LFC vector also carries some global information about the image I, and the weights of local information and global information are automatically determined by training SVM. Finally, feature vector $lfc(I, z)$ is normalized to one in the l^2 norm.

There are a few free parameters in LFC. The first one is the reference point z. Usually we sample reference $n \times n$ points in the image, so there are n^2 reference points for LFC. The second one is the maximal radius of LFC. We set it to λ times diagonal length of the image, where λ ranges in $[1/4, 1/2]$.

LFC is a robust, location-aware image representation method, since it is stable to substantial reference point drift. The robustness is due to it is statistical model (it is based on *bag-of-feature*) and the spatial uncertainty design. Benefiting from the robustness, we can set fewer reference points when mining object locations, which makes our method more efficient. To the contrary, other image representation methods, such as HOG [33] and GIST [34], are sensitive to location drift. Other location-aware image representations, *eg.* the *object-graph* descriptor [21], can also be used in our framework. It is composed of many superpixels described by local features. Compared to LFC, the *object-graph* method is more complex and its parameter expressed as the number of neighboring segments is hard to set.

3.2   Latent SVM

The original formulation of latent SVM is given by Felzenszwalb *et al* in [11] for learning deformable part models. In this section, we present a version of latent SVM for location-aware image classification.

In the training process of image classification, given a training image I, we learn a classifier that scores image I with a function of the form

$$f_\beta(I) = \max_{z \in Z(I)} \beta \cdot lfc(I, z). \tag{3}$$

Here, β is a vector of SVM model parameters, and z the latent variable that represents locations (center points) of LFC descriptors. $Z(I)$ is the set of possible locations of LFC descriptors in image I.

In analogy to classical SVMs, we train β from labeled images $D = (< I_1, y_1 >$ $, \cdots, < I_n, y_n >)$, where $y_i \in \{-1, 1\}$, by minimizing the objective function,

$$L_D(\beta) = \frac{1}{2}\|\beta\|^2 + C\sum_{i=1}^{n} max(0, 1 - y_i f_\beta(I_i))^2, \tag{4}$$

where $max(0, 1 - y_i f_\beta(I_i))^2$ is the standard l^2 hinge loss and the constant C controls the relative weight of the regularization term.

Let Z_p specify a latent value for each positive example in a training set D. We define an auxiliary objective function $L_D(\beta, Z_p) = L_{D(Z_p)}(\beta)$, where $D(Z_p)$ is derived from D by restricting the latent values for positive examples according to Z_p. That is, for positive examples we set $Z(I_i) = \{z_i\}$ where z_i is the latent value specified for I_i by Z_p. Note that

$$L_D(\beta) = \min_{Z_p} L_D(\beta, Z_p). \tag{5}$$

In particular, $L_D(\beta) \leq L_D(\beta, Z_p)$. The auxiliary objective function bounds the LSVM objective. This justifies training a latent SVM by minimizing $L_D(\beta, Z_p)$.

Then we use a "coordinate descent" approach proposed in [11] to minimize $L_D(\beta, Z_p)$:

– *Relabel positive examples:* Optimize $L_D(\beta, Z_p)$ over Z_p by selecting the highest-scoring latent value for each positive example

$$z_i = \arg \max_{z \in Z(I_i)} \beta \cdot lfc(I_i, z).$$

– *Optimize beta:* Optimize $L_D(\beta, Z_p)$ over β by solving the convex optimization problem defined by $L_{D(Z_p)}(\beta)$. We solve this problem using a fast SVM solver: LibLinear [35].

Both steps always improve or maintain the value of $L_D(\beta, Z_p)$. After convergence we obtain a very discriminative SVM classifier as well as positions most likely to contain objects in each training image.

Initialization of β is very important when training latent SVM. If β is not correctly initialized, $L_D(\beta, Z_p)$ is likely to fall into a local optimum. We initialize

β by training a linear SVM using all LFCs whose reference points are in the centers of images. Therefore, we assume that there are many objects are located near to the centers of images. However, because LFC is robust to substantial location drift, small offsets of objects from centers of images do not lead bad initialization.

We do not fix the latent value for negative examples, so all LFCs in negative training images are used for training. To handle the very large number of negative examples, we use the bootstrapping method: first train a model with an initial subset of negative examples, and then collect hard negative examples that are incorrectly classified by this initial model to form a set of hard negatives. A new model is trained with the hard negative examples and the process is repeated a few times.

In the testing phase, inference is simple, as we score a testing image I_t by

$$score(I_t) = \max_{z \in Z(I_t)} \beta \cdot lfc(I_t, z). \tag{6}$$

At the same time, the location of detected object z_t^* in I_t is also obtained as

$$z_t^* = \arg \max_{z \in Z(I_t)} \beta \cdot lfc(I_t, z). \tag{7}$$

4   Experiments

We first describe our experimental setup in Sect. 4.1, then give our experimental results in Sect. 4.2.

4.1   Experimental Setup

Dateset: We use the standard benchmark PASCAL VOC 2007 dataset [15] to test our algorithm, and compare our algorithm to other methods on this dataset. The PASCAL VOC 2007 dataset consists of a total of 9,963 images divided in 20 categories. It is split into fixed training/validation (5,011 images) and testing (4,952 images) sets. In all of our experiments, we use the training+validation sets for training, and the testing set for testing. The main challenge of this dataset is the significant intra-class variation due to changes in scale, texture, position of objects, background clutter, partial visibility, and inclusion of deformable objects such as people and animals.

Evaluation Metrics: To evaluate the performance of image classification we use *average precision* (AP) measure, which is defined as the average of precisions computed at the point of each of the relevant documents in the sequence of images ranked by decreasing classification score. The mean average precision (mAP) is the mean of AP computed over all 20 classes. We also evaluate our weakly-supervised object localization performance. As our method outputs the center point but not the bounding box of the localized object, we use the *bounding box hit rate* (BBHR) measure [36]. For each image I_t, we obtain z_t^* from

Eq. 7. If z_t^* is inside of a ground truth bounding box for a given class, we declare it as a "hit". Hence the definition of BBHR is the number of hits divided by the total number of object instances in the positive images.

Compared Methods: We compare our method to the global image representation method BoF and SPM in [5], and the *Region Based Latent SVM* (RBLSVM) method in [1]. When comparing to BoF and SPM, we use the same local descriptor, the same codebook, and the same coding method to encode local descriptors, and the same SVM options. Details are given in the following paragraphs. As the RBLSVM method uses a different image descriptor and a different coding method, we do not directly compare mAP to it. We compare the improvement ratios over BoF scores of the two methods. To see the performance benefit from latent SVM, we also report the performance of *center feature context* (CFC), CFC uses only one feature context, of which the reference point is in the center of image to represent the whole image, which is fed to linear SVM for image classification. Notice that in our approach, if the iteration of latent SVM is 0, the LFC with latent SVM method is equal to CFC with linear SVM. We do not compare our method with the image classification method in [10]. Since the method in [10] is aimed at giving a precise global image presentation, it cannot deal with the problem caused by object position variation.

Implementation Details: Referring to the image classification system in [5], we use a single feature, the dense SIFT feature implemented by [37], as the local descriptor. Dense SIFT features are extracted from patches densely located every 4 pixels on the image, under three scales, 16×16, 24×24, and 32×32, respectively. About 1 million dense SIFT features are randomly selected from training images to generate a codebook with 1000 codewords using k-means algorithm. All dense SIFT features are encoded using the *Locality-constrained Linear Coding* (LLC) method in [5]. For LLC, the number of nearest neighbors is set to 5. As a pooling method for BoF, SPM and FC, we use max pooling. We noticed that max pooling works better than average pooling with LLC in the experiments. The best setting for SPM on PASCAL VOC dataset is dividing the image in 1×1, 1×3 (three horizontal stripes), and 2×2 (four quadrants) grids, for a total of 8 regions, as suggested in [38].

When computing LFC, n is 6, we sample 6×6 points in image as reference points, for each feature context, local region is divided into 2×6 bins shown in Fig. 1 and $\lambda = 1/3$, spatial uncertainty parameter s is set to 0.05λ times diagonal length of image. We use LibLinear [35] software as the classifier, and ''-s 2 -c 1'' option is used when training a SVM classifier. The training process of latent SVM will stop if the declined value of $L_D(\beta)$ is smaller than 0.5 or the maximal number of iterations (we set it to 10) is reached.

4.2 Experimental Results

There are four methods we have evaluated, BoF with linear SVM, SPM with Linear SVM, CFC with linear SVM, and the proposed method LFC with latent

SVM. Respectively, we call them BoF-SVM, SPM-SVM, CFC-SVM and LFC-LSVM for convenience. The per-category AP and mAP of BoF-SVM, SPM-SVM, CFC-SVM and LFC-LSVM is reported in Table 1.

First of all, BoF-SVM works much worse than the other three methods. This is due to the fact that BoF model discards all spatial layout information of local descriptors. Our LFC-LSVM method gets the best results, outperforming BoF-SVM and CFC-SVM on all 20 categories, and outperforming SPM-SVM on 15 out of 20 categories. LFC-LSVM fails to improve SPM-SVM on the `bottle`, `cow`, `dog`, `plant` and `tv` categories. This is probably due to the fact that objects in the five categories are relatively small, and hence they are hard to detect. Overall, LFC-LSVM improves SPM-SVM by 2.8 % mAP, from 43.7 % to 46.5 %. Notice that the SPM-SVM method in our experiments is exactly the same method as described in [5]. However, in our experiments the codebook size is 1,000, while in [5] the codebook size is larger than 10,000. As compared to image center based classification of CFC-SVM, the proposed latent SVM improves performance on all 20 categories, and mAP increases by 3.9 %, from 42.6 % to 46.5 %, it clearly shows that mining object location using latent SVM can boost image classification performance.

As reported in [1], using their image descriptor and coding method, mAP of BoF method is 40.2 %, mAP of RBLSVM method is 42.3 %. The improvement ratio is hence 0.052. We list all improvement ratios in Table 2. Our improvement ratio of mAP is 0.356. It is obvious that the performance of our method is significantly better than RBLSVM.

Table 1. Comparison of AP (in %) on 20 classes and mAP (in %) for the four methods, BoF with linear SVM, SPM with linear SVM, CFC with linear SVM, and the proposed LFC with latent SVM on PASCAL VOC 2007 dataset.

Object class	aero	bicyc	bird	boat	bottle	bus	car	cat	chair	cow	
BoF-SVM	59.4	41.9	25.3	44.5	10.7	26.0	60.2	29.5	35.2	19.3	
SPM-SVM	66.3	50.5	30.6	52.1	**14.4**	39.5	68.5	36.8	44.0	**27.8**	
CFC-SVM	67.4	50.0	28.7	47.3	11.3	39.4	68.4	40.3	42.4	25.5	
LFC-LSVM	**69.3**	**55.2**	**30.8**	**52.7**	13.8	**47.4**	**71.5**	**43.9**	**46.6**	27.1	

Object class	table	dog	horse	moto	person	plant	sheep	sofa	train	tv	mAP
BoF-SVM	13.4	30.9	46.1	31.3	67.7	11.0	20.6	27.3	46.6	38.4	34.3
SPM-SVM	30.5	**34.1**	68.1	47.3	75.5	**13.4**	28.0	43.8	59.9	**42.1**	43.7
CFC-SVM	28.3	31.1	68.4	46.2	74.8	11.9	31.5	40.7	62.1	35.7	42.6
LFC-LSVM	**38.0**	32.2	**73.7**	**52.3**	**78.0**	12.1	**34.7**	**46.7**	**64.2**	40.5	**46.5**

Table 2. The improvement ratio of RBLSVM, SPM-SVM, CFC-SVM and LFC-LSVM methods against BoF-SVM method on mAP.

RBLSVM	SPM-SVM	CFC-SVM	LFC-LSVM
0.052	0.274	0.242	**0.356**

Table 3. The bounding box hit rates (BBHR) (in %) on training set (train+val) and testing set (test) on PASCAL VOC 2007 dataset.

Object class	aero	bicyc	bird	boat	bottle	bus	car	cat	chair	cow	
train+val	82.8	56.4	70.6	53.6	20.5	69.4	61.3	71.2	28.3	70.9	
test	80.0	61.2	67.1	54.0	20.0	75.4	63.1	74.4	25.7	66.1	
Object class	table	dog	horse	moto	person	plant	sheep	sofa	train	tv	average
train+val	43.0	75.5	85.4	75.1	67.2	29.0	55.2	66.4	88.1	32.0	60.1
test	44.9	76.7	89.2	78.1	65.5	30.7	63.3	63.7	79.9	36.9	60.8

In order to evaluate the object localization accuracy, we list BBHR of our method in Table 3. The average BBHR of our method on training and testing sets are about 60 %. We obtain very good BBHR of above 80 % on aeroplane, horse, train categories. According to our BBHR criteria, the chance of randomly hitting the true object location has average BBHR of 31.9 % and 32.0 % for training and testing set respectively, here randomly hitting means randomly choosing a position in image. Comparing to randomly hitting, average BBHR is

Fig. 2. Some objects are very small, heavy occluded, or on a very cluttered background in PASCAL VOC 2007 dataset. Objects of interest in the images of first row: sofa, train, sheep; second row: cat, person, car.

approximately doubled by our method. Some of the objects are hard to detect, in particular, if the objects are very small, on very cluttered background, or heavy occluded, *eg.* images in Fig. 2. Object localization results for all 20 categories are illustrated in Fig. 3, where detected objects are marked with red log-polar coordinates of their LFC descriptors. We observe that many objects are correctly localized, although their positions vary a lot, *eg.* the boats, motorcycles, airplanes, cars, and sheep. When multiple instances of the same category are present, we localize only one of them, since we focus here on image classification. However, in our framework it would also be possible to identify the other instances.

Time Complexity: As we only set 36 reference points per image to extract LFC, and we use a very efficient linear SVM software (LibLinear), testing speed is very fast. It takes about 20 h to train 20 latent SVM models using a computer with 3.40 GHz CPU and 16 GB memory, and most of the 20 h is consumed on loading features from hard disk to memory. So if the training is performed on a computer with larger memory, training time should decrease significantly. Our time complexity is higher than SPM-SVM method, but much lower than the RBLSVM method.

Fig. 3. The object localization results on PASCAL VOC 2007 dataset. The red log-polar coordinates show the positions of detected objects (Color figure online).

5   Conclusions

In this paper we propose a location-aware image classification framework. The whole framework is composed of a robust local image representation and a latent SVM algorithm. The learning process allows us to iteratively optimize object location and train more discriminative local image classifiers. The testing procedure is extremely fast. The proposed framework clearly demonstrates that by localizing objects in test images, the image classification accuracy can be significantly improved. One of the key advantages of the proposed method is the fact that it can be trained in the standard image classification setting, where images are labeled but the location of the target objects in images is unknown. This is possible because we utilize the latent SVM framework to infer the object locations as part of the learning phase. Therefore, the proposed framework can be used for learning image classifiers from Internet images, where training images are collected using keyword-based search engines.

Acknowledgments. This work was primarily supported by National Natural Science Foundation of China (NSFC) (No. 61503145). This material is also based upon work supported by the NSF under Grants No. IIS-1302164 and OIA-1027897.

References

1. Yakhnenko, O., Verbeek, J., Schmid, C.: Region-based image classification with a latent SVM model. Research report RR-7665, INRIA (2011)
2. Sivic, J., Zisserman, A.: Video google: a text retrieval approach to object matching in videos. In: Proceedings of ICCV, pp. 1470–1477 (2003)
3. Lazebnik, S., Schmid, C., Ponce, J.: Beyond bags of features: spatial pyramid matching for recognizing natural scene categories. In: Proceedings of CVPR (2006)
4. Grauman, K., Darrell, T.: Pyramid match kernels criminative classification with sets of image features. In: ICCV (2005)
5. Wang, J., Yang, J., Yu, K., Lv, F., Huang, T., Gong, Y.: Locality-constrained linear coding for image classification. In: Proceedings of CVPR (2010)
6. Song, Z., Chen, Q., Huang, Z., Hua, Y., Yan, S.: Contextualizing object detection and classification. In: Proceedings of CVPR (2011)
7. Xie, L., Tian, Q., Wang, M., Zhang, B.: Spatial pooling of heterogeneous features for image classification. IEEE Trans. Image Process. **23**, 1994–2008 (2014)
8. Qi, G.J., Hua, X.S., Rui, Y., Tang, J., Zhang, H.J.: Image classification with kernelized spatial-context. IEEE Trans. Multimedia **12**, 278–287 (2010)
9. Dietterich, T.G., Lathrop, R.H., Lozano-Perez, T.: Solving the multiple instance problem with axis-parallel rectangles. IEEE Trans. Pattern Anal. Mach. Intell. **89**, 31–71 (1997)
10. Wang, X., Bai, X., Liu, W., Latecki, L.J.: Feature context for image classification and object detection. In: Proceedings of CVPR (2011)
11. Felzenszwalb, P., Girshick, R., McAllester, D., Ramanan, D.: Object detection with discriminatively trained part based models. IEEE Trans. Pattern Anal. Mach. Intell. **32**, 1627–1645 (2010)
12. Belongie, S., Malik, J., Puzicha, J.: Shape matching and object recognition using shape contexts. IEEE Trans. PAMI **24**, 509–522 (2002)
13. Andrews, S., Tsochantaridis, I., Hofmann, T.: Support vector machines for multiple-instance learning. In: Proceedings of Advances in Neural Information Processing Systems (2003)
14. Hong, R., Wang, M., Gao, Y., Tao, D., Li, X., Wu, X.: Image annotation by multiple-instance learning with discriminative feature mapping and selection. IEEE Trans. Cybern. **44**, 669–680 (2014)
15. Everingham, M., Van Gool, L., Williams, C.K.I., Winn, J., Zisserman, A.: The PASCAL Visual Object Classes Challenge (VOC2007) (2007), Results. http://www.pascal-network.org/challenges/VOC/voc2007/workshop/index.html
16. Wang, M., Li, G., Lu, Z., Gao, Y., Chua, T.S.: When amazon meets google: product visualization by exploring multiple web sources. ACM Trans. Internet Technol. (TOIT) **12**, 12 (2013)
17. Wang, M., Li, H., Tao, D., Lu, K., Wu, X.: Multimodal graph-based reranking for web image search. IEEE Trans. Image Process. **21**, 4649–4661 (2012)
18. Wang, X., Feng, B., Bai, X., Liu, W., Latecki, L.J.: Bag of contour fragments for robust shape classification. Pattern Recogn. **47**, 2116–2125 (2014)

19. Zhu, J., Wu, T., Zhu, J., Yang, X., Zhang, W.: Learning reconfigurable scene representation by tangram model. In: 2012 IEEE Workshop on Applications of Computer Vision (WACV), pp. 449–456. IEEE (2012)
20. Fergus, R., Perona, P., Zisserman, A.: Object class recognition by unsupervised scale-invariant learning. In: Proceedings of the IEEE Conference on Computer Vision and Pattern Recognition (2003)
21. Lee, Y.J., Grauman, K.: Object-graphs for context-aware category discovery. IEEE Trans. Pattern Anal. Mach. Intell. TPAMI **34**, 346–358 (2011)
22. Yuan, J., Wu, Y.: Spatial random partition for common visual pattern discovery. In: Proceedings of ICCV (2007)
23. Zhu, L.L., Lin, C.X., Huang, H., Chen, Y., Yuille, A.L.: Unsupervised structure learning: hierarchical recursive composition, suspicious coincidence and competitive exclusion. In: Forsyth, D., Torr, P., Zisserman, A. (eds.) ECCV 2008, Part II. LNCS, vol. 5303, pp. 759–773. Springer, Heidelberg (2008)
24. Zhu, J., Zou, W., Yang, X., Zhang, R., Zhou, Q., Zhang, W.: Image classification by hierarchical spatial pooling with partial least squares analysis. In: BMVC, pp. 1–11 (2012)
25. Khan, I., Roth, P.M., Bischof, H.: Learning object detectors from weakly-labeled internet images. In: OAGM Workshop (2010)
26. Alexe, B., Deselares, T., Ferrari, V.: What is an object? In: Proceedings of CVPR (2010)
27. Vijayanarasimhan, S., Grauman, K.: Keywords to visual categories: multiple-instance learning for weakly supervised object categorization. In: Proceedings of CVPR (2008)
28. Pandey, M., Lazebnik, S.: Scene recognition and weakly supervised object localization with deformable part-based models. In: Proceedings of ICCV (2011)
29. Russakovsky, O., Lin, Y., Yu, K., Fei-Fei, L.: Object-centric spatial pooling for image classification. In: Fitzgibbon, A., Lazebnik, S., Perona, P., Sato, Y., Schmid, C. (eds.) ECCV 2012, Part II. LNCS, vol. 7573, pp. 1–15. Springer, Heidelberg (2012)
30. Harzallah, H., Jurie, F., Schmid, C.: Combining efficient object localization and image classification. In: International Conference on Computer Vision (2009)
31. Lowe, D.G.: Distinctive image features from scale-invariant keypoints. Int. J. Comput. Vis. **60**, 91–110 (2004)
32. Yang, J., Yu, K., Gong, Y., Huang, T.: Linear spatial pyramid matching using sparse coding for image classification. In: Proceedings of CVPR (2009)
33. Dalal, N., Triggs, B.: Histograms of oriented gradients for human detection. In: CVPR (2005)
34. Oliva, A., Torralba, A.: Modeling the shape of the scene: a holistic representation of the spatial envelope. Int. J. Comput. Vis. **42**, 145–175 (2001)
35. Fan, R.E., Chang, K.W., Hsieh, C.J., Wang, X.R., Lin, C.J.: LIBLINEAR: a library for large linear classification. J. Mach. Learn. Res. **9**, 1871–1874 (2008)
36. Quack, T., Ferrari, V., Leibe, B., Gool, L.V.: Efficient mining of frequent and distinctive feature configurations. In: International Conference on Computer Vision (ICCV 2007) (2007)
37. Liu, C., Yuen, J., Torralba, A., Sivic, J., Freeman, W.T.: SIFT flow: dense correspondence across different scenes. In: Forsyth, D., Torr, P., Zisserman, A. (eds.) ECCV 2008, Part III. LNCS, vol. 5304, pp. 28–42. Springer, Heidelberg (2008)
38. Chatfield, K., Lempitsky, V., Vedaldi, A., Zisserman, A.: The devil is in the details: an evaluation of recent feature encoding methods. In: Proceedings of the British Machine Vision Conference (BMVC) (2011)

Enhancement for Dust-Sand Storm Images

Jian Wang[✉], Yanwei Pang, Yuqing He, and Changshu Liu

School of Electronic Information Engineering, Tianjin University, Tianjin 300072, China
jianwang@tju.edu.cn

Abstract. A novel dust-sand storm image enhancement scheme is proposed. The input degraded color image is first convert into CIELAB color space. Then two chromatic components ($a*$ and $b*$) are combined to perform color cast correction and saturation stretching. Meanwhile, the fast Local Laplacian Filtering is employed to lightness component ($L*$) to enhance details. Experimental results illustrate that enhanced images have natural colors, more clear details and better visual effect than original degraded images.

Keywords: Dust-sand storm images · Color cast correction · Saturation stretching · Local Laplacian filter

1 Introduction

Most outdoor vision applications, such as intelligent surveillance, autonomous navigation and object tracking, require input images have clear visibility. Unfortunately, under bad weather conditions (e.g. haze, fog, or dust-sand storm), the contrast and colorfulness of images are drastically degraded [1]. Therefore, it is imperative to remove weather effects from images in order to make vision systems more reliable.

Recently, there has been an increased works on the issue of bad weather degraded image processing. The previous methods can be divided into two categories: physics model based methods and model-free based methods [2].

Most physics models proposed by researchers [3–5] require many parameters, such as angle of sun shine, location of sky region and horizon, to describe complicated phenomena. Therefore, these methods are limited and do not suitable for general outdoor scenes, which present different levels of bad weather conditions. Model-free based methods normally combine a series of image enhancement techniques to improve the visibility of the degraded images. So they also called enhancement-based method. These techniques include adaptive histogram equalization (AHE) [6], contrast stretch [7], multi-scale retinex with color restoration (MSRCR) theory [8], image fusion [9], and so on.

However, most previous works mainly related to haze, fog, and rain/snow situations. And these methods focus on denoising, contrast stretching or detail enhancement. Compared with those weather types mentioned above, images captured in dust-sand storm situation appear severe color cast. In addition, previous methods are implemented in RGB color space directly. The same procedure is performed in all three color components separately, which may results in serious color distortion.

© Springer International Publishing Switzerland 2016
Q. Tian et al. (Eds.): MMM 2016, Part I, LNCS 9516, pp. 842–849, 2016.
DOI: 10.1007/978-3-319-27671-7_70

In this paper, we propose an enhancement scheme for improving the visibility of degraded dust-sand storm images. The main contributions of our work include two aspects. Firstly, as one kind of typical bad weather conditions in Northeast Asia, the issue of dust-sand storm image enhancement is studied for the first time. Secondly, input degraded images are converted into chromatic and lightness components separately, and then they are employed sequentially for color cast correction, saturation stretching and detail enhancement.

2 Proposed Method

Figure 1 shows the block diagram of our method. The input degraded images are converted into CIELAB color space. Then lightness and chromatic components are separated. Next, color cast correction and saturation stretching procedures are performed successively in the joint $a*$–$b*$ channels. Meanwhile, image details are enhanced in the lightness channel ($L*$ component) by using Local Laplacian Filter (LLF) technology. Finally, three components after processing are combined and transform into RGB color space for storage and displaying. More details are given as follows.

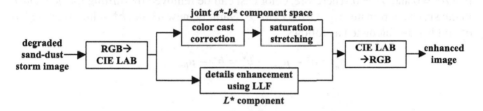

Fig. 1. Block diagram of the proposed algorithm.

2.1 Color Space Transformation

Our enhancement scheme is performed in the CIE 1976 $L*a*b*$ color space, which is also named as CIELAB color space. The CIELAB is a perceptually uniform color space in which it is possible to effectively quantify color differences as seen by the human eye [10]. Three coordinates of CIELAB represent the lightness of the color ($L* = 0$ yields black and $L* = 100$ indicates white), its position between red/magenta and green ($a*$, negative values indicate green while positive values indicate magenta) and its position between yellow and blue ($b*$, negative values indicate blue and positive values indicate yellow).

In our scheme, the lightness component ($L*$ component) and two chromatic components ($a*$ and $b*$) are processed separately. The original RGB values of each pixel in input images are transformed into CIE XYZ space at first, and then been mapped into the CIELAB color space. The chromaticity adjustment procedure is performed by only using chromatic information, while lightness information is solely employed for detail enhancement.

2.2 Color Cast Correction

Degraded dust-sand storm images usually suffered from bad color cast due to the attenuation of the propagated light. Figure 2(a) gives an example of a typical degraded dustsand storm image. It can be seen that the scene looks very yellowish and dim. Hence, it is highly desired to remove the cast effect and restore the natural chromaticity, which is called color cast correction in our method. The cast correction technique we used is similar with the technology proposed by Gasparini [11], which is carried out in the joint $a*–b*$ chromatic space. Two normalized histograms of $a*$ and $b*$ component are computed, which are denoted as $H(a*)$ and $H(b*)$, respectively. Then the chromatic center $O(a*, b*)$ of the original $a*–b*$ chromatic space is obtained as follows:

$$\mu_c = \sum_{k=-127}^{128} k \cdot H_c(k) \tag{1}$$

where subscript $c = \{a*, b*\}$, k represents the serial number of $a*$ or $b*$ bins. The distance between O' and the origin point $O(0,0)$ is computed to measure the degree of color cast. It can be assumed that outdoor images captured in good weather satisfy the condition that $a* = 0$ and $b* = 0$. Therefore, color cast can be removed by shifting the new chromatic space coordinate origin from O to O'. Namely, new $a*$ and $b*$ values (denoted as a_n^* and b_n^*) are calculated as follows:

$$a_n^* = a^* - \mu_{a*}, \quad b_n^* = b^* - \mu_{b*} \tag{2}$$

Fig. 2. Steps of proposed enhancement methods: (a) input image; (b) $L*$ component; (c) color cast correction result; (d) saturation stretching result; (e) detail enhancement result; (f) final result (Color figure online).

2.3 Saturation Stretching

After color cast correction, the ranges of a_n^* and b_n^* maybe are quite narrow, which leads to dim scenes. In the proposed method, we adopt a nonlinear mapping based saturation stretch technology to make images look more colorful.

By consider the ranges of two chromatic components after color cast correction, we find that most of a_n^* and b_n^* values are close to 0. So, we use the sigmoid function to stretch low level range meanwhile compress high level range. The nonlinear mapping function is defined as

$$S(x) = \frac{1 - e^{-\alpha x}}{1 + e^{-\alpha x}} \tag{3}$$

In Eq. (3), parameter α is a positive constant that is used to adjustment saturation stretching level. The larger α is, the higher saturation obtained. According to simulation results, we find that too high value may lead to unnatural colors. So α is set in the range of [0.01, 0.03] in our scheme.

2.4 Detail Enhancement

In original dust-sand storm images, image details (such as leaves and buildings) are blurred or disappeared, which decrease the image quality heavily. Furthermore, we find that the quality degradation of lightness component is worse than chromatic components. So, we employ an effective spatial filter technology named Local Laplacian Filter (LLF) is employed to L^* component to enhance details and contrast of lightness component.

LLF is a well-known edge-aware filter proposed by Paris [12]. They demonstrated that these filters generate high-quality results for detail enhancement. However, these filters are very slow in their original form. Aubry [13] developed a fast LLF algorithm (FLLF) on gray-scale images that is about 50 times faster than the original LLF method. In practice, they pre-computed a small set of pyramids. And they achieved the new coefficients by interpolating the coefficients of pre-computed pyramid j and $j + 1$ directly, instead of remapping the image and computing a new pyramid, which is very expensive. More details about FLLF algorithm can be found in Ref. [13].

In our proposed method, three components after processing are combined and converted into RGB color space for storage and displaying. Figure 2 give a complete procedure of proposed method, where (a) is a input degraded image, (b) is the lightness component, (c) and (d) is the color cast correction and saturation stretch result without detail enhancement process, (e) is the detail enhancement result of (b), and (f) is the final result.

3 Experimental Result and Comparison

Computer simulations are conducted to validate the effectiveness of the proposed algorithm. The test dust-sand storm image dataset include more than 50 images with different scenes and degradation degrees.

Figure 3 show parts of results using our method. First two rows are dust-storm case, while last two rows represent sand-storm case. It can be seen that our approach can recover the vivid color information and unveil the detail in both dust and sand storm cases. However, some false contour artifacts appear in flat regions (such as sky regions), which are caused by LLF steps.

Fig. 3. Dust-sand storm image enhancement results using proposed method.

In order to evaluate the performance of the proposed method, we compared it with three existing conventional color image enhancement algorithms. Method 1 and method 2 both are model-free algorithms based on AHE [6] and MSRCR technology (denoted as MSRCR) [8]; Method 3 is a model based algorithm which using classical dark channel prior hypothesis together with contrast stretching technology (denoted as DARK) [7].

Because weather effects are volumetric, classical image noise or degradation evaluation methods are not suitable for evaluating weather degraded images [14]. Thus we make use of three different metrics present in literature to test the performance of our algorithm and compare it with other three methods. These metrics are described as follows:

- Descriptor *CNT* [14] computes the geometric mean of the ratios of the visible edges, that is, expresses the quality of the contrast restoration by enhancement procedure. This metric is calculated in L^* channel. Higher values mean better contrast restoration.
- Descriptor *COF* [15] calculates the relative variance and mean over the chromatic space in the logarithmic domain. This metric is calculated in CIELAB color space. Higher value means more colorful in the restored image.
- Descriptor *CCR* [16] measures the color cast level, which represents the distribution of a_n^* and b_n^* components in the chromatic space. The higher the *CCR* value, the more severe the color cast is.

Table 1 gives image quality assessment result of the application of the aforementioned metrics for all 50 test images and four algorithms, where numbers in bold denote the best value of each row. The proposed method gets the best performance for all metrics. For AHE method and DARK method, the Descriptor *COF* and *CCR* both are similar with the original images. The Descriptor CNT of DARK method is close to 1, which means nearly no contrast enhanced. For MSRCR method, the values of Descriptor *CNT* and *COF* are both similar with our method, but the *CCR* value is higher than our method.

Table 1. Image quality assessment of different algorithms

Metric	Input images	AHE result	MSRCR result	DARK result	Our result
CNT	-------	1.105	1.201	1.018	**1.436**
COF	0.634	0.973	3.495	0.756	**3.504**
CCR	8.632	5.681	0.872	7.569	**0.299**

Figure 4 gives some comparison results of different methods, where our results are shown in the left column. The dark channel prior hypothesis is false completely for dust-sand storm weather condition, thus DRAK method is almost invalid. Although AHE results obtain higher contrast than original images, their color cast ratio is also very high. Compared with DARK and AHE results, MSRCR results have higher contrast and lower color-cast ratio. But the color distortion problem is very obvious. Furthermore, the chromaticity of flat region (such as sky and road) is quite dim, which make scene looks under saturated. Compared with other three methods, our method leads to a more natural color and rich details. Especially the saturation is increase significantly, which make scenes look very colorful and bright.

(a)   (b)   (c)   (d)   (e)

Fig. 4. Performance comparison. (a) dust-sand images, (b) AHE results, (c) MSRCR results, (d) DARK results, (e) our results.

4 Conclusion

A novel dust-sand storm image enhancement algorithm is proposed. Our strategy is a single image approach that does not require specialized parameters or knowledge about the light conditions or scene structure. Simulation results demonstrate that dust-sand storm scenes with different degradation levels are all improved significantly by using proposed algorithm. Although the proposed method works well, it still shows the potential for further improvement. Future work main focus on developing a model-based restoration method for dust-sand storm images.

Acknowledgment. This work was supported in part by the National Basic Research Program of China 973 Program (Grant No. 2014CB340400) and the National Natural Science Foundation of China (Grant Nos. 61172121, 61472274, 61271412, 61002030 and 61222109).

References

1. Narasimhan, S.G., Nayar, S.K.: Contrast restoration of weather degraded images. IEEE Trans. PAMI **5**(6), 713–724 (2003)
2. Mukhopadhyay, S., Tripathi, A.K.: Combating Bad Weather. Morgan and Claypool Publishers, San Rafael (2015)
3. Tan, R.T.: Visibility in bad weather from a single image. In: IEEE Conference Computer Vision and Pattern Recognition (CVPR), pp. 1–8. IEEE Press, New York (2008)
4. Barnum, P.C., Narasimhan, S.G., Kanade, T.: Analysis of rain and snow in frequency space. Int. J. Comput. Vision **86**, 256–274 (2010)
5. He, K.M., Sun, J., Tang, X.O.: Single image haze removal using dark channel prior. IEEE Trans. PAMI **33**(12), 2341–2353 (2011)
6. Jia, Z., Wang, H.C., Caballero, R., et al.: Real-time content adaptive contrast enhancement for see through fog and rain. In: IEEE International Conference on Acoustics Speech and Signal Processing (ICASSP), pp. 1378–1381. IEEE Press, New York (2010)
7. Li, B., Wang, S.H., Zheng, J., et al.: Single image haze removal using content-adaptive dark channel and post enhancement. IET Comput. Vision **8**(2), 131–140 (2014)
8. Gao, Y., Yun, L.J., Shi, J.S., et al.: Enhancement MSRCR algorithm of color fog image based on the adaptive scale. In: 6th International Conference on Digital Image Processing (ICDIP 2014), SPIE Press, New York (2014)
9. Ancuti, C.O., Bekaert, P.: Effective single image dehazing by fusion. In: IEEE International Conference on Image Processing (ICIP), pp. 26–29. IEEE Press, New York (2010)
10. http://en.wikipedia.org/wiki/Lab_color_space
11. Gasparini, F., Schettini, R.: Color balancing of digital photos using simple image statistics. Pattern Recogn. **37**, 1201–1217 (2004)
12. Paris, S., Hasinoff, S.W., Kautz, J.: Local Laplacian filters: edge-aware image processing with a Laplacian pyramid. ACM Trans. Graph. **30**(4), 1–12 (2011)
13. Aubry, M., Paris, S., Hasinoff, S.W., et al.: Fast local Laplacian filters: theory and applications. ACM Trans. Graph. **33**(5), 167 (2014)
14. Hautière, N., Tarel, J.P., Aubert, D., Dumont, E.: Blind contrast enhancement assessment by gradient ratioing at visible edges. Image Anal. Stereol. J. **27**, 87–95 (2008)

15. Karen, P., Gao, C., Agaian, S.: No reference color image contrast and quality measures. IEEE Trans. Consum. Electron. **59**(3), 643–651 (2013)
16. Li, F., Wu, J., Wang, Y., et al.: A color cast detection algorithm of robust performance. In: IEEE International Conference on Advanced Computational Intelligence (ICACI), pp. 662–664. IEEE Press New York (2012)

Using Instagram Picture Features to Predict Users' Personality

Bruce Ferwerda[⊠], Markus Schedl, and Marko Tkalcic

Department of Computational Perception,
Johannes Kepler University, Altenberger Str. 69, 4040 Linz, Austria
{bruce.ferwerda,markus.schedl,marko.tkalcic}@jku.at
http://cp.jku.at

Abstract. Instagram is a popular social networking application, which allows photo-sharing and applying different photo filters to adjust the appearance of a picture. By applying these filters, users are able to create a style that they want to express to their audience. In this study we tried to infer personality traits from the way users manipulate the appearance of their pictures by applying filters to them. To investigate this relationship, we studied the relationship between picture features and personality traits. To collect data, we conducted an online survey where we asked participants to fill in a personality questionnaire, and grant us access to their Instagram account through the Instagram API. Among 113 participants and 22,398 extracted Instagram pictures, we found distinct picture features (e.g., relevant to hue, brightness, saturation) that are related to personality traits. Our findings suggest a relationship between personality traits and these picture features. Based on our findings, we also show that personality traits can be accurately predicted. This allow for new ways to extract personality traits from social media trails, and new ways to facilitate personalized systems.

Keywords: Instagram · Personality · Photo filters · Picture features

1 Introduction

Instagram is a popular mobile photo-sharing, and social networking application, with currently over 300 million active users a month, over 70 billion pictures shared, with an average of 70 million new pictures a day.[1] Instagram is interconnected with an abundance of social networking sites (e.g., Facebook, Twitter, Tumblr, and Flickr) to let its users share their pictures on. In addition, it encourages users to apply filters to modify the color appearance of their pictures. At this moment Instagram offers 25 predefined photo filters that allow users to customize and modify their pictures to create the desired visual style.

The ease with which a photo filter can be applied allow users to express a personal style and create a seeming distinctiveness with the customized pictures.

---

[1] https://instagram.com/press/ (accessed: 08/07/2015).

© Springer International Publishing Switzerland 2016
Q. Tian et al. (Eds.): MMM 2016, Part I, LNCS 9516, pp. 850–861, 2016.
DOI: 10.1007/978-3-319-27671-7_71

Through the shared content and the way of applying filters, users are able to reveal a lot about themselves to their social network. With that, the question arises: What do Instagram pictures tell about the user? Or more specifically: What do Instagram pictures say about the personality of the user?

Personality traits have shown to consist of cues to infer users' behavior, preference, and taste (e.g., [9,26,28]). Hence, knowing one's personality can provide important information for systems to create a personalized user experience. It can provide systems with estimations about user preferences, and avoid the use of extensive questionnaires or observations.

There is an increasing interest in implement personality in systems (e.g., [8,11,29]), and the implicit acquisition of personality from online behavior trails (e.g., Facebook [2,5,13,24,27], Twitter [12,25], Flickr [7], video blogs [3,4]). In this work we join the personality extraction research. We specifically focus on the relationship between the personality of Instagram users and the way they manipulate their pictures by using photo filters, in order to create a visual style.

Our work makes several contributions. We contribute to personality research, by showing relationships between personality traits and the visual style of users' Instagram pictures. Additionally, we contribute to new ways to extract personality from social media (i.e., Instagram). To the best of our knowledge, we are the first to investigate the relationship between personality traits and how users try to create a visual style by applying photo filters.

An online survey was conducted where we: (1) asked participants to fill in the widely used Big Five Inventory (BFI) personality questionnaire, and (2) grant us access to the content of their Instagram account. We extracted 22,398 Instagram pictures of 113 users, and analyzed them on several color-centric picture features (e.g., related to hue, saturation, value). We found distinct correlations between personality traits and picture features, and show that personality can be accurately predicted from the picture features.[2]

In the remainder of the paper we will continue with related work, materials, features, results, discussion, and conclusion.

2   Related Work

Personality has shown to be an enduring factor that can be related to a person's taste, preference, and interest (e.g., [9,26,28]). For example, Rawlings and Ciancarelli found relationships between personality traits and music genre preferences [26], while Tkalcic et al. found relationships between personality and classical music [28]. These relationships indicate that personality information can be used to create useful proxy measures for applications to cater to a more personalized service (e.g., [8,11,29]). For example, Tkalcic et al. propose to use personality to enhance the nearest-neighborhood measurement for overcoming the cold-start problem (i.e., recommending items to new users) in recommender systems [29]. Ferwerda et al. provide a way to use personality for adjusting the user interface of music applications to fit a user's music browsing style [11].

---

[2] Our preliminary results of this work can be found in [10].

In order to measure personality, several models have been developed. The five-factor model (FFM) is the most well known and widely used one in the computing community [32], and categorizes personality into five general dimensions (traits), that describe personality in terms of: openness to experience, conscientiousness, extraversion, agreeableness, and neuroticism [21]. However, unless using extensive, and time-consuming questionnaires, acquiring personality traits is still a challenging task.

There is an emergent interest in how to implicitly acquire personality traits based on behavioral data (for an overview see [32]). Research has shown it is feasible to compute personality from behavioral data such as mobile phone usage (e.g., [6,23]), or with acoustic and visual cues through cameras and microphones (e.g., [1,19,22]). With the increasing connectedness of people, recent research has started to focus on personality acquisition from online behavior trails, such as, video blogs (vlogs) [3,4], Facebook behavior (e.g., [2,13,24,27]) and profile pictures [5], Twitter behavior (e.g., [12,25]), and Flickr pictures [7].

Little work has been done on personality extraction from pictures. Celli et al. focused on the content of Facebook profile pictures (e.g., facial close-ups, facial expressions, alone or with others) to extract personality [5]. Other work of Cristani et al. showed that personality can be extracted from the visual features of Flickr pictures [7]. Flickr attracts a lot of advanced photographers as an image hosting platform, and thereby consist of more serious and higher quality pictures. Instagram, on the other hand, targets snapshot pictures taken with the mobile phone, and puts emphasis on applying predefined filters. The different usage and interactions on Flickr and Instagram attract different audiences, and therefore personality prediction may be based on different cues. As Instagram is known for its photo filters to create certain effects, we decided to focus more on color-properties; how users manipulate their pictures with help of the filters to achieve a certain expression, rather then picture content.

3   Materials

To investigate the relationship between personality traits and picture features, we asked participants to fill in the 44-item BFI personality questionnaire (5-point Likert scale; Disagree strongly - Agree strongly [16]). The questionnaire include questions that aggregate into the five basic personality traits of the FFM. The distribution of each personality trait can be found in Fig. 1. Additionally, we asked participants to grant us access to their Instagram account through the Instagram API, in order to crawl their pictures. From hereon, we define the picture-collection term as *all* the Instagram pictures of a single user.

We recruited 126 participants through Amazon Mechanical Turk, a popular recruitment tool for user-experiments [18]. Participation was restricted to those located in the United States, and also to those with a very good reputation ($\geq$95 % HIT approval rate and $\geq$1000 HITs approved)[3] to avoid careless

---

[3] HITs (Human Intelligence Tasks) represent the assignments a user has participated in on Amazon Mechanical Turk prior to this study.

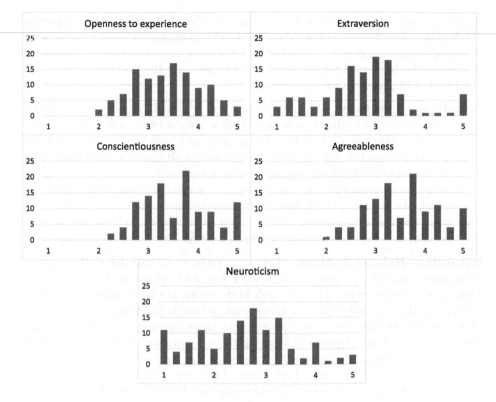

Fig. 1. Collected personality distribution for each personality trait.

contributions. Several comprehension-testing questions were used to filter out fake and careless entries. The Mahalanobis distance was calculated to check for outliers. This left us with 113 completed and valid responses. Age (18–64, median 30) and gender (54 male, 59 female) information indicated an adequate distribution. Pictures of each participant were crawled after the study. This resulted in a total of 22,398 pictures.

4   Features

As the goal in this study was to see how Instagram users manipulate their pictures with photo filters, we extracted mainly color related picture features. Based on the assumption that Instagram users' personality is manifested through the way photo filters are applied, a set of features that relate to color were selected based on the work of Machajdik and Hanbury [20]. For the color-centric features, the color space that is most closely related to the human visual system was selected. That is, the Hue-Saturation-Value (HSV) color space [30]. The H parameter describes the hue (i.e., the color quality of each pixel) from orange through yellow, green, blue, violet to red, on a scale from 0 to 1. The S parameter describes how saturated the color is. That is, the share of white in a picture (high share of white means low saturation). The parameter V (value) represents the brightness of the color. Beside the color-centric features, we additionally performed basic picture content analysis, by counting the number of faces and the number of people in each picture.

Hue-Related Features. We divided the range of the H parameter into intervals that correspond to the hues: orange, yellow, green, blue, violet, and red. For each of these intervals we counted the number of pixels in an image that fall into this interval and divided it with the number of all pixels in the image. This yielded the share of the image surface that a hue covers.

Furthermore, we merged the cold colors (i.e., green, blue violet) and the warm colors (i.e., orange, red, yellow) into the respective shares across an image. On a user level, the features: orange, yellow, green, blue, violet, red, warm and cold, are the average values of the color shares among all the pictures of a user.

Saturation-Related Features. For each picture we calculated the average saturation and the variance. Images with low average saturation tend to be bleak and colorless, while pictures with high saturation have more vivid colors. Pictures with a high saturation variance tend to have both bleak and vivid colors. Here, we also divided the saturation axis into three equally spaced intervals and calculated the share of pixels that fall into each interval (low, mid, and high saturation). Pictures that have a high value in the *low saturation* tend to have more bleak colors, those with a high value in the *mid saturation* feature tend to have neither bleak nor vivid colors, and those that have a high value in the *high saturation* feature tend to have vivid colors across most of the image area.

Value-Related Features. For each image we calculated the average value (*value mean*) and variance (*value variance*) across all the pixels in the image. These features represent how light or dark a picture is and how much contrast it reveals, respectively. Pictures that have a high variance tend to have both dark and light areas, whereas pictures with a low variance tend to be equally bright across the image. Furthermore, we divided the value axis into three equally spaced intervals and counted the share of pixels that fall into each of these intervals (*value low, mid, and high*).

Pleasure-Arousal-Dominance (PAD). As the filters Instagram users can apply to their pictures, are intended to create certain expressions, we adopted the PAD model of Valdez and Merhabian [31], which contains general rules of the expression of pleasure, arousal, and dominance in a picture, and models these as a combination of brightness (value) and saturation levels:

1. Pleasure $= .69$ Value $+ .22$ Saturation
2. Arousal $= -.31$ Value $+ .60$ Saturation
3. Dominance $= -.76$ Value $+ .32$ Saturation

Content-Based Features. In addition to the color-centric features, we computed two extra features for each user: (1) the average number of human faces across all the images of a user, and (2) the average number of full people bodies across all the images of a user.

To extract the number of faces in a picture, we used the Viola-Jones algorithm [33]. We trained the algorithm to recognize the number of faces in an image, by using the Haar-like features and the AdaBoost classifier. For extracting the number of full bodies we used the Histogram of Oriented Gradient (HOG) features with a Support Vector Machine (SVM) classifier. To achieve this, we employed Matlab's Computer Vision System Toolbox.[4]

5   Results

We divided the results section into two parts: the first part discusses the correlations we found, and the second part we discuss our personality regressor to predict personality traits based on the picture features.

5.1   Correlations

Although the main goal of this study was to investigate the relationship between personality and picture features, we decided to explore whether we could find correlations with personality and the usage of certain filters. We shortly discuss our results on the latter, and then continue with the correlations of the picture features.

Instagram Filters. Besides crawling the pictures, we also crawled descriptives about the filters each participant applied to their pictures, in order to explore whether certain personality traits are related to a more frequent use of some filters. Of the collected 22,398 pictures from 113 participants, 1487 unique filters were applied in total (this also include the "Normal" filter which means that no filter is applied). This brings that on average participants use 13 (13.15) different photo-filters to the pictures they upload to their picture-collection. Given that Instagram offers 25 different filters to apply, participants seem to use a whole range of filters. This is not so surprising, as the result of applying a filter depends on how the original picture looks like. So it is possible that the visual characteristics of a picture is eventually the same, but achieved by using different filters.

With using Pearson's correlation ($r \in [-1,1]$) to indicate the linear relationship between personality and photo filters, we found the following correlations: a positive correlation between the conscientiousness personality traits and the "Kelvin" filter ($r = .203$, $p = .044$), a negative correlation between the agreeableness personality trait, the "Crema" ($r = -.205, p = .042$) and the "Gotham" ($r = -.204, p = .042$) filter. Additionally, we found a positive correlation between the neuroticism personality traits and the "Hudson" ($r = .224$, $p = .026$) filter. These results imply that in general conscientiousness participants make more use

---

[4] http://www.mathworks.com/help/vision/index.html.

of the Kelvin filter, agreeable participants used less the Crema and Gotham filter, and neurotic participants applied more the Hudson filter. Besides these four significant correlations, there were no statistical significant correlations found with the remaining 21 photo filters. Given the low number of significant correlations that we found, and that the applied filter does not tell anything about how the end result looks like, we decided to not further pursue this direction, and continue our analyses focusing on the picture features.

Picture Features. For each picture from a crawled picture-collection (i.e., all the pictures of a participant's Instagram account), we extracted all the features that are described in Sect. 4 (Features). As the features in participants' picture-collection show a normal distribution, we calculated mean values for each feature to create a measurement of central tendency, to represent the whole picture-collection of each participant. The mean values of the features were used to calculate the correlation matrix (see Table 1). Pearson's correlation ($r \in [-1,1]$) is reported to indicate the linear relationship between personality and picture features. The correlation matrix shows several features related with personality traits.[5] We discuss the results related to each personality trait below.

Openness to Experience: Openness to experience was found to correlate with the *saturation mean*. This indicates that the pictures of open participants consist of more saturated, vivid colors. We also observed a positive correlation with the feature *saturation variance*, which means that open participants share pictures that have both vivid and bleak colors. Furthermore, a negative correlation with the feature *value mean* was found, indicating that open participants tend to share pictures that are low on brightness. This was further confirmed by the positive correlation on *brightness low*, and the negative correlation on *value high*. These correlations show that the pictures of open participants show more dark areas, and less bright areas. Also, a correlation was observed for the warm and cold features. Pictures of more open participants contained less warm colors (i.e., red, orange), but more cold colors (i.e., blue, green). Also, their pictures tend to express less *pleasure*, but more *arousal* and *dominance*. Additionally, their pictures consist in general of fewer faces and people.

Conscientiousness: A positive correlation was found between the *saturation variance* feature and conscientiousness. This indicates that conscientious participants more frequently shared pictures consisting of bleak and vivid colors.

Extraversion: We found correlations between the picture features and extraversion. Extraverts tend to create pictures with less *red* and *orange*, but with more *green* and *blue* tones. Additionally, their pictures tend to be darker (*brightness low*), but consist of both vivid and bleak colors (*saturation variance*). Also the emotion that the pictures of extraverts consist, are low on *pleasure*, but high on *dominance*.

---

[5] The magnitude of the correlations are commonly found in relationships between personality traits and social media trails (e.g., [2,12,13,24,25,27]).

Table 1. Correlation Matrix of the picture features against the personality traits: (O)penness, (C)onscientiousness, (E)xtraversion, (A)greeableness, (N)euroticism.

	O	C	E	A	N
Red	−0.06	0.02	**−0.17^**	−0.05	0.03
Green	**0.17^**	0.14	**0.23^^**	0.03	−0.12
Blue	−0.01	0	**0.17^**	0.02	−0.01
Yellow	0.01	0.04	0.01	0.14	−0.07
Orange	−0.03	−0.07	**−0.16^**	−0.02	0.06
Violet	0	−0.06	−0.09	−0.07	0.06
Saturation mean	**0.16^**	0.06	0.03	−0.04	0
Saturation variance	**0.20^^**	**0.16^**	**0.19^^**	0.10	−0.05
Saturation low	−0.08	−0.02	0.02	0.07	0.01
Saturation mid	0.08	−0.09	0.02	0.07	0.01
Saturation high	0.13	0.10	0.04	−0.01	0.01
Value mean	**−0.25***	−0.10	**−0.19^**	−0.07	**0.22^**
Value variance	0.06	0	0	−0.07	0.05
Value low	**0.28****	0.09	**0.16^**	−0.05	**−0.16^**
Value mid	−0.09	0.06	0.04	**0.15^**	−0.06
Value high	**−0.20^**	−0.12	**0.18^**	−0.08	**0.21^**
Warm	**−0.05^^**	−0.04	−0.20	0	0.03
Cold	**0.05^^**	0.04	0.20	0	−0.03
Pleasure	**−0.19^^**	−0.08	**−0.18^**	−0.09	**0.22^^**
Arousal	**0.23***	0.09	0.10	0	−0.08
Dominance	**0.28****	0.11	**0.17^**	0.05	**−0.18^^**
# of faces	**−0.16^**	0.03	0.11	−0.11	−0.03
# of people	**−0.22^^**	−0.05	−0.07	−0.01	0.07

Note. ̂p < 0.1, ^^p < .05, *p < .01, **p < .001

Agreeableness: A positive correlation was found between agreeableness and the *brightness mid* feature. This means that the pictures of agreeable participants do not show emphasized bright or dark areas, but are more in between.

Neuroticism: A correlation was found on *value mean, value low*, and *value high*. The positive correlation with *value mean* indicate that participants scoring higher on the neurotic trait tend to share pictures that are high on brightness. This is also reflected in the *value low* (negative correlation) and *value high* (positive correlation) features. Additionally, correlations were found in the emotion expression of the pictures of extraverts. Result show that they adjust their pictures to express more pleasure but less dominance.

5.2 Personality Regressor

Given the correlations that were found, we developed a personality regressor based on the features reported in Table 1. As prior work on prediction personality traits from Twitter behavior uses the same method of analyses [25], we compare our performance against their results. We trained our predictive model with several classifiers in Weka, with a 10-fold cross-validation with 10 iterations. For each classifier we used, we report the *root-mean-square error* (RMSE) in Table 2, to indicate the root mean square difference between predicted and observed values. The RMSE of each personality trait relates to the [1,5] score scale.

Table 2. Comparison of different classifiers to predict personality prediction compared to prior work of Quercia et al. [25]. Numbers in bold represent the results that outperform prior work. Root-mean-square error (RMSE) is reported ([1,5]).

Personality traits	RMSE			
	Radial basis function network	Random forests	M5 rules	M5 rules Quercia et al. [25]
Openness to experience	**0.68**	0.71	0.77	0.69
Conscientiousness	**0.66**	**0.67**	**0.73**	0.76
Extraversion	0.90	0.95	0.96	0.88
Agreeableness	**0.69**	**0.71**	**0.78**	0.79
Neuroticism	0.95	1.01	0.97	0.85

In line with prior work of Quercia et al. [25], we started to train our predictive model with the *M5' rules* [34]. Although our results do not outperform prior work on most facets, we do find similar trends: extraversion and neuroticism are the hardest personality traits to predict. As applying M5' rules did not result in any improvement, we applied the *random forests* classifier. Random forests are known to have a reasonable performance when the features consist of high amounts of noise [15]. The results show slight improvement over M5' rules in general, but for the neuroticism personality trait, the prediction got worse. Finally, we tried using the radial basis function (RBF) network, which is a neural network that has shown to work well on smaller datasets [17]. Results show that the RBF network outperforms the M5' rules, as well as the random forest classifier. Compared to prior work, the RBF network outperforms prediction of openness to experience, conscientiousness, and agreeableness. Also with the RBF network, predictions is most difficult for extraversion and neuroticism. Although we do not outperform prior work on extraversion and conscientiousness, our results do not differ much. In general, we show that with picture features we can achieve better personality prediction than prior work on Twitter data.

6   Discussion

We found Instagram picture features to be correlated with personality. A summary and interpretation of the picture features can be found in Table 3. We found

Table 3. Interpretation and summary of the correlations found between personality traits and picture properties. The properties apply for the pictures of participants who score high in the respective personality trait.

Personality	Picture properties
Openness to experience	More green tones, lower in brightness, higher in saturation, more cold colors, fewer faces and people
Conscientiousness	Mix of saturated and unsaturated colors
Extraversion	More green and blue tones, lower in brightness, mix of saturated and unsaturated colors
Agreeableness	Fewer dark and bright areas
Neuroticism	Higher in brightness

that most correlations appear in the openness to experience personality trait. Even though, less and weaker correlations were found for the other personality traits, we were still able to observe distinct correlations.

Based on the identified correlations between personality and picture features, we created a personality regressor. The results that we achieved with our prediction model show that personality can be accurately predicted from picture features on Instagram. The results of the personality regressor show similar patterns as prior work on personality extraction from social media (i.e., Twitter) [25], and were able to outperform in predicting most of the personality traits. We found that the easiest and most successful prediction are for the openness to experience, conscientiousness, and agreeableness personality traits, whereas the more difficult traits are extraversion and neuroticism.

7   Future Work and Limitations

Our study contains limitations that need to be considered. Although we were able to obtain a fair amount of Instagram pictures ($n = 22{,}398$), our personality measurement was limited to 113 participants. Given that we only had personality information of 113 participants to find relationships with picture features, results would benefit from a bigger sample size.

In this study we solely focused on participants based in the United States. However, color interpretation and meaning could be influenced by cultural factors. Therefore, cultures could engage in different behavior of picture taking [14] and applying filters. Future work should address this. Furthermore, we decided to mainly focus on the color-centric features of the pictures we crawled, as Instagram's main focus is on applying photo filters. However, content analyses of the pictures would be an interesting next step to conduct.

8   Conclusion

With this study we show that personality of Instagram users can be accurately predicted from color-centric features of the pictures they post. Being able to implicitly extract personality from social media trails, gives possibilities to facilitate systems in order to provide a personalized experience. For example, it can help recommender systems to overcome the cold-start problem.

Acknowledgments. This research is supported by the Austrian Science Fund (FWF): P25655; and by the EU FP7/'13-'16 through the PHENICX project, grant agreement no. 601166.

References

1. Aran, O., Gatica-Perez, D.: Cross-domain personality prediction: from video blogs to small group meetings. In: Proceedings of the 15th ACM on International Conference on Multimodal Interaction, pp. 127–130. ACM (2013)
2. Back, M.D., Stopfer, J.M., Vazire, S., Gaddis, S., Schmukle, S.C., Egloff, B., Gosling, S.D.: Facebook profiles reflect actual personality, not self-idealization. Psychol. Sci. **21**, 372–374 (2010)
3. Biel, J.I., Aran, O., Gatica-Perez, D.: You are known by how you vlog: personality impressions and nonverbal behavior in youtube. In: ICWSM. Citeseer (2011)
4. Biel, J.I., Gatica-Perez, D.: The youtube lens: crowdsourced personality impressions and audiovisual analysis of vlogs. IEEE Trans. Multimedia **15**(1), 41–55 (2013)
5. Celli, F., Bruni, E., Lepri, B.: Automatic personality and interaction style recognition from facebook profile pictures. In: Proceedings of the ACM International Conference on Multimedia, pp. 1101–1104. ACM (2014)
6. Chittaranjan, G., Blom, J., Gatica-Perez, D.: Mining large-scale smartphone data for personality studies. Pers. Ubiquit. Comput. **17**(3), 433–450 (2013)
7. Cristani, M., Vinciarelli, A., Segalin, C., Perina, A.: Unveiling the multimedia unconscious: implicit cognitive processes and multimedia content analysis. In: Proceedings of the 21st ACM International Conference on Multimedia (2013)
8. Ferwerda, B., Schedl, M.: Enhancing music recommender systems with personality information and emotional states: a proposal. In: Proceedings of the EMPIRE Workshop (2014)
9. Ferwerda, B., Schedl, M., Tkalcic, M.: Personality & emotional states: understanding users' music listening needs. In: UMAP 2015 Extended Proceedings (2015)
10. Ferwerda, B., Schedl, M., Tkalcic, M.: Predicting personality traits with instagram pictures. In: Proceedings of the 3rd Workshop on Emotions and Personality in Personalized Systems 2015, pp. 7–10. ACM (2015)
11. Ferwerda, B., Yang, E., Schedl, M., Tkalcic, M.: Personality traits predict music taxonomy preferences. In: CHI 2015 Extended Abstracts. ACM (2015)
12. Golbeck, J., Robles, C., Edmondson, M., Turner, K.: Predicting personality from twitter. In: IEEE Third Conference on Social Computing, pp. 149–156 (2011)
13. Gosling, S.D., Gaddis, S., Vazire, S., et al.: Personality impressions based on facebook profiles. In: ICWSM (2007)
14. Huang, C.M., Park, D.: Cultural influences on facebook photographs. Int. J. Psychol. **48**(3), 334–343 (2013)
15. Humston, E.M., Knowles, J.D., McShea, A., Synovec, R.E.: Quantitative assessment of moisture damage for cacao bean quality using two-dimensional gas chromatography combined with time-of-flight mass spectrometry and chemometrics. J. Chromatogr. A **1217**(12), 1963–1970 (2010)

16. John, O.P., Donahue, E.M., Kentle, R.L.: The Big Five Inventory: Institute of Personality and Social Research. UC Berkeley, CA (1991)
17. Khot, L.R., Panigrahi, S., Doetkott, C., Chang, Y., Glower, J., Amamcharla, J., Logue, C., Sherwood, J.: Evaluation of technique to overcome small dataset problems during neural-network based contamination classification of packaged beef using integrated olfactory sensor system. LWT-Food Sci. Technol. **45**(2), 233–240 (2012)
18. Kittur, A., Chi, E.H., Suh, B.: Crowdsourcing user studies with mechanical turk. In: Proceedings of the SIGCHI Conference on Human Factors in Computing Systems, pp. 453–456. ACM (2008)
19. Lepri, B., Subramanian, R., Kalimeri, K., Staiano, J., Pianesi, F., Sebe, N.: Connecting meeting behavior with extraversion—a systematic study. IEEE Trans. Affect. Comput. **3**(4), 443–455 (2012)
20. Machajdik, J., Hanbury, A.: Affective image classification using features inspired by psychology and art theory. In: Proceedings of the International Conference on Multimedia, pp. 83–92. ACM (2010)
21. McCrae, R.R., John, O.P.: An introduction to the five-factor model and its applications. J. Pers. **60**(2), 175–215 (1992)
22. Mohammadi, G., Vinciarelli, A.: Automatic personality perception: prediction of trait attribution based on prosodic features. IEEE Trans. Affect. Comput. **3**(3), 273–284 (2012)
23. de Montjoye, Y.-A., Quoidbach, J., Robic, F., Pentland, A.S.: Predicting personality using novel mobile phone-based metrics. In: Greenberg, A.M., Kennedy, W.G., Bos, N.D. (eds.) SBP 2013. LNCS, vol. 7812, pp. 48–55. Springer, Heidelberg (2013)
24. Park, G., Schwartz, H.A., Eichstaedt, J.C., Kern, M.L., Kosinski, M., Stillwell, D.J., Ungar, L.H., Seligman, M.E.: Automatic personality assessment through social media language (2014)
25. Quercia, D., Kosinski, M., Stillwell, D., Crowcroft, J.: Our twitter profiles, our selves: predicting personality with twitter. In: IEEE Third Conference on Social Computing, pp. 180–185 (2011)
26. Rawlings, D., Ciancarelli, V.: Music preference and the five-factor model of the neo personality inventory. Psychol. Music **25**(2), 120–132 (1997)
27. Ross, C., Orr, E.S., Sisic, M., Arseneault, J.M., Simmering, M.G., Orr, R.R.: Personality and motivations associated with facebook use. Comput. Hum. Behav. **25**, 578–586 (2009)
28. Tkalčič, M., Ferwerda, B., Hauger, D., Schedl, M.: Personality correlates for digital concert program notes. In: Ricci, F., Bontcheva, K., Conlan, O., Lawless, S. (eds.) UMAP 2015. LNCS, vol. 9146, pp. 364–369. Springer, Heidelberg (2015)
29. Tkalcic, M., Kunaver, M., Košir, A., Tasic, J.: Addressing the new user problem with a personality based user similarity measure. In: DEMRA (2011)
30. Tkalcic, M., Tasic, J.: Colour spaces: perceptual, historical and applicational background. In: IEEE EUROCON, Computer as a Tool, pp. 304–308 (2003)
31. Valdez, P., Mehrabian, A.: Effects of color on emotions. J. Exp. Psychol. Gen. **123**(4), 394 (1994)
32. Vinciarelli, A., Mohammadi, G.: A survey of personality computing. IEEE Trans. Affect. Comput. **5**(3), 273–291 (2014)
33. Viola, P., Jones, M.J.: Robust real-time face detection. Int. J. Comput. Vis. **57**(2), 137–154 (2004)
34. Witten, I.H., Frank, E.: Data Mining: Practical Machine Learning Tools and Techniques. Morgan Kaufmann, San Francisco (2005)

Extracting Visual Knowledge from the Internet: Making Sense of Image Data

Yazhou Yao[1,2]($\boxtimes$), Jian Zhang[1], Xian-Sheng Hua[3], Fumin Shen[4], and Zhenmin Tang[2]

[1] Advanced Analytics Institute, University of Technology, Sydney, Australia
[2] Nanjing University of Science and Technology, Nanjing, China
yazhou.yao@student.uts.edu.au
[3] Alibaba Group, Hangzhou, China
[4] University of Electronic Science and Technology of China, Chengdu, China

Abstract. Recent successes in visual recognition can be primarily attributed to feature representation, learning algorithms, and the ever-increasing size of labeled training data. Extensive research has been devoted to the first two, but much less attention has been paid to the third. Due to the high cost of manual data labeling, the size of recent efforts such as *ImageNet* is still relatively small in respect to daily applications. In this work, we mainly focus on how to automatically generate identifying image data for a given visual concept on a vast scale.

With the generated image data, we can train a robust recognition model for the given concept. We evaluate the proposed webly supervised approach on the benchmark *Pascal VOC 2007* dataset and the results demonstrates the superiority of our method over many other state-of-the-art methods in image data collection.

Keywords: Webly supervised · Image data generation · Data purifying · CNN

1 Introduction

With the development of the Internet, we have entered the era of big data. It is consequently a natural idea to leverage the large scale yet noisy data on the Internet [9,10]. Methods of utilizing these data for visual recognition have recently become a hot topic, with the convergence of computer vision, pattern recognition and machine learning being collectively known as *"Internet vision"* [9].

When generating massive identifying image data, it is important to ensure that the data contains sufficient representative visual patterns. Search engines (e.g., Google Images) and social media portals (e.g., Flickr) have been created to obtain the candidate image data [20,25]. For example, we can submit a certain concept as a query to Google Images or Flickr which will return a large number of images based on that query. However, the images returned by search engines or social media portals are usually attractive and representative, but they incorporate fewer visual patterns. For example, if we submit the term "tiger" to Google

© Springer International Publishing Switzerland 2016
Q. Tian et al. (Eds.): MMM 2016, Part I, LNCS 9516, pp. 862–873, 2016.
DOI: 10.1007/978-3-319-27671-7_72

Images, most of the returned images are of tiger faces. To build a high-quality training dataset, however, we need to collect a large number of Internet images which contain different views of an image as visual patterns. To address this problem, we propose a two-step approach in this paper, as follows: We first find the useful related word variations to enrich the given concept from a text perspective, which is known as label purification; we then use these selected word variations to collect images and run a further image clean process, which is known as image purification. Our goal is to achieve a certain level of match between the image labels and their dominant contents to narrow the semantic gap.

Separating Effective Data from Noise. Search results returned from Google Books Ngrams corpora are usually very noisy. For example, word variations returned from Google Books Ngrams corpora for the query "horse" contains not only different "visual patterns" of a horse such as a "jumping horse" and a "rearing horse", but also a large number of different word variations associated with "horse" such as "horse boy" and "horse stealer". Our starting point in this step is to find key "visual patterns" of horse. Therefore, "horse boy", "horse stealer" etc. should be removed from the word variations list.

In addition, images returned from the Google search engine also contain noise. For example, it may contain some clipart images even at the top of the return list. In order to build a large scale robust image dataset, noisy images such as clipart images should be filtered out without too much manual effort.

Leveraging the Massive Amount of Data. By searching in the Google Books Ngrams corpora and Google images, it is easy to get over 1000 related word variations and 10000 images of the given concept. However, how to effectively leverage this massive and noisy data to build a robust image dataset remains a huge challenge. In this paper, we argue that combining word variation purifying and image purifying is a more effective way to use this massive amount of data.

Main Contributions. Through our framework and outcomes, our main contributions are: (a) our method is the first to combine related word variation purifying and image purifying to build the image dataset; (b) our method is able to build the image data for any given visual concept; and (c) our method is capable of training a robust recognition model without manual intervention.

2   Related Work

Recent work has mainly focused on learning from large datasets for recognition and classification [4,14,22,26–28]. To our knowledge, there are three principal methods of building the database: manual annotation, active learning, and using Internet data.

Manual Annotation. Manual annotation has a high level of accuracy but is resource-intensive, so the scale of the dataset is relatively small (both the numbers of categories and images) in the early years. The method of building ImageNet is using manual annotations. It firstly download images from the search engines using different languages for the given concept and then label these images by the power of crowds [4].

Active Learning. To reduce the cost of manual annotation, recent work has also focused on active learning (a special case of semi-supervised learning) [23, 29–31], which selects label requests. [23] introduces an approach for learning object detectors from real-world web videos. The limitation is these web videos should only contain objects of a target class. [29] build a contextual object recognition model by using active learning. It needs the user's answer to update the existing object recognition model. [30] learn a detector for a specific object class using weakly supervised learning and need a initial annotation. Both manual annotation and active learning require pre-existing annotations, which results in one of the biggest limitations to building a large scale dataset.

Using Internet Data. Recent methods propose the use of Internet data to build the training data for visual concept [1, 5, 10, 14, 25]. The general method is to automatically collect images using search engines or social networks to build a training set, and then to re-rank these images using visual classifiers [5, 10] or some form of clustering in visual space [25]. [1] extracts visual knowledge from the pool of visual data on the web, mainly focusing on finding labeled segments/boundaries and relationships between objects. [14] uses iterative methods to automatically collect object datasets from the web and to incrementally learn object category models. This approach has some uncertainty, since it depends on the seed images. In addition, using this method to collect images of objects may contain fewer visual patterns. However, the use of "re-ranking", "clustering" and the other methods mentioned above does not tackle the problem of insufficient visual patterns in these training data. To address this problem, we propose our method for building a high-quality training dataset in the next section.

3   System Framework and Methods

Due to the amount and complexity of Internet data, in order to build a high-quality training dataset, we must separate noisy data from useful data automatically. Figure 1 shows how we purify noisy data and train a recognition model for a given concept. As shown in the diagram, we first use Google Books Ngram corpora to obtain all the related word variations modifying the given concept; second, we use Normalized Google distance (NGD), linear SVM and visual distance to prune these word variations; third, we download images according to the filtered word variations, use exemplar-LDA and progressive CNN to purify these noisy images; lastly, we put all the purified images based on the various

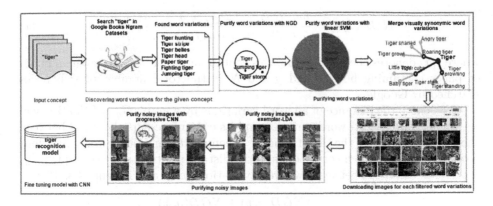

Fig. 1. The proposed framework.

word variations together and fine-tune a robust model for object recognition. The following subsections describe the details of our process.

3.1 Discovering Word Variations for the Given Concept

As mentioned above, we use Google Books Ngrams corpora to generate "visual patterns" on the Internet scale for the given concept [16]. These corpora cover almost all related word variations for any concept at the text level and are much more general and richer than WordNet or Wikipedia [21,32]. We use Google Books Ngrams corpora to discover related word variations for the given concept with parts-of-speech (POS), specifically with NOUN, VERB, ADJECTIVE and ADVERB. Using Google Books Ngrams data helps us cover all variations of any concept the human race has ever written down in books [5]. In addition, using the POS tag can help us to partially purify word variations.

3.2 Purifying Noisy Word Variations

In our experiments, we found that not all the retrieved word variations are relevant to our given visual concept, e.g., "tiger sharks", "tiger belles", etc. Downloading images with these noisy word variations of the concept are harmful for training our target model "tiger"; therefore, we need to purify these noisy word variations and set up key word variations. We developed a fast NGD method to address this issue from the perspective of text semantics.

Purifying Based on NGD. Words and phrases acquire meaning from the way they are used in society, from their relative semantics to other words and phrases. For computers, the equivalent of "society" is "database", and the equivalent of "use" is "a way to search the database" [3]. Normalized Google distance constructs a method to automatically extract similarity distance from the World Wide Web (WWW) using Google page counts [2]. For a search term x and search term y (just the name for an object rather than the object itself), Normalized Google distance is defined by (1):

$$NGD(x,y) = \frac{max\{logf(x), logf(y)\} - logf(x,y)}{logN - min\{logf(x), logf(y)\}} \qquad (1)$$

where $f(x)$ denotes the number of pages containing x, $f(x,y)$ denotes the number of pages containing both x and y and N is the total number of web pages searched by Google.

We denote the distance of all word variations by a graph $G_g = \{N, D\}$ where each node represents a word variation and its edge represents the NGD between the two nodes. We set the target concept as center (d_0) and other word variations have a score (d_x) which corresponds to the distance to the target concept, (d_{xy}) represents the NGD between two word variations x, y, and is defined as (2):

$$d_{xy} = \frac{NGD(x,y) + NGD(y,x)}{2} \qquad (2)$$

By setting the threshold d_x to a value (0.5) we can successfully remove most of the noisy word variations while keeping the vast majority of useful word variations. Then we set d_{xy} to a value (0.1) to merge semantic synonyms. This step can be viewed as part of a cascade strategy that purifies irrelevant word variations, so we set the value of d_x and d_{xy} to ensure the vast majority of useful word variations are passed on to the next stage. Table 1 shows the results of purifying noisy word variations using NGD for the concept "tiger" and "horse".

Table 1. variations using NGD for "tiger" and "horse"

Concept	Found word variations			After NGD filtering			
	Correct	Noisy	Precision	Correct	Noisy	Precision	False pos
Horse	132	460	22.3 %	124	237	34.3 %	8
Tiger	108	536	16.8 %	102	287	26.2 %	6

By experiments, we found that there are still lots of visual non-salient word variations, e.g., "tiger shooting", "social tiger", etc. and we cannot purify these noisy word variations using NGD alone.

Purifying Based on Linear SVM. From the visual perspective viewpoint, we want to identify visual salient word variations and eliminate visual non-salient word variations in this step. The intuition is that visual salient word variations should exhibit predictable visual patterns that are accessible to classifiers [5]. We use the image-classifier based purifying method.

For each filtered word variation, we directly download 100 images from the Google image search engine as positive images; then randomly split these images into a training set (75 images) and validation set (25 images) $I_i = \{I_i^t, I_i^v\}$, we gather a random pool of negative images (50 images) and split them into a training set (25 images) and validation set (25 images) $\overline{I} = \{\overline{I}^t, \overline{I}^v\}$; We then train a linear SVM C_i with I_i^t and $\overline{I}^t$, using dense HOG features and then use $\{I_i^v, \overline{I}^v\}$ as validation images to calculate the classification results S_i [7,15]. Table 2 shows the results of the experiments on the concepts "tiger" and "horse":

Table 2. Classification results of the concepts "tiger" and "horse"

Concept:	S_i	Precision rate:	Recall rate:
Horse	0.7	0.8978	0.9278
	0.68	0.9408	0.8833
	0.66	0.9615	0.8332
Tiger	0.7	0.9045	0.9344
	0.68	0.9523	0.8932
	0.66	0.9731	0.8374

where S_i denotes the percentage of correctly classified images. We declare an word variation i to be visually salient if the classification results S_i give a relatively high score (0.7) as we want to purify visual non-salient word variations as much as possible.

Purifying Based on Visual Distance. By filtering the noisy word variations based on NGD and linear SVM, we obtain a relatively clean word variations set. We found some word variations share similar visual patterns, e.g., "tiger cubs, small tiger, baby tiger, little tiger". To avoid collecting too many visually similar images as training data, we group these word variations [18]. We represent the distance of all word variations or labels by a graph $G_v = \{V, E\}$ where each node represents a word variation and each edge represents the distance between two variations. Each node has a score V_i which corresponds to the classifier C_i on its validation data $\{I_i^v, \overline{I}^v\}$. As previously mentioned, the edge weights E_{ij} from the visual perspective correspond to the distance between two word variations (labels) i, j and are measured by the score of the jth word variation classifier C_j on the ith word variation validation set $\{I_i^v, \overline{I}^v\}$. We group these word variations based on their distances to satisfy (3):

$$E_{ij} + 0.1 \geqslant V_j \tag{3}$$

Table 3 shows some examples of word variations merged by our method:

Table 3. Some examples of the merged word variations

Concept:	Merged word variations:
Horse	{horse cabs vs horse carriages vs horse carts vs horse buses vs horse trams}
Dog	{horse cabs vs horse carriages vs horse carts vs horse buses vs horse trams}
Bus	{double bus vs london bus vs open bus}

By using the above procedures, we obtain relatively clean word variations to represent different "visual patterns" for the given concept. Algorithm 1 shows the process of purifying these irrelevant word variations.

Algorithm 1. Word variation purifying algorithm

Input:
  $X = \{x_0\}$, a concept for image label
1: Discover word variations in Google Books Ngrams corpora with POS and get $X = \{x_0, x_1, x_2 .... x_n\}$
2: Calculate NGD d_i between x_0 and x_i, delete x_i from X if $d_i \geqslant 0.5$
3: Calculate NGD d_{ij} between x_i and x_j $(i, j \geqslant 0)$, merge x_i and x_j if $d_{ij} \leqslant 0.1$
4: Delete visually non-salient x_i from X if $S_i \leqslant 0.7$
5: Merge visually similar x_i and x_j if $E_{ij} + 0.1 \geqslant V_j$
Output:
  a relatively clean word variation X for the given concept

We use these filtered word variations as queries to download the top 120 images from Google image for each word variation. Then we put these images together as an initial image dataset for the given concept. Although we get the initial images from the Internet similar to [1,5,9,10,14,19,24,25], the difference is we firstly get lots of related word variations representing different "visual patterns" for the given concept. Downloaded images using these related word variations are much more ample than only using the given concept. This is also our advantage over other methods. Table 4 shows the relatively clean word variations found by our "purifying word variations":

Table 4. Word variations found on Pascal VOC 2007

Concept:	Bottle	Train	Cat	Cow	Dog	Horse	Sheep	Plane	Bus	Car
vars:	128	47	174	136	145	116	123	135	59	125
Concept:	Bike	Boat	Sofa	Bird	Mbike	Plant	Table	Chair	Pers	Tv
vars:	43	207	38	226	33	5	437	345	209	36

The Limitation of Our Method. From our experiments, we found our method is not able to remove these noisy word variations thoroughly. Also, some positive word variations may be filtered out incorrectly. Filtering our word variations using the previous steps results in an average (for PASCAL Visual Object Classes) of 3.24 % noisy word variations and an average 2.84 % positive word variations being filtered out for the given concept. Using these word variations may result in noisy images to our initial image dataset for the given concept (type 1 noisy images). We found these word variations are a very small number respect to the correct word variations. These few noisy images caused by noisy word variations can be effectively filtered out by the next image purifying steps.

There are other types of noisy images which result from correct word variations in our initial image dataset for the given concept. Although the Google image search engine ranks the returned images, some noisy images are still included (type 2 noisy images). The reason for this is that the Google image search engine is a text based image search engine. To build a high-quality image dataset, both of these two types of noisy images should be removed from the initial image dataset.

3.3   Purifying Noisy Images

Most current approaches handle these problems via clustering [13,17,33]. Clustering can help assist with handling visual diversity and can reject outliers based on their distance from cluster centers. However, clustering presents a scalability issue for our problem. Since our images are sourced directly from the Internet and have no bounding boxes, every image creates millions of data points, the majority of which are outliers. Recent work has suggested that K-means is not scalable [6]. Instead, we propose to use a two-step approach to purify these noisy images.

Purifying Based on Exemplar-LDA. To purify these type 2 noisy images caused by Google image search engine, we download a set of images (120) from the Google image search engine for each selected word variation such as "fighting tiger". The image set is used to train a detector using exemplar-LDA [11], and these detectors are then used for dense detections on the same image set. We select the top 100 images with high scores from multiple detectors for the next step. This method assists us to prune those images which relate less well to the word variations, e.g., in Fig. 2, three of the images are not related to our word variations for the scene concept "fighting tiger". This type of noisy images will be filtered in this step.

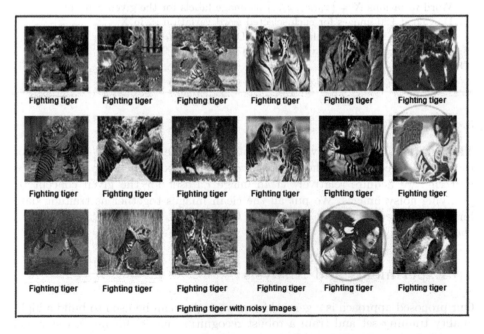

Fig. 2. Images returned by the search engine for "fighting tiger"

Purifying Based on Progressive CNN. To reduce the influence of type 1 noisy images, we use a purifying method similar to that proposed by [34]. The difference is that we do not train a CNN from the beginning; instead, we directly fine-tune a CNN on filtered images with a trained model *"bvlc_reference_caffenet"* [12]. We then use the probabilistic sampling algorithm to select the new training samples according to the prediction score of the fine-tuned model on the training data itself. The intuition is we want to keep images with distinct sentiment scores between the two classes with high probability, and remove images with similar sentiment scores for both classes with high probability. Let $S_i = (S_{i1}, S_{i2})$ be the prediction sentiment scores for the two classes of instance i. We choose to remove the training instance i with probability P_i given by (4):

$$P_i = max(0, 2 - exp(|S_{i1} - S_{i2}|))$$ (4)

The training instance will be kept in the training set if the predicted scores of one training instance are large enough. Otherwise, the smaller the difference between the predicted scores, the large the probability that this instance will be removed from the training set. Type 1 noisy images can be effectively filtered by this step. The reason for this is that the number of this type of noisy images is relatively small in the whole image dataset for the given concept. Algorithm 2 shows the detailed process for purifying noisy images.

Algorithm 2. Image purifying algorithm

Input:
    Word variations $X = \{x_0, x_1, x_2...\}$ as image labels for the given concept
1: Download 120 images for each selected word variation x_i in X
2: Select top 100 images with exemplar-LDA for each x_i in X
3: Purify the remaining noisy images with progressive CNN
Output:
    a relatively clean image dataset for the given concept

3.4   Model Learning

Through the above steps, we firstly obtain the different "visual patterns" for the given concept from the perspective of the text. Then for each "visual pattern", we acquire relatively clean image dataset for the given concept by purifying noisy images. We put all the clean images together as training data for the given concept, and fine-tune a CNN model with a pre-trained model *"bvlc_reference_caffenet"* [12].

4   Experiments and Analysis

Our proposed approach is a general framework that can be used to build a high-quality training set and train a robust recognition model for any given visual concept. To quantitatively evaluate the performance of our approach, we choose

Table 5. Results (average precision) on Pascal VOC 2007 (test) object detection

Method	Supervised	Bottle	Train	Cat	Cow	Dog	Horse	Sheep	Plane	Bus	Car
[30]	Weak	0	**34.2**	7.1	9.3	1.5	29.4	0.4	13.4	31.2	**43.9**
[5]	Web	**9.2**	23.5	8.4	17.5	12.9	30.6	**18.8**	14.0	35	35.9
Our	**Web**	8.7	22.5	**10.3**	**18.7**	**13.6**	**32.7**	13.7	13.8	**36.5**	35.6
[8]	Full	26.6	45.2	22.5	24.3	12.6	56.5	20.9	33.2	52.0	53.7
Method	Supervised	Bike	Boat	Sofa	Bird	Mbike	Plant	Table	Chair	Pers	Tv
[30]	Weak	**44.2**	3.1	3.8	3.1	**38.3**	0.1	**9.9**	0.1	4.6	0
[5]	Web	36.2	10.3	10.3	**12.5**	27.5	1.5	6.5	10.0	6.0	16.4
Our	**Web**	37.1	**12.3**	**11.2**	11.7	26.2	**1.9**	6.4	**12.9**	**7.2**	**20.3**
[8]	Full	59.3	15.7	35.9	10.3	48.5	33.2	26.9	20.2	43.3	16.4

the *Pascal VOC 2007* test set 20 categories for testing. Table 5 displays the results obtained using our algorithm and compares them with state-of-the-art baselines [5, 8, 30].

[30] is trained on *Pascal VOC 2007* training data with image-level labels. [5] is the state-of-the-art results for webly-supervised detection. [8] is the state-of-the-art results for fully-supervised detection.

Compared to [30] which uses weak supervision and [8] which uses full supervision, our method uses web-supervision as even the training set does not need to be labeled manually. Nonetheless, our results substantially surpass the previous best results in weakly supervised object detection. Compared to [5] which also uses web supervision, our method surpasses their results in most of the cases. The main reason for this is that our training data generated from the Internet contains much richer and accurate visual patterns in images. By observing the binding data in Tables 4 and 5, we found that those concepts which have good performance tend to have sufficient word variations for the concept (with the exception of "table"). In other words, our approach discovers concepts that have much more useful linkages to the visual patterns in the corresponding image set.

Lastly, we reach an important conclusion: a good training set should achieve successful results both in scale and in quality.

5   Conclusion and Future Work

We presented a fully automated approach to the generation of high-quality training data for any given visual concept. Our aim was to remove the need to laboriously annotate training datasets and train a robust recognition model with the generated training data for the given concept. Through our experiments on the benchmark *Pascal VOC 2007* test set, we found our approach surpasses most of the previous best result in weakly supervised and webly supervised object detection.

Using related word variations which link to the different "visual patterns" of images and then building the training image set for a concept or query according to these word variations is our first attempt to make use of textual metadata to build the training set. There is still room to improve our approach, for example, we can potentially use more sophisticated approaches to purify noisy images downloaded from the Internet and this will be the focus of our future work.

References

1. Chen, X., Shrivastava, A., Gupta, A.: Neil: extracting visual knowledge from web data. In: 2013 IEEE International Conference on Computer Vision (ICCV), pp. 1409–1416. IEEE (2013)
2. Cilibrasi, R.L., Vitanyi, P.M.: The google similarity distance. IEEE Trans. Knowl. Data Eng. **19**(3), 370–383 (2007)
3. Collosal, C.: How well does the world wide web represent human language. The Economist (2005)
4. Deng, J., Dong, W., Socher, R., Li, L.-J., Li, K., Fei-Fei, L.: Imagenet: a large-scale hierarchical image database. In: IEEE Conference on Computer Vision and Pattern Recognition, CVPR 2009, pp. 248–255. IEEE (2009)
5. Divvala, S.K., Farhadi, A., Guestrin, C.: Learning everything about anything: webly-supervised visual concept learning. In: 2014 IEEE Conference on Computer Vision and Pattern Recognition (CVPR), pp. 3270–3277. IEEE (2014)
6. Doersch, C., Singh, S., Gupta, A., Sivic, J., Efros, A.: What makes paris look like paris? ACM Trans. Graph. **31**(4), 101:9 (2012)
7. Fan, R.-E., Chang, K.-W., Hsieh, C.-J., Wang, X.-R., Lin, C.-J.: Liblinear: a library for large linear classification. J. Mach. Learn. Res. **9**, 1871–1874 (2008)
8. Felzenszwalb, P.F., Girshick, R.B., McAllester, D., Ramanan, D.: Object detection with discriminatively trained part-based models. IEEE Trans. Pattern Anal. Mach. Intell. **32**(9), 1627–1645 (2010)
9. Fergus, R., Fei-Fei, L., Perona, P., Zisserman, A.: Learning object categories from internet image searches. Proc. IEEE **98**(8), 1453–1466 (2010)
10. Fergus, R., Perona, P., Zisserman, A.: A visual category filter for google images. In: Pajdla, T., Matas, J.G. (eds.) ECCV 2004. LNCS, vol. 3021, pp. 242–256. Springer, Heidelberg (2004)
11. Hariharan, B., Malik, J., Ramanan, D.: Discriminative decorrelation for clustering and classification. In: Fitzgibbon, A., Lazebnik, S., Perona, P., Sato, Y., Schmid, C. (eds.) ECCV 2012, Part IV. LNCS, vol. 7575, pp. 459–472. Springer, Heidelberg (2012)
12. Jia, Y., Shelhamer, E., Donahue, J., Karayev, S., Long, J., Girshick, R., Guadarrama, S., Darrell, T.: Caffe: convolutional architecture for fast feature embedding. In: Proceedings of the ACM International Conference on Multimedia, pp. 675–678. ACM (2014)
13. Kankanhalli, M.S., Mehtre, B.M., Wu, R.K.: Cluster-based color matching for image retrieval. Pattern Recogn. **29**(4), 701–708 (1996)
14. Li, L.-J., Fei-Fei, L.: Optimol: automatic online picture collection via incremental model learning. Int. J. Comput. Vis. **88**(2), 147–168 (2010)
15. Lin, Y., Lv, F., Zhu, S., Yang, M., Cour, T., Yu, K., Cao, L., Huang, T.: Large-scale image classification: fast feature extraction and svm training. In: 2011 IEEE Conference on Computer Vision and Pattern Recognition (CVPR), pp. 1689–1696. IEEE (2011)
16. Lin, Y., Michel, J.-B., Aiden, E.L., Orwant, J., Brockman, W., Petrov, S.: Syntactic annotations for the google books ngram corpus. In: Proceedings of the ACL 2012 System Demonstrations, pp. 169–174. Association for Computational Linguistics (2012)

17. Lucchi, A., Weston, J.: Joint image and word sense discrimination for image retrieval. In: Fitzgibbon, A., Lazebnik, S., Perona, P., Sato, Y., Schmid, C. (eds.) ECCV 2012, Part I. LNCS, vol. 7572, pp. 130–143. Springer, Heidelberg (2012)
18. Malisiewicz, T., Efros, A., et al.: Recognition by association via learning per-exemplar distances. In: IEEE Conference on Computer Vision and Pattern Recognition, CVPR 2008, pp. 1–8. IEEE (2008)
19. Mezuman, E., Weiss, Y.: Learning about canonical views from internet image collections. In: Advances in Neural Information Processing Systems, pp. 719–727 (2012)
20. Michel, J.-B., Shen, Y.K., Aiden, A.P., Veres, A., Gray, M.K., Pickett, J.P., Hoiberg, D., Clancy, D., Norvig, P., Orwant, J., et al.: Quantitative analysis of culture using millions of digitized books. Science 331(6014), 176–182 (2011)
21. Miller, G.A.: Wordnet: a lexical database for english. Commun. ACM 38(11), 39–41 (1995)
22. Perona, P.: Vision of a visipedia. Proc. IEEE 98(8), 1526–1534 (2010)
23. Prest, A., Leistner, C., Civera, J., Schmid, C., Ferrari, V.: Learning object class detectors from weakly annotated video. In: 2012 IEEE Conference on Computer Vision and Pattern Recognition (CVPR), pp. 3282–3289. IEEE (2012)
24. Rubinstein, M., Joulin, A., Kopf, J., Liu, C.: Unsupervised joint object discovery and segmentation in internet images. In: 2013 IEEE Conference on Computer Vision and Pattern Recognition (CVPR), pp. 1939–1946. IEEE (2013)
25. Schroff, F., Criminisi, A., Zisserman, A.: Harvesting image databases from the web. IEEE Trans. Pattern Anal. Mach. Intell. 33(4), 754–766 (2011)
26. Shen, F., Liu, W., Zhang, S., Yang, Y., Shen, H.: Learning binary codes for maximum inner product search. In: The IEEE Conference on Computer Vision (ICCV), December 2015
27. Shen, F., Shen, C., Liu, W., Tao Shen, H.: Supervised discrete hashing. In: The IEEE Conference on Computer Vision and Pattern Recognition (CVPR), pp. 37–45, June 2015
28. Shen, F., Shen, C., Shi, Q., Van Den Hengel, A., Tang, Z.: Inductive hashing on manifolds. In: 2013 IEEE Conference on Computer Vision and Pattern Recognition (CVPR), pp. 1562–1569. IEEE (2013)
29. Siddiquie, B., Gupta, A.: Beyond active noun tagging: modeling contextual interactions for multi-class active learning. In: 2010 IEEE Conference on Computer Vision and Pattern Recognition (CVPR), pp. 2979–2986. IEEE (2010)
30. Siva, P., Xiang, T.: Weakly supervised object detector learning with model drift detection. In: 2011 IEEE International Conference on Computer Vision (ICCV), pp. 343–350. IEEE (2011)
31. Vijayanarasimhan, S., Grauman, K.: Large-scale live active learning: training object detectors with crawled data and crowds. Int. J. Comput. Vis. 108(1–2), 97–114 (2014)
32. Völkel, M., Krötzsch, M., Vrandecic, D., Haller, H., Studer, R.: Semantic wikipedia. In: Proceedings of the 15th International Conference on World Wide Web, pp. 585–594. ACM (2006)
33. Wang, W., Song, H.: Cell cluster image segmentation on form analysis. In: Third International Conference on Natural Computation, ICNC 2007, vol. 4, pp. 833–836. IEEE (2007)
34. You, Q., Luo, J., Jin, H., Yang, J.: Robust image sentiment analysis using progressively trained and domain transferred deep networks. In: The Twenty-Ninth AAAI Conference on Artificial Intelligence (AAAI) (2015)

Ordering of Visual Descriptors in a Classifier Cascade Towards Improved Video Concept Detection

Foteini Markatopoulou[1,2]([✉]), Vasileios Mezaris[1], and Ioannis Patras[2]

[1] Information Technologies Institute (ITI), CERTH, 57001 Thermi, Greece
{markatopoulou,bmezaris}@iti.gr
[2] Queen Mary University of London, Mile End Campus, London E14NS, UK
i.patras@qmul.ac.uk

Abstract. Concept detection for semantic annotation of video fragments (e.g. keyframes) is a popular and challenging problem. A variety of visual features is typically extracted and combined in order to learn the relation between feature-based keyframe representations and semantic concepts. In recent years the available pool of features has increased rapidly, and features based on deep convolutional neural networks in combination with other visual descriptors have significantly contributed to improved concept detection accuracy. This work proposes an algorithm that dynamically selects, orders and combines many base classifiers, trained independently with different feature-based keyframe representations, in a cascade architecture for video concept detection. The proposed cascade is more accurate and computationally more efficient, in terms of classifier evaluations, than state-of-the-art classifier combination approaches.

Keywords: Concept detection · Video analysis · Cascade architecture · Classifier ordering

1 Introduction

Video concept detection is a popular research topic that aims to annotate video fragments (e.g. keyframes) with semantic concept labels (e.g. sky, people etc.). Large-scale semantic concept detection systems mainly follow a process where a video is initially segmented into meaningful fragments, called shots; each shot is represented by e.g. one or more characteristic keyframes; and, several features (e.g. different local visual descriptors) are extracted from the keyframes (or any other chosen representation) of each shot. Given a ground-truth annotated video training set, supervised machine learning algorithms are used for building multiple independent base classifiers (concept detectors), using different types of features, for the same concept; the outputs of them are combined by means of late fusion. This ensemble-based approach is more accurate than using a single base classifier, trained on a single type of features (e.g. SIFT only). In this work we

© Springer International Publishing Switzerland 2016
Q. Tian et al. (Eds.): MMM 2016, Part I, LNCS 9516, pp. 874–885, 2016.
DOI: 10.1007/978-3-319-27671-7_73

propose an improved way of ordering and combining independently trained concept detectors using a cascade. The proposed cascade combines handcrafted (e.g. SIFT) and deep convolutional neural network (DCNN) features, and is computationally more efficient and more accurate than other combination approaches by adjusting the required processing (i.e. evaluate fewer classifiers) based on the input video fragment.

The rest of this paper is organized as follows: Sect. 2 reviews related work on learning and combining concept detectors. Section 3 introduces the proposed cascade architecture. Section 4 presents the experimental results and, finally, Sect. 5 presents conclusions.

2   Related Work

State of the art feature extraction methods are often based on local descriptors (e.g. SIFT [7], SURF [2]) in combination with encoding approaches (e.g. VLAD [5], FV [12]) in order to extract global, feature-based keyframe representations. In the last few years, features extracted with the use of pre-trained deep convolutional neural networks (DCNN) have also shown excellent results [15]. Using any of the above features, concept detection is typically treated as multiple independent binary classification problems (one per concept). That is, given the feature-based keyframe representations and also the ground-truth concept annotations for each keyframe, any supervised machine learning algorithm that solves classification problems, such as Support Vector Machine (SVM), can be trained in order to learn the relations between the low-level keyframe representations and the high-level semantic concepts.

It has been shown that combining many different keyframe representations (e.g. SIFT, RGB-SIFT, DCNN) for the same concept, instead of using a single feature (e.g. only SIFT), improves the concept detection accuracy [8]. The typical way of combining multiple features is to train several supervised classifiers for the same concept, each trained separately on a different feature. When all the classifiers give their decisions, a fusion step computes the final confidence score (e.g. by averaging); this process is known as late fusion. Hierarchical late fusion [16] is a more elaborate approach; classifiers that have been trained on more similar features (e.g. SIFT and RGB-SIFT) are firstly fused together and then, more dissimilar classifiers (e.g. DCNN-based) are sequentially fused with the previous groups. A second category of classifier combination approaches performs ensemble pruning to select a subset of the classifiers prior to their fusion. For example, [14] uses a genetic algorithm to automatically select an optimal subset of classifiers separately for each concept. Finally, there is a third group of popular ensemble-based algorithms, namely cascade architectures, that have been used in various visual classification tasks for training and combining detectors [3,4,9,10,18]. In a cascade architecture the classifiers are arranged in stages, from the less computationally demanding to the most demanding ones (or may be arranged according to other criteria such as their accuracy). A keyframe is classified sequentially by each stage and the next stage is triggered only if the previous one returns a positive prediction (i.e. that the concept or object appears in the keyframe). The rationale behind this is to rapidly reject keyframes that clearly do not match the classification criteria and focus on those keyframes that are more difficult and more likely to depict the sought concept or object. Cascades of classifiers have been mainly used in object detection tasks [18], however they have also been briefly examined for video/image concept detection [9,10].

Cascades developed for object and face detection are mainly boosting-based [1,3,4,10,18]. Each stage of the cascade is build using a boosting algorithm such as AdaBoost. Such approaches require the presence of a big pool of weak features (e.g. Haar-like features) in order to combine them and build a strong classifier. In contrast, video concept detection systems utilize a different kind of features, visual local descriptors encoded into global image representations or DCNN-based features that alone can build strong classifiers without boosting. For example, [9] presents a cascade with fixed ordering of the stages in terms of classifier accuracy and a simple threshold selection strategy that selects one rejection threshold per stage on the probability output of a classifier. The authors of [9] use the above cascade to combine binary, non-binary and DCNN-based features. In the present work, the proposed cascade is also developed so as to combine similar types of features, however in contrast to [9] which is based on a fixed ordering of the cascade stages, the proposed algorithm dynamically selects, orders, and combines a larger number of pre-trained base classifiers. This leads to both concept detection effectiveness and computational efficiency gains.

3   Cascade Construction with Pre-trained Classifiers

3.1   Cascade Architecture Overview

Figure 1 shows a cascade architecture suitable for combining many base classifiers that have been trained for the same concept [9]. Each stage j of the cascade encapsulates a stage classifier D_j that either combines many base classifiers $(B_1, B_2, \ldots, B_{f_j})$ that have been trained on different types of features or contains only one base classifier (B_1) that has been trained on a single type of features. In the first case, the output of f_j base classifiers is combined in order to return a single stage output score $D_j(I) = \dfrac{1}{f_j} \sum_{i=1}^{f_j} B_i(I)$, $f_j \geq 1$ in the [0,1] range. The second case is a special case where $f_j = 1$. Let I indicate an input keyframe; the classifier D_{j+1} of the cascade will be triggered for it only if the previous classifier does not reject the input keyframe I. Each stage j of the cascade is associated with a rejection threshold, while a stage classifier is said to reject an input keyframe if $D_j(I) < \theta_j$. A rejection indicates the classifier's belief that the concept does not appear in the keyframe. Given a set of pre-trained classifiers, we will present an algorithm that sets the ordering of cascade stages (i.e. the ordering of stage classifiers) and assigns thresholds to each stage in order to instantiate the above cascade.

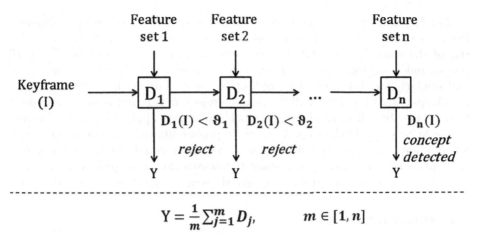

$$Y = \frac{1}{m}\sum_{j=1}^{m} D_j, \qquad m \in [1, n]$$

Fig. 1. Block diagram of a cascade architecture for one concept.

3.2   Problem Definition and Search Space

Let $D = \{D_1, D_2, \ldots, D_n\}$ be a set of n independently trained classifiers for a specific concept. Let $\mathbf{S} = [s_1, s_2, \ldots, s_n]^\top$ denote a vector of integer numbers in $[-1, 0) \cup [1, n]$. Each number represents the index of a classifier from D and appears at most once. The value -1 indicates that a classifier from D is omitted. Consequently, $\mathbf{S}$ expresses the ordering of the pre-trained classifiers $D_1, \ldots, D_n$. For example, given a pre-trained set of 4 classifiers $D = \{D_1, D_2, D_3, D_4\}$, the solution $\mathbf{S} = [2, 1, 3, -1]^\top$ denotes the cascade $D_{2,1,3,-1} : D_2 \rightarrow D_1 \rightarrow D_3$, where stage classifier D_4 is not used at all. In addition, let $\boldsymbol{\theta} = [\theta_1, \theta_2, \ldots, \theta_n]^\top$ denote a vector of rejection thresholds for the solution $\mathbf{S}$ and let $T = \{x_i, y_i\}_{i=1}^M$, where $y_i \in \{\pm 1\}$, be a finite set of annotated training samples for the given concept (x_i being the feature vectors and y_i the ground-truth annotations). The problem we aim to solve is finding the pair of the index sequence $\mathbf{S}$ (that leads to the cascade $D_\mathbf{S} : D_{s_1} \rightarrow D_{s_2} \rightarrow \cdots \rightarrow D_{s_n}$) and the vector of thresholds $\boldsymbol{\theta} = [\theta_1^\star, \theta_2^\star, \ldots, \theta_n^\star]^\top$ that maximizes the expected ranking gain on the finite set T. The implied optimization problem is given by the following equation:

$$(\mathbf{S}^\star, \boldsymbol{\theta}^\star) = \underset{(\mathbf{S}, \boldsymbol{\theta})}{\mathrm{argmax}}\{F(D_\mathbf{S}, T, \boldsymbol{\theta})\}, \tag{1}$$

where the ranking function $F(D_\mathbf{S}, T, \boldsymbol{\theta})$ can be defined as the expected ranking gain of $D_\mathbf{S}$ on T, that is

$$F_{AP}(D_\mathbf{S}, T, \boldsymbol{\theta}) = AP@k(rank(y), rank(D_\mathbf{S}(T, \boldsymbol{\theta})),$$

where, $rank(y)$ is the actual ranking of the samples in T (i.e., samples with $y_i = 1$ are ranked higher than samples with $y_i = -1$), and $rank(D_\mathbf{S}(T, \boldsymbol{\theta}))$ the predicted ranking of the samples of cascade $D_\mathbf{S}, \boldsymbol{\theta}$ on T. $AP@k$ is the average precision in the top k samples.

Let $l \leq n$ refer to the number of variables $s_j \in \mathbf{S}$ whose value is different from -1 (i.e., l is the number of cascade stages that solution $\mathbf{S}$ implies). The size of the search space related to the ordering of cascade stages is $\sum_{l=1}^{n} \binom{n}{l} l!$ (i.e. all index sequences for $l = 1$, all permutations of index sequences for $l = 2$, and similarly for all higher values of l, up to $l = n$). Furthermore, $\Theta \subset \mathbb{R}^n$ is the search space that consists of all the possible rejection thresholds for each stage of the cascade. To collect candidate threshold values, we apply each stage classifier on the finite set T. Each of the M returned probability output scores constitutes a candidate threshold. The size of the search space equals to M^n. Considering that this is a large search space, exhaustive search cannot be practically applied. To solve the problem we propose the greedy search algorithm described below.

3.3   Problem Solution

Our algorithm finds the final solution by sequentially replacing at each iteration a simple solution (consisting of a cascade with a certain number of stages) with a more complex one (consisting of a cascade with one additional stage). Algorithm 1 presents the proposed greedy search algorithm that instantiates the proposed cascade (Fig. 1). Let $\mathbf{S} = [s_1, s_2, \ldots, s_n]^\top$, and $\boldsymbol{\theta} = [\theta_{s_1}, \theta_{s_2}, \ldots, \theta_{s_n}]^\top$, represent a solution. Each variable $s_1, s_2, \ldots, s_n$ can take n possible values, from 1 to n or the value -1 which indicates that a stage is omitted. Each variable $\theta_{s_1}, \theta_{s_2}, \ldots, \theta_{s_n}$ can take M possible values. Initially we set, $s_j = -1$ for $j = 1, \ldots, n$ and $\boldsymbol{\theta} = [0, 0, \ldots, 0]^\top$ where $|\boldsymbol{\theta}| = n$. In the first step the algorithm optimizes $\mathbf{S}$ with respect to s_n (Algorithm 1: States 1–3) in order to build the solution:

$$\mathbf{S}_0 = [-1, -1, \ldots, s_n]^\top, \boldsymbol{\theta}_0 = [0, 0, \ldots, 0]^\top,$$

where according to (1),

$$s_n^\star = \underset{s_n}{\operatorname{argmax}}\{F_{AP}(D_{\mathbf{S}_0}, T, \boldsymbol{\theta}_0)\}. \tag{2}$$

and $\boldsymbol{\theta}_0^\star = [0, 0, \ldots, \theta_{s_n}^\star]$, $\theta_{s_n}^\star = 0$. This can be interpreted as the optimal solution of $l = 1$, that maximizes (1). Then the algorithm, in iteration j (Algorithm 1: States 4–7), assumes that it has solution with $l = j$, that is:

$$\mathbf{S}_{j-1}^\star = [s_1^\star, s_2^\star, \ldots, s_{j-1}^\star, -1, -1, \ldots, s_n^\star]^\top,$$

$$\boldsymbol{\theta}_{j-1}^\star = [\theta_{s_1}^\star, \theta_{s_2}^\star, \ldots, \theta_{s_{j-1}}^\star, 0, 0, \ldots, \theta_{s_n}^\star]^\top,$$

and finds the pair of $\mathbf{S}_j$ and $\boldsymbol{\theta}_j$ in one step as follows. It optimizes the pair of $\mathbf{S}_{j-1}^\star$ and $\boldsymbol{\theta}_{j-1}^\star$ with respect to s_j and θ_j, respectively, in order to find the solution:

$$\mathbf{S}_j = [s_1^\star, s_2^\star, \ldots, s_{j-1}^\star, s_j, -1, -1, \ldots, s_n^\star]^\top,$$

$$\boldsymbol{\theta}_j = [\theta_{s_1}^\star, \theta_{s_2}^\star, \ldots, \theta_{s_{j-1}}^\star, \theta_{s_j}, 0, 0 \ldots, \theta_{s_n}^\star]^\top.$$

According to (1):

$$(s_j^\star, \theta_{s_j}^\star) = \underset{(s_j, \theta_{s_j})}{\operatorname{argmax}}\{F_{AP}(D_{\mathbf{S}_j}, T, \boldsymbol{\theta}_j)\}. \tag{3}$$

Algorithm 1. Cascade stage ordering and threshold search

Input: Training set $T = \{x_i, y_i\}_{i=1}^{M}$, $y_i \in \{\pm 1\}$; n trained classifiers $D = \{D_1, D_2, ..., D_n\}$

Output: (i) An index sequence $\mathbf{S}^*$, of the ordering of cascade stages: $D_\mathbf{S}^* : D_{s_1}^* \rightarrow D_{s_2}^* \rightarrow ... \rightarrow D_{s_n}^*$. (ii) A vector of thresholds $\boldsymbol{\theta}^* = [\theta_{s_1}^*, \theta_{s_2}^*, ..., \theta_{s_n}^*]^\top$

Initialize: $\mathbf{S} = [s_1, s_2, ..., s_n]^\top$, $s_j = -1$, $j = 1, ..., n$, $\boldsymbol{\theta} = [0, 0, ..., 0]^\top$, $|\boldsymbol{\theta}| = n$,

1. $s_n^* = \mathrm{argmax}_{s_n}\{F_{AP}(D_{\mathbf{S}_0}, T, \boldsymbol{\theta}_0)\}$ (1),
   $\mathbf{S}_0 = [-1, -1, ..., s_n]^\top$, $\boldsymbol{\theta}_0 = [0, 0, .., 0]^\top$

2. $maxCost = F_{AP}(D_{\mathbf{S}_0^*}, T, \boldsymbol{\theta}_0^*)$,
   $\mathbf{S}_0^* = [-1, -1, ..., s_n^*]^\top$, $\boldsymbol{\theta}_0^* = [0, 0, ..., \theta_{s_n}^*]^\top$, $\theta_{s_n}^* = 0$

3. $\mathbf{S}^* = \mathbf{S}_0^*$, $\boldsymbol{\theta}^* = \boldsymbol{\theta}_0^*$

for $j = 1$ to $n - 1$ **do**

4. $(s_j^*, \theta_{s_j}^*) = \mathrm{argmax}_{(s_j, \theta_{s_j})}\{F_{AP}(D_{\mathbf{S}_j}, T, \boldsymbol{\theta}_j)\}$ (1),
   $\mathbf{S}_j = [..., s_j, -1, ..., s_n^*]^\top$, $\boldsymbol{\theta}_j = [..., \theta_{s_j}, 0, ..., \theta_{s_n}^*]^\top$

5. $cost = F_{AP}(D_{\mathbf{S}_j^*}, T, \boldsymbol{\theta}_j^*)$,
   $\mathbf{S}_j^* = [..., s_j^*, -1, ..., s_n^*]^\top$, $\boldsymbol{\theta}_j^* = [..., \theta_{s_j}^*, 0, ..., \theta_{s_n}^*]^\top$

   if cost>maxCost **then**

   6. maxCost = max(cost, maxCost)
   7. $\mathbf{S}^* = \mathbf{S}_j^*$, $\boldsymbol{\theta}^* = \boldsymbol{\theta}_j^*$

   end if

end for

The algorithm finds the pair of (s_j, θ_{s_j}) that optimizes (1). The complexity of this calculation equals to $(n - j) \times M$. This corresponds to $n - j$ possible values that variable s_j can take in iteration j and M possible threshold rejection values that variable θ_{s_j} can take for every different instantiation of s_j. Finally, the optimal sequence $\mathbf{S}^*$ equals to

$$\mathbf{S}^* = \mathop{\mathrm{argmax}}_{\mathbf{S} \in \{\mathbf{S}_0^*, \mathbf{S}_1^*, ..., \mathbf{S}_{n-1}^*\}} \{F_{AP}(D_\mathbf{S}, T, \boldsymbol{\theta})\}, \tag{4}$$

which is the sequence that optimizes (1) within all the iterations of the algorithm (Algorithm 1: States 6–7). The optimal threshold vector $\boldsymbol{\theta}^*$ is the vector connected to the optimal sequence $\mathbf{S}^*$. Our algorithm focuses on the optimization of the complete cascade and not the optimization of each stage separately from the other stages. This is expected to give a better complete solution. Furthermore, the algorithm can be slightly modified to make the search more efficient. For example, at each iteration we can keep the p best solutions. However, this would increase the computational cost.

4   Experiments

4.1   Dataset and Experimental Setup

Our experiments were performed on the TRECVID 2013 Semantic Indexing (SIN) dataset [11], which consists of a development set and a test set (approximately 800 and 200 h of internet archive videos for training and testing, respectively). We evaluated our system on the test set using the 38 concepts that were evaluated as part of the TRECVID 2013 SIN Task [11]. The video indexing problem was examined; that is, given a concept, we measure how well the top retrieved video shots for this concept truly relate to it. For experimenting with all methods, one keyframe was extracted for each video shot. Regarding local feature extraction, we followed the experimental setup of [8]. More specifically, we extracted nine local descriptors, presented in Table 1. All the local descriptors were compacted using PCA and were subsequently aggregated using the VLAD encoding. The VLAD vectors were reduced to 4000 dimensions. In addition, we used features based on three different pre-trained convolutional neural networks: (i) The 16-layer deep ConvNet network provided by [15], (ii) the 22-layer GoogLeNet network provided by [17], and (iii) the 8-layer CaffeNet network described in [6]. We applied each of these networks on the TRECVID keyframes and we used as a feature (i) the output of the last hidden layer of ConvNet (fc7), which resulted to a 4096-element vector, (ii) the output of the last fully-connected layer of CaffeNet (fc8), which resulted to a 1000-element, (iii) the output of the last fully-connected layer of GoogLeNet (loss3). We refer to these features as CONV, CAFFE and GNET in the sequel, respectively.

To train our base classifiers, for each concept, a training set was assembled that included all negative annotated training examples for the given concept and three copies of each positive training sample (in order to account for the most often limited number of the latter). Then the positive and negative ratio of training examples was fixed by randomly rejecting any excess negative samples, to achieve an 1:6 ratio. This is important for building a balanced classifier. Given the twelve different types of feature vectors described above, for each concept we trained twelve different base-classifiers, using linear SVMs. In all cases, the final step of concept detection was to refine the calculated detection scores by employing the re-ranking method proposed in [13].

We compared the proposed cascade (Sect. 3) with five different ensemble combination approaches: (i) Late-fusion with arithmetic mean [16]. (ii) The ensemble pruning method proposed by [14]. (iii) The cascade proposed by [9]. In this case we only applied the thresholding strategy of [9], that is, we did not perform any retraining, which appeared to be less accurate and computationally more expensive. We refer to this method as cascade-thresholding. (iv) A cascade with fixed ordering of the stages in terms of classifier accuracy, and the offline dynamic programming algorithm for threshold assignment proposed by [3]. In contrast to [3] that aims to improve the overall classification speed, we optimize the overall detection performance of the cascade in terms of AP. We refer to this method as cascade-dynamic in the sequel. (v) A boosting-based approach (i.e., the multimodal sequential SVM [1]). We refer to this method as AdaBoost. Both for the proposed and also for the cascade-dynamic method we used quantization to ensure that the optimized cascade generalizes well to unseen samples. In these lines, instead of searching for candidate thresholds on all the M examples of a validation set, we sorted the score values in descending order and split at every M/Q example (Q was set to 32). For all the methods, except for the Late-fusion that does not require this, the training set was also used as the validation set.

Table 1. Performance (MXinfAP, %) for each of the stage classifiers that we used in the experiments. For stage classifiers that are made of more than one base classifiers, we report in parenthesis the MXinfAP for each of these base classifiers.

Stage classifier	MXinfAP	Base classifiers
ORBx3	17.91 (12.18,13.81,14.12)	ORB, RGB-ORB, OpponentORB
SURFx3	18.68 (14.71,15.49,15.89)	SURF, OpponentSURF, RGB-SURF
SIFTx3	20.23 (16.55,16.73,16.75)	SIFT, OpponentSIFT, RGB-SIFT
CAFFE	19.80	Last fully-connected layer of CaffeNet
GNET	24.36	Last fully-connected layer of GoogLeNet
CONV	25.26	Last hidden layer of ConvNet

Table 2. Performance (MXinfAP, %) for different classifier combination approaches.

RunID	Stage classifiers	M1	M2	M3	M4	M5	M6
		Late-fusion [16]	Ensemble pruning [14]	Cascade-thresholding [9]	Cascade-dynamic [3]	AdaBoost [1]	Cascade-proposed
R1	ORBx3; SURFx3;CAFFE SIFTx3	**24.97**	23.63	24.96	**24.97**	24.14	23.68
R2	ORBx3; SURFx3; SIFTx3;GNET	27.72	28.47	27.69	27.7	27.69	**28.52**
R3	ORBx3; SURFx3; SIFTx3;CONV	28.14	28.6	28.25	28.08	28.08	**28.84**
R4	ORBx3; SURFx3;CAFFE; SIFTx3;GNET; CONV	29.84	29.74	29.79	29.84	29.70	**29.96**

With respect to the proposed method we calculated the AP for each candidate cascade at three different levels (i.e., for $k = 50,100$ and equal to the number of the training samples per concept) and we averaged the results.

4.2   Experimental Results

Tables 1 and 2 present the results of our experiments in terms of the Mean Extended Inferred Average Precision (MXinfAP) [19], which is an approximation of the Mean Average Precision suitable for the partial ground-truth that accompanies the TRECVID dataset [11]. Table 1 presents the MXinfAP for the different types of features that were used by the algorithms of this study. Each line of this table was used as a cascade stage for the cascade-based methods (Table 2: M3, M4, M6). Specifically, stages that correspond to SIFT, SURF and ORB consist of three base classifiers (i.e. for the grayscale descriptor and its two color variants), while the stages of DCNN features (CAFFE, CONV, GNET) consist of one base classifier each. For the late fusion methods (Table 2: M1, M2) and the boosting-based method (Table 2: M5), the corresponding base classifiers per line of Table 1 were firstly combined by averaging the classifier output scores and then the combined outputs of all lines were further fused together. We adopted this grouping of similar base classifiers as this was shown to improve the performance for all the methods in our experiments, increasing the MXinfAP by $\sim 2\%$. For M2 we replaced the genetic algorithm with exhaustive search (i.e. to evaluate all $2^6 - 1$ possible classifier subsets) because this was more efficient for the examined number of classifiers.

Table 2 presents the performance of the proposed cascade-based method and compares it with other classifier combination methods. The second column shows the stage classifiers that were considered in each run. Runs R1-R3 encapsulate nine types of features from local descriptors and only one type of DCNN features; ultimately, R4 refers to the systems that utilize six stage classifiers and all twelve types of features. The best results were reached by the proposed cascade in R4, where it outperforms all the other methods reaching a MXinfAP of 29.96%. Compared to the ensemble pruning method (M2) the results show that exploring the best ordering of visual descriptors on a cascade architecture (M6), instead of just combining subsets of them (M2), can improve the accuracy of video concept detection. In comparison to the other cascade-based methods (M3, M4) that utilize fixed stage orderings and different algorithms to assign the stage thresholds, the proposed cascade (M6) also shows small improvements in MXinfAP. These can be attributed to the fact that our method simultaneously searches both for optimal stage ordering and threshold assignment. These MXinfAP improvements, of the proposed cascade, although small, are accompanied by considerable improvements in computational complexity, as discussed in the following section.

4.3   Computational Complexity

We continue the analysis of our results with respect to the computational complexity of the different methods compared in this study during the training and classification phase. Table 3 summarizes the computational complexity during the training phase. Let us assume that n stage classifiers need to be learned, M training examples are available for training the different methods and Q is the quantization value, where $Q \leq M$. The late-fusion approach [16], which builds n models (one for each set of features), is the simplest one. Cascade-thresholding [9] follows, which evaluates n cascade stages in order to calculate the appropriate thresholds per stage. Cascade-dynamic [3] works in a similar fashion as the Cascade-thresholding, requiring a little higher number of evaluations. Cascade-proposed is the next least complex algorithm, requiring $Q(n(n+1)/2)$ classifier evaluations. Ensemble pruning [14] follows, requiring the evaluation of $2^n - 1$ classifier combinations. Finally, only AdaBoost requires the retraining of different classifiers, which depends on the complexity of the base classifier, in our case the SVM, making this method the computationally most expensive.

Table 4 presents the computational complexity of the proposed cascade-based method for the classification phase, and compares it with other classifier

Table 3. Training complexity: (a) Required number of classifier combinations during the training of different classifier combination approaches. (b) Required number of classifiers to be retrained.

		Required classifier evaluations	Number of classifiers to be retrained
M1	Late-fusion [16]	-	-
M2	Ensemble pruning [14]	$(2^n - 1)M$	-
M3	Cascade-thresholding [9]	$\sum_{j=0}^{n} M_j, M_j \subseteq Mj - 1$	-
M4	Cascade-dynamic [3]	$(n - 2)Q^2$	-
M5	AdaBoost [1]	$M(n(n + 1)/2)$	$n(n + 1)/2$
M6	Cascade-proposed	$Q(n(n + 1)/2)$	-

Table 4. Relative amount of classifier evaluations (%) for different classifier combination approaches during the classification phase.

RunID	Stage classifiers	M1 Late-fusion [16]	M2 Ensemble pruning [14]	M3 Cascade-thresholding [9]	M4 Cascade-dynamic [3]	M5 AdaBoost [1]	M6 Cascade-proposed
R1	ORBx3; SURFx3;CAFFE SIFTx3	83.33	55.92	66.17	77.69	83.33	**53.50**
R2	ORBx3; SURFx3; SIFTx3;GNET	83.33	55.70	66.98	77.95	83.33	**52.74**
R3	ORBx3; SURFx3; SIFTx3;CONV	83.33	57.68	66.98	78.54	83.33	**54.32**
R4	ORBx3; SURFx3;CAFFE SIFTx3;CONV; GNET	100	66.67	74.94	92.38	100	**62.24**

combination methods. We observe that the proposed algorithm reaches good accuracy while at the same time is less computationally expensive than the other methods. Specifically, the best overall accuracy reached in R4 achieved 37.8 % and 32.6 % relative decrease in the amount of classifier evaluations compared to the late fusion alternative (Table 4: R4-M1) and the cascade-dynamic alternative (Table 4: R4-M4), respectively, which are the two most accurate methods after the proposed-cascade. Figure 2 presents the complexity of the proposed cascade-based method and compares it with other classifier combination methods, separately for each target concept. We can observe that the proposed method is computationally less expensive for 26 out of the 38 concepts.

To sum up, according to Tables 2 and 4, the three best-performing methods are the proposed-cascade, the late fusion [16] and the cascade-dynamic [3]. With respect to runs R2-R3 the proposed-cascade outperforms the two other methods, while it is always computationally more efficient during classification. When the number of features/stage classifiers increases (R4) the proposed-cascade performs

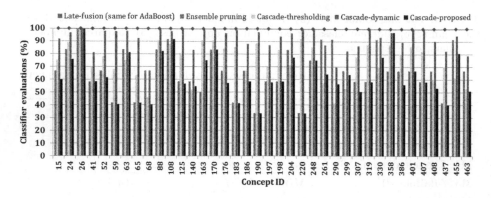

Fig. 2. Relative amount of classifier evaluations (%) per concept for R4 of Table 4.

slightly better in terms of MXinfAP compared to the late fusion and cascade-dynamic method, achieving 0.4 % relative improvement, for both cases. At the same time it is computationally less expensive during classification. Only for R1, which uses a small number of stage classifiers, the proposed-cascade presents lower accuracy than the other two best performing methods; however, it remains computationally less expensive. Finally, we should note that the training of the proposed cascade is computationally more expensive than the training of the late fusion and the cascade-dynamic methods. However, considering that training is performed offline only once, but classification will be repeated many times for any new input video, the latter is more important and this makes the reduction in the amount of classifier evaluations that is observed in Table 4 for the proposed cascade very important.

5   Conclusions

In this work we presented an improved way of ordering and combining independently trained base concept detectors using a cascade. A search-based algorithm that finds the optimal stage ordering and rejection thresholds was presented and evaluated. The resulting cascade-based concept detection method is computationally more efficient, in terms of classifier evaluations, and more accurate than other state-of-the-art approaches.

Acknowledgements. This work was supported by the European Commission under contract FP7-600826 ForgetIT.

References

1. Bao, L., et al.: CMU-Informedia@TRECVID 2011 semantic indexing. In: TRECVID 2011 Workshop, Gaithersburg, MD, USA (2011)
2. Bay, H., et al.: Speeded-up robust features (surf). Comput. Vis. Image Underst. **110**(3), 346–359 (2008)

3. Chellapilla, K., Shilman, M., Simard, P.Y.: Combining multiple classifiers for faster optical character recognition. In: Bunke, H., Spitz, A.L. (eds.) DAS 2006. LNCS, vol. 3872, pp. 358–367. Springer, Heidelberg (2006)
4. Cheng, W.C., Jhan, D.M.: A cascade classifier using adaboost algorithm and support vector machine for pedestrian detection. In: IEEE International Conference on SMC, pp. 1430–1435 (2011)
5. Jegou, H., et al.: Aggregating local descriptors into a compact image representation. In: IEEE Conference on Computer Vision and Pattern Recognition (CVPR 2010), San Francisco, CA, pp. 3304–3311 (2010)
6. Krizhevsky, A., Ilya, S., Hinton, G.: Imagenet classification with deep convolutional neural networks. In: Advances in Neural Information Processing Systems, vol. 25, pp. 1097–1105. Curran Associates, Inc., Red Hook (2012)
7. Lowe, D.G.: Distinctive image features from scale-invariant keypoints. Int. J. Comput. Vis. **60**(2), 91–110 (2004)
8. Markatopoulou, F., Pittaras, N., Papadopoulou, O., Mezaris, V., Patras, I.: A study on the use of a binary local descriptor and color extensions of local descriptors for video concept detection. In: He, X., Luo, S., Tao, D., Xu, C., Yang, J., Hasan, M.A. (eds.) MMM 2015, Part I. LNCS, vol. 8935, pp. 282–293. Springer, Heidelberg (2015)
9. Markatopoulou, F., Mezaris, V., Patras, I.: Cascade of classifiers based on binary, non-binary and deep convolutional network descriptors for video concept detection. In: IEEE International Conference on Image Processing (ICIP 2015). IEEE, Canada (2015)
10. Nguyen, C., Vu Le, H., Tokuyama, T.: Cascade of multi-level multi-instance classifiers for image annotation. In: KDIR 2011, pp. 14–23 (2011)
11. Over, P., et al.: Trecvid 2013 - an overview of the goals, tasks, data, evaluation mechanisms and metrics. In: Proceedings of TRECVID 2013. NIST, USA (2013)
12. Perronnin, F., Sánchez, J., Mensink, T.: Improving the fisher kernel for large-scale image classification. In: Maragos, P., Paragios, N., Daniilidis, K. (eds.) ECCV 2010, Part IV. LNCS, vol. 6314, pp. 143–156. Springer, Heidelberg (2010)
13. Safadi, B., Quénot, G.: Re-ranking by local re-scoring for video indexing and retrieval. In: 20th ACM International Conference on Information and Knowledge Management, pp. 2081–2084. ACM, NY (2011)
14. Sidiropoulos, P., Mezaris, V., Kompatsiaris, I.: Video tomographs and a base detector selection strategy for improving large-scale video concept detection. IEEE Trans. Circ. Syst. Video Technol. **24**(7), 1251–1264 (2014)
15. Simonyan, K., Zisserman, A.: Very deep convolutional networks for large-scale image recognition. arXiv technical report (2014)
16. Strat, S.T., Benoit, A., Bredin, H., Quénot, G., Lambert, P.: Hierarchical late fusion for concept detection in videos. In: Fusiello, A., Murino, V., Cucchiara, R. (eds.) ECCV 2012 Ws/Demos, Part III. LNCS, vol. 7585, pp. 335–344. Springer, Heidelberg (2012)
17. Szegedy, C., et al.: Going deeper with convolutions. In: CVPR 2015 (2015). http://arxiv.org/abs/1409.4842
18. Viola, P., Jones, M.: Rapid object detection using a boosted cascade of simple features. In: CVPR (2001), vol. 1, pp. 511–518 (2001)
19. Yilmaz, E., Kanoulas, E., Aslam, J.A.: A simple and efficient sampling method for estimating AP and NDCG. In: 31st ACM SIGIR International Conference on Research and Development in Information Retrieval, pp. 603–610. ACM, USA (2008)

Spatial Constrained Fine-Grained Color Name for Person Re-identification

Yang Yang[1], Yuhong Yang[1,2]($\boxtimes$), Mang Ye[1], Wenxin Huang[1], Zheng Wang[1], Chao Liang[1,2], Lei Yao[1], and Chunjie Zhang[3]

[1] National Engineering Research Center for Multimedia Software,
School of Computer, Wuhan University, Wuhan 430072, China
yangyang0518@whu.edu.cn
[2] Collaborative Innovation Center of Geospatial Technology,
Wuhan University, Wuhan, China
ahka_yang@yeah.net
[3] University of Chinese Academy of Sciences, Beijing, China

Abstract. Person re-identification is a key technique to match different persons observed in non-overlapping camera views. It's a challenging problem due to the huge intra-class variations caused by illumination, poses, viewpoints, occlusions and so on. To address these issues, researchers have proposed many visual descriptors. However, these visual features may be unstable in complicated environment. Comparatively, the semantic features can be a good supplement to visual feature descriptors for its robustness. As a kind of representative semantic features, color name is utilized in this paper. The color name is a semantic description of an image and shows good robustness to photometric variations. Traditional color name based methods are limited in discriminative power due to the finite color categories, only 11 or 16 kinds. In this paper, a new fine-grained color name approach based on bag-of-words model is proposed. Moreover, spatial information, with its advantage in strengthening constraints among features in variant environment, is further applied to optimize our method. Extensive experiments conducted on benchmark datasets have shown great superiorities of the proposed method.

Keywords: Person re-identification · Color name · Bag-of-words · Spatial constraint

1 Introduction

Person re-identification is an important topic of video surveillance in multimedia community [1]. Its task is to identify specific person over disjoint camera views. The task remains a challenging problem due to huge intra-class variations caused by different viewpoints, illumination, poses, etc. [2]. To tackle these challenges, many researchers have proposed different approaches, the focus of which can be roughly divided into feature representation methods and metric learning methods [3–6]. Feature representation methods refer to seeking stable and

© Springer International Publishing Switzerland 2016
Q. Tian et al. (Eds.): MMM 2016, Part I, LNCS 9516, pp. 886–897, 2016.
DOI: 10.1007/978-3-319-27671-7_74

distinctive appearance descriptions which can easily separate different persons in various cameras [8–12], while metric learning methods refer to learning a distance metric or projecting features from different views into a common space to suppress inter-camera variations [13–16]. The metric learning methods rely too much on the size of training data, however, the practical condition is that the large scale labelled data are difficult to achieve [20]. Consequently, we focus on seeking stable and distinctive feature representation for person re-identification in this paper.

Texture and color are the most commonly used visual features to describe person appearance for person re-identification [1]. Visual texture descriptors such as LBP and Gabor filter [10], have been widely applied to address the problem. In comparison with texture information, color information, however, seems to be another important cue due to the fact that low-resolution images are always obtained [17]. Considering the influence of environmental variations, traditional visual color information, like color histograms calculated in different color spaces separately or jointly, is put forward. Whereas performance by means of traditional color histograms would not be so satisfactory [18]. All these visual feature representation methods are relatively limited. As a supplement, semantic features are more robust in complicated environment [7,19,21]. Therefore, as a typical semantic feature, color name is another tentative way to optimize the representation of person appearance.

Color name is an abstract concept derived from the real word. It's a linguistic label that human utilize to describe the visual feature. Examples of color name are "*red*", "*black*", and so on. More importantly, color name is found to be a more stable semantic feature [7,21]. Taking "*red*" for example, multiple shades of red are all likely shaped to the same color name "*red*". Van de Weijer *et al.* [22] proposed a method to learn the eleven basic color names from Google images automatically, and the result of this learning is a partition of the color space into eleven regions. Kuo *et al.* [2] employed semantic color names, which were learned in [22] to describe colors and achieve improvements over the state-of-art methods in person re-identification. SCNCD was proposed in [17], which employed 16 kinds of color names from 16-color palette in RGB color space. However, two aspects are ignored or insufficiently considered, which can help to obtain a more stable and distinctive feature.

First, existing approaches employ only 11 or 16 kinds of color names extracted from RGB color space, which we called coarse-grained color names in this paper. Discriminative information is obliterated due to the limited color categories. In order to reduce information loss, we propose a framework which embeds bag-of-words model into color name. In our model, local color features are quantized into certain combination of weighted color words by a pre-trained codebook, which we define as fine-grained color names. Consequently, color word features formed by this multiple color words make it possible that images can be represented by fine-grained color names, which retain more subtle information and thus own more discriminative power. Moreover, following the bag-of-words model, instead of performing brute-force feature-feature matching [20], local color features are aggregated into a global vector, leading to higher computational efficiency [23].

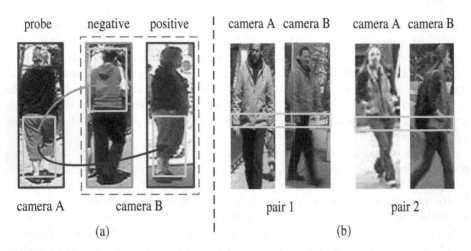

probe   negative   positive | camera A   camera B   camera A   camera B

camera A   camera B | pair 1   pair 2

(a) | (b)

Fig. 1. Illustration of spatial constraints. (a) and (b) are two query samples and their corresponding candidate sets from VIPeR dataset. The left is a query sample observed in camera A. The right are samples of two corresponding candidate persons observed in camera B. We can see that image with red window indicates the truly matched persons, while image with green window indicates the false matched person. Color components between the probe and image with green window are almost identical, both with black and red color. However, influenced by illumination, the color information between the same person has changed. Taking this situation into account, the space information is deemed as an important cue in our method. In addition, as shown in (b), unexpected spatial displacements are caused by changing environment. It's necessary to reasonably handle these displacements (Color figure online).

Second, spatial information is another consideration utilized in this paper to optimize our fine-grained color name model. Influenced by changes in person poses, camera viewpoints and lighting conditions between different cameras, large variations of features between the same person would produce. While other persons with similar features would have a priority to rank higher in the person similarity ranking list. As shown in Fig. 1(a), in view of this situation, spatial distribution information is deemed as an important cue in our approach. Moreover, affected by changing environment, feature displacements often emerge, as shown in Fig. 1(b). All of these motivate us to introduce spatial constraints to optimize our model, which refer to considering the spatial distribution restrictions of various features. Specifically, the spatial constraint is divided into three aspects. First, various feature weights are obtained according to the locations in the person image. The latent assumption is that features near to center of body are always more reliable than others, while features on the edge of body are inclined to be impacted by backgrounds and should be allocated lower weights. Second, while color components of the probe and the gallery person image with green window are more identical, as shown in Fig. 1(a), the color distribution is completely different. Calculating color information at stripe level is what we employ here to describe spatial constraints in a person image. Finally, in order to match the real features and minimize the influence of changing environment, a subwindow, of which the step size is smaller than the its height, is utilized to produce overlapping stripes. In a brief summary, the spatial constrained process can reduce the influence of feature displacements.

In this paper, we propose a new spatial constrained fine-grained color name method to achieve stable and distinctive features. Based on the bag-of-words model, each patch with particular weight is represented by multiple color name. The subwindow whose height is larger than its step size is utilized to form overlapping stripes, which takes the feature displacements into account. The overlapping stripes are then represented by histogram bins chosen from each patch. With the above processing, images are represented by multiple color name features. And simultaneously, the spatial information in person images is specifically utilized to optimize our person re-identification approach.

The main contribution of this paper can be classified into two merits: (1) Based on the bag-of-words model, we introduce the fine-grained color name, which can be utilized to represent the person image. The fine-grained color name feature representation owns more discriminative power and does not rely on complicated online feature extraction process. (2) We introduce spatial constraints to our model, which considers not only what the color is but also where the color is. The remainder of the paper is organized as follows, in Sect. 2, we describe our method in details. Experiments are shown in Sect. 3. Section 4 is the conclusion.

2   Approach

In this section, we first introduce our new fine-grained color name model, which focus on improving the discriminative power of traditional color name-based feature representation. After that, the spatial constraints are presented, which aim at improving robustness of the fine-grained color name method. Furthermore, a combination with visual features makes our method compare favorably with existing state-of-the-art feature representation methods. The details are discussed in the following.

2.1   Fine-Grained Color Name

The bag-of-words model is one of the most popular representation methods in image search. Considering the "*query − search*" mode, person re-identification is potentially compatible with the image search. In off-line processing, we partition person images into multiple patches, several orderless local patches are therefore extracted from images of different categories as candidates for basic elements. Then feature descriptors are computed to represent the patches with numerical vectors. The final step is to convert such vectors to "*codewords*" to generate a codebook using clustering methods. The number of the clusters is the codebook size. Thus, each patch in an image can be mapped to certain code words based on the pre-trained codebook and the image can eventually be represented by the code word histograms.

Feature extraction. Given a person image derived from dataset normalized to 128×48 pixels, and then patches of size 4×4 are densely sampled. For each patch, we calculate semantic features, namely, color name [22,24]. The mean vector is computed as the descriptor of this patch.

Codebook. The codebook is pre-trained on the independent TUD-Brussels dataset offline [9]. Standard k-means is used to quantize the descriptors into clusters which are called color words for person identification, and the codebook size is k. The number of clusters is chosen empirically to maximize identification results. Based on this codebook, the new person image can be represented by multiple code words during online process.

Quantization. In this paper, we employ Multiple Assignment (MA) to find neighbors under Euclidean distance in the codebook [25], while the rest of the value is set to zero entries. MA is set to 10. Each patch is then represented by 10 color words. After that, the bag-of-words based color name (BCN) features is achieved, which we call fine-grained color names.

2.2 Spatial Constrained Fine-Grained Color Name

We have achieved the features of each person image represented by code words in above section, while the spatial information of features is ignored. Spatial constraints are then introduced to the above fine-grained color name model and described in the following. First, this paper adopts a priori knowledge on the position of person, which assumes that the person lies in the center of the image, and therefore patches should be given a weight according to its distances from the center point. Furthermore, by parting the person image into horizontal stripes, the feature processing here considers the spatial relationships in vertical direction. Finally, feature displacements are taken into account by employing the overlapping stripes.

Location weights. Based on the priori assumption that features belong to the center should be assigned a higher weight, while the feature near the edge of the image should be suppressed. A simple way by exerting a 2-D Gaussian template on the image is applied in this paper. Specifically, the Gaussian template function [27] takes on the form of $N(\mu_x, \sigma_x, \mu_y, \sigma_y)$, of which μ_x, μ_y are horizontal and vertical Gaussian mean values, while σ_x, σ_y are horizontal and vertical Gaussian standard variances, respectively. (μ_x, μ_y) is set to the image center, and $(\sigma_x, \sigma_y) = (1, 1)$ is set to all the experiments.

Location constraints. We consider the spatial information as an important factor. As illustrated in Fig. 2, this paper suggests to equally divide a person image into M horizontal stripes. Histogram is computed in each stripe. For each patch, the histogram is denoted as $h = (h_1, h_2, \ldots, h_{MA}, \ldots, h_k)$, where MA is the quantized value chosen from the code words with Euclidean distance, k is the codebooke size, and h represents the occurrence probability of one pattern in a patch. We check all patches at the same horizontal location, and add up the local occurrence of each pattern (i.e. the same histogram bin) among the

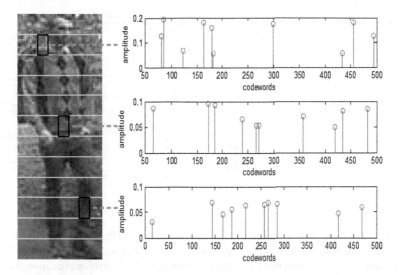

Fig. 2. Person image is divided into M horizontal stripes which composed of numerous patches. And the right part shows the fine-grained color name results.

horizontal patches. The stripe is then denoted as $d^m = (d_1^m, d_2^m, \ldots, d_k^m)^T$, and consequently, the feature vector for the input image can be represented as $f = (d_1, d_2, \ldots, d_M)^T$, which is the concatenation of vectors from all stripes. In this way, the calculation of similarity is performed at stripe level, which considers not only features of the given images, but the location that calculated features belong to.

Displacements corrections. In person re-identification, displacements corrections are scarcely considered. Based on the assumption that person displacements always emerge in the vertical direction. Overlapping stripes are utilized to decrease the influence of vertical displacements caused by changing poses or viewpoints. A subwindow is utilized to locate the overlapping stripes, of which the size and step are 20×48, 8 pixels respectively.

2.3 Combination with Visual Features

As described above, our spatial constrained fine-grained color name model is complementary to existing visual feature representation approaches. Therefore, we combine our method with the visual based features, namely LOMO [26] and wHSV [3]. In order to match the features in different scales, feature distance is utilized to combine different features. Euclidean distance of all features are calculated, and subsequently the distances are normalized and combined with different weights:

$$d_{eSBCN}(x_p^A, x_g^B) = \sum_i \beta_i \cdot d_i(f_i(x_p^A), f_i(x_g^B)) \qquad (1)$$

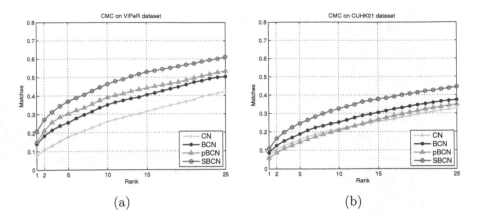

(a)   (b)

Fig. 3. Different feature components on two datasets, VIPeR and CUHK01 respectively. *"CN"* refers to basic feature representation method which utilizes only 11 kinds of color names. *"BCN"* is a method which introduces the bag-of-words model into the basic *"CN"*, multiple color words are employed for stripes representation. *"pBCN"* means to represent the person image by patches in the framework of *"BCN"*, which considers the feature location in horizontal and vertical directions. The final *"SBCN"* represents an additional consideration of the feature displacement caused by changing poses or viewpoints.

where d_{eSBCN} denotes the final distance, d_i means the distance of i-th features, such as LOMO, wHSV or SBCN in this paper. x_p^A and x_g^B describe two person images captured from camera A and camera B respectively. $f_i(\cdot)$ is i-th feature of an image while β_i expresses the corresponding weight.

3   Experiment

3.1   Dataset and Evaluation Protocol

We compare our approach with baseline methods and a number of state-of-the-art methods on publicly available dataset, the VIPeR dataset [28] and the CUHK01 dataset [30]. The widely used VIPeR dataset consists of 632 pedestrians image pairs. The CUHK01 dataset is a larger dataset recently released by Wang *et al.* [30] and contains 971 identities from two disjoint camera views. For each person on these two datasets, a pair of images are taken from cameras with widely inconsistent views. Viewpoint changes of 90 degrees or more as well as huge lighting variations make these datasets as one of the most challenging datasets for person re-identification. Following the methodology used in [26], the group of persons in each dataset are randomly split into two halves, one for training and the other for testing. To reduce the bias, we repeated the whole procedure 10 times, the average of the results is given as the final performance. In the evaluation, Cumulative Matching Characteristic (CMC) curve [29], which represents the expectation of finding the true match, is used as the main measurement.

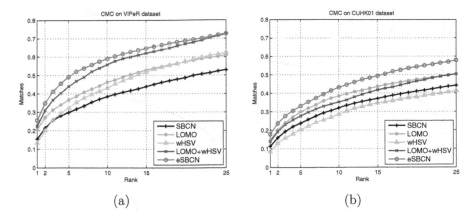

Fig. 4. Different feature representation methods on two datasets, VIPeR and CUHK01 respectively. *LOMO* [26] and *wHSV* [3] are two visual feature representation methods. *"LOMO+wHSV"* is a combination of the two methods. The *"eSBCN"* is a combination of our *"SBCN"* method with the above *"LOMO+wHSV"*. The *"eSBCN"* achieves better results than the direct combination of *"LOMO+wHSV"* in this experiment, which means that semantic features can improve the visual feature representation results.

3.2   Experimental Settings

In our experiments, we normalize all images to 128×48 pixels. Patches of size 4×4 are densely obtained. The sampling step is 4 pixels, so there is no overlapping between patches. For each patch, CN descriptors of all pixels are calculated [14, 22], the mean vector is taken as the description of this patch, as shown in Fig. 2. The codebook size is set to 500. We split the person image into 16 stripes in our experiments. A subwindow, of which the size is 20×48, while the step size is 8 pixels is utilized to form the overlapping stripes.

3.3   Evaluation

Effectiveness of every step. On two datasets, VIPeR and CUHK01, we present CMC results obtained by Euclidean distance, while four different features representation methods are utilized. As shown in Fig. 3. The first baseline feature is *CN*, which utilizes only 11 kinds of color names to represent the person image [22]. Then, the second feature called *BCN* is the combination of the bag-of-words model and the baseline *CN*. In this approach, multiple color words are employed to represent the person image, which enhances the discriminative power of the person image. The third feature is *pBCN*, which is a model in the framework of *BCN*. The differences are that features are represented and matched both by patches in *pBCN*. Spatial constraints referring to features in horizontal and vertical directions are considered in this way. *SBCN* is the final person representation method, which considers the feature displacements by representing and matching overlapping stripes.

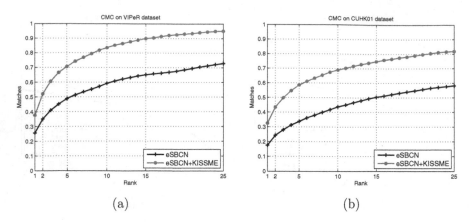

(a)   (b)

Fig. 5. CMC curves on the VIPeR dataset and CUHK01 dataset. The *"eSBCN"* is measured by Euclidean distance, while *"eSBCN+KISSME"* is measured by KISSME. We observed a higher result with a metric learning.

As can seen from Fig. 3, on the datset VIReR, the baseline *CN* produces a relatively low accuracy, the rank-1 matching rate $= 7.88\%$. The *BCN* presents an impressive promotion in re-identification accuracy, with the rank-1 matching rate rising from 7.88% to $13.01\%(+5.13\%)$. The *pBCN* is observed $+1.90\%$ in rank-1 matching rate, and a greater promotion can be observed in the subsequent matching rate. Our final *SBCN* achieves a remarkable promotion in rank matching rates. Similar improvements can be seen on the dataset CUHK01. Need to be noticed that the viewpoint changes of 90 degrees in CUHK01 dataset always occurs, which may lead to a discontented result of the *pBCN*. The reason for these promotions can be summarized in the following aspects. On one hand, by utilizing the fine-grained color name model, multiple color words are used to represent the person image, which improve the discriminative power of these features. On the other hand, two aspects of spatial information which refer to feature constraints in both vertical and horizontal directions are considered, which improves the robustness of our method in the variant poses and viewpoints.

Effectiveness of semantic-based *SBCN*. We present CMC results obtained by Euclidean distance on two datasets. As shown in Fig. 4. *LOMO* [26] and *wHSV* [3] are two feature representation methods based on visual feature. Visual color and texture information are utilized in the LOMO, while wHSV only employs the color information. *LOMO+wHSV* is a combination of the above two methods, β_{LOMO} and β_{wHSV} are set to 0.4, 0.3 respectively. The *eSBCN* is formed by combining our *SBCN* method with the above *LOMO+wHSV*, with the β_{SBCN}, β_{LOMO}, β_{wHSV} is 0.3, 0.4, 0.3 respectively. The *eSBCN* achieves a better result than the direct *LOMO+wHSV*. Results on both datasets are consistent with the previous analysis that semantic features and visual features can be complementary. Simple combination of visual features can't be enough for the feature representation as shown in Fig. 4(b), while a significant promotion can be obtained when the semantic features are combined.

Table 1. VIPeR dataset: top ranked matching rates in [%] with 316 persons.

Methods	$r = 1$	$r = 2$	$r = 5$	$r = 10$	$r = 15$	$r = 25$
eSBCN + KISSME	37.69	**52.12**	**70.51**	**83.51**	**89.87**	**93.10**
ELF [10]	12.08	17.00	31.28	41.00	54.00	65.00
SDALF [3]	19.87	25.20	38.89	49.37	58.22	70.00
PRDC [13]	15.66	22.80	38.42	53.86	64.00	72.78
PCCA [14]	19.27	29.10	48.89	64.91	72.48	82.78
KISSME [16]	22.63	32.72	50.13	63.73	71.65	82.12
SDC [23]	23.32	31.27	43.73	54.05	59.87	68.45
SalMatch [20]	30.16	39.28	52.00	65.00	74.00	*
SCNCD [17]	37.8	50.10	68.5	81.2	87.0	92.7
LOMO+XQDA [26]	**38.23**	50.95	68.71	80.51	87.22	91.21

3.4 Person Matching with Metric Learning

Finally, the popular metric learning method KISSME [11] is utilized to measure the person features. An impressive improvement can be observed on both VIPeR dataset and CUHK01 dataset, as shown in Fig. 5. Specifically, the improvements of the matching rates at rank 1 are 13 % and 15 % respectively. Finally, we compare the performance of *eSBCN* with state-of-the-art methods reported on VIPeR dataset, as illustrated in Table 1, after introducing the metric learning method, our results are competitive.

4   Conclusion

In this paper, we propose a robust feature representation method by considering two items which are simple but easily being overlooked. Firstly, instead of traditional 11 or 16 kinds of color names, fine-grained color name model which embeds bag-of-words model into color name is proposed to represent the person image. Described by multiple color names, the person feature owns more discriminative power. Second, the spatial information is deemed as an important cue in our approach. We solve it by three aspects. On one hand, assigning a higher weight to features near to center of body, and vice versa. On the other hand, parting the body into numerous horizontal stripes. Moreover, we dispose the person image by overlapping stripes which take the spatial displacements into account. Extensive experiments compared to the original baseline methods and existing state-of-the-art methods have been conducted. Effectiveness of every key step is also testified on the benchmark datasets. In the future, we'll apply our approach to some real applications.

Acknowledgement. The research was supported by National Nature Science Foundation of China (61303114, 61231015, 61170023), the Specialized Research Fund for the

Doctoral Program of Higher Education (20130141120024), the Technology Research Project of Ministry of Public Security (2014JSYJA016), the China Postdoctoral Science Foundation funded project (2013M530350), the major Science and Technology Innovation Plan of Hubei Province (2013AAA020), the Guangdong-Hongkong Key Domain Break-through Project of China (2012A090200007), and the Special Project on the Integration of Industry, Education and Research of Guangdong Province (2011B090400601). Nature Science Foundation of Hubei Province (2014CFB712). Jiangxi Youth Science Foundation of China (20151BAB217013).

References

1. Li, W., Zhao, R., Wang, X.: Human reidentification with transferred metric learning. In: Lee, K.M., Matsushita, Y., Rehg, J.M., Hu, Z. (eds.) ACCV 2012, Part I. LNCS, vol. 7724, pp. 31–44. Springer, Heidelberg (2013)
2. Kuo, C.H., Khamis, S., Shet, V.: Person re-identification using semantic color names and rankboost. In: Workshop on Applications of Computer Vision (WACV), pp. 281–287 (2013)
3. Farenzena, M., Bazzani, L., Perina, A, et al.: Person re-identification by symmetry-driven accumulation of local features. In: Computer Vision and Pattern Recognition (CVPR), pp. 2360–2367 (2010)
4. Leng, Q., Hu, R., Liang, C., et al.: Person re-identification with content and context re-ranking. Multimedia Tools Appl. (MTAP) **74**, 1–26 (2014)
5. Wang, Z., Hu, R., Liang, C., Leng, Q., Sun, K.: Region-based interactive ranking optimization for person re-identification. In: Ooi, W.T., Snoek, C.G.M., Tan, H.K., Ho, C.-K., Huet, B., Ngo, C.-W. (eds.) PCM 2014. LNCS, vol. 8879, pp. 1–10. Springer, Heidelberg (2014)
6. Ye, M., Chen, J., Leng, Q., Liang, C., Wang, Z., Sun, K.: Coupled-view based ranking optimization for person re-identification. In: He, X., Luo, S., Tao, D., Xu, C., Yang, J., Hasan, M.A. (eds.) MMM 2015, Part I. LNCS, vol. 8935, pp. 105–117. Springer, Heidelberg (2015)
7. Liu, Y., Zhang, D., Lu, G., et al.: Region-based image retrieval with high-level semantic color names. In: Multimedia Modelling Conference (MMM), pp. 180–187 (2005)
8. Wang, Y., Hu, R., Liang, C., et al.: Camera compensation using a feature projection matrix for person reidentification. Circ. Syst. Video Technol. (TCSVT) **24**, 1350–1361 (2014)
9. Wojek, C., Walk, S., Schiele, B.: Multi-cue onboard pedestrian detection. In: Computer Vision and Pattern Recognition (CVPR), pp. 794–801 (2009)
10. Gray, D., Tao, H.: Viewpoint invariant pedestrian recognition with an ensemble of localized features. In: Forsyth, D., Torr, P., Zisserman, A. (eds.) ECCV 2008, Part I. LNCS, vol. 5302, pp. 262–275. Springer, Heidelberg (2008)
11. Ma, B., Su, Y., Jurie, F.: Local descriptors encoded by fisher vectors for person re-identification. In: Fusiello, A., Murino, V., Cucchiara, R. (eds.) ECCV 2012 Ws/Demos, Part I. LNCS, vol. 7583, pp. 413–422. Springer, Heidelberg (2012)
12. Liu, C., Gong, S., Loy, C.C., Lin, X.: Person re-identification: what features are importantly specific fusion for image retrieval. In: Fusiello, A., Murino, V., Cucchiara, R. (eds.) ECCV 2012 Ws/Demos, Part I. LNCS, vol. 7583, pp. 391–401. Springer, Heidelberg (2012)

13. Zheng, W.S., Gong, S., Xiang, T.: Person re-identification by probabilistic relative distance comparison. In: Computer Vision and Pattern Recognition (CVPR), pp. 649–656 (2011)
14. Mignon, A., Jurie, F.: PCCA: a new approach for distance learning from sparse pairwise constraints. In: Computer Vision and Pattern Recognition (CVPR), pp. 2666–2672 (2012)
15. Hirzer, M., Roth, P.M., Köstinger, M., Bischof, H.: Relaxed pairwise learned metric for person re-identification. In: Fitzgibbon, A., Lazebnik, S., Perona, P., Sato, Y., Schmid, C. (eds.) ECCV 2012, Part VI. LNCS, vol. 7577, pp. 780–793. Springer, Heidelberg (2012)
16. Koestinger, M., Hirzer, M., Wohlhart, P., et al.: Large scale metric learning from equivalence constraints. In: Computer Vision and Pattern Recognition (CVPR), pp. 2288–2295 (2012)
17. Yang, Y., Yang, J., Yan, J., Liao, S., Yi, D., Li, S.Z.: Salient color names for person re-identification. In: Fleet, D., Pajdla, T., Schiele, B., Tuytelaars, T. (eds.) ECCV 2014, Part I. LNCS, vol. 8689, pp. 536–551. Springer, Heidelberg (2014)
18. Kviatkovsky, I., Adam, A., Rivlin, E.: Color invariants for person reidentification. Trans. Pattern Anal. Mach. Intell. (TPAMI) 35, 1622–1634 (2013)
19. Van De Weijer, J., Schmid, C.: Applying color names to image description. In: Transactions on Image Processing (ICIP), pp. 493–496 (2007)
20. Zhao, R., Ouyang, W., Wang, X.: Person re-identification by salience matching. In: International Conference on Computer Vision (ICCV), pp. 2528–2535 (2013)
21. Benavente, R., Vanrell, M., Baldrich, R.: Parametric fuzzy sets for automatic color naming. JOSA A 25(10), 2582–2593 (2008)
22. Van De Weijer, J., Schmid, C., Verbeek, J., et al.: Learning color names for real-world applications. Trans. Image Process. (ICIP) 18, 1512–1523 (2009)
23. Zhao, R., Ouyang, W., Wang, X.: Unsupervised salience learning for person re-identification. In: Computer Vision and Pattern Recognition (CVPR), pp. 3586–3593 (2013)
24. Arandjelovi, R., Zisserman, A.: Three things everyone should know to improve object retrieval. In: Computer Vision and Pattern Recognition (CVPR), pp. 2911–2918 (2012)
25. Jegou, H., Douze, M., Schmid, C.: Hamming embedding and weak geometric consistency for large scale image search. In: Forsyth, D., Torr, P., Zisserman, A. (eds.) ECCV 2008, Part I. LNCS, vol. 5302, pp. 304–317. Springer, Heidelberg (2008)
26. Liao, S., Hu, Y., Zhu, X., et al.: Person re-identification by local maximal occurrence representation and metric learning. In: Computer Vision and Pattern Recognition (CVPR), pp. 2197–2206 (2015)
27. Zheng, L., Shen, L., Tian, L., et al.: Person re-identification meets image search. arXiv preprint arXiv:1502.02171 (2015)
28. Gray, D., Brennan, S., Tao, H.: Evaluating appearance models for recognition, reacquisition, and tracking. In: IEEE International Workshop on Performance Evaluation of Tracking and Surveillance (2007)
29. Wang, X., Doretto, G., Sebastian, T., et al.: Shape and appearance context modeling. In: International Conference on Computer Vision (ICCV), pp. 1–8 (2007)
30. Li, W., Wang, X.: Locally aligned feature transforms across views. In: Computer Vision and Pattern Recognition (CVPR), pp. 3594–3601 (2013)

Dealing with Ambiguous Queries in Multimodal Video Retrieval

Luca Rossetto[(⊠)], Claudiu Tănase[(⊠)], and Heiko Schuldt[(⊠)]

Databases and Information Systems Research Group, Department of Mathematics
and Computer Science, University of Basel, Basel, Switzerland
{luca.rossetto,c.tanase,heiko.schuldt}@unibas.ch

Abstract. Dealing with ambiguous queries is an important challenge in
information retrieval (IR). While this problem is well understood in text
retrieval, this is not the case in video retrieval, especially when multimodal queries have to be considered as for instance in Query-by-Example
or Query-by-Sketch. Systems supporting such query types usually consider dedicated features for the different modalities. This can be intrinsic object features like color, edge, or texture for the visual modality or
motion for the kinesthetic modality. Sketch-based queries are naturally
inclined to be ambiguous as they lack specification in some information
channels. In this case, the IR system has to deal with the lack of information in a query, as it cannot deduce whether this information should be
absent in the result or whether it has simply not been specified, and needs
to properly select the features to be considered. In this paper, we present
an approach that deals with such ambiguous queries in sketch-based multimodal video retrieval. This approach anticipates the intent(s) of a user
based on the information specified in a query and accordingly selects the
features to be considered for query execution. We have evaluated our
approach based on Cineast, a sketch-based video retrieval system. The
evaluation results show that disregarding certain features based on the
anticipated query intent(s) can lead to an increase in retrieval quality of
more than 25 % over a generic query execution strategy.

1 Introduction

Ambiguity is a property of most kinds of expression. While the problem of query
ambiguity in text-based queries is well known, it differs from the situation found
when dealing with multimodal queries. In text retrieval the difficulties lie primarily in the context-sensitivity of certain terms. Multimodal queries in contrast –
such as the ones produced by query paradigms like Query-by-Sketch (*QbS*) or
Query-by-Example (*QbE*) – additionally can have different meaning depending
on the interpretation of the presence or absence or even the amount of information contained within each individual mode of expression. This is especially true
in a QbS scenario where based on the query alone it is not possible to determine
if a user did intend for a certain modality to contain no or limited information
because the desired result should also lack this sort of content — or if the user

© Springer International Publishing Switzerland 2016
Q. Tian et al. (Eds.): MMM 2016, Part I, LNCS 9516, pp. 898–909, 2016.
DOI: 10.1007/978-3-319-27671-7_75

was just not able to provide the information due to the lack of appropriate input devices, lack of artistic skills, or if he simply could not remember this aspect of the piece of content in question.

This problem is particularly relevant in video retrieval as video intrinsically comes with multiple *modalities* (e.g., visual, aural, text, kinesthetic) that are hardly provided jointly in a single query. Moreover, several of these modalities have different *information channels*, such as color, edge, or texture in the visual modality, or pitch, volume, speech, and music in the aural modality. Retrieval is based on inherent features characterizing these information channels individually; this can be done either by a single feature per information channel, or by the combination of several different features for one information channel (e.g., color histograms and color moments for the color information channel). For retrieval purposes, the lack of specification of information for one or several of the information channels or, even worse, the lack of specification for an entire modality has immediate consequences on the required selection and combination of features and thus on the retrieval results. If the wrong set of features is considered (e.g., features for which no or only poor information is given in the corresponding information channel of the modality), the query result is negatively impacted as these features might not all have the same selectivity. While the (amount of) information present in each channel is an inherent property of the content, the absence of information in a channel of the query could be due to the lack of care, memory or artistic skill of the user or because of unsuited means of input. It is therefore not possible to exactly determine the user's search intent based on the query and the amount of information present within its different channels. Hence, the absence of some information channel does not mean that this channel (and the features describing this channel) also does not have to be considered for query execution.

Query-by-Sketch video retrieval focuses on the two modalities that can be provided by users by means of sketches: the visual modality and the kinesthetic modality. For the visual part, users may provide line/contour sketches, color sketches, or a combination thereof via a sketch canvas. The kinesthetic modality can be specified in the form of flow fields via the same canvass to represent motion across frames. Alternatively, other user interfaces as for instance gesture-based UIs [8] can be used. In this paper, we will focus on all information that can be specified via a single sketch canvass (color, edges, and flow fields).

Figure 1 illustrates the problem of ambiguous queries resulting from sketch-based video queries. In Fig. 1a, the query only contains a line-sketch of a car on a white background. It is therefore not clear whether the author of the query was explicitly looking for a white car on a white background or just did not care what color the car has. Similarly, Fig. 1b contains color information, but without crisp edge information. Figure 1c lacks a specification of the kinesthetic modality (motion) in the sketched query. In this case, it is not obvious whether the query asks for a non-moving object such as the statue, or a video of a soccer scene in which motion plays an important role.

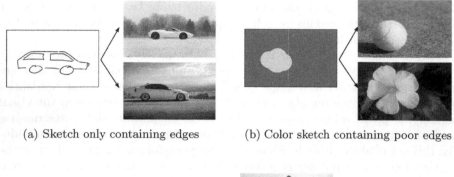

(a) Sketch only containing edges   (b) Color sketch containing poor edges

(c) Sketch without specified motion

Fig. 1. Examples of ambiguous sketch-based video queries

In this paper, we present an approach to disambiguate sketch-based queries in video retrieval, thereby solving the problem that inherently comes with this type of access where not all modalities and/or not all information channels within a modality are specified in a query. This is based on the consideration of the most likely user intent(s) that are anticipated by considering the modalities and information channels on which information is provided (or unintentionally absent) in the sketch and the combination of the retrieval results for these intents. We have evaluated the disambiguation strategy on top of Cineast, a novel multimodal video retrieval engine. The evaluation results show an increase in retrieval quality of more than 25 %, measured in terms of the inverted retrieval rank, over a generic baseline approach which anticipates all possible query intents, independent of the information provided in the query sketch.

The contribution of the paper is twofold: First, we provide a detailed analysis of the problem of ambiguous queries in multimodal (sketch-based) video retrieval, especially with regard to the absence of certain modalities in the query (visual or motion), or the absence of information channels within a modality (e.g., colors or edges in the visual modality). Second, we present an approach to anticipate the possible query intent(s) of a user based on the information provided in the sketch or unintentionally left out to overcome these limitations in ill-posed queries. The evaluation results show that the consideration of only a subset of the available features, tailored to the user intent(s), leads to a significant improvement of the retrieval results compared with a generic combination of features, independent of the query content.

The remainder of the paper is organized as follows: Sect. 2 discusses the problems of ambiguity in multimodal video retrieval. In Sect. 3, we introduce the Cineast multimodal video retrieval system which is used as basis for the evaluation provided in Sect. 4. Section 5 surveys related work and Sect. 6 concludes.

2   Dealing with Ambiguous Queries

In this section, we describe the method we use to deal with ambiguous multimodal queries in general and our concrete case with sketch-based video retrieval in particular. In our case, queries typically contain two modalities, the visual modality with the actual sketch and the motion modality represented by a flow field. The motion modality contains one information channel while the edge- and color information of the visual modality are treated differently, therefore resulting in two information channels for this modality.

2.1   Considering Multiple Intents

If the search intent of a user was known, a retrieval system could usually be optimized to best serve it by for example changing the individual parts used for measuring similarity or by adjusting their influence during the compilation of the final result list. Since determining the search intent for a general query on an unrestricted dataset is generally not reliably doable, we propose a different strategy. Even though one can usually not say with certainty what a user meant by a particular query, it is in most cases feasible to enumerate all possible intents and, depending on the dataset, even order them by expected probability. During query evaluation, a system can then perform different instances of the same similarity search optimized for different query intents and combine the individual result lists accordingly.

In our case with three information channels in the context of (sketch-based) video retrieval, we first have to analyze the query for its colorfulness, its crispiness (i.e., the amount of edge information contained within the query image) and its amount of movement. This analysis is done as a binary decision but could also be done in a more fine-gained way. The result of this analysis returns the possible query intent(s) of a user, called anticipated intent(s). These anticipated intent(s) are ordered by their expected probability, determined empirically by the analysis of actual queries, as shown in the second column in Table 1 (with the anticipated intents ordered with descending probability per cell). Each of these intents leads to one or multiple combinations of feature groups (again ordered per cell with decreasing probability in the rightmost column) which produces appropriate results for a given query.

Table 1. Anticipated intents for different query types, leading to different combinations of feature groups

Query Type			Anticipated Intent(s)	Combination of Feature Groups
colorful	crisp	moving	colorful crisp moving shot	(color, edge, motion) (color, edge) (color) (edge)
		still	colorful crisp still shot colorful crisp moving shot	(color, edge) (color, edge, motion) (color) (edge)
	blobby	moving	colorful blobby moving shot colorful crisp moving shot	(color, motion) (color, edge, motion) (color) (edge)
		still	colorful blobby still shot colorful crisp still shot colorful blobby moving shot colorful crisp moving shot	(color) (color, edge) (color, edge, motion)
colorless	crisp	moving	colorless crisp moving shot colorful crisp moving shot	(color, edge, motion) (edge, motion) (color, edge) (edge)
		still	colorless crisp still shot colorful crisp still shot colorless crisp moving shot colorful crisp moving shot	(edge) (color, edge) (edge, motion) (color, edge, motion)
	blobby	moving	colorless blobby moving shot colorful blobby moving shot colorless crisp moving shot colorful crisp moving shot	(motion)

2.2   Combining Intents

Compared to the overall retrieval time for multiple different similarity measures, the duration of the result combination is usually negligible. For combination strategies which allow sub-queries to be performed in parallel, which is the case for late-fusion approaches, multiple combinations can be jointly considered without significant extra costs, because the sub-queries were performed without interdependency resulting in valid input for any number of combinations. In our case, the used combinations are listed in Table 1.

In what follows, we present three different ways, illustrated in Fig. 2, in which this last combination step can be performed, together with a generic baseline approach we have used for the evaluation. The only commonality of the three proposed approaches is that duplicates are removed during combination.

Equal-Length Sub-lists (EL). In this combination scheme, each result list for the different anticipated intents gets an equal portion of the overall result

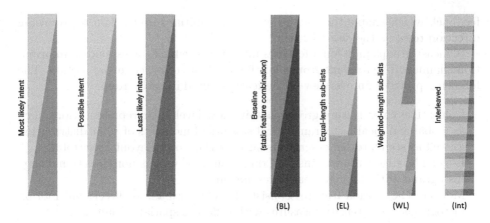

Fig. 2. Illustration of the different combination methods

list. The order of the sub-list is pre-determined based on which intent seems most likely given the query properties.

Weighted-Length Sub-lists (WL). In contrast to the previous evaluation scheme, the final list is comprised of the individual list in a ratio of $1 : 2 : 3$ and so on with the most likely query intent getting the largest part and the least likely intent getting the smallest. The ordering is the same as in the previous case.

Interleaved (Int). Other than in the previous two approaches, this combination method does not preserve the order of the results within the individual lists but rather interleaves the lists. In the Int approach, similar to EL, all lists contribute the same number of elements to the final result.

Baseline (BL). The easiest way to combine results from the individual features is to do so in a generic way, independently of the query. In this case it is beneficial to optimize the combination weights for the most common query type and use them for all queries.

3   Cineast

Cineast [12,13] is a content-based video retrieval engine focused on Query-by-Sketch. It uses many different features in parallel to perform retrieval on different information channels such as color, edge, motion and text (in the form of subtitles if available) as well as meta-information. The results from these modules are combined using a score-based late fusion approach to construct a final result list. Due to its modular architecture, feature modules can be added or removed easily enabling Cineast to use as many information channels as possible given a certain data set. Cineast also supports Query-by-Example as well as relevance feedback which enable the user to expand an initial result set in a promising direction towards the desired result.

Cineast groups individual features into groups which correspond to information channels. The relevant groups for this work are color, edge and motion. The following provides an overview of the features used in each group.

- Color: The color feature group contains low-level color representations such as global and local histograms, local statistical moments of color distribution as well as standardised descriptors such as the Color Layout Descriptor.
- Edge: The edge features consist of directional- as well as non-directional edge histograms with different spatial resolution.
- Motion: The motion features consist of local normalized directional motion histograms and intensity measures with different spatial resolutions.

4   Evaluation

We evaluate our approach on the OSVC1 dataset [14] consisting of 200 creative-commons web-videos with a wide range of content.

4.1   Evaluation Procedure

To evaluate this approach, $N = 100$ query shots q out of the roughly 30'000 shots of the evaluation collection were randomly selected. From each of these shots, seven queries were created containing different combinations of information. Our performance metric is the Mean Inverted Rank (MIR), computed as the mean of the inverse rank value of the ground truth document across all queries $MIR = \frac{1}{N} \left(\sum_q \frac{1}{R_q} \right)$.

In a late score fusion scheme, the relevance score of a retrieved document d is estimated to a weighted sum of relevance scores from individual feature estimators $score(d) = \sum_i w_i \cdot score(d, i)$. In order to learn optimal weights w_i we apply an optimization step of the weight vector w_i with the MIR as objective function, using the Sequential Least Squares Quadratic Programming method. In our experiments, the MIR has been optimized to a mean value of $MIR = 0.9288$ across 10 folds, after less than 8 iterations in each fold.

Weights are not changed across combination methods for the feature groups that are considered in a query.

In our case, a query has three information channels, color, edge, and motion, each of which can be omitted based on the user's intent or ability of expression. This leads to eight possible combinations if the decision whether or not to consider a channel of information is modeled as a binary one as we do in this evaluation. One of the eight possible combinations contains no information at all and can therefore be omitted.

For each of the 100 shots we generate seven queries by omitting different information channels. To remove color information, the representing image is

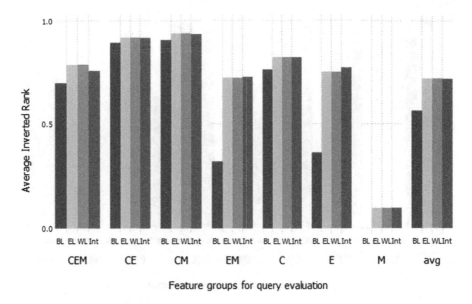

Fig. 3. Average inverted ranks for top 50 results

converted to monochrome and edge information is removed by low-pass filtering the image. Every information channel which is not artificially removed is designated with a character in the query name. A query where no information channel has been removed is designated as *CEM* (for containing *C*olor, *E*dge and *M*otion), while for example a query where the color information has been removed is designated as *EM*. It is important to note that due to the random selection, some of the shots do not contain certain information channels and therefore artificially removing them has no effect e.g., a grayscale shot will remain unchanged in *EM* mode. Similarly, removing the information channels present and keeping those originally without content results in empty or near-empty queries. For this evaluation, these combinations are not manually excluded to avoid the introduction of biases. As a consequence, ground truth is one unique result per query, making this a case of a known-item search.

4.2 Measurements

We compare the three described methods of combining multiple intents, EL, WL, and Int, against the baseline method BL which uses a static combination of weights set to produce generally good results. We measure for each of the seven variants of the 100 queries the number of correctly retrieved results as well as the average inverted rank for a result list with length 50. The number of results returned by the individual modules was always capped at twice the number of overall results.

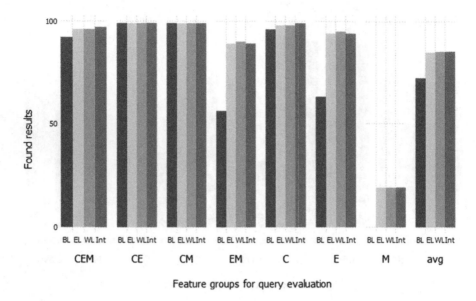

Fig. 4. Number of found correct results in the top 50

4.3 Results

Figures 3 and 4 show the average inverted rank and number of found results out of the 100 possible per method and query type respectively. It can be seen in both figures that the BL method never outperforms any of the three proposed methods EL, WL, and Int. It can also be seen that EL, WL, and Int produce consistently high results for all but one query type. The comparatively bad results for the queries only containing motion is due to the low overall information contained in this channel, which also explains the fact that the BL method was unable to produce a single correct result for this query type. On average, it appears to make no difference if the result lists for the different possible intents are concatenated (EL and WL) or interleaved (Int), as long as they are considered at all. For some query types, some combination methods appear to perform slightly better than others but this can as well be due to artefacts in the query generation process or due to inherent properties of the used dataset. The inverted rank improves on average from 0.56 to 0.71 which is a relative improvement of 26.8 % and the average number of correctly retrieved results increases from 72 to 85 out of 100. Perhaps counter-intuitively, results confirm that by intelligently selecting features, thus disregarding certain feature information, the retrieval performance in terms of both MIR and the number of correctly retrieved objects actually increases.

5   Related Work

While ambiguous queries in search engines are a commonly understood issue, most of the papers that have dealt with quantitative measurements or disambiguation strategies [3,11,16–18] have focused exclusively on web search/ text retrieval. The disambiguation of text searches is achieved using external data such as ontologic information [11] which cannot be easily transferred to video search, as semantic video indexing is still an open problem in itself. A user-centric disambiguation method is presented in [19], where visual examples presented as the user formulates his query, help a content-based image retrieval system better understand the intent. Similarly, in [1] mutual relevance feedback from the user to the system and vice-versa contribute to refining the retrieval results in a multimodal context.

Ranking results is a key problem of information retrieval systems. In the traditional IR view, ranking is achieved by sorting results by their estimated relevance to the query. In later years, machine learning techniques have been used to learn better ranking models from training data. Techniques based on SVMs [2] and neural networks [6] have been applied to learn rankings of documents based on a corpus of queries with corresponding relevances manually rated or derived from external measurements (e.g., click-through rate for web search engine results [7]). The problem of optimally selecting a subset of a large or dynamically growing set of features has been studied by the IR literature as feature selection. There is a significant body of work dedicated to selecting features in order to improve ranking [5,9]. These selectors however work in a query-independent way and thus cannot decide which features are relevant on a specific query.

Content-based video retrieval systems rely on multiple (typically in the tens) heterogeneous features to index the video content. Although more advanced methods exist [15], late score fusion is a simple yet effective multimodal feature fusion method widely used in practical systems because of its scalability to huge datasets, as shown in multimedia information retrieval challenges such as TRECVID [10] and MediaEval [4]. Essentially, late fusion takes estimated scores for a document and applies a set of precomputed weights.

6 Conclusion

In this paper, we have introduced an approach that is able to deal with ill-specified and thus ambiguous queries in sketch-based multimodal video retrieval. We focus on the visual and the kinesthetic modality, which are specified via sketches in a sketch canvas by means of edges, color, and flow fields. The absence of one or several of these information may be intentional and thus should be respected in the result list (e.g., no motion expected in result) – or it may simply be a result of lack of memory, artistic skills, or other reasons. In the latter case, the result should not be impacted by this lack of information. We have presented an approach that first anticipates the query intent(s) of a user based on the information specified in a query sketch. Based on the anticipated user intent(s), the necessary feature groups are chosen for query execution. In an evaluation with the sketch-based video retrieval engine Cineast, we have shown the effectiveness of this approach which significantly improves retrieval quality over a baseline approach agnostic to the query intent at negligeable extra computational cost.

While we have concentrated in this paper on the visual and kinesthetic modalities, the approach can be easily extended to other modalities and the information channels characterizing these modalities. In the IMOTION project, we are currently working on the combination of hand-drawn sketches and spoken queries which will add support for the aural modality. We will also evaluate further approaches to specify information on the motion modality beyond flow fields. In addition, we plan to analyze the impact of a more fine-grained differentiation of the modalities and information channels beyond a binary selection. Similarly, further differentiations within a channel will be considered. This will allow to apply individual weights for features and feature groups dependent on the amount and/or quality contained within an information channel.

Acknowlegements. This work was partly supported by the Swiss National Science Foundation in the context of the CHIST-ERA (European Coordinated Research on Long-term Challenges in Information and Communication Sciences & Technologies ERA-Net) project IMOTION (Intelligent Multi-Modal Augmented Video Motion Retrieval System), contract no. 20CH21_151571.

References

1. Amir, A., Berg, M., Permuter, H.: Mutual relevance feedback for multimodal query formulation in video retrieval. In: Proceedings of the 7th ACM SIGMM International Workshop on Multimedia Information Retrieval, pp. 17–24. ACM (2005)
2. Burges, C., Shaked, T., Renshaw, E., Lazier, A., Deeds, M., Hamilton, N., Hullender, G.: Learning to rank using gradient descent. In: Proceedings of the 22nd International Conference on Machine Learning, pp. 89–96. ACM (2005)
3. Cronen-Townsend, S., Bruce Croft, W.: Quantifying query ambiguity. In: Proceedings of the Second International Conference on Human Language Technology Research, pp. 104–109. Morgan Kaufmann Publishers Inc. (2002)
4. Eskevich, M., Aly, R., Racca, D., Ordelman, R., Chen, S., Jones, G.J.F.: The search and hyperlinking task at mediaeval 2014 (2014)
5. Geng, X., Liu, T.-Y., Qin, T., Li, H.: Feature selection for ranking. In: Proceedings of the 30th Annual International ACM SIGIR Conference on Research and Development in Information Retrieval, pp. 407–414. ACM (2007)
6. Herbrich, R., Graepel, T., Obermayer, K.: Large margin rank boundaries for ordinal regression. Advances in Neural Information Processing Systems, pp. 115–132 (1999)
7. Joachims, T.: Optimizing search engines using clickthrough data. In: Proceedings of the Eighth ACM SIGKDD International Conference on Knowledge Discovery and Data Mining, pp. 133–142. ACM (2002)
8. Kabary, I.A., Schuldt, H.: Using hand gestures for specifying motion queries in sketch-based video retrieval. In: de Rijke, M., Kenter, T., de Vries, A.P., Zhai, C.X., de Jong, F., Radinsky, K., Hofmann, K. (eds.) ECIR 2014. LNCS, vol. 8416, pp. 733–736. Springer, Heidelberg (2014)

9. Novaković, J., Štrbac, P., Bulatović, D.: Toward optimal feature selection using ranking methods and classification algorithms. Yugoslav J. Oper. Res. **21**(1) (2011) ISSN: 0354-0243 EISSN: 2334-6043

10. Over, P., Awad, G., Michel, M., Fiscus, J., Sanders, G., Kraaij, W., Smeaton, A.F., Quénot, G.: Trecvid 2014- an overview of the goals. Tasks, data, evaluation mechanisms, and metrics. In: Proceedings of TRECVID (2014)

11. Qiu, G., Liu, K., Bu, J., Chen, C., Kang, Z.: Quantify query ambiguity using odp metadata. In: Proceedings of the 30th Annual International ACM SIGIR Conference on Research and Development in Information Retrieval, pp. 697–698. ACM (2007)

12. Rossetto, L., Giangreco, I., Heller, S., Tănase, C., Schuldt, H.: Searching in video collections using sketches and sample images - the Cineast system. In: Tian, Q., Sebe, N., G.J., Qi, Huet, B., Hong, R., Liu, L. (eds.) MultiMedia Modeling. LNCS, vol. 9516, pp. 336–341. Springer, Heidelberg (2016)

13. Rossetto, L., Giangreco, I., Schuldt, H.: Cineast: a multi-feature sketch-based video retrieval engine. In: 2014 IEEE International Symposium on Multimedia (ISM), pp. 18–23. IEEE (2014)

14. Rossetto, L., Giangreco, I., Schuldt, H.: OSVC - Open Short Video Collection 1.0. Technical report (CS-2015-002), University of Basel (2015)

15. Snoek, C.G.M., Worring, M., Smeulders, A.W.M.: Early versus late fusion in semantic video analysis. In: Proceedings of the 13th Annual ACM International Conference on Multimedia, pp. 399–402. ACM (2005)

16. Song, R., Luo, Z., Wen, J.-R., Yu, Y., Hon, H.-W.: Identifying ambiguous queries in web search. In: Proceedings of the 16th International Conference on World Wide Web, pp. 1169–1170. ACM (2007)

17. Stojanovic, N.: On analysing query ambiguity for query refinement: the librarian agent approach. In: Song, I.-Y., Liddle, S.W., Ling, T.-W., Scheuermann, P. (eds.) ER 2003. LNCS, vol. 2813, pp. 490–505. Springer, Heidelberg (2003)

18. Weinberger, K.Q., Slaney, M., Van Zwol, R.: Resolving tag ambiguity. In: Proceedings of the 16th ACM International Conference on Multimedia, pp. 111–120. ACM (2008)

19. Zha, Z.-J., Yang, L., Mei, T., Wang, M., Wang, Z., Chua, T.-S., Hua, X.-S.: Visual query suggestion: towards capturing user intent in internet image search. ACM Trans. Multimedia Comput. Commun. Appl. (TOMM) **6**(3), 13 (2010)

Collaborative Q-Learning Based Routing Control in Unstructured P2P Networks

Xiang-Jun Shen[1]([✉]), Qing Chang[1], Jian-Ping Gou[1], Qi-Rong Mao[1],
Zheng-Jun Zha[2], and Ke Lu[3]

[1] School of Computer Science and Telecommunication Engineering,
Jiangsu University, Zhenjiang 212013, China
xjshen@ujs.edu.cn
[2] Institute of Intelligent Machines, Chinese Academy of Sciences, Hefei, Anhui, China
[3] University of Chinese Academy of Sciences, Beijing, China

Abstract. Query routing among peers whilst locating required resources is still an acute issue discussed P2P networking, especially in unstructured P2P networks. Such an issue becomes worse when there is frequent in and out movement of the peers in the network and also with node failures. We propose a new method to assure alternative routing path to balance the query loads among the peers under higher network churns. The proposed collaborative Q-learning method learns the networks parameters such as processing capacity, number of connections, and number of resources in the peers, along with their state of congestion. By this technique, peers are avoided to forward queries to the congested peers. Our simulation results show that the required resources are located more quickly and queries in the whole network are also balanced. Also our proposed protocol exhibits more robustness and adaptability under high network churns and heavy workloads than that of the random walk method.

Keywords: Peer-to-Peer Networks · Q-Learning · Routing control

1 Introduction

Peer-to-Peer (P2P) networks are growing in popularity and are being employed in a wide range of popular Internet applications [1], such as content delivery, file sharing and multimedia streaming etc. P2P networking span across two main types such as structured networks and unstructured networks, and query load balancing is one of the most prevalent issues in both the two types, especially in unstructured P2P networks. This is because such networks forward queries to distant and adjacent peers since the structure of the networks is unknown to the peers. Furthermore, such forwarding strategies often lead to load balancing issues [2] in P2P networking.

Some research works used super peers [3,4] to control queries broadcasted over the network. But this approach might suffer from single-point-of-failure issues. The scalability of the networks will also be limited with the departure

© Springer International Publishing Switzerland 2016
Q. Tian et al. (Eds.): MMM 2016, Part I, LNCS 9516, pp. 910–921, 2016.
DOI: 10.1007/978-3-319-27671-7_76

of the super peers [5]. It is observed that identification of the free riding of uneven popular files over the peers could transform a random network into a star topology [6, 7] in flat networks. Thus, dynamically adapting network overlay topologies. These dynamically adapting network topology strategy are found in some of the recent works. Merino et al. [8], whom introduced the DANTE network, employing a reconnection mechanism to form a balanced cluster overlay topology. And Pournaras et al. [9] proposed ERGO network following a rewiring strategy. It uses virtual servers to monitor the workload of the peers. When overloaded peers are found in the network, ERGO rewires some of the incoming links to the under-loaded peers. AVMON [10] is another network that monitors content discovery and collusion. Small world overlay topologies are introduced in the works of [11–13]. Merugu et al. [11] used two types of links: the short links connecting close peers and the long links connecting randomly chosen peers. Wu et al. [13] proposed a similar mechanism to maintain a state-based shortcut lists in the peers. Efficient search is thus achieved by using links or state information.

In this paper, we propose a new method for unstructured P2P networks in order to overcome the issues of query routing overheads and congestions in peers through congestion control. The proposed collaborative Q-learning learns the networks parameters such as processing capacity, number of connections, and number of resources in the peers, along with their state of congestion. By this technique, peers are avoided to forward queries to the congested peers. Our simulation results show that the desired resources are located more quickly and also queries in the whole network are balanced, by our proposed protocol. Also our proposed protocol exhibits more robustness and adaptability under higher network churns and heavy workloads than that of the random walk method.

The rest of the paper is organized as follows: Sect. 2 formulates our proposed protocol of load balanced routing based on congestion control in P2P networks. The simulation results are presented in Sect. 3. Finally, we draw conclusions in Sect. 4.

2   Proposed Method

In order to identify the congested peers and avoid forwarding queries to them, Q-learning method, which is a method of Reinforcement Learning, is applied to monitor the state of the peers in the network. In this approach of Reinforcement Learning (RL) [14], RL agents learn by interacting with their environment and also by observing the results of these interactions. And Q-learning [15] is an Off-Policy algorithm of RL for temporal difference learning. It uses an action-value function Q to directly approximate the optimal action-value function for an arbitrary target policy. The one-step Q-learning model is defined as follows:

$$Q^{local}(s,a) = R(s) + \gamma max_{a'} Q(s',a')$$
$$Q^{new}(s,a) = Q(s,a) + \alpha Q^{local}(s,a) \tag{1}$$

where $Q(s,a)$ is an action-value function. $R(s)$ is the reward. α is the learning rate, which is set between 0 and 1. γ is the discount factor, also set between

0 and 1. And this parameter of γ considers that future rewards are worth less than the immediate rewards. And $max_{a'}$ is the maximum reward that can be achieved in the next state.

To monitor the state of the peers in the network, state information relevant for the routing process such as processing capacity, number of connections and number of resources is considered. The parameters encoded in the $R(s)$ function reflect the basic state of the peers in the network. And $R(s)$ function is defined as follows:

$$R(s) = \sum_{i=0}^{\infty} \gamma^i \frac{AP_s}{N_s}$$

$$AP_s = C_s \times \chi(s, k_c) \qquad (2)$$

$$\chi(s, k_c) = \sum_{h=1}^{k_c} \frac{N(s, h)}{h^\sigma}$$

where $\chi(s, k_c)$ denotes the degree of peer connectedness of P_s. $N(s, h)$ denotes that the counting degrees of neighboring peers are h hops away from P_s. k_c is the radius parameter for counting P_s's neighbor peers. The role of the control parameter σ is to control the value of h^σ, which is a weight to control the peers at different hop distances away from P_s. C_s is the maximum number of queries that P_s can process per micro-second. In this way, AP_s denotes the positive attractiveness of P_s. The large value of AP_s reflects the higher processing capacity of P_s along with its higher peer degrees. And finally, N_s is the number of resources contained in the peer P_s. N_s is a negative factor in the formula. More the resources in P_s, more is the forwarding queries required to pass through. This implies that for a single resource, less is the bandwidth allocated to it.

From the above formula, it is understood that more number of rewards give larger value of AP_s to the peers. Also peers with more number of neighbors and connections, can process more forwarding queries than others. On the other hand, such peers with more connections can easily for congestion in the network. In order to balance this effect, we regularize the basic Q-learning model by adding the parameter of congestion level (CL), which is defined in formula 3. The modified Q-learning model is as follows:

$$Q^{new}(s, a) = Q(s, a) + \alpha Q^{local}(s, a) +$$
$$\beta \times I(U - CL_s(t)) \times CL_s(t)$$
$$CL_s(t) = \frac{1 + Q_s(t)}{C_s} \qquad (3)$$

where $Q_s(t)$ denotes the number of queries waiting in the input queue of P_s at time t. When a peer is processing a query, other following forwarding queries are put into the waiting queue of that corresponding peer. $CL_s(t)$ reflects the waiting time that a query would spend if it is forwarded to P_s. The bigger the value of $CL_s(t)$, higher is the congestion state of a peer. We use a threshold U to assert a peer to be congested. And when $CL_s(t)$ is bigger than U, the peer

Algorithm 1. Our proposed Q-learning Algorithm

1: Initialize $Q(s,a)$ arbitrary
2: **for** each episode **do**
3:    Initialize s
4:    **repeat**
5:        Choose a from s using policy derived from Q
6:        Take action a, observe $R(s)$, s', $CL_s(t)$
7:

$$Q^{new}(s,a) = Q(s,a) + \alpha Q^{local}(s,a) + \\ \beta \times I(U - CL_s(t)) \times CL_s(t)$$

8:        $s \leftarrow s'$, $a \leftarrow a'$
9:    **until** s is terminal
10: **end for**

is considered to be over-loaded. Finally $I(x) = \begin{cases} +1, & x > 0 \\ -1, & x \leq 0 \end{cases}$ is an indicator function. This function gives a positive sign when a peer is in normal state, and a negative sign when the peer is over-loaded. In this way, our proposed model incorporates the effect of congestion state of the peers.

From formula 3, it is observed that the computation of the Q-values of the peers considers the processing capacity, number of connections and number of resources, along with the congestion state of the peers. In this way, query routings are controlled by collaborative Q-learning among peers. Our proposed Q-learning method is updated as shown in Algorithm 1.

3   Simulation Results

Our simulation environment is composed of 10,000 peers, which are developed based on Gnutella in Python 2.6. Every peer in the network is assigned with 10 neighbors on average. The heterogeneity characteristics of the P2P networks in terms of the processing capacity are obtained from the measure of Gnutella reported in [16]. 1,000 object resources are used in the network, with each object having duplicates. The number of object duplicates is determined by the popularity of the objects having a Zipflike distribution property [8]. And we randomly distribute the objects into different peers. The TTL of a query is set to 8. Since the network load is being caused by the queries of the resources launched by the peers, we assume that for all the peers, the time interval between two successive queries is equal and is termed as the time between searches (tbs), and $tbs = 5s$ implies $2,000$ queries are generated per second in the system on average.

Parameter settings in our proposed formula are as follows: The two parameters k_c and σ are set to 2 and 1, respectively. The parameter U is set to 1.1, which means that the peers are allowed to have 10 % more queries than their

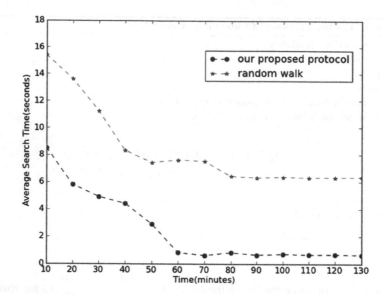

Fig. 1. Timely behavior of average search time is evaluated in our proposed protocol and random walk method in primitive networks.

actual capacity of processing maximum number of queries per micro-second. γ is set to 0.3, α is set to 0.3 and β is 0.5, respectively.

In order to evaluate the performance of our proposed protocol, we simulate the network under two kinds of network settings. And we evaluate the efficiency of our proposed protocol against the random walk method. In Subsect. 3.1, the basic characteristics of our proposed protocol are evaluated. In Subsects. 3.2 and 3.3, high churns and heavy workloads are simulated to test the efficiency of our proposed protocol in more realistic networks.

3.1   Performance Evaluation in Primitive Networks

In this subsection, the network assumes to be static, which means no peers enter or depart the network. In such settings, primitive network characteristics are evaluated as follows.

Firstly, search performances based on the average search time involved in locating the resources are evaluated. Figure 1 shows the timely behavior of the network search performance of our proposed protocol and the random walk method from the 10th min to 130th min. From the figure, it illustrates that the average search time of our proposed protocol falls from 8.7 s to about 1 second, but the fall is from 15.4 s to 7.2 s in the random walk method. Thus the search efficiency of our proposed protocol is improved significantly as a result of query routing optimization.

Secondly, the congestion rates of the peers in the whole network are evaluated. From Fig. 2, it can be observed that the peer congestion rate of our proposed

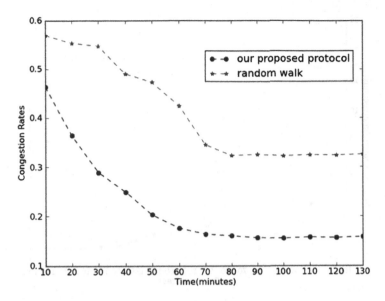

Fig. 2. Timely behavior of peer congestion percentage is evaluated in our proposed protocol and random walk method in primitive networks.

network decreases from 0.47 to 0.16. But the peer congestion rate of the random walk method decreases from 0.57 to 0.33. And this demonstrates that our proposed query routing control method by collaborative Q-learning is more effective than the random walk method. More queries are forwarded to the un-congested peers by our proposed protocol, which decreases the average search time of the network, as illustrated in Fig. 2.

From the above figures, we can conclude that average search time is improved much while peer congestion rates decrease a lot through our routing control strategy. In the next subsections, characteristics of dynamic networks under high churns and heavy workloads are evaluated.

3.2 Network Performance Evaluation Under High Churns

In this subsection, network performances under higher churns are evaluated. In order to observe the network performances under higher churns, peers with a processing capacity of 1,000 are set to leave the network as they more links and forwarding queries. Thereby, in our simulation network settings, all the peers of processing capacity 1,000 leave the network at the 60th min simultaneously and their links are redirected randomly to establish connections with other peers in the network. And these peers re-enter the network again at the 120th min. During the time from 60th min to 120th min, network performances under higher churns are observed.

Firstly, Fig. 3 illustrates the average search time change of our proposed protocol and the random walk method, when the peers of processing capacity 1,000

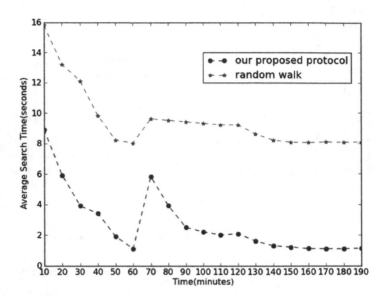

Fig. 3. Timely behavior of average search time is evaluated in our proposed protocol and random walk method under high churns.

leave the network simultaneously at 60th min and enter the network again at 120th min. Between 60th min and 70th min, the average search time of both the methods increases. And the average search time of our proposed protocol increases more than the random walk method. This is because the peers of higher processing capacity of our proposed protocol are prone to be clustered, and the clustered peers have greater impact on the forwarding routings. While, the random walk method shows a little increase in the average search time since the forwarding routings are sent randomly. While between 70th min and 120th min, the average search time of both the methods is observed to be decreasing again. This implies that our proposed protocol dynamically adapts routings when the peers of higher processing capacity leave the network. After 120th min, peers of processing capacity 1,000 enter the network again, and now the average search time of both the methods continues to decrease. This is because the network becomes stable again, and our proposed protocol can learn this network topology change to decrease the average search time to a lower level than that of the random walk method. From the figure, it demonstrates that under higher network churns, our proposed protocol shows high adaptability to the changes in the network topology, and the average search time of our proposed protocol is achieved much quickly than the random walk method during at all times.

Secondly, the congestion rate of the peers in the whole networks is evaluated, as shown in Fig. 4. When the peers of processing capacity 1,000 leave the network simultaneously at 120th min, the peer congestion rates of both the methods are observed to be increasing between 60th min and 70th min, especially in our proposed protocol. This demonstrates that the high processing capacity peers have

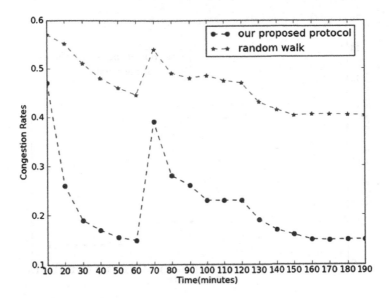

Fig. 4. Timely behavior of peer congestion percentage is evaluated in our proposed protocol and random walk method under high churns.

greater impact on the forwarding routings than that of the random walk method. After 70th min, peer congestion rates of both the methods are decreasing, especially in our proposed protocol. This illustrates that our proposed protocol can adapt to the changes in the network topology. Our routing control method by Q-learning reduces the peer congestion rates of our proposed protocol than that of the random walk method. Thus in our method, more routing messages are forwarded to the un-congested peers. This demonstrates that our network achieves load balancing among peers when high churns happen.

3.3 Network Performance Evaluation Under Heavy Workloads

In this subsection, network performances under heavy workloads are evaluated, In our settings, the workloads in networks are increased at the 60th min and decreased to the normal workload state at the 120th min. During this time from 60th min to 120th min, network performances under heavy workloads are observed. And in our simulations, the heavy workloads are evaluated with parameter $tbs = 0.5$, which means the network generates 10 times more queries in the networks than queries generated in the normal workloads of $tbs = 5$.

Firstly, the search performances based on average number of hops and average search time under heavy workloads are evaluated in Fig. 5. As the first 60 min no heavy workload is imposed in the network, the average search time is observed to decrease in both methods, especially in our proposed protocol. By the time at 60th min, our proposed protocol achieves much lower average search time than the random walk does. And between 60th min and 120th min, ten times

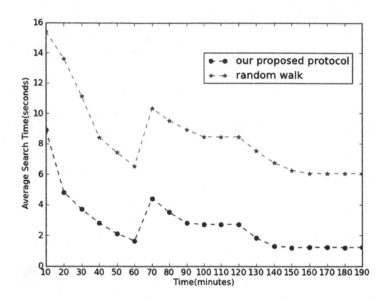

Fig. 5. Timely behavior of average search time is evaluated in our proposed protocol and random walk method under heavy workloads.

queries are imposed in the network and the average search time of both the methods are observed to increase until the 70th min, and decreasing thereafter. During this time, the average search time of our proposed protocol is observed to be recovered more quickly and be much lower than those of the random walk method.

Finally, the congestion rates of the peers in the whole networks are evaluated, as shown in Fig. 6. Similar behavior is also observed in this figure just as the congestion rates change in the network of high churns. As the first 60 min no heavy workload is imposed in the network, the congestion rates are observed to be decreased in both the methods, especially in our proposed protocol. The reason in it is that our proposed protocol can not only change the network topology dynamically, but also can guide forwarding messages to those un-congested peers. While during the time between 60th min and 120th min that the network suffers high workloads, the congestion rates of both the methods are observed to increase until the 70th min, and decrease thereafter. And the congestion rates of our proposed protocol are observed to be recovered more quickly and be much higher than that of the random walk method.

Otherwise, if we compare all figures in the two scenarios of high churns and high workloads in time between 60th min and 70th min, the changing rates of our proposed protocol can be observed. For example, if we observe Figs. 4 and 6 in time between 60th min and 70th min that the networks are suffering high churns and high workloads, the changing of congestion rates of our proposed protocol in Fig. 6 is observed more sharply and highly than the changing of congestion rates of our proposed protocol in Fig. 4. Other figures show same characteristics that

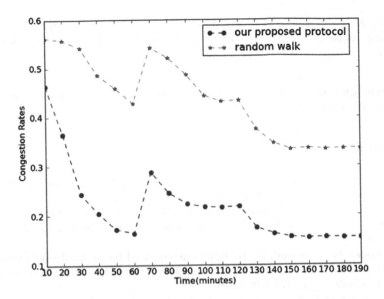

Fig. 6. Timely behavior of congestion rates of the peers is evaluated in our proposed protocol and random walk method under heavy workloads.

the changing rates of our proposed protocol under high churns are more sharp and high than the rates of our proposed protocol under high workloads. They demonstrate that, since those peers of processing capacity $1,000$ form clustered network, the network under such condition of high churns is seen to be more affected than the network under high workloads is.

From the two scenarios of high churns and high workloads, we can conclude that our proposed protocol can adapt to the changes of high churns and high workloads. Our routing control method is also demonstrated more effective than the random walk method in two scenarios.

4 Conclusion

P2P networks have been witnessed in many applications in the past few decades. Load balancing and decentralized resource locating approaches in such networks still suffer various limitations. In this paper, a new method is proposed to assure alternative routing path to balance the query loads among the peers under high network churns. Load balanced routing is achieved by collaborative Q-learning among the peers. Our proposed collaborative Q-learning method not only learns from the network parameters as processing capacity, number of connections and the number of resources in the peers, but also learns from their state of congestion. By this technique, queries are guided to avoid being forwarded to the congested peers. Thus, query routings are forwarded and balanced among the peers. Our simulation results show that the desired resources are located more quickly and also queries in the whole network are balanced, by our proposed protocol. Also our proposed protocol exhibits more robustness and adaptability under high network churns and heavy workloads than that of the random walk method.

Acknowledgments. This work was funded in part by the National Natural Science Foundation of China (No. 61572240).

References

1. Risson, J., Moors, T.: Survey of research towards robust peer-to-peer networks: search methods. Comput. Netw. **50**(17), 3485–3521 (2006)
2. Lua, E.K., Crowcroft, J., Pias, M., Sharma, R., Lim, S.: Survey of research towards robust peer-to-peer networks: search methods. Commun. Surv. Tutorials **7**(2), 72–93 (2005)
3. Khataniar, G., Goswami, D.: HUP: an unstructured hierarchical peer-to-peer protocol. In: Proceedings of the International MultiConference of Engineers and Computer Scientists, pp. 671–676 (2010)
4. Li, J.-S., Chao, C.-H.: An efficient super-peer overlay construction and broadcasting scheme based on perfect difference graph. IEEE Trans. Parallel Distrib. Syst. **21**(5), 594–606 (2010)
5. Fakasa, G.J., Karakostas, B.: An efficient super-peer overlay construction and broadcasting scheme based on perfect difference graph. Inf. Softw. Technol. **46**(6), 423–431 (2004)
6. Xu, Z., Bhuyan, L.N.: Effective load balancing in p2p systems. In: IEEE International Symposium on Cluster Computing and the Grid, pp. 81–88 (2006)
7. Sreenu, G., Dhanya, P.M., Thampi, S.M.: Enhancement of bartercast using reinforcement learning to effectively manage freeriders. In: Advances in Computing and Communications, pp. 126–136 (2011)
8. Merino, L.R., Anta, A.F., Lópze, L., Cholvi, V.: Self-managed topologies in p2p networks. Comput. Netw. **53**(10), 1722–1736 (2009)
9. Pournaras, E., Exarchakos, G., Antonopoulos, N.: Load-driven neighbourhood reconfiguration of gnutella overlay. Comput. Commun. **31**(13), 3030–3039 (2008)
10. Morales, R., Gupta, I.: AVMON: optimal and scalable discovery of consistent availability monitoring overlays for distributed systems. IEEE Trans. Parallel Distrib. Syst. **20**(4), 446–459 (2009)
11. Merugu, S., Srinivasan, S., Zegura, E.: Adding structure to unstructured peer-to-peer networks: the use of small-world graphs. J. Parallel Distrib. Comput. **65**(2), 142–153 (2005)
12. Liu, L., Antonopoulos, N., Mackin, S., Xu, J., Russell, D.: Efficient resource discovery in self-organized unstructured peer-to-peer networks. Concurrency Comput. Pract. Experience **23**(2), 159–183 (2009)
13. Wu, K., Wu, C., Liu, L.: State-based search strategy in unstructured p2p. In: Proceedings of 13th IEEE International Symposium on Object/Component/Service-Oriented Real-Time Distributed Computing, pp. 381–386 (2010)
14. van Hasselt, H.: Reinforcement learning in continuous state and action spaces. In: Wiering, M., van Otterlo, M. (eds.) Reinforcement Learning. ALO, vol. 12, pp. 205–248. Springer, Heidelberg (2012)

15. Gheshlaghi Azar, M., Munos, R., Ghavamzadaeh, M., Kappen, H.J.: Speedy Q-Learning. In: Advances in Neural Information Processing Systems, (NIPS 24), pp. 2411–2419 (2011)
16. Sarolu, S., Gummadi, P.K., Gribble, S.D.: A measurement study of peer-to-peer file sharing systems. In: Proceedings of Multimedia Computing and Networking (2002)

Author Index

Printed in the United States
By Bookmasters